Lecture Notes in Networks and Systems 620

The series "Lecture Notes in Networks and Systems" publishes the latest developments in Networks and Systems—quickly, informally and with high quality. Original research reported in proceedings and post-proceedings represents the core of LNNS.

Volumes published in LNNS embrace all aspects and subfields of, as well as new challenges in, Networks and Systems.

The series contains proceedings and edited volumes in systems and networks, spanning the areas of Cyber-Physical Systems, Autonomous Systems, Sensor Networks, Control Systems, Energy Systems, Automotive Systems, Biological Systems, Vehicular Networking and Connected Vehicles, Aerospace Systems, Automation, Manufacturing, Smart Grids, Nonlinear Systems, Power Systems, Robotics, Social Systems, Economic Systems and other. Of particular value to both the contributors and the readership are the short publication timeframe and the world-wide distribution and exposure which enable both a wide and rapid dissemination of research output.

The series covers the theory, applications, and perspectives on the state of the art and future developments relevant to systems and networks, decision making, control, complex processes and related areas, as embedded in the fields of interdisciplinary and applied sciences, engineering, computer science, physics, economics, social, and life sciences, as well as the paradigms and methodologies behind them.

Indexed by SCOPUS, INSPEC, WTI Frankfurt eG, zbMATH, SCImago.

All books published in the series are submitted for consideration in Web of Science.

For proposals from Asia please contact Aninda Bose (aninda.bose@springer.com).

Bahaaeddin Alareeni · Allam Hamdan ·
Reem Khamis · Rim El Khoury
Editors

Digitalisation: Opportunities and Challenges for Business

Volume 1

Springer

Editors
Bahaaeddin Alareeni
Middle East Technical University, Northern Cyprus Campus
Kalkanlı, Güzelyurt, KKTC via Mersin 10, Turkey

Reem Khamis
University College of Bahrain
Manama, Bahrain

Allam Hamdan
College of Business and Finance
Ahlia University
Manama, Bahrain

Rim El Khoury
Lebanese American University
Beirut, Lebanon

ISSN 2367-3370 ISSN 2367-3389 (electronic)
Lecture Notes in Networks and Systems
ISBN 978-3-031-26952-3 ISBN 978-3-031-26953-0 (eBook)
https://doi.org/10.1007/978-3-031-26953-0

This Springer imprint is published by the registered company Springer Nature Switzerland AG
The registered company address is: Gewerbestrasse 11, 6330 Cham, Switzerland

Preface

We are delighted to write this Foreword for the International Conference on Business and Technology (ICBT'22) proceedings. I deeply believe in the role of such a conference and other similar scientific forums in bringing together leading academicians, scholars, and researchers to share their knowledge and new ideas as well as to discuss current developments in the fields of economics, business, and technology. ICBT'22 provides a valuable window on the implementation of technology such as artificial intelligence, IoT, and innovation in business development. For two days, a large number of distinguished researchers and guest speakers discussed many contemporary issues in business and technology around the world. We have a strong faith that this book will be of great benefit for many parties, especially those aspiring to develop buoyant strategies that will lead to positive impact on any future endeavors. Finally, I hope that the ICBT'22 continues as a destination for researchers, postgraduate students, and industrial professionals.

Bahaaeddin Alareeni
Allam Hamdan

Contents

Marketing, E-commerce and Digitalization

Digital Economy, Business Innovation, Technology and Covid-19

Building Information Modeling for Risk Management: A Literature Review

Lorena Ortiz-Mendez(✉), Alberto de Marco, and Gabriel Castelblanco

Politecnico di Torino, Turin, Italy
lorena.ortizmendez@polito.it

Abstract. Building Information Modeling (BIM) has become more relevant to the construction industry in recent years due to almost all of the biggest industries use BIM as a tool to improve the integration process and risk management. Although there is significant literature on risk management and BIM, the relationship between both of them has not been covered in previous research. This study provides a thorough explanation of the relationship between the interaction with risk management and BIM, as well as how it has evolved. Following a screening procedure, 190 peer-reviewed papers were pulled from the Scopus database. Findings showed that the introduction of risk management into BIM is still in an incipient phase within the construction project management body of knowledge. Overall, three developed nations—the USA (with 30 documents), Australia (21) and China (21)—have steered this research agenda, while a developing country—Malaysia (17)—is an outsider gaining relevance as the fourth contributor to this topic. Five clusters were identified by the network representation, these clusters include risk and project management areas that constitute the research paths to be advanced in the next few years.

Keywords: BIM · Risk · Building information modeling

1 Introduction

The risk in project management has been studied in the construction industry for a long time [1–5]. Proper risk management is essential for preventing renegotiations, identifying sustainability challenges, and addressing issues between stakeholders during the project's lifecycle [6–12]. However, as Building Information Modeling (BIM) is gaining relevance, the whole construction process has increased their interaction with this tool. By strengthening the integrative approach necessary for process groups within project management, BIM capabilities hold the key to overcoming the traditional obstacles in construction project management [13]. This paper aims to explore the implementation of BIM in risk management and its evolution. The integrated analysis of both -the risks in projects and the use of BIM- allows for improving understanding through uncovering conceptual trends and relationships.

The following are some crucial practical concerns with the application of BIM models in the building sector:

B. Alareeni et al. (Eds.): ICBT 2022, LNNS 620, pp. 3–10, 2023.
https://doi.org/10.1007/978-3-031-26953-0_1

- Inefficient methods for managing group projects
- Failure to successfully manage BIM model conceptualization issues and changes,
- Trouble communicating and keeping track of pertinent changes with appropriate BIM engineers
- Difficulty in effectively managing self-inspections and discoveries during BIM model creation
- A lack of effective control on BIM model construction versions during the process when the BIM model must be regularly updated and altered
- There is a paucity of studies on BIM collaboration management (CM), particularly for BIM model production, despite significant research and development efforts in the academic and professional BIM literature.

The paper is structured as followed:

Section 2, Research methodology, Sect. 3, Results and Discussions, in this section the analysis been done the chronology, geographic evolution, and Network Analysis, Sect. 4, the relevance of the BIM information, and Sect. 5, conclusions.

2 Research Methodology

The authors conducted a thorough literature review to determine how risk management is incorporated into BIM. The resulting papers were then utilized to create networks to examine relationships and trends, in accordance with methodological approaches recommended by multiple authors [14–17]. The networks and bibliometric data were then examined.

In the first filter done the query included the following terms: ("Risk") AND ("BIM" OR "Building Information Modelling") including title, abstract, author, and keywords. This initial search resulted in 5268 document types, as shown in Table 1. In the second filter done the query included the following terms: ("construction project") AND ("Risk") AND ("BIM" OR "Building Information Modelling") including title, abstract, author, and keywords. This initial search resulted in 246 document types, as shown in Table 1. And In the third filter done the query included the following terms: ("construction project") AND ("Risk") AND ("BIM" OR "Building Information Modelling") AND (LIMIT-TO (SUBJAREA, "engi")) including title, abstract, author, and keywords. This initial search resulted in 190 document types, as shown in Table 1.

2.1 Network Development and Analysis

By optimizing the objective function of the Euclidean distances between pairs of nodes in VOSviewer, bibliometric networks generated from similarity matrices for mapping data co-occurrence were built and visualized. This method was selected to evaluate and visualize the overlap of several related subjects in the scientific literature.

Table 1. Cluster interpretation

Selected String	Papers
TITLE-ABS-KEY (bim OR "BUILDING INFORMATION MODELING" OR "BUILDING INFORMATION MODELLING") AND risk	5268
TITLE-ABS-KEY ((**bim** OR "building information modeling" OR "building information modelling") AND risk AND ("construction project"))	246
TITLE-ABS-KEY ((**bim** OR "building information modeling" OR "building information modelling") AND risk AND ("construction project")) AND (LIMIT-TO (SUBJAREA , "engi"))	190

3 Results and Discussions

3.1 Chronologic and Geographic Evolution

Figure 1 depicts how risk implementation in BIM has steadily increased since 2015. It's interesting to note that the first article to examine BIM and RISK was released in 2009. In addition, 84% of the research has been published in the past seven years, showing that this field of study is still developing and that more research is anticipated in the future.

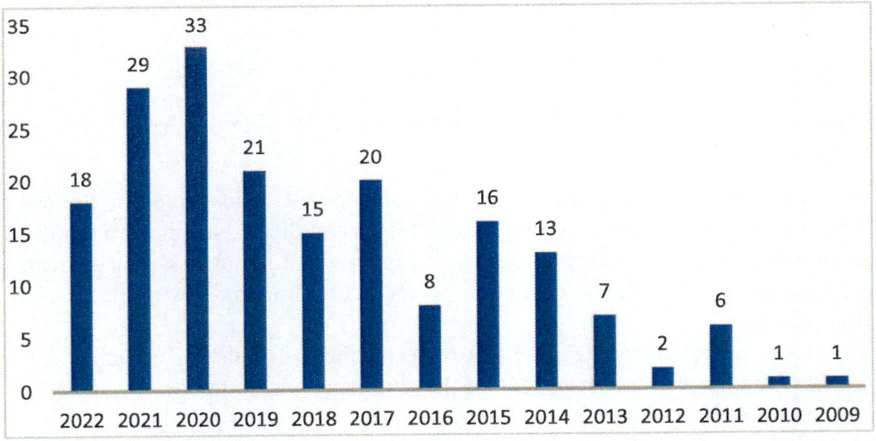

Fig. 1. Chronological evolution.

3.2 Network Analysis

The network study is concentrated on the evolution between 2015 and 2022 due to the considerable concentration of research produced over the previous seven years. Beginning in the United States and Taiwan, the research method that incorporates risk management in the application of BIM then made great strides in Australia, and in the last three years, it has been further developed in China and Malaysia (Figs. 2 and 3).

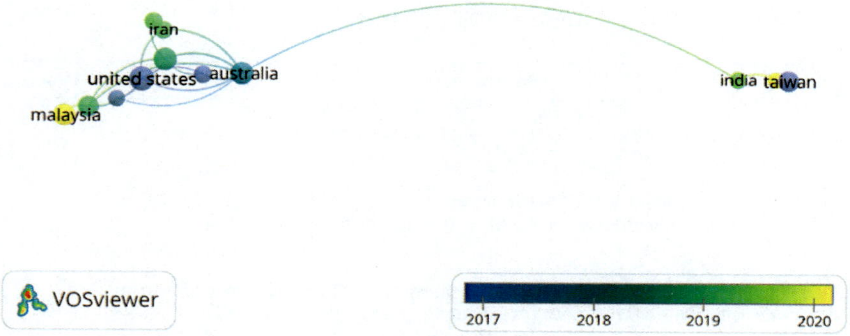

Fig. 2. Countries' network evolution

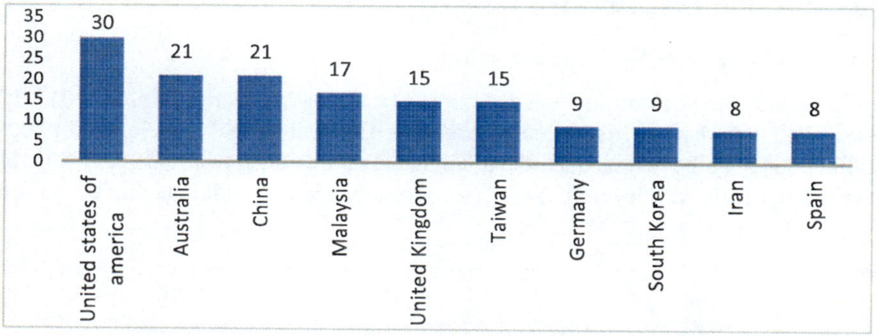

Fig. 3. Top 10 countries with the most articles published on the subject

The evolution of the research process that includes risk management in the use of BIM is presented in Fig. 4, regarding research topics had their first approach in the part of construction processes and plan designs, today we talk about risk management in projects that use BIM to define cost issues, work scheduling and cost projections and execution times.

The cluster representation (Fig. 5) represents the thematic clusters by each of the six (6) colors of the nodes. Moreover, each of the clusters is related to one or more subject groups defined in ISO 21500 [3], as shown in Table 1. The most populated cluster (red color) has reference to the construction safety risk, showing that using BIM is possible to reduce accidents in the construction industry, because is possible to plan strategies to guarantee a correct function in civil construction (Table 2).

The second cluster (green color) is related to risk identification, in which the aim is to identify prospective risk events and their characteristics that, if they take place, would have an impact on the project's goals favorably or negatively. The third cluster (blue color) is driven by BIM a tool that permits doing a Risk Analysis using all the information of the projects in the different models.

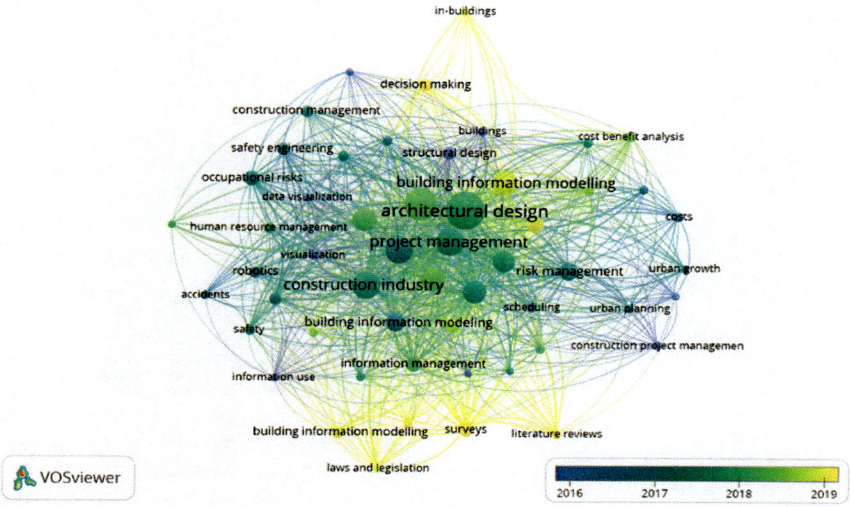

Fig. 4. Topic's network evolution

Table 2. Cluster interpretation

Color	Subject group	Example of keywords
Red	Construction safety risk	Accident prevention, Buildings, Construction Safety, Human resource management, Occupational risks, Safety Engineering, Safety management
Green	Risk identification	Information management, Risks analysis, Information use
Blue	Risk analysis	Cost, economics, cost benefits analysis, risk assessment
Yellow	Risk control	Conceptual framework, construction process, life cycle, scheduling
Violet	Risk treatment	Architectural design, building information modeling, decision-making, risk
Cyan	Legal risk	Laws and legislation

The fourth cluster (yellow) is focused on risk Control whose purpose is to determine whether the risk responses are carried out and whether they have the desired impact to minimize disruption to the project.

The fifth cluster (violet) is centered to improve possibilities and lessen threats to project objectives, treat risks and aims developing options, and deciding on actions.

And the last cluster (Cyan) is set in the legal risk, which includes all the documentation required that must be done to do the project.

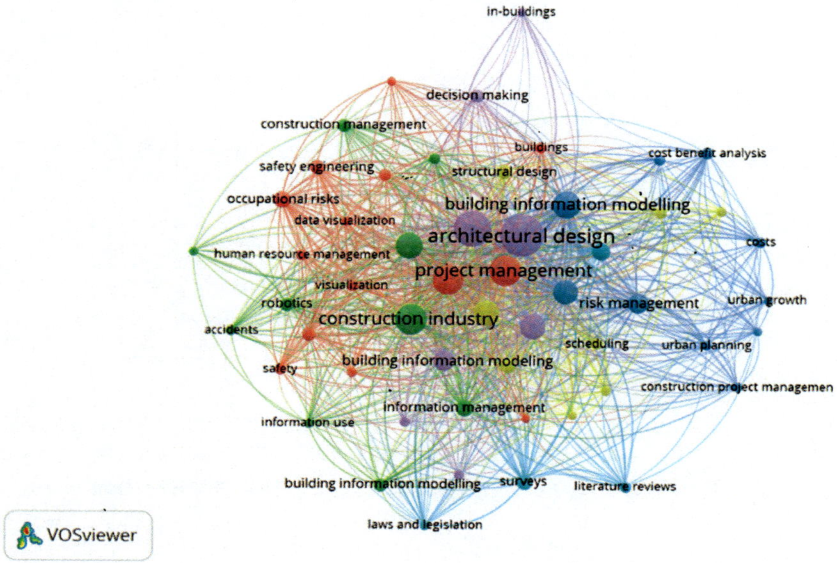

Fig. 5. Most co-occurrence terms of all keywords

4 The Relevance of the BIM Formation

- Greater cooperation and planning between disciplines during the design phase.
- Constantly coordinated, updated, and secured documentation.
- Understanding of 3D information visualization software.
- Ability to improve the level of precision, rigor, and detail while shortening design timeframes.
- increases the degree of certainty in the work's measurement and quantification.
- Include all deadlines, maintenance instructions, and file information in all of the construction equipment's parts.
- To enable sophisticated building management, integrate sensors into the equipment.
- 3D simulations should be used to train operators and prevent occupational risk.
- Give each agent working on project transparency.

5 Conclusion

This study gives a thorough understanding of the development of the relationship between BIM and risk management.

By undertaking a virtual construction of the project, a well-implemented BIM technology in occupational risk management enables risks to be combated from their inception or design phase, facilitating the detection, mitigation, and/or elimination of risks generated in the project's design phase.

Is demonstrated that the use of BIM reduces the risk of lost money in a project of construction, some of the reasons have been written in the chapter "The relevance of the BIM formation" It is demonstrated by the rise in publications linking BIM to risk analysis and prevention, starting with the USA and also involving countries like Malaysia that was not included in this research topic before.

Future research should focus on presenting specifics of technology implementation processes that demonstrate the usage of BIM as a tool to enhance risk management as this work is confined to evaluating the Scopus database.

References

1. Castelblanco, G., Guevara, J., Mesa, H., Flores, D.: Risk allocation in unsolicited and solicited road public-private partnerships: sustainability and management implications. Sustainability **12**, 1–28 (2020)
2. Castelblanco, G., Guevara, J.: Risk allocation in PPP unsolicited and solicited proposals in Latin America: pilot study in Colombia. In: Construction Research Congress 2020, pp 1321–1329 (2020)
3. Marcellino, M., Castelblanco, G., De Marco, A.: Multiple linear regression model for project's risk profile and DSCR. In: IOP Conference Series: Materials Science and Engineering (2022)
4. Castelblanco, G., Fenoaltea, E.M., De Marco, A., Demagistris, P., Petruzzi, S., Zeppegno, D.: Integrating risk and stakeholder management in complex mega-projects: a multilayer network analysis approach. Megaprojects Research Interdisciplinary Team, pp. 1–17 (2022)
5. Castelblanco, G., Mesa, H., Serra, L.: Risk analysis in private building projects: a pilot study in Chile. Megaprojects Research Interdisciplinary Team, pp 1–8 (2022)
6. Castelblanco, G., Guevara, J., Mendez-Gonzalez, P.: PPP renegotiation flight simulator: a system dynamics model for renegotiating PPPs after pandemic crisis. In: Construction Research Congress 2022 (2022)
7. Castelblanco, G., Guevara, J., Mendez-Gonzalez, P.: In the name of the pandemic: a case study of contractual modifications in PPP solicited and unsolicited proposals in COVID-19 Times. In: Construction Research Congress 2022 (2022)
8. Rojas, R., Bennison, G., Gálvez, V., Claro, E., Castelblanco, G.: Advancing collaborative water governance: unravelling stakeholders' relationships and influences in contentious river basins. Water (Switzerland) **12**, 1–25 (2020)
9. Castelblanco, G., Guevara, J., Mesa, H., Hartmann, A.: Social legitimacy challenges in toll road PPP programs: analysis of the Colombian and Chilean cases. J. Manag. Eng. **38**, 1–15 (2022)
10. Castelblanco, G., Guevara, J., Rojas, D., Correa, J., Verhoest, K.: Environmental impact assessment effectiveness in public-private partnerships: study on the Colombian road program. J. Manag. Eng. (2022)
11. Castelblanco, G., Guevara, J.: Building bridges: unraveling the missing links between public-private partnerships and sustainable development. Proj Leadersh. Soc. **3**, 1–10 (2022)
12. Marcellino, M., Castelblanco, G., De Marco, A.: Contract renegotiation in PPPs: evidence from Italy. In: IOP Conference Series: Materials Science and Engineering (2022)
13. Marcellino, M., Castelblanco, G., De Marco, A.: Building information modeling for construction project management: a literature review. In: IOP Conference Series: Materials Science and Engineering (2022)
14. Castelblanco, G., Guevara, J., Mendez-Gonzalez, P.: Sustainability in PPPs: A network analysis. In: Interdisciplinary Civil and Construction Engineering Projects. ISEC-11, pp 1–6. ISEC Press, Fargo (2021)
15. Castelblanco, G., Guevara, J., Mesa, H., Sanchez, A.: Semantic network analysis of literature on public-private partnerships. J. Constr. Eng. Manag. **147**, 1–16 (2021)

16. Castelblanco, G., Guevara, J.: Crisis driven literature in PPPs: a network analysis. In: IOP Conference Series: Earth and Environmental Science, Melbourne, Australia (2022)
17. Castelblanco, G., Guevara, J., Salazar, J.: Remedies to the PPP crisis in the Covid-19 pandemic: lessons from the 2008 global financial crisis. J. Manag. Eng. **38**, 1–18 (2022)

Knowledge and Preferences of Urban Population of Bengaluru: Fortified vs. Non-fortified

Jaspreet Kaur(✉) and B. Subha

Department of Management, Kristu Jayanti College, Autonomous, Bengaluru 560077, India
{jaspreetkaur,subha}@kristujayanti.com

Abstract. The study is carried out to find out preferences for food items in Bangalore's urban Population. The study focuses on the purchasing choices of the population varying between fortified and non-fortified food items. Primary data is collected from male and female respondents of various age groups, belonging to various professional backgrounds. The study aimed to find out connect between preferences for fortified food and professional backgrounds. Additionally, data is collected and divided into various age groups to further understand age-related preferences for fortified food. Convenience sampling is used for data collection. Data are analysed using statistical techniques. Results showed a connection between professional backgrounds and a preference for fortified food.

Keywords: Fortified food · Food preferences · Bangalore urban Population · Non-fortified food

1 Introduction

The fortification of food is the practice of adding vitamins and minerals to food commonly consumed by people. The additives are added during processing to enhance their nutritional value. It is a proven strategy to improve diets and control and prevent micronutrient deficiencies. It is a cost-effective and safe way of improving dietary intake, Olson et al. (2021). The human body requires small amounts of vitamins and micronutrients, having a critical impact on its health. Any deficiency of these nutrients can cause serious and even life-threatening conditions. The most common deficiencies across the world are Vitamin A and iron. It is prevalent among children and pregnant women. Deficiencies of these micronutrients can bring about reduced energy levels along with a lack of mental clarity, WHO.

The most common deficiencies among the population of low and middle-income countries include iron, zinc and vitamin A. These deficiencies can be of one or more micronutrients, Olson et al. (2021). Fortification of food is classified as commercial and industrial fortification (wheat flour, corn meal, cooking oils), Biofortification (breeding crops to increase their nutritional value, which can include both conventional selective breeding, and genetic engineering) and Home fortification (for example vitamin D drops), Wikipedia (2022).

B. Alareeni et al. (Eds.): ICBT 2022, LNNS 620, pp. 11–16, 2023.
https://doi.org/10.1007/978-3-031-26953-0_2

Micronutrients are important for the growth and development of the body. Any deficiencies of these micronutrients lead to disease or improper development, Wikipedia (2022). Common nutrients which are more commonly of concern include Vitamin A, B6, B 12, C, and D, Calcium, Folate, iodine, Iron, Magnesium and Zinc. Lack of sufficient intake of these nutrients by the general population can lead to nutrient inadequacy. Nutrient deficiencies vary with age, gender, race and ethnicity, https://health.mo.gov/. About 6000 children under the age of five years die in India daily. Malnutrition, mainly deficiency of vitamin A, iron, Zinc and folic acid is the major cause of more than half of these deaths. Vitamin A deficiency is found in about 57% of children less than the age of 6 and also amongst mothers of these children Kotecha (2008). Essentially, there are six micronutrients. They have a specific role in the human body. Iron is important for motor and cognitive development. Any deficiency of iron causes anemia.40% of children under the age of 5 years and 30% of pregnant women globally are affected by anaemia. Women have a high risk of death due to Anaemia and it also causes a low birth rate of the newborn. Vitamin A is essential for eyesight and other immune functions. Any deficiency of this vitamin can lead to blindness. Its deficiency can be life-threatening with infections like measles and diarrhoea. 190 million young children are affected by vitamin A deficiency. Vitamin D helps in stronger bones by assisting in the absorption of calcium. A deficiency of vitamin D results in rickets among children and causes osteomalacia among adults. There is a widespread deficiency of Vitamin D globally. Iodine is essential for healthy growth amongst infants and their cognitive development. 1.8 billion are affected by insufficient intake of iodine. Folate (Vitamin B9) helps in making new cells. It is essential for the healthy brain and spine development of the foetus. Zinc helps in immune functions and resistance to infectious diseases. This nutrient helps in the prevention of diseases like diarrhoea, pneumonia and malaria. 17.3% of the world population is at risk of Zinc deficiencies, (https://www.cdc.gov/). Consuming natural foods like fruits, vegetables, meat and dairy products is the best way to have these micronutrients, (https://www.webmd.com). However due to poverty, lack of access to a variety of food, insufficient knowledge of correct dietary practices and due to high rate of infectious diseases micronutrient malnutrition is prevalent, (www.webmd.com). Fortification of food is an essential strategy identified by the WHO to combat nutrient deficiencies at the global level WHO/Wikipedia. There is a need to address malnutrition in all forms. Fortification of food items with micronutrients seems to play a very important role in this regard, Guarantee (2022).

Significant effects were observed in haemoglobin levels, serum ferritin levels and anaemia prevalence for preschool and school-going children, as a result of micronutrient fortification, Das et al. (2013). Inefficiency in the absorption of calcium and other minerals results from a deficiency of vitamin D. One of the major causes of this deficiency is limited dietary intake of vitamin D, Calvo and Whiting (2013). Participants from the study are less aware of dietary sources of vitamin D and few could state fortified products. There was a favourable attitude towards fortification, Clark et al. (2019). A study showed that about55% of households are aware of fortified sugar and this awareness is higher in urban consumers, Pambo (2013) Residents of rural areas have less knowledge about fortified food, (Mabaya et al. 2010). It is found that if a consumer is found to have higher knowledge about fortified food, the likelihood of consumption of the same is higher, Pounis et al. (2011). Willingness to pay higher for bio-fortified food is found to be higher amongst higher-income groups, Meier et al. (2020).

2 Objectives of the Study

- To understand preferences of fortified food items among different age groups of Bangalore's urban Population.
- To find the relation between professional background and fortified food purchase.

3 Materials and Methods

First-hand data is collected from 270 respondents, belonging to different professional backgrounds and age groups. Among the respondents, 50 were working professionals, 100 were students, 70 were housewives and 50 were self-employed or running small businesses. Social-demographic details are collected from respondents belonging to different professional backgrounds in different age groups. Respondents belong to the city of Bengaluru, Karnataka. Further data is arranged as per the age of the respondents to test age-related preferences for fortified food. Data is collected using convenience sampling. The questionnaire is used as a tool to collect data. Questions about age, professional background, and awareness about various aspects of fortified food and buying preferences and habits were included in the questionnaire. The population of Bengaluru urban is chosen for the survey. Data is analysed using one-way ANOVA.

4 Results

(See Table 1.).

Table 1. Demographic characteristics of study participants

	Total N = 270 Frequency (%)	Working Professionals n = 50 Frequency (%)	Students n = 100 Frequency (%)	Housewives n = 70 Frequency (%)	Self-employed/small business n = 50 Frequency (%)	p-value
Age (in Years)						
>20	80(29.6)	–	80 (80)	–	–	0.8729
20–40	90 (33.3)	19 (38)	20 (20)	14 (20)	37 (74)	
40–60	92 (34.0)	31 (62)	–	48 (68.5)	13 (26)	
60–80	8 (2.9)	–	–	8 (11.4)	–	
Gender						
Male	91 (33.7)	22 (44)	45 (45)	–	24 (48)	0.7351
Females	159 (58.8)	28 (56)	55 (55)	70 (100)	26 (52)	

(continued)

Table 1. (*continued*)

	Total N = 270 Frequency (%)	Working Professionals n = 50 Frequency (%)	Students n = 100 Frequency (%)	Housewives n = 70 Frequency (%)	Self-employed/small business n = 50 Frequency (%)	p-value
Family income (in Rupees)						
10,000–25000	33 (12.2)	–	–	–	33 (66)	0.8421
25,000–50,000	27 (10)	10 (20)	–	–	17 (34)	
50,000–1,00,000	95 (35.1)	13 (26)	66 (66)	16 (22.8)	–	
>1,00,000	115 (42.5)	27 (54)	34 (34)	54 (77.1)	–	

There is a significant difference between the ages of the respondents ($p \geq 0.05$). From a total sample size of 270, 29.6% of the respondents are below the age of 20 years, 33.3% are in the age group of 20–40, 34% fall in the age group of 40–60 and 2.9% are from the 60–80 age group. The student group is the youngest in terms of age, the entire group falls under 40 years. A significant difference is observed between the monthly family incomes of the respondents ($p \geq 0.05$).12.2% of the respondent's family income falls between Rs. 10,000 to 25,000. 10%, 35.1% and 42.5% of the respondents fall in the income group of Rs 25,000–50,000, Rs.50,000–1,00,000 and above Rs. 1,00,000 respectively. 100% of self-employed respondents' family income is less than 50,000 Rs. a month. Whereas 100% of the housewives fall above a family income of Rs. 50,000 or more. The majority of the housewives (77.1) belong to the monthly income group of Rs.1,00,000 and above. Working professionals participating in the study worked as professionals in IT firms, college professors, doctors and the like. Self-employed/ small business owners include people working in shops, houses or owned small shops and the like. 33.7% of male and 58.8% of female respondents participated in the study. A significant difference is found in the gender ($p \geq 0.05$). The Housewives group constitutes a female sample (Table 2.).

Table 2. Awareness and preferences of fortified food

	Total N = 270	Working Professionals n = 50	Students n = 100	Housewives n = 70	Self-employed/small business n = 50	p-value
Awareness about the availability of Fortified food						
Yes	140 (51.8)	26 (52)	60 (60)	49 (70)	5 (10)	0.554
No	130 (48.1)	24 (48)	40 (40)	21 (30)	45 (90)	
Awareness about the health benefits of Fortified food						
Yes	135 (50)	28 (56)	58 (58)	42 (60)	7 (14)	0.4023

(*continued*)

Table 2. (*continued*)

	Total N = 270	Working Professionals n = 50	Students n = 100	Housewives n = 70	Self-employed/small business n = 50	p-value
No	135 (50)	22 (44)	42 (42)	28 (40)	43 (86)	
Preference to purchase Fortified food knowingly						
Yes	136 (50.3)	30 (60)	54 (54)	45 (64.2)	7 (14)	0.4164
No	134 (49.6)	20 (40)	46 (46)	25 (35.7)	43 (86)	
Awareness about food fortification logo in India						
Yes	119 (44.0)	24 (48)	51 (51)	38 (54.2)	6 (12)	0.3426
No	151 (55.9)	26 (52)	49 (49)	32 (45.7)	44 (88)	

*P-value for differences in frequencies between the groups
**$p < 0.05$

51.8% of respondents are aware of the availability of fortified food in the market. 48.15 of the respondents are not aware of the availability of fortified food. Amongst the respondents, housewives have the highest level of awareness (70%), followed by students (60%) and working professionals (52%). A significant difference is found between the groups regarding the awareness of the availability of fortified food in the market ($p \geq 0.05$. Whereas the $p = 1$ while comparing the working professional group with the self-employed group. There is a difference between the awareness about the availability of fortified food among working professionals and the self-employed group of respondents.

When speaking about the awareness of health benefits gained by consuming fortified food the response is 50–50. Awareness about the benefits is found to be highest amongst housewives (60%), followed by students (56%) and working Professionals (56%). Awareness about the benefits of fortified food is found to be the least among self-employed and owners of small businesses (10%). However, no significant difference is observed ($p \leq 0.05$) among the groups regarding awareness of the health benefits of fortified food. However, a significant difference is observed between working professionals and self-employed groups ($p \leq 0.05$). There is a difference between the awareness about the health benefits of fortified food among working professionals and the self-employed group of respondents. $P = 1$ between these groups.

50.3% of the respondents preferred fortified food while shopping for food/ groceries. Amongst these 64.2% are housewives, 60% are working professionals, 54% are students and 14% are self-employed. No significant difference is observed among the groups for preference to purchase fortified food ($p \leq 0.05$). There is a difference between the purchase behaviour of fortified food among working professionals and a self-employed group of respondents. $P = 1$ between these groups.

Overall only 44% of respondents are aware of the logo of fortified food in India. 66% of the respondents are not aware of the food fortification symbol. Awareness is highest among housewives (54.2%) followed by students (51%), working professionals (48%) and self-employed (12%). No significant difference is observed among the groups regarding the awareness of the logo of fortified food ($p \leq 0.05$).

5 Summary and Conclusion

Housewives are found to have the highest awareness about the availability and benefits of fortified food. They are also the highest purchasers of fortified food amongst another group of respondents. A good number of working professionals and college students are also well informed about the availability and benefits of fortified food and also prefer to purchase based on fortification. The awareness level about fortification amongst self-employed/ small business owners is found to be less. Hence the majority of them do not look for buying fortified food.

References

Olson, R., Gavin-Smith, B., Ferraboschi, C., Kraemer, K.: Food fortification: the advantages, disadvantages and lessons from sight and life programs. Nutrients **13**(4) (2021)

Kotecha, P.V.: Micronutrient malnutrition in India: let us say "no" to it now. Indian J. Commun. Med. **33**(1), 9 (2008)

Das, J.K., Salam, R.A., Kumar, R., Bhutta, Z.A.: Micronutrient fortification of food and its impact on woman and child health: a systematic review. Syst. Rev. **2**(1) (2013)

Pambo, K.O.: Analysis of consumer awareness and preferences for fortified sugar in Kenya. Thesis, University of Nairobi Research Archive (2013)

Calvo, M.S., Whiting, S.J.: Survey of current vitamin D food fortification practices in the United States and Canada. J. Steroid Biochem. Mol. Biol. **136**, 211–213 (2013)

Clark, B., Hill, T., Hubbard, C.: Consumers' perception of vitamin D and fortified foods. British Food J. **121**(9), 2205–2218 (2019)

Mabaya, E., Jordaan, D., Malope, P., Monkhei, M., Jackson, J.: Attribute preferences and willingness to pay for fortified cereal foods in Botswana. Sabinet Afr. J. **49**(41), 459–483 (2010)

Pounis, G.D., et al.: Consumer perception and use of iron fortified foods is associated with their knowledge and understanding of nutritional issues. Food Qual. Prefer. **22**(7), 683–688 (2011)

Meier, C., et al.: Are non-farming consumers willing to pay "a good market price" for iron-biofortified finger millet? Evidence from experimental auctions in Karnataka, India. Emerald insight, 2044–0839 (2020)

World Health Organization and Food and Agriculture Organization of the United Nations Guidelines on food fortification with micronutrients. Wayback Machine (2006). https://www.who.int/health-topics/micronutrients#tab=tab_1. Accessed 26 Dec 2016

https://www.cdc.gov/nutrition/micronutrient-malnutrition/micronutrients/index.html

https://www.webmd.com/diet/what-to-know-about-micronutrients#:~:text=These%20fruits%2C%20vegetables%2C%20meats%2C,greens%2C%20fish%2C%20and%20lean%20meats

https://www.thehmt.com/importance-fortification-fortified-foods-combat-global-burden-disease/

Economic and Social Challenges of Dialysis During the COVID-19 Pandemic

Jaspreet Kaur[1](✉) and C. H. Madhavi Latha[2]

[1] Department of Management, Kristu Jayanti College, Autonomous, Bengaluru 560048, Karnataka, India
jaspreetkaur@kristujayanti.com

[2] Department of Professional Accounting and Finance, Kristu Jayanti College, Autonomous, Bengaluru 560048, Karnataka, India
madhavi@kristujayanti.com

Abstract. Covid-19 is reported to have originated in Wuhan, China in December 2019. It slowly spread through the world causing global health problems. Having a high transmission rate the infection spread across the population affecting the most vulnerable. The worst affected world is those who had high-risk factors such as age hypertension, diabetes, chronic respiratory diseases, cancer, cardiovascular diseases and the like. Patients with end-stage renal disease were hit hard. These patients world vulnerable to covid-19 they are older and also have comorbidity. Their immune system is weak which makes them more vulnerable to infection. Moreover, patients with end-stage kidney disease need to visit the dialysis centres three times a week for around 4 h. Exposure to the dialysis centre made them more exposed to the infection. Many patients are treated simultaneously at these dialysis centres. The non-availability of an antiviral drug for covid-19 it makes more important to prevent the disease. The infection can be prevented by limiting exposure to the infected areas. Many guidelines have been issued by various bodies for the prevention and containment of the disease in these hemodialysis centres. European dialysis working group has published guidelines to prevent the spread of infection in hemodialysis centres. Despite the availability of vaccines, the immune system in the patient having end-stage kidney failure diseases responds poorly to the vaccine. Poor Immunity of dialysis Patients and the emergence of a variant of SARS-COV-19 calls for a booster dose in all the patients undergoing dialysis. Due to Covid-19 stress levels and anxiety levels among the patients increased. However, not much information is available on the mental health of the patients undergoing Haemodialysis during the Pandemic. This study is carried out to study the economic impact of the COVID-19 outbreak on patients undergoing haemodialysis in the city of Bengaluru, India.

Keywords: End-stage renal disease · Haemodialysis · COVID-19 outbreak · Economic challenges of dialysis

1 Introduction

The spread of SARS covid infection is very quick among dialysis patients. Studies suggest a fourfold increase in mortality among patients on dialysis as compared to other

B. Alareeni et al. (Eds.): ICBT 2022, LNNS 620, pp. 17–25, 2023.
https://doi.org/10.1007/978-3-031-26953-0_3

patients. These patients have comorbidities like hypertension diabetes obesity old age cardiovascular diseases etc. Also, so many patients belong to low social economic status. These problems result in a low-grade immune system. Patients on dialysis have a low time gap between the appearance of symptoms and death. It was suggestive of a lack of infection control during the early phase of the disease. 10–50% of the patients on dialysis had asymptomatic infections. The emergence of variants of the virus is characterized by increased transmissibility also it escapes the acquired immune responses. VOCs-Variants of concern are B.1.1.7 (Alpha), B.1.351 (Beta), P.1 (Gamma), and B.1.617.2 (Delta). Omi Karan is the most recent of these variants and has a high potential rate of infection. Not much data is available on the long-term consequences of covid-19 in patients on dialysis. Symptoms of covid-19 include fatigue symptoms include fatigue, dyspnea, cardiac involvement, muscle ache, headache, joint pain, or neuropsychological disorders new. These symptoms are those which are common among patients on dialysis hence the prevalence of covid-19 is difficult to assess among patients. Most of the 2 to 3 million patients are treated with dialysis worldwide, (Lancet 2017). A nephrologist is the main care provider for such kidney dialysis patients they depend upon dialysis for their survival. There is a Surge in infection due to the non-maintenance of social distancing. Mortality among patients receiving dialysis who have COVID-19 was approximately 20%, (Weiner 2020). This mortality figure is confirmed by Jager et al. (2020)., 7 who, reporting from the European Renal Association—European Dialysis and Transplant Association (ERA-EDTA) Registry, note an approximately 20% mortality rate due to COVID-19 among both patients receiving dialysis and kidney transplant recipients, a rate that is dramatically higher than the estimated 4% mortality rate overall in Europe among people diagnosed with COVID-19 (Hsu 2020).

2 Review of Literature

Due to the outbreak of Covid-19, there is a decline in Non-Covid-19 care units. This has impacted health services. The government imposed Lockdown policies limiting outpatient visits to the hospitals. Even after the relaxation by the government, the outpatient count did not reach the original level. Social distancing proved helpful in controlling the pandemic. However, which policies can be used by the government to produce social distance at the lowest costs are not clearly stated. The nations are trying to mitigate the spread of COVID-19 to save the healthcare system from being overwhelmed with increasing cases and also a slower spread of COVID could help to save lives, Gupta et al. (2020).

The disease is primarily transmitted through social interactions. To protect the patients with ESKD and the staff of the dialysis center, it is important to take safety measures. Social Distancing alone reduced mortality by 70% Gregor et al. (2020). Patients undergoing haemodialysis are at risk of developing further complications if they contract COVID-19. If patients are given two sessions of HD instead of three, there is no effect on them... This is an important way to reduce contact with others and stay protected from COVID-19 Lodge et al. (2020).

A study co-authored by Indiana University researchers indicates that the use of non-COVID-19 care declined during this period as people either deferred or skipped care, which may have important implications for their current and future health, Simon (2020). Patients at End stage renal disease are highly dependent on dialysis three times a week. It is a non-elective service as dialysis is a lifesaver in case of renal failure. Dialysis can not be delayed whether a patient is infected with Covid-19 or not. Dialysis Patient infected with Covid-19 is a high risk for other patients on dialysis and healthcare personnel. Patients, due to age and other health-related factors are at high risk of death related to Covid-19. The median age of patients on dialysis is 62 years, National Center for Immunization and Respiratory Diseases March 24 (2021). Another problem which was faced by ESKD patients infected with COVID-19 is circuit clotting. Continuous kidney replacement therapy CKRT circuit clotting is more common in COVID-19 patients. Routine use of anticoagulation amongst COVID-19 patients should be considered, Khoo et al. (2021). Patients with COVID-19 are frail as compared to patients without COVID-19. ESKD patients have a high mortality rate from COVID-19. This is because most of the patients on dialysis are of a higher age group and have co-morbidities. The need to attend frequent HD sessions exposes these patients to a high risk of contracting COVID-19. Real-time reverse transcription polymerase chain reaction (RT-PCR) is used for the diagnosis of COVID-19. The accuracy of the test depends upon the sufficiency of the sample via respiratory track or missing the window period of viral replication, Wickens et al. (2021). COVID-19 patients with ESRD had high mortality as compared to those without ESKD. Also, older patients have a high mortality rate, Rastad (2021). Due to the onset of the COVID-19 pandemic, the number of transplants has dropped considerably, increasing the risk for patients, Tuschen et al. (2021). Patients need to stay on dialysis during this period. It is important to stay shielded from COVID infection, Andersen et al. (2021). ESKD patients additionally experience fatigue, depression and reduced sleep quality during the COVID-19 pandemic, Naamani et al. (2021).

CKD and dialysis patients are at high risk of COVID-19 infection and poor outcomes. Mitigation strategies to reduce rates of infection in this population remain essential. Whether COVID-19 will increase the risk of CKD long-term and increase the demand for maintenance dialysis needs to be observed and investigated further. The cumulative reports of long-standing post-infectious symptoms and lingering organ damage after COVID-19 suggest that this will be important to monitor also in the dialysis population, Smolander et al. (2021). The impact of COVID-19 on patients with end-stage kidney disease (ESKD) on dialysis is substantial. Dialysis patients are especially vulnerable to COVID-19 because of their significant comorbidities, impaired immune function, and frequent face-to-face interactions as part of their life-sustaining therapy. Consistent with this premise, dialysis units are prone to COVID-19 outbreaks, and ESKD patients with COVID-19 experience higher morbidity and mortality. John M. Conlyeneral population, with a reported case fatality rate of 20% to 30%, Qirjazi et.al. (2021).

3 Research Methodology

Primary data is collected from 11 dialysis centres in Bangalore. The study is conducted to study the impact of the Covid-19 pandemic on the patients availing of haemodialysis services in Bangalore. The data is collected from the centres to check the economic and social impact of the pandemic and lockdown imposed by the government, on availing the services. Data collected, is analysed to measure its impact on end-stage kidney disease patients availing dialysis.

3.1 Statement of the Problem

During the outbreak of the COVID-19 Pandemic, patients with end-stage renal disease were hit hard. They experienced social and economic trauma, like many others. This study is conducted to measure the economic and social impact of COVID-19 on these patients. The study focuses on the rise in the basic cost of dialysis services, the increase in spending on dialysis due to additional cost of safety equipment like PPE kits, gloves, masks etc., and added costs due to mandatory RTPCR tests as suggested by many hospitals/dialysis centres.

3.2 Objectives

1. To study the price rise of Dialysis during Covid-19
2. To study the impact of Covid-19 on Dialysis treatment.

4 Data Analysis and Interpretation

Patients visiting every month for Haemodialysis before and during Pandemic (Table 1).

Table 1. Additional charges levied for safety equipment per session

Number of Patients visiting dialysis centres before Pandemic	Several hospitals responded	Percentage	Number of Patients visiting during Pandemic	Percentage	Additional charges levied for safety equipment per session	Hospitals responded	Percentage
50–100	3	27%	5	45.4%	None	5	45.4%
100–150	2	18%	1	9%	250–500 Rs.	3	27%
150–200	2	18%	2	18%	500–750 Rs.	1	9%
200–250	1	9%	1	9%	750–1000 Rs.	2	18%
250–300	1	9%	2	18%			
More than 300	2	18%	0	0%			

Source: Primary data

Out of 11 hospitals, eight of them made RTPCR Test Mandatory for patients. Three of them did not emphasize the test. 27.27% of respondents did not emphasize the test. A majority of hospitals favoured RTPCR. It was mandatory for 72.72% of respondents. By making the diagnosis test mandatory, these hospitals can screen patients who contracted Covid-19 infections. This becomes the first step towards the top of the spread of infection. Additional charges are levied for safety equipment per session. About 45% of the hospitals/Centres did not charge any extra amount from the patients for safety

Table 2. RTPCR requirement

RTPCR requirement	Several hospitals responded	Percentage
Yes	8	72.72%
No	3	27.27%

equipment. 27% of respondents charged Rs. 250–500 extra per session of dialysis, 9% charged about Rs. 500–750 extra for each session and 18% of respondents charged Rs. 750–1000 extra from the patients. Percentage of Patients contacting Covid-19/Patients missing more than one dialysis during the lockdown. A dip in the number of patients visiting the hospitals is seen in some of the hospitals. This dip is because some centres did not entertain patients with Covid-19 infection. These patients moved to hospitals that were open to Covid-19 infected patients (Table 2).

Table 3. Patient information during pandemic

Percentage of Patients contacting Covid -19	Hospitals responded	Percentage	How many patients missed at least one dialysis during the lockdown	Hospitals responded	Percentage	How many patients missed more than one dialysis during the lockdown	Number of Hospitals responded	Percentage
Up to 20%	9	81.8%	0–10%	9	81.8%	0–10%	9	81.8%
20–40%	3	27.2%	10–20%	1	9%	10–20%	1	9%
40–60%	0	0%	20–30%	1	9%	20–30%	1	9%
60–80%	0	0%						
80–100%	0	0%						

Source: Primary data

An average of 1065 patients visited these hospitals during the Pandemic. Less than 20% of the patients visiting the hospitals during the Pandemic contracted Covid-19 as responded by 9 respondents. As to the three respondents, the infection rate was up to 40% among haemodialysis patients. In about 81.81% of the respondents, less than 10% of patients missed more than one dialysis during the weeks of lockdown. As per 9% of respondents less than 20% of patients missed more than one dialysis and another 9% responded that less than 30% of respondents missed the same (Table 3).

81.81% of the respondents responded that less than 10% of their patients missed more than one dialysis during the lockdown. As per another 9%, about 20% and 30% of their patients did not have access to more than one dialysis during the lockdown. A small percentage of haemodialysis patients missed around two dialysis a week due to lockdown situations. The majority of the patients had access to dialysis centres during the lockdown. A small number of patients missed around one dialysis during the lockdown. Dialysis centres were kept open during the lockdown. There has been a reduced number of patients visiting the hospitals during the pandemic, especially during the lockdown. One of the reasons for this reduction in the number in some hospitals is that some centres did not provide dialysis service to the patients infected with COVID- 19 (Table 4).

Table 4. Sanitization measures/dialysis services provided to COVID-19 patients during pandemic

Additional sanitizing measures were taken by the dialysis centre to prevent Covid-19	Percentage	Safety equipment like gloves, masks, PPE kits etc.was used by the dialysis centre to prevent Covid-19 from spreading	Percentage	Safety equipment like gloves, masks, PPE kits etc.was made mandatory for patients to prevent Covid-19 from spreading	Percentage
10	90.9%	11	100%	6	54.5%
1	9%	0	0%	5	45.45%

Source: Primary data

To Prevent the spread of Covid-19 infection additional measures like sanitizing the centres were taken up by the centres. Equipment, like PPE kits, gloves, and masks were made mandatory for the staff dealing with haemodialysis patients. Around 55% of the hospitals had made it mandatory for patients to make use of Masks, PPE kits etc. Haemodialysis was performed at isolated units for patients infected with Covid-19 (Table 5).

Table 5. Dialysis services provided to COVID-19 patients during pandemic

Isolation units used for performing dialysis on patients infected with Covid-19	Percentage	Increase in the cost of haemodialysis during the Pandemic	Percentage
11	100%	9	81.8%
0	0%	2	18.1%

Source: Primary data

7 out of 11 respondents confirmed that Dialysis services were not available for COVID- 19 Patients during the Pandemic. Those centres that provided services to haemodialysis Patients, did so in the isolation wards. 81.8% of the hospitals in the city did not increase the basic cost of haemodialysis. 18% of the hospitals increased basic dialysis costs. This increase in the cost of haemodialysis was in the range of Rs.500 per dialysis during the lockdown.

5 Results and Discussion

72.7% of hospitals made RTPCR tests mandatory for patients. More than a quarter did not emphasize the test. The majority of the hospitals used this method to screen patients with Covid-19 infection, hence the first step towards prevention of the spread of infection. There has been a reduced number of patients visiting the hospitals during the pandemic, especially during the lockdown. One of the reasons for this reduction in the number in some hospitals is that some centres did not provide dialysis service to the patients infected with COVID- 19. Less than 20% of the patients visiting the hospitals during the Pandemic contracted Covid-19 as responded by 81.8% of respondents. As per 18% of respondents, the infection rate was up to 40% among haemodialysis patients. 63.6% of respondents confirmed that dialysis services were not available for COVID- 19 patients during the Pandemic. Those centres that provided services to haemodialysis patients, did so in the isolation wards. The majority of the patients had access to dialysis centres during the lockdown. Less than 10% of patients missed at least one dialysis during the lockdown. Dialysis centres were kept open during the lockdown. A small percentage of haemodialysis patients missed around two dialysis a week due to lockdown situations. This number was less than 10%. To prevent the spread of Covid-19 infection, additional measures like sanitizing the centres were taken up by the centres. Equipment, like PPE kits, gloves, and masks were made mandatory for the staff dealing with haemodialysis patients. Around 55% of the hospitals had made it mandatory for patients to make use of masks, PPE kits etc. Haemodialysis was performed at isolated units for patients infected with Covid-19. This ensured the safety of other patients and staff. The majority of the hospitals in the city did not increase the basic cost of haemodialysis. Less than a fifth of hospitals studied increased the basic cost of haemodialysis. The increase in the cost of haemodialysis was in the range of Rs.500 per dialysis during the lockdown. About 45% of the hospitals/Centres did not charge any extra amount from the patients for safety equipment. 27% of respondents charged Rs. 250–500 extra per session of dialysis, 9% charged about Rs. 500–750 extra for each session and 18% of respondents charged Rs. 750–1000 extra from the patients.

6 Summary and Conclusion

More than a quarter of the hospitals/dialysis centres did not make RTPCR tests mandatory for patients. Making this test a must for all patients can prove helpful in checking the spread of infection.COVID-19 affected the entire society. Those with End stage kidney failure were hit hard as they were the ones with co-morbidities. Many patients on dialysis missed out on dialysis during the lockdown. Though the majority of the hospitals/Centres did not increase the basic cost of Haemodialysis, still, there was an increase in the cost for patients in the form of increased charges towards safety equipment like PPE kits, gloves, masks etc., mandatory RTPCR, haemodialysis charges (in some cases). Moreover, many Dialysis centres/Hospitals did not provide services to COVID-19-positive patients. A formal setup keeping in mind the plight of dialysis patients can be enhanced to meet their requirements in challenging times.

References

Cockwell, P., Fisher, L.-A.: The global burden of chronic kidney disease. Lancet **395**(10225), 662–664 (2020)

Hsu, C.M.: COVID-19 in dialysis patients: outlasting and outsmarting a pandemic. National Libr. Med. **98**(6), 1402–1404 (2020)

Gupta, S., Simon, K.I.: Mandated and voluntary social distancing during the Covid-19 epidemic: a review (2020)

Lodge, M.D.S. Abeygunaratne, T.: Safely reducing haemodialysis frequency during the COVID-19 pandemic. National Libr. Med. **21**(1) (2020)

Da Silva, M., Lodge, T.A., et al.: Safely reducing haemodialysis frequency during the COVID-19 pandemic. BMC Nephrol. **21**, 532 (2020). https://doi.org/10.1186/s12882-020-02172-2

Lim, R.S., Goh, S.M., Yeo, S.C.: Renal outcomes in immunoglobulin a nephropathy following COVID-19 vaccination: a retrospective cohort study (2022)

Wickens, O., Chinnadurai, R., et al.: Investigating the utility of COVID-19 antibody testing in end-stage renal disease patients receiving haemodialysis: a cohort study in the United Kingdom. Litterature scientifique international sur la maladie a coronavirus 2019, **22**(1), 154 (2021)

Rastad, H.: Risk and predictors of in-hospital mortality from COVID-19 in patients with diabetes and cardiovascular disease. Food Agric. Organ. United Nations **1**(12), 261–272 (2020)

Tuschen, K., et al.: Renal transplantation after recovery from COVID-19 - a case report with implications for transplant programs in the face of the ongoing corona-pandemic. National Libr. Med. **22**(1), 251 (2021)

Al Naamani, Z., Gormley, K., Noble, H.: Fatigue, anxiety, depression and sleep quality in patients undergoing haemodialysis. Re. Gate **22**(1) (2021)

Iqbal, S., Iqbal, A., Blair, K.A.A., et al.: Challenges faced by the patients on dialysis treatment in COVID-19 era and the possible solutions. Biomed. J. Sci. Techn. Res. **36**, 28279–28282 (2021)

Clin Invest, J.: Kidney diseases in the time of COVID-19: major challenges to patient care. J. Clin. Investig. **130**(6), 2749–2751 (2020)

Geetha, S., Guganathan, M., Giridharan, G., et al.: Challenges faced by nursing and dialysis staffs during COVID-19 pandemic - tanker foundation a model organization. J. Biomed. Res. **2**(6), 529–531 (2021)

Suri, R.S., Antonsen, J.E., et. al.: Management of outpatient hemodialysis during the COVID-19 pandemic: recommendations from the canadian society of nephrology COVID-19 rapid response team. Can. J. Kidney Health Dis. (2020)

Al Amin, S., Morrison, S.D., et al.: Challenges for Non-COVID patients with chronic kidney disease in Bangladesh: an observation during coronavirus disease pandemic, INQUIRY. J. Health Care Organ. Provision Financing 58 (2021)

Yu, X.,·Jha, V., et al.: Should more patients with kidney failure bring treatment home? What we have learned from COVID-19 (2022)

Malo, M.F., Affdal, A., et al.: Lived experiences of patients receiving hemodialysis during the COVID-19 pandemic: a qualitative study from the Quebec renal network. Kidney360, **3**(6), 1057–1064 (2022)

Kliger, A.S., Silberzweig, J.: COVID-19 and dialysis patients: unsolved problems in early 2021. J. Am. Soc. Nephrol. **32**(5), 1018–1020 (2021)

Chan, A.S.W., Ho, J.M.C., Li, J.S.F., et al.: Impacts of COVID-19 Pandemic on psychological well-being of older chronic kidney disease patients. Frontiers **8**, 666973 (2021)

Mahalingasivam, V., Su, G., Iwagami, M., Davids, M.R., Wetmore, J.B., Nitsch, D.: COVID-19 and kidney disease: insights from epidemiology to inform clinical practice. Nat. Rev. Nephrol. **18**, 485–498 (2022)

Alkhunaizi, A.M., Al-Mueilo, S.H., Atiyah, M.M., Alnasrallah, B.: Hemodialysis during lockdown due to coronavirus disease 2019 pandemic in Eastern Saudi Arabia. Saudi J. Kidney Dis. **32**(3), 794–797 (2021)

Simon, K.: Non-COVID-19 health care visits declined dramatically as pandemic hit. News at IU Bloomington (2020). https://news.iu.edu/stories/2020/08/iub/releases/03-non-covid-19-health-care-declines-during-pandemic.html

Weiner, D.E., Watnick, S.G.: Hemodialysis and COVID-19: an Achilles' heel in the pandemic health care response in the United States. Kidney Med. **2**(3), 227–230 (2020)

Jager, K.J., et al.: Results from the ERA-EDTA Registry indicate a high mortality due to COVID-19 in dialysis patients and kidney transplant recipients across Europe. Kidney Int. **98**(6), 1540–1548 (2020). https://doi.org/10.1016/j.kint.2020.09.006

Qirjazi, E., et al.: SARS-COV-2 shedding in dialysis patients with covid-19. Kidney Inte. Rep. (2021). https://www.ncbi.nlm.nih.gov/pmc/articles/PMC8393503/. Accessed 16 Feb 2023

Features of the Selection of Foreign Securities for Investment Activities

P. Reznik Nadiia[1(✉)], V. Dolynkyi Serhii[2], V. Miroshnychenko Oleksandr[3], Alieksieiev Ihor[4], Yarmoliuk Anatoliy[1], and Svitlyshyn Ihor[5]

[1] Department of Management, National University of Life and Environmental Sciences of Ukraine, Kyiv, Ukraine
nadya-reznik@ukr.net
[2] Department of Economics and Management, Carpathian Institute of Entrepreneurship, Open International University of Human Development Ukraine, Khust, Ukraine
[3] National University of Life and Environmental Sciences of Ukraine, Kyiv, Ukraine
[4] Department Finance, Lviv Polytechnik National University, Lviv, Ukraine
ihor.v.alieksieiev@lpnu.ua
[5] Department of Management, Business and Marketing Technologies, Zhytomyr Polytechnic State University, Zhytomyr, Ukraine

Abstract. The article examines the modern investment policy of Ukraine, as well as the regulation of investment activity. Such an important issue as the state's guarantee of investment protection regardless of the forms of ownership, as well as foreign investment, is indicated. Illuminated ways of regulating the conditions of investment activities. Some types of financial are considered investment and their regulation by the state.

Currently, there are not enough investment tools in Ukraine. Recently popular methods of saving money either no longer work properly (deposits), or are too expensive for the majority of the population (real estate), or are little known (US stocks). Today, the process of investing in foreign shares is quite complicated, and therefore inaccessible to a wide range of consumers. In addition, the threshold for entering international investment markets is too high for the average Ukrainian.

Measures are certain on the increase of interest of investors to the fund market of Ukraine. The analysis of above all tasks of development of internal imperious investors is conducted.

Article reveals the essence of investments, the main ways of attracting capital, the main measures to stimulate investment. In macroeconomic policy, emphasis is placed on creating prerequisites for investment growth – weakening inflation, ensuring optimal interest rates on deposits and investments, reducing interest rates on loans, etc.

Keywords: Regulation of investment activities · Corporate rights · Guarantees · Securities market · Foreign investments

1 Introduction

Selecting shares for investment is a responsible process, which may seem like a game of chance only at first glance. In general, the process of investing is the purchase of

B. Alareeni et al. (Eds.): ICBT 2022, LNNS 620, pp. 26–34, 2023.
https://doi.org/10.1007/978-3-031-26953-0_4

financial instruments with the aim of long-term ownership of them. Famous investor Warren Buffett said: «Buy stocks only if you are willing to hold them for 10 years or more. If you are not sure about the company's prospects for the next decade, you should not spend your capital».

A practical recommendation for establishing criteria for selecting securities for one's investment portfolio, in particular for investors operating on the foreign stock market.

In the conditions of the market transformation of the economy of Ukraine, the need for significant foreign investments is very acute. For most countries with a transition economy, effectively used foreign capital becomes a key factor in their development. Of course, the attraction of foreign investments plays an important role in the structure of priorities of the Ukrainian economy.

The current state of economic development requires an active policy of attracting foreign direct investment. In Ukraine, there is a legislative framework in the field of regulation of investment activities, which is gradually being improved with the aim of achieving a greater inflow of foreign investments and increasing the efficiency of their use.

Today, the state of production, the level of technical equipment of enterprises of the national economy, the ability of structured restructuring of the economy, and the solution of social and environmental problems depend on the effectiveness of the investment policy.

As long as innovative tools are at the stage of development, the interest of Ukrainians in investments is growing. Those who do not want to wait for launches can already work directly with foreign brokers. Here, after signing the contract, you need to withdraw money abroad through the SWIFT system within the framework of e-limits and carry out transactions with securities already there.

This method is not cheap and easy, and for the long-term investor it has more disadvantages than advantages. In the case of small amounts of investment, the cost of a SWIFT transfer makes it economically unreasonable. In addition, there are high inheritance taxes here, and the investor has to report to the Ukrainian tax office on his own.

Another way of investing in foreign securities, which you can use right now, is cooperation with Ukrainian securities traders (both brokers and banks with the appropriate license can act as brokers). A trader can provide the opportunity for Ukrainians to invest in foreign shares here in Ukraine in two ways:

Introduction of foreign securities to Ukraine with their further sale in the domestic jurisdiction. In this case, all further operations with shares of foreign companies will take place by analogy with Ukrainian assets. Due to the creation by a Ukrainian trader of an omnibus account abroad with a foreign broker.

It is about the fact that the right to the client's assets, which are kept in the investment firm, is confirmed by the Ukrainian broker, and the right to them to the Ukrainian broker is confirmed by the foreign investment firm. The longer the chain, the greater the risk that one of the links will break. In the event that something happens to a Ukrainian or foreign broker, it will become more difficult to prove ownership of your securities abroad.

2 Materials and Methods

State regulation of investment activity includes management of state investments, as well as regulation of the conditions of investment activity and control over its implementation by all investors and participants of investment activity. Investment protection is a set of organizational, technical and legal measures aimed at creating conditions that contribute to the preservation of investments, the achievement of the goal of investment, the effective operation of investment and reinvestment facilities, the protection of legal rights and interests, including the right to receive profit (income) from investments. The state guarantees the protection of investments regardless of the forms of ownership, as well as foreign investments.

Economists H. Markowitz, J. Keynes, R. Piotroski, R. Shiller, U. Sharp devoted their works to the study of investment activity and related processes. Contribution to the study of problems related to investment activity was also made by domestic scientists: I.A. Blank, V.M. Gridasov, A.A. Peresada, N.P. Reznik, V.G. Fedorenko, O.M. Tsarenko.

3 Results and Discussion

Based on the studied sources and studies, we have identified the main criteria by which stocks should be selected for the investment portfolio:

1) presence of a trend;
2) financial stability of companies;
3) liquidity of shares;
4) historical trends;
5) correct allocation of capital.

The presence of a trend can be monitored using Bollinger lines. This indicator is based on one of the basic indicators of technical analysis - the average moving price for a specific period of time. After constructing the moving average, it is necessary to calculate the standard deviation of quotes from its moving average for a specific period of time. It is worth noting that 1 standard deviation is the range in which the price was approximately 2/3 or 68% of the time of the analyzed period. Bollinger lines are the magnitude of deviations from the moving average up and down. For example, I took a chart of the broad market index S&P500 for the last 5 years and recreated the indicator myself. If the price goes beyond the upper line, this indicates a strong upward trend in the stock. If the price falls below the lower deviation, this indicates a bearish trend. If the price is in a channel between two deviations, this indicates a neutral trend and high uncertainty. Of course, as you can see on the chart, Bollinger does not work with 100% probability. As with any technical indicator, it has its shortcomings, but in combination with other factors, it is a good tool for determining the presence or absence of a price trend (Fig. 1).

Fig. 1. Line graph of the S&P500 index

Financial stability of the company. To determine financial stability, I borrowed one of the criteria for selecting stocks according to the S&P500 index, which is conducted quarterly by S&P Global Inc. The company must show a total net profit for the last 4 quarters, and the last quarter must be profitable [1]. This stock selection policy looks for a company that has proven financial results as well as successful operations. Recently, the Tesla company reported for the 2nd quarter of 2020 and, despite the loss-making

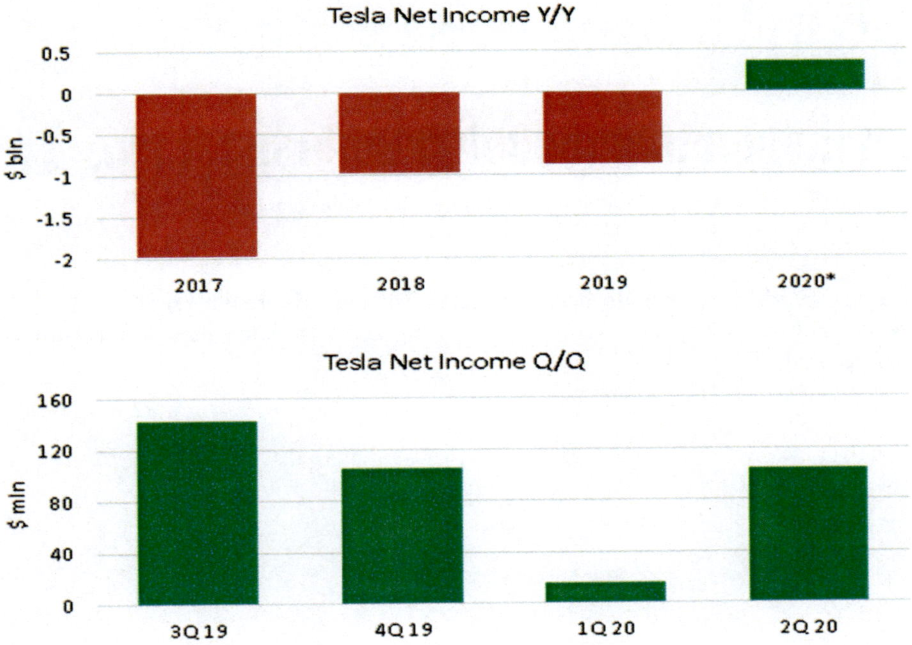

Fig. 2. Tesla's net income

previous years, showed growth for the last 4 quarters [2]. In protest, the committee decided not to include the company in the index. However, that doesn't convey the fact that the company will finally enter its first profitable year, and that quality management will continue to fuel investor appetite for Tesla stock (Fig. 2).

Liquidity of shares. In fact, liquidity is one of the most underestimated criteria, especially among lay investors. In simple words, it is the ability to quickly buy or sell a security with a minimal difference in price. The greater the number of trading participants, the greater the liquidity of the financial instrument [7]. One of the liquidity indicators trade volumes appear. If you look at the shares of McDonalds, you can see that the average daily trading volume of approximately 2.5 million contracts per 1 trading session is a high indicator that allows the investor to be more mobile and flexible in terms of managing his capital [4]. In case of urgent need, he can always immediately sell his shares and release the required amount of funds with minimal commission costs (Fig. 3).

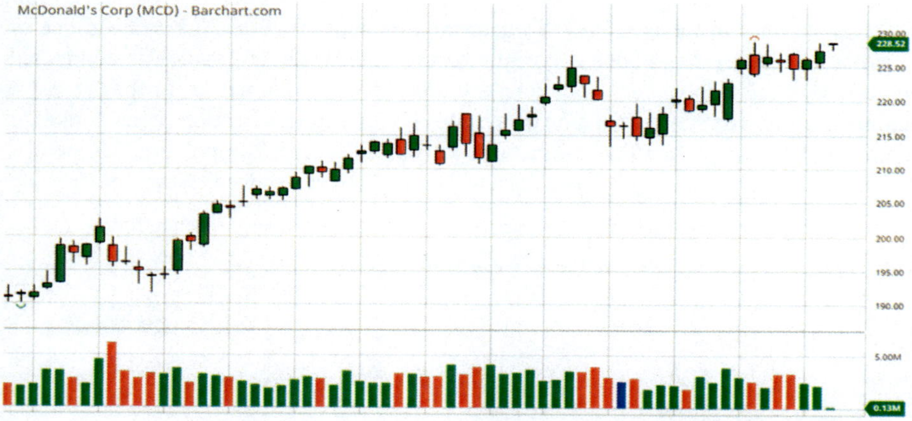

Fig. 3. Festive schedule of the McDonalds company

Historical trends. It is generally accepted that seasonal trends are inherent only in commodity markets, where, for example, the sowing and harvesting periods are clearly separated (Fig. 4).

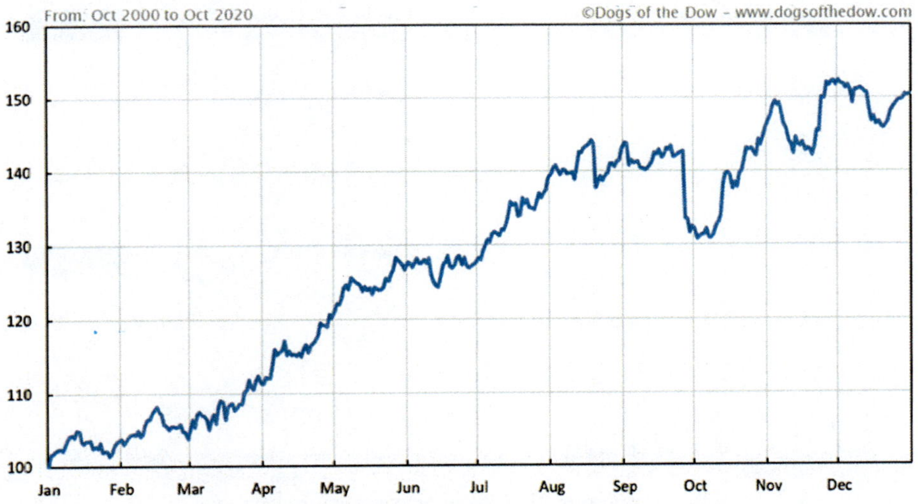

Fig. 4. Apple's seasonal schedule

However, there are seasonal trends in the financial markets, including the stock market. When choosing stocks for your investment portfolio, you should pay attention to how it behaved in the past, which periods were mostly unprofitable, and which were the most profitable. Reporting periods, vacations, New Year's holidays, and other corporate events – all this is reflected in stock quotes and repeats with a certain frequency. For example, consider the seasonal chart of Apple shares over the past 20 years. In general, a pronounced upward trend, however individual months from year to year show corrections and weakening of quotations. This can be important when choosing an entry point or an exit period [4].

Distribution of capital. One of the main rules in the formation of an investment portfolio is diversification, that is, the distribution of capital among several low-correlated instruments. This makes it possible to even out the yield curve of the portfolio, when during the decline of one asset, the other shows growth and vice versa [5]. For a clear example, the article shows how sectors of the US economy behaved during the sharp fall of the stock market in 2020. The economy came to a standstill during the March crisis, but the technology sector has fully recovered as of October and even showed year-to-date growth, the financial sector is somewhat weaker, although it has recovered from the crisis. Companies in the energy sector (mostly oil and gas corporations) have lost 50% of their value since the beginning of the year. This clearly emphasizes the need for diversification. It is best to form a portfolio from shares of those sectors that are not related to each other or have a minimal number of business touch points [8] (Fig. 5).

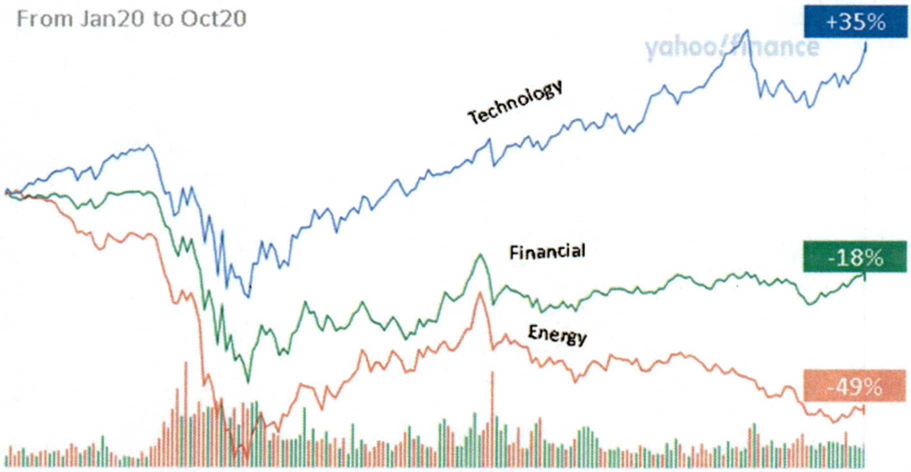

Fig. 5. Line graph of SPDR sector exchange-traded funds

For example, the technology sector and telecommunications have a lot in common, but the healthcare and energy sectors have a low correlation, which makes them more attractive in terms of capital allocation in a portfolio.

To date, in our country there is a problem of insufficient use of investment potential, which is associated with political instability, excessive state interference in investment regulation, constant changes in current legislation, the absence of a single central body in Ukraine for state investment management, insufficient development of small and medium entrepreneurship, the presence of a large number of «shadow capitals». As for foreign investment, it will have positive effects for us effects. With the help of foreign investments, it is possible to modernize the production base, create new jobs, develop important sectors of the economy, etc.

Thus, it is necessary to revise the investment policy directions of Ukraine and develop a single clear strategy for attracting foreign investments, since attracting foreign investors with the purpose of investing money in the state's economy is the main organic part of the investment policy of any country.

The development of measures to increase investors' interest in the stock market involves changing the nature of the market from an insider to an efficient and transparent one; strengthening the effectiveness of control and audit activities and law enforcement, developing measures to prevent fraud and market manipulation; development of relevant normative legal acts regulating the procedure for obtaining income by investors; improvement of corporate management of companies by issuers; ensuring information transparency in financial and economic activities of issuers.

Implementation of measures to protect the issuer's rights provides for the regulation of the processes of buying up large and controlling stakes; disclosure of information about investors.

Measures to protect the interests of issuers should to be implemented, first of all, in the following groups of issuing enterprises:

1. Enterprises of strategically important industries.
2. Enterprises recognized as monopolists.
3. Banks, joint investment institutes and insurance companies.

4 Conclusion

This is far from the entire list of factors for selecting stocks for forming one's own investment portfolio. Of course, for a more detailed familiarization and analysis of shares, it is worth using deep and point analytics, paying attention to trends in the middle of the sector, company news, fundamental indicators, price, membership in the index, the country of origin of the company, etc. [6]. However, these basic 5 criteria will be more than enough for the primary analysis and filtering of a sample of shares. Any investor will always find the following resources that have organized stock screeners or useful information on the stock market in general useful: Finviz, Yahoo Finance, Guru Focus, Trading View, Investing, Morning Star, Seeking Alpha, Investor's business daily, Wall Street Journal, Barron's.

References

1. S&P Global [Electronic resource] - Resource access mode: https://www.spglobal.com/en/
2. Yahoo! Finance [Electronic resource]. Tesla Inc.: Resource access mode (2020). https://finance.yahoo.com/quote/TSLA/financials?p=TSLA
3. Barchart [Electronic resource]. McDonalds Inc.: Resource access mode (2020). https://www.barchart.com/stocks/quotes/MCD/interactive-chart
4. Dogs of the Dow [Electronic resource]. Apple Inc.: Resource access mode (2020). https://www.dogsofthedow.com/aapl-chart-today.htm
5. Murphy, J.: Cross-market analysis. Principles of interaction of financial markets. John J. Murphy, USA, 212 p. (1991)
6. Guru focus [Electronic resource]. Stock screener: Resource access mode (2020). https://www.gurufocus.com/screener
7. Bespalov, V.M., Yu, A., Vakula, O.M., Gostrik, K.: Informatics for economists: teaching. In: Manual for Students of Higher Educational Institutions of Economic Specialties. TsUL, 788 p. (2003)
8. Islam, J.S., Islam, M.R., Islam, M., Mughal, M.A.: Economics of Sustainable Energy. Hoboken, 627 p. (2018)
9. Abuselidze, G., Reznik, N., Slobodianyk, A., Prokhorova, V.: Global financial derivatives market development and trading on the example of Ukraine. In: SHS Web of Conferences, vol. 74, p. 05001. EDP Sciences (2020). https://doi.org/10.1051/shsconf/20207405001
10. Reznik, N.P., Dolynskyi, S.V., Voloshchuk, N.Y.: Retrospective analysis of basic risk as a part futures trading in Ukraine. Int. J. Sci. Technol. Res. **9**(1), 3419–3423 (2020)
11. Reznik, N.P., Gupta, S.K., Sakovska, O.M., Ostapchuk, A.D., Demyan, Y.Y.: A research of state regulation of stock exchange in Ukraine: significance and growth for economic development. Int. J. Eng. Adv. Technol. **8**(6), 3851–3857 (2019)
12. Reznik Nadiia, P., Slobodianyk Anna, M., Blahodatnyi Andrii, S.: The use of speculative operations in the capital market and their importance. In: Alareeni, B., Hamdan, A., Elgedawy, I. (eds.) ICBT 2020. LNNS, vol. 194, pp. 1487–1495. Springer, Cham (2021). https://doi.org/10.1007/978-3-030-69221-6_110

13. Olena, A., Reznik, N., Dmytro, S.: Organizational and economic mechanism of the development of foreign economic activity of industrial production. In: Alareeni, B., Hamdan, A., Elgedawy, I. (eds.) ICBT 2020. LNNS, vol. 194, pp. 1011–1022. Springer, Cham (2021). https://doi.org/10.1007/978-3-030-69221-6_77
14. Slobodianyk, A.N., Reznik, N.P., Abuselidze, G.D.: The analysis of hedging instruments on the exchange commodity market of Ukraine. In: Bogoviz, A.V. (ed.) The Challenge of Sustainability in Agricultural Systems. LNNS, vol. 205, pp. 379–385. Springer, Cham (2021). https://doi.org/10.1007/978-3-030-73097-0_42

Efforts to Increase Core Capital for Core Capital Bank Group Base on Regulation

Nur Ellyanawati Esty Rahayu and Dessy Isfianadewi(✉)

Department of Management, Faculty of Business and Economics, Universitas Islam Indonesia, Yogyakarta, Indonesia
dessy.isfianadewi@uii.ac.id

Abstract. This study aims to analyze the efforts made by the management of the core capital bank group 1 (KBMI 1) in Indonesia to increase the bank's core capital. OJK Regulation Number 12/POJK.03/2020 concerning Consolidation of Commercial Banks in Indonesia stipulates a minimum bank core capital of IDR 3 trillion as of December 31, 2022. There are 30 commercial banks with core capital below IDR 3 trillion in 2021. This study uses the CAMEL and RGEC methods to analyze the bank's health. The study results indicate that banks with a soundness predicate will easily attract investors, while banks with a relatively soundness and unsoundness predicate must work harder to ensure that core capital is met. The bank cannot meet the OJK provisions and will be downgraded to a BPR or BPRS. Efforts to increase capital can be made by offering to the parent company, not the parent company, and looking for investors. Companies already listed on the stock exchange can issue new shares or rights issues.

Keywords: Consolidation of commercial banks · Bank soundness · CAMEL · RGEC

1 Introduction

The Financial Services Authority has determined that all commercial banks in Indonesia, by December 31, 2022, must have a minimum core capital of IDR 3 trillion. This provision is stated in OJK Regulation Number 12/POJK.03/2020, dated March 16, 2020, concerning the Consolidation of Commercial Banks, which has been effective since its promulgation on March 17, 2020. For commercial banks included in the core capital bank group (KBMI) 1, which initially had to have a core capital of IDR 1 trillion, then with the new provisions, they must increase their core capital to a minimum of IDR 3 trillion.

If, in the initial provisions, large banks are only allowed to have one sharia bank subsidiary or one combined bank, then with the issuance of POJK Regulation Number 12/POJK.03/2020, banks are given the convenience of being able to have several bank subsidiaries with a consolidation scheme through mergers, consolidation/integration, as well as the establishment of a Bank Business Group [1]. On the one hand, the issuance of this regulation will put tremendous pressure on KBMI 1, especially those who still have

B. Alareeni et al. (Eds.): ICBT 2022, LNNS 620, pp. 35–45, 2023.
https://doi.org/10.1007/978-3-031-26953-0_5

capital below IDR 3 trillion. On the other hand, immense pressure occurred because even though some of these banks have been operating for decades, it turns out that their core capital is still below IDR 3 trillion.

The bank's business performance can be seen from the published annual financial reports, and from these financial statements, it can be seen that the business performance and health of the bank can be seen. Ratios from financial statements can even be used to predict financial difficulties and company bankruptcy with the Z-Score model formula [2]. In addition, the Z-Score model can measure strengths and weaknesses and help companies have sound financial turnover [2]. In this study, only analyzing bank soundness with the CAMEL and RGEC models and the results of the predicate level of bank soundness can be used as a basis for considering future bank predictions [3] and as material for evaluating strategies that must be taken by bank management in achieving business goals and expected financial performance [2, 4]. KBMI 1 can use the health level predicate from the CAMEL and RGEC formulas to set the strategy that must be taken until December 31, 2022, to meet IDR's minimum core capital of 3 trillion.

2 Literature Review

2.1 Financial Services Authority

Otoritas Jasa Keuangan issued OJK Regulation Number 12/POJK.03/2020 dated March 16, 2020, concerning the Financial services authority, which has been in effect since its promulgation on March 17, 2020, which regulates the minimum core capital of commercial banks of IDR 3 trillion on December 31, 2022 [1]. This regulation on bank consolidation was issued as an effort to strengthen the structure, resilience, and competitiveness of the banking industry to support national economic stability and growth, as well as an attempt to encourage the banking industry to reach a more efficient level toward higher economies of scale [5, 8].

The current provisions that require a single presence policy through merger/consolidation are also inflexible and limit the Controlling Shareholder (PSP) from taking over banks to empower small banks (in large bank groups); bank takeovers in helping rescue troubled banks [1]. Through bank consolidation, it is hoped that it will create banks that can face the challenges and demands of technology-based product and service innovation so that they have more extraordinary adaptability and can respond to various challenges in global economic conditions [6]. Bank consolidation encourages national banks to be competitive in the regional and international scope [7, 9].

2.2 Bank Soundness, CAMEL, RGEC

Bank soundness interests various parties, including owners, bank management, customers, Bank Indonesia and OJK as the banking supervisory authority, and the government. The assessment of the level of the soundness of commercial banks is regulated in BI Regulation No.13/1/PBI/2011 and OJK Regulation No.4/POJK.03/2016. In addition, bank health must be maintained or improved so that public confidence in the bank is maintained [12]. Meanwhile, for investors, welfare increases if the bank's condition is soundness [13, 17].

CAMEL

Assessment or measurement of the soundness of commercial banks in Indonesia using CAMEL analysis has been used since the enactment of BI Regulation Number 6/10/PBI/2004 for commercial banks and BI Regulation Number 9/1/PBI/2007 for commercial banks based on sharia principles. CAMEL analysis also functions to assess performance and detect problems that have the potential to disrupt the smooth operation of a bank [14]. CAMEL can be used as a benchmark for determining a bank's level of soundness and performance [14]. Bank research using CAMEL was also carried out by comparing banks in Malaysia and Indonesia. The results stated that CAMEL could be used significantly to assess banks' profitability [15], and CAMEL is essential because it describes bank performance and overall bank soundness [18]. The ratios in CAMEL are as follows [19]:

Capital

The soundness level of the bank in terms of capital shows the bank's ability to operate its capital to stop the decline in assets caused by losses.

$$CAR = \frac{\text{CAPITAL}}{\text{ATMR}} \times 100\%$$

Asset

Productive asset quality describes the financial performance of banking companies.

$$\text{ASSET} = \frac{\text{Classified earning assets}}{\text{Total productive asset}} \times 100\%$$

Management

This level can be seen in the management's ability to maintain, demonstrate, measure, and control the risk of daily activities in the company.

$$\text{Management} = \frac{\text{Income Operating}}{\text{Income Net}} \times 100\%$$

Earnings

The bank's ability to earn income or profit also shows its level of health. The greater the income obtained illustrates that the bank's performance is also getting better so that its financial condition is getting healthier.

$$ROA = \frac{\text{Earning Before Tax}}{\text{Total Assets}} \times 100\%$$

$$BOPO = \frac{\text{Operation Cost}}{\text{Operation Income}} \times 100\%$$

Liquidity
This ratio measures the level of the company's ability to meet its obligations in the short term or due date (Table 1).

$$\text{Cash Ratio} = \frac{\text{Current Assets}}{\text{Current Assets}} \times 100\%$$

$$\text{LDR} = \frac{\text{Loan}}{\text{Deposit}} \times 100\%$$

Table 1. Bank rating credit value based on CAMEL method

Rating credit	Description
81–100	Soundness
66–< 81	Quite soundness
51–< 66	Less soundness
0–< 51	Not soundness

Source: Bank Indonesia, 2007.

RGEC

The provisions for assessing the health of commercial banks in addition to using CAMEL analysis, namely by using the RGEC analysis as stipulated in Bank Indonesia Regulation Number 13/1/PBI/2011 concerning the assessment of the soundness of commercial banks. The assessment indicators contained in the RGEC are as follows [20]:

Risk
The Risk Profile factor assesses the inherent risk and the quality of implementing Risk Management in the bank's operational activities.

Risk Management Components measured = Credit Risk, Market Risk, Liquidity Risk, Operational Risk, Legal Risk, Strategic Risk, Compliance Risk, Reputational Risk, Return Risk, and Investment Risk.

Good Corporate Governance (GCG)
The Good Corporate Governance factor for Commercial Banks is an assessment of the quality of bank management. Good Corporate Governance is implementing 5 (five) principles of Good Corporate Governance. The principles are accountability, responsibility, transparency, fairness, and professionalism. The principles of Good Corporate Governance and the focus of assessment on the implementation of the principles of Good Corporate Governance are guided by the provisions of Good Corporate Governance that apply to Commercial Banks by taking into account the characteristics and complexity of the bank's business.

Earnings

The assessment of Profitability factors includes an evaluate the performance of profitability, sources of profitability, sustainability of profitability, management of profitability, and implementation of social functions.

$$\text{Earning component} = \text{ROA, ROE, and current profit growth}$$

Capital

Capital factor assessment includes the evaluation of capital adequacy and adequacy of capital management. Calculating capital refers to the applicable provisions regarding the minimum capital adequacy requirement for commercial banks (Table 2).

$$\text{Capital component} = \text{CAR ratio (Capital Adequacy Ratio)}$$

Table 2. Rating of commercial banks' soundness

No	Composite rating	Weight (%)	Description
1	CR 1	86–100	Very soundness
2	CR 2	71–85	Soundness
3	CR 3	61–70	Quite soundness
4	CR 4	41–60	Less soundness
5	CR 5	< 40	Not soundness

Source: Bank Indonesia 2011

3 Research Methods

This research is a descriptive comparative by grouping KBMI 1 banks in Indonesia whose core capital is still less than IDR. 3 trillion, then analyzing the bank's financial statements to measure the soundness of the bank based on the analysis of the CAMEL and RGEC models. The population in this study focused on all KBMI 1 Commercial Banks in Indonesia, both conventional and Islamic banks. The sampling technique used is purposive sampling. The samples taken in this study were all KBMI 1 Commercial Banks with core capital still below IDR. 3 Trillion until December 31, 2022. A sample of 30 banks was obtained. The research period is one year, namely in 2021 (Table 3).

3.1 Result

KMBI 1

Table 3. KBMI 1 - Capital below IDR 3 Trillion (IDR Million)

No	Bank Name KBMI 1	December 31, 2021
1	Bank Mega Syariah	1.869.586
2	Bank BCA Syariah	2.792.291
3	Bank Nationalnobu	1.628.300
4	Bank Index Selindo	1.472.559
5	Bank Sahabat Sampoerna	2.053.586
6	Bank Mas	2.709.297
7	Bank Ina Perdana	2.322.502
8	Bank Seabank Indonesia	2.358.707
9	Bank IBK Indonesia	2.902.185
10	Bank CTBC Indonesia	2.868.608
11	Bank Neo Commerce	2.754.751
12	Bank MNC International	2.041.755
13	Bank J.Trust Indonesia	2.208.402
14	Bank Panin Dubai Syariah	2.082.126
15	Bank Resona Perdana	2.167.057
16	Bank Victoria International	2.339.061
17	Bank Raya	2.083.285
18	Allo Bank Indonesia	1.274/748
19	Bank Bisnis	2.067.802
20	Bank Oke Indonesia	2.881.666
21	Bank Jasa Jakarta	2.084.788
22	Bank Bumi Arta	2.211.485
23	Bank SBI Indonesia	2.109.069
24	Bank Mayora	1.139.309
25	Prima Bank	289.464
26	Bank Ganesha	2.072.676
27	Bank Victoria Syariah	260.291
28	Bank Amar Indonesia	260.291
29	Bank Fama International	1.921.694
30	Bank Aladin Syariah	1.038.915

Source: Financial Bank Report, 2022.

CAMEL and RGEC Model Analysis

The conclusion of the CAMEL and RGEC model analysis for 2021as follows (Table 4).

Table 4. The Conclusion of the CAMEL and RGEC 2021

No	Bank Name in KBMI 1	Total Value CAMEL	Total Value RGEC
1	Bank Mega Syariah	88	81
2	Bank BCA Syariah	88	73
3	Bank Nationalnobu	90	74
4	Bank Index Selindo	81	79
5	Bank Sahabat Sampoerna	87	78
6	Bank Mas	85	81
7	Bank Ina Perdana	81	72
8	Bank Seabank Indonesia	74	68
9	Bank IBK Indonesia	71	66
10	Bank CTBC Indonesia	72	66
11	Bank Neo Commerce	74	65
12	Bank MNC International	71	71
13	Bank J.Trust Indonesia	65	59
14	Bank Panin Dubai Syariah	58	64
15	Bank Resona Perdana	57	65
16	Bank Victoria International	56	66
17	Bank Raya	53	64
18	Allo Bank Indonesia	97	86
19	Bank Bisnis	89	84
20	Bank Oke Indonesia	83	71
21	Bank Jasa Jakarta	77	82
22	Bank Bumi Arta	76	74
23	Bank SBI Indonesia	69	73
24	Bank Mayora	69	71
25	Prima Bank	61	69
26	Bank Ganesha	61	68
27	Bank Victoria Syariah	59	70
28	Bank Amar Indonesia	56	60
29	Bank Fama International	56	65
30	Bank Aladin Syariah	60	65

Source: Processed Data, 2022.

4 Discussion

Analyzing the bank's soundness using the CAMEL and RGEC model analysis to determine the bank's predicate. The results show that the assessment predicate with the CAMEL and RGEG models has different effects. We can say a predicate of 4 when the CAMEL weight value is divided into four and a predicate of 5 when the RGEC value is divided by five. The CAMEL weight value tends to be lower than RGEC, so the results of the bank predicate, as measured by RGEC, will look better. Here we show the difference between CAMEL and RGEC predicates (Table 5):

Table 5. CAMEL and RGEC Predicate

Predicate	CAMEL	RGEC
Very Soundness	–	86–100
Soundness	81–100	71–85
Quite Soundness	66–<81	61–70
Less Soundness	51–<66	41–60
Not Soundness	0–<51	<40

Source: [11, 16]

Banks must continue to improve their financial performance in the third and fourth trimesters so that the development of bank financial performance is getting better, so that it is easier to attract investors or the parent bank (company) [12] and are willing to place additional core capital before December 31, 2022. Assessment of bank soundness using the CAMEL and RGEC methods can help core capital bank group 1 to increase core capital. Investors can use it to determine investment decisions in the core capital bank group 1. Investors have minimal risk of loss if they invest because the analysis of the bank's soundness indicates that the bank's condition is soundness or very soundness. Banks with soundness or even very soundness predicate will easily attract investors. For the group of core capital banks with the predicate relatively fit and not soundness, they must work harder to achieve a minimum core capital of IDR 3 trillion until December 31, 2022, because if the bank is unable to meet the provisions in question, it will be designated as a BPR or BPRS based on a decision. Financial Services Authority [1] can say that the commercial bank will be downgraded to a BPR or BPRS.

Offer to the parent company (bank) and other companies (not the parent bank) [21] for additional core capital. In line with the statement [22] that the subsidiary's positive performance is related to the possibility that the bank subsidiary receives more capital from the parent company. Banks appointed as implementing companies for holding companies, as referred to in POJK Number 12/POJK.03/2020, are banks that have total assets greater than banks or have a better bank soundness level. Looking for investors to increase the company's capital [6]. In line with the statement [23] that these investors can come from domestic or foreign investors if no domestic investors are willing. Companies already listed on the stock exchange can increase their capital by issuing new shares or rights issues [24]. The scheme for adding core capital based on the literature reference that the author obtained can be implemented through the following methods [1, 10]: a. Merger, fusion, or integration; b. The takeover is followed by a merger, consolidation, or integration; c. Establishment of a bank business group against a bank that is already owned; d. Establishment of a bank business group due to the separation of sharia business units; or e. Establishment of a bank business group due to take over.

5 Conclusion

Fulfilment of bank core capital in accordance with POJK Number 12/POJK.03/2020 concerning the consolidation of commercial banks in Indonesia can be done through: 1) Assessment of bank soundness using the CAMEL and RGEC methods can help banks determine efforts to increase core capital of at least IDR 3 trillion to December 31, 2022, while for investors it can be used to determine decision making related to investment; 2) Bank management must continue to make efforts to improve financial performance in the third trimester and fourth trimester it can be seen that the development of bank financial performance is getting better; 3) For a company with a soundness predicate, it will be easy to attract investors to invest by investing their funds, while for a KBMI 1 bank with a relatively soundness and unsoundness predicate, they must work harder to ensure that their core capital is fulfilled because if the bank cannot meet the OJK regulations, it will be downgraded to BPR or BPRS; 4) Efforts to increase capital can be done through the following sources, namely offering to the parent company and other companies (not the parent company), looking for investors to increase the company's capital, and for companies that are already listed on the stock exchange, they can increase capital by issuing shares. New or rights issue; 5) The core capital increase scheme can be implemented through a merger, consolidation, or integration scheme; takeover followed by merger, consolidation, or integration; establishment of a bank business group against a bank that is already owned; establishment of a bank business group due to the separation of sharia business units; or the formation of a bank business group due to take over.

Based on the above conditions, it will be more difficult for a bank to achieve a healthy predicate using CAMEL analysis because to get a soundness predicate, it must achieve a score of 81, while with an RGEC analysis, it only needs a score of 71. However, the RGEC study can reach a score of 86, so banks with good financial performance will quickly achieve very soundness if assessed using RGEC analysis. Suggestions for further research in evaluating the soundness of banks should still use CAMEL and RGEC. CAMEL purely uses financial ratios that can be obtained from the annual report, while in the RGEG analysis, there are components of risk assessment and organizational management, which are represented by self-assessment of Good Corporate Governance assessment. Thus, the criteria used can be complete to assess the bank's health.

References

1. Keuangan, O.J.: Peraturan Otoritas Jasa Keuangan Nomor 12/POJK.03/2020 about Financial services authority, *Ojk*, p. 26 (2020)

2. Altman, E.I.: Applications of distress prediction models: what have we learned after 50 years from the z-score models? Int. J. Financ. Stud. **6**(3), 70 (2018). https://doi.org/10.3390/ijfs6030070
3. Higgins, H.N.: Banks and corporate decisions: evidence from business groups. Finance. Manag. **47**(3), 679–713 (2018). https://doi.org/10.1111/fima.12214
4. Hamilton-Hart, N.: How well is Indonesia's financial system working? J. Southeast Asian Econ. **35**(2), 125–142 (2018). https://doi.org/10.1355/ae35-2b
5. Fich, E.M., Nguyen, T., Officer, M.: Large wealth creation in mergers and acquisitions. Financ. Manag. **47**(4), 953–991 (2018). https://doi.org/10.1111/fima.12212
6. Mohd, I.M., Y., and Abdul Rashid, M. Z. H.: Regulatory capital funds and risk-sharing behavior in distressed financial conditions: an empirical analysis on Islamic banks in Malaysia. J. Financ. Report. Account. **16**(1), 197–216 (2018). https://doi.org/10.1108/JFRA-06-2015-0066
7. Nurfalah, I.: Early warning to banking crises in the dual financial system in Indonesia: the markov switching approach. J. King Abdul Aziz Univ. **31**(2), 133–156 (2018). https://doi.org/10.4197/Islec.31-2.10
8. Bank Indonesia.: Peraturan bank Indonesia nomor 14/26/PBI/2012 tentang kegiatan usaha dan jaringan kantor berdasarkan modal inti bank. Bank Indones. **1**(1), 24 (2012)
9. Martínez-Campillo, A., Fernández-Santos, Y., del Pilar Sierra-Fernández, M.: How well have social economy financial institutions performed during the crisis period? exploring financial and social efficiency in Spanish credit unions. J. Bus. Ethics **151**(2), 319–336 (2016). https://doi.org/10.1007/s10551-016-3192-9
10. Çelik, M., Tekşen, O.: Does it matter how to fund?: a research on Turkish deposit banks. Istanbul Bus. Res. **50**(2), 359–383 (2021). https://doi.org/10.26650/ibr.2021.50.861522
11. Fadrul, F., Ridawati, R.: Analysis of method used to predict financial distress potential in pulp and paper companies of Indonesia. Int. J. Econ. Dev. Res. **1**(1), 57–69 (2020). https://doi.org/10.37385/ijedr.v1i1.29
12. Budiman, T., Herwany, A., Kristanti, F.T.: An evaluation of financial stress for Islamic banks in Indonesia using a bankometer model. GATR J. Financ. Bank. Rev. **2**(3), 14–20 (2017). https://doi.org/10.35609/jfbr.2017.2.3(3)
13. Muzakir, M.F.A.: "Struktur Kepemilikan, Kebijakan Deviden, Kebijakan Hutang dan Nilai Perusahaan di Sektor Perbankan", *EKOMBIS Rev*. J. Ilm. Ekon. dan Bisnis **10**(1), 13–24 (2022). https://doi.org/10.37676/ekombis.v10i1.1687
14. Bank Indonesia, "Peraturan Bank Indonesia No. 6/10/PBI/2004 tentang Sistem Penilaian Tingkat Kesehatan Bank Umum," *Peratur. bank Indones.*, vol. 1, no. 1, pp. 1–23, 2004, [Online]. Available: www.bi.go.id
15. Badrul Munir, M., Bustamam, U.S.A.: Camel ratio on profitability banking performance (Malaysia versus Indonesia). Int. J. Manag. Innov. Entrep. Res. **3**(1), 30–39 (2017). https://doi.org/10.18510/ijmier.2017.314
16. AL-Najjar, D., Assous, H.F.: Key determinants of deposits volume using CAMEL rating system: the case of Saudi banks. PLoS One, **16**(12), 1–16 (2021). https://doi.org/10.1371/journal.pone.0261184
17. Arafa, S., Pastory, D.: Effect of treasury single account (TSA) on the financial performance of commercial banks in Tanzania. Int. J. Res. Bus. Soc. Sci. **11**(2), 172–182 (2022). https://doi.org/10.20525/ijrbs.v11i2.1639
18. Quoc Trung, N.K.: Determinants of bank performance in Vietnamese commercial banks: an application of the camels model. Cogent Bus. Manag. **8**(1), 1–36 (2021). https://doi.org/10.1080/23311975.2021.1979443
19. Bank Indonesia, Peraturan Bank Indonesia Nomor 9/1/PBI/2007 Tentang Sistem Penilaian Tingkat Kesehatan Bank Umum Berdasarkan Prinsip Syariah, p. 26 (2007)

20. Bank Indonesia, Peraturan Bank Indonesia No.13/1/PBI/2011 Tentang Penilaian Tingkat Kesehatan Bank. *Peratur. Bank Indones.*, pp. 1–31 (2011)
21. Kuehn, J.: Spillovers from entry: the impact of bank branch network expansion. RAND J. Econ. **49**(4), 964–994 (2018). https://doi.org/10.1111/1756-2171.12258
22. Dahl, D., Coster, D.C.: Subjective assessment of managerial performance and decision-making in banking. Fed. Reserve. Bank St. Louis Rev. **104**(3), 210–223 (2022). https://doi.org/10.20955/r.104.210-23
23. Delice, G., Karadas, H.A.: KüreselEkonomiKrizlerin Katilim Bankalarinin Performans ÜzerindekiEtkileri: Covid-19 SalginiÖrneği, no. 62 (2022)
24. Fahlenbrach, R., Prilmeier, R., Stulz, R.M.: Why does fast loan growth predict poor performance for banks? Rev. Financ. Stud. **31**(3), 1014–1063 (2018). https://doi.org/10.1093/rfs/hhx109

Cyclicality as a Manifestation of the Volatility of Economic Systems at the Sectoral Level

Anton Moholivets[1], Olena Molodid[1], Yuliia Zapiechna[1], Mykola Stetsko[2](✉), Danylo Bohatiuk[1], and Iryna Fedun[3]

[1] Kyiv National University of Construction and Architecture, Kyiv, Ukraine
{moholivets.aa,dubinin_dv,zapiechna.iuo,bahatiuk_dv}@knuba.edu.ua

[2] Department of Financial Technologies and Banking Business, West Ukrainian National University, Ternopil, Ukraine
nsteco6691@gmail.com

[3] Department of Entrepreneurship, Cooperation, and Agro-Industrial Integration, National Scientific Centre "Institute of Agrarian Economics", Kyiv, Ukraine

Abstract. An urgent problem is the search for indicators of development for various sectors of the economy and the study of their volatility at individual phases of the economic cycle, as well as the determination of the most suitable for detecting the impact of cycles of different depth and duration. The article examines the performance indicators of the construction sector of Ukraine in 2003–2022, namely: Changing the value of 1 sq.m of housing, Shareholders Equity ratio, Current ratio and Long-term Debt to equity. Cyclical fluctuations of the specified indicators during the studied period were revealed, which coincide with the cycles of Kitchin J. and Juglar C. Changing the physical volume of GDP was also investigated. Fluctuations of the mentioned indicator also have signs of cyclicity.

The gradual decrease in the financial stability of the sector was the result of the fact that the equity capital of construction companies has been steadily decreasing since 2002 and has been close to zero for the last few years. This happened thanks to the unprofitability of the sector's enterprises almost throughout the analyzed period, with the exception of the last few years.

In moments of reduced business activity in the country and in the industry, the financial stability of the sector increases, which can act as a kind of lever for market stabilization of crisis phenomena.

Keywords: Construction · Economic cycle · Indicators · Business activity · Financial stability

1 Introduction

Economic systems are characterized by the organic interconnection of all constituent elements, the integrity and interdependence of the system of industrial relations, the process of functioning of productive forces, methods of coordination, regulation in accordance with the existing economic system.

B. Alareeni et al. (Eds.): ICBT 2022, LNNS 620, pp. 46–55, 2023.
https://doi.org/10.1007/978-3-031-26953-0_6

The market economic system is characterized by cyclical fluctuations of supply and demand, employment, the number of economic entities and the wave-like development of business activity. Sometimes the variability of the key parameters of the system is so great that it leads to its transformations, and sometimes the sharp fluctuations of the parameters of development during a certain period of time decrease and the system again acquires a state of equilibrium.

Purpose of research to determine a set of indicators of changing phases of economic cycles separately for different sectors of the economy, which will increase the accuracy of forecasting. An important direction of state regulation is smoothing the negative impact of such fluctuations on the development of the economy of countries, sectors of the economy, individual subjects of market relations, as well as strengthening positive trends. And if at the macroeconomic level, methods of regulating and smoothing out sharp fluctuations, indicators of the financial stability of the system are quite well researched, then at the sectoral level, additional research is needed.

2 Literature Review and Methodology

In modern economic science, completely different interpretations of the reasons for the volatility of the development of economic systems and the occurrence of cyclical fluctuations at the level of the world economy, the economies of individual countries, and individual sectors have developed. In particular (Kitchin 1923; Burns and Mitchell 1946; Juglar 1982; Schumpeter 1939) factors of consumption (saturation or undersaturation of demand, propensity of economic subjects to consume and save), factors of supply (inelasticity of aggregate supply leading to a crisis of overproduction), volumes of capital accumulation, amount of money in circulation are attributed to the causes of fluctuations in economic activity, as well as non-economic factors (scientific and technical progress, natural climatic and meteorological conditions, etc.).

The main causes of fluctuations at the level of economic systems are (Karpenko 2008):

- technical innovations that affect investments and consumer spending, and, accordingly, on production, employment, and the price level;
- political and random events;
- changes in credit and monetary policy (fluctuations in the volume of the money supply);
- changes in economic conditions, material resources, etc.

Considering the given classification, one can support the opinion regarding the inclusion of random events in the list of causes of volatility of economic systems. Random events such as hostilities, the global epidemic of Covid-19 and other diseases, violent or carried out by democratic methods of destroying the existing borders of countries, cannot initiate the cyclicality of the economy, but can lead to significant fluctuations in economic activity, shift the time of the cycle phase change and affect the intensity of fluctuations.

After the identification of economic cycles that differ in intensity, duration and degree of influence on the development of various sectors of the economy and the study of their effects on economic and organizational systems, work has begun in many countries to improve the assessment of the stability of financial systems and to determine the impact of fluctuations in economic activity on the financial stability of various sectors.

Indicators that determine the volatility of economic systems at the sectoral level were studied in works (Sorokina et al. 2017; Bielienkova 2011).

Financial stability as a lever for regulating fluctuations in business activity was studied in works (Shpakov et al. 2021; Bielienkova et al. 2021; Stetsenko et al. 2021; Stetsenko et al. 2020). Procyclical indicators of the development of enterprises, sectors of the economy in the long-term or short-term perspective were studied in their works (Akselrod 2022; Bozhanova et al. (2022); Shpakov et al. 2021; Bielienkova etal. 2021; Reznik et al. 2022; Zeltser et al. 2019).

Today, there is a need to determine a set of indicators of changing phases of economic cycles separately for different sectors of the economy, which will increase the accuracy of forecasting.

3 Methodology

Discussions are currently underway and research on choosing financial stability indicators at different sectors of the economy is underway. The composition of indicators and detail of the calculations differs for different countries at the level of institutions, markets and relationships between them. But usually the basis of the financial stability of individual sectors of the economy is the data of the aggregate balance and the statement of financial results, as well as individual indicators and indicators of expectations of business activity.

In Ukraine, there is currently developed system of monitoring and state regulation of the financial stability of the banking sector, although there is no single system of indicators in accordance with the international methodology for the real sector of the economy.

Using the development of the International Monetary Fund (Financial Stability Review International Monetary Fund 2019) and the experience of researchers (Bielienkova 2011; Bielienkova and Antropov 2013) to determine the impact of economic cyclicality on the financial stability of different sectors of the set of indicators, the procedure for calculating in the construction sector of the economy, were selected to determine the financial stability (Table 1).

Using the traditional approach to the decomposition of time series, which is used in scientific works (Sorokina et al. 2017; Bielienkova et al. 2022), and which allows each time series to be imagined as the result of the action of an evolutionary, cyclical and random component:

$$y = f(t) + c(t) + \varepsilon t \qquad (1)$$

where f(t) is the evolutionary component (characterizes the trend), c(t) is a cyclic component, εt is a random component that reflects the influence of unpredictable factors that are not explicitly taken into account when making a forecast, and may reflect an error.

The impact of cyclicality on indicators characterizing business activity and financial stability of construction and economy of Ukraine is identified.

Table 1. Indicators of development and determination of volatility of economic systems at the sectoral level

№	Indicator	Output data/Calculation procedure	Conventional designation
1	Changing the physical volume of GDP, %	Change in the volume of gross domestic product (The State Statistics Service of Ukraine 2022)	Δ GDP
2	Changing the value of 1 sq.m of housing, %	Calculated by the authors based on the portal data www.svdevelopment.com	Δ VH
3	Shareholders Equity ratio (Shareholders Equity/Total Assets)	Calculated by the authors based on the State Statistics Service of Ukraine (2022)	SER
4	Current ratio (Current Assets/Current Liabilities)		CR
5	Long-term Debt and Equity to Total Assets Ratio (Long-term Debt + Equity/Total Assets		LTED

4 Research Results

During the analyzed period, the indicators of the sector underwent quite significant fluctuations (Table 2). The largest increase in the price of 1 square meter was observed in 2004. At that time, the growth of prices in the real estate market reached 97% per year, while in 2015 there was a drop in prices by 35%.

Table 2. Development indicators of the construction sector in 2003–2022

	Year									
	2003	2004	2005	2006	2007	2008	2009	2010	2011	2012
Δ GDP	9.5	12.1	3	7.4	7.6	2.3	−14.8	4.1	5.5	0.2
Δ VH	x	97	49	38	39	24	−45	11%	−3	5
SER	0.51	0.39	0.37	0.38	0.32	0.25	0.26	0.23	0.20	0.18
CR	1.05	1.07	1.06	1.16	1.18	1.18	1.20	1.10	1.08	1.08
LTED	0.56	0.46	0.46	0.50	0.50	0.51	0.51	0.47	0.43	0.41

(continued)

Table 2. (*continued*)

	Year									
	2003	2004	2005	2006	2007	2008	2009	2010	2011	2012
	Year									
	2013	2014	2015	2016	2017	2018	2019	2020	2021	2022*
Δ GDP	0	−6.6	−9.8	2.3	2.4	3.5	3.2	−3.8	3.4	−33.4
Δ VH	6	−1	−35	−4	−3	−4	−3	−5	3	5
SER	0.17	0.08	0.02	0.02	0.00	−0.01	0.02	0.03	0.11	0.05
CR	1.07	1.03	0.96	0.94	0.89	0.92	0.89	0.87	1.04	0.98
LTED	0.40	0.37	0.29	0.29	0.23	0.22	0.18	0.17	0.21	0.16

*Forecast values were used for 2022

Fluctuations in real estate prices coincide with fluctuations in Ukraine's GDP (Fig. 1), therefore, price fluctuations in the real estate market can be used as an indicator showing the growth or decline of the economy.

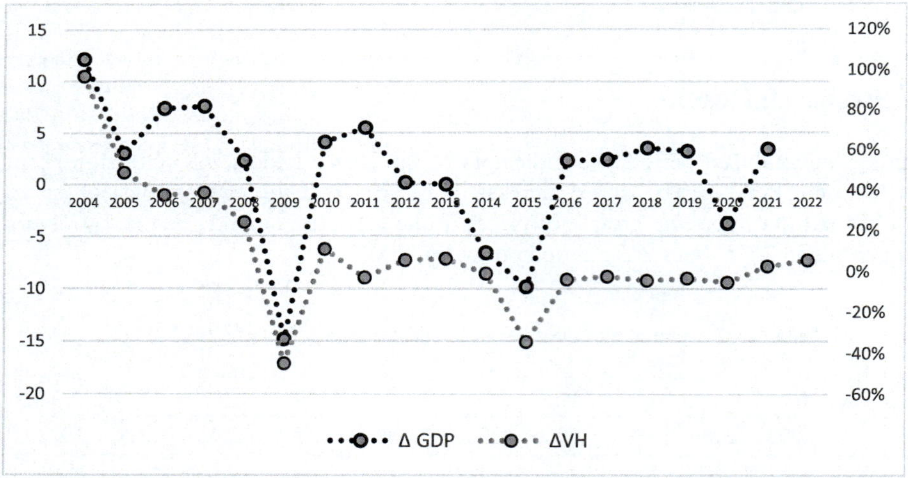

Fig. 1. Fluctuations in GDP and the price of 1 square meter of real estate in 2003–2022

The dynamics of SER, CR, LTED indicators (Fig. 2, 3, 4) show that the equity capital of construction enterprises has been steadily decreasing since 2003, and has been close to zero for the last few years. This happened due to the unprofitability of enterprises in the sector during the analyzed period. In accordance with the methodology given in (Sorokina et al. 2017), we will analyze the specified coefficients in order to identify the tendency of their development to isolate cyclical fluctuations without taking into account the general trend.

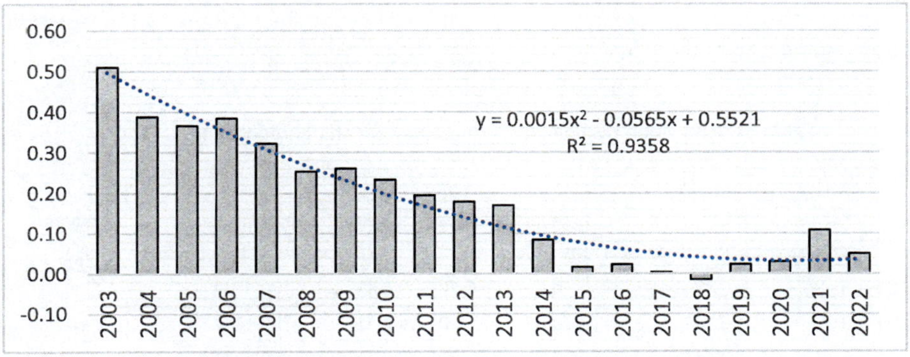

Fig. 2. Fluctuations in shareholders equity ratio in 2003–2022

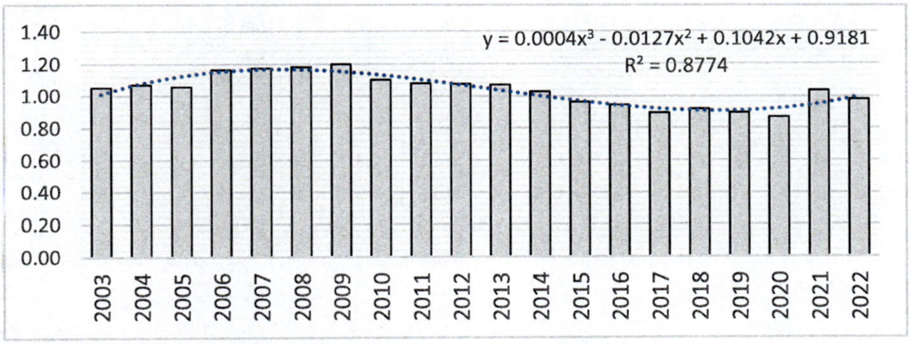

Fig. 3. Fluctuations in current ratio in 2003–2022

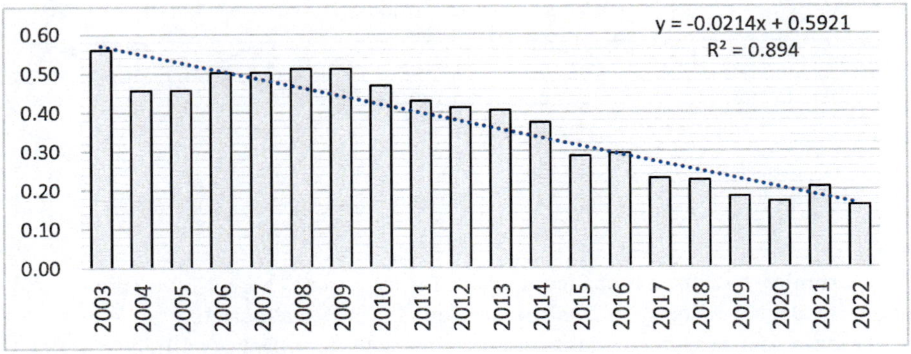

Fig. 4. Fluctuations in long-term debt and equity to total assets ratio in 2003–2022

The Shareholders Equity ratio has a sharp downward trend (Fig. 2). During the analyzed years, its value decreased from 0.5 to values close to zero, which is a negatives sign and a consequence of the unprofitability of the construction sector for a long time.

The next stage of the research is the isolation of cyclic fluctuations by the isolation of the general trend (Table 3, Fig. 5).

Table 3. Allocation of the cyclic component for shareholders equity ratio

Year	Actual values	Trend	Cyclical component	Year	Actual values	Trend	Cyclical component
	SERfactual	SERtheoretical	Δ SER		SERfactual	SERtheoretical	Δ SER
2002	0.48	0.4731	0.01	2012	0.48	0.5232	−0.04
2003	0.51	0.4563	0.05	2013	0.51	0.4742	0.04
2004	0.39	0.4335	−0.05	2014	0.39	0.4276	−0.04
2005	0.37	0.4059	−0.04	2015	0.37	0.3834	−0.02
2006	0.38	0.3747	0.01	2016	0.38	0.3416	0.04
2007	0.32	0.3411	−0.02	2017	0.32	0.3022	0.02
2008	0.25	0.3063	−0.05	2018	0.25	0.2652	−0.01
2009	0.26	0.2715	−0.01	2019	0.26	0.2306	0.03
2010	0.23	0.2379	0.00	2020	0.23	0.1984	0.03
2011	0.20	0.2067	−0.01	2021	0.20	0.1686	0.03
2012	0.18	0.1791	0.30	2022	0.05	0.1412	0.04

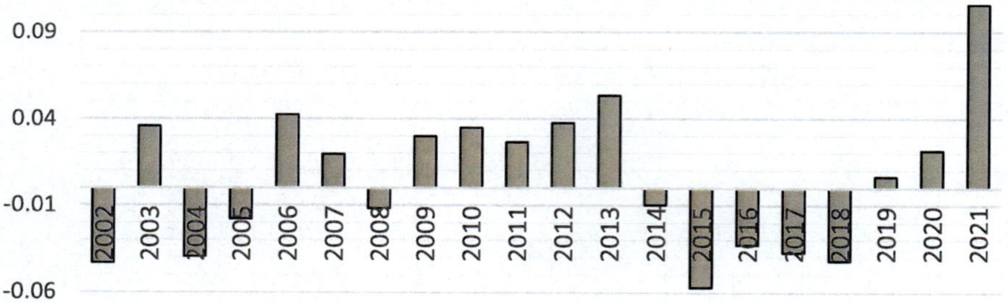

Fig. 5. Allocation of the cyclic component for Shareholders Equity ratio

In a similar way, the cyclical component for indicators Current ratio (Fig. 6) and Long-term Debt and Equity to Total Assets Ratio (Fig. 7) is allocated.

Three- to four-year fluctuations of indicators Long-term Debt and Equity to Total Assets Ratio, Current ratio, Shareholders Equity ratio can be observed during the analyzed period, which coincides with Kitchin's three- to four-year cycles. Also, the graphs clearly trace the cycles of the Juglar, which have a period of 7–11 years.

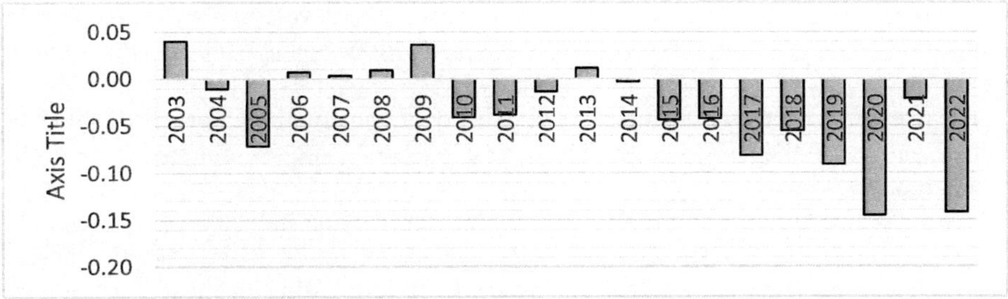

Fig. 6. Allocation of the cyclic component for current ratio

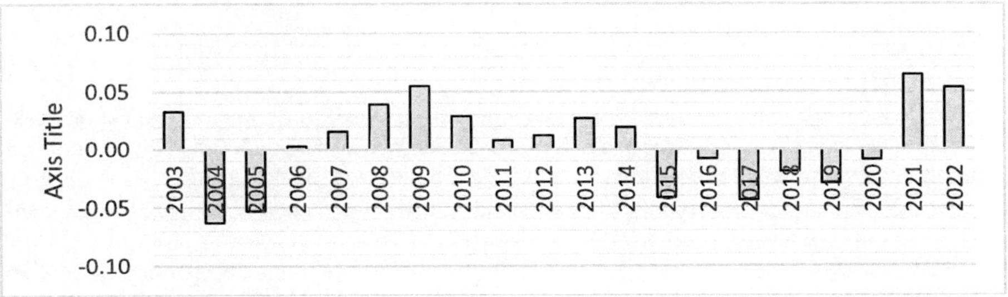

Fig. 7. Allocation of the cyclic component for long-term debt and equity to total assets ratio

5 Conclusions

The current state and development trends of the construction sector in Ukraine were studied, and the cyclical nature of its development was revealed. It was established that among the fluctuations of business activity, two- to four-year Kitchin cycles can be quite clearly distinguished, as well as less clearly - Juglar cycles, which have a period of fluctuations of seven to eleven years. According to the results of the analysis, the hypothesis about the volatility of indicators of the development of the construction sector was confirmed, as well as about the possibility of using the volatility of the price of 1 square meter of housing as an indicator of financial stability, since the percentage change in the GDP of Ukraine repeats the trend of changes in prices for residential real estate. Based on the analysis of trends in their development, the presence of periodic fluctuations during the analyzed period was revealed.

The influence of economic cycles on the financial stability of construction, which during 2003–2022 was constantly decreasing due to the decrease in the equity of the sector, was comprehensively assessed. It was determined that during the downward phase of the economic cycle and the reduction of business activity, the values of the coverage coefficients increase and the values of the indicators of autonomy and long-term financial independence decrease. That is, during an economic boom, the financial stability of construction decreases, and during a recession it increases somewhat, which is due to the reorientation of the goals of the sector's enterprises from profit maximization to survival.

It was revealed that one of the important factors in the development of construction during 2003–2022 is the decrease in the size of construction capital, which is a threatening sign of the tendency to decrease financial stability.

The need to regulate the financial sustainability of construction in order to influence business activity in the construction sector and as a prerequisite for influencing the GDP of construction is proven.

Given the unstable economic situation in the construction industry, today there is a need to develop a mechanism for anti-cyclical management of the construction enterprise. What is in demand is a toolkit designed for forecasting cyclical fluctuations, smoothing out the negative factors of the volatility of the sector's development and strengthening the positive ones.

References

Akselrod, R.B.: Integral model for choosing an alternative for an intra-company project of transformational changes in a construction enterprise. Středoevropský věstník Pro vědu a výzkum, 9 (2022). http://czvestnic.info/ojs/index.php/cz_ojs/article/view/156

Arefieva, O.V.: Ekonomichna stiikist pidpryiemstva: sutnist, skladovi ta zakhody yii zabezpechennia. Aktualni Problemy Ekonomiky **8**, 79–85 (2012)

Bielienkova, O.: Assessment of financial stability of small construction enterprises of Ukraine in 2000–2008. Ekonomika i Derzhava. **1**, 56–60 (2011)

Bielienkova, O., Novak, Y., Matsapura, O., Zapiechna, Y., Kalashnikov, D., Dubinin, D.: Improving the organization and financing of construction project by means of digitalization. Int. J. Emerg. Technol. Adv. Eng. **12**(8), 108–115 (2022). https://doi.org/10.46338/ijetae0822_14

Bielienkova, O., Izmailova, K., Zapiechna, Y., Loktionova, Y.: Dependence of competitiveness on the level of business confidence of the enterprise. Sci. J. Astana IT Univ. **7**(7), 4–14 (2021)

Bielienkova, O., Moholivets, A., Loktionova, Y.: Measurement of the level of economic security of enterprises by the indicator of the attractiveness for hostile takeovers. Baltic J. Economic Studies **7**(5), 29–40 (2021)

Bozhanova, V., et al.: Green enterprise logistics management system in circular economy. Int. J. Math. Eng. Manag. Sci. **7**(3), 350–363 (2022)

Burns, A.F., Mitchell, W.C.: Measuring Business Cycles, Studies in Business Cycles. NBER (1946)

Fedun, I.L., Novikova, I.V., Igor, L.V., Marina, M.K., Marta, O.G.: Concept of innovative marketing in management of enterprise. Concept Innovative Mark. Manag. Enterp. **12**(7), 352–358 (2020)

Juglar, C.: Des Crises Commerciales Et de Leur Retour Périodique En France, An Angleterre Et Aux Etats-Unis. Guillaumin, Paris (1862)

Karpenko, O.A.: Modeli ta stratehii funktsionuvannia avtotransportnykh pidpryiemstv za umov tsyklichnosti: NTU. K. (2008)

Kitchin, J.: Cycles and Trends in Economic Factors. Rev. Econ. Stat. **5**(1), 10–16 (1923)

Mohey-ud-din, G., Siddiqi, M.W.: GDP fluctuations and long-run economic growth: a study of selected south Asian countries. Pakistan J. Appl. Econ. **27**(1), 41–66 (2017)

Nadiia, R., Yusuf, I., Yaroslav, K.-S., Nataliia, B., Mykola, S., Olha, B.: Systems thinking to investigate the archetype of globalization. In: Alareeni, B., Hamdan, A. (eds.) Financial Technology (FinTech), Entrepreneurship, and Business Development: Proceedings of The International Conference on Business and Technology (ICBT 2021), pp. 123–140. Springer International Publishing, Cham (2022). https://doi.org/10.1007/978-3-031-08087-6_9

Shpakov, A., Stetsenko, S., Shpakova, H., Sorokina, L., Akselrod, R.: Assessment of the influence of adaptability factors on the effectiveness of managing changes in enterprises by fuzzy logic. Sci. Horiz. **24**(10), 72–82 (2021)

Shumpeter, J.A.: Business Cycles. McGraw-Hill, New York (1939)

Sorokina L.V., Hoiko A.F., et al.: Ekonometrychnyi instrumentarii upravlinnia finansovoiu bezpekoiu budivelnoho pidpryiemstva: monohrafiia. KNUBA, Kyiv (2017)

Sorokina, L.V.: Improving the monitoring procedures for processes of capital cost accumulation in building & construction works. Actual Probl. Econ. **124**(10), 113–124 (2011)

Stetsenko, S., Bolila, N., Sorokina, L., Tsyfra, T., Molodid, O.: Monitoring mechanism of resilience of the anti-crisis potential system of the construction enterprise in the long-term period. Econ. Finan. Manag. Rev. **3**, 29–40 (2020)

Stetsenko, S., Sorokina, L., et al.: Model of a company competitiveness control by means of artificial intelligence tools. Int. J. Emerg. Trends Eng. Res. **9**(2), 60–65 (2021)

Stetsenko, S.P., et al.: The interrelation of digital technologies and organizational and economic mechanisms in construction: adaptation to change management. Int. Rev. **1 Part I**, 21–31 (2021)

The cost of apartments (housing prices). Consulting companiy SV Development statistics. http://surl.li/dblwo

The state statistics service of Ukraine (2022). https://www.ukrstat.gov.ua/

Zeltser, R.Y., et al.: Digital transformation of resource logistics and organizational and structural support of construction. Sci. Innov. **15**(5), 38–51 (2019)

Conditional Conservatism in Islamic Banks During the COVID-19 Pandemic

Zuhair Barhamzaid(✉)

Palestine Technical University – Kadoorie, Tulkarm, Palestine
z.barham@ptuk.edu.ps

Abstract. This study explores conditional conservatism (CC) in listed Islamic banks (IB) during the COVID-19 pandemic. The author collects data manually over the period from 2019 to 2020. In order to capture CC, the author uses the C_score measurement in the main model. As predicted, the author finds an increase in CC level within IB during the COVID-19 pandemic. This finding enriches the current literature on CC, IB, and the economic outcomes of the COVID-19 pandemic. On the other hand, decision-makers (investors, creditors, etc.) can benefit from the governance role of CC experienced in IB during crises such as the COVID-19 pandemic.

Keywords: Islamic banks · Conditional conservatism · The COVID-19 pandemic

1 Introduction

Researchers have shown a considerable influence of religion on firm behavior, such as corporate ethical codes [1], corporate social performance [2], and corporate philanthropy [3]. Moreover, religion mitigates other business practices such as tunneling practice [4], financial restatement, misrepresentation of financial statements, tax sheltering [5], financial reporting irregularities, accruals manipulation [6], and risk exposure [7], etc.

One form of the application of religiosity is the employment of the Islamic financial system, which has been applied currently in the financial sector, especially IB, in many countries around the world. This system is continuously developing over time [8]. At the present time, IB play a vital role in the global economy and are not simply considered alternative banks [9]. Therefore, It is crucial for IB to be efficient in order to withstand financial crises and competitive pressures [10].

Accounting conservatism (AC) in Islamic banking is a vital issue [9]. So far, AC is in line with Islamic values [11]. Prior studies prove a positive relationship between AC and Islamic banking [12–14]. In effect, AC plays a role in mitigating the opportunistic behavior of managers [15] and information asymmetry [16], suggesting that AC plays an essential role in mitigating agency problems [17]. Thus, the interaction between IB and AC is a considerable subject to be investigated.

Studies on IB are limited in the literature, and the Arabian Gulf countries have been promoting Islamic banking [18]. In addition, research on the COVID-19 crisis in the

B. Alareeni et al. (Eds.): ICBT 2022, LNNS 620, pp. 56–65, 2023.
https://doi.org/10.1007/978-3-031-26953-0_7

financial sector is rare [19]. Thus, the author is motivated to explore CC within IB in the Arabian Gulf countries during the COVID-19 pandemic. However, studying such a rare topic is expected to enrich the literature on CC, IB, and the economic outcomes of the COVID-19 pandemic. In addition, exploring the CC level within IB is an essential issue for users since CC can play a governance role against agency problems [20].

According to the author's knowledge, this is the first paper that explores CC in IB during the COVID-19 pandemic. In a recent study, Rinaldi [21] reviews accounting studies conducted during the COVID-19 pandemic and only captures one paper conducted by Cui, Kent [22] that explores the role of CC on firm performance during the pandemic.

The author finds an increase in CC within IB during the pandemic. This finding enriches the current accounting research concerning CC, IB, and the outcomes of the COVID-19 pandemic. However, decision-makers (investors, creditors, etc.) can benefit from the governance role of CC experienced in IB during the pandemic.

This study is structured as follows: The relevant literature is reviewed in the next section. The third section develops hypotheses. In the fourth section, the main model is explained. The fifth section describes the data. The sixth section reports the main results. The seventh section represents the robustness tests. An additional test is reported in the eighth section. Finally, the ninth section concludes the current study.

2 Literature Review

The impact of religion on AC is explored in the literature. Almost researchers find a positive association between religion and AC. Ma, Zhang [23] study the impact of religion on AC. They observe higher degrees of AC for listed firms in areas with higher degrees of religiosity. Bjornsen, Do [24] confirm that firms in more religious societies provide more conservative accounting numbers. Swenson [25] finds higher levels of AC in a larger presence of Western religion.

In particular, most studies also report a positive association between the Islamic religion and AC. For example, AlAbbad [12] finds that IB are more conservative than conventional banks. Matching IB with non-Islamic banks, Almutairi and Quttainah [13] find that IB are more likely to adopt AC. Fariz, Mohammed [26] show a positive connection between AC and financial performance among Malaysian Shariah-compliant firms. Abdul Rahman, Omar [14] concur that Shariah firms represent more conservative reporting than other firms. They conclude that religious values improve accounting quality. Rudiawarni, Sulistiawan [11] observe that AC decreases the crash risk for Sharia stocks. On the other hand, their results show no effect of AC on crash risk for non-Sharia stocks. Rosdini and Fildzah [8] find that IB are less conservative compared to conventional banks in Indonesia. In contrast, they document that Malaysian IB show a higher degree of conservatism than conventional banks.

The economic outcomes of financial crises on AC are also investigated in the literature. Almost, past research reveals that AC is increased during financial crises. For instance, Herrmann, Pornupatham [27] show that AC is increased following the financial crisis that occurred in Asia. Francis, Hasan [28] find that AC positively affects company performance during the global financial crisis. Al-Hroot, Al-Qudah [29] indicate that the degree of AC is significantly improved during the financial crisis (2009–2011) in

Jordanian commercial banks. Balakrishnan, Watts [30] document that companies with low AC experienced a reduction in investment following the financial crisis. They conclude that AC reduces the likelihood of underinvestment during information frictions. Cerqueira and Pereira [31] analyze the role of IFRS adoption on CC, surrounding the global financial crisis (2007–2008). They find that CC is increased after the crisis period. Kim and Shawn [32] show that firms experience more conservative practices during the financial crisis (2007–2008). Moreover, conservative firms outperform in the post-crisis periods. On the other hand, some studies find a low degree in AC in financial crises. For example, Vichitsarawong, Eng [33] find that AC is low during the 1997 Asian financial crisis, but enhanced and becomes greater after the crisis period. Gul, Srinidhi [34] consider Ac and audit fees during the financial crisis that occurred in Asia. According to their results, the crisis is correlated with a decrease (increase) in AC (audit fees). In the Netherlands, Leune [35] tests the role of the financial crisis on AC. The investigation shows that Dutch-listed firms are not conservative before and during the crisis. However, some researchers fail to capture a significant relationship between AC and crises. As an example, Heijden [36] show no differences in AC before or during the credit crisis across seven European countries.

Rinaldi [21] reviews accounting research conducted during the COVID-19 pandemic. He shows that accounting research during the pandemic focused on five main areas: accounting education, public budget response, the public sector, financial markets, and corporate disclosure. Concerning research in AC during the pandemic, he only captures one paper conducted by Cui, Kent [22] that explores the role of CC on company performance during the pandemic, addressing a significant role of CC in mitigating declines in company performance during the pandemic.

3 Hypotheses Development

In general, past studies suggest a positive connection between AC and religions [23–25], including Islamic religion [12–14]. In effect, religion has a positive connection with ethical norms [37] and helping others [38], suggesting that religious values are expected to mitigate managerial opportunism behavior [14] and enhance AC as a result. As a governance mechanism, conservatism is not in line with the opportunistic behavior of managers [20], precisely the conditional form of AC [15]. Thus, religious values and conservatism are expected to be in the same direction.

On the other hand, most studies confirm that the demand for conservatism increases during crises [see, e.g., 28, 29, 31, 32]. A high level of risk is expected to drive greater demand for AC [39], as AC is a governance tool to lessen information risk [28]. Based on the litigation explanation explained by Watts [20], the financial crisis is expected to increase AC. Indeed, litigation risk is expected to be higher during financial crises since litigation is expected if share prices drop sharply [40]. Thus, crises and conservatism are expected to be connected.

In sum, prior studies suggest that IB experience high levels of AC [12–14], and AC is found to be increased during crises [29, 31, 32]. In close, the current study argues that CC is expected to be increased in IB during the COVID-19 pandemic. Accordingly, the author adopts the following alternative hypothesis:

H1: CC is significantly increased in IB during the COVID-19 pandemic.

4 Research Model

In the accounting literature, C_score, suggested by Khan and Watts [41], is widely used to capture CC [42]. Therefore, the author relies on C_score, which is a firm-specific version of the model. The following equation represents the Basu [43] cross-sectional regression:

$$E_i = \beta_1 + \beta_2 D_i + \beta_3 R_i + \beta_4 R_i D_i + e_i \tag{1}$$

where E_i is net income before extraordinary items divided by lagged market value. R_i is a yearly share return ending three months after the fiscal year-end. D_i is a dummy variable that equals one where R_i is negative and zero otherwise.

Khan and Watts [41] suggest that CC is affected by company size (SIZE), leverage (LEV), and market-to-book ratio (MTB). Thus, they add SIZE, LEV, and MTB to Basu [43] cross-sectional regression, which becomes as follows:

$$\begin{aligned} E_i = &\beta_1 + \beta_2 D_i + \beta_3 R_i(\mu_1 + \mu_2 SIZE_i + \mu_3 MTB_i + \mu_4 LEV_i) \\ &+ \beta_4\, R_i D_i(\lambda_1 + \lambda_2\, SIZE_i + \lambda_3\, MTB_i + \lambda_4\, LEV_i) \\ &+ \delta_1 SIZE_i + \delta_2\, MTB_i + \delta_3\, LEV_i \\ &+ \delta_4\, D_i SIZE_i + \delta_5\, D_i MTB_i + \delta_6 D_i LEV_i + \varepsilon_i \end{aligned} \tag{2}$$

where $SIZE_i$ is the natural log of a firm's equity market value. MTB_i is the ratio of market-to-book. LEV_i is Leverage, defined as total liabilities divided by the market value of a firm's equity.

Using the estimates ($\lambda_1, \lambda_2, \lambda_3, \lambda_4$) from Eq. (2) , they compute the C_score based on the following equation:

$$C\_score = \beta_4 = \lambda_1 + \lambda_2 SIZE_i + \lambda_3 MTB_i + \lambda_4 LEV_i \tag{3}$$

To explore the CC level in IB during the COVID-19 pandemic, the author regresses the C_score on CO19 (a dummy variable) along with the determinants of C_score (SIZE, LEV, and MTB) as control variables, following Khan and Watts [41]. Accordingly, the study model is constructed as follows:

$$C\_score_{it} = \beta_0 + \beta_1 CO19_{it} + \beta_2 SIZE_{it} + \beta_3 LEV_{it} + \beta_4 MTB_{it} + e_{it} \tag{4}$$

where C_score is the measurement of CC suggested by Khan and Watts [41]. The variable CO19 is a dummy variable representing the COVID-19 period that equals 1 for 2020 and 0 for 2019. All other variables are defined above.

The coefficient β_1 can determine the level of CC during the COVID-19 pandemic. A significantly negative (positive) value of β_1 specifies that CC is positively decreased (increased) during the COVID-19 pandemic.

5 Data

5.1 Data Collection

The author manually collects data from financial reports of IB listed on the financial markets in the Arabian Gulf countries (Kuwait, Saudi Arabia, the United Arab Emirates, Qatar, Oman, and Bahrain) from 2019 to 2020. In addition, share prices are collected manually from the websites of financial markets, Yahoo Finance or Investing.com. The Arabian Gulf countries are chosen since they account for the largest proportion of the global Islamic financial assets, about 37.6% [44], and the banking sector assets of the Arabian Gulf countries represent 38% of the total Islamic banking assets [19]. The study sample includes 30 listed Islamic banks, and the total firm-year observations with complete data for analysis are 60 observations. Table 1 provides a distribution of IB by country. As can be seen, most observations are from the AUE (10 banks), as two financial markets are represented in the UAE, namely the Dubai Financial Market (9 banks) and Abu Dhabi Securities Exchange (1 bank). The lowest number of observations is from Oman (1 bank).

Table 1. Distribution of IB by country

No	Country	No. of IB	Observations (Final sample)
1	Kuwait	4	8
2	Saudi Arabia	4	8
3	UAE	10	20
4	Qatar	6	12
5	Oman	1	2
6	Bahrain	5	10
Total		30	60

5.2 Data Description

Table 2 shows descriptive statistics for all variables used in the main model. The mean (median) C_Score is 1.5578 (0.0755) and ranges from −12.3596 to 11.5396. The average (median) SIZE and LEV are 21.4019 (21.1618) and 9.5719 (5.3184), respectively. The average (1.5803) of MTB is greater than one, suggesting that IB experience a high level of AC. AC results in overstating the market value of a firm's equity relative to its book value of equity, which, in turn, firms using AC should have MTB ratios greater than one [45]. The median (1.0697) of MTB is also greater than one, suggesting that half of IB, at least, experience a high level of AC.

Table 2. Descriptive statistics

Variables	N	Mean	Std. Dev	Min	Median	Max
C_Score	60	1.5578	4.2228	−12.3596	0.0755	11.5396
SIZE	60	21.4019	2.2053	17.6716	21.1618	25.4684
LEV	60	9.5719	14.9695	0.2052	5.3184	83.6593
MTB	60	1.5803	1.9585	0.2253	1.0697	14.1389

6 Main Results

Prior to the analysis, the author clusters standard errors at the firm level, as suggested by Petersen [46], to mitigate the potential effect of homoscedasticity and autocorrelation. Table 3 reports the estimated results of the main model, Eq. (4), to explore the CC level in IB during the COVID-19 pandemic. As predicted, results show that IB experience a significant increase in CC level during the COVID-19 pandemic (see estimated coefficient CO19 = 5.0166, significant at less than 1%, one-tailed).

Table 3. Results of the main model

Variables C_score	Coefficients	Standard errors	P-value
CO19	5.0166***	0.9548	0.000
SIZE	0.3848***	0.0448	0.000
LEV	−0.1164***	0.0270	0.000
MTB	0.2742***	0.0578	0.000
Constant	−8.5047***	1.2598	0.000
Observations	60		
Number of clusters	30		
R-squared	0.6434		
F-statistic	138.22***		
p-value	0.000		

7 Robustness Tests

As in the current study, most studies dealing with the Basu model estimator of CC adopt the OLS model. In this regard, Ball, Kothari [47] recommend using the FEM. As a robustness check, the author also tries the FEM and the REM and finds no significant differences from the main model (the OLS).

On the other hand, winsorization is widely used in the literature. As a robustness check, the author winsorizes all continuous variables used in the main model at 1% and 5%. The untabulated results show similar findings.

In addition, this study computes the returns (R_{it}) using the simple formula for calculating returns, $(P_{it+1} - P_{it})/P_{it}$, where P_{it+1} is the share price of $firm_i$ on March 31 in $year_{t+1}$, P_{it} is the share price of $firm_i$ on March 31 in $year_t$. Following Barth, Landsman [48], and others, the author recalculates R_{it} based on the natural logarithm of the share price ratio, $Ln[P_{it+1}/P_{it}]$, but these alternative definitions yield similar results.

8 Additional Test - The Basu (1997) Model

The Basu [43] model is also widely used by researchers to measure CC [42]. Therefore, the author uses the Basu [43] model as an additional test. The Basu [43] model is conducted as follows:

$$E_{it} = \beta_0 + \beta_1 D_{it} + \beta_2 R_{it} + \beta_3 R_{it} D_{it} + e_{it} \tag{5}$$

To explore CC during the COVID-19 pandemic, the author transforms the Basu (1997) model by adding the $CO19_{it}$ variable and its interaction effects with share returns. This method is suggested by Ball and Shivakumar [49] who transform the model in order to compare differences between subsamples. However, this approach is followed by scholars in accounting such as André, Filip [39] and Barhamzaid [50]. Therefore, the author transforms the Basu (1997) model as follows:

$$\begin{aligned} E_{it} = {} & \beta_0 + \beta_1 D_{it} + \beta_2 R_{it} + \beta_3 R_{it} D_{it} \\ & + \beta_4 CO19_{it} + \beta_5 CO19_{it} D_{it} + \beta_6 CO19_{it} R_{it} + \beta_7 CO19_{it} D_{it} R_{it} + e_{it} \end{aligned} \tag{6}$$

In Eq. (6) A significantly negative (positive) value of β_7 reveals that CC is decreased (increased) in IB during the COVID-19 pandemic.

As can be seen in Table 4, Results show an increase in CC level within IB during the COVID-19 pandemic. The modified Basu (1997) model shows consistent results with the initial results of the main model (see estimated coefficient CO19_D_R = 2.1517, significant at less than 5%, one-tailed).

Table 4. Results using the modified Basu (1997) model

Variables	Coefficients	Standard errors	P-value
E			
D	0.2053	0.1524	0.188
R	0.1237	0.5275	0.816
D_R	0.2228	0.5510	0.689
CO19	0.0974	0.1455	0.509
CO19_D	0.1437	0.2700	0.599
CO19_R	-0.1554	0.5324	0.772
CO19_D_R	2.1517**	1.0544	0.025
Constant	−0.0421	0.1451	0.774
Observations	60		
Number of clusters	30		
R-squared	0.3474		
F-statistic	15.98***		
Prob > F	0.000		

9 Conclusion

This study explores CC in IB during the COVID-19 pandemic. A sample of 60 firm-year observations is analyzed from 2019 to 2020. The current study finds a significant increase in CC levels within IB during the COVID-19 pandemic. This finding enriches the research on CC, IB, and the economic outcomes of the COVID-19 pandemic. However, decision-makers (investors, creditors, etc.) can benefit from the governance role of CC experienced in IB during the pandemic. The current study also has some limitations. First, it explores the conditional form of AC, not the unconditional form of AC, in IB during the COVID-19 pandemic. Second, the study sample is drawn only from the Arabian Gulf countries. Therefore, the generalization of the results may be limited to this region.

Acknowledgments. The author is very thankful to the Palestine Technical University –Kadoorie for funding the present study.

References

1. Weaver, G.R., Agle, B.R.: Religiosity and ethical behavior in organizations: a symbolic interactionist perspective. Acad. Manag. Rev. **27**(1), 77–97 (2002)
2. Casselman, R.M., Sama, L.M., Stefanidis, A.: Differential social performance of religiously-affiliated microfinance institutions (MFIs) in base of pyramid (BoP) markets. J. Bus. Ethics **132**(3), 539–552 (2014). https://doi.org/10.1007/s10551-014-2360-z

3. Du, X., Jian, W., Du, Y., Feng, W., Zeng, Q.: Religion, the nature of ultimate owner, and corporate philanthropic giving: evidence from China. J. Bus. Ethics **123**(2), 235–256 (2013). https://doi.org/10.1007/s10551-013-1804-1
4. Du, X.: Does religion mitigate tunneling? evidence from Chinese buddhism. J. Bus. Ethics **125**(2), 299–327 (2013). https://doi.org/10.1007/s10551-013-1917-6
5. Dyreng, S.D., Mayew, W.J., Williams, C.D.: Religious social norms and corporate financial reporting. J. Bus. Financ. Acc. **39**(7–8), 845–875 (2012)
6. McGuire, S.T., Omer, T.C., Sharp, N.Y.: The Impact of religion on financial reporting irregularities. Account. Rev. **87**(2), 645–673 (2012)
7. Hilary, G., Hui, K.W.: Does religion matter in corporate decision making in America? J. Financ. Econ. **93**(3), 455–473 (2009)
8. Rosdini, D., Fildzah, N.: Comparison of conservatism in Islamic and conventional banks in Indonesia and Malaysia. Jurnal Manajemen **23**(2), 239–256 (2019)
9. Ardiansyah, M.: Accounting conservatism in the perspective of positive accounting theory: a study of Islamic banking in Indonesia. Asian Econ. Financ. Rev. **12**(6), 380–396 (2022)
10. AlAbbad, A.: Essays on the performance, disclosure, and corporate governance of Islamic banks. Rutgers University-Graduate School-Newark (2015)
11. Rudiawarni, F.A., Sulistiawan, D., Sergi, B.S.: Is conservatism good news? The case of stocks of Jakarta Islamic index. Heliyon **8**(4), e09292 (2022)
12. AlAbbad, A.: Accounting conservatism in islamic banking. Int. Rev. Account. Bank. Finan. **8**(1), 32–53 (2016)
13. Almutairi, A.R., Quttainah, M.A.: Corporate governance and accounting conservatism in Islamic banks. Thunderbird Int. Bus. Rev. **61**(5), 745–764 (2019)
14. Abdul Rahman, R., Omar, N., Rahman, A.: Shariah Governance And Accounting Conservatism. In: Proceedings of the 36th International Business Information Management Association Conference, IBIMA. Granada, Spain: IBIMA (2020)
15. García Lara, J.M., García Osma, B., Penalva, F.: Conditional conservatism and the limits to earnings management. J. Account. Public Policy **39**(4), 106738 (2020)
16. LaFond, R., Watts, R.L.: The information role of conservatism. Account. Rev. **83**(2), 447–478 (2008)
17. Ahmed, A.S., Duellman, S.: Evidence on the role of accounting conservatism in monitoring managers' investment decisions. Account. Finan. **51**(3), 609–633 (2011)
18. Elbadri, M., Bektaş, E.: Dynamic relationship among the bank stability, oil, and gold prices: evidence from the Islamic banks operating in the Gulf cooperation council countries. Int. J. Finan. Econ. **27**(2), 2153–2168 (2022)
19. Abdulla, Y., Ebrahim, Y.: Effect of COVID-19 on the performance of Islamic and conventional GCC banks. Rev. Finan. Econ. **40**(3), 239–258 (2022)
20. Watts, R.L.: Conservatism in accounting part i: explanations and implications. Account. Horiz. **17**(3), 207–221 (2003)
21. Rinaldi, L., Accounting and the COVID-19 pandemic two years on: insights, gaps, and an agenda for future research. In: Accounting Forum, pp. 1–32 (2022)
22. Cui, L., et al.: Accounting conservatism and firm performance during the COVID-19 pandemic. Account. Finan. **61**(4), 5543–5579 (2021)
23. Ma, L., et al.: The Effect of religion on accounting conservatism. Eur. Account. Rev. **29**(2), 383–407 (2020)
24. Bjornsen, M., Do, C., Omer, T.C.: The influence of country-level religiosity on accounting conservatism. J. Int. Account. Res. **18**(1), 1–26 (2019)
25. Swenson, L.: The impact of world religions on conservatism. J. Int. Account. Res. **19**(2), 197–218 (2020)
26. Fariz, N.M., et al.: Accounting conservatism and financial performance: accountability of Shariah compliant companies in Malaysia. Humanit. Soc. Sci. Lett. **8**(3), 280–297 (2020)

27. Herrmann, D.R., Pornupatham, S., Vichitsarawong, T.: The impact of the asian financial crisis on auditors' conservatism. J. Int. Account. Res. **7**(2), 43–63 (2008)
28. Francis, B., Hasan, I., Wu, Q.: The benefits of conservative accounting to shareholders: evidence from the financial crisis. Account. Horiz. **27**(2), 319–346 (2013)
29. Al-Hroot, Y., Al-Qudah, L.A.M., Alkharabsha, F.I.A.: The effect of the global financial crisis on the level of accounting conservatism in commercial banks: Evidence from Jordan. Int. J. Bus. Manage. **12**(2), 151–159 (2017)
30. Balakrishnan, K., Watts, R., Zuo, L.: The effect of accounting conservatism on corporate investment during the global financial crisis. J. Bus. Financ. Acc. **43**(5–6), 513–542 (2016)
31. Cerqueira, A., Pereira, C.: The effect of economic conditions on accounting conservatism under IFRS in Europe. Rev. Econ. Perspect. **20**(2), 137–169 (2020)
32. Kim, T., Shawn, H.: Conservative financial reporting and resilience to the financial crisis. Sustainability **14**(14), 8535 (2022)
33. Vichitsarawong, T., Eng, L.L., Meek, G.K.: The impact of the Asian financial crisis on conservatism and timeliness of earnings: evidence from Hong Kong, Malaysia, Singapore, and Thailand. J. Int. Financ. Manage. Account. **21**(1), 32–61 (2010)
34. Gul, F.A., Srinidhi, B., Shieh, T.: The Asian financial crisis, accounting conservatism and audit fees: Evidence from Hong Kong. SSRN Electron. J. (2002)
35. Leune, H.G.: The impact of the financial crisis on accounting conservatism in the Netherlands, in Erasmus School of Economics. Erasmus University Rotterdam (2014)
36. Heijden, R.J.J.V.D.: Accounting Conservatism and the crisis: What was the impact of the credit crisis on conservatism?. Erasmus University Rotterdam (2011)
37. Conroy, S.J., Emerson, T.L.N.: Business ethics and religion: religiosity as a predictor of ethical awareness among students. J. Bus. Ethics **50**(4), 383–396 (2004)
38. Einolf, C.J.: The link between religion and helping others: the role of values, ideas, and language. Sociol. Relig. **72**(4), 435–455 (2011)
39. André, P., Filip, A., Paugam, L.: The effect of mandatory IFRS adoption on conditional conservatism in Europe. J. Bus. Financ. Acc. **42**(3–4), 482–514 (2015)
40. Jenkins, D.S., Kane, G.D., Velury, U.: Earnings conservatism and value relevance across the business cycle. J. Bus. Finan. Account. **36**(9–10), 1041–1058 (2009)
41. Khan, M., Watts, R.L.: Estimation and empirical properties of a firm-year measure of accounting conservatism. J. Account. Econ. **48**(2), 132–150 (2009)
42. Zhong, Y., Li, W.: Accounting conservatism: a literature review. Aust. Account. Rev. **27**(2), 195–213 (2017)
43. Basu, S.: The conservatism principle and the asymmetric timeliness of earnings. J. Account. Econ. **24**(1), 3–37 (1997)
44. Miah, M.D., Uddin, H.: Efficiency and stability: A comparative study between Islamic and conventional banks in GCC countries. Future Bus. J. **3**(2), 172–185 (2017)
45. Beaver, W.H., Ryan, S.G.: Biases and lags in book value and their effects on the ability of the book-to-market ratio to predict book return on equity. J. Account. Res. **38**(1), 127–148 (2000)
46. Petersen, M.A.: Estimating standard errors in finance panel data sets: comparing approaches. Rev. Finan. Stud. **22**(1), 435–480 (2009)
47. Ball, R., Kothari, S.P., Nikolaev, V.V.: On Estimating conditional conservatism. Account. Rev. **88**(3), 755–787 (2013)
48. Barth, M.E., Landsman, W.R., Lang, M.H.: International accounting standards and accounting quality. J. Account. Res. **46**(3), 467–498 (2008)
49. Ball, R., Shivakumar, L.: Earnings quality in UK private firms: comparative loss recognition timeliness. J. Account. Econ. **39**(1), 83–128 (2005)
50. Barhamzaid, Z.A.A.: Unconditional conservatism under the Chinese version of IFRS. China J. Account. Res. **12**(4), 395–409 (2019)

An Assessment of Corporate Zakat Payment During Covid-19 Pandemic

Dodik Siswantoro[1] (✉), Mohamad Soleh Nurzaman[1], Sri Nurhayati[1], Abdul Ghafar Ismail[2], and Syed Musa Bin Syed Jaafar Alhabshi[3]

[1] Universitas Indonesia, Depok, Indonesia
dodik.siswantoro@ui.ac.id
[2] Universiti Kebangsaan Malaysia, Selangor, Malaysia
[3] International Islamic University Malaysia, Selangor, Malaysia

Abstract. This study aims to assess corporate zakat payment during the Covid-19 Pandemic Period. This is intriguing due to the fact that Covid-19 caused lockdown, recession, and financial crisis that could affect the company. Unlike business activities and performance which are significantly affected by locked down policies due to the Covid 19 Pandemic charitable activities including Corporate Zakat exhibit a more positive resilient response to continuously support those in need and affected by covid 19 during the pandemic period. This study employs a qualitative method based on the data provided. The outcome indicates that Covid-19 pandemic period had no impact on corporate zakat payment, as some businesses continued to pay zakat during the pandemic. This demonstrates that Covid-19 had no direct impact on zakat payments made by corporations.

Keywords: Corporate · Zakat · Covid-19 · Impact · Pandemic

1 Introduction

Zakat is required of Muslims who meet the criteria for obligatory zakat. If a Muslim does not meet the criteria for paying zakat, they are not required to do so. In Quran, the word for paying zakat is وَآتُوا الزَّكَاةَ, where the word آتُواis a passive command verb. It carries the meanings to derive, turn up, bring, come about, originate, develop, take place and occur. On the other hand, it may not belong to Muslim, zakat should be given to needy. Zakat is the right of other people.

The same treatment standards apply to corporate zakat. If the company fits the conditions for zakat payment, it must pay zakat. Corporate zakat is essentially zakat levied on the assets of individuals who invest in the firm and the company's earnings. The subject of zakat is the owner of the company who is required to pay zakat. Only in Saudi Arabia and Kuwait is zakat seen as obligatory for local businesses, therefore zakat payment is required. On the other hand, zakat is not mandatory in Malaysia and other Muslim countries.

The outbreak of the Covid-19 pandemic between 2020 and 2021 had a detrimental influence on the world economy, including Muslim nations. Aside from the health and

B. Alareeni et al. (Eds.): ICBT 2022, LNNS 620, pp. 66–71, 2023.
https://doi.org/10.1007/978-3-031-26953-0_8

communication industries, where personnel are most needed during the pandemic, other economic sectors are experiencing a slowdown, including retail, transportation, consumer, tourist, hotel, and energy. People cannot leave their homes and engage in activities, especially during lockdown periods. Decreased demand for goods and services disrupts the production process and reduces the company's earnings.

There was no studies found that explored the influence of Covid-19 on corporate zakat. Some publications focused solely on the influence of Covid-19 on an individual basis. Therefore, the objective of this paper is to analyze the impact of Covid-19 on corporate zakat payment. The paper's structure is as follows: introduction, literature review, research technique, and analysis. It ends with conclusion and recommendation.

2 Literature Review

There were no studies found on the influence of Covid-19 on corporate zakat. As a result, this study will use individual zakat payments for comparison purposes, because corporate zakat is based on an individual basis and is charged to the company's owner.

Some attempts and techniques were made to collect zakat during Covid-19 [1]. Furthermore, despite the fact that some tactics have been implemented, Covid-19 has an impact on zakat collection [2, 3]. As a result, zakat institutions should be vigorous in encouraging Muslims to pay zakat.

Covid-19 had no effect on the quantity of zakat paid by Muslims; rather, the amount grew [4]. In addition, It is discovered that awareness of zakat keeps Muslims paying zakat despite the Covid-19 pandemic [5, 6]. Therefore, the government should facilitate zakat payment for Muslims. Similarly, it is discovered that direct and indirect tactics boosted zakat collection [7]. Direct techniques include zakat collection and zakat counter services, whilst indirect strategies include Internet-based digital fundraising, crowd funding, and program marketing via Facebook and Instagram [8]. It is discovered a similar result when zakat collection rose in 2020 by employing techniques such as the food program, school collaboration, and Friday program [9]. The digitalization of zakat, government laws, and awareness of zakat payments are other determining factors [10]. In order to collect zakat funds, a zakat institution must have an effective program and strategy, particularly during economic downturns and crises.

Some initiatives in Malaysia could reduce the impact of Covid-19 [11]. Covid-19 may have a greater influence in other countries that have many lockdowns and closed economic activity than in countries that have not implemented such regulations.

3 Research Method

This study applies a qualitative research methodology to analyze the company's financial data. Only countries where zakat is discretionary would be included in the analysis. Hence, Saudi Arabia and Kuwait, where zakat is obligatory are removed from the analysis whereas Malaysia, Qatar, Sudan, Bahrain, and Egypt that do not mandate businesses to collect zakat are included in the sampling frame. The final sample consists of 38 companies. One company is excluded owing to a lack of information.

Data description	Total
Total Companies paid zakat	346
Compulsory payment	(307)
Voluntarily payment	39
Incomplete data	(1)
Final data	38

The analysis covers the descriptive analysis, total zakat payment (both obligatory and optional), and total corporate voluntary zakat. The study then specifies which companies paid zakat and which did not.

4 Analysis

The analysis begins with descriptive statistics, with 2021 having the highest average zakat payment, followed by 2018. 2020 had the highest median, followed by 2019. The largest standard deviation occurred in 2018, followed by 2021. This phenomenon revealed that the most diverse data was collected in 2018. It was also demonstrated by the highest kurtosis and skewness, as well as the range (see Table 1.).

Table 1. Descriptive statistics

Item	2017	2018	2019	2020	2021
Mean	945,370	1,593,994	1,131,753	1,349,709	1,736,381
Standard Error	227,207	537,965	288,735	304,301	402,723
Median	247,774	475,838	602,225	608,085	577,276
Standard Deviation	1,400,598	3,316,238	1,779,880	1,875,836	2,482,549
Sample Variance (million)	1,961,674	10,997,436	3,167,974	3,518,761	6,163,047
Kurtosis	3	17	14	8	6
Skewness	2	4	3	3	2
Range	5,854,599	17,990,315	9,640,743	9,308,209	11,546,350
Minimum	-	-	-	-	-
Maximum	5,854,599	17,990,315	9,640,743	9,308,209	11,546,350
Sum	35,924,058	60,571,764	43,006,601	51,288,952	65,982,493
Count	38	38	38	38	38

Total zakat payment, which includes compulsory payment from countries like Saudi Arabia and Kuwait, fell in 2019, then grew until 2021. (see Fig. 1). Hence, prior to the occurrence of Covid-19, overall zakat payment was already reduced in 2019. Covid-19 occurred in 2020, resulting in numerous lockdowns and economic downturns. However, total zakat payments increased in 2020 and 2021 in comparison to 2019.

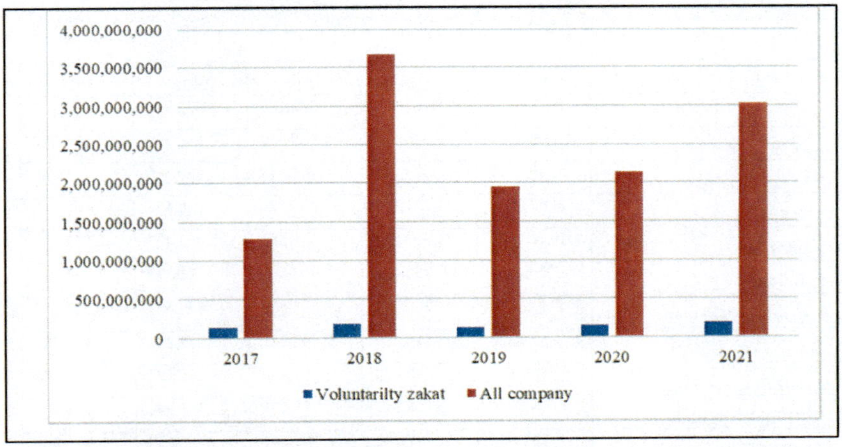

Fig. 1. Total zakat payment (in USD)

A similar pattern emerged in voluntary zakat payments, which excluded enterprises from Saudi Arabia and Kuwait (see Fig. 2). The overall voluntary zakat payment is less than the total zakat payment which includes Saudi Arabia and Kuwait. When compared to 2018, total zakat payment fell in 2019. This shows that before Covid-19 in 2020, most companies had smaller zakat payment in 2018 and 2020.

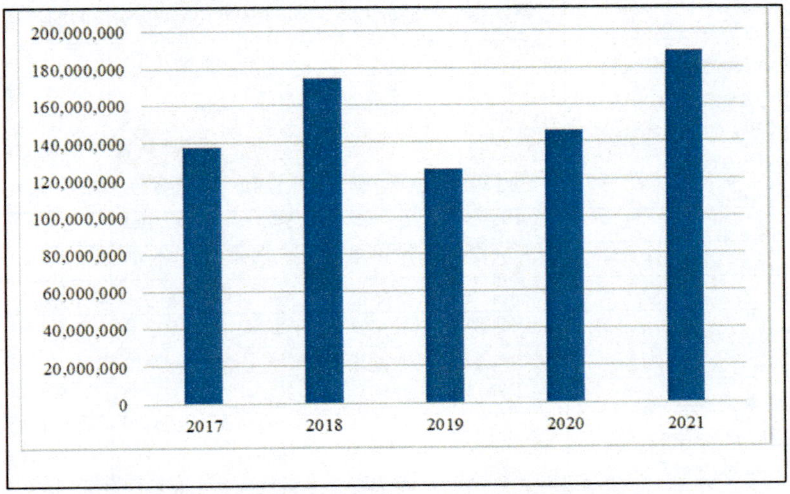

Fig. 2. Voluntarily zakat payment (in USD)

In 2020, there are two companies who paid zakat at the first time, which are Bursa Malaysia Bhd and Boustead Plantations. This demonstrated that the Covid-19 outbreak had no effect on corporations' refusal to pay corporate zakat. Four enterprises, Fima Corporation Bhd, Dagang NeXchange, AWC Bhd, and Kumpulan Fima Bhd, paid zakat

Table 2. Zakat payment

Company Name	2017	2018	2019	2020	2021
United Gulf	1,373,378	1,484,830	302,171	2,492,923	3,409,710
Ajwa for Food	2,052,957	2,340,775	1,108,933	2,543,234	3,144,949
UEM Edgenta Bhd	404,550	690,073	749,572	588,806	169,789
UEM Sunrise Bhd	678,536	1,261,501	490,584	268,657	0
UMW Holdings Bhd	1,710,930	1,307,022	568,599	2,063,433	2,062,440
CIMB Group Holdings	309,100	750,605	904,867	721,393	1,215,178
Boustead Holding Bhd	0	266,344	635,852	870,647	6,556,196
Bank Islam Malaysia	3,575,420	3,556,659	2,998,288	2,927,861	2,947,887
AMMB Holdings Bhd	526,670	588,480	701,832	245,597	515,822
MNRB Holdings Bhd	145,779	166,912	298,168	311,942	272,900
Syarikat Takaful	265,826	247,215	307,899	318,657	316,042
Public Bank Bhd	64,293	62,954	73,368	300,000	171,230
Fima Corporation Bhd	0	0	0	0	25,696
Dagang NeXchange	98,063	110,296	0	0	144,803
Symphony Life Bhd	0	0	193,833	147,165	0
Gas Malaysia Bhd	865,480	847,458	855,955	870,647	840,538
Telekom Malaysia Bhd	2,200,791	1,573,850	1,834,189	1,542,289	1,705,091
Axiata Group Bhd	229,723	112,107	422,842	440,050	316,282
FGV Holdings Bhd	1,250,247	4,504,358	1,564,441	4,054,726	4,180,115
Malayan Banking Bhd	5,854,599	9,908,475	9,640,743	9,308,209	11,546,350
Tenaga Nasional Bhd	0	17,990,315	4,695,525	5,671,642	5,979,827
MSM Malaysia	0	363,196	3,424	369,652	240,154
AWC Bhd	0	0	0	0	27,236
Kumpulan Fima Bhd	0	0	0	0	99,453
Malaysia Building	2,935,460	3,147,700	219,858	90,299	34,102
Pharmaniaga Bhd	148,368	259,322	547,811	627,363	5,781,220
Bursa Malaysia Bhd	0	0	0	165,920	189,001
Company Name	2017	2018	2019	2020	2021
Malaysia Airports	1,100,396	1,357,869	1,497,921	371,891	319,404
Boustead Plantations	0	0	0	226,866	772,574
Kumpulan Perangsang	39,812	121,065	273,661	378,109	600,384
Ranhill Utilities Bhd	0	0	1,659,819	2,240,547	127,762
Kenanga Investment	0	0	91,220	110,697	89,337
RHB Bank Bhd	0	0	1,222,793	1,017,662	1,409,222
Affin Bank Bhd	833,581	995,884	1,352,409	1,857,960	1,541,306
Zad Holding Company	4,077,765	4,094,398	4,183,266	4,033,647	4,917,333
Barwa Real Estate Company QPSC	0	0	1,099,176	973,077	554,167
Sudatel Telecom	3,469,125	1,532,676	1,559,852	2,160,133	2,968,883
Al Salam Bank Sudan	1,713,209	929,426	947,729	977,253	790,110
Total	35,924,058	60,571,764	43,006,601	51,288,952	65,982,493

for the first time in 2021. UEM Sunrise Bhd and Symphony Life Bhd were the only two companies that did not pay zakat in 2021. However, total zakat payments did not fall during the Covid-19 pandemic. This is consistent with individual zakat payment, as corporate zakat is essentially also predicated on the company's owner paying zakat [4, 5] (Table 2).

5 Conclusion

The effect of Covid-19 on corporate zakat payment is not evidence. However, in 2021, two corporations failed to pay corporate zakat. This requires additional investigation and research. The limitation of this study was that it only looked at the overall amount and number of companies that paid zakat. Further research needed to explore how and why the amount of corporate zakat payment increase or decrease. While recommendation of this research is government and zakat institutions must increase the variety of strategies to raise zakat funds.

References

1. Al Anshory, A.C., Adhiningsih, S.M., Saputra, M.I.: Strategy of Zakat Collection during COVID-19 Pandemic: Muzaki Perception (2021)
2. Kamrizal, K., Nenengsih, N.: Strategy of Zakat Collection during the Covid-19 Pandemic at Padang City (Case Study at LAZ Charity Risalah). Int. J. Adv. Sci. Educ. Relig. **4** (2021). https://doi.org/10.33648/ijoaser.v4i1.101
3. Nurhidayat, N.: Strategi Fundraising Zakat Pasca Pandemi Covid-19. SALAM J. Sos. dan Budaya Syar-i **7** (2020). https://doi.org/10.15408/sjsbs.v7i8.16553
4. Napitupulu, R.M., Hayati, L.R., Sapna, F.: Perilaku masyarakat dalam menunaikan zakat di masa pandemi COVID-19. Jeiei. **7** (2021)
5. Hakimi, F., Widiastuti, T., Al-Mustofa, M.U., Al Husanaa', R.: Positive effect of attitude, peer influence, and knowledge zakat on zakat compliance behavior: update In Covid 19. J. Islam. Econ. Laws **4** (2021). https://doi.org/10.23917/jisel.v4i2.13859
6. Zulfikar, H., Kamiluddin, K.: Digital zakat fundraising in times of Covid-19 pandemic: evidence from BAZNAS, indonesia. J. Islam. Econ. Bus. Res. **1** (2021). https://doi.org/10.18196/jiebr.v1i2.1
7. Ahyani, F., Syahbudin, F.: ZIS fund-raising strategy in the time of the Covid-19 Pandemic (Case Study of LAZ Nahwa Nur). Focus (Madison) **1** (2021). https://doi.org/10.37010/fcs.v1i2.353
8. Fitriani, E.S., Agrosamdhyo, R., Mansur, E.: Strategi Penghimpunan dan Penyaluran Zakat, Infak, dan Sedekah (ZIS) dalam Program Sebar Sembako pada Masa Pandemi Covid-19 di Badan Amil Zakat Nasional (BAZNAS) Provinsi Bali. Widya Balina. **5** (2020). https://doi.org/10.53958/wb.v5i1.52
9. Khaerunisa, S., Yayuli, Y., Febriandika, N.R.: Analysis of the Impact of Covid-19 on the Collection of Zis Funds (Zakat, Infaq and Shodaqoh) in Lazismu Surakarta in 2020. Proc. B. Int. Conf. Islam. Econ. Islam. Financ. Islam. Law 2021. **1** (2021)
10. Dafiq, M.I., Rohim, A.N., Priyatno, P.D.: Strategy analysis of ZIS fund collection during the Covid-19 pandemic at baznas DKI jakarta province. El-Barka J. Islam. Econ. Bus. **4** (2021). https://doi.org/10.21154/elbarka.v4i2.3231
11. Asni, F.: Pengurusan kutipan dan agihan zakat oleh maips ketika Covid-19 di negeri perlis: suatu sorotan kajian (Management of Zakat Collection and Distribution By. Int. J. Islam. Econ. Financ. Res. **4** (2021)

The Role of Blockchain Technology in the Management of Waqf

Nur Hidayah Laili, Khairil Faizal Khairi(✉), and Rosnia Masruki

Universiti Sains Islam Malaysia, Bandar Baru Nilai, 71800 Nilai, Negeri Sembilan, Malaysia
khairil@usim.edu.my

Abstract. Waqf has historically been a vital institution in Islam in fulfilling religious needs and improving social welfare. Its effectiveness in medieval and modern Islam is evident. However, this institution is seemingly not without inefficiencies. Public trust and contributions are also low due to the opacity of the accounting and accountability system. Scholars have proposed adopting the blockchain technology to mitigate these issues. The implementation of the blockchain technology is to provide a solution for waqf organisations towards offering transparency to their donors and building trust to gain more support towards the waqf organisation's mission and work. The blockchain technology manages information within the waqf system by enhancing the aspects of transparency and reliability in how funds are collected, managed, and distributed. This study has two objectives: first, to look into the technological role of blockchain in managing waqf assets in ensuring the most efficient way of financing waqf projects. Second, this research is aimed at exploring and understanding the blockchain technology regarding the issues of trust and accountability in waqf organisations in Malaysia. The study methods consisted of content analysis and an interview. The content analysis focused on publicly accessible information online such as home pages, websites, user interface pages, and annual and financial reports. The semi-structured interview was carried out with a representative of a waqf organisation who has extensive knowledge of these technologies. This article is an effort to contribute to the body of knowledge by exploring blockchain technology's potential applications and their limits.

Keywords: Blockchain technology · Waqf · Accountability

1 Introduction

Blockchain captured the world's attention in 2015 by offering a new way of creating, exchanging, and tracking data on a peer-to-peer basis (Yermack 2015). The use of blockchain is limitless: it gives access to cryptocurrencies, tracks and settles digital and mainstream assets in an environment that is cryptographically secure, reduces the complexity and number of intermediaries participating in existing transaction processes, and manages digital risks such as settlement and counterparty risks better (Piazza 2017). Since blockchain engages peer-to-peer operations on a distributed structure made of secure records, no physical entity is required for it to work (Monich 2017).

B. Alareeni et al. (Eds.): ICBT 2022, LNNS 620, pp. 72–81, 2023.
https://doi.org/10.1007/978-3-031-26953-0_9

Muslims' contributions to the waqf assets are estimated to be worth more than US$1 trillion globally (Noordin 2018). Despite the large amount, it is difficult for individuals and institutions to maximise the waqf value. Waqf institutions face several challenges. It has been reported that the database of Islamic Religious Council on waqf assets is not adequate, and detailed and important information is not made available to the stakeholders. This inadequacy has affected the process of developing waqf lands, raised questions on the accountability of waqf organisations, and increased the waqf organisations' risk. Looking at the potential of waqf assets and the risk related to waqf organisations, these organisations need to be transparent and communicate their accountability. Effective and efficient communicated accountability can overcome the issues and attract donors. Thus in the long run, waqf organisations may become sustainable and achieve their missions.

Blockchain is a revolutionary technology that waqf organisations could leverage to exponentially enhance the management of waqf assets and streamline operations. This technology is still at an infancy stage for waqf organisations, with only a few organisations pursuing the blockchain system in managing and investing Islamic assets. A Singapore-based financial technology company, Finterra, has developed a crowdfunding platform that uses digital ledger technology, or blockchain, to create "smart contracts' that would be tied to a specific waqf project (Young 2018). Blockchain makes it possible to coordinate organisational transactional activities within a strong mechanism of trust and transparency (Lehr and Lamb 2018). This technology can provide a more efficient way of raising money and managing and transferring the ownership of Islamic assets for waqf organisations, which receive donations from Muslims to operate social projects such as mosques, schools, and welfare schemes. It is apparent that the characteristics and conditions of blockchain are in alignment with the principles of Islam. More importantly, blockchain is a perfect medium to incorporate and operationalise the Islamic values of justice, equality, trust, and fairness into waqf organisations, which embody the spirit of Shariah (Zainuddin 2017). In this regard, this study first examines the technological role of blockchain in managing waqf assets in the most efficient manner to finance waqf projects. Second, this research seeks to explore and understand the blockchain technology regarding the issues of trust and accountability among waqf organisations in Malaysia.

This paper is divided into four sections. The first section highlights the waqf system in Malaysia. The second section discusses the blockchain mechanism and characteristics and the role of blockchain in waqf management. Then, Sect. 3 discusses the methodology and Sect. 4 discusses the result and conclusion of the study.

2 Literature Review

2.1 Waqf System in Malaysia

Waqf is one of the most important institutions that contribute to the economic growth of the Muslim ummah. It plays similar roles to other charity instruments in Islam like *infaq*, *sadaqah jariyyah*, *hibah*, and others. However, waqf is unique in that for as long as the asset endowed is still in use and benefits other people, the *waqif* will continue to be rewarded. From the socioeconomic aspect, waqf can be used as a capital resource to fund economic activities such as businesses and investments. The capital resource from waqf can produce high profits if managed in efficient and structured manners. Besides, waqf practices can also inculcate responsible members of society who help one another (Ramli et al. 2014).

In Malaysia, waqf organisations have been growing with their role expanding to be more holistic. In managing waqf, State Islamic Religious Councils (SIRCs) play the most important governance role relating to Islamic matters that fall under the states' responsibilities. There are 14 SIRCs in Malaysia, and each is aimed at making the administration of waqf systematic and effective for the Muslim ummah. The SIRC is the sole trustee or also known as *nadzir* or *mutawalli* for all matters pertaining to waqf and waqf funds for each state (Sulaiman 2012). As a nadzir, the SIRC is also responsible for investing a portion of the waqf profits and channelling the collected waqf funds to charitable projects that help waqf beneficiaries, particularly in terms of education and health. Excellent management and governance of waqf organisations is an indicator of economic achievement that would strengthen the evidence that waqf is a tool that can help boost the economy and the social development of Muslims.

2.2 Blockchain Mechanism and Characteristics

Blockchain is simply a decentralised network of nodes—servers, laptops, and computers connected to the blockchain—that verify, record, and store data blocks that are interlinked by chains. Each block contains multiple transaction records. Each record is identified with a hash, a fixed-length alphanumeric string generated from a string of characters using a mathematical function. This hash details the specific record for each transaction, including the sender, recipient, other participants, date, time, and other details as coded into the blockchain algorithm (Brito and Castillo 2013; Salmon and Myers 2019). After the block has been forged, it is also "stamped" with a hash, which connects it to the preceding block to form the "chain". This hash also stores the transaction hashes recorded inside the block. Each block is added to the chain subsequent to the verification by the network participants of its authenticity, which is proven using a consensus mechanism (Zīle and Strazdiņa 2018). This entire process continues indefinitely, and theoretically, there is no limit to the number of blocks on the network.

Blockchain is considered immutable and secure because any alteration to a block would require the rebuilding and rehashing of the subsequent blocks (Hileman and Rauchs 2017). For instance, slightly modifying the information, even a character, contained in the seventh block of a 10-block network would produce a new hash for the altered block. This will cause the block to disconnect from the original eighth block, since the latter identifies the original seventh block by its hash. Effectively, modifying the seventh block transforms the network into a seven-block chain. To cover this clandestine deed, one must update the remaining three blocks so that they will generate new hashes that can continue this new chain. Failure to do so, or if the majority of the nodes in the network have verified the original chain to be the true chain that connects to the genesis block (the first block in a chain), this altered seventh block becomes an extinct or orphan block, and this new chain ends there (Decker and Wattenhofer 2013). Thus, two forks are created in the blockchain network, with one leading to the now abandoned seventh block and the other being the consensually agreed correct path. Without the collusion from the majority of the nodes—an expensive and resource-intensive undertaking—any attempt to alter information in a blockchain is effectively unfeasible (Bonneau 2019).

2.3 The Role of Blockchain in Waqf Management

Waqf organisations rely on contributions from the public and the trust placed in them by the wider community. The advantage of combining the technological concept of blockchain with the historical and traditional concept of waqf is that both of these innovations deal with maintaining a sense of public trust for the benefit of society as a whole. Waqf organisations are now turning to blockchain solutions as a way to improve social and economic outcomes. One of the many potential blockchain applications is in supporting transactions involving cryptocurrencies such as bitcoins. Bitcoin is a new form of currency and blockchain is the database that enables its secure transaction by providing transparency, enabling efficiency, and reducing costs (Tomasicchio 2016).

Transparency is another benefit of using blockchain with donations being recorded immutably and transparently. Givetrack is an organisation rooted in blockchain technology that allows donors to donate bitcoins to their favourite project and track its progress in real time, such as Crowdfunding 2.0 (Lamb 2018). The potential of blockchain for Islamic organisations is focused more on the financial industry. Blockchain's strength has attracted hundreds of financial institutions across the Organisation of Islamic Cooperation (OIC) countries to pursue solutions relating to the blockchain system. Dubai had unveiled its plan to become the leader of blockchain utilisation across its private and public sector by 2020 (Zainuddin 2017). The first movers in applying the blockchain technology in Islamic finance include Blossom Finance, EthisCrowd, and OneCoin Ltd, a cryptocurrency company that is the first to launch a certified Shariah-compliant product. Other first movers include Finocracy, a Dubai-based fintech start-up (Sundararajan 2017).

This technology is still in its early stage among waqf organisations, with only a few organisations pursuing the blockchain technology in managing and investing Islamic funds. A Singapore-based financial technology company, Finterra, has developed a crowd-funding platform that uses digital ledger technology, or blockchain, to create "smart contracts' that would be tied to a specific waqf project (Young 2018). Thus, waqf organisations need to be forward thinking and adoptive in order to incorporate new technologies and practices such as artificial intelligence, blockchain, and Internet of Things (IOT) into their strategies and workforces.

The application of blockchain technology offers numerous opportunities for waqf organisations by reestablishing trust in the sector, where every transaction must be verified through a process known as consensus that requires multiple system participants to independently verify authenticity (Piazza 2017). It also reduces transaction costs, as participants can avoid using expensive third-party services to facilitate transactions. With traditional donations, money goes through various middlemen such as government agencies and financial institutions, and fees are charged by credit or debit card as well as by clearing houses. Blockchain donations are executed directly, reaching the charity instantaneously and at a fraction of the cost (Lamb 2018). By leveraging the blockchain technology into waqf organisations, donors can transfer and track their money and obtain permanent records of financial transactions across the globe from beginning to end. No record on the blockchain can be erased, thereby providing traceable accountability by showing the exact flow of any coin in the ecosystem (Campbell 2016).

The blockchain technology implementation is to provide a solution for waqf organisations to enable them to offer transparency to their donors and build trust with the expectation of gaining more support towards the waqf organisations' mission and work. There is a smart contract managing the transactions, the beneficiaries, the waqifs, and the recorded amounts of materials in secure data that are private and confidential. Regarding the reporting function, the blockchain technology would bring a major transformation by making information available in real time. Information would post to the blockchain automatically if the organisation used digital currency or digital tokens (Hans 2017). Real time updating would also help the regulators. Since the transactions would be governed by smart contracts, functions could be built into the contracts to ensure an organisation was referred to a regulator if certain events occurred (Mark et al. 2016).

3 Methodology

This study employed a qualitative research approach to gain insights into waqf organisations on their views concerning the use of blockchain for waqf management. In gathering their views, a semi-structured interview was conducted focusing on the role of blockchain in managing waqf organisations. The interview was guided by an interview protocol and probing questions were asked to pursue issues and to enhance the researchers' understanding of the responses, while at the same time improving the validity and reliability of the interview conducted (Yin 2003). The interview responses provide an understanding of the role of blockchain in waqf from the managerial perspective. Hence, the research findings from the study provide a comprehensive and realistic input regarding the potential of blockchain as a mechanism for waqf management and offer a rich understanding of the context in which the technology can be deployed.

The Head of Asset Development & Waqf Properties Unit of Majlis Agama Islam Pulau Pinang (MAINPP) has been interviewed for this study. The interviewee has extensive knowledge about the operation and management of waqf due to the interviewee's experience in different roles in the unit, i.e., reviewing and planning waqf property development for future investments as well as maintaining waqf property data and information. MAINPP was chosen due to its focus on enhancing transparency in waqf management and its involvement with donors. Insights gained from the waqf organisation help improve the researcher's understanding of donors' attitudes and roles in blockchain-based projects. MAINPP has a property management and waqf division which is responsible for managing waqf. It plays an important role in protecting waqf properties in Penang. The division aims to maintain and administer waqf properties properly in accordance with the function of MAINPP as a trustee of all waqf in Penang that collects information and data on waqf assets, identifies the waqf properties that are not registered and enrolled under the name of MAINPP, and ensures that waqf properties are well-governed for the benefit of Penang's Muslim community. In 2018, MAINPP established Wakaf Pulau Pinang Sdn. Bhd. to assist in the management of waqf properties under its administration.

4 Result and Discussion

4.1 The Practices of Waqf System in Malaysia

The interviewee was asked regarding the waqf system practices in MAINPP. At the same time, the interviewee was also asked about the challenges faced by MAINPP in managing the waqf fund. Below are some of the responses gathered from the interview:

In MAINPP, waqf is divided into two types: general and specific waqf. General waqf refers to any form of waqf property or cash made for charitable or general purposes without specifying any specific beneficiary (whether individual, organisation or institution) or specific purpose. Waqf from this category can be applied to matters or purposes of public welfare. The waqf does not impose any conditions on the endowed property and can be used for the community's best interest for the benefit of Islam. Cash waqf is a common type of fund contributed by the waqif. This form of waqf must be managed wisely until it grows and generates income for social welfare. MAINPP is responsible for managing waqf assets including cash waqf. Cash waqf application in Pulau Pinang began in 1994. It constitutes money collected for waqf purposes for charity in accordance with Islamic law. The waqif can contribute cash via three options: Internet banking by transferring to MAINPP's waqf trust account, salary deduction, and walk-in to MAINPP's office. In MAINPP, general waqf refers to cash waqf that is donated for general purposes. The *nadzir* is responsible for managing the waqf fund as long as the objectives of the waqf are achieved. The nadzir will invest the collected fund in various investment portfolios to buy assets. The nadzir also channels funds to identified projects, and any balance will be transferred back to the waqf fund for future undertakings. The gains or profits from investments are channelled to waqf programmes.

The second type of waqf is contributed by the waqif for a specific purpose and intention on the usage of the waqf property. In specific waqf, the donor declares the purpose of the waqf in advance, for example, to build a mosque or a school or for any other purpose that the donor specifies. The purpose will then be specified in a waqf deed and the manager (mutawalli) will be entrusted to administer the waqf as specified in the deed. In MAINPP, specific waqf which is accepted for a specific purpose is classified as assets waqf. The waqif may specify the beneficiaries, the purpose of use, or the purpose for which the waqf is created in the foundation's deed.

4.2 The Role of Blockchain in Waqf Management

Based on the interviewee's response to the first question concerning the current waqf management system at MAINPP, the researchers proposed a blockchain system to overcome the challenges faced in waqf management.

One of the special features of the waqf blockchain technology is the total transparency offered in the distributed ledger, where every transaction is recorded publicly and authorised individuals can view the transactions. The waqif who donates on the waqf EOS platform can track their contribution from the moment it is gifted until the beneficiaries have reaped the rewards. Moreover, using the smart contract, which is a computer program or algorithm that automatically executes itself when the predefined conditions

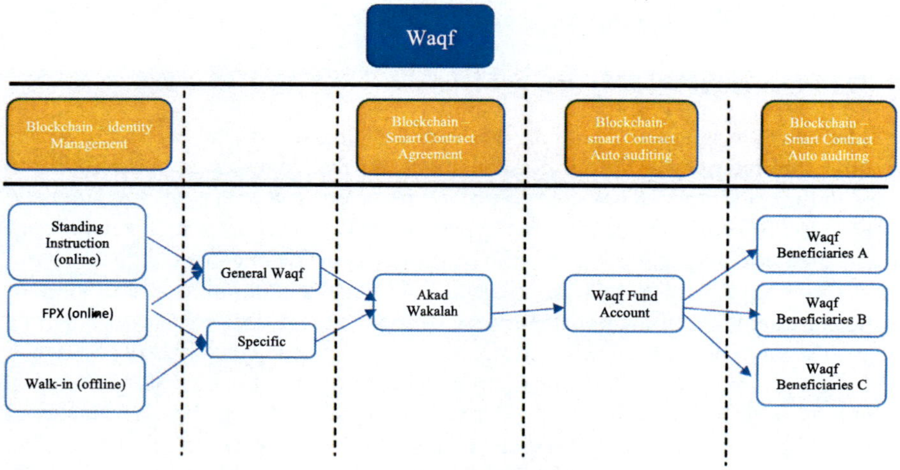

Fig. 1. Waqf Blockchain system

are met, could further enhance the performance and efficiency of waqf management while also increasing the security, thus providing transparency.

Figure 1 above shows the proposed waqf blockchain that solely focuses on one of the blockchain features, "smart contract auto auditing", which is important for a waqf institution. The main advantage of this system is it can reduce the operation cost, documentation, and manpower due to the auto-auditing function. Traditionally, the cost of managerial operation is high due to several layers of auditing documentation. If the smart contract is introduced into waqf, the cost of operation and monitoring can be reduced significantly.

The interviewee was asked on the potential role of the proposed blockchain technology in the development and management of cash and assets waqf. Then, the interviewee was asked about which blockchain characteristics are seen to be most relevant for the development of cash and assets waqf. Below are some of the responses given by the interviewee:

There is a need to improve the transparency, security, and accountability of the waqf fund and the spending in order to improve public trust and confidence. The record keeping technology offered by blockchain's smart contract auto auditing records every transaction that has been verified and approved and keeps track of every transaction broadcasted from the waqif all the way to the waqf beneficiaries. In this instance, there is a verification that the waqf fund has been transferred to the intended beneficiary. By monitoring the entire sequence of transactions, the waqif is assured as to where the waqf fund is and who receives it, and that the fund is properly managed in ensuring it is effectively and efficiently used for socioeconomic development.

The Asset Development & Waqf Properties Unit of MAINPP is responsible for three main duties: planning for the accumulation of waqf fund, waqf investment, and waqf fund distribution. Therefore, it requires competent personnel with knowledge and skills to manage and report the data and the waqf fund for project planning to generate income and highly productive investments. In exercising these responsibilities, the ability to collect and report data on the waqf fund is critical for planning and analysing the waqf projects and for making productive investments. Apart from the cash and assets waqf, the information on waqif that MAINPP receives via the waqf fund platform is extremely valuable. The waqf fund report should have information on the waqf management's overall performance as well as the specific performance of each waqf fund to which the beneficiaries are associated. This waqf fund report should provide appropriate checks and balances and can serve as an official record of the progress and performance of waqf projects.

The technology offered by blockchain (identity management, smart contract agreement, and smart contract auto auditing) will allow the waqf management team to account for their activities. The real-time information on the distribution of the waqf collection will track the passage of funds from the waqif to the beneficiary, and the waqf fund data can be automatically imported into MAINPP's database. The database is important for monitoring and controlling the waqf fund and the progress and performance of waqf projects. The technology also offers transparent accounting of the waqf fund route, enabling MAINPP to track the logistics and operations and to produce a specialised report of the amount that the waqif contributed, the percentage of completion of the waqf project, and the total amount raised for the waqf project.

Under the administration of MAINPP, institutions such as *maahad* or religious institutions have to register as beneficiaries and have to obtain permission to collect donations from the public. This rule is to prevent the public from being deceived by syndicates that collect donations. However, a lack of enforcement and poor monitoring by the government have resulted in an increase in the number of unlicensed donor collectors. Thus, by providing the beneficiaries, i.e. maahad and religious institutions, with identity management, they will be equipped with an identity tool as registered beneficiaries and entitled to receive the waqf fund or assets contributed by the waqif. The identity management function would allow MAINPP to digitally register entitlements and accumulate transactional data of the beneficiaries, thereby capturing records permanently and preventing identity theft. Thus, this technology can be used to tackle the issue of unregistered and unlicensed donor collectors in ensuring that the waqf fund is provided to the intended beneficiaries and assuring the waqif that their waqf funds or assets have reached the desired destinations.

5 Conclusion

The blockchain technology is developed mainly to help the management of waqf solve financial constraints in a sustainable way. The institution of waqf has always played a pivotal role in ensuring the sustainable economic development of the Muslim society. This paper has identified the challenges faced in waqf management in the context of tracking the waqf fund transactions to the beneficiaries, reporting data from the waqf fund, measuring the performance of waqf projects, and addressing the issues concerning unregistered waqf fund collection. Based on the existing shortcomings, this study has explored ways that technology can help in the management of waqf. We examined how a blockchain solution consisting of identity management, a smart contract agreement, and smart contract auto auditing might overcome several challenges currently associated with the management of waqf. The blockchain innovation has substantial potential and will be beneficial to the management of waqf in the hope of providing a more effective way to manage the waqf fund and increasing public trust in waqf organisations.

This research exploring the blockchain technology in regards to the issues of trust and accountability in Wakaf Pulau Pinang Sdn. Bhd managed by MAINPP, hence it is impossible to generalize the result of this study to all Majlis Agama Islam in Malaysia. Thus, this study recommends future research to explore other Majlis Agama Islam in Malaysia in an effort to contribute to the blockchain technology's potential applications and their limits.

Acknowledgements. This research was supported by Fundamentals of Research Grant Scheme (FRGS) provided by the Ministry of Education of Malaysia with grant code: FRGS/1/2019/SS01/USIM/02/7 for the research titled "Developing A Blockchain-Based Governance Framework For Waqf Organisations Toward A Sustainable Economics Wellbeing". Special thanks to the staff and the Head of Asset Development & Waqf Properties Unit under Majlis Agama Islam Pulau Pinang (MAINPP) for their full cooperation and commitment in supporting this study.

References

Bonneau, J. Hostile Blockchain Takeovers (Short Paper). Berlin, Heidelberg (2019)

Brito, J., Castillo, A.: Bitcoin: A Primer for Policymakers. Mercatus Center, Arlington (2013).

Campbell, R.: BitGive Launches GiveTrack, a Real-Time Donation Tracking System (2016). https://bitcoinmagazine.com/articles/bitgive-launches-givetrack-a-real-time-donation-tracking-system-1482179875/

Ecker, C., Wattenhofer, R.: Information Propagation in the Bitcoin Network. In: Paper presented at the IEEE P2P 2013 Proceedings (9–11 Sept. 2013)

Hans, B.: Blockchains, Real-Time Accounting and the Future of Credit Risk Modeling. Lund University Working Paper 2016, p. 4 (2017)

Hileman, G., Rauchs, M.: Global blockchain benchmarking study. Cambridge Centre for Alternative Finance, Cambridge (2017)

Lamb, P.: Transforming the Social Sector: Bitcoin and Blockchain for Good (2018). https://www.huffingtonpost.com/entry/transforming-the-social-sector-bitcoin-andblockchain_us_59c169e3e4b0f96732cbc9c7

Lehr, D., Lamb, P.: Digital Currencies and Blockchain in the Social Sector (2018). https://ssir.org/articles/entry/digital_currencies_and_blockchain_in_the_social_sector1

Mark, C., Snagg, F., Claire, H.: Distributed ledgers: a future in financial services? J.I.B.L.R. **31**(5), pp. 246–247 (2016)

Monich, I.: Blockchain - A revolutionary solution to problems of modern corporate governance. Economics Journal of Transbaikal, State University, vol. 15 (2017)

Noordin, K.A.: Islamic Finance: Using Blockchain to Solve Waqf Challenges. The Edge Malaysia (2018)

Piazza, F.S.: Bitcoin and the blockchain as possible corporate governance tools: strengths and weaknesses. Penn State Journal of Law & International Affairs, Is. 2. vol. 5 (2017)

Ramli, R.N.A., Hamid, N.A.: Kelestarian Wakaf Dalam Membangunkan Institusi Pendidikan: Kajian Awal. In: Proceeding of International Conference on Postgraduate Research (ICPR 2014). Kuala Lumpur, Malaysia (2014)

Salmon, J., Myers, G.: Blockchain and Associated Legal Issues for Emerging Markets (2019). Retrieved from https://openknowledge.worldbank.org/handle/10986/31202

Sundararajan, S.: Islamic Development Bank to Research Sharia-Compliant Blockchain Products (2017). https://www.coindesk.com/islamic-development-bankresearch-sharia-compliant-blockchain-products/

Sulaiman, S.: Isu Pembangunan Wakaf Menggunakan Struktur Amanah Pelaburan Hartanah Islam di Malaysia: Satu Tinjauan (2012). Accessed on 25 Januari 2017. http://e-muamalat.islam.gov.my/sites/default/files/kertas_ilmiah/2016/01/pembangunan_wakaf_menerusi_pendanaan_kerajaan_dan_kerjasama_institusi_kewangan_dan_korporat_hala_tuju_cabaran_dan_harapan.pdf

Tomasicchio, A.: Italian Startup Creates Bitcoin Platform for Transparent Natural Disaster Donations (2016). https://cointelegraph.com/news/italian-startup-creates-bitcoin-platform-for transparent-natural-disaster-donations

Yermack, D.: Corporate Governance and Blockchains. Nber Working Paper Series. National Bureau of Economic Research (2015). http://www.nber.org/papers/w21802

Yin, R.K.: Case study research design and methods, 3rd ed. Sage Publications, Thousand Oaks, CA (2003)

Young, M.J.: Islamic Endowments to be Modernized by Blockchain (2018). https://www.newsbtc.com/2018/02/12/islamicendowments-modernized-blockchain/

Zainuddin, A.: Blockchain. Fintech Industry, Islamic Finance - Blockchain in Islamic Finance (2017). https://www.ethiscrowd.com/blog/blockchain-islamic-finance/

Zīle, K., Strazdiņa, R.: Blockchain use cases and their feasibility. Applied Computer Systems **23**(1), 12–20 (2018). https://doi.org/10.2478/acss-2018-0002

Prospects of Using Digital Technologies in the Activities of Agricultural Enterprises

Novykova Innola[1] (✉), Lynovytska Olesya[2], Nykytiuk Oleksandr[3], Marchenko Svitlana[4], Pysklyvets Vitalii[5], and Fedun Iryna[5]

[1] Economics and Management Department, Institute of Innovation Education, Kyiv National University of Construction and Architecture, Separated Structural Subdivision, Kyiv, Ukraine
innolanovykova@gmail.com

[2] Conservative Dentistry Department, Bogomolets National Medical University, Kyiv, Ukraine
q45ua@ukr.net

[3] Department of Agronomy, Kyiv Agrarian University of the National Academy of Agrarian Sciences of Ukraine, Kyiv, Ukraine

[4] Kyiv National University of Construction and Architecture, Kyiv, Ukraine
svitlanamarchenko@ieu.edu.ua

[5] Department of Entrepreneurship, Cooperation, and Agro-Industrial Integration, National Scientific Centre «Institute of Agrarian Economics», Kyiv, Ukraine

Abstract. The purpose of the article is to reveal the modern significance of digitalization and digital platforms for the economic development of the subjects of the agricultural sector of the economy.

The following methods were used in the research process, in particular: abstract-logical method (when studying the theoretical foundations of digitalization of the economy), monographic (to clarify the essence of the relevant categories), comparative analysis, economic-statistical (in identifying trends and patterns of development of the state economy), illustrative (for a visual representation of the digital platform of agriculture), as well as methods of systematic approach, grouping and generalization, other methods in the field of economic research.

The need to use a digital platform, especially for small and medium-sized agricultural enterprises, has been identified. In particular, this system will allow: public authorities (to make a plan of visits and yields; to report on the allocation of subsidies, as well as their use; to conduct a financial audit for each enterprise; to assess the competitive advantages of each enterprise; enterprises (to analysed data from weather stations; to study information on the condition of soils; choose crops for sowing; assess the risks of crop loss; obtain data on climatic risks).

Theoretical provisions on the impact of digitalization of the economy on agricultural production and the need for digitalization of production and economic processes have been further developed.

As a result of the implementation of the proposed digital platform in agriculture, the following benefits will be obtained: increased crop yields; the field planning system will improve; reduce production costs based on efficient use of resources and science-based approaches.

Keywords: Digitalization; digital platform; agriculture; innovations · Innovative activity · Agricultural enterprises · Production · Digital technologies · Efficiency · Productivity

B. Alareeni et al. (Eds.): ICBT 2022, LNNS 620, pp. 82–93, 2023.
https://doi.org/10.1007/978-3-031-26953-0_10

1 Introduction

The introduction of digital technologies in agricultural production is one of the most important elements of strategic development in this area. Bio and nanotechnologies, the use of genetic development, the ability to adapt agricultural products to the needs of specific categories of buyers are important factors in increasing the competitiveness of the industry, but without the active use of digital innovation technologies can not quickly turn the domestic agricultural sector into a hightech industry.

Given the current state of functioning of domestic agriculture and the acceleration of the processes of change in the external and internal environment, the digital transformation of agricultural production is especially relevant because it is a significant source for significant economic growth.

In our opinion, digitalization means transformations caused by the mass introduction of digital technologies that generate, process, exchange and transmit information. In economics, it allows you to automate business operations, which increases their efficiency and productivity, opens new opportunities, affects the development of entrepreneurship, promotes the formation of new products and management models. In agriculture, a comprehensive digital transformation can provide a significant increase in productivity, reduce non production costs, improve the quality of agricultural products.

The changes brought about by digitalization and informatization of agriculture are fundamental, as they affect all areas of organization of agricultural enterprises, ways of participation of economic entities in production, innovation, commodity and economic processes, opening new opportunities and qualitatively transforming the nature of interaction, social integration and communication in general. Alternative use of potentially released financial resources through the widespread introduction of information methods of doing business, stimulating domestic consumer demand, is an incentive that encourages producers to effectively use material and labour resources, to develop competitive production of goods and services to meet the diverse needs of customers.

Thus, with the help of digitalization of agriculture can provide a significant economic effect by increasing productivity, land use efficiency, monitoring the quality of machinery, crops, automation, transparency and controllability of processes, reducing costs of farmers.

For the first time in history, people are fighting not only in the physical plane, but also in the digital one. The front in our laptops has become almost as important as in the East of Ukraine. In the last two and a half years, Ukraine has been creating a digital country. And it turned out that its products and approaches can be easily adapted to modern realities.

2 Literature Review

Problems of digitalization of agrarian sector are constantly in the field of view of theorists. In particular, various aspects of the general process of digitalization have been fundamentally explored in the work of the Negroponte N. (1995), who was the first to use the word "digitalization" and to give definition to it.

A number of Ukrainian researchers have recently become devoted to the problems of digitalization in different branches and sectors of the domestic economy. In particular, attempt to find a creative solution to the problem of forecasting the new workplaces in the industrial sector of Ukrainian economy was done in the work of Kozhemiakina S., Cherkasov A., Reznik N., Zhuravka O., Mazurov S. (2018) [18]. Digitalization of economic relations as a factor of improving the business processes at enterprises was investigated by L. Lazebnyk (2018) [19].

Attention of the Ukrainian scholars is also being kept by innovation and digitalization problems of concrete directions of the agrarian production within the sector. For instance, the challenges for growth of oilseeds and the following exchange of them at the world market have been studied by N. Reznik, S. Gupta, O. Sakovska, A. Ostapchuk, R. Levkina (2019) [20].

Despite the importance of the research that has already been carried out by scientists there are some aspects of digitalization of economy left without attention. That is why the topic of the current research was considered to be relevant and actual and the challenges of digitalization of agrarian sector in terms of human development have been considered in the article.

The most important tasks of the article are substantiation of the digitalization itself in current conditions, its impact on the human development; demonstrating the crucial role of the agrarian sector in the Ukrainian economy and outline the present achievements in digitalization of it; and working out the main solutions to the challenges that arise in the process of digitalization of the agrarian sector.

3 Methodology

To achieve the formulated goal of the research and to fulfill the stated tasks a set of well-known scientific methods and techniques have been used. The method of logic synthesis has been applied in order to well-ground theoretically the importance of studying the challenges of digitalization of the agrarian sector in terms of human development. Using the methods of analysis and synthesis allowed showing the importance the agrarian sector for the economy of Ukraine and the present achievements in the digitalization of it. The method of finding logical connections was used to demonstrate the necessity of simultaneous providing innovations and preparing workers for manipulating them. Method of building schemes and models has been used for a visual representation of results of the research and their schematic representation.

4 Results

«ДІЯ Diia» (Action), one of the best government apps in the world. More than 17.4 million Ukrainians use it now. All the necessary documents are at hand. You can pay your taxes in just a few clicks. Now – watch TV and pay for the army.

All documents are not only digital, but it is also possible to conveniently share them via a QR code.

Photocopies and waiting are a thing of the past. Checking into a hotel is now a matter of two minutes. Get help – too. How to issue a new card at the bank.

If you need to sign something instantly – we have a unique «ДІЯ Diia» (Action. Signature). You can get it simply in your smartphone. Without queues and visits to state institutions. It is already used by 8 million Ukrainians.

The coronavirus pandemic pushed us to eSupport. This is a unique project: never before have Ukrainians received financial assistance so quickly, in just a few clicks. We also developed a high-speed Internet network – more than a million people in 3,000 villages across Ukraine were able to study, conduct business, and simply communicate.

And last year, more than 1 million Ukrainians learned basic digital skills in the format of educational miniseries. Almost like Netflix, only about digital education.

At the same time, Ukraine's economy was transformed all this time. Virtual assets were legalized and IT education was reformed in Ukraine. A special legal and tax regime for technology companies was launched – Diia City. These are the best tax conditions for IT companies in Europe. Hundreds of companies have already joined Diia City.

Today, total digitization is not a fad, but actual necessity. Computer technologies have become an integral part of any sphere of social life, including the sphere of public administration. Today we see that almost every government body, service or agency has its own websites and databases.

In general, the implementation of the «state in a smartphone» concept began in 2019 and has shown its effectiveness in a year of active use. In fact, a new ministry was formed in our country – the Ministry of Digital Transformation of Ukraine. According to the Regulation on the Ministry of Digital Transformation of Ukraine, approved by the Resolution of the Cabinet of Ministers of Ukraine dated September 18, 2019 No. 856, this body is responsible for digital transformation all spheres of public life and optimization of public administration through the introduction of electronic document management systems and improvement of electronic identification means [1].

However, is it worth implementing digital ideas in the field of agro-industry? Ukraine is an agrarian country – 65% of the country's territory is occupied by chernozems, so it is logical that grain agricultural production takes the first place in terms of profitability. Given its specificity, it must be said that even recognizing the need for digitalization, it is worth understanding that such a procedure in agro-industry will have its own characteristics.

The digital transformation of agro-industrial production is considered as one of the main ways of diversifying the national economy, its reorientation from the raw material export model to the supply of products with high added value. Reducing costs for the production of agricultural products, increasing their quality and competitiveness based on the effective use of resources and science-based approaches is the main task of digitalization of agriculture [2, p. 127]. Currently, such solutions are popular on the agricultural market in Ukraine, such as: the introduction of precision farming systems, aerial surveying to control the quality of crops, keeping a history of fields for choosing the optimal crop, laboratory soil research to obtain information about the biochemical composition. These technologies make it possible to increase productivity and reduce the cost of production due to the reduction of costs for fuel, seeds and fertilizers [3, p. 311].

However, this is only one of the components of complex digitalization, in addition, it is necessary to consider its legal component. Thus, the Concept of the Development of Digital Economy and Society 2018–2020 operates in Ukraine today, it, in particular, provides for the development of so-called digital infrastructures. The main goal of their development is that all citizens of Ukraine without restrictions and difficulties of a technical, organizational and financial nature (in particular, socially vulnerable sections of the population) could use digital opportunities regardless of their location or residence and not be in the «digital divide» segment [4]. The goal is very timely and attractive to the population, but in practice we see a lack of means of its implementation, so all the above «sowing digital measures» can be implemented by only 12% of all agricultural enterprises in Ukraine.

In addition, there are no financial support programs at the state level for agricultural companies that have decided to implement digitalization. Therefore, it is simply not profitable for a Ukrainian agricultural producer to implement digital technologies in their production, since the equipment and its software for agricultural needs in Ukraine it is not actually produced, and it is very expensive abroad.

Examples of digitalization of infrastructure maintenance are the launch of an electronic farm register, the pilot project of which was created on the basis of Decree of the President of Ukraine No. 837/2019 of 08.11.2019 [5], the purpose of which is to introduce state support for increasing the financial capacity of farmers and agricultural producers. In the following, the circle of participants of such a register was expanded to include all producers of agricultural products [6].

Later, this register was reformatted into the electronic farm register «State Agrarian Register», the purpose of which is to identify market subjects and establish support [7].

The next step is the legislative regulation of the functioning of the State Agrarian Register, for which draft law No. 3295 [8] was developed and submitted to the Verkhovna Rada of Ukraine for consideration. The draft law is aimed at mandatory interaction of all electronic registers. Also, the bill envisages corresponding changes in the Law of Ukraine «On State Support of Agriculture of Ukraine» [9]. Although, in our opinion, maintaining such a register is not fully perceived as any help to agricultural producers.

Such actions should be qualified as an ordinary means of economic law regulation of certain relations of the farmer with subjects and participants of the agricultural market.

The next example of special digitization is keeping a register of agricultural receipts. The procedure for conducting which is defined in Art. 10 of the Law of Ukraine «On Agrarian Receipts» [10] and the «Procedure for maintaining the Register of Agrarian Receipts», approved by the CMU No. 665 dated August 17, 2013. Ensuring the openness and stability of the market of agricultural products and ensuring the rights of the participants of the relevant legal relations should be considered a positive factor in the development of such a register. The legal protection of the interests of the participants in the legal relationship when using these agricultural receipts is additionally provided by entering a pledge of the future harvest into the register of movable property encumbrances in accordance with Art. 7 of the Law of Ukraine «On Agrarian Receipts» and to the Law of Ukraine «On Securing Creditors' Claims and Registration of Encumbrances» dated 11.18.2003 No. 1255-IV [11], as it was correctly noted in scientific sources [12, p. 232].

In addition to digitization, Ukraine is also pursuing a decentralization policy, which involves the deconcentration of power and resources from the centre to individual regions. At the same time, it is necessary to redistribute and delegate the functions of users of the relevant registers to rural united territorial communities, which will be able to provide more economic activity in their territory, the need for which is emphasized by scientists [13]. Maybe that's what will happen the key to solving the problems of digitalization of the agricultural sector, since it will be easier for the local administration to analyse the specifics of production in its region and accordingly distribute budget funds so that a certain part of them goes to the modernization of a certain branch of agriculture, which is the most developed in the region.

In the process of research in the field of implementation of the digital transformation of agriculture, a number of economic problems were identified [14]:

- unavailability of modern technologies and techniques for small agricultural enterprises due to high cost;
- low level of mechanization in agricultural enterprises, especially in medium and small ones;
- the presence of cheap and low-quality goods in the consumption structure;
- low integration with all business processes of the enterprise;
- lack of a ready-made comprehensive solution that could ensure automation and digitization.

Among the existing barriers in organizational, information and hardware support, the following can be distinguished:

- the need for import substitute components;
- the need to use a single software platform and exchange protocol between different software agents and structural units;
- insufficient staffing of measures for digital registration at the regional level;
- the need to provide stable navigation and mobile signals throughout the territory of Ukraine.

To solve these problems, a system should be developed that allows to increase the level of digitization of enterprises. The system envisages the creation of a digital platform, the purpose of which is to implement a system of digital technologies in the agricultural sector to improve the efficiency of enterprises.

For the functioning of the platform, information providers, which serve the state authorities, are necessary. They can provide remote sensing data of lands (DZZ): satellites and drones, information on stimulation methods, as well as, at the initial stage, the largest agricultural holdings that have data from drones, sensors and sensors, video surveillance systems and other devices.

Small and medium-sized enterprises do not have sufficient funds to purchase a number of modern technologies and techniques due to the high cost. As part of the implementation of the proposed model, a free connection to the general network is offered for obtaining data on climate, soils, productivity, seeds, etc.

There are various services, sites, events that provide information about the modern digital world and help to increase agricultural IQ. The most common option for filling the lack of professional information and knowledge is taking online courses. You can choose a convenient platform for distance learning on the Internet. Examples of such websites for taking agrarian courses are:

1) EdX is a world-renowned distance learning platform;
2) Coursera – a website that has general and specialized modules on agriculture;
3) Udemy – a site with lectures on current and modern technologies and implementation of the agricultural sector;
4) Factor Academy - a platform that has an online course on the peculiarities of accounting in agriculture;
5) «Business Incubator» from the Ukrderzhfond – a course on how to get a return from the state for a farmer financial assistance;
6) Webinars of the Ukrainian Fruit and Vegetable Business Development Project (UHBDP).

Also, agricultural online games and simulators, such as:

1) FarmForesight is an agribusiness planning tool that has software that models the business process of agricultural production;
2) Farming Simulator – a series of video games simulating agriculture;
3) Pure Farming – a «farmer» simulator in which you can engage in agriculture, build farms, grow plants, etc. in different cities of America and Europe.

We offer the use of a subsidy system for small agricultural enterprises. The amount of subsidies received will directly depend on the amount of information received. Otherwise, small and medium-sized enterprises must provide information about their activities, in particular, the use of data obtained from various devices: drones, agricultural machinery, sensors, sensors, etc. Thanks to this approach, these small and medium-sized enterprises will increase their competitive advantage over others.

Robotic greenhouses and animal husbandry complexes can also be introduced into the practice of agricultural production, which will allow significant savings due to more efficient use of water, fertilizers and plant protection products. In addition, such technologies will allow to optimize the number of personnel engaged in the care of the frame cult, as well as to minimize losses caused by the human factor.

The use of special sensors and software monitoring tools included in the complex of operations of so-called animal husbandry complexes helps to increase the productivity of animals and, as a result, to improve the quality of final products due to the introduction of automated cow feeding and milking systems in agricultural enterprises. The use of software and hardware for monitoring the health of cows contributes to an increase in milk yield by up to 40%. The ultimate level of automation of agricultural activity, increasing productivity and quality of final products is the key goal of robotic agriculture.

The joint work of developers and integrators of IT solutions in agriculture, investors, representatives of the expert community and authorities will contribute to overcoming the existing obstacles on the way to the implementation of digital technologies in the activities of farms of the corporate sector of the agrarian economy, accelerating the transfer of the industry to a new technological level of development. The world has already entered the era of the information method of production. In 2025, 50% of the world economy will transition to the implementation of digitalization technologies that allow business to work efficiently. Developed countries are rapidly developing innovative technologies dominated by digital platforms, artificial intelligence, and robotics [15]. The economy of Ukraine as a whole and the agricultural sector in particular must be included in this process to increase efficiency agricultural production and conservation environment. According to scientists' calculations, the comprehensive digitalization of agricultural production allows to reduce costs in the industry by 23%. At the same time, the entire range of technologies that are used in the economy today can be used in this area. The creation of data arrays will make it possible to ensure remote sensing of the earth, hyper-spectral, with maximum efficiency well, aerial photography, use hydrometeorological data, will increase the quality of precision farming, plant protection, provision of the seed fund, maintenance of the machine and tractor fleet, etc. According to the calculations, only land use using GPS navigation technologies provides an average cost savings of 11–14%, differentiated application of mineral fertilizers – 8–12%, use of parallel driving systems – 8–13%.

With the help of the introduction of digital technologies in the field of agricultural production, it is possible to significantly reduce the retail price of agricultural products: since in the structure of price formation up to 80% falls on the payment of numerous intermediary structures, due to its cross-cutting nature, IT technologies allow connecting the needs of specific consumers of products with the capabilities of agricultural producers, eliminating the need to pay for redundant operations and eliminating their duplication.

Additional elements of the mentioned platform are the decision support system. In particular, this system will allow:

1. To state authorities:

   - draw up a visit and productivity plan;
   - submit a report on the allocation of subsidies, as well as their use;
   - conduct a financial audit of each enterprise;
   - assess the competitive advantages of each enterprise.

2. Enterprises:

   - analyse data from weather stations;
   - study information about soil condition;
   - choose an agricultural crop for sowing;
   - assess the risks of yield loss;
   - to obtain data on climate risks.

     As a result of the implementation of the specified model in agriculture, the following advantages will be obtained:
   - the yield of agricultural crops will increase;
   - the field work planning system will improve;
   - production costs will decrease based on effective use of resources and scientifically based approaches.

It is possible to highlight a number of problems in enterprises of the agrarian sector related to the transition to digitalization. First, the use of outdated equipment and technologies, particularly in small and medium-sized agricultural enterprises, which in turn reduces their competitiveness in the domestic market. Secondly, the lack of necessary resources for the introduction of modern equipment and technology in small agricultural enterprises. Thirdly, access to specialized information resources, where you can get the necessary information on optimizing your activities, is practically non-existent for small businesses. Fourth, there are several companies that offer their own platforms for data processing and analysis on the market of manufacturers of technological equipment for agriculture. As a result, the user is forced to master several platforms at the same time. Fifth, the existing state and regional support measures for the agricultural sector do not have a targeted focus on the implementation of advanced technologies in the activities of small and medium-sized enterprises.

In order to solve the identified problems, a model of a digital platform for the agricultural sector is proposed, which will allow to involve in the process of digital transformation all small and medium-sized enterprises at the expense of:

- point means of support for small and medium-sized enterprises of the agrarian sector.
  That is, the more technologically the equipment is used, the larger the amount of information and the higher the quality of the data that is transferred to the digital platform, and the available information about business processes facilitates quick decision-making regarding further development;
- high-quality information about the condition of land, crops, transport, and other things, based on aggregated data from various sources (data from DZZ, drones, etc.);
- the possibilities of introducing technological equipment into the activities of agricultural enterprises.

5 Discussion

The digitalization of agrarian sector is the main trend in the current condition, something that cannot be avoided or escaped. The choice not to implement digital technologies into the sector can cause widening of the gap between the Ukrainian economy and economy of the developed countries of the world. Taking into consideration the crucial role of the agrarian sector in the economy of Ukraine the process of implementing digital technologies in it should become the priority.

Implementing the digital technologies into the sector helps to solve not only the economic goals of the agrarian companies in particular and of the sector as a whole but it allows solving the human development problems, raising quality of life of people, improving their mentality.

However fast digitalization causes resistance of people it because of their reluctance to changes, to studying, to new forms of management. That is why practical providing of digital modernization should be well-thought and evolutionary process attended by simultaneous workers' training.

6 Conclusions

The implementation of the proposed digital platform model will significantly speed up the process of digital transformation of agricultural production, which will ultimately strengthen the position of small and medium-sized agricultural producers on the domestic and international markets.

The changes that digitalization and informatization of agriculture will cause are of a fundamental nature, as they affect all spheres of activity of agricultural enterprises, methods of participation of business entities in production, innovation, commodity economic processes, opening up new opportunities and qualitatively transforming the nature of interaction, social integration and communication as a whole. Alternative use will potentially free up financial resources due to the widespread implementation of information methods of conducting business, which stimulate domestic consumer demand, acts as a factor that encourages product manufacturers to effectively use material and labor resources, develop competitive production of goods and servants to meet the most diverse needs of merchants.

Thus, with the help of digitalization of the activities of agricultural enterprises, a significant economic effect will be ensured due to the increase in labor productivity, the efficiency of land use, monitoring the quality of the use of agricultural machinery, ensuring the quality of agrotechnical measures, ensuring automation, transparency and manageability of processes, reducing costs in the corporate sector of the agrarian economy.

Summarizing all of the above, we can say that the implementation of digitization in agro-industry is a very progressive and positive trend, but our country has problems with its implementation: legal, financial and technical. The lack of a nationwide support program makes digitization unprofitable for agricultural users. In addition, one cannot fail to note the lack of preparation for digitalization in rural areas.

In our opinion, the full implementation of decentralization and the adoption of appropriate targeted state support programs will significantly speed up the process of digitalization in the agricultural sector.

References

1. Pytannia Ministerstva tsyfrovoi transformatsii. Postanova Kabinetu Ministriv Ukrainy (2019). https://zakon.rada.gov.ua/laws/show/856-2019-%D0%BF#Text [in Ukrainian]
2. Rudenko, M.V.: Vplyv tsyfrovizatsii na rozvytok ahrosfery. Materialy mizhnarodnoi naukovo-praktychnoi konferentsii. In: Proceedings of the international scientific-practical conference, pp. 127–129 (2019). [in Ukrainian]
3. Sherstiuk, L.M.: Tsyfrove silske hospodarstvo: zarubizhnyi dosvid ta osoblyvosti vprovadzhennia y vykorystannia v Ukraini, pp. 310–318 (2019). [in Ukrainian]
4. Kontseptsiia rozvytku tsyfrovoi ekonomiky ta suspilstva Ukrainy na 2018–2020 roky zatverdzhena Rozporiadzhenniam Kabinetu Ministriv Ukrainy № 67-r (2018). https://zakon.rada.gov.ua/laws/show/67-2018-%D1%80#Text [in Ukrainian]
5. Ukaz Prezydenta Ukrainy № 837/2019 vid 08.11.2019. https://zakon.rada.gov.ua/laws/show/837/2019#Text [in Ukrainian]

6. Ofitsiinykh Veb-sait Ministerstva rozvytku ekonomiky, torhivli ta silskoho hospodarstva Ukrainy. https://www.me.gov.ua/News/Detail?lang=ukUA&id=ef25db5b-94f4-441d-a577-df83426ca104&title=InnaMtlvaProvelaOnla innaraduSchodoStvorenniaElektronnogoFermerskogoRestruZPilotnimiRegionami [in Ukrainian]
7. Oskilska hromada, Kharkivska oblast, Iziumskyi raion. https://oskilskasilrada.gov.ua/news/1599158922/ [in Ukrainian]
8. Normatyvno-pravove rehuliuvannia provedennia pilotnoho proektu zi stvorennia ta funktsionuvannia derzhavnoho ahrarnoho reiestru. http://valky-rda.kh.gov.ua/news/208/78502 [in Ukrainian]
9. Zakonoproekt № 3295 vid 30.03.2019. http://w1.c1.rada.gov.ua/pls/zweb2/webproc4_1?pf3511=68504 [in Ukrainian]
10. Zakon Ukrainy «Pro ahrarni rozpysky». https://zakon.rada.gov.ua/laws/show/5479-17#Text [in Ukrainian]
11. Pro zabezpechennia vymoh kredytoriv ta reiestratsiiu obtiazhen: Zakon Ukrainy vid 18.11.2003 № 1255-IV. https://zakon.rada.gov.ua/laws/show/1255-15#Text [in Ukrainian]
12. Pravove zabezpechennia adaptatsii investytsiinoi modeli rozvytku ekonomiky Ukrainy ta rynkiv finansovykh posluh do prava Yevropeiskoho Soiuzu. In: Hlibko, S.V., Vnukova, N.M., Dmytryk, O.O. (eds.) Kharkiv: Pravo (2017). [in Ukrainian]
13. Stan hospodarskoi diialnosti okremykh obiednanykh terytorialnykh hromad Kharkivskoi oblasti. In: Hlibko, S.V., Heorhiievskii, Yu.V., Vnukova, N.M. (eds.) Kharkiv: NDI PZIR NAPrN Ukrainy (2020) [in Ukrainian]
14. Borodina, O.M., Lopatynskiy, Ju.M.. Prokopa, I.V.: Pidtrymka ahrarnoho sektora v systemi derzhavnoho rehuliuvannia ekonomiky: navch. posib. [Support of the agrarian sector in the system of state regulation of the economy]. In: Borodina, O.M., Lopatynskyi, Yu.M. (ed.) Nats. akad. Nauk Ukrainy, In-t ekonomiky ta prohnozuvannia, Cherniv. nats. un-t im. Yu. Fedkovycha. Chernivtsi: ChNU (2009). [In Ukrainian]
15. Fedun, I.L.: Digitalization as a priority of development of public-private partnerships in the agrarian sector. In: Fedun, I.L., Reznik, N., Ostapchuk, A., Alekseieva, K., Cherkasov, A. (eds.) International Journal of Advanced Science and Technology **29**(8s), 2524–2530 (2020). http://sersc.org/journals/index.php/IJAST/article/view/14755
16. Agricultural sector of Ukraine: Securing the global food supply (2018). file:///C:/Users/User/Downloads/agro-small.pdf [in English]
17. Digital technologies in Agriculture: Briefing paper. Food and Agriculture Organization of the United Nations. Rome (2019). http://www.fao.org/3/ca4887en/ca4887en.pdf [in English]
18. Kozhemiakina, S., Cherkasov, A., Reznik, N., Zhuravka, O., Mazurov, S.: Problems and Perspectives in Management. New workplace forecasting in the industrial sector of the Ukrainian economy**16**(4), 384–394 (2018). https://www2.scopus.com/authid/detail.uri?authorId=57202444035 [in English]
19. Lazebnyk, L.: Didzhytalizatsiia ekononichnykh vidnosyn yak factor udoskonalennia biznes-protsesiv pidpryiemstva. Ekonomichnyi visnyk. Seriia: finansy, oblik, opodatkuvannia (2), 69–74 (2018). http://ojs.nusta.edu.ua/index.php/ojs1/article/view/75 [in Ukrainian]
20. Reznik, N., Gupta, S., Sakovska, O., Ostapchuk, A., Levkina, R.: Ukrainian world exchange market of oilseeds: a research of challenges for growth. Int. J. Eng. Adv. Technol. **8**(6), 3823–3829 (2019). https://www2.scopus.com/authid/detail.uri?authorId=57202444035 [in English]
21. Danyliuk, V., Riepina, I., Shafalyuk, A., Valinkevych, N., Reznik, N.: Methodical approaches to measurement of the technical level of a production enterprise. AIP Conf. Proc. **2413**, 040017 (2022). https://doi.org/10.1063/5.0080013
22. Mamchur Volodymyr, A., Stetsko Mykola, V., Yuriy, T., Irkliienko Ivan, S., Reznik Nadiia, P.: Organizational transformations for the development of agrarian business. AIP Conference Proceedings **2413**, 040006 (2022). https://doi.org/10.1063/5.0079157

23. Andrii, M., Liudmyla, H., Yuliia, M.P.R.N.: Gender Equality in Access to the Profession of Land Surveyor and Geodesist & Land Appraiser in Ukraine: National and Regional Assessment. Int. Trans. J. Eng. Manage. Applied Sci. Technol. **13**(2), 13A2S and 1–8 (2022). http://TUENGR.COM/V13/13A2S.pdf https://doi.org/10.14456/ITJEMAST.2022.40
24. Malakhova, Y., Lozhachevska, O, Ilchenko, V., Smagin, V., Reznik, N.P.: Financial opportunities management of ensuring enterprise investment costsinternational. Trans. J. Eng. Manage. Applied Sci. Technol. **13**(2), 13A2I and 1–10 (2022). http://TUENGR.COM/V13/13A2I.pdf https://doi.org/10.14456/ITJEMAST.2022.30

A Sir Model for Viral Growth of Coronavirus: A System Dynamics Approach

Nadiia P. Reznik[1](✉), Olena M. Sakovska[2], Olexandr Yu. Yemelyanov[3], Kateryna I. Petrushka[3], Ihor M. Petrushka[3], and Krystyna Dramaretska[4]

[1] Department of Commodity Science and Customs Affairs, State University of Trade and Economics, Kyiv, Ukraine
nadya-reznik@ukr.net
[2] Uman National University of Horticulture, Uman, Ukraine
sakovska_lena@ukr.net
[3] Lviv Polytechnic National University, Lviv, Ukraine
{Oleksandr.y.yemelianov,Igor.m.petrushka}@lpnu.ua,
Petrushkat24@ukr.net
[4] National University of Life and Environmental Science of Ukraine, Kyiv, Ukraine

Abstract. In this paper, a system dynamics model depicts the viral growth of COVID-19 at an exponential rate. The outbreak of Corona virus was started from the Feb 26, 2020 when the first case was reported in Pakistan. Local bodies and law enforcing agencies had taken initial preventive measures to restrict the COVID-19 to a particular locality but all in vain. The infected people were increasing every day rising the stocks of recoveries and deaths. Numbers of infected people were alarming and a need was felt to develop the model to calculate the existing reproduction number and transmission rate and highlight its varied values in coming days. People-oriented measures and government-based policies must be explored to fight against this deadly disease. This paper aims the development of epidemic model using the system dynamic framework on simulation software STELLA. The objective of the research is to experiment with the model to replicate the progression of the communicable disease and probe the multiple combinations of the people-based and government-based measures to reduce its spread. The containment measures are of two types; people-based measures and government-based measures and both directly affect the reproduction number and infection growth fraction for mitigating the spread of deadly coronavirus. Combined efforts of public and government can combat this pandemic. Reduced degree of reproduction number and infection growth fraction are the key metrics to judge and evaluate the effectiveness of containment measures.

This research points to more holistic combination of public and government-oriented measures that play the vital role to flatten the curve and reduce its spread affecting the reproduction number. Simulation results have been traced to replicate the real-life settings against four combinations of containment measures in tabular form and graphical form.

Keywords: COVID-19 · Epidemic models · System dynamics · Containment measures · Simulation software · People-based measures · Government-based measures

B. Alareeni et al. (Eds.): ICBT 2022, LNNS 620, pp. 94–105, 2023.
https://doi.org/10.1007/978-3-031-26953-0_11

1 Introduction

COVID-19 is an emerging biohazard which has drastically influenced millions of people globally causing many deaths with its exponential transmission rate.

In our study, we have devised a system dynamic model (Forrester 1958) based on well-known (Susceptible-Infected-Recovered) SIR model (Kermack and McKendrick 1927) which is an epidemiological model that consists of three coupled non-linear equations (Sterman 2001) that represent the real life setting and grounded in the data gathered from the secondary source i.e., government website. In our research, the system dynamics model is developed in computer based simulation software STELLA Professional version 1.1.2 using system dynamics modelling framework (Sterman 2000). Simulated SIR model replicates the real-life data gathered over the period of 55 days and experimentation with the model helps to explore the possibilities of containment measures that flatten the curve and reduce the spread of infectious disease. It also predicts that to draw a comparison between people vs government-based measures.

Outbreak of COVID-19 was reported in late December in 2019 in Wuhan which is a province of China. It is a highly contagious disease with an expeditious transmission rate and because of this, the WHO has announced a Public Health Emergency of International Concern on 31st January 2020 (Wu and McGoogan 2020). On account of its accelerated global spread, it was declared a pandemic, by World Health Organization on 11th March 2020, due to thirteen-fold increase in cases in two weeks. The basic reproduction number for corona virus is estimated to be around 2.2 (range 1.4 to 6.5) (Wu and McGoogan 2020). In Pakistan first case of COVID-19 was reported on 26th February 2020 owing to its geographical association with China and Iran with continuous immigration of people (Waris et al. 2020). Unluckily the first immigrant was from Iran in Sind province. In order to combat COVID-19, preventive measures were enforced by the provincial government and unprecedented measures were taken to delay the epidemic peak. Till March 26, 2020, 1197 cases of COVID-19 were reported with 9 deaths. Lockdown was enacted on march 2020 and it was gradually eased till 9th May 2020 to shore up the economy although the proportion of infected people had a rising trend. Over the period 55 days till April 20, 2020 the total number of infected cases in Pakistan had amounted to 8418 with 1970 recoveries and 176 deaths.

Pakistan is currently fighting two battles; one against corona virus and the other against poverty. This economic meltdown is due to disruption of supply chains, lockdown and sudden closure of businesses leading to decreased livelihood opportunities. In addition to this, money is being consumed for implication of safety measures, PPEs, testing kits and increasing health care facilities. Favero et al. (2020) has categorized the effectiveness of strategic policies for COVID-19 on basis of their costs and fatalities in his research article ''Restarting the economy while saying lives under COVID-19" on 8th May 2020 to optimize the management strategies from the ecological viewpoint.

In another research article "Why is it difficult to accurately predict the COVID-19 epidemic" Roda et al. (2020) have mentioned that detected cases are only a fraction of total infected people using the metaphor of an iceberg since many asymptomatic patients are not even tested making it difficult to predict the disease course. In his article he has mentioned that SIR model is better in predicting the disease progression than SEIR and other complex models. SIR model provides a theoretical verification of effectiveness of the mitigating strategies as strict implication of these can halt the epidemic surge and it can be used to address the risk of second peak if occupational S activities are started during the disease course.

The objective of the research article is to understand what containment measures should be deployed to flatten the curve, push down the infection rate and mitigate the spread of COVID-19.

2 Research Methodology

System Dynamics is a versatile methodology to deal with the non-linear, dynamic and complex problems (Saeed 2014) in industrial, social and in medical science. System Dynamics masterly attempts to combine the key concepts like feedback controls, mutual causality, non-linearity in the functions, cybernetics, complexity, counterintuitive behavior, deviation correcting and deviation amplifying processes to the organizational systems (Sweeney and Sterman 2007).

COVID-19 has enhanced the importance of disease epidemiological modelling. A well-known epidemiological model SIR (Susceptible, infected and resolved) is now a days talk of the town (Kermack and McKendrick 1927). Even the common man is interested to know about the infection rate and resolution rate. There is a need to develop the SIR model using system dynamics modelling framework and determine the containment measures to control the governing factors BETA (Infection rate) and GAMMA (Resolution rate) that cause the spread of this infectious disease.

Figure 1 Indicates the simplified representation of the SIR Epidemic Model. Saturation loop **(S)**, reinforcing loop **(R)** and balancing loops **(B)** exist in the model and generate the model behaviour. The loop dominance may change over the period of time and creating the various modes of behaviour as the time pass on. As the transmission rate increases due to increase in infected people, there is a decline in susceptible people (people who are healthy but are at risk) and it is going to saturate till all the susceptible persons become infected people. Model structure represents the Stock and Flow diagram and symbols used in the model are shown in Appendix A.

The value of the GAMMA recovery rate strength the balancing loop. No doubt the constant value of gamma does not bring significance effect of the resolved people. Changing values of the BETA (infection rate) keep on increasing the stock of infected people and susceptible people becomes the infected at exponential rate (Rahmandad and Lim, 2021) if there is no vaccine for susceptible (Ghaffarzadegan and Rahmandad 2020). The spread of COVID-19 is with a jet speed. Limited testing and screening facilities, non-availability of the testing kits, poor quality of kits and low quality personal protective equipment are the hurdles to catch the viral growth of coronavirus (Table 1).

The model consists of three non-linear differential equations based on Euler method where 't' is the time, Beta is the infection rate and gamma are the resolution rate.

$$\mathbf{dS/dt = -Beta * S * I} \tag{1}$$

$$\mathbf{dI/dt = Beta * S * I - Gamma * I} \tag{2}$$

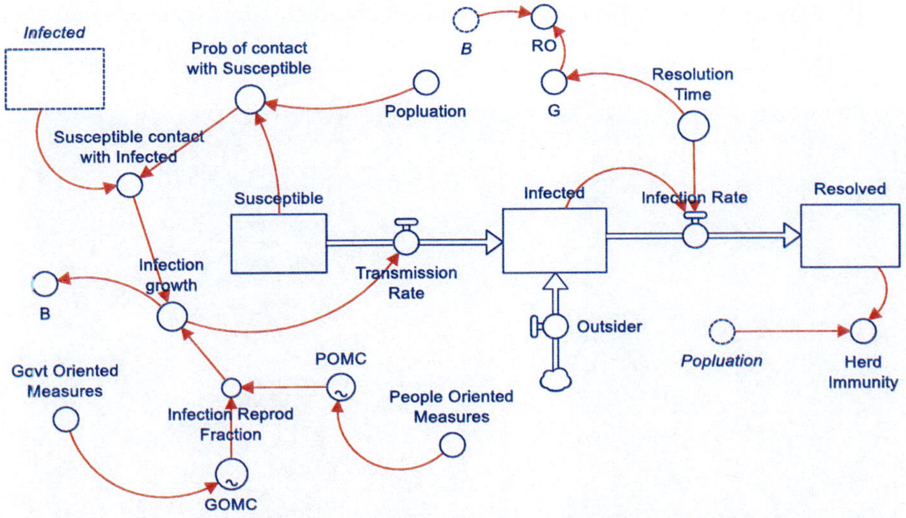

Fig. 1. Stock and Flow Diagram of SIR Epidemic Model

Table 1. Anatomy of SIR model

Symbol	Nomenclature	Description
S	Susceptible	How many available for infection
I	Infected	How many are infected
R	Resolved	How many are resolved either recovered or died (no longer part of total population)
B	Beta	Number of contacts per infected person per day
G	Gamma	Number of recoveries per person per day
R0	Reproduction Number	Reproduction number is used to measure the transmission potential of a disease. It is the average number of secondary infections produced by a typical case of an infection in a population where everyone is susceptible
N	N	Total Population i.e. N = S + I + R

$$\mathbf{dR/dt = Gamma * I} \tag{3}$$

$$\mathbf{N = S + I + R} \tag{4}$$

Beta is the infection rate that is controlling variable that transmission from susceptible to infected and ratio of Beta to Gamma called reproduction number (RO) determine the spread of epidemic COVID-19 (Wu et al. 2020). It is interesting to note that infection rate (Beta) and reproduction number (RO) both exponentially rise. The model is calibrated fifty-five days for value of infection rate (Beta) and resolution rate (gamma) shown in Fig. 2 and Fig. 3.

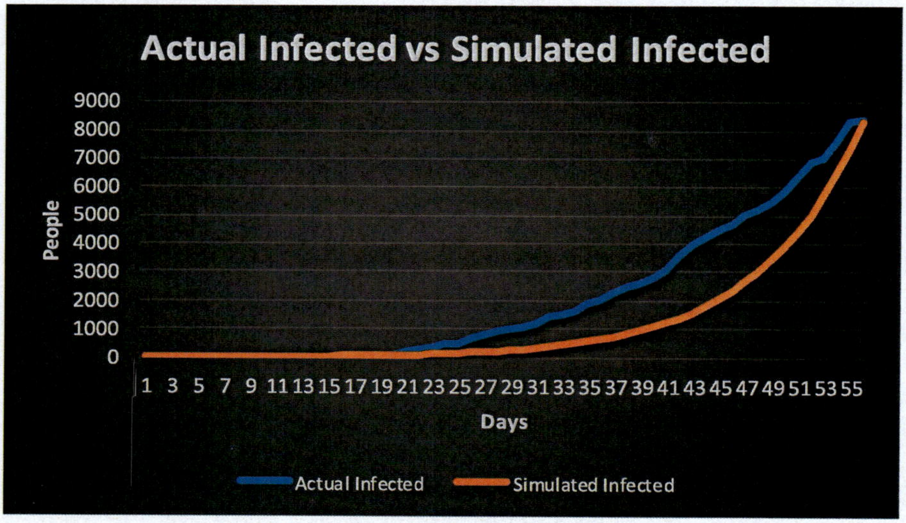

Fig. 2. Actual infected vs simulated infected

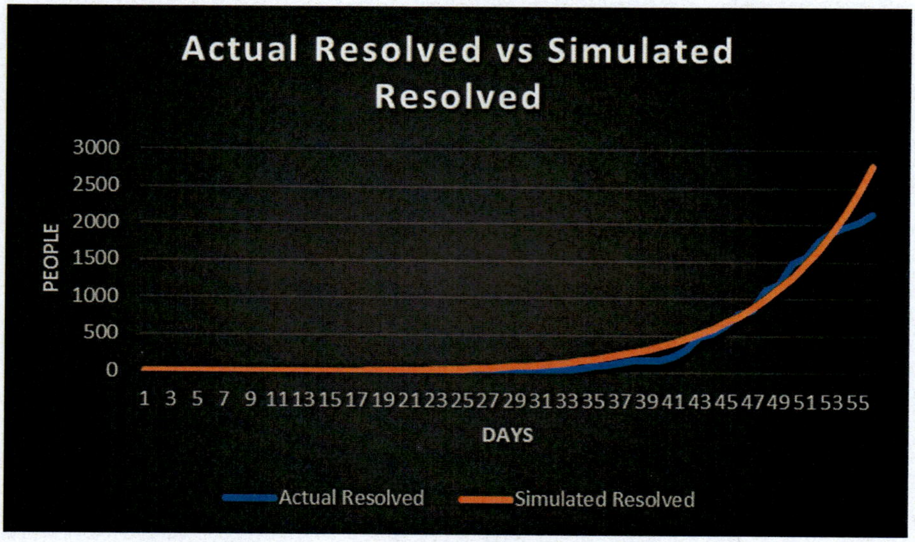

Fig. 3. Actual resolved vs simulated resolved

The days to resolved either recovered or died usually it takes around 23 days. The model is the third order differential equation with associated flows (Forrester 1968). The order of the model depends upon the number of levels and the number of delays (Sushil 1993). The detail listing of the variables of the model are highlighted in Appendix B.

Base Line Model Equations.

SIR model based on Euler Integration that are as under:

Susceptible = INTEGRAL (- Transmission rate) * dt -- (1)
Transmission rate = Infection growth ---------------------------------- ------------------------------(1.1)
Infection growth = Infection reproduction fraction * susceptible contract with infected -------(1.2)
Susceptible contact with infected = prob of contact with susceptible * infected contact -------(1.3)
Infected contact = Contact rate * Infected ---(1.4)
Prob of contact with susceptible = Susceptible / population --(1.5)
B = Infection growth (Beta represents infection rate)---(1.6)

Infected = INTEGRAL (Transmission rate – infection rate + outsider inflow) * dt--------(2)
Infection rate = Infected / Resolution time --(2.1)
Outsider Inflow = One person per day (Constant)---(2.2)

Resolved = INTEGRAL (Infection rate) * dt --(3)
Herd Immunity = Resolved / Population --------------------- -------------------------------------(3.1)
G = 1/resolution time (Gamma represents the resolution rate) ----------------------------------(3.2)
RO= B/G --- (3.3)

The summation of all the stocks and rates are equal to the total population that depicts the dynamic balance (Saeed et al. 2018). The system dynamic model has following assumptions:

1) There is a fixed recovery time that in case of COVID-19
2) People are well mixed and cannot segregate the asymptomatic with people with healthy people.
3) There is no information about the immunity loss period.
4) Infected person is not to be infected again.

On the basis of the actual time series data, the system dynamics model is calibrated for fifty-five days data highlighted in Fig. 2 and Fig. 3 representing the true picture of COVID-19 spread in Pakistan considering the people-oriented measures and government-based measures. The detail of model equations has been shown in Appendix C. The value of Beta (infection rate) and Gamma (resolution rate) derived from the time series data of Pakistan and world statistics on COVID-19 (Cori et al. 2013) and tracing of the behavioral pattern for various simulation periods validated the estimations. The various containment measures as shown in the Table 2 have been proposed for experimentation. The only fifty-five days had passed so far and level of awareness was minimal. If we project the model behaviour for next 365 days the results are shown in Fig. 4 that indicate the peak will appear after 146 days and by that time 80 million people will be infected and around 120 million people will be resolved. The figures are so alarming that compel the researcher to go for experimentation and explore the effectiveness of the containment measures to lower the infection rate. The model consists of positive and negative loops (Lane 2007) and dominance of the polarity and its shift from positive to negative and negative to positive adds complexity (Saeed 1992).

Table 2. Containment measures people based versus government based

People-based measures	Government-based measures
Hand washing	Lock down duration
Social Distancing	Lock down effectiveness
Avoid meetings, get-togethers, parties, festivals	Contact trancing and quarantine people on quarantine centers
Self-isolation and quarantine	Smart Lock Down-Open up with Standard Operating Procedures (SOPs) like temperature gun, PPEs, Face mask, frequent medical checkup
Maintaining the 6 feet distance between two individuals	Usage of Facemask is mandatory for visiting the shops and all public places
Using the hand sanitizers	Walk through Dis-infected tunnels for company employees and entrants in shopping malls, hospitals, factories, airports, retail outlets

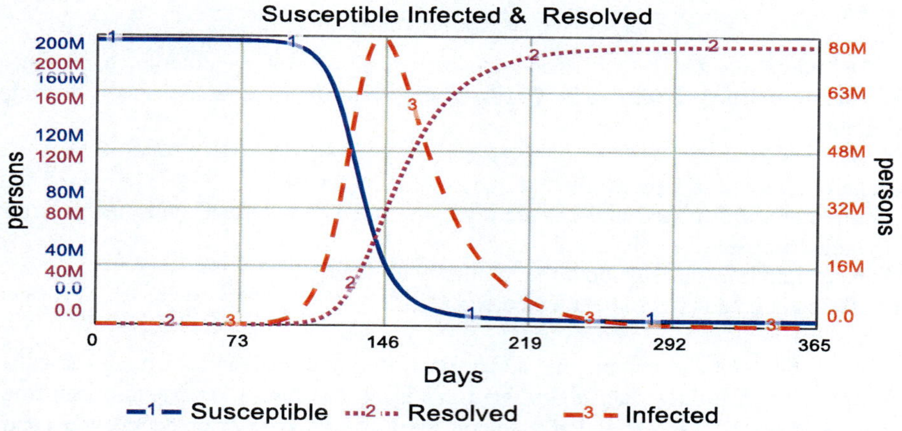

Fig. 4. Susceptibles, infected and resoved during the simulation period 365 days

Graphical functions based on people oriented measures and government oriented measures have been introduced shown in Appendix D and designed to understand the behavioral patterns arising out of the model structure rather than point prediction of the future deaths and recoveries (Saeed et al. 2018).

3 Results and Discussion

People based measures like social distancing is a non-pharmaceutical prevention containment measure that is outcome of people behaviour and attitude (Raouf and Yusuf 2011).

Knowledge, understanding and attitude of medical students (Bostanudin et al. 2020) contribute to run the awareness campaign about wearing face mask, hand washing, usage of sanitizers, maintaining social distance, avoid parties, social get togethers, festivals and funerals. It is all about the people attitude and behaviour of the individuals. Government based measures associated with lock down duration, lock down effectiveness, contract tracing, enhancing health care capacity, Smart lock down with the effective implementation of standard operating procedures (SOPs) and personal protective equipment (PPEs). Lock down duration in days stops everything and other government-oriented initiatives are very much impeded in it as highlighted in Table 2.

The model was calibrated for initial 55 days and then under the government-based measures and people-based measure the actual data was replicated till July 10, 2020 with the simulation numbers indicating the reduced reproduction number. The details are shown in Table 3.

Table 3. Lowering reproduction number shows effectiveness of the containment measures

Period	Days	Actual infected	Simulated infected	Actual resolved	Simulated resolved	Reproduction number	Measures reference	People based measures (%)	Government based measures days
Ist Period	55	8418	8400	2146	2800	3.97	Base Case NO. 1	15	30
2nd Period	85	45898	44000	14086	19000	3.29	Measure No 2	17	55
3rd Period	116	171666	171000	66886	89000	2.92	Measure No. 3	21.5	62
4rth Period	136	243599	255000	154151	152000	2.68	Measure No. 4	24.5	65

The effectiveness of the containment measures people based and government based are equally important and both are significant to reduce the reproduction number shown in Fig. 7 leading to lowering in infected people highlighted in Fig. 5 due to reduced level of Beta-Infection fraction shown in Fig. 8. This diversity of the patterns of behaviors is under the influence of multiple combinations of people based and government based containment measures (Richardson and Pugh III 1981). Measure reference indicates the set of combination (people-based and government-based) of containment measure. The number 1 measure represents the base value and that was the source of model calibration after passing fifty-five days. Reproduction number has gone down from 3.97 to 2.68 that shows the government strategy is at the right track shown in Table 3.

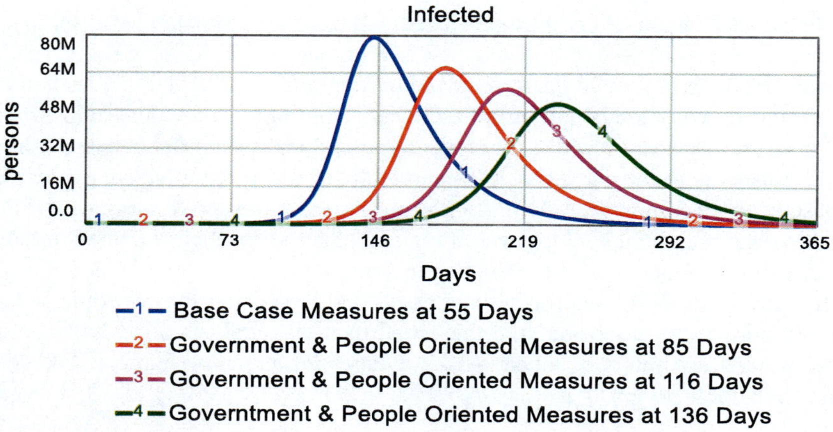

Fig. 5. Affect of containment measures on infected people at varied time interval

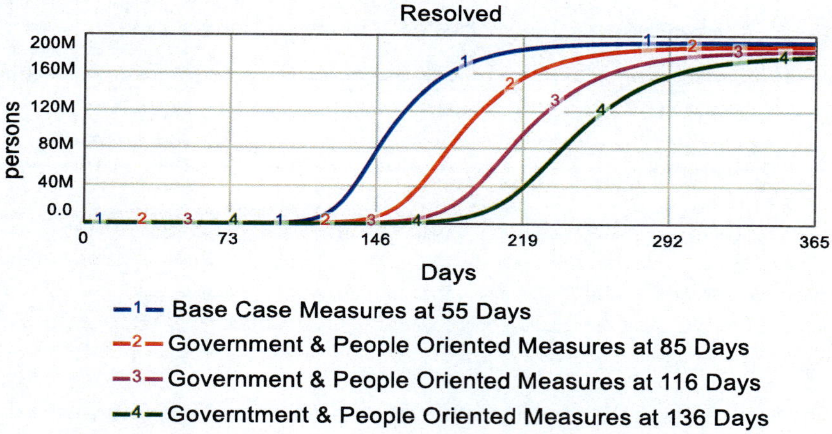

Fig. 6. Affect of containment measures on resolved people at varied time interval

Number of resolved cases have gone down due to decrease in infected people highlighted in Fig. 6. Beta- infection fraction has declined from 0.18 to 0.11 as shown in Fig. 8. This is the result of people attitude and fear due to deaths reported in the media every day.

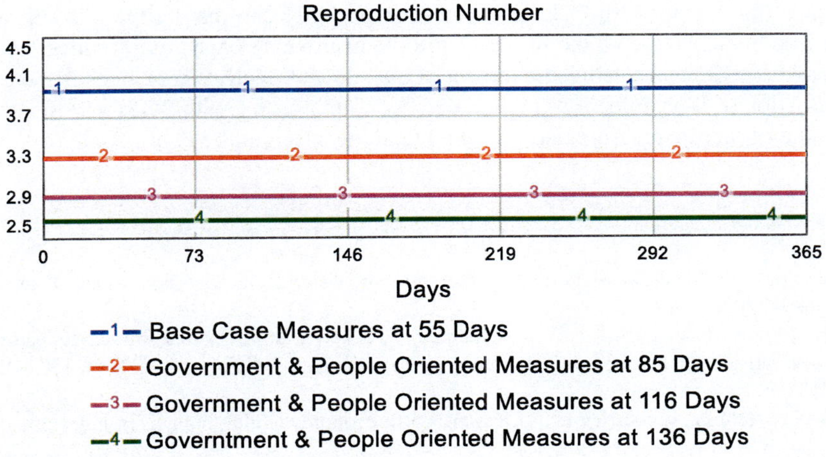

Fig. 7. Affect of containment measures on reproducton numbers at varied time interval

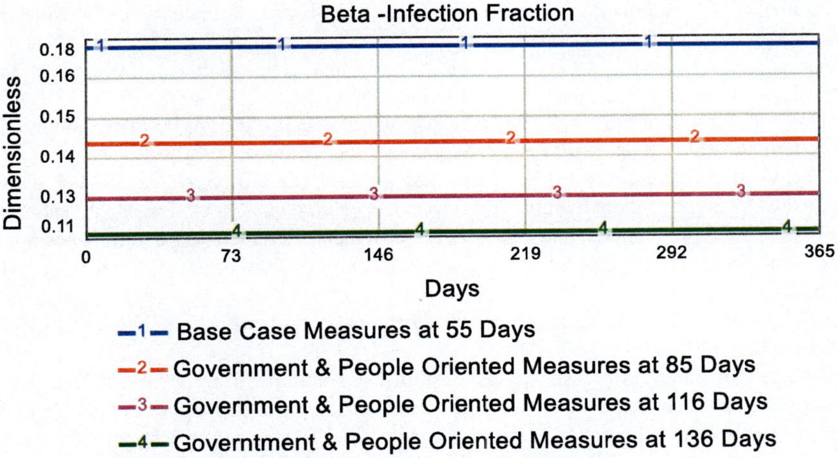

Fig. 8. Affect of containment measures on beta-infection fraction at varied time interval

4 Conclusion

System dynamics model is used to be explored and experimented with various combinations of containment measures (Saeed 2017) and determining the various levels of reproduction number and beta-infection fraction to mitigate the spread of infectious disease COVID-19. It is quite obvious that government-oriented measures are more effective than people-oriented measures. People oriented measures based on education level, effectiveness of the awareness campaign, attitude and mind-set of the social fabric. Measure based on attitude are time depended and cannot be achieved quickly. Government has the power to establish the writ of its government and can implement the policies using stick or carrot. With lock down duration not only can restrict the movement of the people but as well take the time to build the health care capacity to reduce the rising death toll. Contact tracing and territorial lock down effectiveness can contribute for the reduction of Beta-Infection fraction and open up industry with SOPs and PPEs can lessen the economic pressure and give the breathing space to the poor people.

References

Bostanudin, M.F., Al Atassi, S., Fei, T.S.: Epidemiology of Headache Disorders among Pharmacy Students (2020)

Cori, A., Ferguson, N.M., Fraser, C., Cauchemez, S.: A new framework and software to estimate time-varying reproduction numbers during epidemics. Am. J. Epidemiol. **178**(9), 1505–1512 (2013)

Favero, C.A., Ichino, A., Rustichini, A.: Restarting the economy while saving lives under Covid-19 (2020)

Forrester, J.: Industrial dynamics: a major breakthrough for decision makers. Harv. Bus. Rev. **36**(4), 37–65 (1958)

Forrester, J.W.: Industrial dynamics - after the first decade. Manage. Sci. **14**(7), 398–415 (1968)

Ghaffarzadegan, N., Rahmandad, H.: (2020). Simulation-based Estimation of the Spread of COVID-19 in Iran. MedRxiv (2020)

Kermack, W.O., McKendrick, A.G.: A contribution to the mathematical theory of epidemics. Proc. Roy. Soc. London. Ser. A, Containing Pap. Math. Phys. Charact. **115**(772), 700–721 (1927)

Lane, D.C.: The power of the bond between cause and effect: Jay Wright Forrester and the field of system dynamics. Syst. Dyn. Rev.: J. Syst. Dyn. Soc. **23**(2–3), 95–118 (2007)

Nadiia, R., Yusuf, I., Yaroslav, K.S., Nataliia, B., Mykola, S., Olha, B.: Systems thinking to investigate the archetype of globalization. In: Alareeni, B., Hamdan, A. (eds) Financial Technology (FinTech), Entrepreneurship, and Business Development. ICBT 2021. Lecture Notes in Networks and Systems, vol 486. Springer, Cham (2022). https://doi.org/10.1007/978-3-031-08087-6_9

Raouf, A., Yusuf, I.: Quality Management: A customer driven approach (First ed. Vol. First). Lahore: UMT (2011)

Reznik, N.P, Slobodianyk, A.M.: Social Responsibility of Ukrainian Business in the Conditions of the Covid-19 Pandemic: Foreign and Domestic Practice. In: The Importance of New Technologies and Entrepreneurship in Business Development: In The Context of Economic Diversity in Developing Countries, vol. 1, no. 194, pp. 1850–1858 (2021). PMCID: PMC7972443. https://doi.org/10.1007/978-3-030-69221-6_133

Richardson, G.P., Pugh III, A.I.: Introduction to system dynamics modeling with DYNAMO: Productivity Press Inc. (1981)

Roda, W.C., Varughese, M.B., Han, D., Li, M.Y.: Why is it difficult to accurately predict the COVID-19 epidemic? Infectious Disease Modelling (2020)

Saeed, K.: Slicing a complex problem for system dynamics modeling. Syst. Dyn. Rev. **8**(3), 251–261 (1992)

Saeed, K.: Jay Forrester's operational approach to economics. Syst. Dyn. Rev. **30**(4), 233–261 (2014)

Saeed, K.: Circumscribing system dynamics modeling and building confidence in models a personal perspective (2017). Available at SSRN 3093080

Saeed, K., Ryder, E., Manning, A.: Cancer as a system dysfunction (2018).. Available at SSRN 3235843

Gupta, S.K., Dongul, E.S., Reznik, N.P., Lohia, D., Kamboj, N.: The effect of economic pro-growth initiatives on the recent increase in economic development in the USA millennial entrepreneurs. In: AIP Conference Proceedings, vol. 2413, pp. 040011 (2022) https://doi.org/10.1063/5.0079113

Sterman, J.: Business dynamics: systems thinking and modeling for a complex world (2000)

Sterman, J.D.: System dynamics modeling: tools for learning in a complex world. Calif. Manage. Rev. **43**(4), 8–25 (2001)

Sushil: System Dynamics: A Practical Approach For Managerial Problems. Wiley Eastern Limited, India (1993)

Sweeney, L.B., Sterman, J.D.: Thinking about systems: student and teacher conceptions of natural and social systems. Syst. Dyn. Rev.: J. Syst. Dyn. Soc. **23**(2–3), 285–311 (2007)

Waris, A., Khan, A.U., Ali, M., Ali, A., Baset, A.: COVID-19 outbreak: current scenario of Pakistan. New Microbes and New Infections, pp. 100681 (2020)

Wu, J.T., Leung, K., Leung, G.M.: Nowcasting and forecasting the potential domestic and international spread of the 2019-nCoV outbreak originating in Wuhan, China: a modelling study. Lancet **395**(10225), 689–697 (2020)

Wu, Z., McGoogan, J.M.: Characteristics of and important lessons from the coronavirus disease 2019 (COVID-19) outbreak in China: summary of a report of 72 314 cases from the Chinese Center for Disease Control and Prevention. JAMA **323**(13), 1239–1242 (2020)

Rahmandad, H., Lim, T.Y.: Risk driven respones to COVID-19 eliminates the tradeoff between lives and livelihoods. SSRN **3747254**, 2021 (2021)

The Influence of Green Knowledge Sharing and Green Organizational Commitment on Green Competitive Advantage: The Mediating Role of Green Innovation

Nala Tri Kusuma and Muafi Muafi(✉)

Department of Management, Business and Economics Faculty, Universitas Islam Indonesia (UII), Yogyakarta, Indonesia
muafi@uii.ac.id

Abstract. This study has a purpose to examine and analyze the influence of green knowledge sharing and green organizational commitment on green competitive advantage, mediated by green innovation. This study uses census method for the sampling technique, because the number of samples is similar with the population, namely all employees of Giriloyo Batik MSMEs in Giriloyo, Karang Kulon, Wukirsari, Imogiri Districts, Bantul Regency, Daerah Istimewa Yogyakarta, Indonesia. The sample used in this study is 60 employees. The statistical technique used in this study is SmartPLS ver. 3. The findings of this study prove that green organizational commitment has a positive and significant influence on green innovation; and green knowledge sharing has an influence on competitive advantage.

Keywords: Green knowledge sharing · Green organizational commitment · Green competitive advantage · Green innovation

1 Introduction

Currently, the trend that forms the competitive advantage for organization is the environment. When organization has the initiative to implement green practice, they will automatically gain advantage for the organization itself by carrying out such a contrast differentiation compared to other organization [1, 2]. The success or failure of the implementation carried out by the organization is highly depended on the human resource of the organization [3]. Employees who have high productivity and motivation will surely generate benefits for the organization. Employees are expected to put their maximum efforts at work, not only working according to their job desc, but also take part in displaying extra-role behavior and building engagement to their organization [4]. In a similar vein, employees who has pro-environmental behavior will facilitate the organization's goal to carry out green practice [5, 6].

Green service innovation is a discipline to innovate, a management technique to create new ideas and form it into new products or services that are environmentally-friendly

B. Alareeni et al. (Eds.): ICBT 2022, LNNS 620, pp. 106–117, 2023.
https://doi.org/10.1007/978-3-031-26953-0_12

[3]. Green dynamic capability displays the ability of company to integrate, build, and reconfigure their internal and external competencies to overcome environmental changes, such as using green product design [7]. Green dynamic capability is very crucial for the company's success, especially in responding to the increasing demand and shortage of resources and energy [8, 9] suggested that service innovation influences competitive advantage. Green service innovation can significantly act as a criteria in value-added product and services, as well as ensure the high competitiveness and creativity, thus increasing green competitive advantage [9–11].

Green knowledge sharing enables organization to utilize innovation capability that is required for their products and processes [6–8]. Developing green knowledge sharing means to empower organization to exploit their resources and knowledge in order to create innovative products and processes that can reduce the environmental impact [12]. Moreover, currently demand from the market keeps changing, thus the organization must find a technique to manage their resources, expand their business operations, carry out innovations, and be able to survive with their competitors, especially in the new normal era of the Covid-19 pandemic [13].

The study carried out by [14] proved that green practice in the organization will also affect employee attitude such as commitment. Green practice is able to increase employee performance, as well as develop psychological relationship between leaders and followers, or between colleagues. Commitment owned by employees will influence their performance which then assist organization to achieve their competitive advantage [15]. In order to achieve their competitive advantage, organization also must be able to create positive value that exceeds their competitors [11]. In addition, organization needs to have valuable resource, rare and inimitable skills, and have the ability to respond to changes by developing their resource capability [16].

This study uses MSMEs in DIY Province which implement green practices in their operational activities, especially in using natural materials and the waste management process. The author uses Giriloyo Batik as the research object, which is located in Giriloyo Tourism Gazebo, Karang Kulon, Wukirsari, Imogiri District, Bantul Regency, DIY Province.

2 Objective and Structure of Research

The objective of this study is to examine and analyze the influence of green knowledge sharing and green organizational commitment on green competitive advantage, mediated by green innovation. This study has three sub-main part, namely:

- Introduction, which explains the theoretical issue and business phenomenon that is related to green management practices in MSMEs context.
- Literature Review and Methods, which includes theories of green knowledge sharing, green organizational commitment, green competitive advantage, and green innovation.
- Results, discussion, and research limitations.

3 Literature Review

Green Knowledge Sharing

[17] stated that knowledge sharing is a media for receiving or disseminating important information, knowledge, and skills related to work experience. Knowledge sharing includes interaction between individuals and teams to engage and coordinate in work groups [13]. In individual level, an employee who carry out green knowledge sharing with other employees will generate more collaborative knowledge within the organization [4].

The study from [12] showed that knowledge sharing acts as mediator in the relationship between trust and virtual team effectiveness [11], job satisfaction, friendship at work, service innovation [6], servant leadership, and team performance. In HRM research, knowledge sharing is also found to be a significant mediator in distinct relationship, such as between organizational culture and job satisfaction, high-commitment work system and employee service behavior, as well as HRM practice and company entrepreneurship [6].

Green Organizational Commitment

Bhatia & Jakhar [7] mentioned that organizational commitment is a psychological relationship that must be maintained between employees and their organization. [18] identified six approaches used for commitment studies, namely affective, instrumental, normative, sociological, behavioral, and affiliation. From the review carried out by [19], it was emphasized that although there are several concepts of commitment, there are at least three common dimensions: attachment or affection to the organization (affective commitment), perception of loss when leaving the organization (instrumental commitment), and the obligation to remain in the organization (normative commitment).

Affective commitment, the prevailing approach in the organizational commitment literature, is one that defines the concept as an affective or emotional bond to the organization [18]. Individuals who are committed to the organization identified themselves as involved and want to be a part of the organization. On the other hand, normative commitment is conceptualized by [20] as a combination of normative pressures internalized by individual behavior in accordance with organizational goals and interests.

Green Innovation

Green innovation enables firms to have more significant average than their competitors because they exploit their green capital and ability to react quickly and appropriately to consumer demands [8] as well as combine their intangible assets and wealth.

Innovation is the main factor used as one of the strengths in increasing the organization's competitive advantage, with changes in environmental regulations becoming the basis for consideration in product life cycle when organizations make decisions to create these products and processing them [21]. Environmentally-friendly products and processes not only reduce the negative impact on the environment, but it can also increase the company's competitive advantage [22].

Green Competitive Advantage

Competitive advantage is defined as "an effort carried out by an organization where competitors cannot imitate the strategy or product created" so that the organization gets sustainable benefits [23]. Characteristics of resources in an organization that can be utilized to obtain competitive advantage are value, rarity, imitability, and unsubstitutability [11]. The competitive advantage of a company is generally known as low cost advantages, superior managerial capabilities, profitability, and first driving advantage [16]. Organizations that take the initiative in managing their environment will benefit themselves, as they carry out varying differentiation strategies, improve their green image, and eventually increase their competitive advantage [11].

3.1 The Influence of Green Knowledge Sharing on Green Inovation

The study carried out by [6] suggested that organizations that have the ability to increase green knowledge sharing will use their resources more efficiently, thereby supporting more innovative employees and producing better performance. Using the perspective of green knowledge sharing in designing products or services can make a difference with other organizations [17]. Since not all organizations think from an environmental perspective, organizations need to create opportunities, increase employee awareness of environmental issues, educate employees, as well as provide training and bring green ideas to inspire their services [4].

Service innovation shows how organizations are able to focus their resources on producing good results and turning their innovative ideas into reality [4]. What drives sales for green products and services? In order to compete in the international market, it is necessary for organizations to implement a green knowledge sharing system for employees with fellow coworkers to lead to employee knowledge and ability to create green innovation ideas for the products or services they create [8].

H1: Green knowledge sharing has a positive and significant influence on green innovation.

3.2 The Influence of Green Organizational Commitment on Green Innovation

Employees' green commitment is an encouragement in individuals to do something in order to support the organization's success according to its goals and prioritize the interests of the organization [7]. Employee green commitment can be influenced by individual characteritsics, the existence of alternative job opportunities, and organizational treatment [24]. Stated that one of the factors of innovative behavior in employees is employees' green commitment. Organizations help individuals to grow and build commitment by providing needed resources. Employees' green commitment will form elements of innovative green behavior that are in line with expectations for an organization that requires positive behavior from individuals as employees [8]. Based on this, employees' green commitment and innovative behavior certainly have an important relationship with individuals as employees to improve performance and help organizations to develop further [24].

H2: Green organizational commitment has a positive and significant influence on green innovation.

3.3 The Influence of Green Knowledge Sharing on Green Competitive Advantage

Green knowledge sharing is the green knowledge and skills acquired by each individual [25]. The availability of resources and knowledge allows organizations to exploit their ability to design products/services to reflect a green philosophy [8]. Organizations must continuously adapt to the changing market conditions to survive and remain competitive in business [26].

This study defines green competitive advantage as a cruicial factor for organization to increase their achievement of sustainable development [16]. Green knwoledge sharing enables organization to increase the flow and integration of both internal and external green knowledge [23]. Once an organization has a defined goal, it must work towards achieving the goal and create a project plan to achieve it conclusively, thus having a competitive advantage [8, 26].

H3: Green knowledge sharing has a positive and significant influence on green competitive advantage.

3.4 The Influence of Green Organizational Commitment on Green Competitive Advantage

Employee commitment to the organization is a form of organizational HRM results which show that employees are interested and willing to accept the values of the organization to achieve green behavior [7]. In green context, [24] also discussed employee internal motivation, defining employee environmental commitment as internal motivation. A study conducted by [7] supports from organizational leaders increase the successful implementation of green environmental management practices for organizations and employees. It can also encourage employees to think creatively, increase employees' commitment, and enhance innovative practices for the employees itself [8]. Employees play a crucial role in the organization, and when they have high commitment in the organization, they will direct all abilities, increase their motivation, and fulfill the organizational desire for green organizational activities [7].

H4: Green organizational commitment has a positive and significant influence on green competitive advantages.

3.5 The Influence of Green Innovation on Green Competitive Advantage

Organizations need to strive to implement green marketing activities to achieve superior brand differentiation and gain competitive advantage [11]. However, how can organizations increase their competitiveness by creating green innovation? [8] reported that in product aspect, green product innovation combines product modification, repackaging, and creation. On the other hand, in process aspect, green product innovation describes manufacturing processes that reduce negative environmental impacts on processing materials, resources, and knowledge. [16] added that competitive advantage can imply the exploitation of resources which results in an organization's distinctive position compared to its competitors. Developing responsible green relational capital enables managers to strengthen their market leadership in environmental products/services. The availability of resources and knowledge allows organizations to exploit their ability to design products/services to reflect the green philosophy of customers [8].

H5: Green innovation has a positive and significant influence on green competitive advantages.

3.6 The Mediating Role of Green Innovation

Using green knowledge sharing in designing products or services reflects the green benefits of products or services, which can make a difference [6]. Not everyone thinks from an environmental perspective, thus organizations might want to create opportunities to turn environmental principles into real-life business lessons by first collecting and organizing green knowledge [8]. Organizations must have their own green knowledge sharing to create green knowledge from their own experiences and innovations, thus they can create green service innovations [7]. Green knowledge sharing is veru important for organizational methods in identifying, creating, collecting, organizing, storing, disseminating, and applying green knowledge [4]. When an organization implements a different innovation compared to other organizations, they will achieve a competitive advantage [8].

Employees who are pro to innovation that will be carried out by the organization will also devote all their abilities and involve in organizational activities so that they form a commitment to themselves and support the organization to achieve organizational competitive advantage [7].

H6: Green innovation mediates the influence of green knowledge sharing on green competitive advantage.

Competitive advantage can also be seen from innovations made by the organizations, especially in terms of environmental issues, because currently consumers prefer products that are environmentally friendly, thus it becomes a competitive advantage for every company [16]. Organizations that have highly committed employees and have the same perceived value as the organization regarding environmental issues will be more involved in organizational activities, devoting all their abilities so that they are able to achieve competitive advantage, which is the goal of the organization [11].

Thus, employees can find out whether organizational green commitment is part of the company's core values and can produce environmental conservation values, where the next competitive advantage will be achieved [20].

H7: Green innovation mediates green organizational commitment on green competitive advantage.

4 Methodology and Approach

This study uses quantitative approach with survey using questionnaire to collect the data. This study also examines the characteristics or causal relationships between variables without any intervention from the researcher. The relationship studied is the relationship between green knowledge sharing and green organizational commitment on green competitive advantage with the mediating role of green innovation. The positivism approach is used in this study because analyzing causality is done on a value-free basis. The researchers used surveys by giving questionnaires and staying focused on variables related to the conditions of the empirical issues studied [27]. The sample used in this study is 60 employees of the Giriloyo Batik MSMEs which is located at the Giriloyo Tourism Gazebo, Karang Kulon, Wukirsari, Imogiri District, Bantul Regency, DIY Province. The MSMEs in this study have indeed used natural dyes, where the raw materials used in the business process from input to output is natural dyeing process. The indicators of green innovation (GI) is adopted from [13], green competitive advantage (GCA) from [23], green organizational commitment (GOC) from [14], and green knowledge sharing (GKS) from [13]. The score of the variables in this study uses Likert scale with the score 1 (strongly disagree) to 5 (strongly agree). The respondents were asked for statistical techniques using SmartPLS 3.0.

5 Result and Discussion

5.1 Respondents' Characteristics

Table 1 indicates that employees who work at Giriloyo Batik MSMEs are mostly women, have high school education (41 employees), have been working for 1–5 years (26 respondents), and are 31–40 years old (17 respondents).

Table 1. Respondents characteristics

Characteristics		Amount
Gender	Female	60
Education	Junior High School	16
	Senior/Vocational High School	41
	Diploma	1
	Bachelor/Master/Doctor	2
Length of work	1–5 years	26
	6–10 years	10
	11–15 years	1
	>15 years	23
Age	Less than 20 years old	0
	20–30 years old	14
	31–40 years old	17
	41–50 years old	14
	Above 50 years old	15

5.2 Validity and Reliability Test

The item of validity test is valid if the value of r count > r table 0.2108 (2 tail; α = 0.005). The results of the test show r-count green knowledge sharing (0.232 s.d. 0.433), green organizational commitment (0.370 s.d. 0.532), green innovation (0.374 s.d. 0.495), and green competitive advantage (0.270 s.d. 0.413) > 0.2108. In addition, the value of Cronbach's Alpha green knowledge sharing (0.603), green organizational commitment (0.652), green innovation (0.623), and green competitive advantage (0.0.602) > 0.60 or the instruments are reliable.

The value of outer loading test can be seen on Table 2 and Table 3. The results indicate that green knowledge sharing is 0.996, green organizational commitment is 0.756 to 0.746, green innovation is 0.847 to 0.837, and green competitive advantage is 0.759 to 0.865 > 0.70. The AVE test results in green knowledge sharing (0.996 to 0.996), green organizational commitment (0.583), green innovation (0.709), and green competitive advantage (0.663) > 0.50. The value of composite reliability indicates that green knowledge sharing (0.996), green organizational commitment (0.807), green innovation (0.830), and green competitive advantage (0.796) > 0.70. Based on these results, it can be seen that all scores of the latent variables passed the minimum critera.

Table 2. Outer Loading

Green knowledge sharing		Green organizational commitment		Green innovation		Green competitive advantage	
GKH1	0.996	GOC1	0.756	GI4	0.847	GCA1	0.759
GKH2	0.996	GOC2	0.787	GI7	0.837	GCA3	0.865
		GOC4	0.746				

GKH = Green Knowledge Sharing; GOC = Green organizational commitment; GI = Green innovation; GCA = Green competitive advantage
Source: Primary Data, 2022

Table 3. AVE and composite relianility

	AVE	Composite reability
Green Knowledge Sharing	0.992	0.996
Green Organizational Commitment	0.583	0.807
Green Innovation	0.709	0.830
Green Competitive Advantage	0.663	0.796

Source: Primary Data, 2022

5.3 Hypothesis Test

Table 4 indicates the hypothesis test results which concluded that H2 and H3 are accepted, while H1, H4, H5, H6, and H7 is rejected.

Table 4. Hypothesis test results

Hypothesis		Original sample	Standard deviation	T Statistic	P Values
H1 rejected	GKH-GI	– 0.154	0.203	0.757	0.449
H2 accepted	GOC-GI	0.898	0.293	3.071	0.002*
H3 accepted	GKH-GCA	0.962	0.337	2.860	0.004*
H4 rejected	GOC-GCA	– 0.674	0.612	1.102	0.271
H5 rejected	GI-GCA	0.196	0.428	0.458	0.647
H6 rejected	GKH-GI-GCA	– 0.030	0.089	0.338	0.736
H7 rejected	GOC-GI-GCA	0.176	0.432	0.407	0.684

Note * = sign with alpha 0.05
Note: GKH = Green Knowledge Sharing; GOC = Green Organizational Commitment; GI = Green Innovation; GCA = Green Competitive Advantage

The analysis results of mediating role of green innovation in this study does not prove that green innovation is able to mediate the relationship between green knowledge sharing and green organizational commitment on green competitive advantage.

5.4 Discussion

The results of the first hypothesis test (H1) indicate that green knowledge sharing does not have positive and significant influence on green innovation. The results of this study are in accordance with those conducted by [28]. Green knowledge sharing is actually able to strengthen the ability to absorb green knowledge which will increase the ability of employees to innovate, but at Giriloyo Batik MSMEs, it turns out that green knowledge sharing that is carried out needs to be further improved between the organization and its employees. Green knowledge sharing indicators that have low question item scores are found in "Our organization will share distribution knowledge with supply chain partners frequently in environmental collaboration" (mean = 3.92) and "Our organization will share process design knowledge with supply chain partners in environmental collaboration." (mean = 3.93).

The results of the second hypothesis (H2) prove that green organizational commitment has a positive and significant influence on green innovation. The results of this study support [24] who stated that green organizational commitment can increase green innovation. Therefore, to increase employee innovation in a green environment in the organization, it is necessary to have an attitude of commitment that appears in employees. Employees' green commitment will form elements of innovative green behavior that are in line with expectations for an organization [8]. Indicators of green organizational commitment are "Green practice values in the organization and self" (mean = 3.95) and "organizations need to increase the best inspiration in employees, especially in carrying out green practices" (mean = 3.98), while the highest average on the statement item "I am willing to make a lot of effort beyond what has been assigned by the organization so that the organization becomes more successful in implementing green practices" (mean = 4.02).

Furthermore, from the results from the third hypothesis (H3), it was found that green knowledge sharing has a positive and significant influence on green competitive advantage. The results of this study are in accordance with the study from [16] which confirmed that green knowledge sharing has a significant influence on green competitive advantage. The availability of resources and knowledge enables organizations to exploit their ability to design products/services to reflect a green philosophy so that organizations are able to achieve organizational competitive advantage and be able to compete with competitors. The green knowledge sharing indicator that has the lowest question item value is found in "Our organization will share distribution knowledge with supply chain partners often in environmental collaboration" (mean = 3.92) and the highest question item is in "Organization leaders and our supply chain partners will share knowledge purchasing with employees within the organization in terms of environmental collaboration" (mean = 4.03).

On the contrary, the results of the fourth hypothesis test (H4) indicate that green organizational commitment has no positive and significant influence on green competitive advantage. This study supports the research conducted by [7]. The commitment of employees at Giriloyo Batik MSMEs needs to be improved again, so that their working employees feel really involved in the organization and become part of the organization. When they feel involved, employees will give their abilities to create products that are more innovative and achieve organizational competitive advantage.

In addition, the results of the fifth hypothesis (H5) test show that green innovation has no positive and significant influence on green competitive advantage. This study does not support the research conducted by [11, 23] that green innovation has a significant influence on green competitive advantage. Green innovation applied to Giriloyo Batik MSMEs needs to be improved again for the employees there, including the form of training or skill improvement, so that employees are able to develop ideas that can compete with other organizations, and achieve competitive advantage which is the goal of the organization. Question items that need to be improved, and the lowest score is in the questions "The organization where I work uses the smallest amount of materials to make products" (mean value = 3.92), and "By recycling and using natural materials for the production process can reduce water consumption, electricity, or oil" (mean = 3.90).

6 Conclusion

The findings of this study suggested that green organizational commitment has a positive and significant influence on green innovation, while green knowledge sharing has an influence on green competitive advantage. This study provides implications for company leaders to encourage an increase in green innovation and low green competitive advantage. The results of the mediation test also show that green innovation is not able to mediate green knowledge sharing on green competitive advantage, while green innovation mediates green organizational commitment to green competitive advantage. Further research can add other antecedent variables that affect green competitive advantage, for example involving organizational green culture variables [11] and environmental orientation [23].

References

1. Yong, J.Y., Yusliza, M., Ramayah, T., Fawehinmi, O.: Nexus between green intellectual capital and green human resource management. J. Cleaner Prod. **215**, 364–374 (2019). https://doi.org/10.1016/j.jclepro.2018.12.306
2. Renwick, D.W.S., Redman, T., Maguire, S.: Green human resource management: a review and research agenda*. Int. J. Manag. Rev. **15**(1), 1–14 (2013). https://doi.org/10.1111/j.1468-2370.2011.00328.x
3. do Rosário Cabrita, M., Cruz-Machado, V., Matos, F., Safari, H.: Green knowledge: developing a framework that integrates knowledge management and eco-innovation. In: Proceedings of the European Conference on Knowledge Management ECKM, vol. 2016, pp. 127–135 (2016)
4. Shahzad, M., Qu, Y., Zafar, A.U., Rehman, S.U., Islam, T.: Exploring the influence of knowledge management process on corporate sustainable performance through green innovation. J. Knowl. Manag. **24**(9), 2079–2106 (2020). https://doi.org/10.1108/JKM-11-2019-0624
5. Sugandini, D., Effendi, M.I., Thamrin, H.M., Priyadi, U., Muafi: From Environmental Knowledge to Conservation Behaviour. Qual. Access Success, **20**(172), 101–107 (2020)
6. Rubel, M.R.B., Kee, D.M.H., Rimi, N.N.: The influence of green HRM practices on green service behaviors: the mediating effect of green knowledge sharing. Empl. Relat. **43**(5), 996–1015 (2020). https://doi.org/10.1108/ER-04-2020-0163
7. Bhatia, M.S., Jakhar, S.K.: The effect of environmental regulations, top management commitment, and organizational learning on green product innovation: evidence from automobile industry. Bus. Strateg. Environ. **30**(8), 3907–3918 (2021). https://doi.org/10.1002/bse.2848
8. Lin, Y.-H., Chen, Y.-S.: Determinants of green competitive advantage: the roles of green knowledge sharing, green dynamic capabilities, and green service innovation. Qual. Quant. **51**(4), 1663–1685 (2016). https://doi.org/10.1007/s11135-016-0358-6
9. Ardyan, E., Rahmawan, G., Tinggi, S., Ekonomi, I.: of Sustainable competitive advantages and smes marketing. In: International Journal of Civil Engineering and Technology, vol. 8, pp. 1114–1122 (2017)
10. Zhou, M., Govindan, K., Xie, X.: How fairness perceptions, embeddedness, and knowledge sharing drive green innovation in sustainable supply chains: an equity theory and network perspective to achieve sustainable development goals. J. Clean. Prod. **260**, 120950 (2020). https://doi.org/10.1016/j.jclepro.2020.120950
11. Wang, C.H.: How organizational green culture influences green performance and competitive advantage: the mediating role of green innovation. J. Manuf. Technol. Manag. **30**(4), 666–683 (2019). https://doi.org/10.1108/JMTM-09-2018-0314
12. Song, M., Yang, M.X., Zeng, K.J., Feng, W.: Green knowledge sharing, stakeholder pressure, absorptive capacity, and green innovation: evidence from Chinese manufacturing firms. Bus. Strateg. Environ. **29**(3), 1517–1531 (2020). https://doi.org/10.1002/bse.2450
13. Hung, S.W., Chen, P.C., Chung, C.F.: Gaining or losing? The social capital perspective on supply chain members' knowledge sharing of green practices. Technol. Anal. Strateg. Manag. **26**(2), 189–206 (2014). https://doi.org/10.1080/09537325.2013.850475

14. Joong, Y., Gon, W., Choi, H., Phetvaroon, K.: International journal of hospitality management the effect of green human resource management on hotel employees' eco- friendly behavior and environmental performance. Int. J. Hosp. Manag. **76**, 83–93 (2019). https://doi.org/10.1016/j.ijhm.2018.04.007
15. Pham, N.T., Thanh, T.V., Tuckova, Z., Thuy, V.T.N.: The role of green human resource management in driving hotel's environmental performance : interaction and mediation analysis. Int. J. Hosp. Manag. **88**, 1–4 (2020). https://doi.org/10.1016/j.ijhm.2019.102392
16. Nanath, K., Pillai, R.R.: The influence of green IS practices on competitive advantage: mediation role of green innovation performance. Inf. Syst. Manag. **34**(1), 3–19 (2017). https://doi.org/10.1080/10580530.2017.1254436
17. Rajabion, L., Sataei Mokhtari, A., Khordehbinan, M.W., Zare, M., Hassani, A.: The role of knowledge sharing in supply chain success: literature review, classification and current trends. J. Eng. Des. Technol. **17**(6), 1222–1249 (2019). https://doi.org/10.1108/JEDT-03-2019-0052
18. Shoaib, M., et al.: The role of GHRM practices towards organizational commitment: a mediation analysis of green human capital. Cogent Bus. Manag. **8**(1), 1870798 (2021). https://doi.org/10.1080/23311975.2020.1870798
19. Machado, C., et al.: Organizational commitment, job satisfaction and their possible influences on intent to turnover. Rev. Gestão, 1–19 (2018). https://doi.org/10.1108/REGE-12-2017-008
20. Bell, M. Sheridan, A., Bell, M.: Corrigendum to: how organizational commitment influences nurses' intention to stay in nursing throughout their career International Journal of Nursing Studies Advances, 2 (2020). 100,007. Int. J. Nurs. Stud. Adv. p. 100087 (2022). https://doi.org/10.1016/j.ijnsa.2022.100087
21. Ahakwa, I., Asamany, M.: Green human resource management practices and environmental performance in Ghana: the role of green innovation **4**(4), 100–119 (2021). https://doi.org/10.33215/sjom.v4i4.704
22. Muisyo, P.K., Qin, S.: Enhancing the FIRM'S green performance through green HRM : the moderating role of green innovation culture. J. Clean. Prod. **289**, 125720 (2021). https://doi.org/10.1016/j.jclepro.2020.125720
23. Zameer, H., Wang, Y., Yasmeen, H., Mubarak, S.: Green innovation as a mediator in the impact of business analytics and environmental orientation on green competitive advantage. Manag. Decis. **60**(2), 488–507 (2022). https://doi.org/10.1108/MD-01-2020-0065
24. Buhl, A., Blazejewski, S., Dittmer, F.: The more, the merrier: why and how employee-driven eco-innovation enhances environmental and competitive advantage. Sustainability **8**(9), 946 (2016). https://doi.org/10.3390/su8090946
25. Fong, C.Y., Ooi, K.B., Tan, B.I., Lee, V.H., Chong, A.Y.L.: HRM practices and knowledge sharing: an empirical study. Int. J. Manpow. **32**(5), 704–723 (2011). https://doi.org/10.1108/01437721111158288
26. Chiou, T.Y., Chan, H.K., Lettice, F., Chung, S.H.: The influence of greening the suppliers and green innovation on environmental performance and competitive advantage in Taiwan. Transp. Res. Part E Logist. Transp. Rev. **47**(6), 822–836 (2011). https://doi.org/10.1016/j.tre.2011.05.016
27. Simonson, I., Carmon, Z., Dhar, R., Drolet, A., Nowlis, S.M.: Consumer research. In: Search of Identity May 2014 (2001). https://doi.org/10.1146/annurev.psych.52.1.249
28. Song, W.: Effects of green human resource management and managerial environmental concern on green innovation management. Eur. J. Innov. **24**(3), 1–17 (2020). https://doi.org/10.1108/EJIM-11-2019-0315

Examining the Impact of Strategic Thinking on Organizational Innovation: The Moderating Role of Autonomy: A Study at Jordanian Information Technology Companies

Sahar Moh'd Abu Bakir and Motteh S. Al Shibly(✉)

Business Administration Department, Faculty of Business, Amman Arab University, Amman, Jordan

Sh-mottee@aau.edu.jo

Abstract. The study aimed to test the impact of strategic thinking in terms of (systems thinking and future insights) on organizational innovation in terms of (introducing new products and developing the current products): The moderating role of Autonomy at Jordanian IT companies. The quantitative analytical descriptive method was used. 98 IT companies were randomly selected out of 220 companies operating in the Jordanian market. For collecting the needed information and data the questionnaire was used and distributed electronically to 205 managers working at the selected IT companies, 173 questionnaires were retrieved and statistically analyzed. The main results of the statistical analysis revealed that there is a statistically significant impact of strategic thinking on organizational innovation, and Autonomy as a moderator improved this impact by 0.02. The study recommended enhancing the capabilities of all the staff in strategic thinking and empowering them to be able to freely make innovative decisions.

Keywords: Strategic thinking · Organizational innovation · Autonomy · Jordanian IT companies

1 Introduction

There is interest by managers and stakeholders in organizations about strategic thinking. Where it is possible to develop the vision and goals of the organization [1], which confirms the importance of strengthening strategic thinking in organizations because it also has a role in enhancing future insight [2]. This enhances the performance of employees in the organization by generating creativity for them and thus creating an interactive work environment between managers and employees [3]. Strategic thinking develops solutions to face any current problems and enhances facing any future problems and challenges, which is reflected in current and future assumptions. This, in turn, makes strategic thinking positively reflected on the performance of the organizational and functional work of the organization, making it more adapted to the business environment [4,

B. Alareeni et al. (Eds.): ICBT 2022, LNNS 620, pp. 118–128, 2023.
https://doi.org/10.1007/978-3-031-26953-0_13

5], The volatility of the business environment, its changing conditions, and its continuous development increase the importance of strategic thinking that achieves the right response at the right time [6].

Strategic thinking is influenced by employees' perceptions of issues and problems and how to deal with them [7]. To reach organizational innovation through strategic thinking, there are many factors that affect this, including the organization's structure, organizational culture, Autonomy, infrastructure, and its development [8]. The employees in the organization must enjoy Autonomy in order to reach higher flexibility in order to achieve strategic thinking and organizational Innovation. Consequently that they can take appropriate decisions and policies at the right time based on their vision and thinking, and This may add appropriate reactions in the course of work [9].

It is worth noting that there are many studies that have examined strategic thinking and organizational innovation separately [5–10] but there are few studies that dealt with strategic thinking and its relationship with organizational innovation [3]. Except that the study population is in the Jordanian technology companies. What distinguishes this study also is that it deals with Autonomy as a moderating variable. And from here, there is a great tendency among departments in organizations to adopt strategic thinking and reach organizational innovation while granting independence to employees after the Corona pandemic, which affected organizations, especially in developing countries such as Jordan. Especially since the Jordanian economy suffers from many pressures, and it increased after the Corona pandemic, which imposed many restrictions on companies [11]. Even after the end of the pandemic, companies must adopt effective strategies to face the high competition in the Jordanian market, as the business environment is shallow in Jordan, and this forces companies to invest in business strategies to empower themselves in front of competitors and to face any possible fluctuations in the future [12]. Which reinforced the importance of the study with the need to support companies in Jordan as they face many challenges through strategic thinking strategies because of their important role that is reflected on the institution as a whole. And for a sensitive service sector, such as information technology companies, which have been highly relied upon in light of the pandemic by all sectors. Accordingly, the current study aimed to examine the impact of strategic thinking on organizational innovation: the moderating role of autonomy: a study at Jordanian information technology companies.

2 Research Model

This study proposes a framework for studying the impact of strategic thinking on organizational innovation (Fig. 1):

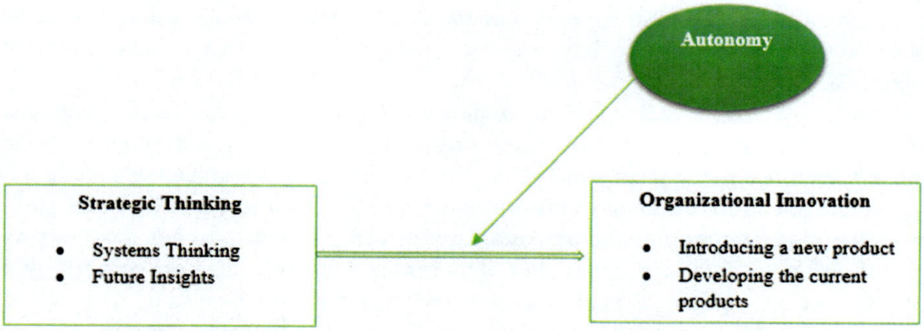

Fig. 1. Research model

2.1 Strategic Thinking with Organizational Innovation

With the dynamic business environment, there are challenges and developments facing business organizations [13], which increases the need for them to work in a harmonious and effective manner in order to be able to keep pace with the environment and adapt to it [14]. Hence, strategic thinking has become important to help organizations face challenges and enhance their stability, which in turn creates organizational innovation. In addition to that, including the ability to analyze problems and face situations in a way that gives autonomy to employees to enhance their ability to make decisions and raise their expertise and skills to reflect on the performance of the work of the organization as a whole, which develops the organization's position among its competitors [3]. According to study of [15], which examined the relationship of strategic planning and strategic thinking with organizational innovation, it revealed that strategic thinking has always been at the core of organizations' work to reach organizational innovation to ensure facing the fluctuations of the business environment, which confirms the relationship between them in the course of work of organizations. Through strategic thinking and innovation, the past and the present can be linked to predict the future. This makes strategic thinking and innovation one of the factors that enhance the company's performance in the future and the present [16].

Strategic thinking is the key to business innovation. This is a very sensitive matter for organizations. This puts strategic thinking and organizational innovation side by side in the business process [17]. According to study of [18], which examined strategic thinking and strategic innovation, it confirmed the existence of a consensual relationship between them to positively influence the performance of the work of organizations. In the same context, [19] study revealed a positive impact of organizational thinking on organizational innovation. Sequence organizational and strategic thinking is considered one of the most important contributors to achieving innovation for the organization. Where organizational thinking directly affects the creation of innovation in organizations [20]. In the same context the advantages of systemic thinking have become one of the most important factors in supporting organizations, as it supports innovation through the system's ability to generate its own ideas in developing existing products and introducing new products. This is the same way that enhances future visions. The course of business and its challenges and potential surprises to develop commodities in line with new desires and developments, in addition to continuing to provide new commodities that are required in the future [21]. According to the discussion of previous studies of the subject of the study and the relationships and variables it dealt with, the following main hypothesis is assumed, and its sub-sections are divided as follows:

H1 There is a statistically significant impact of strategic thinking (systems thinking, future insights) on organizational innovation (introducing new products, developing the current products).

H1-1 systems thinking has a positive impact on introducing new products
H1-2 systems thinking has a positive impact on developing the current products
H1-3 future insights has a positive impact on introducing new products
H1-4 future insights has a positive impact on developing the current products

2.2 Autonomy

The concept of autonomy began in Greece with political entities, as it is said, an independent state, so that this state sets its own laws without imposing on them from a certain party. Hence, independent people, like independent states, conduct their business according to their own principles and rules in which they believe. Independent people are more impulsive and creative [22]. According to study of [23], which examined the role of autonomy and job satisfaction among workers in Greece, it found that there is a modified role of autonomy on employees in their job satisfaction and the extent of their performance. In the same context, [24] study showed that independence has a positive effect on raising the productivity and performance of employees.

Functional autonomy plays an important role on the organizational performance of the organization, and this in turn is linked with organizational thinking and organizational innovation. This makes autonomy important in creating organizational adjustment among employees [25]. Therefore, it must be recognized that independence has a positive impact on organizational innovation [26]. Sequentially study of [27], which examined organizational activities that enhances employees' autonomy, revealed that autonomy played an important role in innovation and the performance of the organization's work. According to study of [28], which examined the behavior of the innovative leader and its impact on employees in terms of management development and access to innovation within the organization, it was found that autonomy affects innovation among employees and that effectively enhances innovation behavior. Based on the discussion of previous studies of the subject of the study and the relationships and variables that they dealt with about the impact of autonomy on organizational thinking and organizational innovation, the following main hypothesis was assumed:

H2-Autonomy has a statistically significant role in improving the impact of strategic thinking on organizational innovation.

3 Research Methodology

The study depends on the quantitative descriptive analytical method, and the questionnaire to collect the needed information and data from the sampling unit [29, 30].

3.1 The Study Population, Sample and Sampling Unit

The study measures relied on the previous studies as follows: For the strategic thinking variable [31] and [32] measures were adapted. Organizational innovation measure was adapted from [33] And [34].The measure of the moderator variable (Autonomy) adapted from [35–37]. Likert 5 scale measurement was used in each part of the questionnaire; to assess the degree of agreement of the participants on the study questions.

3.2 The Study Measure

Cronbach's alpha was calculated to identify the internal consistency of the tool questions. [38–40] commented that if the results of this measurement were 0.70 or more the questions of the variables are internally consistent. The results of organizational justice = 0.811, Organizational trust = 0.84,3 and Job security = 0.798. Consequently the results revealed that the study tool is reliable.

3.3 The Questionnaire Reliability

To identify of the questionnaire's reliability, the researchers employed Cronbach's alpha coefficient measurement for the questions of each variable, the results of all the variables were more than 0.70 the accepted level for internal consistency according to [40, 41].

4 The Statistical Analysis Results

4.1 Descriptive Statistics Analysis

The purpose of this part of the statistical analysis is to identify the degree of applying each of the study variables and sub variables. The arithmetic means and standard deviations were computed for each variable questions. Table 1 illustrated the total means and standard deviations of each variable. The rating scale of the means will be as follows, it will be considered high degree of application if the mean is >3.7, and moderate between 3–3.7, and weak if the mean is <3. When it comes to strategic thinking elements (systems thinking and future insights) both variables get >3.7 means which indicated that these elements dominated managers' way of thinking during work. The same can be said about the two sub variables of organizational innovation (introducing a new product, and developing the current products). These values revealed that the companies' managers are aware of the vital role of novelty, creativity and innovation to survive, grow and achieve sustainable competitive advantage for their organizations. Finally the mean of the moderator variable autonomy = 3.84 which indicated that strategic thinking is

Table 1. The results of descriptive statistics.

Variables	Total Arithmetic Means	Standard deviations	Degree of application
Systems thinking	3.76	.33591	High
Future insights	4.27	.64384	High
Introducing a new product	4.06	.49428	High
Developing the current products	3.86	.37880	High
Autonomy	3.84	.31905	High

accompanied with high level of autonomy and freedom to make decisions as long as these decisions will add value to the company's performance.

The values of standard deviations are low, which revealed that there is no spread of the answers and there is an agreement among the participants of the questions content.

4.2 Hypotheses Testing

H1: There is a statistically significant impact of strategic thinking (systems thinking, future insights) on organizational innovation (introducing new products, developing the current products) to test H1 and its sub hypotheses, multiple regression was calculated with p value = 0.05 the hypothesis will be accepted if t sig value was less than 0.05. Table 2 illustrated the results of the multiple regression as follows: First part is related the results of testing strategic thinking in terms of (systems thinking and future insights) impact on organizational innovation as one variable, The values of R (the correlation between the independent variable/s and the dependent variables/s) respectively = 0.566, 0.581, and 0.450 revealed a moderate correlation between variables that are manifested in Table 1.The values of R^2 = respectively 0.320, 0.314, and 0.202 Indicated that the independent variable/s in each part is responsible by the value of R^2 for the positive variation of organizational innovation and its sub variables respectively. The values of F sig in each part = 0.000, since it is <0.05 revealed that there is at least one independent variable has a statistically impact on the dependent/s. While the values of t sig = (0.000) < 0.05 for all the variables except for introducing new products = 0.56 $>$ 0.05.

Table 2. Multiple regression results

Independent variables	Dependent variables	Model summery		ANOVA		Coefficient		
		R	R Square	F	Sig.	Beta	t	Sig.
Systems thinking	Organizational innovation	.566[a]	.320	63.635	.000[b]	.217	3.994	.000
Future insights						.445	8.177	.000
Systems thinking	Introducing new products	.561[a]	.314	61.911	.000[b]	.105	1.922	.056
Future insights						.512	9.358	.000
Systems thinking	Developing the current products	.450[a]	.202	34.222	.000[b]	.247	4.959	.000
Future insights						.292	4.181	.000

H2: Autonomy has a statistically significant role in improving the impact of strategic thinking on organizational innovation.

To test the moderation impact in the relationship between the independent and dependent variables the researchers computed the hierarchal regression through 3 steps / models. In the first model the impact of strategic thinking as one variable on organizational innovation as one variable was calculated, based on t sig value (0.000) the results revealed that there is a statistically significant impact of strategic thinking on organizational innovation. In the second model the moderator (Autonomy) impact on organizational innovation was tested, with t sig = 0.000 indicated a statistically significant impact of the moderator variable on the dependent.

In the third model the interaction between (strategic thinking and Autonomy impact) on the organizational innovation was tested, t sig value = 0.000 indicated a statistically significant impact of strategic thinking on organizational innovation in the existence of autonomy as a moderator. Table 3 illustrated that the values of R and R^2 were as follows respectively: In the first model = 0.539 and 0.290In the second model = 0.660 and 0.439In the third model 0.673 and 0.453 The results show that there is an increase in the values of R and consequently R^2 The change in $R^2 = .018$ revealed that the impact of strategic thinking on organizational innovation improved by 0.02 with the existence of Autonomy.

Table 3. The second hypothesis results (hierarchical regression results)

Model	R	R^2	R^2 Change	F	Sig	F Change	Beta	T	T sig
1	$.539^a$	.290	.290	110.837	$.000^b$	110.837	.641	7.600	.000
2	$.660^b$	.436	.145	104.195	$.000^c$	69.526	.513	8.369	.000
3	$.673^c$	.453	.018	74.391	$.000^c$	8.779	.296	2.963	.000

5 Results

The results of t sig revealed that there is a statistically significant impact of strategic thinking (systems thinking and future in-sights) on organizational innovation as a single variable, and on developing the cur-rent products. Meanwhile there is no statistically significant impact of strategic thinking on introducing new products. The justification of this results is due to the financial risk, time and efforts devotion, and high cost of introducing new products in the cur-rent market. Based on Beta values future insights has a stronger impact on organizational innovation and its dimensions than systems thinking. Then the results were completed that there is a statistically significant effect of strategic thinking on organizational innovation. In the second model, the influence of the mediator (Autono-my) on organizational innovation was tested, as there is a statistically significant effect of the mediator variable on the dependent. In addition to the existence of a statistically significant effect of strategic thinking on organizational innovation in the presence of independence as a mediator. Thus, the impact of strategic thinking on organizational innovation has improved with the presence of autonomy.

6 Discussions

The results of the analysis discussed earlier revealed several important findings. First, this study of strategic thinking (systemic thinking and looking to the future) emphasized organizational innovation as a single variable, and on the development of existing products. The results showed that, at the same time, there is no statistically significant effect of strategic thinking in introducing new products. These results indicate that the explanatory power was great in the impact of strategic thinking. It has been shown that based on BETA values, future visions have a stronger impact on organizational innovation and its dimensions than systems thinking. Within the integrated model, the last part of the research model showed that the impact of strategic thinking on organizational innovation improved with the presence of independence. So I found it important and the result was consistent with the results of [17, 19, 21, 42]. The importance of independence in achieving innovation was in line with that as well [28].

7 Conclusions and Recommendations

Their rationality is built on their enthusiasm and entrepreneurial tendency to add value and excel in comparison to competitors. Organizations depend on those who are able to read the future and accurately predict changes in the external environment to invest in appropriate opportunities and avoid potential threats, those (strategic thinkers). Therefore, this study aimed to study the impact of strategic thinking in terms of (systemic thinking and future visions) on organizational innovation, taking into account independence as a coordinator. The study targeted Jordanian IT companies. Based on the results of the study, there is a significant statistical effect of strategic thinking and its dimensions on organizational innovation and independence. When it comes to the role of mediator in independence, the results confirmed that independence improved the impact of strategic thinking on organizational innovation by 0.02, and the study emphasized the importance of independence in achieving creativity and innovation. Strategic thinking will be re-constrained with high levels of formality, low empowerment, and low delegation terms. Hence the study has management implications, as it provides insights into the critical role of strategic thinking in organizational creativity and innovation, and in enhancing the capabilities of employees to be able to make decisions freely and rationally. Accordingly, the study recommends providing employees of information technology companies with training courses to enhance their competencies in strategic and critical thinking. In addition to the importance of empowering all employees to give them the opportunity to express their ideas and suggestions freely without restrictions and obstacles. As for future research, it is recommended to conduct more research related to the impact of strategic thinking on other variables, and in sectors different from IT companies. As there are future studies and recommendations, the current study is not devoid of the determinants that have manifested itself in targeting information technology companies in the city of Ma'an only, where if other cities from the north and center were included, we would have more comprehensive results.

References

1. Habes, M., Alghizzawi, M., Elareshi, M., Ziani, A., Qudah, M., Al Hammadi, M.M.: E-marketing and customers' bank loyalty enhancement: Jordanians' perspectives. In: Hamdan, A., Shoaib, H.M., Alareeni, B., Hamdan, R. (eds.) The Implementation of Smart Technologies for Business Success and Sustainability. Studies in Systems, Decision and Control, vol. 216, pp. 37–47. Springer, Cham (2023). https://doi.org/10.1007/978-3-031-10212-7_4
2. Salloum, S.A., Al-Emran, M., Khalaf, R., Habes, M., Shaalan, K.: An innovative study of e-payment systems adoption in higher education: theoretical constructs and empirical analysis. Int. J. Interact. Mob. Technol. **13**(6) (2019)
3. Shahbazzadeh, M., Rahiminik, A., Ardestani, A.S.: Survey of relationship between strategic thinking within deployment of knowledge management and organizational innovation in miandoab NIOPDC. Mediterr. J. Soc. Sci. **7**(4 S1), 294 (2016)
4. Al-Shibly, M.S., Alghizzawi, M., Habes, M., Salloum, S.A.: The impact of de-marketing in reducing Jordanian youth consumption of energy drinks. In: Hassanien, A.E., Shaalan, K., Tolba, M.F. (eds.) AISI 2019. AISC, vol. 1058, pp. 427–437. Springer, Cham (2020). https://doi.org/10.1007/978-3-030-31129-2_39
5. Shahul Hameed, N.S., Salamzadeh, Y., Abdul Rahim, N.F., Salamzadeh, A.: The impact of business process reengineering on organizational performance during the coronavirus pandemic: moderating role of strategic thinking, foresight (2021)
6. Tovstiga, G.: Strategy in Practice: A Practitioner's Guide to Strategic Thinking. John Wiley & Sons, Hoboken (2015)

7. Gross, R.: Situational interview method as a predictor of strategic thinking: a theoretical framework. Glob. Bus. Econ. (2015)
8. Haycock, K.: Strategic thinking and leadership. Libr. Leadersh. Manag. **26**(3/4) (2012)
9. Wynen, J., Verhoest, K., Ongaro, E., Van Thiel, S.: In cooperation with the COBRA network.: innovation-oriented culture in the public sector: do managerial autonomy and result control lead to innovation? Public Manag. Rev. **16**(1), 45–66 (2014)
10. Steptoe-Warren, G., Howat, D., Hume, I.: Strategic thinking and decision making: literature review. J. Strateg. Manag. (2011)
11. Abu-Mater, W., Alsufy, F., Afifa, M.A.: The effect of coronavirus (Covid-19) on the Jordanian economy: a comprehensive analysis of the economy and how to return growth rapidly. J. Account. Financ. Manag. Strateg. **15**(2), 1–30 (2020)
12. de Luna, I.R., Liébana-Cabanillas, F., Sánchez-Fernández, J., Muñoz-Leiva, F.: Mobile payment is not all the same: the adoption of mobile payment systems depending on the technology applied. Technol. Forecast. Soc. Change **146**, 931–944 (2019)
13. Rahi, S., Alghizzawi, M., Ishtiaq, M., Ngah, A., Mehta, A.M.: Examining consumer behaviour towards continuance use of mobile shopping apps with the integration of expectation confirmation theory and flow theory.
14. Alghizzawi, M.: A survey of the role of social media platforms in viral marketing: the influence of eWOM. Int. J. Inf. Technol. Lang. Stud. **3**(2) (2019)
15. Bouhali, R., Mekdad, Y., Lebsir, H., Ferkha, L.: Leader roles for innovation: strategic thinking and planning. Procedia-Soc. Behav. Sci. **181**, 72–78 (2015)
16. AlQershi, N.: Strategic thinking, strategic planning, strategic innovation and the performance of SMEs: the mediating role of human capital. Manag. Sci. Lett. **11**(3), 1003–1012 (2021)
17. Calabrese, A., Costa, R.: Strategic thinking and business innovation: abduction as cognitive element of leaders' strategizing. J. Eng. Technol. Manag. **38**, 24–36 (2015)
18. Kasemsap, K.: Strategic innovation management: an integrative framework and causal model of knowledge management, strategic orientation, organizational innovation, and organizational performance. In: Organizational Culture and Behavior: Concepts, Methodologies, Tools, and Applications, pp. 86–101. IGI Global (2017)
19. Crawford, V.P., Iriberri, N.: Level-k auctions: can a nonequilibrium model of strategic thinking explain the winner's curse and overbidding in private-value auctions? Econometrica **75**(6), 1721–1770 (2007)
20. Olaleye, B.R., Akkaya, M., Emeagwali, O.L., Awwadd, R.I., Hamdane, S.: Strategic thinking and innovation performance; the mediating role of absorptive capabilities. Rev. Argent. Clín. Psicol. **29**(5), 2030–2043 (2020)
21. Urze, P., Abreu, A.: System thinking to understand networked innovation. In: Working Conference on Virtual Enterprises, pp. 327–335 (2014)
22. Miller, S.R.M.: Autonomy. In: The SAGE Encyclopedia of Surveillance, Security, and Privacy, pp. 79–82. SAGE Publishing (2018)
23. Belias, D., Koustelios, A., Sdrolias, L., Aspridis, G.: Job satisfaction, role conflict and autonomy of employees in the Greek banking organization. Procedia-Soc. Behav. Sci. **175**, 324–333 (2015)
24. Langfred, C.W.: To be or not to be autonomous: exploring why employees want more autonomy. N. Am. J. Psychol. **15**(2) (2013)
25. Tabiu, A., Pangil, F., Othman, S.Z.: Does training, job autonomy and career planning predict employees' adaptive performance? Glob. Bus. Rev. **21**(3), 713–724 (2020)
26. Demircioglu, M.A.: Sources of innovation, autonomy, and employee job satisfaction in public organizations. Public Perform. Manag. Rev. **44**(1), 155–186 (2021)
27. Burcharth, A., Knudsen, M.P., Søndergaard, H.A.: The role of employee autonomy for open innovation performance. Bus. Process Manag. J. (2017)

28. Lee, W.R., Choi, S.B., Kang, S.-W.: How leaders' positive feedback influences employees' innovative behavior: the mediating role of voice behavior and job autonomy. Sustainability **13**(4), 1901 (2021)
29. Rahi, S., Othman Mansour, M.M., Alharafsheh, M., Alghizzawi, M.: The post-adoption behavior of internet banking users through the eyes of self-determination theory and expectation confirmation model. J. Enterp. Inf. Manag. (2021)
30. Rahi, S., Othman Mansour, M.M., Alghizzawi, M., Alnaser, F.M.: Integration of UTAUT model in internet banking adoption context. J. Res. Interact. Mark. (2019)
31. Jalali, S.H., Golmohammadi, E.: Role of strategic thinking in increasing the innovation of companies: an emphasis on strategic learning. J. Bus. Adm. Res. **13**(26), 239–266 (2022)
32. Kumar, S., Dhir, A., Talwar, S., Chakraborty, D., Kaur, P.: What drives brand love for natural products? The moderating role of household size. J. Retail. Consum. Serv. **58**, 102329 (2021)
33. Sareen, A., Pandey, S.: Organizational innovation in knowledge intensive business services: the role of networks, culture and resources for innovation. FIIB Bus. Rev. **11**(1), 107–118 (2022)
34. Baumane-Vītoliņa, I., Woschank, M., Apsalone, M., Šumilo, Ē, Pacher, C.: Organizational innovation implications for manufacturing SMEs: findings from an empirical study. Procedia Comput. Sci. **200**, 738–747 (2022)
35. Guo, M., Ahmad, N., Adnan, M., Scholz, M., Naveed, R.T.: The relationship of CSR and employee creativity in the hotel sector: the mediating role of job autonomy. Sustainability **13**(18), 10032 (2021)
36. Habes, M., Elareshi, M., Youssef, E., Ali, S., Qudah, M.: Social impact of videos at new media platforms on the eLearning acceptance during the COVID-19. Inf. Sci. Lett. **11**(3), 913–923 (2022)
37. Habes, M., Ali, S., Pasha, S.A.: Statistical package for social sciences acceptance in quantitative research: from the technology acceptance model's perspective. FWU J. Soc. Sci. **15**(4), 34–46 (2021). https://doi.org/10.51709/19951272/Winter-2021/3
38. Alhammad, K.L., Habes, M., Al Olaimat, F., Haddad, I.: Attitudes of students of the faculty of mass communication at yarmouk university towards using video platforms in distance education. Rev. Int. Geogr. Educ. Online **11**(5), 1041–1052 (2021)
39. Habes, M., Ali, S., Elareshi, M., Tahat, K.M., Ziani, A.: Understanding users' social TV content using artificial intelligence approach: a survey. In: 2021 International Conference of Modern Trends in Information and Communication Technology Industry (MTICTI), pp. 1–7 (2021)
40. Sekaran, U., Bougie, R.: Research Methods for Business: A Skill Building Approach. John Wiley & Sons, Hoboken (2016)
41. Elareshi, M., Habes, M., Ali, S., Ziani, A.: Using online platforms for political communication in Bahrain election campaigns. Soc. Sci. Humanit. **29**(3), 2013–2031 (2021)
42. Bedi, H.S.: Strategic autonomy: a step towards scale development. Int. J. Bus. Glob. **24**(2), 153–173 (2020)

Analysis and Forecasts of the Impact of Non-performing Loans on the Economy in Pandemic Conditions

George Abuseldze(✉)

Batumi Shota Rustaveli State University, Ninoshvili, 35, 6010 Batumi, Georgia
george.abuselidze@bsu.edu.ge

Abstract. Credit risk is particularly important for the banking system and any financial institution. The risk associated with lending affects not only financial institutions but also other sectors with a "domino" effect. Therefore, it is important to assess the impact of loans, including non-performing loans, their degree of risk, and based on their analysis, it is necessary to draw fair conclusions in terms of forecasting. The aim of the paper is to analyze the impact of non-performing loans on the economy. The paper examines foreign trends and Georgia's reality regarding non-performing loans. It has been assessed how risky the credit portfolio of commercial banks in Georgia is. The article assesses the impact of the pandemic on the share of non-performing loans and the financial system. In our research, a separate analysis of the influence of factors affecting non-performing loans and interrelationships between variables is made using a regression model and a correlation matrix. Based on the results, the article draws conclusions and offers appropriate recommendations for maintaining financial sustainability.

Keywords: Non-performing loans · Banks · Banking · Banking sector · Credit risk

1 Introduction

It is practically impossible to carry out banking activities without risk, and this is especially true of lending, since lending always involves inherent credit risk. Even a high level of paying ability is not a guarantee and carries the risk that the loan will not be returned. Accordingly, banks should pay special attention to the share of non-performing loans in the structure of loans. The spread of the Covid pandemic has made it even more urgent and has put on the agenda the policy activation aimed at reducing the share of non-performing loans.

Non-performing loans directly or indirectly affect the overall economic situation of the country. Non-performing loans reduce banks' profitability levels, and each non-performing loan affects the ability to issue new loans. The problems created in the banking sphere find reflection in the related spheres, hinder business development, reduce jobs, etc. Ultimately, the high level of non-performing loans seriously harms the country's economic situation [1].

B. Alareeni et al. (Eds.): ICBT 2022, LNNS 620, pp. 129–143, 2023.
https://doi.org/10.1007/978-3-031-26953-0_14

Taking into account the mentioned circumstances, certain benefits were established, in particular, during 2020, commercial banks offered borrowers a deferral of loan payments:

* In the first stage, which mainly covered March-June, commercial banks offered borrowers a 3-month deferral.
* After the expiration of the mentioned period, the banking sector offered an additional 3-month loan deferment to those people who have reduced or lost their income. The grace period applied to both the loan principal and interest.
* It is worth noting the fact that with the mentioned change, the loans would not be classified as negative, and if the payments continued without problems, the loan would retain the standard category.
* The National Bank did not ask commercial banks to create an additional expected loss reserve for such borrowers [2].

Although the mentioned grace period was a kind of relief for the borrowers in the conditions of the pandemic, however, I consider it turned out to be less effective. The mentioned steps were failing to help the borrowers, as, after the grace period, the burden of the loan became even heavier for them. Since the term of the loan increased due to the grace period, the accrued interest increased as well. After the end of the grace period, the borrowers had to pay the increased interest and principal amount. It should also be noted that the pandemic situation has not ended yet, and the loan postponement for 3 months does not change the situation, because the economic situation of the employed people in the sector vulnerable to the pandemic has not improved. Although the said benefit reduced non-performing loans in a short period, however, after the expiry of the benefit, the said figure increases.

It is fact that when a commercial bank has a high proportion of non-performing loans in its loan structure, it will focus on improving the quality of assets rather than issuing new loans. Accordingly, non-performing loans reduce the available resources needed for lending, which ultimately affects the bank's profitability.

2 Methodological Bases

The quantitative research method is used in the paper, which is based on the annual or quarterly reports of commercial banks of Georgia, as well as the data of the National Statistics Service of Georgia. In this paper, the commercial banks in Georgia are the research objects for credit risk assessment. Secondary data collected from each bank's annual or quarterly reports are used. The relationship between different factors is evaluated based on correlation analysis with individual calculations. A regression model is used to evaluate the impact.

3 Results and Discussion

3.1 General Overview and Foreign Trends

Immediately after the spread of the pandemic, the imposition of certain restrictions was on the agenda, which certainly affected the country's economic situation, including the banking sector. The restriction of economic activity led to the fact that many people remained unemployed, which in turn was reflected in the repayment of loan obligations.

Non-performing loans and its management are a challenge for any country. Its proper management and non-performing loan policies differ from country to country [3–13], but all of them aim to reduce the share of non-performing loans (Fig. 1).

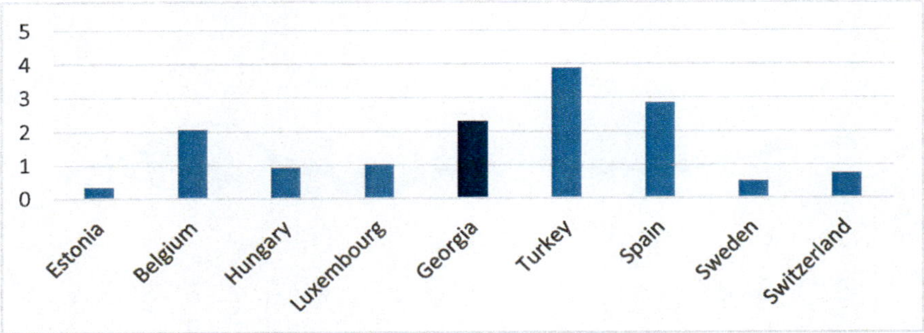

Fig. 1. Share of non-performing loans in total loans in different countries.

As we can see, European countries manage to maintain a low share of non-performing loans in pandemic conditions by 2020. To avoid the risk caused by non-performing loans, the supervision of the European Central Bank annually checks how commercial banks manage the level of non-performing loans, and whether they have appropriate strategies and governance structures (SREP). To reduce the number of non-performing loans, the European Central Bank issues the following recommendations [3–16]:

- Commercial banks should grant loans only to those customers who are likely to repay them;
- Banks should monitor information on loans to identify insolvent borrowers at an early stage and correct the situation;
- They should engage in the loan restructuring process on time;
- They should "supply with" provisions in time to ensure adequate loss coverage. Provisioning means that the bank recognizes a loss on the loan ahead of time.

However, the European Central Bank also points out that caution is needed in order not to support undesirable borrowers by preventing their loans from being classified as non-performing loans. Different countries applied for benefits of the borrowers affected by the pandemic [14, 17–19]. For example, the Belgian government has offered households and companies affected by the crisis to postpone their obligations to banks or the insurance sector for 9 months until the end of June 2021. In Poland, the banking association recommended deferring loan payments for a period of up to 3 months, which would be voluntary. Banks in Poland have also increased credit card limits. In addition, the Romanian government issued a law extending the repayment period of loans for households and businesses affected by covid 19 by 9 months - effective until March 2021 [17–19].

In foreign practice, there are frequent cases when the grace period is much longer than 3 to 6 months. As we can see, the share of non-performing loans increases significantly after the expiry of the mentioned grace period, which is logical, because the borrowers were unable to make payments during the crisis period.

Consider the ratio of non-performing loans to total loans in Georgia by years, allowing us to assess how risky the credit portfolio in commercial banks of Georgia is (Fig. 2).

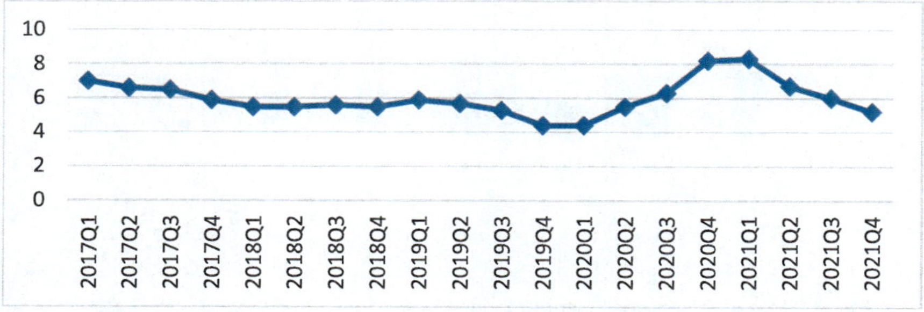

Fig. 2. Share of non-performing loans in total loans in Georgia. Source: Pillar 3 Quarterly Report [20]

Figure 2 shows the share of non-performing loans in the banking sector as a whole, but it is interesting to see what the share of non-performing loans is by individual banks. Figure 3 shows the share of non-performing loans over the last 5 years in systemically important banks such as Bank of Georgia and TBC Bank.

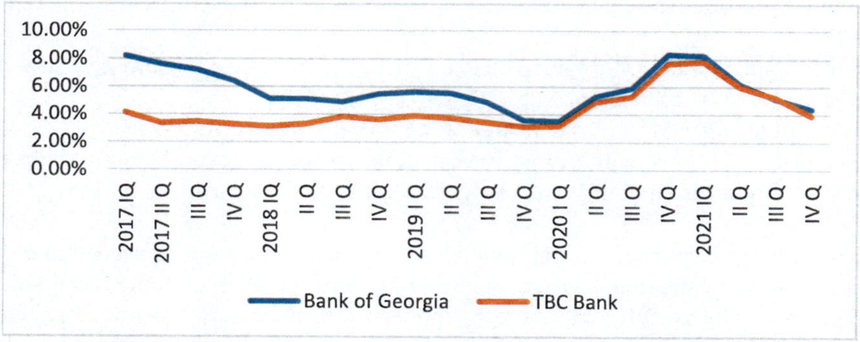

Fig. 3. Share of non-performing loans in Bank of Georgia and TBC Bank. Source: Pillar 3 Quarterly Report [21, 22]

Despite the fact that the average rate of total loans of TBC Bank in the mentioned period is higher than that of the Bank of Georgia, non-performing loans have a smaller share in the total loans issued by TBC Bank. However, if we look at the graph, we can easily notice that the share of non-performing loans increased significantly in the pandemic conditions - in the case of Bank of Georgia, it reached 8.35%, and in the case of TBC Bank, it reached 7.81%. In the analysis of non-performing loans, it is important to analyze the non-performing loans of such systemically important banks and reduce them, since and because the number of total loans issued by TBC Bank and Bank of Georgia is much higher than the loans issued by other banks [23].

According to the share of non-performing loans, Kartu Bank is the leader in the banking sector of Georgia, the data is given in the graph (Fig. 4):

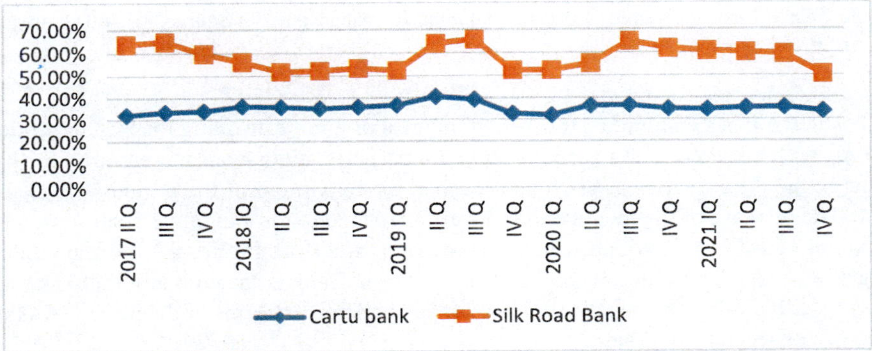

Fig. 4. Share of non-performing loans in total loans in Kartu Bank and Silk Road Bank. Source: Pillar 3 Quarterly Report [24, 25]

The share of non-performing loans in Kartu Bank was not favorable even before the pandemic period, in the last 5 years, the share of non-performing loans reached its maximum value of 40.77% even before the pandemic in the second quarter of 2019. No commercial bank has recorded such a high rate in the banking sector of Georgia. Although during the last 5 years, the maximum loan volume issued by Bank Kartu is 1,119,004,967 GEL per quarter, which is significantly lower than the minimum rate of loans issued by TBC Bank, but the number of non-performing loans with such a high share puts both the bank and the banking sector at risk. Silk Road Bank is distinctive with an equally high rate, although it maintains a downward trend in 2021.

Besides, worth noting is the example of Pasha Bank, which is one of the most visible examples of the negative impact of the pandemic on the share of non-performing loans (Fig. 5).

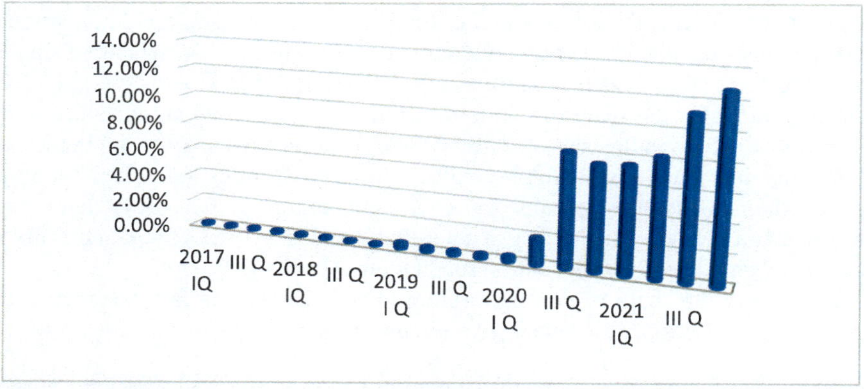

Fig. 5. Share of non-performing loans in total loans in Pasha Bank, %. Source: Pillar 3 Quarterly Report [26].

Pasha Bank has maintained the minimum level of non-performing loans in the entire banking sector for years, however, the share of non-performing loans increased by 6.19% points in the third quarter of 2020 and reached the maximum of the period - 12.7% in 2021. The increase of the mentioned indicator is related to the pandemic conditions and the losses caused by it, which were reflected in non-performing loans after the end of the grace period. Therefore, it can be said that Pasha Bank is the commercial bank that was most affected by the negative impact of the pandemic in terms of the increase in the share of non-performing loans.

As for the rest of the banks, their data is given in the graph (Fig. 6):

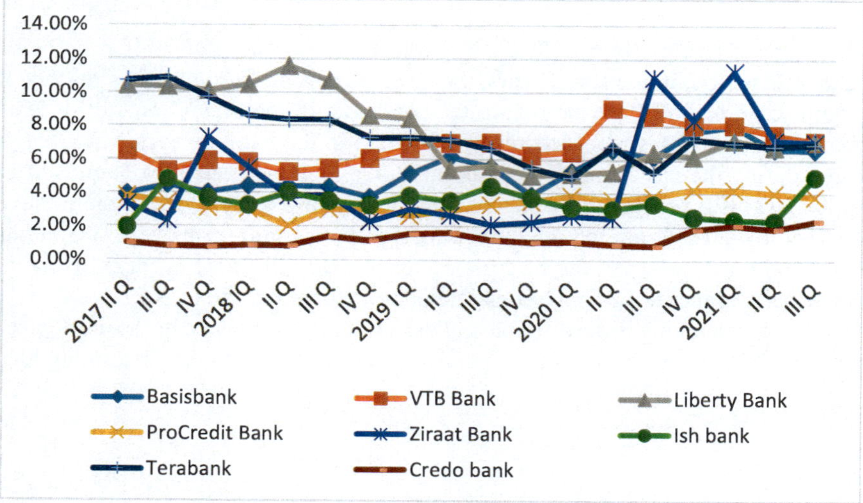

Fig. 6. Share of non-performing loans in total loans in different commercial banks, %. Source: Pillar 3 Quarterly Report of Commercial Banks [27–34].

As we can see in the graph, the trend of growth of non-performing loans can be observed from the first quarter of 2020, which increases significantly in the third quarter. In the pandemic conditions, the share of non-performing loans of Ziraat Bank experienced a very strong growth, which increased by 8.45% points in the second quarter of 2020.

The strength of the financial sector in Georgia depends on the stability of commercial banks. The growth of non-performing loans initially affects individual commercial banks, and in the long run it destroys the financial system. In such a developing country as Georgia, which is still heavily dependent on the banking sector, it is important to study the causes of the increase in the share of non-performing loans and the macroeconomic or specific factors affecting it.

3.2 Correlational Analysis of the Influence of Different Actors on Non-performing Loans and the Economy

Many factors affect non-performing loans and also non-performing loans themselves affect the economy of the country. These factors can be:

Exchange Rate

It is known that the economy of Georgia is characterized by high dollarization, and despite the implementation of dollarization measures over the years, the share of dollars in the structure of loans remains high. That is why, when we talk about the growth of non-performing loans, the GEL-USD exchange rate makes a big contribution, because most of the loans are denominated in dollars. Loans are given on Rafik according to currencies (Fig. 7):

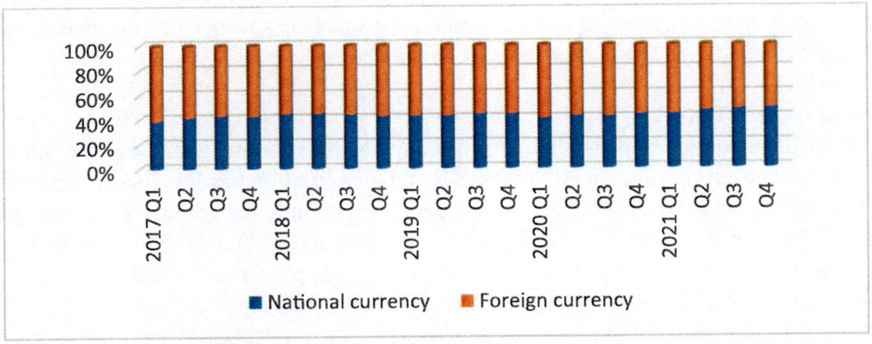

Fig. 7. Loans by currencies (Thousand Gel). Source: National Bank of Georgia [20, 37]

When the exchange rate increases, the amount of GEL equivalent to the dollar increases, so the population has to spend much more to cover the loan obligations that they have taken in dollars. Therefore, the impact of the exchange rate on non-performing loans should be significant.

Income Diversification

Diversification of incomes is considered in the study as the ratio of incomes other than interest incomes on loans (interbank, loans granted to individuals and legal entities) to total incomes. This means that commercial banks receive income from other sources besides interest income from loans. All activities are accompanied by risk, therefore, not only in the banking sector, but in any activity, income diversification is necessary in order to avoid possible losses caused by risks. That is why, when the types of income are diversified in commercial banks, they try less to lend in risky cases, and this of course reduces the credit risk, which in turn is reflected in the reduction of non-performing loans. Therefore, income diversification should be one of the important influencing factors for non-performing loans.

GDP Per Capita

In this work, depending on the specifics of loan repayment, we took into account the volume of GDP per capita, which reflects both the incomes of all members of the economy as a whole, and the distribution of expenditures on goods and services for each person. It is a generally accepted indicator that measures the economic activity of the country. In the case when the population does not have enough income to pay the principal amount of the loans and the accrued interest, the number of non-performing loans increases.

Interest Rate

The interest rate set by commercial banks on loans is a factor with a significant influence, both on the volume of loans in general and on the share of non-performing loans. The higher the interest rate, the higher the cost of the loan.

Unemployment Rate

The level of unemployment is also an important factor, because if the level of unemployment in the country increases, it means that the population has less income to cover obligations. Accordingly, when the level of unemployment in the country increases, the number of non-performing loans also increases. It is also worth noting the fact that the level of unemployment increased significantly during the pandemic, which is one of the important factors in the increase in the share of non-performing loans (Fig. 8).

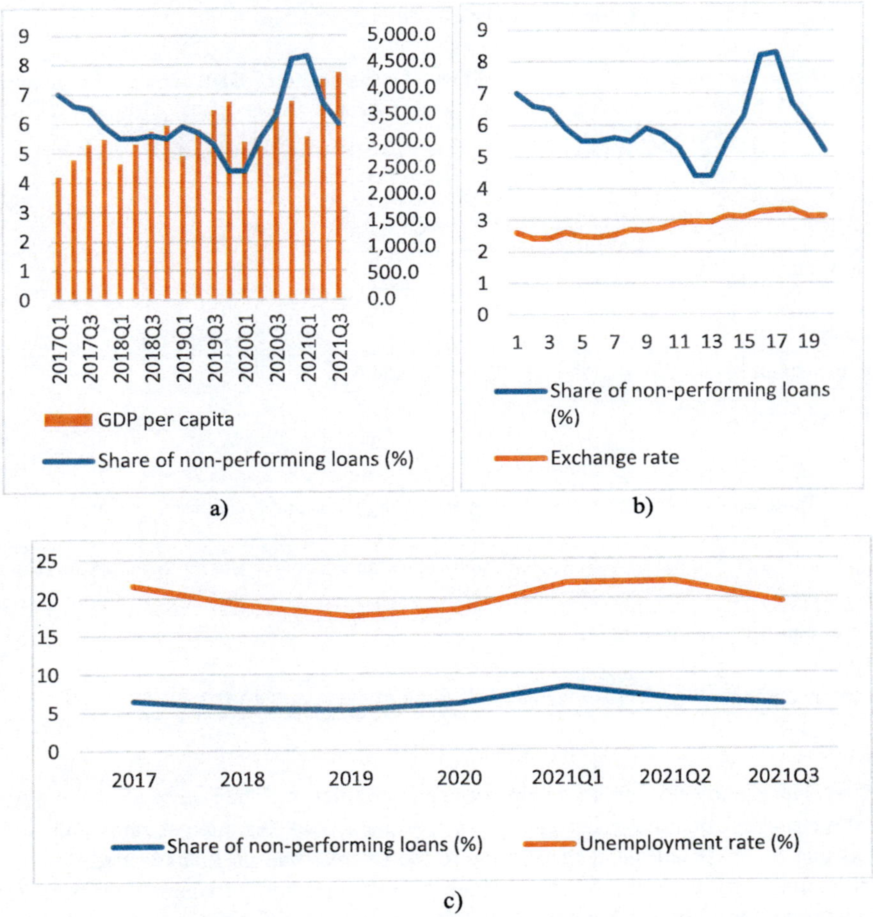

Fig. 8. Dynamics of some factors over time. Source: National Bank of Georgia, National Statistical Service of Georgia [20, 27, 29–31, 35–37].

To determine the relationship between them, we use the correlation coefficient.

The variables are the share of non-performing loans in total loans by quarter, as well as the average exchange rate per quarter, GDP per capita, and average annual interest rates on loans issued by commercial banks. Income diversification is presented quarterly in percentages, and the unemployment rate is presented in annual percentages.

The data for the variables are taken from 2010–2019, as the statistics for 2020–2021 have changed the trend of previous years due to the covid pandemic. Therefore, 40 quarterly data are taken to evaluate the impact. In each case, in the case of the unemployment rate, 10 data per year. However, to take into account worst-case scenarios and the consequences of the pandemic, data from 2017–2021 is also applied (Tables 1 and 2).

Table 1. Correlation matrix of variables

	Share of non-performing loans (%)	Interest rate on loans	Diversification of income	GDP per capita	Exchange rate
Share of non-performing loans (%)	1				
Interest rate	0.831	1			
Diversification of income	−0.544	−0.544	1		
GDP per capita	−0.786	−0.902	0.547	1	
Exchange rate	−0.602	−0.882	0.396	0.87	1

Table 2. Correlation matrix of non-performing loans and unemployment rate

	Share of non-performing loans (%)	Unemployment rate (%)
Share of non-performing loans (%)	1	
Unemployment rate (%)	0.765	1

Correlation shows the relationship between variables and their strength. As we can see, the strongest positive relationship was revealed between interest rates and non-performing loans. It is logical that the more the interest rate on loans issued by commercial banks increases, the more the share of non-performing loans increases, as for businesses or households it becomes difficult to pay the interest accrued on the loan.

As for the relationship between GDP per capita and non-performing loans, it is negative. The negative relationship shows that the higher the number of non-performing loans, the smaller the economy. If the amount of GDP in the country increases, the income of each citizen increases and they have cash funds left to cover their obligations and vice versa. That is why the relationship between the mentioned variables is negative.

As we can see, the correlation between income diversification and the share of non-performing loans is negative, which indicates that the more diversified the sources of income, the smaller the share of non-performing loans. The correlation between them is −0.544, which indicates a rather significant relationship.

One of the strongest positive correlations is the relationship of the unemployment level with the share of non-performing loans. It seems that the level of unemployment is a significant influencing factor for the share of non-performing loans, which is logical.

As we can see, the correlation between the exchange rate and non-performing loans, based on the mentioned data, proves to have a negative sign. It should be assumed that the mentioned connection is due to the small part of dollar-denominated loans in non-performing loans. However, it is interesting how this connection is in the pandemic conditions and what changes it underwent during this period. Therefore, I will discuss the dependence of non-performing loans and exchange rates over the last five years (Table 3):

Table 3. Correlation matrix between non-performing loans and exchange rate, 2017–2021

	Share of non-performing loans (%)	Exchange rate
Share of non-performing loans (%)	1	
Exchange rate	0.267	1

A positive correlation between 2017 and 2021 indicates that an increase in the exchange rate leads to an increase in non-performing loans. As we can see, the obtained coefficient is 0.267, which does not show a strong relationship, probably due to the scarcity of dollar-denominated loans in non-performing loans. However, there is definitely a positive relationship between them, which is noteworthy in the analysis of non-performing loans and its reduction policy. As it was revealed based on the analysis, the share of non-performing loans was sensitive to the exchange rate in crisis conditions, accordingly it is important to pay attention to this factor in pandemic conditions.

3.3 Analysis and Forecast of the Impact on the Economy

We already know the relationships between the variables of empirical research. However, it is interesting how the variable changes when one of them changes. To understand this, we will apply the regression model. In our case, the values of the obtained coefficients are (Table 4):

Table 4. Regression model variables

	X = exchange rate/y = non-performing loan	X = non-performing loan/y = GDP
B_0	16.722	3705.711
B_1	−3.994	−172.785
R_2	36.3%	61.8%

As it is known, in the regression model $Y = B_0 + B_1x$.

It is interesting to see what effect the change in the exchange rate will have on non-performing loans if it reaches the historical maximum since the 2000s of 3.4842 GEL. For this, we can enter the variables we are interested in into the model:

$$Y = 6.722 - 3.994 * 3.4842$$
$$Y = 2.81$$

As it is already mentioned, the maximum of the exchange rate of 3.4842 was recorded in the first quarter of 2020. The actual data for the share of non-performing loans for this period is 4.4%, but based on our results, we conclude that if the pandemic had not spread, the share of non-performing loans would have been 2.81%. As we can see, the difference between the mentioned indicators is quite significant.

Now let's consider what the forecast will be if we take into account the impact of the pandemic (Table 5):

Table 5. Regression model variables

	X = exchange rate/y = non-performing loan	X = non-performing loan/y = GDP per capita
B_0	3.774	3129.27
B_1	0.785	12.651

It will be interesting to see how the share of non-performing loans and, as a result, GDP per capita will change in the future, considering the worse and better scenarios of the indicators recorded during the pandemic. At this stage, we can discuss how the share of non-performing loans will change in the case of maximum and minimum exchange rates under pandemic conditions. The maximum quarterly occurrence is 3.33 GEL. We can enter the variables we are interested in into the model:

$$Y = 3.774 + 0.785 * 3.33$$
$$Y = 6.39$$

We can see that if the exchange rate reaches 3.33 GEL, the share of non-performing loans in total loans will increase to 6.39% points. After that, we can see how this change will affect the economy.

$$Y = 3129.27 - 12.651 * 6.39$$
$$Y = 3048.43$$

We have concluded that if the share of non-performing loans increases to 6.39%, then this will lead to a decrease in GDP per capita to 3048.43 GEL, while according to the data of the third quarter of 2021 it is 4290.4 GEL.

As for the better scenario, the minimum quarterly incidence in 2020–2021 is 2.93 GEL, respectively:

$$Y = 3.774 + 0.785 * 2.93$$
$$Y = 6.07$$

We concluded that if the exchange rate will decrease to 2.93 lari in the future, then the share of non-performing loans will also decrease to 6.07 percent. This will affect the economy as follows:

$$Y = 3129.27 - 12.651 * 6.07$$
$$Y = 3052.48$$

Accordingly, the GDP per capita increased slightly, but still did.

4 Conclusions

The banking sector plays an important role in the economy of Georgia, which is characterized by high profitability. However, it, like most countries in the world, faces the challenge of non-performing loans. Credit risk in the banking sector is vulnerable to various economic and social factors and is significantly influenced by both macroeconomic and microeconomic factors. Studying these factors allows us to maintain financial stability in the banking sector and avoid crises. As we have seen, the share of non-performing loans experienced special changes during the pandemic.

In general, based on empirical data, we can make some recommendations:

* To reduce the share of non-performing loans, first of all, it is important to provide a correct credit policy. The paper identified commercial banks, whose credit portfolio has a particularly high share of non-performing loans; consequently, it is necessary to adopt stricter quantitative or qualitative standards;
* There is a need for timely detection of non-performing loans, so that the commercial bank can identify it in time and protect it from credit risks;
* In Georgia, the terms of loans issued in dollars within the framework of the de-dollarization program should be further tightened, because the rate of dollarization in bank loans is still high. This creates a significant credit risk and, under conditions of exchange rate growth, leads to an increase in the share of non-performing loans;
* Because of the strong correlation between unemployment rate and non-performing loans, it is necessary to establish such a grace period that ensures the easing of the borrower's credit conditions in case of job loss.

Based on empirical data, it can be said that there is not a one-way relationship between non-performing loans and the economy. Non-performing loans are significantly influenced by a lot of economic factors, and non-performing loans themselves affect such an important indicator of the economy as GDP. Accordingly, it is important to maintain a low volume of non-performing loans to maintain the positive trends of economic growth.

Acknowledgement. This study was supported by the Scientific Grant the Batumi Shota Rustaveli State University (Georgia) under the contract No. 01-06/206, 2022. The author would like to thank management of the Batumi Shota Rustaveli State University for their supports.

References

1. Abuselidze, G.: The impact of banking competition on economic growth and financial stability: an empirical investigation. Eur. J. Sustain. Dev. **10**(1), 203–220 (2021)
2. Prevention of the spread of coronavirus in Georgia. Government report Covid 19. https://stopcov.ge/Content/files/Government--report.pdf
3. Serrano, A.S.: The impact of non-performing loans on bank lending in Europe: an empirical analysis. North Am. J. Econ. Financ. **55**, 101312 (2021)

4. Cucinelli, D.: The impact of non-performing loans on bank lending behavior: evidence from the Italian banking sector. Eur. J. Bus. Econ. **8**(16), 59–71 (2015)
5. Rajha, K.S.: Determinants of non-performing loans: evidence from the Jordanian banking sector. J. Financ. Bank Manag. **4**(1), 125–136 (2016)
6. Khairi, A., Bahri, B., Artha, B.: A literature review of non-performing loan. J. Bus. Manag. Rev. **2**(5), 366–373 (2021)
7. Abuselidze, G.: Modern challenges of monetary policy strategies: inflation and devaluation influence on economic development of the country. Acad. Strateg. Manag. J. **18**(4), 1–10 (2019)
8. Bhattarai, B.P.: Effects of non-performing loan on profitability of commercial banks in Nepal. Management **6**(6), 164–170 (2020)
9. Wood, A., Skinner, N.: Determinants of non-performing loans: evidence from commercial banks in Barbados. Bus. Manag. Rev. **9**(3), 44–64 (2018)
10. Mazreku, I., Morina, F., Misiri, V., Spiteri, J.V., Grima, S.: Determinants of the level of non-performing loans in commercial banks of transition countries. Eur. Res. Stud. J. **21**(3), 3–13 (2018)
11. Yurttadur, M., Celiktas, E., Celiktas, E.: The place of non-performing loans in the Turkish banking SECTOR. Proc. Comput. Sci. **158**, 766–771 (2019)
12. Curak, M., Pepur, S., Poposki, K.: Determinants of non-performing loans–evidence from Southeastern European banking systems. Banks Bank Syst. **8**(1), 45–53 (2013)
13. Roman, A., Bilan, I.: An empirical analysis of the macroeconomic determinants of non-performing loans in EU28 banking sector. Rev. Econ. **67**(2), 108–127 (2015)
14. Abuselidze, G., Kizinidze, M.: Influence of central bank regulations on interbank competition in association with EU. In: E3S Web of Conferences, vol. 135, p. 04037 (2019)
15. European central bank. non-performing loans. https://www.bankingsupervision.europa.eu/banking/priorities/npl/html/index.en.html
16. Tarasenko, I., Saienko, V., Kirizleyeva, A., Vozniakovska, K., Harashchenko, L., Bodnar, O.: Comparative characteristics of the banking sector in eastern Europe. IJCSNS Int. J. Comput. Sci. Netw. Secur. **22**(1), 639–649 (2022)
17. Abuselidze, G., Mamaladze, L.: The impact of the COVID-19 outbreak on the socio-economic issues of the black sea region countries. In: Gervasi, O., et al. (eds.) ICCSA 2020. LNCS, vol. 12253, pp. 453–467. Springer, Cham (2020). https://doi.org/10.1007/978-3-030-58814-4_32
18. Abuselidze, G., Slobodianyk, A.: Social responsibility of business and government in the conditions of the COVID-19 pandemic. In: E3S Web of Conferences, vol. 210, p. 15016 (2020)
19. Abuselidze, G., Slobodianyk, A.: Pandeconomic crisis and its impact on small open economies: a case study of COVID-19. Adv. Intell. Syst. Comput. **1258**, 718–728 (2021)
20. National Bank of Georgia. Statistical data. https://nbg.gov.ge/statistics/statistics-data
21. TBC Bank. Financial results. https://www.tbcbank.ge/web/ka/web/guest/results-announcements
22. Bank of Georgia. Financial results. https://bankofgeorgia.ge/ka/about/about-us
23. Abuselidze, G., Sharabidze, M.: Competition in the Georgian banking sector and its impact on the credit policy of commercial banks. In: MATEC Web of Conferences, vol. 342, p. 08003 (2021)
24. Cartu Bank. Pillar 3 quarterly reports. https://www.cartubank.ge/ge/597/pilar-3-is-kvartaluri-angariSebi
25. Silk Road Bank. Financial highlights. https://www.silkroadbank.ge/geo/page/387
26. Pasha Bank. Financial reporting. https://www.pashabank.ge/ge/about-us/financial-reports
27. National Bank of Georgia. Financial Stability Report. https://nbg.gov.ge/publications/financial-stability-reports

28. Basis Bank. Financial highlights. https://bb.ge/ge/about-holding/basisbank/bank/financial-highlights
29. VTB Bank. Financial reporting. https://vtb.ge/ge/about-the-bank/financial-reporting
30. Credo Bank. Financial highlights. https://credobank.ge/finances/
31. Liberty Bank. Financial highlights. https://libertybank.ge/ka/chven-shesakheb/investorebistvis/pinansuri-inpormatsia/regulatory-financial-reporting-to-the-national-bank-of-georgia
32. Procredit Bank. Pillar 3 quarterly reports. https://www.procreditbank.ge/ge/%E1%83%99%E1%83%95%E1%83%90%E1%83%A0%E1%83%A2%E1%83%90%E1%83%9A%E1%83%A3%E1%83%A0%E1%83%98?mid=1104
33. Tera Bank. Financial highlights. https://www.terabank.ge/ge/1245
34. Ziraat Bank. Financial reports. https://www.ziraatbank.ge/ka/financial-reports
35. National Statistics Office of Georgia. Employment, salaries. https://www.geostat.ge/ka/modules/categories/37/dasakmeba-khelfasebi
36. National Statistics Office of Georgia. National accounts. https://www.geostat.ge/ka/modules/categories/22/erovnuli-angarishebi
37. National Statistics Office of Georgia. Monetary statistics. https://www.geostat.ge/ka/modules/categories/92/monetaruli-statistika

Leadership Styles Adopted by Scottish Micro-businesses During the COVID-19 Pandemic

Sayed Gilani[1], Liza Gernal[2](✉), Ansarullah Tantry[2](✉), Naveed Yasin[1], and Rommel Sergio[1]

[1] Canadian University, Dubai, UAE
[2] Westford University College, Sharjah 50325, UAE
{LIZA.G,ansarullah.t}@westford.org.uk

Abstract. The current exploratory study investigated leadership styles adopted by Scottish micro-businesses during COVID-19. A qualitative research approach was employed on 20 owners/managers in Scotland, using semi-structured interviews. The data were analyzed using qualitative thematic analysis. The findings of the study revealed thematic variations across a range of leadership styles as the autocratic style was identified by businesses as the most common approach during the pandemic due to the influence of the external environment on rapid decision-making. In essence, the findings highlight the importance of recognizing the role of leadership approaches under uncertain and volatile market conditions. Based on the qualitative findings of the study, a novel framework presented as the "COVID-19 Leadership Framework" was proposed that addresses the contextualization of the findings to a specific and contemporary context. The results of the study presented theoretical and practical implications for micro-business, policymakers, and Small to Medium Enterprise support services.

Keywords: Leadership framework · Scotland · Micro-businesses · Leadership style

1 Introduction

1.1 Leadership

Leadership is defined as the ability of an individual to lead people (Keizer 2017). A variety of leadership styles are used in businesses of varying sizes depending on the circumstances of the business (Cowen 2018). The link between leadership styles and potential business growth and survival can be applied to Small to Medium Enterprises (SMEs) that are looking to survive or grow (Cowen 2018). Research to identify appropriate leadership styles to ensure SMEs' growth is important for SMEs in the United Kingdom (UK), where SMEs account for half of all businesses.

Micro-businesses make up 99% of businesses in Scotland (9 or fewer employees), and in March 2019, 55.4% of employment in Scotland's private sector was attributed

B. Alareeni et al. (Eds.): ICBT 2022, LNNS 620, pp. 144–156, 2023.
https://doi.org/10.1007/978-3-031-26953-0_15

to micro-businesses (Scottish Government 2020a). The productivity, effectiveness, and growth of micro-businesses in Scotland have been negatively impacted by COVID-19 (Corona Virus Disease 2019) because of the government-imposed restrictions for non-essential businesses during the pandemic, where convenience stores, supermarkets, and pharmacies are considered essential businesses (Scottish Government 2020c).

The importance of micro-businesses justifies the need to investigate leadership styles that ensure the survival/sustainability of businesses during uncertain times (Scottish Government 2020c). There have been studies conducted worldwide investigating the impact of leadership styles (e.g., Song et al. 2021; Vorobeva and Dana 2021), but there is limited research that focused on investigating the impact of different styles on micro-businesses during the pandemic in Scotland. Authors have advocated for additional research into the impact of the pandemic on SMEs in a specific geographic and operational context (e.g., Çera et al. 2021; Griffin and Häyrén 2022).

The purpose of this paper was to investigate the operation of leadership styles used by micro-businesses in Scotland during the pandemic. The remainder of this paper is structured as follows: The second section gives an overview of COVID-19. The third section discusses the impact of SMEs. The fourth section conducts a review of studies investigating the effect of leadership styles on business productivity. The fifth section provides an overview of the research method used in this paper. The sixth section reports, analyses, and discusses the empirical work findings. The seventh section develops a conceptual framework for the research. The paper concludes with a summary of all key findings, limitations, implications, and recommendations.

1.2 COVID-19

COVID-19 was found to attack the victim's lungs and airways (Boyd 2020). The first report of a COVID-19 outbreak was made in late 2019 in China's Wuhan province (Roberts et al. 2020). There had been reported cases globally by late March 2020. (Boyd 2020). To avoid/minimize the spread of COVID-19, all international and domestic travel was banned in March 2020 (GOV.UK 2020).

1.3 Impact of Small to Medium Enterprise

Small to Medium Enterprises (SMEs) are defined by the number of employees and revenue generated by a company, however, the definition varies in different countries (McQuerrey 2019). In Canada, for example, businesses with fewer than 500 employees are classified as SMEs, whereas businesses with fewer than 250 employees are SMEs in the European Union (Percy 2019). In New Zealand, companies with under 19 employees are classified as SMEs (Kirby and Watson 2017). In the UK, an SME is defined as having a turnover rate under 5.6 million and fewer than 50 employees. Kirby and Watson (2017) defined a micro-business as one that employs 0–9 people in the UK. As a result, the definition of business sizes relevant to the UK context informed the Scotland-based research in this paper. In 2019, over 98% of all private-sector businesses in Scotland were classified as small (0 to 49 employees), which accounted for 42.6% of private-sector employment (Scottish Government 2020b). Sole traders account for 69.3% of all private sector businesses in Scotland (Scottish Government 2020b).

1.4 A Review of Studies on Leadership Styles and Productivity

Through a survey in Asia, Raja and Palanichamy (2011) discovered that the respondent's positional identity had a significant impact on the perception of leadership style and organizational commitment, but salary did not appear to make a difference among the sample respondents. Transformational leadership behaviors and performance were suggested to be relevant in this study, with characteristics such as charisma, inspirational motivation, intellectual stimulation, extra effort, and satisfaction (Obiwuru et al., 2011). A positive effect on organizational performance was identified through the adoption of the Laissez-faire leadership style if used in a limited capacity over a short period (Khan and Adnan 2014). Therefore, based on the results and analysis, the researchers suggested the adoption of a hybrid leadership style to ensure optimum performance within a business (Khan and Adnan 2014). Igbaekemen and Odivwri (2015) through a survey in Nigeria also identified the benefits of adopting a hybrid leadership style for improving organizational performance. Like Samad (2012), Hurduzeu's (2015) research in Asia suggested transformational as the best leadership style for improving organizational performance as it inspires and drives employees to work harder and longer. Khademian (2016) and, Kaushal and Mishra (2017) identified that a supportive behavior-based leadership style had a higher probability of influencing more ethical-based practices among employees.

Girling (2018) emphasized that leadership style was not simply an accident based on personality but could be shaped by practice and the need for quality and could have been adapted to fit any situation. Girling (2018) agreed with Amer (2017) that adopting a hybrid leadership style to ensure optimal productivity in an organization was an innovative idea. While agreeing with the points made by the preceding authors, Pringgabayu and Ramdlany (2017) find that leadership, as well as internal business culture, had an impact on knowledge management in government organizations.

Salamzadeh et al. (2021), in their Malaysia-based study, added to Girling's (2018) findings by emphasizing the positive impact of digital-based leadership on areas such as dynamic capabilities, innovation-driven capabilities, and organizational effectiveness.

1.5 Methods

The research in this paper involved semi-structured interviews with a theoretical underpinning provided by the Leadership Interactional Framework (LIF) (Deductive approach). The reason behind the selection of the Deductive approach over the Inductive approach was to provide a route for the analysis of data generated in this research which is focusing on a previously unexplored or under-researched area, e.g., investigating leadership styles adopted in Scottish micro-businesses during the pandemic (Silverman 2017). Therefore, the Deductive approach through the LIF provided a theoretical underpinning for the previously unexplored/unresearched area.

The questions were designed considering the research aim of "investigating the effect of leadership styles on productivity in Scottish micro-businesses during the Coronavirus pandemic". It was decided on the questions to identify leadership style(s) adopted (1) before and (2) during the COVID-19 lockdown where a question looking to identify whether the owner-manager may alter the leadership style (3) after the lockdown was also constructed These questions looked at the potential variation in leadership styles

(1. Before 2. During and 3. After the lockdown) to ensure that the research aim was achieved. The owner-managers were initially sent information on leadership styles via email.

Data were analyzed through thematic analysis which entails separating retrieved data into separate categories via the identification of themes and patterns (Orfanidou et al. 2015). The main motive behind the selection of thematic analysis in this research was its simplicity and user-friendly nature along with its compatibility with a Pragmatic philosophy that supports exploratory research (Groenland and Dana 2020). The sampling method included in this research was the stratified random sampling method. Stratified random sampling involves sampling from a population which consists of dividing the population into subgroups (Saunders et al. 2017). The reason behind the selection of this method was that it enabled the collected samples to be fully represented by the target population (Groenland and Dana 2020). The initial sample of essential and non-essential businesses (56 businesses) was retrieved from yell.com, e.g., a search on yell.com consisted of inputting a postcode in the search field which led to particulars of several businesses appearing on the screen. The empirical research was carried out during the period of June 2021 to September 2021.

The focus was to generate a sample consisting of an equal number of businesses that were regarded by the Scottish Government (2020b) as essential and non-essential businesses during the pandemic. Essential businesses during the pandemic were defined as businesses that sold essential everyday items like food and toiletries (e.g., corner shops and supermarkets), otherwise, businesses were identified as non-essential, e.g., mobile stores, pubs, and clothing shops (Scottish Government 2020b).

It was initially attempted to contact the essential and non-essential businesses to check whether they were micro-businesses (0–9 employees) and if the owner-manager was willing to take part in a semi-structured interview. However, due to no response to the calls or refusal of owners/managers to speak to the researcher, the sample size was reduced to 39 where 23 agreed to participate in follow-up interviews. However, to ensure an equal number of essential and non-essential businesses, the sample was then reduced to 20 which consisted of 10 essential and 10 non-essential businesses. An email was initially sent out to 20 businesses which consisted of information related to the authenticity of the research and researchers to ensure that participants were assured and encouraged to participate in follow-up interviews conducted via telephone. The follow-up interview questions are provided in Table 1.

Table 1. Questions that were used in the telephonic interview.

Q1	When was the business established?
Q2	What is your business specialty, e.g., products/services sold?
Q3	How many workers are employed by the business?
Q4	Based on the 10 leadership styles what style matched your style before the lockdown?

(continued)

Table 1. (*continued*)

Q1	When was the business established?
Q5	Have you ever had to change your leadership style in the past?
Q6	If yes/no to Q5, then how did this impact the productivity of your business? If not, then why have you never changed your leadership style?
Q7	Have you changed your leadership style after the COVID-19 lockdown?
Q8	If yes/no to Q7, then why you did/did not change your leadership style?
Q9	What is your current leadership style in this period of COVID-19 lockdown?
Q10	Will you keep the same leadership style after the COVID-19 lockdown?
Q11	Why are you going to change/not change your leadership style after the lockdown?

1.6 Results and Discussion

The results from the 20 interviews are summarized in Table 2 where 'Q' represents Question.

Table 2. Summary of interview findings

	Q2	Q4	Q7	Q8	Q10	Q11
I1	Pharmacy	Transactional	Yes	I am looking to find ways to keep employees engaged and motivated during these uncertain times which may be achieved from a Transformational style	No	I am looking to work more closely with my colleagues
I2	Convenience store	Transactional and maybe a bit of Transformational sometimes	Yes	I think I can adapt the Transformational style to communicate with lower-level employees a bit more in these uncertain times	No	I think there will not be a requirement to micro-manage as things will go back to normal

(*continued*)

Table 2. (*continued*)

	Q2	Q4	Q7	Q8	Q10	Q11
I3	Travel agent	Transformational style	Yes	I changed my style to an Autocratic style to focus on results where I have had to make major cutbacks after the lockdown as my business has been greatly affected due to restrictions on international travel	Yes	I will until the business recovers back to the same level as before the lockdown
I4	Farm	The Democratic style	No	I have not been directly affected by the lockdown as I work from home in an isolated area	Yes	The style has served me well up to now so I will not change
I5	Convenience store	A mixture of Coach and Pacesetter	Yes	I believe me alternating between the Coach and Pacesetting style will aid the business' growth. However, the lockdown has not greatly influenced this decision	No	I believe by working in a way that considers each employee's capabilities will lead to the firm growth in the future
I6	Newsagent	A mix between a Transformational and a Coach style	Yes	I believe the introduction of a Pacesetter style for short periods will benefit the business during these uncertain times	No	I will revert to the Transformational and Coach style as they were better-suited styles during normal business hours
I7	Pharmacy	Transactional and Bureaucratic styles	No	After the lockdown, I have run my business as normal with some added precautions	Yes	I will keep the same style but remove the precautions adopted during the lockdown

(*continued*)

Table 2. (*continued*)

	Q2	Q4	Q7	Q8	Q10	Q11
I8	Taxi	Pacesetter and Transformational	No	The running of my business has stopped due to the government-imposed lockdown; therefore, I am currently unemployed	Yes	Pacesetter and Transformational styles to ensure that I am pushing the pace to recover the business after it re-opens again
I9	Online services	Transformational leader	No	I have not been greatly affected after the lockdown as mostly I work from home, and I have no employees working for me	Yes	The same style has worked for me up to now regardless of the pandemic
I10	Mechanic	Coach and Autocratic style	No	I have not been able to work due to the lockdown not allowing non-essential businesses to deal with customers face-to-face	No	I will adopt solely an Autocratic style to drive recovery in the business after its re-opens
I11	Mechanic	Autocratic and Coach style	No	I have not been able to work due to the lockdown not allowing non-essential businesses to deal with customers face-to-face	No	I will adopt this hybrid style to ensure recovery and productivity after the lockdown
I12	Bakery	Servant style	No	We have been open throughout the lockdown so there was no need for change	Yes	I have not been affected by the lockdown so the style will remain the same

(*continued*)

Table 2. (*continued*)

	Q2	Q4	Q7	Q8	Q10	Q11
I13	Café/Restaurant	Transactional style	Yes	The working conditions due to the lockdown have changed (e.g., social distancing) so I had to change to a more Transformational leader	Yes	Once we revert to normal working conditions, I will adopt a Transactional style again which benefitted the business greatly before the lockdown
I14	Pharmacy	Autocratic style	No	There has been no change in the way our business operates outside of the social distancing rule which is why I am still an Autocratic leader	Yes	I have not had any issues with this approach before and after the lockdown so I will carry on doing this
I15	Ice cream truck	Autocratic style	No	I did not have to change due to me keeping a distance from customers by serving from the van	Yes	I do not need to change anything as I have not really been affected in any way outside of the social distancing rule
I16	Convenience store	Autocratic style	No	Demand for services has been the same whereas all operations have been the same	Yes	Nothing has changed throughout the pandemic in my case as my business has remained open

(*continued*)

Table 2. (*continued*)

	Q2	Q4	Q7	Q8	Q10	Q11
I17	Gym	Coach and Transformational style	Yes	I cannot run the business during the lockdown, so I must think about outgoing and incoming money	No	I will adopt a Coach and Transformational style as before, but I will also add in phases of adopting the Pacesetter style to trigger recovery along with regular growth
I18	Law	Autocratic style	No	We have not been able to run the practice as normal due to a decrease in activity after the start of the lockdown, e.g., closed courts	Yes	I will adopt the Autocratic style as it was effective before but there may be periods of the Transformational style as well as we will be looking to recover the business back to normal
I19	Joinery	Autocratic style	No	I have not been able to work due to the lockdown so there has been no need to change the style	Yes	I am a sole trader so there will be no need to change anything
I20	Newsagent	Coach and Autocratic style	Yes	I have had to become more results-focused due to the business being partially closed because of the 2-m rule	No	I will go back to the Autocratic and Coach styles as they were effective in the growth and expansion of the business before the lockdown

As highlighted in Table 2, overall, the most popular leadership style adopted by businesses was autocratic. During the pandemic, businesses identified autocratic as the most common leadership style. An autocratic leader is defined as someone who is exclusively focused on results and efficiency where the leader either makes decisions by themselves or with a small group of people (Cowen 2018). Khajeh (2018) compared an autocratic style to military commanders due to their strict approach. Cowen (2018) stipulated that the autocratic style can be beneficial for a business when it is used with employees who need a great deal of supervision, e.g., employees with little to no experience. However, it was argued that the Autocratic style can adversely affect creativity and make employees feel restricted (Khajeh 2018).

The transformational style and coach style were also adopted by businesses where both styles focus on communication, goal setting, and motivation amongst employees (Cowen 2018). However, transformational leaders were driven by a commitment to organizational objectives and standards rather than focusing on each employee's goals (Cowen 2018). The interviewees identified that the adoption of the autocratic style was attributed to the volatile and uncertain nature of the business environment created by the pandemic where they collectively believed that this was appropriate for COVID-19 as it allowed the leader to make immediate decisions.

Most business owners identified that they would remain autocratic leaders, however, some businesses were aspiring to change to a hybrid style where it was believed that the hybrid style might also have involved the autocratic style, as businesses may be anticipating the possibility of further restrictions in the future where such an approach will allow businesses to develop a versatile leadership style which will be applicable in pandemic/non-pandemic settings.

In Sect. 2, a collective consensus amongst the reviewed studies was that the transactional and transformational styles along with a hybrid style might lead to optimum results in terms of productivity within businesses (Choudary et al. 2012; Kaushal and Mishra 2017). However, limited COVID-19 and Scotland-based research were identified in the literature review. Therefore, these identified gaps informed the need for this research. However, as mentioned earlier the interviews in this study identified the popularity of the autocratic style during the pandemic. The findings from this study agreed with the literature review on the effectiveness of a hybrid leadership style. It was noted that the questions in the interviews could have addressed the link between the leadership styles with culture, gender, and innovation which were areas addressed by Pringgabayu and Ramdlany (2017) (culture), Palalic et al. (2017) (gender) and Salamzadeh et al. (2021) (innovation). Therefore, there may be scope for further research.

1.7 Development of Theoretical Frameworks/Models Investigating Leadership

Salamzadeh (2020) highlighted the importance and mandatory requirement of ensuring a theoretical contribution from the output of a research study. Therefore, the Leadership Interactional Framework (LIF) was included in this research. LIF involved an investigation of the interaction between the leader/management, the followers/employees, and the situation within a business (Deng 2017). The LIF is illustrated in Fig. 1.

Figure 1, the LIF consists of 3 components which were the leader, the followers and the situation associated with a business. The adoption of LIF in this study was critical as the current research aimed to explain leadership styles adopted by Scottish

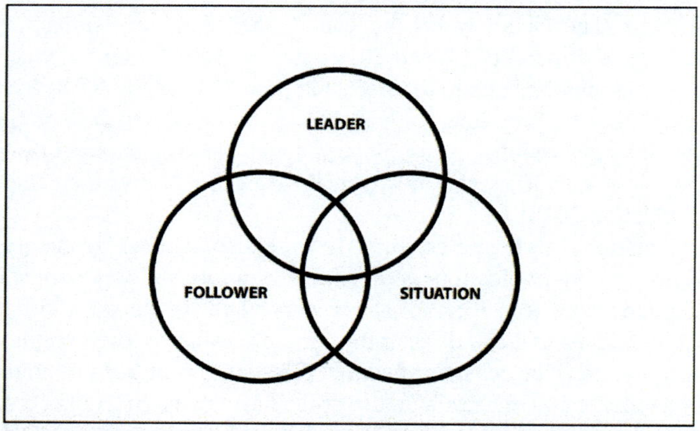

Fig. 1. Leadership Interaction Framework/Interactional Framework. **Source:** Lindsay and Woycheshin (2015)

micro-businesses during COVID-19. Therefore, the identified compatibility and similarities between the LIF and the key findings from this research have led to a theoretical contribution of the COVID-19 Leadership Framework (CLF) (Fig. 2).

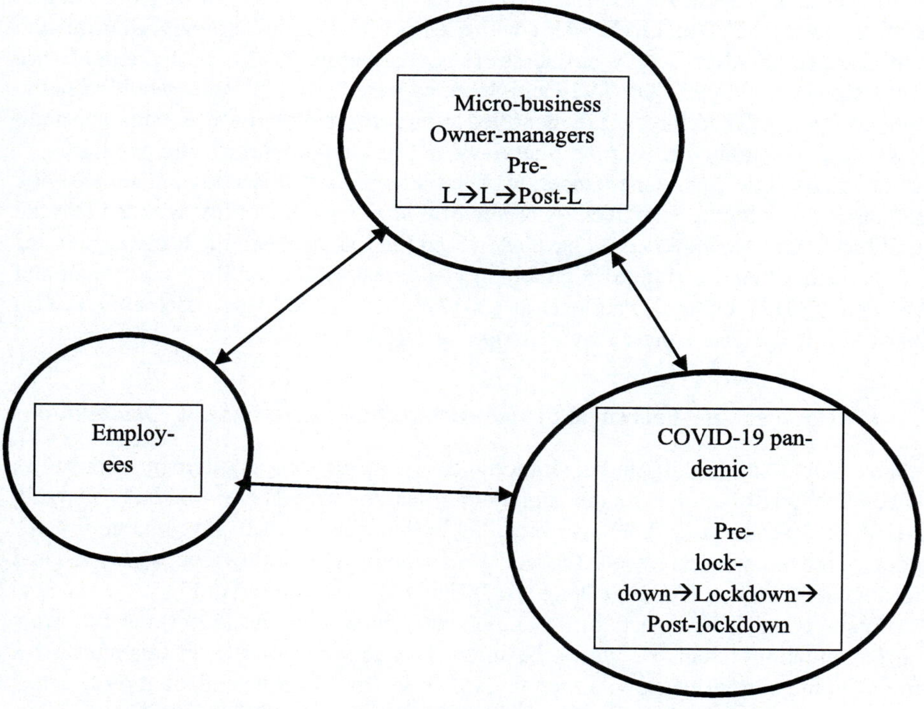

Fig. 2. COVID-19 Leadership Framework (CLF). **Source:** Authors

In Fig. 2, the leaders were the Scottish micro-business owner-managers, the followers were the employees of these businesses, and the situation was the lockdown brought on by the COVID-19 pandemic.

1.8 Conclusions and Recommendations

The review of the literature identified a gap in investigating the role of leadership styles on businesses in Scotland during the pandemic. This paucity informed this research study which involved interviews with 20 businesses. The paper demonstrated findings exclusive to businesses in a Scottish context where owner-managers preferred adopting an autocratic style during the pandemic, however, mostly the literature preferred transformational, transactional and coach styles. However, findings from this paper highlighted that owner-managers from Scotland were considering hybrid leadership styles to adapt between normal and pandemic-like conditions. There may be practical implications from this research for businesses regarding what may be appropriate leadership styles to employ during and outside of the pandemic. Policymakers may be informed by the findings while developing policies to support businesses' survival in turbulent times like the recent pandemic or the economic recession in 2008. This research has had implications on theory as the findings together with the LIF have informed the development of the conceptual framework of the CLF which may be included in future research related to this research area. Despite the key findings and implications of the research, there were problems/limitations encountered during the research.

Limited interview options due to the pandemic-led government restrictions were a research limitation, therefore, most interviews were carried out online. Another limitation was the research did not explore the role of culture, gender, and innovation in the leadership styles of business owners during the pandemic in Scotland. Therefore, the following recommendations were proposed.

- Conducted the same research in settings outside of COVID-19.
- The research explored the role of culture, gender, and innovation on leadership styles in businesses during the pandemic.
- Included and developed the CLF in further research.
- Included government officials in future studies to allow researchers in gaining insight into businesses' survival during the pandemic from the perspective of policymakers.

References

Khajeh, E.H.: Impact of leadership styles on organizational performance. J. Hum. Resour. Manag. Res. **2018**, 1–10 (2018)

Boyd, M.: Coronavirus LIVE updates: UK child, 5, dies as Brits continue to flout lockdown (2020)

Çera, E., Çera, G., Skreli, E.: The relationship between entrepreneurship education and entrepreneurial intention: evidence from a transition country. Int. J. Entrep. Small Bus. **43**(4), 548–569 (2021)

Deng, J.: Leadership in education abroad office: a case study based on the interactional framework of leadership and the transformational-transactional leadership theory (Doctoral dissertation, the State University of New York at Binghamton) (2017)

Girling, N.: The Impact of Leadership Style on Organisational Performance (2018)
Griffin, G., Häyrén, A.: Female entrepreneurship as rupture: becoming a female entrepreneur in Sweden in neoliberal times. Int. J. Entrep. Small Bus. **45**(3), 300–313 (2022)
Groenland, E., Dana, L.P.: Qualitative methodologies, and data collection methods: toward increased rigour in management research (2020)
Hurduzeu, R.E.: The impact of leadership on organizational performance. SEA–Pract. Appl. Sci. **3**(07), 289–293 (2015)
Igbaekemen, G.O., Odivwri, J.E.: Impact of leadership style on organization performance: a critical literature review. Arab. J. Bus. Manag. Rev. **5**(5), 1–7 (2015)
Keizer, P.: A power model of the labour market. Theor. Econ. Lett. **10**(4), 770–802 (2017)
Kirby, D.A., Watson, A.: Small Firms and Economic Development in Developed and Transition Economies: A Reader. Routledge, Milton Park (2017)
Legislation Gov, U. K. Data Protection Act (2018)
Lindsay, L.C.R.D., Woycheshin, C.D.: Overcoming Leadership Challenges: International (2015)
McQuerrey, L.: The economy's effects on small businesses (2019)
Obiwuru, T.C., Okwu, A.T., Akpa, V.O., Nwankwere, I.A.: Effects of leadership style on organizational performance: a survey of selected small-scale enterprises in Ikosi-Ketu council development area of Lagos State, Nigeria. Aust. J. Bus. Manag. Res. **1**(7), 100 (2011)
Orfanidou, E., Woll, B., Morgan, G.: Research Methods in Sign Language Studies: A Practical Guide, 2nd edn, pp. 74–75. Wiley Blackwell (2015)
Palalic, R., Ramadani, V., Dana, L.P., Ratten, V.: Gender entrepreneurial leadership in family businesses: a study from Bosnia and Herzegovina. In: Women Entrepreneurship in a Family Business, vol. 2, no. 3, pp. 208–226 (2017)
Pringgabayu, D., Ramdlany, D.M.A.: Creating knowledge management with the role of leadership and organizational culture. J. Entrepreneurship Bus. Econ. **5**(2), 147–171 (2017)
Salamzadeh, A., Dana, L.: The coronavirus (COVID-19) pandemic: challenges among Iranian startups. J. Small Bus. Entrep. **33**(5), 489–512 (2021)
Salamzadeh, A.: What constitutes a theoretical contribution? J. Organ. Cult. Commun. Conflict **24**(1), 1–2 (2020)
Salamzadeh, A., Kesim, H.K.: The enterprising communities and startup ecosystem in Iran. J. Enterprising Commun.: People Places Glob. Econ. **11**, 456–479 (2017)
Salamzadeh, A., Arasti, Z., Elyasi, G.M.: Creation of ICT-based social start-ups in Iran: a multiple case study. J. Enterprising Cult. **25**(01), 97–122 (2017)
Salamzadeh, Y., Farzad, F.S., Salamzadeh, A., Palalić, R.: Digital leadership and organizational capabilities in manufacturing industry: a study in a Malaysian context. Periodicals Eng. Nat. Sci. **10**(1), 195–211 (2021)
Samad, S.: The influence of innovation and transformational leadership on organizational performance. Procedia – Soc. Behav. Sci. **57**(1), 486–493 (2012)
Scottish Government. Businesses in Scotland- Headline Figures (2020a)
Scottish Government. Coronavirus (COVID-19): returning to work safely (2020b)
Scottish Government. Coronavirus in Scotland (2020c)
Sheiner, L.: How does the coronavirus pandemic compare to the Great Recession, and what should fiscal policy do now? (2020)
Silverman, D.: Doing Qualitative Research, 5th edn, pp. 222–223. Sage (2017)
Song, Y., Dana, L.P., Berger, R.: The entrepreneurial process and online social networks: forecasting survival rate. Small Bus. Econ. **56**(3), 1171–1190 (2021). https://doi.org/10.1007/s11187-019-00261-7
GOV.UK: The number of coronavirus (COVID-19) cases and risk in the UK (2020). https://www.gov.uk/guidance/coronavirus-covid-19-information-for-the-public
Vorobeva, E., Dana, L.P.: The COVID-19 pandemic and migrant entrepreneurship: responses to the market shock. Migration Lett. **18**(4), 477–485 (2021)

COVID-19 and Digitizing Accounting Education: Theory and Literature Review

Hassan Ali Ahmed[1], Zainab Sayed Al Mosawi[1], Qassim Mohamed Shabib[1], Nabaa Qarooni[1], Maryam Mohammed[1], Allam Hamdan[2(✉)], Abdullah Silawi[3], and Esmail Qasem[4]

[1] College of Business and Finance, Manama, Bahrain
[2] Ahlia University, Manama, Bahrain
allamh3@hotmail.com
[3] Jeddah College of Engineering, University of Business and Technology, Jeddah 21448, Saudi Arabia
[4] Islamic University of Gaza, Gaza, Palestine

Abstract. This article critically analyzes previously published literature and discusses the lessons learned in detail. Learning gleaned from the theories has been used to create a conceptual framework. Using data from the published sources, a null hypothesis is developed to suit the current case. The results highlight the importance of digitalization in education in general and the accounting field in particular. This paper also reveals the impact of COVID-19 on education and addresses the main challenges faced by education and the accounting sector after the pandemic.

Keywords: Digitizing accounting education · COVID-19 · Bahrain

1 Introduction

The COVID-19 pandemic has led to a huge change in all sectors worldwide. People's lives were transformed because of the pandemic, and education at all levels was affected intensely. The lockdown that was enforced in most countries led to the immediate closure of universities and college campuses, and all academic activities and services had to be delivered digitally.

Online teaching existed before the COVID-19 pandemic, but it was not popular; however, as COVID-19 spread across the globe, it became mandatory. Therefore, researchers have conducted studies to compare the effectiveness of virtual teaching and traditional teaching.

The COVID-19 pandemic brought about a huge global crisis, especially in the education sector. Educational authorities digitalized all courses, including in accounting education. This meant that various activities that were previously conducted offline were now accessible online. This section reviews the literature on this topic, and a conceptual framework has been developed using models and theories that could be used for analysis.

B. Alareeni et al. (Eds.): ICBT 2022, LNNS 620, pp. 157–165, 2023.
https://doi.org/10.1007/978-3-031-26953-0_16

2 Literature Review

2.1 Critical Analysis of the Change in the Modes of Accounting Education

A survey found that 29.1% of faculties accepted the value and validity of online teaching. A study on the global perception of students during the lockdown period of COVID-19 found that 82.29% of the respondents were willing to shift to virtual learning, and 80.57% determined it would be more suitable during the quarantine (Radha et al. 2020). The students were also asked if they preferred online teaching to traditional learning, and 77.71% of the students responded that they did not.

The new COVID-19 norms meant that modes of education shifted to digital, and the new teaching methods prevented students from maintaining any form of external physical contact. Knudsen (2020) suggests that these changes have encouraged a phase of technological advancements in the field of education, and digitalization has meant that the internet has been used more than it was previously.

The notion that education systems have been disrupted is supported by several individuals. Sarea et al. (2021) state that the COVID-19 pandemic is one of the most disturbing occurrences in the "history of education," as the means of education have been dismantled and systems have been shifted online. For example, methods of examination have moved online, and systems can be used to detect if there is more than one person on the screen, and if there is, the examination can be stopped.

The mode of learning changed to "e-learning," which is the new normal being followed by most universities and colleges. The continuity of academia during the pandemic was possible due to this method of learning. Shahid and Mughal (2020) put forward the idea that the educational sectors came up with immediate damage control, which was then widely used in various educational backgrounds, including accounting. The author argues that this helped communities maintain the normal course of education and prevented extinction at the hands of the virus. Teaching and feedback systems occurred online instead of the usual face-to-face interactions that had previously occurred between the students and the teachers. It was seen that attending online classes became easier for students in the presence of the COVID-19 virus, which prohibited any form of physical contact. Thus, these changes occurred after the digitalization of the accounting education system.

Since online-based learning has been launched in many universities, it has proven to be an effective way of learning, especially during a crisis that prevents faculty members and students from attending classes. It has taken education to higher levels and helped companies when hiring employees that have good backgrounds in how to use digital systems.

Accounting graduates are also more capable when dealing with information technology (IT) developments and have built the skills and knowledge needed to deal with the technology applied in their field of work. The base for any accounting professional is established via academics; therefore, many universities have incorporated IT and information and communication technology (ICT) within their teaching plans.

Moreover, universities all over the world have started to include accounting courses in their study plans to prepare students as data scientists and accounting analysts. Many companies have also moved along with innovations in the accounting field; for example,

KPMG has offered sponsorships for master's degree students to provide them with programs that build and develop new technological and data-related skills.

During the current COVID-19 pandemic, many companies and universities have digitalized their jobs and educational channels (Kergroach et al. 2021). The accounting profession, which normally uses traditional rules, has recently started to respond to technological changes and use digitalization opportunities to equip its students with the new skills needed (Stanciu et al. 2020).

2.2 Critical Analysis of the Digitalization that Occurred Before the Pandemic

It has been noted that several changes occurred in accounting education, which resulted in the development of systems that were present before the pandemic. Borrego et al. (2020) opine that the digitalization of accounting education started before the pandemic in various parts of the world. For example, various modes of digitalization started when internet usage increased and online technologies were being developed that helped students progress through their courses more effectively and efficiently. These technologies also made it easier for teachers. The author believes that these changes started to gain more attention and become more prominent after the worldwide spread of the coronavirus because all modes of physical contact ceased at short notice.

Although the effect of the coronavirus stopped all connections, in reality, these changes were inevitable. Sova and Popa (2020) think that the changes came about because of developments in the field of education, long before they were collectively known as "e-learning." These methods helped students and teachers change the stereotypes and basic norms of educational systems that were sped up by the COVID-19 pandemic.

The historical changes that were taking place in accounting education became more developed in the wake of the pandemic and slowly increased the opportunities that were provided to accountants. Nagari (2021) states that accounting education is important and more dependent on the performance of educators and their competency. The calculational aids and similar systems that were used during the pandemic existed long before the pandemic. According to the author, these methods seem to be gaining more popularity due to the pandemic. Although the current generation has been affected by the pandemic, it has seen advancements that have increased student competencies.

2.3 Critical Analysis of the Benefits of Digitalized Accounting Education

It has been seen that online modes of study have hugely helped students during the pandemic. A primary benefit is that students were still able to study even though physical contact was prohibited. Digitalization due to the pandemic allowed students to attend classes, even if they were not in the same city or university; this also applied to teachers who lived quite a distance away from the university where they taught. Frumușanu et al. (2020) suggest that the different online methods used successfully in education and accounting firms during the pandemic prove that online learning can dominate offline methods of accounting.

The effectiveness of the education system improved because students could attend classes regardless of where they were. Furthermore, Alshurafat et al. (2021) note that students' skill sets also increased as they started to spend their time at home more constructively. This would not have been possible if classes had been taught offline. The author proposes that students' and teachers' technological skills greatly improved, which led to an improvement in technological advances in the education sector.

It is proposed that online classes and the associated student activities became more developed during the pandemic and would probably have ceased if the technologies did not exist. Tartavulea et al. (2020) argue that COVID-19 made the transition from offline to online mode faster than it would have happened in the normal course of development. Although the quality and standard of classes dipped at the start of the pandemic, they slowly improved as teachers and students became more familiar with online teaching and learning systems. The author states that the technologies are huge assets for the opportunities afforded by the pandemic that will help accounting education develop an online base for its courses.

There are many advantages to online learning for accounting and other courses. One advantage is its cost-effectiveness, as teaching via online channels is less costly than teaching in traditional classrooms. Another advantage is easy accessibility, where education is provided to everyone regardless of where they are; they only need internet access (Nguyen 2015). Moreover, student participation is also less intimidating, and the quality and quantity of student interaction increase in an online class. Due to the advancements and improvements in technology, students can view and rewatch lectures at any time.

The COVID-19 pandemic forced accounting education to shift from traditional learning to online teaching, which brought about significant changes and many issues. For example, it was sometimes frustrating when dealing with online technologies, and there were difficulties in making personal connections with students.

Most universities and education establishments had no previous experience with online teaching. Accounting education in universities was faced with the challenge of creating a digital platform for accounting.

Technology is vital to accounting curricula, and many technologies are appropriate for the subject (Chugh 2010). Students appreciate the use of technology and online learning (Helfaya 2021), and the performance of accounting students improved with online exams instead of traditional learning (Aisbitt 2005). A positive and significant relationship was found between the time students spent using online educational platforms and their performance in the final exam of the accounting course (Perera and Richardson 2009). This argument was reinforced by Duncan et al. (2012) who found that accounting students' performance improved in the online examination, particularly in courses that involved synchronal and asynchronous connections among students.

The managerial accounting course was evaluated, as all business majors are required to take this course. In 2020, at the beginning of the semester, the course was delivered for eight weeks via traditional teaching methods, and the remainder of the semester (six weeks) was taught using the online platform. Student performance over the two different teaching methods was compared, and the results showed that the students performed better when learning virtually for six weeks. The performance for the virtual class was 87% and the traditional class was 73%, an increase of 14%.

Universities and educational organizations should enhance the quality of online learning and concurrently consider the elements that have been found to influence the use of online learning systems during COVID-19. Control over students and social experiences will improve their perception of the advantages and ease of using online learning systems.

2.4 Critical Analysis of the Issues in Imparting Accounting Knowledge Online

It has been noted that students faced many problems in the accounting course due to the lack of face-to-face interaction with teachers. Teachers also had difficulties communicating and connecting with their classes on a personal basis, and this also helps students understand the course better. Bourmistrov (2020) found that the goal for universities became how to get students to pass exams rather than educate them as well as how to make students more employable. Thus, this became an issue in teaching, and the course often remained unclear for a lot of students.

Studies became dependent on technologies and online modes, and if the technology went down for any reason, the classes would stop. Sova and Popa (2020) thought that COVID-19 created a crisis in the education system, and students and teachers strived to take various measures that could help the whole of accounting education. When accounting education shifted to an online basis, it affected more introverted students, as they could not consult the teacher in front of the other students.

Additionally, international students who were living far away from home suffered the most due to the quarantine and the online modes of classes that prevented the students from meeting each other. These students could not return to their homes, go out with friends, go to college, or attend university in person. Students were affected on a psychological basis and had difficulties understanding their courses, leading them to develop bad practices. Sangster et al. (2020) found that essential research and various developmental examinations stopped due to the lack of physical contact among accountants, which hampered the better performance of the researchers. Reports from accountants who were researching the new technologies and "new normal" were also delayed because of the never-ending pandemic.

3 COVID-19 and Digitizing Accounting Education: Empirical Evidence from the GCC

The crisis affected Gulf Cooperation Council (GCC) nations as well as the rest of the world. The first shutdown of all educational institutions began in Bahrain on February 25, 2020, followed by the rest of the GCC countries. All GCC higher education institutions switched to an e-learning system and employed learning management systems (LMSs), including Blackboard, Microsoft Teams, Big Blue Button, etc.

According to Sarea et al. (2021), during the crisis, accounting education may have encountered a few problems that could impact the quality of the outputs, including student evaluation procedures, teacher self-efficacy, accounting education digitization, lecture time, and instructional techniques.

3.1 Digitizing Accounting Education

The majority of institutions across the world are currently investing in LMSs. Combining the internet with accounting education is viewed as a beneficial approach for teachers to electronically assess learners and offer e-feedback. This has resulted in an exponential increase in the usage of e-learning (Sarea et al. 2021).

According to Humphrey and Beard (2014), educators may be worried about students' learning and understanding, although digitizing accounting education may provide more freedom. The abrupt shift to e-learning and the lack of an effective learning process may impact the student's future career prospects.

3.2 COVID-19 and the Evaluation Process of Accounting Students

To evaluate whether learners are learning or not, it is critical to have an effective and efficient evaluation process. Summative and formative evaluations are the two most common forms of assessment. In a summative evaluation, students are evaluated to see how far they have progressed toward their learning objectives, whereas a formative assessment is a continual review process by the instructor to understand the requirements of the students. As all educational institutions are currently closed, it is critical to utilize more formative assessments to gain a better understanding of students' learning (Liberman et al. 2020).

3.3 Online Teaching Self-efficacy of Faculty Members

The effective use of technology in the educational process is largely dependent on the user's acceptance and perception of these tools. Teachers' self-efficacy in utilizing the internet in the classroom was investigated by Lee and Tsai (2010) who found that teachers with more web expertise had better self-efficacy.

Although most institutions worldwide have an LMS, faculty members are not prepared enough to offer their courses online. As a result, the rapid transfer of education to online learning may pose a danger to educational quality due to a lack of teaching self-efficacy.

3.4 Lecture Timing During the COVID-19 Pandemic

Van de Vord and Pogue (2012) tracked online and traditional courses and found that traditional teaching requires more time per student than virtual learning. However, the time log reveals that several tasks, such as student work evaluation, recording grades, and technological difficulties, took more time online than face-to-face instruction.

The impact of the virus on teaching methods and the transition to online distance learning are seen positively by accounting faculty members in the GCC. Sarea et al. (2021) found that accounting educators have modified their teaching techniques in response to the COVID-19 pandemic, which has resulted in a major change in the manner of course delivery.

3.5 Insights into Accounting Education in a COVID-19 World

The epidemic has had a significant impact on the higher education sector. The lockdowns in most countries necessitated the immediate closure of university and college campuses, as well as the switch to remote delivery of all academic activities and services.

Sangster et al. (2020) conducted research on COVID-19 and accounting education in 45 nations throughout the world. Their findings suggest that during the crisis, accounting professors felt a great deal of uncertainty and stress about their jobs.

On the one hand, when it comes to a university's accounting curriculum, the online delivery of courses has demonstrated that computational and procedural components may be covered successfully via online training platforms. On the other hand, there are considerable issues that should be investigated concerning online teaching and the assessment of accounting.

When examining the assessment methods, a large percentage of participants said they had to change their planned evaluation techniques. Assessments shifted from closed-book, invigilated exams to open-book, at-home exams in many situations. Moreover, considering the sudden shift from traditional to virtual mode at short notice, the implementation of an assessment methodology was even more challenging, as it required both the reform of testing tools, methods, and content, as well as a change in providing new understanding for teachers and students, including a rethink on how materials were provided, classes were delivered, and practice-based activities were carried out (Sangster et al. 2020).

4 Model

The learning aspect of the various modes of teaching and learning for students needs to be performed and maintained by both teachers and students. Fogarty (2020) suggests that the new systems that have been set up are a huge prospect and are essential for maintaining the various needs of new forms of education.

1. Launch. The launch of new learning systems is a primary step that needs to be taken so that the systems can become familiar to students.
2. Plan. Course planning and adaptive methods must be conducted properly so that students can adopt the "new normal" ways of learning.
3. Research. Various methods need to be researched deeply by learners, and a better outlook needs to be seen for the "new normal."
4. Critique. Critiquing the new accounting functions needs to be managed and analyzed to ensure that the known knowledge is true and apt.
5. Share. The knowledge that is collected in the course of understanding the subject also needs to be shared with other learners.

5 Conclusion

COVID-19 has provided the opportunity for change, not only for students but also for teachers who now have the chance to expand their teaching knowledge and skills. They have the opportunity to build a mix of online-based learning and in-class learning, which is helpful for those students who are working in the accounting field and cannot attend classes.

Furthermore, digitalization and online teaching have opened the door for students to read and obtain more knowledge, as they are required to produce more project-based assessments rather than complete normal paper tests or quizzes, as well as literature reviews. This helps students expand their knowledge as they are reading and learning more (Sangster et al. 2020).

To conclude, even with the rapid changes and new advancements in technologies, educators should remain aware and continue developing courses and skills because such changes can create instability for accounting education. At the same time, teachers should maintain a proper balance between innovation and stability in developing curricula and provide their graduates with new developmental data and skills (Stanciu et al. 2020).

References

Aisbitt, S., Sangster, A.: Using internet-based on-line assessment: a case study. Account. Educ. Int. J. **14**(4), 383–394 (2005)

Alshurafat, H., Al Shbail, M.O., Masadeh, W.M., Dahmash, F., Al-Msiedeen, J.B.: Factors affecting online accounting education during the COVID-19 pandemic: an integrated perspective of social capital theory, the theory of reasoned action and the technology acceptance model. Educ. Inf. Technol. **26**, 1–19 (2021)

Borrego, A. C., Pardal, P., Carreira, F. J. A.: The accountant in the digital era and COVID-19. Repositório Comum 19 (2020)

Bourmistrov, A.: From educating agents to change agents: experience of using foresight in accounting education. J. Account. Organ. Chang. **16**(4), 607–612 (2020)

Chugh, R.: E-learning tools and their impact on pedagogy. In: Ubha, D.S (Ed.), Emerging Paradigms in Commerce and Management Education, pp. 58–81 (2010)

Duncan, K., Kenworthy, A., McNamara, R.: The effect of synchronous and asynchronous participation on students' performance in online accounting courses. Account. Educ. **21**(4), 431–449 (2012)

Fogarty, T.J.: Accounting education in the post-COVID world: looking into the mirror of erised. Account. Educ. **29**(6), 563–571 (2020)

Frumuşanu, N.M., Zgardan, A.M., Martin, M.: The impact of COVID-19 pandemic on the activity of the professional accountant. "Ovidius" University Annals, Economic Sciences Series, 1–10 (2020)

Helfaya, A.: Assessing the use of computer-based assessment-feedback in teaching digital accountants. Account. Educ. **28**(1), 69–99 (2021). https://doi.org/10.1080/09639284.2018.1501716

Humphrey, R.L., Beard, D.F.: Faculty perceptions of online homework software in accounting education. J. Account. Educ. **32**(3), 238–258 (2014)

Kergroach, S., Pionnier, P-A., Fietz, L.: Government support has helped save many SMEs but challenges remain (2021). https://www.oecd.org/newsroom/government-support-has-helped-save-many-smes-but-challenges-remain-oecd.htm

Knudsen, D.R.: Elusive boundaries, power relations, and knowledge production: a systematic review of the literature on digitalization in accounting. Int. J. Account. Inf. Syst. **36**, 100441 (2020)

Lee, M.-H., Tsai, C.-C.: Exploring teachers' perceived self-efficacy and technological pedagogical content knowledge with respect to educational use of the world wide web. Instr. Sci. **38**(1), 1–21 (2010)

Nagari, P.M.: The utilization of ICT in the COVID-19 pandemic: does it optimize the competencies of accounting lecturers? In: Advances in Economics, Business and Management Research, pp. 1 – 8 (2021)

Nguyen, T.: The effectiveness of online learning: beyond no significant difference and future horizons. MERLOT J. Online Learn. Teach. **11**(2), 309–319 (2015)

Perera, L., Richardson, P.: Students' use of online academic resources within a course website and its relationship with their course performance: an exploratory study. Account. Educ. **19**(6), 587–600 (2009)

Radha, R., Mahalakshmi, K., Kumar, V.S., Saravanakumar, A.R.: E-Learning during lockdown of Covid-19 pandemic: a global perspective. Int. J. Control Autom. **13**(4), 1088–1099 (2020)

Sangster, A., Stoner, G., Flood, B.: Insight into accounting education in a COVID-19 world. Account. Educ. **29**(5), 431–562 (2020)

Sarea, A., Alhadrami, A., Al-Madhagy Taufiq-Hail, G.: COVID-19 and digitizing accounting education: empirical evidence from GCC. PSU Res. Rev. **5**(1), 68–83 (2021)

Shahid, R., Mughal, A.M.: E-learning: a way out in COVID-19 crisis. J. Rawalpindi Med. Coll. **24**(3), 180 (2020)

Sova, R.A., Popa, A.F.: Accounting education–between digitalisation and the COVID-19 pandemic crisis. CECCAR Bus. Rev. **1**(10), 59–63 (2020)

Stanciu, V., Punga, I.B., Gheorghe, M.: New coordinates of accounting academic education a Romanian insight. Bucharest Univ. Econ. Stud. **19**(1), 158–178 (2020)

Tartavulea, C.V., Albu, C.N., Dieaconescu, R.I., Petre, S.: Online teaching practices and the effectiveness of the educational process in the wake of the COVID-19 pandemic. Amfiteatru Econ. **22**(55), 920–936 (2020)

Van de Vord, R., Pogue, K.: Teaching time investment: does online really take more time than face-to-face? Int. Rev. Res. Open Distrib. Learn. **13**(3), 132–146 (2012)

Impact of COVID-19 on Knowledge Management: The Double Edged Sword of Big Data

Noor Al Shehab[1](✉) and Salem M. Aljazzar[2]

[1] Ahlia University, Manama, Bahrain
noor_shehab@hotmail.com
[2] Jeddah College of Engineering, University of Business and Technology, Jeddah 21448, Kingdom of Saudi Arabia

Abstract. With a fluctuating status of the global market post to the spread of COVID-19 pandemic, the repeated and disoriented lockdowns lead to unexpected and harmful consequences to many firms around the world. However several businesses try hard to cope with the disaster, they ultimately arrived to a sad conclusion by putting their fundamental projects or functions on hold since they do not acquire convenient plans to be resilient among these unanticipated events. Leaders and successful firms take the advantage of artificial intelligence and the tsunami of data to enhance their efficiency, innovation, supply chain, knowledge management, forecasting, problem solving and decision making. The phenomenon of Big Data has opened new horizons in the field of research and currently been used to track the nature of COVID-19 virus. In addition, Big Data is employed nowadays to capture the consumers' behavior as a result of the increased online transactions in COVID times. Furthermore, it can be useful to make wise decisions during merge and acquisition practices where knowledge and experience are massively transferred between parties. Oppositely, Big Data bears some drawbacks which are linked to procuring outstanding computational skills, pleasing infrastructure, privacy and security concerns, effect on scientific research and more.

This paper aims to highlight the concept of Big Data in literature and focuses on its double-edged sword. Beside this, it discusses whether Big Data breeds excellent output at all times or this may depend upon other factors. Moreover, it explores how Big Data and Knowledge Management are related to each other and where they meet exactly. Likewise, it provides some insights about the role of Big Data during the Pandemic and lastly, the paper offers a number of recommendations for future directions in the background of Big Data.

Keywords: Artificial intelligence · Big data · Knowledge management · Business research · Consumer behavior · Forecasting · Pandemic

1 Introduction

It is almost a decade when both terms Artificial Intelligence (AI) and Big Data have considerably occurred in the business field promising innovative insights and solutions

B. Alareeni et al. (Eds.): ICBT 2022, LNNS 620, pp. 166–174, 2023.
https://doi.org/10.1007/978-3-031-26953-0_17

to all societies. The significant presence of big data allows predicting that by 2020, foremost research firms around the world would be replaced by Google and Facebook which controlled the core behavioral data of millions of consumers. Nowadays, data analysis is increasingly automated by different AI approaches despite the fact that several business firms are still employing the conventional methods in analyzing and forecasting. Amid COVID times, it is clear that a superior understanding of the pivotal role of Big Data has been pronounced especially in health and finance sectors alike. Absolutely, the option to use Big Data analytics is subject to the assessment of management since many firms are not yet ready to be involved in such intentions. In fact, companies focus on delivering their products and services to their customers with specific criteria and seek for a combination of sustainability and profitability. No matter which data analysis is being implemented, the most important thing is making the right and intelligent decisions that drive for a long lasting success by meeting the initial desired goals.

In the event that the world continues to open up in terms of technology changes, the market opportunities are bringing new types of competition that force countless business firms to review their plans and processes. Likewise, this has contributed to increase the importance of knowledge management where adequate data and information handling pave the way for better efficiency and sustainable economic development. It is extraordinary that statistics exhibited that by 2020, every online individual may generate nearly 1.7 MB of fresh data every second (SAS Institute 2017).

Big Data in 2019 has been defined by Wibisono et al. as "a large volume of information sets in terabytes or exabytes which is resulted from the web, financial, administrative and other records." Normally, those data are unstructured, complex and need to be captured, stored, managed, analyzed and distributed (Chen et al. 2012). Today, Big Data is a trendy concept and practice that could provide actionable understandings and competitive advantage for numerous business environments (Salehan and Kim 2016). Back to 1962, Phillips was the first scholar to remark the essential need for the Big Data in Macroeconomic studies. He claimed that policymakers could find it challenging to design a proper economic policy deprived of quantitative knowledge and measurable economic factors such as inflation.

Because of the great pressure of competition, the supply chain becomes more multifaceted and interlinked. Thus, companies pay substantial attention to gain a deeper understanding of their customers through leveraging Big Data in order to generate more sophisticated output. It is apparent that marketing and production departments have led businesses to adopt modern technologies for which Big Data is harnessed to maximize the innovation in business models, predictive capabilities and opportunities (Tan and Zhan 2017). This means that there is a correlation between Big Data and Knowledge that is resulted from business' daily activities. As future filled with ambiguity, Big Data yields impressive knowledge inventory especially in Finance industry where banks and financial firms unceasingly produce a vast number of financial Big Data around the globe. The new concept of "Big Data Finance" refers to all financial services that are carried out by employing Big Data technology. With this in mind, the financial sector under the umbrella of Big Data is going through a major era of innovation and transformation.

The trend now of exploring Big Data will continue to strengthen the correlation between academic research and financial applications. It is important to realize that knowledge-based economies are on the rise because of the high global demand on data. The phenomenon of Big Data mandates a quick and ample response to some business subjects that are getting the power in business analytics and architecture such as the Machine Learning (ML), Data Mining (DM), Cloud analytics, Internet of Things (IoT) and others. So far, the usage of mobile applications arrives at a stable and more matured era of development than other technologies.

1.1 Can Big Data Always Help?

The rise of Artificial Intelligence accelerates the construction of data power in many countries such as the United States which in 2015 promoted its national strategy by improving the field of Big Data. The strategy basically concentrated on how to process and analyze huge information excellently and how to extract value from such resources. Indeed, Big Data along with business analytics are currently important resources and been classified as assets for several big companies (Xie et al. 2016). It is worth mentioning that open data portals deliver several remunerations such as improving efficiency and quality of public administration. Besides, open data portals offer higher transparency of public services and allow for more innovation. The European Commission projected that by 2020, the open data initiatives would participate by more than EUR 739 billion which is equivalent to around to 4% of the EU GDP (Open Data Barometer 2016). Srujana et al. in 2016 clarified that the perception of "Data Democratization" supports the accessibility of data by people who required them at all organizational levels. Conversely, some financial and non-financial firms have the tendency to limit the full access to databases to the senior management and ICT department only.

Despite the fact that Big Data has a weighty impact on the field of research, scholars admit that there is a great need to understand the phenomenon of Big Data in the organizations and societies and what its concerns are. Moreover, deep investigations and observations about the practices and applications of Big Data should be conducted in order to develop the field of theories, methods and frameworks. It is true that Big Data participates in managerial revolution for decision making, problem solving, competitive strategy and formulation. To put it another way, Big Data approaches and analysis used to predict or explain the factors behind certain results. It seems that Big Data by itself is not a major issue. Yet, the way of handling and managing the numerous formless data accelerate the associated risks since it requires distinct environment to breed fruitful results. Therefore, it is crucial to mention both the advantages and disadvantages of Big Data in order to obtain a better thoughtful as stated below:

Advantages of Big Data during the Pandemic:

- **Fighting COVID Virus:** As revealed previously, Big Data at the present is commonly used for more advanced research especially in health sector. The spread of COVID-19 enters its second year producing many catastrophic effects. The good part of it is that the pandemic has generated extraordinary volume of varied data which can be connected to improve the understanding of this particular type of virus's nature. With the purpose of protecting the public health, Big Data technology helps to store massive amount of infected people's data such as their names, ages, blood type, duration of infection, symptoms, history of other diseases, recoveries, contact numbers, locations, travelling records, and other relevant data.

Moreover, in several countries such as the United Arab Emirates, the government created a distinct database to store the data of the health workers who serve in the front line to be appreciated through offering them with exceptional merits in housing, schooling, job professionalism and furthermore. It has been noted that Big Data analysis provides possibilities to track and control the virus's situation globally and to foster innovative remedies in the medical field. Analytics improve the development of vaccine as it builds more knowledge to predict the effective cure for the virus. At the moment, more researches are undergoing to identify the infection mechanism of the virus and scientists are still examining how COVID-19 can be slowed or banned.

- **Capturing Consumer's Behavior:** Big Data has been widely seized to estimate stock prices, purchase manners and voting intents (Pappas et al. 2018). Apart from the medical arena, COVID-19 pandemic rehabilitated the consumer behavior when social distancing and home quarantines were imposed worldwide. Stores and retailers which are technologically capable have shifted to online selling and delivering. Sheth in 2020 acknowledged that this alteration in shopping patterns may be temporary, but it could sustain in critical stages also. Because of the frequent lockdowns during the epidemic, firms which acquire a solid online presence are doing well. Nevertheless, many others are struggling to cope with the New Normality.

  In 2018, Zeng and Glaister found that companies which are armed with Big Data setup are more agile, flexible and decidedly responsive to different business needs and unanticipated opportunities. To illustrate, the knowledge given by Big Data enhances the understanding of market behavior and characteristics of customers. It provides excellent learning capabilities to support the business strategies as well (Johnson et al. 2019). What is more is that it pertains accelerating the development of new products, cultivating the utilization of existed assets, increasing the supply chain efficiencies and reinforce innovation (Trabucchi and Buganza 2019).

  On the top of this, banks, financial firms, insurance companies, and telecommunication agencies around the world take the advantage of Big Data to formulate specific knowledge profile for each client for the aim of delivering precise offers according to his/her interests and lifestyle.
- **Embracing Knowledge Management Phases:** The entire world is fronting a tsunami of data generating from dissimilar sources in different formats. Big Data is an integral part of the Knowledge Management cycle. Lately, Knowledge Management has taken the ground due to the fact that organizations affirm that it is exceptionally vital in making decisions, solving problems, improving efficiencies, delivering innovation and enhancing the overall performance (Jose Duarte 2016). Knowledge Management is shaped through collecting, capturing, organizing, storing and retrieving the critical information via specific designed platforms. Correspondingly, Big Data serves as the initial stage in Knowledge Management that would be aggregated, processed, arranged, saved and visualized subsequently.

In like manner, Big Data can discover the hidden knowledge, build more adaptive know-hows and deliver a superior competitive advantage among others (Khan and Vorley

2017). Here again, the knowledge extracted from the Big Data helps firms to diminish uncertainty and make well-informed decisions as mentioned by Rust and Huang in 2014.

Likewise, Big Data facilitates communication and knowledge sharing within firms. For example, employees attain an improved knowledge about another firm at the time of merge and acquisition due to the fact that their data becomes accessible and transferable. All transactions and undertakings can be communicated among parties. For this, it is advisable that both better data management and noble scenario planning can lead to increase the correctness in predictions. At the time of takeovers and mergers which has been also increased during the pandemic, firms should take care of working with the same vision and avoid any lack of synergy which can damage the value and reputation of business. It is essential to plan what will happen next in terms of integrating systems and consolidating data. Even the most careful estimation can be thrown out by unforeseen occasions. As a consequence of the global COVID crisis, organizations had to embrace change. Otherwise, they would face collapse. Business winners would deal proactively with chaos, seeing it as an opportunity to learn, innovate, and resilience.

Visibly, governments and organizations employ communication channels and platforms to allow for sharing ideas and thoughts with their audiences such as Instagram and Facebook. Unlimited feedback and comments by online users yield Big Data and through this approach, several business entities undertake serious arrangements to develop their products and services. Business firms opt to ensure efficiency and effectiveness in all of their operations by purifying and processing data through an advanced computational analysis (Tan and Zhan 2017). Coupled with this, meaningful data allows for better predicting of the purchasing patterns, allocating future budgets, addressing probable risks and setting the best pricing for products and services (Fernando et al. 2018). As knowledge is rapidly produced from available data, the impact of it on the overall performance of the firm is valuable. Thus, firms should consider the following issues while applying Big Data analysis (Saltz 2015):

Disadvantages of Employing Big Data:

- **Changing the Fashion of Handling Scientific Research:** Due to the "New Normality" post to the pandemic, the changes in global economies, the emergence of 5G and Internet of things, the research modes and innovation have undergone irregular ride which massive business firms found themselves obliged to alter their methods to gather, manipulate and protect data. Large and leading firms anchor the foundation of replacing traditional qualitative and quantitative methods by the Big Data analysis. They believe that approaches such as fuzzy cognition, data mining, deep learning, machine learning, intelligent computing, algorithm, and so on could offer better and concise answers. In fact, this kind of research method is vividly demonstrated in technology firms such as Facebook. Some researchers pointed that classical methods cannot capture the speed of knowledge and the stream of information.

  A foremost challenge in Big Data analysis is how to harness and suitably interpret the huge volumes of data which lack appropriate structure and coherent. A lot of questions lead to a main debate about deploying Big Data in research field. Researchers and editors enquire what Big Data contains? Is it the end of theory and scientific method? And how Big Data research differs from conventional ones? It seems that Big Data analysis poses access to a huge data source for a certain phenomenon as the first step in conducting research whereas the scientific research started with a theory (Johnson et al. 2019).

Moreover, Big Data analysis appreciates distinct computational and programming skills in order to collect, interpret, store and visualize data. Yan et al. in 2015 indicated that unstructured data is produced daily in millions of files at the original work environment although the structured data is stored in databases. Normally, those data are unstructured and not necessarily meet the research questions and constructs. Thus, the researcher should arrange the available data sets to serve the desired aim of the research. It is worthwhile that financial firms should benefit from the Big Data by not only deploying its own data, but also by importing other data from external institutions to breed superior results.

Most of the time, researchers may have to offer additional justification for all types of data, variables and constructs collected. Additionally, it is equally important to highlight the reliability and validity of variables that minimize the margin of errors in order to avoid misleading conclusions.

- **Requiring Appropriate Environment and Skills:** It cannot be overlooked that dealing with Big Data demands specific computational, statistical and informational knowledge which is usually not conquered by many enterprises. In the United States alone, there is a shortage of 190,000 of "Data Scientist" job (Srujana et al. 2016). Beside this, Big Data analysis mandates computing power and advanced infrastructure to provide great consequences.
- **Bearing more Complexities:** Silver in 2012 pointed out that Big Data obtainability is important but neither sufficient nor precise to improve the business model's predictions. By the same token, Bańbura and Modugno in 2014 observed that Big Data lead to more complex forecasting due to the large size of databases. However it might be true that Big Data by itself is not a problem, the way of dealing and processing the huge databases could bear complications and ambiguities for many.

  Albeit Big Data goes hand in hand during the pandemic to support medical researches, others initiated the debate that there are several challenges facing Big Data that deals with COVID-19. Firstly, Big Data should be accumulated globally to identify the hotspots and make predictions accordingly. Although this might be true, various countries reject to provide data on this regards due to some reasons which contribute to strict the data gathering process. Secondly, Big Data experts somehow forget the target of their research and try to answer countless questions about the disease as a result of overvaluing their capacity. Thirdly, there would be billions of pairs exhibiting numerous correlations between the countless given variables of COVID-19. This drives for more complications and vague. Fourthly, as this kind of disease is totally new, no proper prediction models and dynamics have been explored yet. Finally, the limitation of employing Big Data in COVID-19 researches stated that labs should be equipped with the latest computational software packages in order to process, analyze and visualize the big volume of collected data (Business World 2020).
- **Privacy and Security Troubles:** Under the background of Big Data, the main issue confronts societies is the personal information privacy especially the financial ones where many have been victims of hackers. In the view of improper legislation of

personal information protection, the abuse of Big Data in unauthorized activities stimulates a number of conspicuous problems. The heavy reliability on modern technology and Internet could generate adverse output. Therefore, Big Data acquisition should be properly handled by trustworthy and wise teams. Otherwise, the improper use of confidential data could be extremely harmful. In the event of personal data leakage or illegal trade of personal information, this would lead to serious problems that disturb the rights and interests of numerous financial clients. At present, there are endless attacks on both commercial and governmental websites from different parties such as foreign governments, illegal groups and even individuals who could threat all types of financial and confidential information. Of course, it is difficult to completely avoid such technical risks. For this, it is deemed necessary for enterprises to enhance and review their process of Knowledge Management on periodic basis to maintain adequate practices.

Webber During the pandemic in 2020 declared that data protection regulations became harder at the time of aggregating ethnicity data which are needed to analyze the mortality rates of COVID-19. He argued that it appears inevitable that more data protection regulations could increase the cost and complexity of data assembly. With no doubt, greater freedom in data collecting and transferring could help in solving issues related to COVID with paying attention to the fact that some countries prohibit the dissemination of information during crises.

2 Recommendation and Conclusion

To sum up, this paper pointed the rise of Big Data during the last decade and how the literature is still on the rise at this particular topic. There are several gaps to be bridged by business researchers who have to empower their coordination with other financial business firms and network information institutions alike. The second argument was whether Big Data brings fruitful results at all times or not due to the fact that it may breed more complexity to business analytics. Others claim that Big Data approaches lead to higher time and budget consumption. Above all, it requires skillful specialists and excellent infrastructure in Internet networks that can bear the heavy extraordinary processes and storage tasks. It briefly discussed the differences between the Big Data research and the traditional scientific research. Then, it mentioned the role of Big Data during the pandemic and how firms take the advantages of it. Thirdly, the paper described how Big Data acts as the primary stage in the Knowledge Management for which meaningful information is employed to make well-informed decisions afterwards. Finally, below are some recommendations that may enhance the use of Big Data to generate knowledgeable firms and economies:

a. More researches have to be conducted in the field of Big Data in order to enrich the literature with new theories, frameworks and strategies.
b. Embrace the connection between researchers, Information institutions, universities, artificial intelligence academies, financial firms to aggregate data in proper settings and create value out of it to serve societies.

c. Increase the awareness of individuals of the proper and improper use of data since illegal practices yield serious and harmful consequences. However, there are no unique remedies regarding the level of protecting confidential information.
d. Design adequate regulations which belong to personal data protection.
e. Not all data are supposed to be obtainable.
f. In financial firms, executive management and IT departments are responsible for personal data leakage. Therefore, a careful supervision must be undertaken constantly.
g. In the event that Big Data is a multifarious environment, several firms are not yet in an actual need to tap into this type of analysis. They can utilize their traditional ways to collect and analyze data to make their correct decisions.
h. Successful firms pay considerable attention to review their knowledge management systems on constant basis to ensure that Data are well captured, saved, scrutinized, examined and distributed.

References

Business World. Five reasons why the COVID-19 pandemic is a failure of big data analytics. Available at: http://www.businessworld.in/article/Five-reasons-why-the-Covid-19-pandemic-is-a-failure-of-Big-DataAnalytics/ (2020). 03-11-2020-339005. Accessed 28 Jan 2021

Chen, H., Chiang, R.H., Storey, V.C.: Business intelligence and analytics: from big data to big impact. MIS Q **36**(4), 1165–1188 (2012)

Fernando, Y., Chidambaram, R.R.M., WahyuniTD, I.S.: The impact of Big Data analytics and data security practices on service supply chain performance. Benchmarking: An International Journal **25**(9), 4009–4034 (2018). https://doi.org/10.1108/BIJ-07-2017-0194

Johnson, D.S., Muzellec, L., Sihi, D., Zahay, D.: The marketing organization's journey to become data-driven. J. Res. Interact. Mark. **13**(2), 162–178 (2019)

Duarte, J.: Factors in the effective transfer of knowledge from multinational enterprises to their foreign subsidiaries: a Mozambican case study: a Mozambican case study. Int. J. Train. Dev. **20**(3), 224–237 (2016). https://doi.org/10.1111/ijtd.12077

Khan, Z., Vorley, T.: Big data text analytics: an enabler of knowledge management. J. Knowl. Manag. (2017). https://doi.org/10.1108/JKM-06-2015-0238

Open Data Barometer. https://opendatabarometer.org/?_year=2017&indicator=ODB (2016)

Pappas, I.O., Mikalef, P., Giannakos, M.N., Krogstie, J., Lekakos, G.: Big data and business analytics ecosystems: paving the way towards digital transformation and sustainable societies. IseB **16**(3), 479–491 (2018). https://doi.org/10.1007/s10257-018-0377-z

Rust, R.T., Huang, M.H.: The service revolution and the transformation of marketing science. Mark. Sci. **33**(2), 206–221 (2014). https://doi.org/10.1287/mksc.2013.0836

Salehan, M., Kim, D.J.: Predicting the performance of online consumer reviews: a sentiment mining approach to big data analytics. Decis. Support Syst. **81**, 30–40 (2016). https://doi.org/10.1016/j.dss.2015.10.006

Saltz, J.S.: The need for new processes, methodologies and tools to support big data teams and improve big data project effectiveness. In: IEEE Computer Society (ed.) 2015 IEEE international conference on big data: 2066–2071. IEEE Computer Society, Los Alamitos, CA (2015)

SAS Institute Inc.: Becoming a Data-Driven Organization. IFL Science (2017)

Sheth, J.: Impact of Covid-19 on consumer behavior: will the old habits return or die? J. Bus. Res. **117**, 280–283 (2020)
Silver, N.: The signal and the noise: why so many predictions fail-but some don't. Penguin Publishing Group (2012)
Srujana, H.M., Sharma, S.S., Amitava, D.: Democratization of analytics: new frontier of data economy. Analytics, March/April **2016**, 42–50 (2016)
Tan, K.H., Zhan, Y.: Improving new product development using big data: a case study of an electronics company. R&D Management **47**(4), 570–582 (2017). https://doi.org/10.1111/radm.12242
Trabucchi, D., Buganza, T.: Data-driven innovation: switching the perspective on Big Data. Eur. J. Innov. Manag. **22**(1), 23–40 (2019). https://doi.org/10.1108/EJIM-01-2018-0017
Xie, K., Wu, Y., Xiao, J., Hu, Q.: Value co-creation between firms and customers: the role of big databased cooperative assets. Inf. Manag. **53**(8), 1034–1048 (2016)
Yan, J., Yu, W., Zhao, J.L.: How signaling and search costs affect information asymmetry in P2P lending: the economics of big data. Financ. Innov. **1**(1), 1–11 (2016). https://doi.org/10.1186/s40854-015-0018-1

Impact of Job Crafting on Employee Performance While Working-From-Home

Isa Abdulla Mustafa[1], Allam Hamdan[1](✉), Muneer Al-Mubarak[1], and Megren Altassan[2]

[1] Ahlia University, Manama, Bahrain
allamh3@hotmail.com
[2] College of Business Administration, University of Business and Technology, Jeddah, Kingdom of Saudi Arabia

Abstract. The current study is focused on Job crafting is a phenomenon that is spread widely all across the globe in different occupations including childcare educators, special education teachers and political advocacy employees. The job crafting process provides the employees with a major role of redesigning their jobs in such a way within certain limits that they can work satisfactorily while happily getting engaged in their jobsx. Job crafting is always done by the employees to feel comfortable in the environment and pays more attention to their job. Job crafting within certain limits can be done in three different ways. The Covid-19 epidemic has affected countries throughout the world, exposing hundreds of millions of people and claiming many lives. Governments in several nations have implemented lockdown measures, one of which is a Working from Home (WFH) policy, in which employees are not required to report to work every day.

Keywords: Worker's performance · Job crafting · Relational crafting · Task crafting · Covid 19 pandemic · Work-from-home

1 Introduction

Changing competition led to uncertainty in the organizations which eventually increased the short-term employees and reduced the security associated with jobs thus making the employees more responsible in managing their own careers. For increasing well-being and better performance, the employees need to engage actively and enthusiastically in the task they are supposed to perform i.e. through job crafting activities (task crafting, relational crafting, and cognitive crafting). The job crafting process involves basically the changes in jobs or tasks done by employees either physically or cognitively within their certain limits [1]. These changes help the employees in performing the task with their own goals and preferences [2].

Job crafting involves altering the work in such a way that the employees can easily perform their tasks while being satisfied with their job and are ready to put extra effort into their jobs. Jobs basically involve different tasks and relationships among the different employees and other people around them in their workplace. In job crafting either the

B. Alareeni et al. (Eds.): ICBT 2022, LNNS 620, pp. 175–182, 2023.
https://doi.org/10.1007/978-3-031-26953-0_18

employers or managers go for making a feasible environment for their employees to work or the employees themselves go for changing their job within their domain to customize the job in such a way that they can easily put extra effort and perform their task without having conflicts between their goals and the organizational goals. The people involved in the job crafting are called job crafters who eventually go for customizing their jobs either through changes in the tasks/roles and responsibilities to be completed or changing the connections with other employees and people around [3].

The job crafters can simply change their jobs by increasing or decreasing the number of tasks they are supposed to perform and the scope of their tasks, likewise changing the methods of doing those tasks and the time and efforts given to them which is called task crafting. Similarly, the job crafters can go for changing the relationships with the people involved in that task or work within the organization so that they can have better social relationships and learn new things from them and likewise teach them new things as well which is called relational crafting. The last one is cognitive crafting done by the job crafters which includes the changes in the perceptions of the employees regarding the jobs and task they are performing or are supposed to perform and the relationships with other employees and similarly focusing as a whole on the job instead of the individual tasks [4].

The pandemic created by Covid 19 has drastically altered the business world as well as the working conditions of people thus making several people lose their jobs and shifting several businesses from traditional to digitalized ones. Due to the virus transmission, there were several threats attached with the job security as the business working was not in the condition of paying salaries to every employee and the main focus of companies was on retaining the most important employees in order to ensure their successful survival [5]. Within such a situation, several employees of different companies went for altering the ways in which they were doing their work previously to bring innovation in their work and ensure that they don't lose their jobs in such a situation. There has been various research done on evaluating the role of Covid 19 in the changes in work engagement as well as the job crafting activities of employees during the pandemic situation, but there arises a question that has Covid 19 did pandemic bring any changes in the job crafting and the performance of employees of telecommunication companies?

2 Literature Review

2.1 Job Crafting: A Conceptual Introduction

The job crafting basically is focused on reshaping the jobs of employees by either adding more tasks in their jobs, by changing the relationships between the different employees involved in different projects and ultimately changing the perception of the employees towards their job and relationship with employees [6]. In job crafting, the structure is changed in such a way that the employees feel comfortable in that environment and put their complete effort to get the desired organizational goals. The job crafting is always done by the employees to feel comfortable in the environment and pays more attention to their job. Job crafting is a phenomenon that is spread widely all across the globe in different occupations including childcare educators, special education teachers, and political advocacy employees [7].

The job crafting process leads to positive outcomes in the employee's behavior and their performance including job satisfaction, wellbeing, creativity, organizational commitment [8–10]. The job crafting process provides the employees with a major role of redesigning their jobs in such a way within certain limits that they can work satisfactorily while happily getting engaged in their jobs thus coming up with better results. Job crafting within certain limits can be done in three different ways including task crafting activities (amending the work they are supposed to do), relational crafting activities (amending the relations with people they are supposed to work), and cognitive crafting activities (amending the personal point of view about work they are supposed to do). Job crafting activity is a little difficult task in order to achieve meaningfulness in the work as the employee's working point of view at a certain job detail is common now but the changes in the economy and the technology demand from the employees a proactive behavior where the employees can put their full efforts happily to achieve their task-related goals without ignoring their personal life goals [11]. Job crafting constitutes of task crafting activities, relational crafting activities, and finally cognitive crafting activities which are as discussed below:

Task Crafting Activities: Task crafting activities involves a job crafting form in which the employees go for changing the task they are supposed to do in such a manner that they can efficiently and effectively perform that certain task. During task crafting the employees of the firm either leave the other tasks, add more, change the timing and effort level to different tasks, provide more priority to certain tasks as compared to others to achieve the organizational goals as well as self-satisfaction, physical and psychological both [6].

Relational Crafting Activities: Relational crafting activities involves a job crafting form in which the employees go for improving their relations with other people at their workplace in such a manner that they can work with them and take their help in efficiently and effectively completing their tasks. While doing relational crafting, the employees either mend the relations with other people they are working in a group or prioritize their relations on the basis of work they are supposed to do in order to complete their work efficiently [6].

Cognitive Crafting Activities: Cognitive crafting activities involves a job crafting form in which the employees go for bringing changes in their perception towards their job in order to increase their motivation level for completing their work at the right time. While doing cognitive crafting, the employees go for identifying what is their work at the job place, what is not for them actually, and why are they important at the workplace [6].

2.2 Previous Literature on Job Crafting and Its Influence on Employees

A study was conducted by Guan and Frenkel [12] to find how HR practices as well the work engagement (WE) of employees and job crafting brings changes in the performance of employees. The research explored the mediating role of WE and job crafting among the perception of HR practices and employee' working output while using a survey questionnaire. Data was gathered from 455 workers working in 5 Chinese' manufacturing companies. The study found that when there are strong HR management practices there is an increase in work engagement (WE) as well as job crafting among the employees which eventually leads to an increase in employee performance. The study likewise illustrated that job crafting individually as well as with work engagement, is directly linked with the increase in employee performance.

Liu, Wan, and Fan [13] focused on exploring the effect of remote work on worker's performance while considering the mediating role of job crafting while moderating the role of performance goal orientation (PEO). The study used 1309 questionnaire responses gathered using the survey method revealed that there is a noteworthy effect of remote work on worker's performance and a noteworthy mediating role of job crafting among the telework and worker's performance. The study illustrated that when people started working online, there were more changes in the working of people and their ways of collaborating with others eventually leading to an increase in the performance of employees.

A study was done by Saragih, Margaretha, and Anantyanda [14] focused on exploring the changes in job autonomy as well as job crafting along with the wellbeing of employees during remote work. Job crafting was used as a mediator between the job autonomy and the wellbeing of employees and conducted an online survey using 427 responses and evaluated that there is no significant impact of job autonomy on the well-being of employees however there is a noteworthy effect of job autonomy on job crafting and job crafting on workers' wellbeing.

Kim and Beehr [15] stated that the task crafting technique basically includes the molding of the jobs of the employees by them in such a way that enhances their ability to perform efficiently and effectively thus putting full energy in their task and eventually enhancing their performance in an organization. The job crafting process is considered to be the cause of improving the working environment and conditions, filling somehow difference between job demand and resources which eventually increases the performance of the employees [8].

2.3 Relationship Between Job Crafting and Job Embeddedness

Previous study has shown that crafting activities improve the likelihood of employees staying with a company. Allowing employees to develop or remodel the way they perform their work responsibilities offers them control over the job, and they will alter the job to fit their unique features and quirks. JC practices result in a better fit in the workplace between the person and the job, as well as enhanced work engagement, because the task will be given more meaning by the employees responsible. According to research, employees may make little changes to their work environment on a regular basis, which can boost job performance and retention. According to Slemp and Vella-Brodrick [16], task, relational, and cognitive crafting has a positive link with work satisfaction and organizational citizenship behavior. According to past study, JC activities improve employee well-being and self-image. On a personal level, personal motivations to maintain a positive self-image, deepen job meaning, or increase one's well-being and performance are hypothesized to cause JC. Employees with a positive self-image are more likely to want to stay longer since they risk losing it if they leave. Job engagement aims to achieve, among other things, a better fit in the workplace, an exponential degree of work meaning, an enhanced work identity, stronger work-related well-being, and superior job performance [6, 17].

2.4 Summary of Literature

The above-mentioned literature reveals that overall job crafting is focused on bringing improvement in the working of the employees in order to bring improvement in their performance along with efficiency and effectiveness while making the employees more engaged in their work and better mentally at their workplace.

2.5 Literature Gap

Previously, analysis has been done on exploring the job crafting activities and its elements considering the general well-being of workers by Tims et al. [17], which found a noteworthy effect of job crafting activities on wellbeing of employees whereas several other studies by Guan and Frenkel [12], Saragih et al. [14] and many others used job crafting as a mediating variable among the different situations, however, very rare research is available on the effect of job crafting activities on worker's performance while considering the pandemic situation of Covid 19 which led to work-from-home, which is the literature gap the current study has identified. Similarly, there has been no research yet done on exploring the job crafting and employee performance during Covid 19 pandemic in Bahraini companies which is a population gap identified by the current study.

2.6 Theoretical Foundation

The current study is based on job crafting theory by Wrzesniewski and Dutton [6] which illustrates that there are 3 main elements of job crafting including task crafting activities, relational crafting activities, and cognitive crafting activities where task crafting activities focuses on the changes in work of the employees, relational crafting activities focuses on changes in relations at work, and cognitive crafting activities relates to changes in perception of employees related to their workplace. Job crafting theory states that when an employee crafts his task, relations, or perception at the workplace, it brings efficiency in the working of employees which reduces the burnout and stress at the workplace eventually leading to better performance.

3 Work from Home (WFH)

Since 2020, the Covid-19 epidemic has affected countries throughout the world, exposing hundreds of millions of people and claiming many lives. To combat the spread of the virus, governments in several nations have implemented lockdown measures, one of which is a Working from Home (WFH) policy, in which employees are not required to report to work every day [18]. WFH is being held until the Covid-19 epidemic is believed to have passed, allowing offices to reopen and return to normal. Work is carried out through WFH using digital technologies and internet network services. Workers in the public sector, often known as the Civil State Apparatus, operate in a variety of ministries, institutions, and other government entities. To limit the spread of Covid-19, has been forced to do work from home activities since March 2020. WFH is not a new concept, as freelancers, translators, SEO professionals, youtubers, resellers, and others have demonstrated [18]. The Covid-19 outbreak has ravaged nations all over the globe since 2020, exposing hundreds of millions of people and taking countless lives. To counteract the virus' spread, governments in some countries have enacted lockdown measures, one of which is a Working from Home (WFH) policy, which allows workers to work from home on a daily basis [18]. WFH will be held until the Covid-19 pandemic has passed, enabling offices to reopen and resume regular operations. WFH uses digital technology and internet network services to carry out its activities. The Civil State Apparatus, or public sector workers, work in a range of ministries, institutes, and other government agencies. Since March 2020, has been obliged to undertake work from home activities in order to minimize the spread of Covid-19. As freelancers, translators, SEO specialists, youtubers, resellers, and others have proved, WFH is not a new notion [18]. WFH, on the other hand, would be required to work from home for the first five days of the week, as per the office's timetable. WFH, of course, involves a shift in working habits. Employees who are used to working in a clean, fresh, and pleasant environment with air conditioning, digital technology, a free and powerful internet network, access to critical data and information, and good social contact and mutual support with coworkers, among other things, are relocating to a home with a different work environment, owning digital technology, and incurring their own expenses, among other things. As a consequence, WFH is thought to affect personal and professional behavior, particularly in terms of emotional mental health, psychological well-being, work performance, and job satisfaction, all of which influence employee job expectations [18].

Employees' mental and emotional states are considered to be harmed by WFH, particularly emotional symptoms, behavioral issues, hyperactivity/inattention, peer connection concerns, and prosocial conduct [18]. Employees' emotions deviate from their typical patterns and routines, resulting in emotional symptoms. Depression or anxiety, fury, irritation, and physical symptoms such as stomach discomfort, headaches, or nausea are all examples of emotional symptoms [18]. Workplace actions that are improper, unpleasant, ugly, uncomfortable, or incorrect are known as behavioral challenges [19]. Hyperactivity/inattention is a sort of suppressed response that results in a lack of self-control, a diminished capacity to meet job goals, and difficulties transitioning to a new work environment [18]. Emotional, cognitive, and interpersonal behavioral obstacles contribute to problems in colleague relationships. Relationships with coworkers are crucial for a number of reasons, including employment, status, creating friends, and sharing sentiments [18]. The suspension of voluntary activity to aid, support, and benefit colleagues is known as prosocial behavior [18]. Interruptions in prosocial conduct occur when numerous components of task are not completed, such as sharing, amusing, speaking, assisting, and so on. Employees' psychological well-being is considered to be affected by WFH. Individuals with psychological well-being have a good attitude toward themselves and others, organize and manage their environment to satisfy their requirements, have positive relationships with others, try to find and develop their potential, and accept and govern their own behavior [18].

Autonomy, environmental mastery, personal progress, meaningful relationships, life objectives, and self-esteem are all factors that contribute to psychological well-being [18]. The autonomy indicator assesses a person's capacity to control their actions autonomously, confidently, and responsibly. The capacity to successfully manage the environment by adjusting it to fulfill the job's objectives and expectations demonstrates environmental mastery. The capacity to build self-actualization potential is a sign of personal development. Negative relationships have closed, less caring, less sensitive, and less controlled environments, but good relationships have the capacity to develop connections with other people based on trust, care, empathy, and a grasp of the principles of mutual acceptance and giving. The life purpose indicator evaluates how the workplace may help people live better lives and give their lives greater significance in the future. Working from home should not lead to disappointment, discontent, or the suffocating of personality aspects owing to incapacity to make the required self-changes, according to self-acceptance indications. WFH is also thought to have an effect on public employees' job performance, namely as a demonstration of skill, employment prospects, time [18].

4 Conclusion

Finally, the current research yields encouraging empirical findings that demonstrate the interconnectedness of perceived organizational support and job crafting in enhancing teachers' career satisfaction, as well as the critical mediating role of work engagement during such a critical time of COVID-19-related job crafting. Our findings add to the body of knowledge by explaining how instructors use their job resources to deal with constantly changing job demands. Our findings demonstrate that organizational support and job crafting may assist teachers in performing well in the classroom while also emphasizing work engagement.

References

1. Van Wingerden, J., Bakker, A.B., Derks, D.: Fostering employee well-being via a job crafting intervention. J. Vocat. Behav. **100**, 164–174 (2017)
2. Tims, M., Bakker, A.B., Derks, D.: Development and validation of the job crafting scale. J. Vocat. Behav. **80**(1), 173–186 (2012)
3. Berg, J.M., Dutton, J.E., Wrzesniewski, A.: Job crafting and meaningful work (2013)
4. Peral, S., Geldenhuys, M.: The effects of job crafting on subjective well-being amongst South African high school teachers. SA J. Ind. Psychol. **42**(1), 1–13 (2016)
5. Vyas, L., Butakhieo, N.: The impact of working from home during COVID-19 on work and life domains: an exploratory study on Hong Kong. Policy Design And Practice **4**(1), 59–76 (2021)
6. Wrzesniewski, A., Dutton, J.E.: Crafting a job: revisioning employees as active crafters of their work. Acad. Manage. Rev. **26**(2), 179–201 (2001)

7. Lyons, P.: The crafting of jobs and individual differences. J. Bus. Psychol. **23**(1–2), 25–36 (2008). https://doi.org/10.1007/s10869-008-9080-2
8. Bakker, A.B., Oerlemans, W.G.M.: Daily job crafting and momentary work engagement: a self-determination and self-regulation perspective. J. Vocat. Behav. **112**, 417–430 (2019). https://doi.org/10.1016/j.jvb.2018.12.005
9. Wang, H., Demerouti, E., Bakker, A.B.: A review of job crafting research. Proactivity at work: Making things happen in organizations **77**, 95–122 (2016)
10. McClelland, G.P., Leach, D.J., Clegg, C.W., McGowan, I.: Collaborative crafting in call center teams. J. Occup. Organ. Psychol. **87**(3), 464–486 (2014)
11. Lee, J.Y., Lee, Y.: Job crafting and performance: Literature review and implications for human resource development. Hum. Resour. Dev. Rev. **17**(3), 277–313 (2018). https://doi.org/10.1177/1534484318788269
12. Guan, X., Frenkel, S.: How HR practice, work engagement and job crafting influence employee performance. Chinese Manag. Stud. **12**(3), 591–607 (2018)
13. Liu, L., Wan, W., Fan, Q.: How and when telework improves job performance during COVID-19? job crafting as mediator and performance goal orientation as moderator. Psychol. Res. Behav. Manag. **Volume 14**, 2181–2195 (2021). https://doi.org/10.2147/PRBM.S340322
14. Saragih, S., Margaretha, M., Anantyanda, L.: Job autonomy, job crafting and employees' well-being during working from home. Jurnal Managements Dan Kewirausahaan **23**(2), 177–185 (2021)
15. Kim, M., Beehr, T.A.: Can empowering leaders affect subordinates' well-being and careers because they encourage subordinates' job crafting behaviors? J. Leadersh. Organ. Stud. **25**(2), 184–196 (2018)
16. Slemp, G.R., Vella-B., D.A.: The job crafting questionnaire: a new scale to measure the extent to which employees engage in job crafting. Int. J. Wellbeing **3**(2) (2013)
17. Tims, M., Bakker, A.B., Derks, D.: The impact of job crafting on job demands, job resources, and well-being. J. Occup. Health Psychol. **18**(2), 230 (2013)
18. Thamrin, S., Sariwulan, T., Suryatni, M., Ridlo, M., Qamarius, I., Capnary, M.C.: The impact of work from home (WFH) during covid-19 pandemic period on job expectations: the case of the state civil apparatus. J. Manag. Inf. Decis. Sci. (25) (2022)
19. https://www.honestdocs.id/masalah-perilaku. Las accessed 13 Sep 2022

Managing Small and Medium Enterprises (SMEs) During Unexpected Situations: Strategies for Overcoming Challenges

Ahlam Mahmood[1], Allam Hamdan[2](✉), Lamea Al Tahoo[2], and Hatem Akeel[3]

[1] College of Business and Finance, Manama, Bahrain
[2] Ahlia University, Manama, Bahrain
Allamh3@hotmail.com
[3] Finance Department, College of Business and Administration (CBA), University of Business and Technology (UBT), Jeddah 21448, Kingdom of Saudi Arabia

Abstract. Crises are part of life, no one has not gone through a crisis or may go through it. Recently, it has been noticed that many crises occur globally and locally rapidly and increasingly. In this research, the focus will be on small and medium enterprises (SMEs), on the way they deal with difficult situations, in addition to the strategies and processes that they follow during crisis and post crisis. Moreover, it will shed light on the role of innovation in reducing risks associated with crises, the effects and influences that are related to the survival of the enterprises and the main causes of success during these difficult times; in turn, the main causes of failure. It will be a comprehensive analysis of crisis management and overcoming challenges during and after crises. The occurrence of the unexpected and unplanned situations is inevitable at any time, however the challenge is how to turn this matter into an opportunity, and what is the importance of planning and full readiness for such situations, as well as predicting and finding solutions before the problem arises, as in emergency situations often thinking becomes harder and narrower.

Keywords: Crises · SMEs · Challenges · Survival · Crisis management · Innovation · Unexpected situations

1 Introduction

Every circumstance that a business goes through will add more experience to it to deal with similar crises in the future; there are many businesses that have turned crises into opportunities, where crises create innovation. While some managements spend time blaming and despairing of a situation out of control, there are those who invest this time thinking and finding a way out of a more difficult situation. Therefore, preparing plans and strategies aimed at addressing these problems is important in these situations.

This research examines the effects of unexpected circumstances in depth like pandemics and abrupt changes on Small and medium enterprises SMEs. COVID19 has put a considerable halt to global economic growth; firms have been forced to implement new management standards in order to adapt to the severe conditions and thrive in this

B. Alareeni et al. (Eds.): ICBT 2022, LNNS 620, pp. 183–192, 2023.
https://doi.org/10.1007/978-3-031-26953-0_19

new position (Carracedo et al. 2021). Oftentimes perfect planning coupled with innovation is the best way to face any challenge; where managers in various sectors resort to brainstorming when sudden problems occur, as one idea may change the business for the better. There are many abrupt changes that businesses are witnessing at the present time, such as the rapid change in prices globally, the change in the tax rate, and even the change in procedures and digital transformation as a result of the conditions the world is witnessing.

SMEs are considered the largest source of innovation; furthermore, it is one of the largest attractors of labor which has a great potential to provide job opportunities, therefore, it is considered the mainstay of the economy of any country (Al Qubtan et al. 2021); as many countries seek to enhance the contribution and performance of the SMEs sector through initiatives, programs and systems which helps in financing and ensuring its continuity.

Many businesses have closed from the first stumble, while there are many businesses that turn obstacles into challenges that must be faced and transformed into opportunities. It is difficult to identify companies that are able to survive in light of rapid changes in the economy and successive crises, however it is possible to determine vulnerability, measuring and evaluating expected risks; when the administrations realize the importance of future planning for any abrupt change, they will come out of any crisis with a positive outcome, as quick decisions without prior planning will be lacking in information and inaccurate, moreover, it may cause disastrous results.

In fact, SMEs may suffer from many problems that may affect their continuity, especially with the current challenges of digital transformation and changes in the economic environment in addition to unexpected crises, thus, there is currently some reluctance to invest in SMEs because of the tough challenges they face, which forcing many to leave the market. One of the problems that is important to mention is that SMEs depend on the usual systems with a simple electronic intervention, are difficult to adapt to the massive digital transformation (Prasanna et al. 2019). The study begins with the literature review that explain the abrupt change during crisis and how it can be managed. Also, how crisis can cultivate innovation and how government gave their full support during hard times to overcome those challenges. Moreover, the literature review will illustrate how the crisis have accelerated global digital transformation trends and helped in the recovery. This is followed by the conclusion and future research.

2 Literature Review

2.1 Abrupt Change

The economic crisis is a state of difficulty that countries are going through, as a result of an extraordinary condition of unforeseen events in the financial system and its components, which has a negative impact on the economy (Hertati et al. 2020). This crisis occurs as a result of an economic imbalance due to pandemics, natural disasters, technological changes and political changes.

Change Occurrence

Pedersen et al. (2020) state that when a crisis occurs change occurs, the company must launch crisis responses. Depending on the nature of the crisis, the organization may take several actions. Even though the situation is unpredictable and dynamic, decision-makers must stick to logical patterns, which can be challenging when evidence is few or contradictory. Furthermore, decision-making speed is frequently critical, implying that many judgments must be taken on the fly during a crisis. This isn't to say that judgments aren't made after careful consideration. The use of simple cost–benefit assessments, effect models, stakeholder analyses, and trade-off models is common. As a result, in this stage of a crisis, the necessity for strong leadership tends to be more obvious (Bajaba et al. 2021).

Types of Changes During a Crisis
There are many crises that have faced and are still facing SMEs and most notably are the following:

- Financial implications: The results of the research conducted by Bartik (2020) indicate that Just a few weeks after Covid-19 commencement, and before the support from various sides became available, it had already caused tremendous disruption among small enterprises. Across the whole sample, 43% of firms had temporarily shuttered and their financial system is disrupted. Reduced demand and staff health concerns were the main causes for temporary closures, with supply chain disruptions playing a smaller role. Since January, employers have reported reducing active employment by 39% on average. The decrease was most pronounced in the Mid-Atlantic area, where 54% of businesses shuttered and 47% of jobs were lost.
- Legal Effects: Past and current economic crises have caused many legal challenges for companies due to accumulated debts, deferred or canceled contracts, unpaid bills, closings, bankruptcy and many others. This led to amendments and changes in some commercial laws in some countries during crises and disasters; The companies that had sufficient legal knowledge at all levels were able to avoid any legal obligations against them, while the companies that did not plan for the legal implications on business operations were negatively affected.
- Organizational Change: It refers to the continuous transformations that occur within the institution, starting with the appointment of employees or the dismissal and resignation of others, in addition to continuous changes in market conditions in general, or changing the suppliers that the organization deals with. Moreover, making some modifications to the ways of performing the work. During crises, many SMEs are forced to modify the framework for the implementation of activities and the relationship between them within the organization. The most prominent example of this is when many organizations tended to shift towards online working, cancel many previous tasks and activities, and dispense with many roles during last pandemic. As the leaders' awareness of all the concepts of adjusting the organizational structure will help them to know the best way to ensure the continuity of work and its management and directing employees towards success in achieving strategic goals.

2.2 Crisis Management

Crisis management is how to handle crises and adapt to changes, and how to overcome crises using a variety of scientific and administrative techniques, while avoiding and exploiting their bad aspects. It is critical to protect the organization's assets and property, as well as its capacity to create money when the revenue of most of the similar businesses declines. In addition to protecting employees and workers from various risks, and striving to minimize or lessen the impact of prospective risks on the business when all previous methods are not practicable (Trachsler and Jong 2020).

Crises must happen at any time, no one can guarantee that a crisis will not occur at a specific time. Some businesses that do well throughout the crisis may gain new customers, while others appear to be doomed. Responses and reactions of businesses during the crisis will, at least in part, decide whether or not they survive. As during the economic crisis caused by COVID-19, where the huge lockdown of enterprises triggered a recession, with significant unemployment, more bankruptcies and long-term spending reductions in consumer investment markets (Pedersen et al. 2020).

Ramesh Rajasingham state that it is critical to understand that no single organization can deliver a full crisis response. Many individuals working together and bringing their varied sources of experience, resources, and talents to the table are required for a successful effort. Under the leadership of national authorities, coordination is critical to make the collective worldwide effort function (Moonen 2021).

Crisis Management Phases

It is clear that planning before crises occur is one of the basics in SMEs, as this guarantees their survival and success. In this regard, crisis management consists of five phases:

1. Pre-crisis: Organizations can try to prevent or prepare for it if feasible. In some circumstances, organizations have the power to proactively avoid a crisis; some even argue that a pre-crisis, preventative phase should involve organizational readiness, changes to structure, and stakeholder relationships to avert system failures. Where strong relationship with stakeholder can help prevent a crisis mitigate the effects of the crisis. The outcome can be evaluated using probabilistic outcomes at this phase; however, uncertainty is difficult to measure, therefore its evaluation cannot be based on probabilistic basis.
2. Emergence of Crisis: At this stage, the crisis has not yet begun, nevertheless its signs have become clearer. Depending on the pace, stakeholders still have a chance to prepare and possibly postpone the onset of the crisis. With regard to previous crises, many countries prepared for a crisis before it occurred by raising readiness and full preparedness and taking all necessary measures, and other organizations took other measures to postpone the crisis or try to mitigate its effects.
3. Crisis Occurrence: When a crisis occurs, the company must launch crisis responses, which are often tactical in character and involve actions, communication and behaviors. The organization may take various actions depending on the sort of crisis. Even though the situation is unpredictable and dynamic, decision makers must stick to logical patterns, which can be challenging when evidence is few or contradictory.
4. Crisis aftermath: Following the crisis, there is a period of time dedicated to repair damage and making up on delayed or disturbed job processes. Extraordinary actions precede the new normality in this era. Recovery and remediation are two of the most important managerial tasks.

5. Post-crisis: The organization is back to business as normal in this phase. However, the situation is no longer the center of management's attention, however it still has to be addressed. Damage repairs and bridging the gaps may continue or begin during this period.

It is not correct to deal with crises with hasty and confused reactions, and accordingly there must be planning, organization and follow-up processes, which lead to carefully studied decisions and ideal results, and here the administration must continuously monitor what is happening around, and follow up on any emergency situation that is expected to occur (Pedersen et al. 2020).

Effective Use of Crisis Management
Appropriate conditions in the organization must be created for the conception, execution, and effective application of the crisis management process. Above all, this need a diverse set of skills and knowledge from numerous fields, as well as suitable infrastructure. These prerequisites, on the other hand, can be met if the management is dedicated to foresee a crisis and is ready to take proactive actions. Top management's job is to aggressively support this choice. The essential knowledge and abilities for coping with crisis situations can then be determined. The interconnection of these factors will have an impact on the successful handling of possible crises (Mikušová and Horváthová 2019).

2.3 Innovation

How do crises create innovation? Crises creates unique conditions that make innovation the best solution to bring about rapid and impactful change. This will open up opportunities for business owners and managers to put forward the most innovative ideas. Organizations have been forced to innovate in limited ways in the past. (Davis et al. 2021).

In-spite-of the fact that the COVID19 problem is incomparable, we may draw on previous crises such as the 2008 financial crisis or natural crisis such as humanitarian disasters to inform our thinking on innovation in crisis situations. Typically, crises tend to have a detrimental impact on total innovation activity in economies, as the COVID-19 crisis is expected to demonstrate. Crises, on the other hand, present opportunities for new entrants to meet new requirements with novel solutions (Ebersberger and Kuckertz 2021).

During crises, there are several changes that may occur to organizations to make positive changes and encourage innovation and creativity (Adam and Alarifi 2021). Obviously, one of the most important challenges facing SMEs is to stimulate performance and encourage innovation to achieve enterprise goals and ensure survival. Employees often increase their motivational energy during crises for fear of losing their job. Hence the role of the administration to guide towards a clear goal to develop radical solutions to the crisis.

2.4 Support and Guidance

In Bahrain, companies find support directly from the government during any unexpected event. Since the beginning of the pandemic, The government in Bahrain has started to support many affected companies from all sectors to ensure survival, continuity and prosperity.

During the COVID-19 pandemic, unemployment support and wage subsidies in most countries have helped maintain jobs or living standards. Cash transfers have been particularly useful in supporting self-employed and those who have lost their jobs in the private sector. Boosting corporate liquidity prevented a wave of defaults and mass layoffs. This is especially important for SMEs that contribute a large share of employment (Brülhart et al. 2020).

2.5 Overcoming Challenges

There is no fixed rule for overcoming the challenges facing SMEs, as this is a broad concept that includes many aspects (Eggers 2020). During the unexpected changes that the world witnessed recently such as economic changes, many companies relied on developing plans and strategies based on innovation and diversification to meet the challenges. One of the biggest challenges facing SMEs is the inability to predict the economic changes that may affect them.

During COVID-19 pandemic, it was critical to recovery to have a post crisis organizational learning experience. SMEs are creative and eager to learn from disasters. These businesses could take part in business development courses and networking events or special lectures to learn from people who have overcome obstacles (Engidaw 2022). The process of crisis assessment helps to understand the dimensions of the crisis as required. For example, their current employment status, sources of funding and support, spending channels and levels of development helps in optimal planning to get out of crises and face challenges efficiently (Mehr and Jahanian 2016).

2.6 Crises in Light of Digital Transformation

Recent crises such as the pandemic led to high prices due to disruption of supply chains in all countries. This trend accelerated global digital transformation, as indicated by greater digital technology used in industry and growth of digital infrastructure (Rha and Lee 2022). Although the pandemic had a severe impact on many enterprises, it has also opened up new business prospects. For example, it has promoted digital entrepreneurship, reflecting shifting customer behavior during and after the epidemic. Digitalizing becoming more common, and even minor modifications can result in significant efficiency advantages. Organizations might connect teams and build closer working ties between headquarters and subsidiaries using more advanced new technology. As the cost of communications, data storage, and gadgets has decreased while their capabilities have expanded tremendously, businesses have acquired new prospects for digitalization (Amoah et al. 2021). To prosper in competitive circumstances, organizations must integrate efficiently. Digital procedures and collaboration technologies are the only ways to accomplish effective integration (Kraus et al. 2021). Organizations seek to strengthen competitiveness in light of emerging challenges in the global market. One of the most important modern software that is used is the Enterprise Resource Planning (ERP) system. ERP systems facilitate decision-making by collecting all company data, and making them available to managers as usable information by providing an integrated software suite to process information requests in the organization (Marsudi and Pambudi 2021).

2.7 Recovery Phase

Governments acknowledge that SMEs will be critical to the rebuilding of the economy following the crisis. In reaction to the epidemic, several of them have already taken action by introducing different stimulus packages and activities (Thukral 2021). Every organization should have a bootstrapping strategy in place not just for the start-up phase, but also throughout the downturn and recovery phases. The subject of how to run a firm sustainably that it can function in whatever economic scenario should be on every entrepreneur's mind. As we have already seen, covid-19 will lead many firms to fail (Arkadiusz 2021).

After the crisis, organizations will attempt to continue the business as usual. In a simplified classification of crisis outcomes, the organization may be unable to restore to its original position, may revert to its original position, or may emerge stronger from the crisis in some way. Different systems, such as organizations, networks, or nations, are likely to influence the result. Systems that deteriorate after a crisis are susceptible. These systemic impacts are also linked to how well-prepared organizations were prior to the crisis and how they responded during the three main crisis periods (Pedersen et al. 2020).

3 Conclusion and Future Work

Considering all of the facts, it is clear that there is a strong relationship between pre-planning for crises and the ability to survive in SMEs. In addition, the performance of SMEs during the crisis determines their viability and success; as when crises occur, events accelerate incredibly, where the smallest impact or problem becomes uncontrollable, and thus the leadership in this unexpected situation undergoes administrative pressure resulting from trying to absorb all the successive effects. Crises are complicated, and their consequences are felt not just instantly but also over time. Innovation has been acknowledged as a major factor for SMEs' organizational resilience in times of crisis. in fact, this was the most impacted sector during most crisis. Innovative activity may help SMEs in mitigating the consequences of crises, and innovation provides a survival advantage. From this perspective, innovation may be viewed as a means of escaping a crisis and as a tool for improving SMEs performance and competitiveness. The combined consequences of innovation are critical to the success and survival of SMEs in global marketplaces. Entrepreneurs in highly competitive SMEs must create relevant innovations to maintain their financial performance and gain a competitive edge (Auken et al. 2021). Until now, research on SMEs in times of crisis has mostly concentrated on macro-perspectives, that is, how crises impacted the economy and companies, the tactics enterprises took in times of crisis, and government policy responses. The influence of crises on entrepreneurial activity and success, the entrepreneurial finance ecosystem, and the impact of government policy responses on small enterprises have all been studied by researchers (Antonarakis et al. 2022). Although academics have begun to investigate how entrepreneurs navigate and function in times of crisis, the majority of research on small enterprises in times of crisis, including that undertaken during the COVID-19 epidemic, has taken an organizational-level approach. Assuredly, the entrepreneur of SMEs may react differently to a crisis, or not at all, depending on how the impacts of the crisis are cognitively understood. Individual and corporate-level reactions may include tangible entrepreneurial efforts and business decisions (Newman et al. 2022). SMEs are affected by crises more and faster, for example, during crises it is easy to decrease the percentage of sales and consequently the profit margin in addition to the shrinking of the customer base (Martínez et al. 2021), which leads to the emergence of difficult challenges, as it was previously mentioned that many companies that may be well-known were also closed during the Covid-19 crisis, while some organizations realized that successive crises had guided them to crisis management, as this term was not very common, or taken seriously.

Finally, the importance of crisis management lies in responding to unexpected situations and getting out of them with the least possible damage, as the ability to optimally communicate during crises is an effective element during a crisis and helps to understand and clarify the situation in an easier way. Moreover, planning before the occurrence of the situation which called crisis management, helps to make an integrated analysis and a holistic view of the expected risk before it occurs, which contributes to reducing the consequences of it when it actually occurs, as these plans are used to manage critical moments and develop exit plans of critical emergencies and crises. In addition, the crisis management helps in training to deal with unexpected risks and ensure the smooth running of work during the crisis, in addition to activating advanced technology to take advantage of modern technologies and social media and use them in effective communication during crises.

Future research may focus on economic crises, and what are the effects of them, moreover, more research can be conducted to investigate these effects and their actual impact on the workforce and jobs availability.

References

Kraus, S., Jones, P., Kailer, N., Weinmann, A., Banegas, N., Tierno, N.: Digital transformation: an overview of the current state of the art of research. SAGE Open (2021). https://doi.org/10.1177/21582440211047576

Adam, N.A., Alarifi, G.: Innovation practices for survival of small and medium enterprises (SMEs) in the COVID-19 times: the role of external support. J. Innov. Entrepreneurship **10**(1), 1–22 (2021). https://doi.org/10.1186/s13731-021-00156-6

Al Qubtan, T., Gan, P., Hadi, A.P., Jalil, A.N., Rambeli, N.: Practical risk management approaches among small and medium enterprises. TEM J. **10**(2), 1004 (2021). https://doi.org/10.18421/TEM102-65

Amoah, J., Khan, Z., Wood, G., Knight, G.: COVID-19 and digitalization: The great acceleration. J. Bus. Res. **136**, 602–611 (2021). https://doi.org/10.1016/j.jbusres.2021.08.011

Antonarakis, A.S., Pacca, L., Antoniades, A.: The effect of financial crises on deforestation: a global and regional panel data analysis. Sustain. Sci. **17**, 1–21 (2022). https://doi.org/10.1007/s11625-021-01086-8

Arkadiusz, A.: How can businesses thrive in the post-COVID-19 world. Am. J. Manag. **21**(2) (2021). https://doi.org/10.33423/ajm.v21i2.4230

Auken, H., Ardakani, M., Carraher, S., Avorgani, R.: Innovation among entrepreneurial SMEs during the COVID-19 crisis in Iran. Small Bus. Int. Rev. **5**(2), e395 (2021). https://doi.org/10.26784/sbir.v5i2.395

Bajaba, A., Bajaba, S., Algarni, M., Basahal, A., Basahel, S.: Adaptive managers as emerging leaders during the COVID-19 crisis. Front. Psychol. **12**, 661628 (2021). https://doi.org/10.3389/fpsyg.2021.661628

Bartik, A., Bertrand, M., Cullen, Z., L. Glaeser, E., Luca, M.: The impact of COVID-19 on small business outcomes and expectations. Nat. Acad. Sci. **117**(30), 17656-17666 (2020). https://doi.org/10.1073/pnas.2006991117

Brülhart, M., Lalive, R., Lehmann, T., Siegenthaler, M.: COVID-19 financial support to small businesses in Switzerland: evaluation and outlook. Swiss J. Econ. Stat. **156**(1), 1–13 (2020). https://doi.org/10.1186/s41937-020-00060-y

Carracedo, P., Puertas, R., Marti, L.: Research lines on the impact of the COVID-19 pandemic on business. A text mining analysis. J. Bus. Res. **132**, 586–593 (2021)

Davis, K., Snider, A., Archibald, T., Grove, B., Babu, S.: Organizational innovation in times of crises: the case of extension and advisory services. J. Int. Agric. Extension Educ. **28**(1) (2021). https://doi.org/10.5191/jiaee.2021.28101

Ebersberger, B., Kuckertz, A.: Hop to it! The impact of organization type on innovation response time to the COVID-19 crisis. J. Bus. Res. **124**, 126–135 (2021). https://doi.org/10.1016/j.jbusres.2020.11.051

Eggers, F.: Masters of disasters? Challenges and opportunities for SMEs in times of crisis. J. Bus. Res. **116**, 199–208 (2020). https://doi.org/10.1016/j.jbusres.2020.05.025

Engidaw, A.E.: Small businesses and their challenges during COVID-19 pandemic in developing countries: in the case of Ethiopia. J. Innov. Entrepreneurship **11**(1), 1–14 (2022). https://doi.org/10.1186/s13731-021-00191-3

Hertati, L., Widiyanti, M., Desfitrina, Syafarudin, A., Safkaur, O.: The effects of economic crisis on business finance. Int. J. Econ. Financ. Issues, **10**(3) 236 (2020). https://doi.org/10.32479/ijefi.9928

Marsudi, A., Pambudi, R.: The effect of enterprise resource planning (ERP) on performance with information technology capability as moderating variable. J. Econ. Bus. Accountancy **24**(1), 1–11 (2021)

Martínez, U., Mediano, J., Rodríguez, A.: The impact of the COVID-19 crisis on consumer purchasing motivation and behavior. Eur. Res. Manag. Bus. Econ. **27**(3), 100166 (2021). https://doi.org/10.1016/j.iedeen.2021.100166

Mehr, M., Jahanian, R.: Crisis management and its process in organization. Mediterr. J. Soc. Sci. **7**(5), 143 (2016). https://doi.org/10.5901/mjss.2016.v7n51p143

Mikušová, M., Horváthová, P.: Prepared for a crisis? Basic elements of crisis management in an organisation. Econ. Res.-Ekonomska Istraživanja **32**(1), 1844–1868 (2019). https://doi.org/10.1080/1331677X.2019.1640625

Moonen, G.: Solidarity in the face of a clear and present danger. Disasters Crisis Manag. (2021)

Newman, A., Obschonka, M., Block, J.: Small businesses and entrepreneurship in times of crises: the renaissance of entrepreneur-focused micro perspectives. Int. Small Bus. J. **40**(2), 119–129 (2022)

Pedersen, C., Ritter, T., Benedetto, C.: Managing through a crisis: managerial implications for business-to-business firms. Ind. Mark. Manag. **88**, 314–322 (2020). https://doi.org/10.1016/j.indmarman.2020.05.034

Prasanna, R., Jayasundara, J., Gamage, S., Ekanayake, E., Rajapakshe, P., Abeyrathne, G.: Sustainability of SMEs in the competition: a systemic review on technological challenges and SME performance. J. Open Innovation: Technol. Market Complex. **5**(4), 100 (2019). https://doi.org/10.3390/joitmc5040100

Rha, J., Lee, H.-H.: Research trends in digital transformation in the service sector: a review based on network text analysis. Serv. Bus. **16**(1), 77–98 (2022). https://doi.org/10.1007/s11628-022-00481-0

Thukral, E.: COVID-19: Small and medium enterprises challenges and responses with creativity, innovation, and entrepreneurship. Strateg. Chang. **30**, 153–158 (2021). https://doi.org/10.1002/jsc.2399

Trachsler, T., Jong, W.: Crisis management in times of COVID-19: Game, set or match? J Contingencies Crisis Manag. **28**(4), 485–486 (2020). https://doi.org/10.1111/1468-5973.12306

Impact of FinTech on the Sustainable Development of Bahrain During Covid-19 Pandemic

Isa Abdulla[1], Latifa Khaled[1], Khaled Mohd[1], Allam Hamdan[2(✉)], and Hatem Akeel[3]

[1] College of Business and Finance, Manama, Bahrain
[2] Ahlia University, Manama, Bahrain
allamh3@hotmail.com
[3] College of Business Administration, University of Business and Technology, Jeddah 21448, Kingdom of Saudi Arabia

Abstract. Fintech involves advanced technology-based solutions for the customers while not being dependent on the banking system and rather focuses on the other virtual currency options. The emergence of these methods especially cryptocurrencies and blockchains after the financial crisis of 2008–9 is more important for the sustainability and development of a country since Fintech is also an advancement in mediums of money flow. The present research has explored the effect of Fintech on sustainable development (economic, society, and environment) while focusing on the existing literature. The previous literature and theories represent that there is a significant influence of Fintech on sustainable development however, the influence differs in different countries and situations. Based on this, the study has identified that there has been no research on the role of the Covid 19 pandemic among Fintech and sustainable development of Bahrain and suggested quantitative research for the future.

Keywords: Fintech · Cryptocurrency · Blockchain · Sustainable development · Economic · Society · Environment · Bahrain

1 Introduction

Covid 19 pandemic has emerged as a result of a recent virus that emerged from Wuhan China and got transmitted to the whole world from one person to another while creating significant problems for the people [1]. Due to the emergence of the Covid 19 pandemic, people were no longer able to go out of their houses (the virus was transmitting and spreading through physical contact). Due to this, the majority of countries eventually led the whole world to shift toward lockdown as well as curfews in several areas across the globe [2]. The virus led to disruption in the lives of people as well as in the economic conditions of the country where measurement of all these consequences was considered to be hard due to unexpected changes caused by the happenings due to Covid 19 pandemic [3].

B. Alareeni et al. (Eds.): ICBT 2022, LNNS 620, pp. 193–202, 2023.
https://doi.org/10.1007/978-3-031-26953-0_20

In such a situation, there has been a great level of uncertainty in the environment as well as quite uneven influences, where the recovery of the countries from all of these uneven influences is presently a question mark. When it comes to the economic influences, the banks along with the financial institutions have been influenced by the epidemic situation to great extent [4]. Since the banking sector is termed as the backbone of the economy of a country, the uneven influences on the banking sector due to any epidemic situation are critical. The banking sector helps the economy of a country by ensuring the flow of money in the country while providing the residents of a country with credit as well as managing the money and business markets' sustainability and ensuring greater liquidity level in the country [3].

Since there have emerged several economic as well as political challenges to the countries to ensure the survival as well as the development of all the sectors which influence the successful survival of individuals [5]. Due to this reason, the sustainable development of countries is turning out to be more crucial while emerging as a most important strategic choice for the countries to ensure the successful survival of their people [6]. Where, sustainable development is defined as bringing innovative solutions/development which competitive advantage help in meeting all the desired needs and wants without putting the ability of future generation at stake to meet their needs and demands [7].

The global financial crisis emerged in 2008–9 as a result of the problems in the banking system at a global level which eventually not just influenced a single bank or a company but the stock exchanges and companies at a global level while becoming the reason for a financial emergency [7]. However, the financial crisis led to the emergence of innovative integration-based technologies including the emergence of Fintech (cryptocurrencies & blockchain) [8]. Nevertheless, the Fintech innovation is something different from the other innovative solutions which have emerged so far as Fintech is a far deeper innovative solution (cryptocurrency; blockchain; digital advisory & trading systems; digital payment systems & much more) [9].

Presently, previous researchers have explored the role of Fintech in the economies of several countries including the previous researches of Malmendier [10]; Fuster et al. [11]; Beck et al. [12]; Popescu and Popescu [13]; Haddad and Hornuf [14] and many other researchers. These researchers in their studies represented that there is a significant role of Fintech in the sustainable development of different countries [7]. However, the problem has emerged with the arrival of the Covid 19 pandemic which has led to no more physical contact among people whiles stressing more on staying at home rather than going out of their homes without all the required precautionary measures. Since Fintech included different payment and money exchange methods available to the people in such a pandemic situation, the effect on sustainable development can differ compared to that of the influence in other situations as represented in the study of Deng et al. [7] that Covid 19 pandemic has uneven results to great extent.

Bahrain is a country that is considered the financial hub of GCC economies and has proficiency in the banking and financial services sector [15]. During Covid 19 pandemic, in order to meet the desires and needs of people in a difficult time, the banks of Bahrain focused on switching to remote work in order to continue meeting the changing needs and demands of people. Moreover, Bahrain is working on ensuring the needs and demands of its people while switching from its only dependence on natural resources towards development in other sectors in order to ensure a boost in its economy. Since Covid 19 pandemic has collapsed the smooth working of economies as well as the lives of common people at a global level, there is a need to understand the role of the Covid 19 pandemic among the influence of Fintech on the sustainable development of Bahrain.

The present research has identified the need of conducting research on finding the impact of Fintech on the sustainable development of Bahrain during the Covid 19 pandemic since the Covid 19 pandemic emerged as the biggest problem after the financial crisis of 2008–9 which led to the emergence of Fintech. Since the great financial crisis led to Fintech's emergence, the role of an unfavorable situation is crucial and needs to be explored in order to find the outcomes or the role of uneven influences or uncertainties. While considering the role of the Covid 19 pandemic in Fintech and sustainable development, the present study can analyze when there are unexpected situations, how can different financial technological methods be utilized, and which not in order to bring improvement in the sustainable development of a country. Moreover, the present research can be useful in finding the disparity in the previous research results where some studies represent significant while the rest represents no significant results as discussed in the Sect. 2-literature review.

2 Literature Review

The modern monetary theory by Reynolds [17] focused on Fintech while illustrating that electronic money as well as the monetary policy on the overall economic condition of a country. The theory illustrates that the economic condition of the country is gets deteriorated due to the changes in the flow of money in the country as well as the mediums available for the flow of money in the country's economy. As per the modern monetary theory, the economic condition and the sustainability of a country are influenced to great extent due to the changes in mediums available for the flow of money. Similarly, another theory named resource-based view theory by Wernerfelt [18] represented that when the companies go for availing and utilizing efficient resources, there are increasing chances of companies achieving as well as maintaining a comparative edge in the market. As per the research of Melville et al. [19], it is stated that technological resources are also termed tangible assets whereas the competency of the companies is considered an intangible resource. The Management of technology theory by Roger et al. [20] illustrates that when the technology is created by the individual, modified, and improved for the purpose of adoption, it helps in achieving a competitive edge in the market. This eventually creates competition in the market and leads to more innovative solutions which eventually leads to development in the country [16].

The sustainability development theories also play a key role in the sustainable development of a country while incorporating the economic, social as well as environmental levels [21]. There are reported to be several drivers of the businesses in the Fintech businesses as well which involves the technology as well as the businesses and the flow of money in the country as Fintech is involved in improving the ability of the companies to meet the financial needs of the companies in a different manner [16]. The business drivers lead to Fintech dimensions which might include the development of Fintech, improvement in the Fintech, or the application of technology in Fintech for the purpose of improving the quality of the services being provided to the clients [16]. This eventually results in the sustainable development of the economy while bringing improvements at the social level, economic level as well as environmental level [16]. These theories represent that Fintech leads to improvement in the sustainable development of the economy of a country.

2.1 Stimulus Organism Response Theory

Stimulus organism response theory was developed by Mehrabian and Russell [22] which illustrates that every single reaction is based on a specific situation where the respondent provides certain feedback or reaction to certain stimuli or external action. The organism is the inner state of the respondent whereas the response is the reaction that the respondent gives in that situation. When there occurs any situation or change in the environment, influences the overall working of a system while bringing changes in the behavior and reaction of respondents within that specific situation [22].

Considering the above-mentioned theories, the present study focuses on exploring how the change in the overall environment, society, and economy due to Covid 19 pandemic has influenced the effect of Fintech on sustainable development when there were strict barriers to physical interactions in Bahrain.

2.2 Emergence of Fintech

2.2.1 Actor-Based Evolutionary Approach Towards Fintech

The emergence & development of Fintech has been classified into 3 different stages named Fintech 1.0; 2.0; and 3.0 by Amer, Barberis & Buckle [23] as shown in the figure below.

The very first stage of Fintech was from the tenure 1866 till 1987 since it's not a new concept and rather in practice since 1866 [24]. In all this evolutionary tenure of Fintech, there were several physical tele-communicatory infrastructures laid down at a global level while facing several problems and challenges at a global level due to the difficulties faced in installing the transatlantic cables. The first stage of Fintech 1.0 played a key role in the development of the banking system while increasing the connection among the banks and other financial institutions at the global level [23]. The integrated system that emerged as a result of Fintech 1.0 is presently even being practiced by the banks at a global level in order to provide more valid and reliable services to clients. If innovation would not be done, the outcomes would not have been present in the form of innovative Fintech solutions [24].

The second stage of Fintech named Fintech 2.0 emerged in 1987 and ended in 2008 due to the arrival of a financial crisis. However, this overall tenure enhanced the working of the financial sector while improving the ability of the banks to provide more reliable services to clients. In this time of century, there was significant digitalization in the banking sector while bringing advanced and innovative IT infrastructures which helped in improving the quality of services provided to the clients [16]. The emergence of ATMs as well as several new payment methods have been the result of the digitalization of financing 2.0 in all this tenure. Along with that, in all this time there were several other innovations including the stock exchanges as well as the central clearing houses or the international spread of the banking system. Based on this, several banking regulatory systems were also generated which gave way to modern banking which is currently being used by several banks at the global level [24].

The present stage of Fintech named Fintech 3.0 is the presently ongoing technological advancement of the financial industry at a global level which evolved after the financial crisis of 2008–9. As per the study of Amer et al. [23], there was more than 12 billion US dollars investment done in the startups of the Fintech 3.0 as of 2014 when the Fintech 2.0-based institutions also spent more than 197 US billion dollars in order to ensure their competitive position in the market [23]. The revolutionary model represents that Fintech has changed to a great extent in all these three stages while bringing significant results in the banking systems as well as the workings of the financial institutions. Moreover, the recent development after the 2008–9 financial crisis brought totally different techniques and technologies including cryptocurrency and blockchain-based payment methods which have changed the financing industry to a significant level [24].

2.2.2 Resource-Based Approach towards Fintech

The evolutionary approach can be diversified on the basis of logic stating that the Fintech-based financial sector is influenced more by the value-added design rather than the technological origin [24]. There are three layers in which Fintech has evolved over time and has supported sustainable development in the countries as shown in the Table 1 below.

Table 1. Evolutionary layers of Fintech [24]

Evolutionary layers of Fintech	Key drivers of the evolutionary Fintech over time
Bottom or first layer – development of the Ecosystem	The emergence of cheap mobiles along with the access to internet The emergence of cheap system hardware and software The emergence of a global level telecommunication-related infrastructure
Middle or the second layer – pioneering services	The emergence of highly rapid scalable services Highly innovative approaches for the purpose of providing reliable services The emergence of diverse business models for innovation
Top or third layer – human-focused design	Consumers' needs and demands with changing times Utilization of data analytics Better quality of user experience to clients Experimental approach for bringing innovative solutions

The very first innovative layer of Fintech focused on the development of the infrastructure for the purpose of developing a base for development in the technological setup which can help in improving the quality of services to the client [16]. There was the emergence of various mobile phones as well as the system hardware as well as software for the purpose of providing the clients with advanced technologies which can make their lives more reliable and efficient. The hardware incorporated computers as well as mobiles and tablets along with laptops which made the entry into the market and ensured the attention of clients easier [24]. There were several telecommunications-related setups installed in various cities and countries at a global level in order to ensure the interaction of people in the virtual or digital space easier. Along with this, Fintech helped in improving the banking systems while providing them with digitalized payment systems while replacing the previously existing systems, and ensuring the lives of clients of banks are more flexible and reliable. However, the lack of agility, as well as the lack of quick-to-market processes, led to a lack of efficient response from the banking sector [24].

The second layer focused on the development of pioneering services which involved the development of innovative solutions while incorporating creativity-based methods and approaches. All the business models designed in the second layer were focused on describing how to create and provide reliable and creative value to the clients while improving the performance of the businesses [25]. This layer also focused on the development of innovative solutions including cheap-priced mobiles and computers which have enabled the interaction among people easier and more reliable on the basis of internet services being provided to the clients [7]. The Fintech services led to an increase in the digital services being available to clients at a global level, especially to adults since the banking system started providing digitalized services to clients compared to the regular services previously being provided. In the financial sector, there was the introduction of credit cards as well as fund transfers, saving accounts, international remittances, and several advanced loan methods available to the clients as a result o the second layer of Fintech innovation [24].

The third layer of Fintech innovation focused on the development of advanced and innovative solutions while being primarily focused on the clients. Here the focus of the Fintech creators is on the development of the toolset as well as the experimental frameworks which can help the consumers in order to provide innovative solutions to businesses [7]. There were several Fintech innovation solutions that emerged in the Fintech third layer including the new financial products as well as services that were more focused on non-banks. The non-bank products and services emerged as a result of the decrease in the trust of the clients in the banking system due to the financial crisis of 2008–9 incorporating the lack of transparency as well as great misconduct within the banks. This was the prime reason for the decrease in the trust of clients in the banks and they started looking at the other factors and options available [7]. Cryptocurrency and blockchain-based financial services emerged as a result of the lack of trust in the banking system. Similarly, several online or mobile payment systems also emerged as a result of the lack of trust in the banking systems [24].

2.3 Fintech and Sustainable Development

Previous research by Deng et al. [7] focused on Fintech and sustainable development while gathering information in China where the study focused primarily on the peer-to-peer companies from the 31 provinces of China. The study results represented that there is a U-shaped connection between Fintech and the sustainable development of China, especially in the eastern and central regions whereas there is no significant connection between Fintech and sustainable development in the western region of China. The study represented that increasing Fintech opens the path for sustainable development in the economic as well as social and environmental circumstances of China especially in the central region and somehow in the eastern region [7].

Research by Legowo et al. [16] focused on the sustainable development in a country on the basis of the emergence of Fintech in the financial as well as the banking industry of a country while considering the mix-method approach. They focused on Indonesia in order to find that either the emergence of Fintech has an adverse influence on the sustainable development of the country or it has paved the path to the development at the social, economic, and environmental levels. The study represented that there is a statistically significant role of Fintech in the sustainable development of the banking sector of Indonesia as well as the financial services sector [16].

Zhang et al. [26] focused on the role of Fintech in the sustainable development of China during the age of digitalization while focusing on the environmental development incorporating land and forest restoration in China. The study represented that the ant forest Fintech activity of Alibaba is playing a key role in the forest and land restoration as well as development in China while bringing a significant reduction in carbon emission in the region as well as the reduction in poverty in China. Moreover, there have been several other environmental influences including the changes in the health of the people of China which was also termed environmental improvement [26].

The research was done by Xu and Xu [27] which focused on the utilization of Fintech for the purpose of bringing sustainable development in the People's Republic of China while working on the financial risks generated by the rise of Fintech. For the purpose of reduction of the risk, there have been several applications in Fintech incorporating the improvement in peer-to-peer lending as well as the payment through the third party, regulations for the blockchain-based currencies incorporating cryptocurrencies, and much more. It was represented in the research that Fintech is playing a sufficiently great role in improving the sustainable development of China where all the advanced Fintech methods with proper regulations are playing a key role while reducing all the possible risks involved in the process of achieving sustainable development through Fintech [27].

Research by Al Hammadi and Nobanee [28] focused on Fintech and sustainability in order to find the role of the emergence of Fintech and innovative solutions on the sustainability of the country while gathering information from the previously existing literature. The study focused on the 9 previous research articles which focused on the Fintech and sustainability of the country. The study exploration represented that the emergence of innovation in Fintech for the purpose of sustainable money flow has a significant influence on the sustainability of the countries. The study represented that improving the Fintech innovations leads to improvement in the performance of companies that incorporate Fintech while helping them in gaining an edge in the market.

3 Conclusions

3.1 Summary of Thesis

The previous research exploration can be summarized as increasing the Fintech innovations leads to improvement in the sustainable development of a country. The modern monetary theory, as well as the resource-based view along with management of technology and sustainability development theory, illustrates that the need for technological advancements as well as the financial technological acts involving creation or improvement leads to improvement in the development of a country especially on the level of the economy as the well social and environmental situation of a country. The results of previous research were in line with these results representing that Fintech plays a key role in the sustainable development of a country. However, the results of some studies found different results in different areas or situations especially the study of Deng et al. [7] which represented that influences differed in the different regions.

3.2 Conclusion

The present research was focused on studying the connection between Fintech and the sustainable development of a country while exploring the previously existing literature on the connection between Fintech and sustainable development. The present study found that Fintech has emerged in three layers in 3 centuries from the 18th century till today, where, the biggest changes have emerged after the financial crisis of 2008–9 when the trust and dependence of people on the banking and other financial sectors reduced due to lack of transparency. On the basis of previously existing literature as well as the theories, it can be stated that increasing Fintech application in the countries can help in ensuring the sustainable development of a country in the social, economic, and environmental sectors. Along with this, the previous research suggests that an unexpected situation like Covid 19 pandemic has brought uneven influences. Based on this, the study concludes that quantitative research can be very useful in finding how the changes in Fintech during the Covid 19 pandemic brought changes in the sustainable development of a country, especially Bahrain.

3.3 Implications

The present study has found from the previous theories and literature that there is a statistical connection between Fintech and sustainable development where increasing Fintech activities brings an increase in sustainable development. However, the influences have differed also based on which the study implies that more focused research rather than generalized is required in order to explore the influence of Fintech on sustainable development in Bahrain. Moreover, the study implies that an uncertain situation also plays a key role in influencing the impact of one variable on another.

3.4 Limitations of the Study

The present research was limited in several aspects as discussed below:

i. The present research focused on Fintech and sustainable development on a specific aspect while gathering information from the previously existing research in order to gain insight into the connection among two variables to identify the gap in previous research. Due to this reason the present research only gathered the previously existing research on Fintech and sustainable development.
ii. The present research has focused on Fintech in a generalized way while incorporating all the types of Fintech rather than focusing on a specific form of financial technology.

3.5 Suggestions for Future Research

Based on the analyzed literature and existing research along with theories, the present research suggests the following directions for future research considered as a gap in previous research:

i. Research can be done to a narrow extent rather than generalizing the influences in a different situation on all the situations.
ii. Research can be done on the quantitative-based approach in order to find how Fintech influences sustainable development.
iii. Research can be done while selecting one Fintech and finding its effect on the sustainable development of a country.
iv. Research can also be done on the comparison of different Fintech's influence on sustainable development while doing a comparison in different regions or countries.

References

1. Hellewell, J., et al.: Feasibility of controlling COVID-19 outbreaks by isolation of cases and contacts. Lancet Glob. Health **8**(4), e488–e496 (2020)
2. Xu, S., Li, Y.: Beware of the second wave of COVID-19. Lancet **395**(10233), 1321–1322 (2020)
3. Rabbani, M.R., et al.: Exploring the role of Islamic FinTech in combating the aftershocks of COVID-19: the open social innovation of the Islamic financial system. J. Open Innov. Technol. Mark. Complex. **7**(2), 136 (2021)
4. Mooney Jr, C.W.: Fintech and secured transactions systems of the future. Law Contemp. Probs. **81**, 1 (2018)
5. Caiado, R., Nascimento, D., Quelhas, O., Tortorella, G., Rangel, L.: Towards sustainability through green, lean and six sigma integration at service industry: review and framework. Technol. Econ. Dev. Econ. **24**(4), 1659–1678 (2018)
6. Deng, X., Cheng, X.: Can ESG indices improve the enterprises' stock market performance?—An empirical study from China. Sustainability **11**(17), 4765 (2019)
7. Deng, X., Huang, Z., Cheng, X.: FinTech and sustainable development: evidence from China based on P2P data. Sustainability **11**(22), 6434 (2019)
8. Shim, Y., Shin, D.H.: Analyzing China's Fintech industry from the perspective of actor–network theory. Telecommun. Policy **40**(2–3), 168–181 (2016)
9. Schindler, J.W.: FinTech and financial innovation: drivers and depth (2017)
10. Malmendier, U.: Law and Finance at the origin. J. Econ. Lit. **47**(4), 1076–1108 (2009)

11. Fuster, A., Plosser, M., Schnabl, P., Vickery, J.: The role of technology in mortgage lending. Rev. Financ. Stud. **32**(5), 1854–1899 (2019)
12. Beck, T., Chen, T., Lin, C., Song, F.M.: Financial innovation: the bright and the dark sides. J. Bank. Financ **72**, 28–51 (2016)
13. Popescu, C.R.G., Popescu, G.N.: An exploratory study based on a questionnaire concerning green and sustainable finance, corporate social responsibility, and performance: evidence from the Romanian business environment. J. Risk Financ. Manag. **12**(4), 162 (2019)
14. Haddad, C., Hornuf, L.: The emergence of the global Fintech market: economic and technological determinants. Small Bus. Econ. **53**(1), 81–105 (2019)
15. Al Shehab, N., Hamdan, A.: Artificial intelligence and women empowerment in Bahrain. In: Hamdan, A., Hassanien, A.E., Khamis, R., Alareeni, B., Razzaque, A., Awwad, B. (eds.) Applications of Artificial Intelligence in Business, Education and Healthcare. Studies in Computational Intelligence, vol. 954, pp. 101–121. Springer, Cham (2021). https://doi.org/10.1007/978-3-030-72080-3_6
16. Legowo, M.B., Subanidja, S., Sorongan, F.A.: Model of sustainable development based on FinTech in financial and banking industry: a mixed-method research. In: 2020 3rd International Conference on Computer and Informatics Engineering (IC2IE), pp. 194–199. IEEE, September 2020
17. Reynolds, G.: Electronic money and monetary policy. In: Southwestern Economic Proceedings, pp. 47–55 (2004)
18. Wernerfelt, B.: A resource-based view of the firm. Strateg. Manag. J. **5**(2), 171–180 (1984)
19. Melville, N., Kraemer, K., Gurbaxani, V.: Information technology and organizational performance: an integrative model of IT business value. MIS Q. 283–322 (2004)
20. Rogers, E.M., Singhal, A., Quinlan, M.M.: Diffusion of Innovations. In: An Integrated Approach to Communication Theory and Research, pp. 432–448. Routledge, Abingdon-on-Thames (2014)
21. Arowoshegbe, A.O., Emmanuel, U., Gina, A.: Sustainability and triple bottom line: an overview of two interrelated concepts. Igbinedion Univ. J. Account. **2**(16), 88–126 (2016)
22. Mehrabian, A., Russell, J.A.: An Approach to Environmental Psychology. The MIT Press, Cambridge (1974)
23. Arner, D.W., Barberis, J., Buckley, R.P.: The evolution of Fintech: a new post-crisis paradigm. Geo. J. Int'l L. **47**, 1271 (2015)
24. Varga, D.: Fintech: supporting sustainable development by disrupting finance. Bp. Manag. Rev. **8**(11), 231–249 (2018)
25. Boons, F., Wagner, M.: Assessing the relationship between economic and ecological performance: distinguishing system levels and the role of innovation. Ecol. Econ. **68**(7), 1908–1914 (2009)
26. Zhang, Y., et al.: The contribution of Fintech to sustainable development in the digital age: ant forest and land restoration in China. Land Use Policy **103**, 105306 (2021)
27. Xu, Z., Xu, R.: Regulating Fintech for sustainable development in the People's Republic of China (2019)
28. Al Hammadi, T., Nobanee, H.: Fintech and sustainability: a mini-review (2019). SSRN 3500873

Artificial Intelligence AI, TechManagement, Entrepreneurship and Development

Gender Divergence on Entrepreneurial Proclivity – An Empirical Analysis of Polytechnic Diploma Holders

T. K. Murugesan[1], Madhu Druva Kumar[1], K. P. Jaheer Mukthar[1(✉)], Guillermo Pelaez-Diaz[2], Julián Pérez-Falcón[2], and Jorge Castillo-Picon[2]

[1] Kristu Jayanti College, Autonomous, Bengaluru, India
jaheer@kristujayanti.com
[2] Universidad Nacional Santiago Antúnez de Mayolo, Huaraz, Peru

Abstract. This study primary aims to throw a light on the entrepreneurial proclivity of polytechnic diploma holders on the basis of gender divergence to derive the demand-driven actions for imbibing an entrepreneurial eco-system and culture among polytechnic diploma holders in select higher educational institutions in Tamilnadu. This empirical study also investigates the substantial obstacles faced by the study respondents on the basis of gender alone that might prevent them from intent of becoming the entrepreneurs. This study was effectively administered on 374 polytechnic diploma holders judgementally sampled and drawn on the basis of purposive-cum-area sampling technique. The outcome of this empirical study clearly revealed that male polytechnic diploma holders have demonstrated higher proclivity for entrepreneurship than their female counterparts. Furthermore, the male diploma holders considered the lack of capital, the difficulty in accessing capital for the new start-up idea and the dearth of break-through support from Governments as higher obstracles to embrace the notion of the entrepreneurship, whereas female diploma holders regarded the lack of encouragement and support from the family, the lack of capital and the lack of knowledge about how to kick start a venture as higher obstracles towards becoming future entrepreneurs.

Keywords: Entrepreneurial proclivity · Entrepreneurship image · Gender divergence · Obstacles · Polytechnic diploma holders

1 Introduction

Today, Indian economy is emerging as one of the leading economies in the world as the Honorable Prime Minister of India has launched national level campaign called "Make in India" to encourage entrepreneurs to make the innovative products in India. In India, the progress of entrepreneurial culture has become the national agenda in both UPA and NDA Governments. The pressing issues of unemployment and the assertiveness of the current graduates who are totally dependent on the private, public and government organizations for their employment were envisaged to be a major concern in the developing countries

B. Alareeni et al. (Eds.): ICBT 2022, LNNS 620, pp. 205–219, 2023.
https://doi.org/10.1007/978-3-031-26953-0_21

like India. This has paved the clear-cut way for modern researchers to reconnoiter whether the students fraternity are more inclined about becoming entrepreneurs.

Generally, entrepreneurship would contribute to the nation's economic growth by encouraging innovation, stimulating competition, generating employment and thus contribute to the nation's economic wealth and spending power (Holmgren and From 2015). Entrepreneurship was observed to be a dynamic and vibrant process whereby entrepreneurs can discover, evaluate and exploit the business opportunities in and around world to create the start-ups to meet the future demands of the country (Shane and Venkataraman 2010). Entrepreneurs must have the access to the indispensable resources in order to exploit the business opportunities and transform their intents into the actions of creating new start-ups.

In the today's world, entrepreneurship awareness must be created and promoted among Indian universities and affiliated college students so that the entrepreneurial challengers have more choices and alternative options upon their graduation. As the notion of entrepreneurship has been acknowledged as the potential vehicle and prospective incubator for the technological innovation, product innovation, and the market novelty and development (Mueller and Thomas 2012), the researchers believed these benefits would help augment nation's economic advancement as well.Over a couple of periods, the research on entrepreneurship took extensively by the research scholars in the pursuit of research and education, particularly at educational institution level (Rushing 2014). Today, the higher education can play a dynamic role in generating the number of employed graduates in numerous countries, which seek to the entrepreneurs and new business start-ups as a genuine and profitable career option (Nabi and Holden 2008). Generally, the educators and academicians have a clear intent to better concoct their students for a constantly mutable market by encompassing entrepreneurship education apart from imparting core subjects in polytechnic educational institutions (Shinnar et al. 2009).

Entrepreneurship has apprehended considerable attention of both policy makers and academic scholars during a couple of the previous decades. Moreover, one of the key reasons of this apprehension is the emergent need for the entrepreneurs to accelerate and speed-up the economic growth via spawning new innovative ideas and transfiguring them into profitable and sustainable business ventures. Thus, entrepreneurial undertakings are not only the start-ups of commercial and technological innovation; they also offer huge employment opportunity and enhance great amount of competitiveness within the national players also global players (Reynolds 1998 and Zahra 1999).Entrepreneurs are generally called as the engine of the nation's economic growth and development. They support the nation's economic growth and societal development by bringing massive positive contributions in the business world. Among those, the most imperative contributions are found to be the innovation and the job creation. The entrepreneurial intention should be investigated on the incorporation of various insights from the psychological and behavioural approach (Karimi et al. 2014).

Training programs on entrepreneurship are initiated to prepare and educate the students fraternity everywhere in the world towards the entrepreneurship. Having assessed the impact of education and training on their entrepreneurial tendency of the participants, we will apply the Planned Behaviour Theory (PBT), originally propound by Azjen

(1991). The theory of planned behaviour advocates that the human social behaviour is planned, controlled, organized and reasoned in the manner that it takes into consideration the prospective consequences of the desired behaviour (Ajzen and Fishbein 2016). The underlying model has been applied for the prediction of many types of human behaviours. This model offers a vibrant framework to analyze how the education and training programmes can effectively influence its abettors regarding their entrepreneurial & risk-taking behaviour.

The entrepreneurship education and training can effectively contribute to the progress of entrepreneurial traits, skills and qualities for new corporate start-up (Kolvereid and Moen 1997). Developing entrepreneurial traits, skills and qualities are highly recognized that the students can acquire the real-time business experience (Gibb 1996). The resultant outcomes of these studies are highly consistent with the conventional theories of learning, where it is evidently implicit that the real-time learning and conceptual learning on the entrepreneurship enables the students fraternity to explore into the new business world.

As per the study conducted by Nabi and Holden (2008), the primary intention of the entrepreneurship education and training is to foster the entrepreneurial spirits and cultures in the minds of the students community, which clearly defines the reciprocal interaction between the graduate students as the product of the higher educational institutes and their keenness to pursue their professional career as the future entrepreneurs. Today, the most of the graduates were eyeing for better employment in the government organizations and private corporates even after they underwent a couple of courses on entrepreneurship and ventureship. Some of the graduates irrespective of the gender would like to end-up with entrepreneurs because the policy makers have indicated that the entrepreneurs are considered to be the central engine of the economic growth of the country.

2 Literature Review

Over a past few periods, education and research on the ventureship and entrepreneurship has been constantly growing (Alstete 2012; Klapper 2014; Gurol and Atsan 2016). Moreover, entrepreneurial spirit has been substantially growing due to the significance of the entrepreneurship in driving dynamic economic development and growth of both developed and developing nations (Gormanet al. 1997). In few decades, there has been substantial and mounting interest in entrepreneurs and entrepreneurship at both domestic and global levels because it symbolizes innovation and a dynamic economy (Klapper 2014). In the global era, the term entrepreneurship has received a snowballing response from India and other countries in light of creating and generating new business start-ups and small-, medium- and large-scale ventures for the nation's overall economic progress and societal development (Acs et al. 2015). Not only the notion of the entrepreneurship promotes the nation's overall economic progress and employment generation, but it is gradually recognized as backbone strength of the nation's overall well-being (Acs 2006). The significant benefits of the entrepreneurship are highly leveraged by the corporate sectors in and around abroad. Nevertheless, entrepreneurial potentials, traits, talents and skills among the fraternity of graduate students based on the gender exclusively remained intact in many contexts of entrepreneurship and ventureship (Audretsch 2012).

The reassurance of the 'entrepreneurial spirit' among graduating students was labelled as a pre-condition for success in employment generation, economic growth, market competitiveness and business innovation (European Communities Commission 2006). The entrepreneurship education is deemed to be one of the vibrant paradigms of the socio-economic sciences and it has created a great deal of interest not only in the scientific and academic community but also in the political sphere in the couple of few decades. Thus, the entrepreneurship education was considered to be the highest priority on the state and central political agenda and currently a driving force for the countries worldwide to be survival (Mitra and Matlay 2004).

An in-depth examination of literature reviews comprehensively discloses a quite amount of the empirical research studies in the last few decades, which have thrown a lime light on the entrepreneurial proclivity. Nevertheless, most of these empirical research studies were administrated in developed countries (Koh 2010; Wang and Wong 2004; Veciana Aponte and Urbano 2005). There was confined research on the area of the entrepreneurial proclivity of college students from developing nations like India. This research study made an attempt to close this gap by offering some meaningful insights into entrepreneurial intention of the student's fraternity in the developing country like India. In the last few years, an extensive investigation of literature reviews by the researchers indicates that the quite amount of empirical and pragmatic research studies done on the area of entrepreneurial prolivity. However, most of the empirical studies were conducted in the developed countries (Peterman and Kennedy 2003; Guerrero et al. 2008).

Veciana et al. (2005) carried out a widespread empirical study of entrepreneurial proclivity of the university students in Puerto Rico (435 university students) and Catalan (837 university students). Their studies have thrown a lime light on the antecedents that are closely associated with the intention of becoming successful entrepreneurs and setting up new ventures. For the survey conducted by Puerto Rico, nearly 90% of graduating students exhibited high level of appetite of setting up a new venture. For the survey conducted by Catalan, approximately 74% of the graduating students exhibited a high level of entrepreneurial proclivity to set up a firm. This result corroborated the findings from other studies in Catalan where more than 70% of the students showed high entrepreneurial proclivity (Guerrero et al. 2008).

With respect to the rational relationship between the demographic factors of the sample respondents and entrepreneurial proclivity, it is evident from the outcome of the studies that the entrepreneurial proclivity of the sample respondents significantly differs among the demographic factors. The demographic factor "Gender" was found to have a positive correlation with entrepreneurial appetite in most of the studies and the male students have tended to exhibit the higher proclivity levels than the female counterparts (Veciana et al. 2005). However, a pragmatic study of the engineering and technical students was conducted in Russia with a sample size of 512 and the outcome of this study was not consistent with the previous research findings. Here, gender was found not to have a positive correlation with the entrepreneurial tendency (Tkachev and Kolvereid, 1999). Nevertheless, the female students' entrepreneurial tendency was found to be higher in another study administered in Spain (Guerrero et al. 2008).

There is a wide-ranging unanimity as to the roles played by current education system of the nation in overall development of entrepreneurial cultures (Lundstrom and Stevenson 2002). There have been high deliberations and arguments that the training and education for the entrepreneurship amount students' community should originate as early as possible (Collins and Moore 1994). In most of the studies on entrepreneurship, the stages of infancy & adolescence were often acknowledged as the congenial and amicable periods for developing positive cultures and attitudes with regard to entrepreneurship and the procurement of basic knowledge on the theme of entrepreneurship (Parker 2014).

The demographic physiognomies such as those relating to the age group, the area of residence, the gender, and the educational background, can be applied as a differenciate factors to explore the budding or prospective entrepreneurs on the basis of the psychological traits. Conversely, majority of these research variables were found to have no influence or little influence on the person's predisposition and successtability for entrepreneurship, nor can they be applied as the underlying predictors of such the lifestyle choice or the professionl career (Hatten and Ruhland 1995).

With regard to cogent relationship between the demographic physiognomies and entrepreneurial traits and behaviours, the study results have been varied and inconclusive. The entrepreneurial traits were significantly varied among the sample respondents based upon the demographic characteritis. With respect to the the categorical variable "Gender", the male students have exhibited extra entrepreneiral traits and behaviours in terms of becoming entrepreneirs than female students (Kolvereid and Moen 1997; Wang and Wong 2004; Veciana et al. 2005).

3 Research Gap

The empirical studies on entrepreneurship carried out in a couple of recent years clearly revealed that entrepreneurs have substantial and significant roles in economic progress of the country. Today, a study on entrepreneurship has become a topic of town extensively by many academic researchers due to its prominence to the progress of an economy by way of job creation and wealth creation. The major chunk of recent research studies on the entrepreneurship primarily threw a light on the tendency of the students toward becoming entrepreneurs in the developed and developing nations. There is a narrow research on proclivity of the students for entrepreneurship on the bases of gender alone in developing and developed nations. This paper is exclusively designed to minimize this research gap by providing meaningful insights and acumens into the proclivity of polytechnic diploma holders for entrepreneurship on the basis of their gender divergence.

4 Research Objectives

To explore the proclivity of polytechnic diploma holders for entrepreneurship on the basis of gender divergence, the researchers have framed the following two broad research objectives:

a) To highlight the underlying antecedents and image attributes associated with the proclivity of the male and female polytechnic diploma holders for entrepreneurship.
b) To explore the stumbling blocks that might prevent polytechnic diploma holders on the basis of gender from the intention of becoming successful entrepreneurs.

5 Research Methodology and Design

The methodology of this research study entails concise research framework, hypotheses formation, research instrument & reliability and sampling frame & design.

5.1 Research Framework

As stated earlier, the crux of this research is to analyze the target group's divergence on underlying antecedents and image attributes that are associated with the entrepreneurship. These antecedents are required to derive demand-oriented eco-system for imbibing the entrepreneurial cultures in higher educational institutions. This study also explores the degree to which the obstacles have become ubiquitous on the basis of gender from the intention of becoming entrepreneurs. In order to realize the concrete objective of this study, the concise research framework was effectively designed by the researchers as presented in Fig. 1. The research framework depicted below is a modest linear research model that designates significant relationships to test the proposed hypotheses of this study.

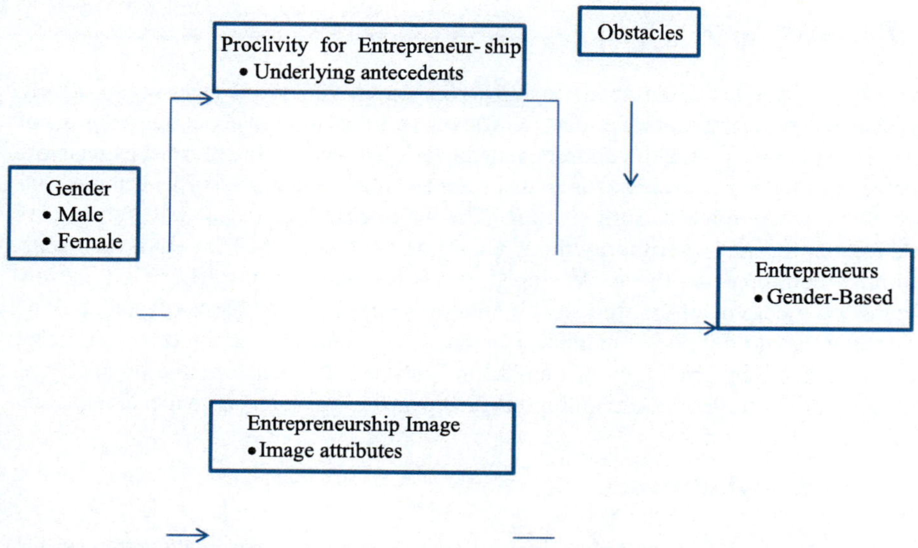

Fig. 1. Gender divergence on entrepreneurial proclivity, entrepreneurship image & obstacles to become successful entrepreneurs

5.2 Hypotheses Formulation

In this study, the following hypotheses were framed to find out whether there is any significant divergence in the polytechnic diploma holders on the basis of gender from the intention of becoming entrepreneurs.

- H1: The degree to which the polytechnic diploma holders on the basis of gender differ in their likelihood of embracing the notion of entrepreneurship.
- H2: To analyze for the significant difference in the extent of attributes associated with entrepreneurship image on the basis of gender divergence.
- H3: polytechnic diploma holders by and large differ in their perspectives on a set of obstacles that prevent them from the propensity of becoming entrepreneurs.

5.3 Research Instrument and Reliability

The research instrument used in this study was primarily developed on the basis of new measurement scales underpinning the proclivity of polytechnic diploma holders for entrepreneurship because the researchers were not able to figure out any past studies directly or indirectly addressing gender-based issues in determining the entrepreneurial proclivity of the students. However, and wherever it is possible to validate the scales, the researchers have applied validated measures properly that have been earlier applied by other scholars. The validity and reliability of the underlying research constructs and scale measurement items applied in the survey instrument were effectively tested and well verified through the pilot survey and the Cronbach's Alpha.

6 Data Analysis, Results and Discussions

The data analysis, survey results and managerial discussions of the study are summarized in the following section.

6.1 Demographic Profile of Study Respondents

The first and foremost facet of this study is to explore the demographic characteristics of the sample respondents taken for the survey. With regard to gender, more than half of the sample respondents (57.90%) are male and 42.10% are female. With respect to the type of polytechnic diploma courses undergone by the sample respondents, 64 (17.11%) are pursuing diploma in electronics and communication engineering, 58 (15.51%) are studying diploma in civil engineering, 55 (14.71%) are undergoing diploma in computer science and engineering, 54 (14.44%) are doing diploma in mechanical engineering, 51 (13.64) are pursuing diploma in electrical engineering, 45 (12.03%) are pursuing diploma in petroleum engineering, 31 (8.29%) are doing diploma in fashion engineering and remaining 16 (4.28%) are studying other polytechnic diploma courses like textile, areopace, mining & biotehcnology.

With regard to the number of training programs on entrepreneurship undergone, more than half of the sample respondents (50.10%) have attended more than 10 training & education programs, 28.40% have attended 5 – 10 training & education programs, and remaining 21.50% have attended less than 5 training & education programs. It was also found from the survey that more than three-fourth of the sample respondents (75.10%) don't have entrepreneurial family background and nearly quarter of the sample respondents (24.90%) have entrepreneurial family background.

6.2 Independent Sample T-Test on Entrepreneurial Proclivity on the Basis of Gender

The independent sample t-test was conducted to test the first hypothesis (H_1) of this study. This hypothesis was framed to identify whether the mean responses of the polytechnic diploma holders on various antecedents associated with entrepreneurial proclivity differ in terms of their gender and the results of this analysis were summarized in the Table 1.

It was evident from the outcome of the Table 1 that the p-value for the mean responses of the male and female polytechnic diploma holders on various antecedents of entrepreneurship were found to be highly significant at 5% ($p < 0.05$), the assumed level of significance. Hence, we have enough evidence to accept the alternative hypothesis. These results also suggest that there was a significant difference in the responses of the both male and female polytechnic diploma holders on various antecedents related to entrepreneurship.

It was also observed from the Table 1 that the mean scores of male polytechnic diploma holders on various triggers of entrepreneurship were found to be high as compared to that of female counterparts taken for the survey. These results also inferred that the male polytechnic diploma holders are more inclined to become entrepreneurs than the female polytechnic diploma holders.

Table 1. Independent sample t-test results of gender of polytechnic diploma holders and their proclivity on entrepreneurship

Sl. No.	Entrepreneurial Proclivity (EP)	Gender	N	Mean	SD	df	F-value	t-value	p-value	Result
1	I have a tendency to become an entrepreneur	Male	431	3.95	1.039	742	24.132	9.694	0.000*	Sig.
		Female	313	3.13	1.275					
2	My professional object is to become an entrepreneur in my field of study	Male	431	4.00	0.986	742	72.466	8.542	0.000*	Sig.
		Female	313	3.27	1.325					

(*continued*)

Table 1. (*continued*)

Sl. No.	Entrepreneurial Proclivity (EP)	Gender	N	Mean	SD	df	F-value	t-value	p-value	Result
3	I have a desire to be an entrepreneur rather than an employee in the company	Male	431	4.04	0.977	742	60.596	5.603	0.000*	Sig.
		Female	313	3.59	1.243					
4	I am willing to take a high risk to become an entrepreneur	Male	431	4.01	0.926	742	36.923	8.981	0.000*	Sig.
		Female	313	3.32	1.174					
5	I prefer to put every effort and time to kick-start my own business	Male	431	4.11	1.128	742	55.953	8.717	0.000*	Sig.
		Female	313	3.26	1.535					
6	I am completely determined to start new venture in the future	Male	431	3.99	0.991	742	56.084	9.005	0.000*	Sig.
		Female	313	3.23	1.303					
7	I consider entrepreneurship as a highly desirable career option for diploma holders	Male	431	4.06	0.924	742	45.632	10.406	0.000*	Sig.
		Female	313	3.21	1.308					
8	I always think of entrepreneurship as my career choice	Male	431	4.10	0.941	742	54.705	6.144	0.000*	Sig.
		Female	313	3.61	1.217					

* Significance at 5% ($p < 0.05$), SD = Standard Deviation, SEM = Standard Error Mean & df = degree of freedom

6.3 Chi-Square Test of a Set of Attributes Associated with Entrepreneurship Image on the Basis of Gender

The second hypothesis (H_2) of this empirical research also focused on the degree of agreement the sample polytechnic diploma holders have shown on a set of attributes associated with entrepreneurship image on the basis of their gender. A Chi-square test was conducted to determine whether the observed means of the underlying attributes associated entrepreneurship image on the basis of gender of the sample respondents are significantly different or not. The resultant outcomes of this test were summarized in Table 2. According to Table 2, the p-value for all the attributes associated with entrepreneurial image was found to be highly significant at 5% level of significance ($p < 0.05$) with regard to the gender of polytechnic college. Therefore, there was a sufficient evidence to accept the alternative hypothesis (H_a) for all the attributes on entrepreneurship image on the basis of gender. Hence, we can conclude that the mean responses of the sample respondents for all the attributes on entrepreneurship image significantly very at 5% ($p < 0.05$) with regard to the gender of the polytechnic diploma holders. It also suggests that there was a significant difference in the perceptions of male and female polytechnic diploma holders on the entrepreneurial image. It is obvious from the descriptive statistics illustrated in the Table 2 that male polytechnic diploma holders have shown stronger image with entrepreneurship than their female counterparts. The Table 2 gives the descriptive statistics, Pearson Chi-square values, p-values, degree of freedom, significance level and 95% confidence interval for the mean.

6.4 One-Sample T-Test for Obstacles on the Basis of Gender Towards Becoming Entrepreneurs

A one-sample t-test was applied to test third hypothesis (H_3) of this study and find whether the observed means of the obstacles from becoming entrepreneurs were significantly different on the basis of gender from the mid-value 3.0 as indicated in the Table 3. The resultant outcomes of this test were categorically presented in Table 3. As seen from the Table 3, the ensuing results were observed to be significantly different from the mid-value 3.0 at 5% level of significance ($p < 0.05$). Having described the entrepreneurial proclivities of the polytechnic diploma holders, the next part of the analysis involved the indispensable obstacles that might put off the polytechnic diploma holders to become entrepreneurs, which is the focus of this survey. Table 3 shows a summary of the mean score of each resisting obstacle that prevents male and female polytechnic diploma holders to become entrepreneurs. As can be seen, the mean score ranges from 11.571 to 33.232 for male and 15.111 to 32.127 for female, which is obviously higher than the mid-point value 3.0.

Table 2. Descriptive statistics and Pearson Chi-Square test results for the independence of the attributes associated with entrepreneurship image by the gender of polytechnic diploma holders

S. No.	Attributes of entrepreneurship image	N	Gender					
			μMale	μFemale	χ2	df	p-value	Sig.
1	Entrepreneurship is about job creation	744	4.56	3.10	10.084	2	0.008	Sig.
2	Entrepreneurship is a prospect for higher income	744	4.55	3.29	15.406	2	0.003	Sig.
3	Entrepreneurs are job providers rather than job seekers	744	4.16	3.58	19.292	2	0.000	Sig.

(continued)

Table 2. (*continued*)

S. No.	Attributes of entrepreneurship image	N	Gender					
			μMale	μFemale	χ2	df	p-value	Sig.
4	Entrepreneurship is an honourable & respectable profession	744	4.55	3.28	16.989	2	0.001	Sig.
5	I respect people who are entrepreneurs	744	4.54	3.31	15.843	2	0.002	Sig.
6	I always admire those who succeed in their own business	744	4.53	3.24	18.996	2	0.000	Sig.
7	Entrepreneurs have the scope to achieve social status easily	744	4.59	3.24	15.342	2	0.003	Sig.
8	Entrepreneurs have a social image in the society	744	4.20	3.63	14.573	2	0.004	Sig.

* Significant at 5% ($p < 0.05$), ** Not Significant at 5% ($p > 0.05$), Pearson Chi-square (χ^2) & df = degree of freedom

Table 3. One-sample test for obstacles perceived by male and female towards becoming entrepreneurs

Obstacles	Male				Female			
	T	df	Sig. (2-tailed)	Mean Difference	T	df	Sig. (2-tailed)	Mean difference
1. Lack of capital	33.232	743	0.000*	0.617	31.991	743	0.000*	1.050
2. Lack of innovative business ideas	12.427	743	0.000*	0.786	22.573	743	0.000*	0.931
3. Difficulty in accessing capital for new start-up idea	26.025	743	0.000*	0.642	25.765	743	0.000*	0.882
4. Lack of managerial & business skills	19.931	743	0.000*	0.702	22.451	743	0.000*	0.780

(*continued*)

Table 3. (*continued*)

Obstacles	Male				Female			
	T	df	Sig. (2-tailed)	Mean Difference	T	df	Sig. (2-tailed)	Mean difference
5. Lack of knowledge about how to kick start a venture	19.818	743	0.000*	0.539	26.911	743	0.000*	1.019
6. Lack of training and education	17.342	743	0.000*	0.672	22.683	743	0.000*	0.591
7. Fear of starting a venture due to high risk	11.571	743	0.000*	0.368	17.482	743	0.000*	0.571
8. Lack of exposure on economic & market environment	16.629	743	0.000*	0.473	20.855	743	0.000*	0.352
9. Dearth of break-through support from Governments	24.528	743	0.000*	0.599	15.111	743	0.000*	0.816
10. Lack of encouragement and support from the family	22.527	743	0.000*	0.664	32.127	743	0.000*	0.788

* Significance at 5% ($p < 0.05$)

Of these 10 tenacious obstacles, the most significant stumbling barrier that might prevent the polytechnic diploma holders from becoming entrepreneurs was found to be 'Lack of capital' perceived by male with a highest mean score of 33.232 and 'Lack of encouragement and support from the family' perceived by female with a highest negative mean score of 32.127. Linked to this, the other significant barriers that might prevent male polytechnic diploma holders from becoming entrepreneurs found at 5% level of significance were 'Difficulty in accessing capital for new start-up idea' (mean score = 26.025), 'Dearth of break-through support from Governments' (mean score = 24.528), 'Lack of encouragement and support from the family' (mean score = 22.527), 'Lack of knowledge about how to start a venture' (mean score = 19.818), 'Lack of managerial & business skills' (mean score = 19.931), 'Lack of training and education' (mean score = 17.342), 'Lack of exposure on economic & market environment' (mean score = 16.629), 'Lack of innovative business ideas' (mean score = 12.427), and 'Fear of starting a venture due to high risk' (mean score = 11.571).

In contrast, the other significant barriers that might prevent female polytechnic diploma holders from becoming entrepreneurs found at 5% level of significance were 'Lack of capital' (mean score = 31.991), 'Lack of knowledge about how to start a venture' (mean score = 26.911), 'Difficulty in accessing capital for new start-up idea' (mean score = 25.765), 'Lack of training and education' (mean score = 22.683), 'Lack of innovative business ideas' (mean score = 22.573), 'Lack of managerial & business skills' (mean score = 22.451), 'Lack of exposure on economic & market environment' (mean score = 20.855), 'Fear of starting a venture due to high risk' (mean score = 17.482), and 'Dearth of break-through support from Governments' (mean score = 15.111).

7 Conclusion and Managerial Implications

The rationale of this study is to present a detailed empirical investigation of the underlying antecedents and image attributes that are closely oriented with entrepreneurship on the basis of gender divergence of polytechnic diploma holders. It was concluded from the study that there was a higher level of concurrence found for male polytechnic diploma holders on most of the underlying antecedents and image attributes triggered with the intention of becoming entrepreneurs than female counterparts. The study also clearly revealed that the stronger entrepreneurial spirit has been embedded in the minds of male polytechnic diploma holders because of the economic necessity and unemployment conditions persistent in the state. In contrast, the female polytechnic diploma holders acknowledged getting suitable corporate jobs as more important than venturing into an entrepreneurial career.

The empirical study has come to conclusion that there was a significant difference in the perspectives of the male and female polytechnic diploma holders on various underlying antecedents and image attributes triggered with the notion of entrepreneurship. Furthermore, the male students regarded lack of finance, difficulty in obtaining finance for the business idea and lack of support from Government as highest barriers to embrace entrepreneurship, whereas female students regarded lack of support and encouragement from family, lack of finance and lack of knowledge about how to start a venture as highest barriers towards becoming entrepreneurs. Altogether, both student groups should be imparted as an interdisciplinary approach on the specific knowledge of new start-ups as well as entrepreneurial skills during the tenure of their studies. It was also evident from the study that entrepreneurial appetites were commonly found among the polytechnic diploma holders and this has policy implications for both Government and educational institution to design appropriate policies and programs for imbibing entrepreneurial spirit and culture among the students fraternity.

References

Acs, Z.: How is entrepreneurship good for economic growth? Innovations **2**, 97–107 (2006)

Acs, Z., Arenius, P., Hay, M., Minniti, M.: Global Entrepreneurship Monitor – Exclusive Report. London Business School, London and Babson College, Babson Park (2015)

Ajzen, I.: The theory of planned behavior. Organ. Behav. Hum. Decis. Process. **50**, 179–211 (1991)
Ajzen, I., Fishbein, M.: Attitudes and the attitude-behavior relation: reasoned and automatic processes. Eur. Rev. Soc. Psychol. **4**, 28–29 (2016)
Alstete, J.: On becoming an entrepreneur: an evolving typology. Int. J. Entrep. Behav. Res. **8**, 222–234 (2012)
Audretsch, D.: Entrepreneurship research. Manag. Decis. **5**, 755–764 (2012)
Collins, O.F., Moore, D.G.: The Enterprising Man. Michigan State University Press, East Lansang (1994)
Commission of the European Communities. Communication from the commission to the council, the European Parliament, The European Economic and Social Committee and the Committee of the regions, implementing the Community Lisbon Programme: Fostering Entrepreneurial Mindsets Through Education and Learning. COM, Brussels (2006)
Gibb, A.A.: Entrepreneurship and small business management: can we afford to neglect them in the twenty-first century business school? Br. J. Manag. **7**, 309–324 (1996)
Gorman, G., Hanlon, D., King, W.: Some research perspectives on entrepreneurship education. Enterprise education and education for small business management: a ten-year literature review. Int. Small Bus. J. **15**(3), 56–77 (1997)
Guerrero, M., Rialp, J., Urbano, D.: The impact of desirability and feasibility on entrepreneurial intentions: a structural equation model. Int. Entrep. Manag. J. **4**(1), 35–50 (2008)
Gurol, Y., Atsan, N.: Entrepreneurial characteristics amongst university students: some insights for entrepreneurship education and training in Turkey. Educ. + Train. **48**(1), 25–38 (2016)
Hatten, T.S., Ruhland, S.K.: Student attitude toward entrepreneurship as affected by participation in an SBI program. J. Educ. Bus. **70**(4), 224–227 (1995)
Holmgren, C., From, J.: Taylorism of the mind: entrepreneurship education from a perspective of educational research. Eur. Educ. Res. J. **4**(3), 382–390 (2015)
Karimi, S., Biemans, H.J.A., Lans, T., Chizari, M., Mulder, M.: The impact of entrepreneurship education: a study of Iranian students' entrepreneurial intentions and opportunity identification. J. Small Bus. Manage. **6**(3), 23–28 (2014)
Klapper, R.: Government goals and entrepreneurship education—an investigation at Grande Ecole in France. Educ. Train. **46**(3), 127–137 (2014)
Koh, H.C.: Testing hypotheses of entrepreneurial characteristics: a study of Hong Kong MBA students. J. Manag. Psychol. **11**(3), 12–25 (2010)
Kolvereid, L., Moen, O.: Entrepreneurship among business graduates: does a major in entrepreneurship make a difference? J. Eur. Ind. Train. **21**(4), 23–31 (1997)
Kothari, H.C.: Impact of Contextual Factors on Entrepreneurial Intention. Int. J. Eng. Manag. **3**(2), 76–82 (2013)
Lundstrom, A., Stevenson, L.: On the road to entrepreneurship policy. Swedish Foundation for Small Business Research: Stockholm (2002)
Mitra, J., Matlay, H.: Entrepreneurial and vocational education and training: lessons from eastern and central Europe. Ind. High. Educ. **18**(1), 53–69 (2004)
Mueller, S.L., Thomas, A.S.: Culture and entrepreneurial potential: a nine country study of locus of control and innovativeness. J. Bus. Ventur. **16**(4), 51–75 (2012)
Nabi, G., Holden, R.: Graduate entrepreneurship: intentions, education and training. Educ. + Train. **50**(7), 545–551 (2008)
Parker, S.C.: The Economics of Self-employment and Entrepreneurship. Cambridge University Press, Cambridge (2014)
Peterman, N., Kennedy, J.: Enterprise education: influencing students' perceptions of entrepreneurship. Entrep. Theory Pract. **28**(2), 129–144 (2003)
Reynolds, P.D.: New firms societal contribution versus survival potential. J. Bus. Ventur. **2**(1), 231–246 (1998)

Rushing, F.W.: Entrepreneurship and education. In: Kent, C.A. (ed.) Entrepreneurship Education: Current Developments. Future Directions & Westport CT: Quarum Books, pp. 29–40 (2014)
Shane, S., Venkataraman, S.: The promise of entrepreneurship as a field of research. Acad. Manag. Rev. **25**(4), 217–226 (2010)
Shinnar, R., Pruett, M., Toney, B.: Entrepreneurship education: attitudes across campus. J. Educ. Bus. **84**(3), 151–159 (2009)
Tkachev, A., Kolvereid, L.: Self-employment intentions among Russian students. Entrep. Reg. Dev. **3**, 269–280 (1999)
Veciana, J.M., Aponte, M., Urbano, D.: University students' attitudes toward entrepreneurship: a two countries comparison. Int. Entrep. Manag. J. **4**(1), 165–182 (2005)
Wang, C.K., Wong, P.K.: Entrepreneurial interest of university students in Singapore. Technovation **24**(4), 161–172 (2004)
Zahra, S.: The changing rules of global competitiveness in the 21st century. Acad. Manag. Execut. **13**(1), 36–42 (1999)

Mediating Role of Business Tactics on the Relationship Between Entrepreneurial Resilience and Business Survival – A Study Across Micro Entrepreneurs in Bangalore

CH. Madhavi Latha[1](✉), Jaspreet Kaur[2], Gokilavani S[1], and Vanlalhlimpuii[1]

[1] Department of Professional Accounting and Finance, Kristujayanti College, Bengaluru 560077, India
{madhavi,gokilavani}@kristujayanti.com
[2] Department of Management, Kristujayanti College, Bengaluru 560077, India
jaspreetkaur@kristujayanti.com

Abstract. "The Small and Dynamic" Small micro enterprises are mostly labour intensive, more change susceptible and highly influenced by socio-economic conditions. Many studies have discussed the antecedents of microentreprenuers successful survival. Most of the research is based on theories, strategies and personal traits.But these studies did not aim at the combined influence of business tactics and personal traits on business survival. The present study aims at studying the mediating effect of business tactics on entrepreneurial resilience and business survival. The study evidenced that business tactics have mediating effect on the relationship of Microentreprenuers resilience and Business survival. For mediation analysis Andrew hayes process model 4 was used.

Keywords: Entrepreneurial resilience · Business tactics · Business survival · Micro entrepreneurs

1 Introduction

Micro enterprise is mostly carried by the owner itself. We can say it's a one man show. The owner he himself will carry on all the activities of the business by himself. The scope of operation of micro enterprises is generally limited to local and regional demands when compared to large business units. Micro enterprises will have less gestation period after which the return on investment starts. Studies revealed that 70% of micro enterprise businesses are unsuccessful after two years. Micro enterprises are mostly labour intensive with small capital. But at the same time with minimum capital lot of business opportunities would be generated compared to large enterprises i.e., nearly 20 times more than the big enterprises. They are open to new business models, new products and strategies too. Small enterprises get small finances from various sources posing challenges to the performance of small entrepreneurs. Considering the challenges that microenterprenuers face in the 21st century, it is vital to analyze and understand the

B. Alareeni et al. (Eds.): ICBT 2022, LNNS 620, pp. 220–229, 2023.
https://doi.org/10.1007/978-3-031-26953-0_22

factors contributing for the business survival of micro entrepreneurs. Various studies have revealed that both business strategies and personal traits have a profound impact on the success of the micro enterprises. But there is a necessity to understand how these influence individually, together and what mediates what. Thus there exists a necessity to study the mediating effect of business tactics on Entrepreneurial resilience and business survival.

2 Review of Literature

Various studies have been undertaken to understand the influencing factors of microenterprenuers success. Jamak (2014) through his intensive review of literature identified some enablers which are positive influencers for the business survival and some disablers known as negative influencers which need to be overcome by the microenterprenuers in order to sustain. The positive influencers are clear objectives, Human capital, entrepreneur readiness, technical skills, marketing skills and financial literacy. According to his study, the negative influencers are lack of business network, lack of finance, tremendous competition and lack of managerial skills. Microentreprenuers if they overcome the disablers and adopt enablers, then survival rate of microentreprenuers will increase beyond critical point of life cycle. Perceived customer service and experience are the factors which contribute to the success of the micro entrepreneurs. Bureaucratic hurdles and environmental uncertainties are the hurdles for the growth of microentreprenuers Hussain (2010).Microenterprenuers need to do four things in order to cope up with the changing environment.They need to perform external environment analysis, internal environment analysis, need to have a proper plan and should have a good network Gosenpud (2011).Green practices pay a way to new business models and lot of avenues are ahead for the micro enterprenuers to succeed in their business ventures Yaacob (2010). The five types of skills namely leadership skills, communication skills, human relation skills, technical skills and inborn aptitude, have been identified from previous literatures and which have relationship with the success of the microenterprenuers.Personal characters, decision making, geographical location intrinsic characters, education, experience are all the factors which contribute to the success of the micro enterprenuers Chatterjee (2016). Entrepreneurial success and individual entrepreneurial spirit can be measured by accumulating and evaluating the activities conducted in the course of setting up a business Davidsson and Honig (2003). Depending on which human resource is considered important, successful entrepreneurs analyze the relationship between human capital and different factors of success (Unger, Andreas, Michael, and Nina, 2011).Some individuals have inherent instinct characters which will make them successful enterprenuers whereas these characters cannot be adopted Farmer (2011).In a family run business, the influence of family experience, knowledge passed on from generation to generation will have a profound influence on the success of the microentreprenuer business (Karpak and Topcu, 2010).From the intensive literature it is evident that both business tactics and entreprenurial resilience contribute for the survival of microenterprenuers.

Objectives

- To understand the importance of Business tactics and Entrepreneurial resilience for Business Survival.
- To study the mediating role of Business tactics on the relationship between Entrepreneurial resilience and Business survival.

Hypothesis
H(1): Business Tactics mediates the relationship between Entrepreneurial resilience and Business survival.

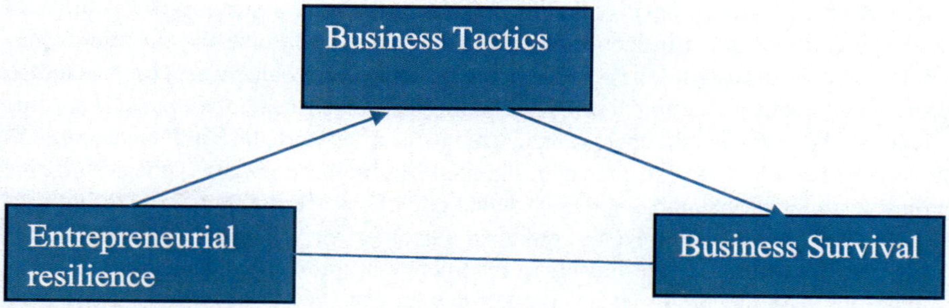

Research Methodology: A descriptive study is undertaken to understand the mediating role of Business tactics on Entrepreneurial resilience and Business Survival. A convenience sample of 120 microentreprenuers are considered in the city of Bangalore for the study. Primary data is collected from the entrepreneurs through repeated casual interviews. Data is analysed using SPSS software. For mediation analysis Andrew hayes process model 4 was used.

Data Collection Tool
The questionnaire was prepared through extensive review of literature and consulting the subject experts in the field, the questionnaire was subjected to testing the scale validity and reliability and the results are as follows (Table 1):

Table 1. Data collection tool- scale validity and reliability

Source				Reliability and validity
Demographic variable	5	Multiple choice	Nominal scale	
Business Tactics	5	Likert scale	Ordinal scale	CA = 0.812/CR = 0.875/AVE = 0.621

(*continued*)

Table 1. (*continued*)

Source				Reliability and validity
Demographic variable	5	Multiple choice	Nominal scale	
Entrepreneurial Resilience	5	Likert scale	Ordinal Scale	CA = 0.897/CR = 0.899/AVE = 0.613
Business survival	5	Likert scale	Ordinal Scale	CA = 0.836/CR = 0.911/AVE = 0.533

All statistics are within the acceptable range CA- Cronbach Alpha should be > 0.700, AVE >0.50 AND CR – Composite reliability > 0.90. Therefore, the questionnaire is considered to adhere to all criteria of scale validity and reliability.

Sample Design: Keeping in mind the Cochran formula of an unknown population with a ten percent margin of error and ninety-five percent confidence, a sample size of one hundred people who participated in the survey would be perfect for the investigation. The number of Micro business owners in Bangalore city is unknown. However, Glen d. Isreal recommended that an extra 30% may be included to account for odd answers. As a result, 130 questionnaires were circulated, and after removing responses that were determined to be outliers, 120 legitimate responses were selected for the study. The sample was selected by the use of the convenience sampling method in each of the city of Bangalore's four distinct zones: north, south, east, and anekal.

Plan of Analysis – Simple frequency and percentage age analysis and descriptive statistics was carried out through SPSS Software, the scale validity and reliability was tested using the same software. For mediation analysis Andrew hayes process model 4 was used (Fig. 1).

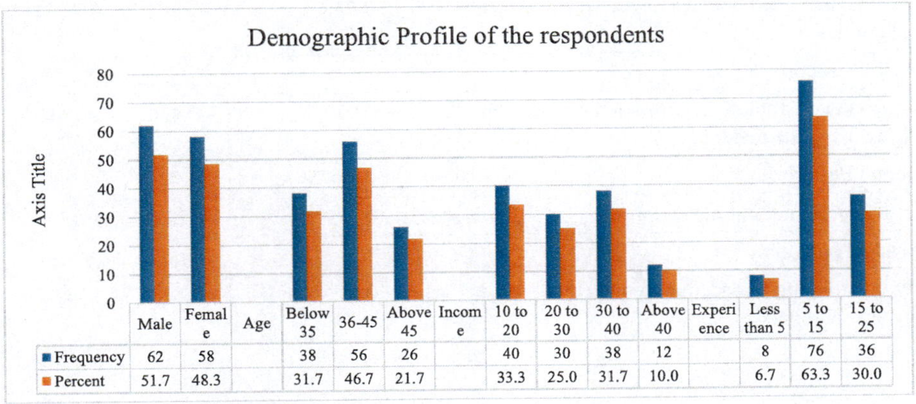

Fig. 1. Demographic variables of the respondents

3 Discussion and Results

3.1 Demographic Profile of the Respondents

In the current study, there are [n = 62] 51.7% male micro entrepreneurs and [n = 58] 48.3% female micro entrepreneurs, according to the demographic profile of the respondents. [n = 38] 31.7% are under the age of 35, [n = 56] 46.7% are between the ages of 36 and 45, and 21.7% are above 45. When asked about their monthly income, [n = 40] 33.3% earned between Rs 10,000 and Rs 20,000, and 25% earned between Rs 20,000 and Rs30,000. A total of 41.7% of respondents earned more than Rs 30,000 per month from their present job. The micro entrepreneurs had a lot of experience, with 63.3% having between 5 and 15 years of experience and 30% having more than 15 years of experience. Babysitters, Bakers, Beauticians, Cab and auto drivers, carpenters, Dairy Products sellers, electricians, Flower vendors, Food Mess owners, fruit vendors, gardeners, grocery Shop owners, Medical Shop owners, milk vendor, Petty Shop owners, Snacks makers, and vegetable vendors are just a few of the micro entrepreneurs (Tables 2 and 3).

Descriptive Statistics

Table 2. Descriptive statistics for Entrepreneurial resilience

Entrepreneurial Resilience	Mean	Std. Deviation	Skewness	Kurtosis
I could continue my business/services during covid period	4.60	0.492	−0.413	−1.860
Determination				
I am patient. No matter how long it is going to take. I am going to do it	4.20	0.751	−0.348	−1.151
Eventual Experience				
I learned from the failures of others	3.98	0.809	−0.356	−0.516
Positivism				
I believe that I can achieve anything with hope and confidence	4.27	0.658	−0.344	−0.731
Societal Support				
My friends and relatives supported me in all the ways during the pandemic	3.85	0.932	−0.456	−0.613
Mental Equilibrium				
Criticism did not put me down	4.25	0.677	−0.682	0.709
Physical condition				
My inner strength made me to overcome physical ailments	4.12	0.611	−0.515	1.549

Table 3. Descriptive statistics for Business tactics.

Business Tactics E-Technology	Mean	Std. Deviation	Skewness	Kurtosis
I learned about various modes of digital payment during pandemic	4.25	0.748	– 0.690	– 0.068
CRM				
Prompt/convenient home delivery or services helped me to increase my business	4.07	0.796	– 0.324	– 0.812
Hygiene				
All the time I maintained hygienic ambience to my customers	4.32	0.594	– 0.232	– 0.609
Innovative Offers				
I offered competitive prices to retain customers	4.07	0.985	– 1.208	1.475
Best Quality				
I offered best quality products/services at reasonable prices	4.47	0.501	0.135	-2.016

Through extensive review of literature, the researcher has identified 5 statement in each of the twelve variables which measures business survival. 7 measures - Survival Parameters, Determination, Eventual experience,Positivism, Societal support,Mental equilibrium and Physicalcondition are factors influencing Entrepreneurial resilience and 5 factors influencing business tactics are – E-technology, CRM, hygiene, innovative offers and best quality. The Micro entrepreneurs were given the questions in form of likert scale in which 1denotes – strong disagreement and 5 denotes strong agreement. The results of the descriptive statistics arranged in its variable form.

The mean scores for all constructs is above 4.00 which indicate agreement for these 12 parameters to enhance business survival in opinion of Micro entrepreneurs.

When it comes to standard deviation, it is a measure which shows how far or how near the responses of the respondents are to its mean. In the current study, all 12 constructs measuring business survival have standard deviation below 1.00 indicating that majority of respondents have agreed to this statement.

The skewness is the measure of how the responses are distributed and Kurtosis measures the shape of the present curve in comparison to the normal distribution. As per (Hair and et al. 2007) the accepted range of Skewness is − 1 to + 1 and kurtosis is − 3 to + 3. Negative skewness indicates that more responses are arranged towards the right. In addition, positive skewness indicates responses arranged towards the left. In the current study, items the skewness values are Positive, fall within the acceptable limit, and tailed towards the left indicating that more responses are towards higher raking. The Kurtosis is also within the adequate limits indicating nearness to the Normal Distribution.

Testing of Hypothesis

Hypothesis – Business Tactics mediates the relationship between Entreprenuerial resilience and Business survival.

Model : 4						
Y : Business Survival						
X : Entrepreneurial Resilience						
M : Business tactics						
Sample						
Size: 120						
**						
OUTCOME VARIABLE:						
Business tactics						
Model Summary						
R	R-sq	MSE	F	dfl	df2	p
.4480	.2007	.8526	7.0300	1.0000	28.0000	.0130
Model						

	coeff	se	t	p	LLCI	ULCI
constant	1.1610	.4498	2.5812	.0154	.2396	2.0824
Entrepreneurial Resilience	.8729	.3292	2.6514	.0130	.1985	1.5473

****

OUTCOME VARIABLE:

Business survival

Model Summary

R	R-sq	MSE	F	df1	df2	p
.6389	.4082	.6202	9.3134	2.0000	27.0000	.0008

Model

	coeff	se	t	p	LLCI	ULCI
constant	.1649	.4268	.3863	.7023	-.7110	1.0407
Entrepreneurial Resilience	.1860	.3141	.5923	.5586	-.4584	.8305
Business tactics	.5733	.1612	3.5568	.0014	.2426	.9040

************************* Σ EFFECT MODEL ****************************

OUTCOME VARIABLE:

Business survival

Model Summary

R	R-sq	MSE	F	df1	df2	p
.3619	.1310	.8783	4.2203	1.0000	28.0000	.0494

Model

	coeff	se	t	p	LLCI	ULCI
constant	.8305	.4565	1.8192	.0796	-.1047	1.7657
Entrepreneurial Resilience	.6864	.3341	2.0543	.0494	.0020	1.3709

************** Σ, DIRECT, AND INDIRECT EFFECTS OF X ON Y **************

The Σ effect of X on Y

Effect	se	t	p	LLCI	ULCI	c_ps	c_cs
.6864	.3341	2.0543	.0494	.0020	1.3709	.6949	.3619

Direct effect of X on Y

Effect	se	t	p	LLCI	ULCI	c'_ps	c'_cs
.1860	.3141	.5923	.5586	-.4584	.8305	.1883	.0981

(a) Relationship between Entrepreneurial Resilience and Business Survival is significant at t(28) = 2.6514, p = .0000, p = 0.0130, the lower-level confidence interval

.1985 and higher-level confidence interval is 1.5473 does not move through 0. Therefore, this relationship is significant. The co-efficient is .8729 indicating that Entrepreneurial Resilience leads to 87.29% positive change in Business tactics.

(b) Relationship between Business tactics and business survival is significant at t(28) = 0.3.5568, p = .000, p = 0.014, the lower-level confidence interval .2426 and higher-level confidence interval is .9040 does not move through 0. Therefore, this relationship is significant. The co-efficient is .5733 indicating that Entrepreneurial Resilience leads to 57.33% positive change in Business survival (Fig. 2).

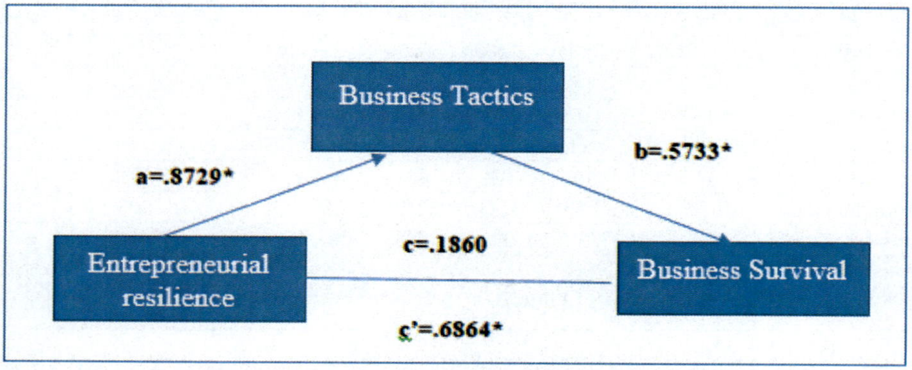

Fig. 2. Business Tactics has a mediating effect on the relationship between entrepreneurial resilience and Business survival

(C) Direct effect of X on Y Relationship between Entrepreneurial Resilience and Business survival is insignificant at t(28) = .1860, p = .0000, p = 0.5586, the lower level confidence interval – 0.4584 and higher-level confidence interval is .8305 moves through 0. Therefore, this relationship is insignificant.

(C1) Σ Effects of X on Y Relationship between Entrepreneurial Resilience and Business survival through Business tactics is significant at t(28) = .6864, p = .0000, p = 0.0494, the lower-level confidence interval .6949 and higher-level confidence interval is .3619 does not move through 0. Therefore, this relationship is significant.

The direct relationship between Entrepreneurial Resilience and business survival is insignificant; the Σ effect relationship is significant showing Entrepreneurial Resilience leads to Business survival through Business tactics.

Therefore, Alternate Hypothesis – Business Tactics Has a Mediating Effect on the Relationship Between Entrepreneurial Resilience and Business Survival is Accepted.

Recommendations: Micro Entreprenuers need to have proper training and understanding of business. They should have financial literacy and also should be able to adopt to the changing technology.Inherent resilience will help them to overcome certain hurdles but at the same time new business tactics will help them in Business survival.

4 Conclusion

In the tremendous competitive world of 21st century, even though certain inherent characters of micro entrepreneurs like Determination, Eventual experience,Positivism, Societal support,Mental equilibrium and Physical condition helps in establishing business, new business tactics like maintaining customer relationship, adopting new technology, better managerial skills, financial literacy,inventing new products and services helps in stabilising business beyond the point of profits in the business and helps in successful business survival. From the study it is evident that business tactics mediates the relationship of entrepreneurial resilience and business survival.

Scope for further research: Further studies can be conducted to analyse the mediating effect of entrepreneurial resilience on business tactics and business survival.

References

Jamak, A.B.S.A., Ali, R.M.M., Ghazali, Z.: A breakout strategy model of Malay (Malaysian indigenous) micro-entrepreneurs. Procedia Soc. Behav. Sci. **109**, 572–583 (2014)

Hussain, D., Yaqub, M.Z.: Micro-entrepreneurs: motivations, success factors, and challenges. Int. Res. J. Financ. Econ. **56**, 22–28 (2010)

Gosenpud, J., Vanevenhoven, J.: Using tools from strategic management to help micro-entrepreneurs in developing countries adapt to a dynamic and changing business environment. J. Appl. Bus. Res. (JABR) **27**(5), 1–14 (2011)

Yaacob, M.R.: A preliminary study of green micro-entrepreneurs in Kelantan, Malaysia. Int. J. Bus. Manage. **5**(3), 81 (2010)

Chatterjee, N., Das, N.: A study on the impact of key entrepreneurial skills on business success of Indian micro-entrepreneurs: a case of Jharkhand region. Glob. Bus. Rev. **17**(1), 226–237 (2016)

Tu, C., Hwang, S.N., Chen, J.S., Chang, F.Y.: The joint effects of personal and relationships characteristics on micro-entrepreneurial success. Procedia Econ. Finan. **4**, 365–372 (2012)

Davidsson, P., Honig, B.: The role of social and human capital among nascent entrepreneurs. J. Bus. Ventur. **18**(3), 301–331 (2003)

Farmer, S.M., Yao, X., Mcintyre, K.K.: The Behavioral impact of entrepreneur identity aspiration and prior entrepreneurial experience. Entrep. Theory Pract. **35**(2), 245–273 (2011)

Karpak, B., Topcu, Y.I.: Small medium manufacturing enterprises in Turkey: an analytic network process framework for prioritizing factors affecting success. Int. J. Prod. Econ. **125**, 60–70 (2010)

A Study on Career Choice as Entrepreneurs Among Undergraduate Students in Bangalore

M. S. Kokila[1], Shubha Chandra[1], and Ch. Raja Kamal[2](✉)

[1] School of Commerce and Management, Garden City University, Bangalore, India
[2] Kristu Jayanti College, Bangalore, India

Abstract. Due to the crisis and high unemployment, the labour market needs diversified skills. Higher education must encourage entrepreneurship. This research examines the impact of entrepreneurial incentives on prospective students' entrepreneurial intentions and the role of entrepreneurship education on entrepreneurship development. Undergraduates' career perspectives were investigated. Surveys collected original data. The study included 250 students, 200 of whom completed the questionnaire. Anova analyses data (single factor). Most responders are risk takers, and parent's occupation influences entrepreneurship choice. Most respondents favour huge companies than entrepreneurship. Family morale and financial support inspired entrepreneurs. Entrepreneurship is associated to risk-taking, student understanding of financing sources, moral and financial support, and parent employment.

Keywords: Career choice · Risk takers · Financial and moral support

1 Introduction

Entrepreneurship creates financial freedom, jobs, innovation, and economic growth. Increasing university entrepreneurship programmes train tomorrow's entrepreneurs. Student entrepreneurship is understudied. This paper investigates undergraduates' entrepreneurial motivations.

This study examines how students see entrepreneurship as a career. It seeks enterprising students. It analyzed risk factors, students' goals, and the effect of curriculum, college aid, and co-curricular and extra-curricular activities. It also looked at the impact of family morale and finances on entrepreneurs.

This study is set out with the main objective to study the career option as an entrepreneur among undergraduate students in Bangalore City. This study is guided with few research questions.

RQ1: what is the perception among undergraduate students towards entrepreneurship as a career choice?
RQ2: Does the course curriculum and educational program help them in making entrepreneur as a career choice?

B. Alareeni et al. (Eds.): ICBT 2022, LNNS 620, pp. 230–239, 2023.
https://doi.org/10.1007/978-3-031-26953-0_23

RQ3: Does moral and financial support from their parents help the students in making entrepreneur as a career choice? This examination is embraced to comprehend the student's perception towards choosing entrepreneurs as a career choice.

2 Review of Literature

Daz-Garca and Jiménez-Moreno (2010) examine the effect of gender in entrepreneurial ambition. The survey found no gender gap in entrepreneurial ambition.

Nabi et al. (2010) wanted to refocus graduate entrepreneurship research, also monitored students' attitudes and intentions to pursue entrepreneurship. Over 8000 European students were sampled. A large proportion of students had significant start-up goals, according to the report, second, higher education's influence on entrepreneurship and venture formation.

Davey et al. (2011) compared African and European students' entrepreneurial ambitions, attitudes, role models, and experience. First-year business students self-administered a questionnaire, and quantitative empirical research methodology was adopted. Students are optimistic about becoming entrepreneurs in the future, according to the study.

de la Cruz Sánchez-Escobedo et al. (2011) studied gender, entrepreneurship, perception, and attitude. It examines gender variations in university students' entrepreneurial beliefs and attitudes. The research used Spanish student data. Bivariate and multivariate analyses showed that gender affects students' entrepreneurial attitudes and intentions.

Majundar (2013) compared male and female entrepreneurship in UAE. UAE Business first-years conducted the research. Statistical significance was tested using multivariate econometric model and logistic regression. Male and female students chose entrepreneurship at the same rate. Women are more risk-taking and cautious. Future entrepreneurs are motivated, creative, and aware, regardless of gender.

Rasil et al. (2013) evaluated entrepreneurial intention and its antecedents among UTM graduates. This research compared job experience, vicarious experience, general attitude, demographic characteristics, and entrepreneurial image to entrepreneurial conviction and ambitions. A crucial conclusion shows that conviction influences entrepreneurial inclinations more than general attitude. Men with job experience are more enterprising.

Barba-Sánchez and Atienza-Sahuquillo (2018) studied entrepreneurship, employment motivation, entrepreneurial aim, and engineering education. This study aims to investigate the influence of entrepreneurial motives on future engineers' entrepreneurial aspirations and the role entrepreneurship education plays in developing engineers' entrepreneurship. The research found that several agents boost the efficiency of firm-creation operations in this region.

González-Serrano et al. (2021) wrote on sports entrepreneurship. Their research attempted to identify the entrepreneurial potential of sports sciences students and investigate the effect of country culture on their entrepreneurial intentions. Data was analyzed using multigrain PLS-SEM. The key results show that sports sciences students have positive views on entrepreneurship and a helpful environment to be an entrepreneur.

3 Objectives of the Study

The study was undertaken with the following objectives.

1. To examine the students perception on inclination towards demographic variables of the study.
2. To identify their students interest in choosing entrepreneur as a career choice.
3. To analyze the risk bearing traits and their plans on taking entrepreneur as a career choice
4. To examine the relationship between inclination towards entrepreneurship, entrepreneurship curriculum, and students participation in extracurricular activities.
5. To identify the students awareness on fund raising for their entrepreneurial activities.

4 Data and Methods

Participants: This research focused on undergraduates in Bangalore. 200 undergraduate and graduate students from Bangalore were surveyed 102 men, 98 women.

4.1 Procedure

Literature review and informal interaction with undergraduate students helped develop organized and unstructured preliminary questions. 20 students pretested, the final questionnaire, and 30 students' comments were used.

4.2 Domain of the Study

The study identified significant informants in Bangalore undergraduate institutions. What Sapp provided key informants the Google form link? After submitting their answers, they snowballed the questionnaire. Since some questions were partly finished, the researcher took the full answer of 200 questionnaires. After 10 days, Google blocked the link. Before delivering their crucial answer, students were given instructions clarifying the research's goal and keeping participant responses anonymous.

4.3 Data Analysis

Data were collected on demographic features, followed with their student's career option towards entrepreneurship, risk traits, Preference of students towards get employed before taking entrepreneur as a career choice etc.....

4.4 Study Instrument

The instrument of the study consists of an online questionnaire; the contents were adapted from the students pursuing undergraduate programs in various educational institutions of Bangalore. The questionnaire has undergone rewording and were designed in 4 sections.

Sec A: Demographic profile
Sec B: Plan, opportunity, risk traits,
Sec C: Course curriculum, participation of towards co-curricular and extracurricular activities
Sec D: Encouragement from college, Moral and financial support from parents

The claims were based on a Literature Review and expert consultations to reduce research bias. To analyses and summaries impression, respondents scored statements on a 5-point Likert scale (1 = least worry, 5 = strong concern).

4.5 Research Gap

Existing literature explored undergraduate students' goals and views of entrepreneurship as a vocation. Many studies have examined how entrepreneurship education affects students' entrepreneurial inclinations. Many studies have examined the link between entrepreneurial intention, attitude, family education, and environment. Entrepreneurship among undergraduates isn't studied. This fills the research void.

4.6 Research Methodology

Nature of research: Descriptive research
Sample size: 200
Sampling method: Non probability (selective sampling)
Statistical tools: Anova (Single factor)

4.7 Hypothesis of the Study

It is hypothesized in the study that:

The following are the hypothesis framed for the study (Fig. 1).

$H0_1$: There is no significance difference in parent's occupation and career option as entrepreneurship
$H0_2$: There is no significant difference in Course curriculum and career option as entrepreneurship
$H0_3$: There is no significant difference in moral and financial support from parents and career option as entrepreneurship
$H0_4$: There is no significant difference in students risk traits and career option as entrepreneurship
$H0_5$: There is no significant relationship in awareness of students in raising funds and career option as entrepreneurship

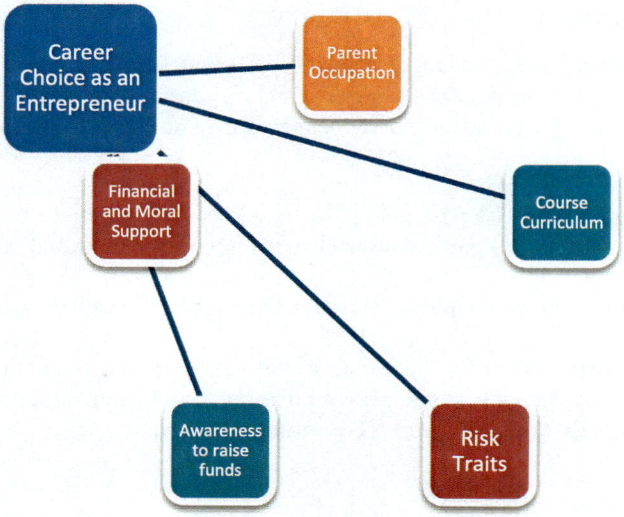

Fig. 1. Career choices as an entrepreneur

4.8 Conceptual Framework

4.8.1 Methodology and Data Analysis

Table 1. Section A Demographic Profile of the respondents

Age	Percentage of respondents	Gender	Percentage of respondents
<21 years	37	Male	51
21–25 years	55.5	Female	49
>25 years	7.5		
Parents occupation	Percentage of respondents	Place of stay	Percentage of respondents
Entrepreneur	52	Rural	24.5
Employed	32.5	Semi urban	22
Both	15.5	Urban	53.5
Career option as entrepreneur	Percentage of respondents	Risk traits	Percentage of respondents
Strongly agree	8.5	Risk taker	58.5
Agree	6.5	Risk averse	7
Neutral	28	Can't stay definitely	34.5

(continued)

Table 1. (*continued*)

Career option as entrepreneur	Percentage of respondents	Risk traits	Percentage of respondents
Agree	36.5		
Strongly agree	20.5		
Section B:: Plan, opportunity, risk traits			
Grabing opportunity to start their venture	Percentage of respondents	Plan to start the venture	Percentage of respondents
Yes	75.5	Yes	50
No	8	No	18
May be	16.5	May be	32
Preference to work in a big organization before start of venture	Percentage of respondents	Preference to be your own boss	Percentage of respondents
Yes	43	Yes	79.5
No	22.5	No	8.5
May be	34.5	May be	12
Section C: Course curriculum, participation of towards co-curricular and extracurricular activities			
Perception of starting venture is risk, hence will not start enterprise	Percentage of respondents	Participation on co- curricular activities	Percentage of respondents
Strongly agree	13.5	Yes	50.5
Agree	19	No	22.5
Neutral	25	May be	27
Agree	26.5		
Strongly agree	16		
Participation in extracurricular activities	Percentage of respondents	Assistance from college to become entrepreneur	Percentage of respondents
Yes	62	Yes	40
No	18.5	No	28.5
May be	19.5	May be	31.5

(*continued*)

Table 1. (*continued*)

Participation in the course curriculum	Percentage of respondents
Strongly agree	42
Agree	39.5
Neutral	14.5
Agree	2.5
Strongly agree	1.5

Section D: Encouragement from college, Moral and financial support from parents

Encouragement from college	Percentage of respondents	Moral support from parents	Percentage of respondents
Yes	61	Yes	63.5
No	15	No	12
May be	24	May be	24.5
Financial support from parents	**Percentage of respondents**	**Awareness to raise funds to start their business**	**Percentage of respondents**
Yes	46	Yes	42
No	22	No	26
May be	32	May be	32

Source: Primary data

4.8.2 Major Findings

111% of respondents are aged 21–25, and 51% are men, according to Table 1. 52% of respondents' parents are entrepreneurs. 52.5% of responders are from Bangalore. A majority of respondents say 36.5% agree and decided to choose entrepreneur as a career choice, and they will grab their opportunity to start a business venture. A majority say 50% has a plan to start a business venture after their undergraduate programme. A majority of respondent 43% from the study revealed their preference to work in a big organization before start of venture. 79.5% are self-employed. 50.5% of respondents participated in entrepreneurial extracurricular. 40% of respondents believe that college fosters entrepreneurship. 63% of 200 respondents gain moral support for choosing entrepreneurship as a profession, while 46% say their parents financially support them. 58.5% are risk takers. 42% of respondents know where to receive startup capital.

4.9 Data Analysis and Interpretation

Table 2. $H0_1$: There is no relationship between parent's occupation and career choice as entrepreneurs

ANOVA						
Source of variation	SS	Df	MS	F	P-value	F crit
Between groups	301.0225	1	301.0225	343.37	0.012	3.864929
Within groups	348.915	398	0.876671			
Total	649.9375	399				

Interpretation: From the above Table 2, Since P = 0.012 is less than 0.05, the null hypothesis is rejected. Parental occupation and entrepreneurship are linked.

Table 3. $H0_2$: There is no relationship between Course Curriculum and career option as entrepreneur.

ANOVA						
Source of variation	SS	df	MS	F	P-value	F crit
Between groups	392.04	1.00	392.04	403.95	0.016	3.86
Within groups	386.27	398.00	0.97			
Total	778.31	399.00				

From the above Table 3 it is revealed that we reject null hypothesis as 0.016 is smaller than 0.05. Co-curricular entrepreneurial activities and entrepreneurship are related.

Table 4. $H0_3$: There is no relationship between family financial support and career option as an entrepreneur

ANOVA						
Source of variation	SS	df	MS	F	P-value	F crit
Between groups	38.44	1	38.44	36.57714	0.012	3.864929
Within groups	418.27	398	1.05093			
Total	456.71	399				

From the above Table 4, it is understood that P = 0.012 0.05, hence the null hypothesis is rejected. There's a link between family support and entrepreneurship.

Table 5. H04: There is no significant relationship between risk traits and career option as entrepreneur.

ANOVA						
Source of variation	SS	df	MS	F	P-value	F crit
Between groups	324.00	1.00	324.00	296.86	0.015	3.86
Within groups	434.39	398.00	1.09			
Total	758.39	399.00				

From the above Table 5 it is revealed that because p = 0.015 0.05. We reject the null hypothesis; risk qualities and entrepreneurship are related.

Table 6. H05: There is no significant relationship in student's awareness to raise funds and career choice as entrepreneurs.

ANOVA						
Source of variation	SS	df	MS	F	P-value	F crit
Between groups	275.56	1	275.56	270.0039	0.004	3.864929
Within groups	406.19	398	1.020578			
Total	681.75	399				

From the above Table 6 it is revealed that Since 0.004 is less than 0.05, we reject the null hypothesis and find a significant link between students' knowledge to generate money and entrepreneurial job choice (Table 7).

Table 7. Summary Table of data analysis

Sl. No	Hypothesis	P value	Significance level 0.05	Accepted or rejected
1	H01	0.012	0.05	Rejected
2	H02	0.016	0.05	Rejected
3	H03	0.012	0.05	Rejected
4	H04	0.015	0.05	Rejected
5	H05	0.004	0.05	Rejected

5 Discussions and Conclusion

The research found that students' exposure to entrepreneurship education influenced their decision to become entrepreneurs. Having an entrepreneur parent is also linked to a student's optimistic attitude, stronger norms, and professional self-efficacy. This confirms Krueger's (1993) conclusion that kids with self-employed dads get early exposure and tacit knowledge of entrepreneurship, which affects their attitudes and self-efficacy towards entrepreneurship as a career option. The data imply students are increasingly aware of sources of startup funding.

This research suggests colleges should encourage students to do internships in new enterprises or establish their own ventures to shape their entrepreneurial attitudes and behaviors. This proved the impact of entrepreneurship education programmes on students' attitudes regarding entrepreneurship as a vocation. Future research should use bigger samples. Future research on this subject should focus on female students' entrepreneur goals and if these lead to entrepreneurial entrance and success.

References

Díaz-García, M.C., Jiménez-Moreno, J.: Entrepreneurial intention: the role of gender. Int. Entrep. Manag. J. **6**(3), 261–283 (2010)

Nabi, G., Holden, R., Walmsley, A.: Entrepreneurial intentions among students: towards a re-focused research agenda. J. Small Bus. Enterp. Dev. **17**(4), 537–551 (2010). https://doi.org/10.1108/14626001011088714

Davey, T., Plewa, C., Struwig, M.: Entrepreneurship perceptions and career intentions of international students. Educ. Train. **53**(5), 335–352 (2011). https://doi.org/10.1108/00400911111147677

de la Cruz Sánchez-Escobedo, M., Díaz-Casero, J.C., Hernández-Mogollón, R., Postigo-Jiménez, M.V.: Perceptions and attitudes towards entrepreneurship. An analysis of gender among university students. Int. Entrepreneurship Manag. J. **7**(4), 443–463 (2011)

Majumdar, S., Varadarajan, D.: Students' attitude towards entrepreneurship: does gender matter in the UAE? Foresight **15**(4), 278–293 (2013). https://doi.org/10.1108/FS-03-2012-0011

Rasli, A., Khan, S.U.R., Malekifar, S., Jabeen, S.: Factors affecting entrepreneurial intention among graduate students of Universiti Teknologi Malaysia. Int. J. Bus. Soc. Sci. **4**(2) (2013)

Barba-Sánchez, V., Atienza-Sahuquillo, C.: Entrepreneurial intention among engineering students: the role of entrepreneurship education. Eur. Res. Manag. Bus. Econ. **24**(1), 53–61 (2018)

Vodă, A.I., Florea, N.: Impact of personality traits and entrepreneurship education on entrepreneurial intentions of business and engineering students. Sustainability **11**(4), 1192 (2019)

Herman, E.: Entrepreneurial intention among engineering students and its main determinants. Procedia Manuf. **32**, 318–324 (2019)

Badri, R., Hachicha, N.: Entrepreneurship education and its impact on students' intention to start up: a sample case study of students from two Tunisian universities. Int. J. Manag. Educ. **17**(2), 182–190 (2019)

Ghatak, A., Chatterjee, S., Bhowmick, B.: Intention towards digital social entrepreneurship: an integrated model. J. Soc. Entrep. 1–21 (2020)

González-Serrano, M.H., González-García, R.J., Carvalho, M.J., Calabuig, F.: Predicting entrepreneurial intentions of sports sciences students: A cross-cultural approach. J. Hosp. Leis. Sport Tour. Educ. **29**, 100322 (2021)

Soomro, B.A., Memon, M., Shah, N.: Attitudes towards entrepreneurship among the students of Thailand: an entrepreneurial attitude orientation approach. Educ. Train. **63**(2), 239–255 (2020). https://doi.org/10.1108/ET-01-2020-0014

Palmer, C., Fasbender, U., Kraus, S., Birkner, S., Kailer, N.: A chip off the old block? The role of dominance and parental entrepreneurship for entrepreneurial intention. RMS **15**(2), 287–307 (2019). https://doi.org/10.1007/s11846-019-00342-7

Big Data in I-O Psychology and HRM: Progress for Research and Practice

Raja Kamal[1] (✉) and M. S. Kokila[2]

[1] Kristu Jayanti College, Bengaluru, India
rajkamalmba2006@gmail.com
[2] New Horizon College, Bengaluru, India

Abstract. Big data and AI may be advantageous for both businesses and consulting organisations when it comes to measuring and analysing workforce-relevant data, as well as interpreting and implementing big data results. Researchers and practitioners in IOP and HRM may provide organisational decision makers, workers, policymakers, and other employment stakeholders significant knowledge. Big data issues in IOP and HRM are studied at the micro and macro levels (e.g., changing nature of big data, developing big data teams, educating professionals and graduate students, ethical and legal considerations). Academics in IOP and HRM will likely play a bigger role in industry-specific big data, AI, and machine learning deployments.

Keywords: Artificial intelligence · Big data · Personnel selection · Talent management

1 Introduction

Big data seems to be everywhere, whether it's a job candidate's virtual reality game, door card readers, or shop traffic footage. Enhanced data collection and storage have contributed to the data explosion. These developments enable complex algorithms to analyse vast and diverse data sources. Big data solutions may be advantageous for both individuals and organisations. The question "What is it?" arises in the majority of narratives on big data. Possible replies include "volume," "variety," "velocity," and "authenticity." The 3–42 Variation in Significant Data Sources (Shaffer 2017), the VS are an effective memory tool for remembering the advantages and disadvantages of big data sets. Regarding big data, businesses require more than what is stated by the Vs.

Integration and aggregation of data are the key subjects of this research. Then, we discuss the finest analytic tools for big data research and possible solutions to the skills gap in data management. In the context of IOP and HRM, technology, data visualisation, analytics, and metrics are then studied. Following an introduction of the technological possibilities afforded by big data, the ethical and legal consequences of this research approach in the context of business are examined.

B. Alareeni et al. (Eds.): ICBT 2022, LNNS 620, pp. 240–251, 2023.
https://doi.org/10.1007/978-3-031-26953-0_24

Table 1. Factors Integrating data sources

Factor	Description
Timing and updating	Information from a variety of sources, compiled and updated at irregular intervals
Samples	Several Hierarchies of Analysis (individual, team, business unit, and organization)
Country, language, and culture	Analogous lengths, populations, and contents are compared
Time intensity	Data may be collected as often as every minute (consumer interactions and purchases) or as seldom as once every two years (team meetings) (traditional performance reviews)
Structure and consistency	Various types of information, both qualitative and quantitative, from unstructured text files to audited financial records
Sources	Data from a variety of sources, including self-report questionnaires, mobile and biometric devices, and human resource information system files

1.1 Big Data's Nature and Management

1.1.1 Need of Big Data IOP and HRM Better Data Management

It is a significant barrier to enhanced data, metric, and analytics use (Harv. Bus. 2014). Moreover, large-scale data analysis may be challenging or impossible due to data peculiarities. One of the most time-consuming operations in analytics is data cleansing (Lohr 2014). The key tasks of a data scientist are data cleansing and organisation (60%) and data collecting (19%). A requirement for data cleaning is comparable to that of a housekeeping service. Inaccurate data might render analyses inaccurate and risk judgments. IOP and HRM are responsible for cleaning up soiled statistics.

1.1.2 Data Linking

Every collection of data that has been integrated or connected must be assigned a unique identification. As firms construct analytics processes, it is anticipated that issuing a unique identification to each employee will become normal practise. You may use a different identification to search for a record in another database, but only if the file is encrypted and safe. Linking IDs across many data sources and formats is difficult. Due to inconsistent attention to detail, links may be missed, ID matches that are missing, duplicated, or incorrect lower the data collection's dependability, predictive capacity, and conclusiveness.

1.1.3 Big Data Variables

When a candidate for a job is evaluated on their level of conscientiousness, their knowledge of the job, and their ability to work in a team, for example, combining item-level results into scale scores for hiring, this is an example of the type of data that is gathered intentionally in typical organisations and in scientific research. Organizational structures that are practical and use standardised measures. IOP and HRM standards are answers to these problems (e.g., SIOP 2018). This is in stark contrast to organisations that receive unanticipated enormous amounts of data in real time, where the interest structure analysis is at best ambiguous, an whole year of real-time collaboration individual, group, and task-focused constructions. Psychometrically defining, scoring, and ranking concepts and actions, given the objectives of the measurement? Top-down identification of such unstructured huge data is nonetheless conceivable (for example, experts may be classed as preparatory, task-based, interpersonal, or adaptive. or regulatory team member behaviours (Rousseau et al. 2006). Using algorithmic grouping and network analysis, it may be feasible to identify variables in huge datasets from the bottom up. This bottom-up big data technique aims to enhance Campbell and Fiske's (1959) multi trait multi method matrix (Cronbach and Meehl 1955).

1.1.4 Assembling Big Data

Due to the difficulty of connecting systems using a broad array of protocols and data types, big data analytics necessitate HR, IOP, and HRM specialists, systematically interconnecting once isolated processes (Ryan and Herleman 2015) of data Integration. Although careful preparation and effort go a long way toward assuring data accuracy, mistakes may occur (Tonidandel et al. 2018). There are deviations from the usual information and analysis. It is important to connect files with identities, locate and reconcile duplicate data, and process raw unstructured data for analysis in order to create huge HR data sets. There are several resources and objectives, thus eliminating duplication might be advantageous (across variables, but potentially across people or cases as well). Multiple files include employee identification and associated information. A near empirical duplicate may occur if two comparable employee engagement surveys are done almost simultaneously. Excluding possible components or information sources is an example of redundancy. People employed in IOP and HRM often make sound judgments.

2 Big Data Infrastructure

Researchers in IOP and HRM need many platforms to analyse the amount, variety, velocity, and authenticity of corporate big data. Human resource information systems (HRIS), corporate data warehouses, data lakes, and cloud-based platforms provide obstacles for business big data analytics (Ryan 2015). To avoid privacy, legal, and ethical difficulties, it is necessary to utilise and store certain technologies and data with discretion (McLean et al. 2016).

IOP and HRM teams must become fluent in a range of computer languages, database lingo, application programming interfaces (APIs), and web scraping methods in order to perform research (Braun et al. 2017). HR analytics experts seek to enhance data-driven decision-making by modelling organisational phenomena such as recruiting, firing, training, and collaboration (e.g., Kozlowski et al. 2016).

2.1 Big Data Skill Gaps

IOP and HRM personnel need training in data collection, administration, and analysis in order for their respective organisations to perform well. According to a research conducted by the Society for the use of HR data for workforce performance prediction, multiyear workforce planning, and data correlation by HR professionals was unprepared (SHRM 2016). This lack of analytic acumen or abilities is considered as one of the greatest barriers to the more effective use of company data, metrics, and analytics.

2.2 Addressing the Skills Gap

Numerous organisations actively recruit and train workers to obtain big data expertise (SHRM Found. 2016). IOP and HRM specialists often have proficiency in R, Python, SQL, Hadoop, Map Reduce, Unix Shell/AWK/Awk, Apache Spark, and Java However, we want to be clear about the abilities required to handle big data-focused organisational research, therefore we will maintain this website for the time being.

3 Big Data Visualization

Despite their origins in statistics and research methods, visualisation techniques have become increasingly relevant and useful due to the requirements of big data. Al indicates that visualisation facilitates management decision-making, Intelligent, interactive data and model visualisation. In data visualisation, Big Data's volume, velocity, diversity, and veracity are brought together (Sinar 2015). Researchers in IOP and HRM, in addition to practitioners and decision-makers, may gain a great deal by visualising their results. As more data becomes accessible, manual techniques of analysing velocity data are become useless.

3.1 Visualizations Can Provide an Audience

Through the use of more aesthetically attractive representations and reports (think Gapminder, the most recent effort of Hans Rosling), Visualisation of HR data for diversity analysis. By highlighting out-of-the-ordinary information, visualisations aid scientists in determining the precision of their data.

3.2 Two Info Graphic Hints

Dimensions (Domingos 2012). Focusing on certain variables or summaries might help to reveal interpretable trends or outliers in data visualisation. Estimations of error are necessary when dealing with huge datasets and outcomes. Those that research IOP and HRM have uncovered AI and ML problems with massive data.

Several articles give examples of human resource information. Comparing categories, analysing hierarchies, and exhibiting temporal changes are just a few of the corporate communication applications cited by Sinar (2015) for visualisation. Visualization of big data and machine learning models applied to large data sets for categorization and prediction are powerful information communication approaches.

4 Big Data Algorithms

Never overlook the larger picture and data analysis. Experts in IOP and HRM use data analysis to advance organisational and personal objectives. Analyses are used to evaluate hypotheses, confirm concepts, and discover workplace difficulties. Using the outcomes of studies, firms may better plan for, react to, and manage the occurrence of diverse organisational events.

Human resources (HR) indicators that may be tracked and recognised using big data include effective recruiting, employee happiness and engagement, and turnover (e.g., division, business unit, region).

Many firms categorise a variety of organisational aspects, including recruitment, applications, workers, departments, and even questionnaire and evaluation items. Frequently, organisations classify professions and job functions according to quantifiable outputs or human attributes, such as knowledge, skill sets, and capacities (KSAOs). Clustering expedites hiring, choosing, staff planning, and performance management. For a deeper understanding of the data, exploratory factor analysis clusters columns (variables, items). This data summary may be used to more effectively communicate survey findings and guide your team's activities.

4.1 Evolving Analytical Methods, Graduate Training, and Big Data

An advanced degree is necessary for organisational analysis. IOP and HRM students are exposed to statistical analysis of organisational data. IOP graduate students study descriptive statistics, psychometrics, factor analysis/clustering, experimental design, ANOVA, and basic GLM variations, with an emphasis on structural equation modelling (SEM) and multilevel models (Aguinis et al. 2009). No courses in econometrics, finance, or corporate strategy are required of students at the Institute of Organizational Performance and HRM.

Using machine learning to bridge the gap between theory and practise would be beneficial for IOP and HRM graduates. Caret gives access to over two hundred distinct categorization, prediction, and grouping techniques. NLP (natural language processing) can quantify job advertisements and interview outcomes (NLP). NHST may benefit from massive-scale machine learning (ANOVA, linear regression).

Induction disregards artificial intelligence, machine learning, and predictive modelling. A noteworthy similarity exists between a big data, machine learning, and inductive analysis for bottom-up prediction organization's structure utilising rich qualitative data. Each hemisphere is used in the examination of an organisation.

4.2 What's New with Big Data Analyses?

Few statistical methods are used in IOP and HRM. For many years, low sample sizes and lack of statistical power have hampered research (Hunter and Schmidt 2015). The restricted study variables are due to the use of historical information. A small sample size is insufficient to prove a difficult theory.

To comprehend the advantages and disadvantages of big data analytics, it is beneficial to consider the findings of publishing deductive theory and hypothesis testing.

What if models used for prediction and explanation included more variables? What if significant empirical and predictive relationships were overlooked in large-scale studies? How might prior knowledge of these relationships benefit theory or writing, and workplace audio/video? When sample numbers are limited, understanding nonlinearities and interactions may benefit from large data.

5 Capitalizing on Unstructured Data

Rarely are statistics used in IOP and HRM like multiple linear regression, ANOVA, SEM, or mixed-effects models. Previous research was hindered by insufficient statistical power and limited sample numbers (Shen et al. 2011). Studies are limited by the past e.g., research conducted prior to the Internet and other technological innovations. Too little of a sample size or too few variables cannot be used to test complicated hypotheses.

5.1 High P-to-N Ratios

With a research and development sample size of 51,060 and a total sample size of 66,732 for statistical significance, we discover that $p > N$. If p is extremely close to N, it is impossible to do linear regression. This problem is not exclusive to text data or repeated-measures data from wearable sensors like (group/team communication or interaction patterns) in the workplace; it exists in other disciplines of research as well e.g., neuroscience and fMRI, or genetics microarray analysis.

5.2 Parsimony, Statistical Power, and Analysis Need P and N

When p is big, PCA or CFA is required (e.g., hundreds of items across numerous measures). After eliminating all probable variables, do research to confirm your hypotheses. Optimize p to N. Elastic nets, ridge regression, and lasso regression are linear model choices for conducting predictor selection and weighting (i.e., giving regression coefficients less than zero to the remaining predictors) and keeping important predictors (Zou and Hastie 2005).

Using big data techniques, the amount of elements to be retrieved is selected based on the desired conclusion like principal components regression (Hastie et al. 2009). Bair (2006) and Lee (2007) used the PLS regression technique (2011). Utilizing a randomised principal component analysis (PCA) and a latent Dirichlet allocation (LDA) may minimise the quantity of text data (Kosinski et al. 2016). There are text reduction approaches that use prediction. Unlike ANOVA and SEM, big p-to-N ratios need model parameter optimization tuning parameters determine model coefficients..

5.3 Identifying Nonlinearity and Interactions

It is standard practise to assume that the unknown functional form of predictor-criterion relationships with vast volumes of data. It would seem that the discovery of nonlinearities and interactions is difficult in the absence of theory and with the ultimate goal being the building of a highly accurate prediction model. In this case, conventional approaches Multiple regression and polynomial regression are unrelated.

6 Acting on Model Selection Uncertainty

Estimating model parameters and their associated sampling error variance-driven uncertainty is key to all traditional inferential statistics. Sampling error variance may be thought of as the differences between a sample and the target population. Researchers in the domains of IOP and HRM have invested a great deal of time and effort to enhance meta-analysis methodologies and gather meta-analytic data in a variety of contexts, but their concerns over the possibility of large inaccuracy in the samples they've utilised are still evident (Hunter and Schmidt 2015). As with other social science academics, IOP and HRM experts have usually ignored model selection uncertainty.

6.1 Considering the Purpose for the Methods Review

Large data sets associated with IOP and HRM are covered here. The bulk of books in this discipline concentrate on practical examples but provide little supplemental resources. We did not restrict our search to graduate programmes in HRM and IOP. Big data investigations are needed when the number of variables or measurements exceeds the sample size. Cross-validation is a technique for identifying complicated and robust machine learning relationships. The methods and instruments for evaluating algorithmic data are explained.

6.2 Measurement Techniques

Big data dominates HR. Traditional HR data sources may also be used yield massive amounts of information when collected from a large number of employees, over a long period of time, and/or in a variety of locations and times. Uncertainty surrounds the legality of using social media and big data in the recruiting process, and novel HR information sources may be superior to conventional sources for predicting organisational outcomes and directing HR decisions (Chamorro-Premuzic et al. 2016). The sources of Big Data relevant to human resources are described below.

6.3 Serious Games and Gamification

Gamification increases candidate engagement, employee engagement, and team engagement. All of the following are desirable: leaderboards that display relative standings, badges or recognitions for completing particular activities or achieving certain objectives, timely feedback, and monetary awards for meeting certain milestones. According to Landers (2014), gamification may not provide the intended results and of course, outcomes should be defined and not assumed.

In order to detect smugglers and terrorists, Airport Scanner analyses several search variations (Mitroff et al. 2015). Three billion trials including twelve million participants are potentially unanalyzable. This aphorism "garbage in, garbage out" is not entirely accurate. Poorly designed video games are neither entertaining nor educational. As advised by the Society of Information Game Professionals (SIOP), people working in the area of serious games must Psychometric dependability, criterion-related validity, group mean differences, unfavourable effect repercussions, and fairness must be considered. In the lack of evidence, it is vital to rely on a test vendor's data-free assurances or create a bespoke game for your company.

6.4 Data from Internet of Things Devices

Digital assistants give essential information. Office and mobile devices are now voice-operable. This programme "hears" and records requests from employees for subsequent use in pattern analysis and big data. If their data is saved in the cloud, light switches that detect motion and doors that scan employee credentials have the potential to generate vast quantities of data. By providing multidimensional psychological data on employee behaviour, intelligent workplaces have the potential to enhance performance management, training, and development, and even expose illegal employee conduct.

6.5 Cameras/Biometric Information

Only automated onsite security and surveillance and business intelligence, including targeted marketing, are two uses for the processing of video data (Gandomi and Haider 2015). Robotic security guards with video cameras are becoming widespread in an increasing number of businesses. Law enforcement officers, delivery drivers, and warehouse employees are increasingly using body-worn cameras to capture their settings and interactions.

6.6 Social Media

According to the media, enormous volumes of data collected from social networking sites are being utilised unethically. The views of employees may be expressed in postings or blogs. Workers might potentially profit from analytical resources. In the next generation of job and occupational analysis, social media postings are used to analyse the KSAOs that are required and desired by employers. Using application programming interfaces (APIs), Web data is Facebook and Twitter are often criticised when utilised in the hiring process, despite their use in social research (Kosinski et al. 2015).

6.7 Text or Sentiment Analysis

Textual information may be discovered in applications, resumes, interviews, and transcripts, among other places. A user-defined dictionary technique may be used to investigate the frequency, categories, and co-occurrences of dictionary words. Personality studies often use dictionary entries. Bottom-up feature extraction and co-occurrence research depend on massive volumes of undefined textual data.

6.8 Mobile Sensors

ID tags, mobile phone applications, and other work-related devices include concealed sensors that monitor the movements, heart rates, and other biometric data of employees.

In this category are RFID, sociometric badges, and Fitbits. Badges that measure sociometry may capture information such as the manner in which group members converse and the frequency with which they shake hands. Individual and team performance, as well as the quality of their connections and the ease with which they change positions, may be monitored in real time (Kozlowski and Chao 2018).

6.9 Public Data Repositories

A lot of academics and institutes make secondary data sets accessible (see Table 1). The Workforce Data Initiative gathers and distributes information on occupational certifications, educational achievement, and current employment status. Data sets may be valuable for IO psychology and management researchers. Data.gov provides access to 100 IOP and HRM data sets, including Survey and the Job Openings and Turnover Survey. Open Science Framework enables access to research materials and data sets e.g., experimental protocols, measures, programming code. Meta BUS has classified 1 million correlation impact sizes using a database of 23 publications and ten years of organisational study (Bosco et al. 2014), Taxonomy of conceptualization, interrelationship, and meta-analysis. The new database search engine from Google can now discover publicly accessible data sets. Add HRM and IOP information.

6.10 Traditional Data on Human Resources and Organizations

We put an emphasis on salary, performance, attendance, and tenure. As it is rare to discover the cause for an employee's leave, turnover rates are often underestimated. Generally, just a yes/no departure code and a voluntary/involuntary turnover code are supplied. If employers performed connection analyses more often than annually during performance reviews, we may be able to better forecast employee departure (e.g.,Wanberg and Banas 2000).

7 Privacy, Ethical, and Legal Considerations

Human resource research using large data sets should include concerns of permission and confidentiality. Acceptance of GDPR in May 2018 is achievable, but training and compliance are essential for its actual implementation have been devoted to the shortcomings of big data to enhance HR and other fields.

Frequently, face recognition research used incorrect photos. Artificial intelligence has progressed to the point that it can now identify full people, entire pieces of clothing, and even minute picture fragments.

7.1 Ethical Codes and Standards

In the domains of IOP and HRM, organisational psychologists are restricted by two moral issues. In the beginning, they are required to adhere to the APA's code of ethics and standards of conduct (APA 2017). (APA 2017). Second, it is essential to conform to national, state, and industry data standards. The government organisations and agencies are covered under CFR 45.46. Numerous psychologists contend that the "Common Rule" has outlived its usefulness and should be discarded. Institutional review boards and informed consent do not apply to the usage of big data.

7.2 Legal Requirements

Protecting Individually Identifiable Data Protecting Private Information Personal information that is not very sensitive may nonetheless identify specific individuals (Rocher et al. 2019). When a person's birthdate and ZIP code are linked in a database, their identity is revealed.

PII includes identification information and other types of data. Impact of Big Data. Big data may raise the accountability of employees (Douthitt and Aiello, 2001). (it lowers morale) According to Kosinski et al., persons who like curly fries have a higher IQ. According to gathered data, Facebook users preferred curly fries (Chamorro-Premuzic et al. 2016). It may be necessary to update the algorithm if it drives corporations and/or workers to alter their behaviour. It is gathered and distributed something which has its own sophisticated administration: massive amounts of data. No? The ownership structure of a corporation may suffer if contracts make it impossible to collect and monitor data. Research and data collection are enhanced by transparency. Big data may be favoured by future workers who gain the benefits of their predecessors' labour.

8 Conclusion

This article discusses information visualisation, technique, measurement, technology, and ethics in human resource management. While the big data science and practise community in IOP and HRM undergoes significant change, Likewise, the technology, data analytics, practises, and policies that will govern the future of big data in organisations are constantly evolving. We hope that the reader finds this essay beneficial in whole or in part, and that the advice given is still valid in light of current changes. Using integrated business systems, employee, team, and unit data are analysed. Collaboration between IOP and HRM scholars capable of bridging the gap between their respective subjects and degrees of analysis would be beneficial to the success of enterprises (Molloy et al. 2011). How sensitive are these forecasts when traditional theory-driven methodologies and data-informed computer simulations of the organisational system (such as firm finances, remuneration, selection ratios, employee turnover, and collaboration; Grand et al., 2016) are taken into account? Big data accurately depicts the dynamic and complicated character of contemporary enterprises. IOP and HRM make it possible to live a more fulfilling life. There are several unexpected commercial uses for enormous data sets.

References

Aguinis, H., Pierce, C.A., Bosco, F.A., Muslin, I.S.: First decade of organizational research methods: trends in design, measurement, and data-analysis topics. Organ. Res. Methods **12**, 69–112 (2009)

Aiken, L.S., West, S.G., Millsap, R.E.: Graduate training in statistics, measurement, and methodology in psychology: replication and extension of Aiken, West, Sechrest, and Reno's 1990 survey of PhD programs in North America. Am. Psychol. **63**, 32–50 (2008)

Al-Kassab, J., Ouertani, Z.M., Schiuma, G., Neely, A.: Information visualization to support management decisions Int. J. Inf. Technol. Decis. Mak. **13**, 407–428 (2014)

Angrave, D., Charlwood, A., Kirkpatrick, I., Lawrence, M., Stuart, M.: HR and analytics: why HR is set to fail the big data challenge. Hum. Resour. Manag. J. **26**, 1–11 (2016)

APA (American Psychological Association): Ethical Principles of Psychologists and Code of Conduct. American Psychological Association, Washington (2017)

Austin, J.T., Scherbaum, C.A., Mahlman, R.A.: History of research methods in industrial and organizational psychology: measurement, design, analysis. In: Rogelberg, S.G. (ed.) Handbook of Research Methods in Industrial and Organizational Psychology, pp. 1–33. Blackwell, Oxford, UK (2002)

Bair, E., Hastie, T., Debashis, P., Tibshirani, R.: Prediction by supervised principal components. J. Am. Stat. Assoc. **101**, 119–137 (2006)

Bal, K.: Building an HR analytics team. Hum. Res. Executive (2016). http://hrearchive.lrp.com/HRE/print.jhtml?id=534360037

Berk, R.A.: An introduction to ensemble methods for data analysis. Sociol. Methods Res. **34**, 263–295 (2006)

Berman, J.J.: Principles of Big Data: Preparing, Sharing, and Analyzing Complex Information. Morgan Kaufmann, Waltham, MA (2013)

Bien, J., Taylor, J., Tibshirani, R.: A lasso for hierarchal interactions. Ann. Stat. **41**, 1111–1141 (2013)

Blei, D.M., Ng, A.Y., Jordan, M.I.: Latent Dirichlet allocation. J. Mach. Learn. Res. **3**, 993–1022 (2003)

Bosco, F.A, Uggerslev, K., Steel, P.: Scientific findings as big data for research synthesis: the metaBUS Project. In: 2014 IEEE International Conference on Big Data, New York, pp. 18–22. IEEE (2014)

Boudreau, J.W., Jesuthasan, R.: Transformative HR: How Great Companies Use Evidence-Based Change for Sustainable Advantage. Jossey-Bass, San Francisco, CA (2011)

Braun, M.T., Kuljanin, G., DeShon, R.P.: Special considerations for the acquisition and wrangling of big data. Organ. Res. Methods **21**, 633–659 (2017)

Breiman, L.: Random forests. Mach. Learn. **45**, 5–32 (2001)

Breiman, L.: Statistical modeling: the two cultures. Stat. Sci. **16**, 199–215 (2001)

Breiman, L., Friedman, J.H., Olshen, R.A., Stone, C.J.: Classification and Regression Trees. Chapman & Hall, New York (1984)

Burnham, K.P., Anderson, D.R.: Model Selection and Multimodel Inference: A Practical Information-Theoretic Approach, 2nd ed. Springer, New York (2002)

Campbell, D.T., Fiske, D.W.: Convergent and discriminant validation by the multitrait-multimethod matrix. Psychol. Bull. **56**(2), 81–105 (1959)

Campbell, J.P., Wilmot, M.P.: The functioning of theory in IWOP. In: Anderson, N., Ones, D.S., Sinangil, H.K., Viswesvaran, C. (eds.) Handbook of Industrial, Work, and Organizational (IWOP) Psychology, Personnel Psychology, vol. 1, pp. 3–37. SAGE (2018)

Campion, M.C., Campion, M.A., Campion, E.D., Reider, M.H.: Initial investigation into computer scoring of candidate essays for personnel selection. J. Appl. Psychol. **101**, 958–975 (2016)

Cascio, W., Boudreau, J., Fink, A.: Investing in People: Financial Impact of Human Resource Initiatives, 3rd ed. Alexandria, VA: Socio Human Resource Management (2019)

Chaffin, D., Heidl, R., Hollenbeck, J.R., Howe, M., Yu, A., et al.: The promise and perils of wearable sensors in organizational research. Organ. Res. Methods **20**, 3–31 (2017)

Chamorro-Premuzic, T., Winsborough, D., Sherman, R.A., Hogan, R.: New talent signals: shiny new objects or a brave new world? Ind. Organ. Psychol. **9**, 621–640 (2016)

Chatfield, C.: Model uncertainty, data mining, and statistical inference. J. R. Stat. Soc. A **158**, 419–466 (1995)

Chen, E.E., Wojcik, S.P.: A practical guide to big data research in psychology. Psychol. Methods **21**, 458–474 (2016)

Connelly, B.S., Ones, D.S.: An other perspective on personality: meta-analytic integration of observers' accuracy and predictive validity. Psychol. Bull. **6**, 1092–1122 (2010)

Cronbach, L.J., Meehl, P.E.: Construct validity in psychological tests. Psychol. Bull. **52**, 281–302 (1955)

da Silva, N., Cook, D., Lee, E.-K.: Interactive graphics for visually diagnosing forest classifiers in R (2017)

Conceptual Principles of Choosing Rational Forms of Labor Organization of Personnel of Motor Transport Enterprises

Nadiia Antonenko[1](✉), Kateryna Kompanets[2], Victoria Ilchenko[1], Nataliia Kovalenko[1], Tetiana Diachenko[1], and Nataliia Kukhtyk[1]

[1] National Transport University of Ukraine, Kyiv, Ukraine
3142206@gmail.com, kovalenko-nv@ukr.net, mart_09@i.ua
[2] State University of Trade and Economics of Ukraine, Kyiv, Ukraine
ket13@ukr.net

Abstract. The implementation of a rational form of labor organization of the company's personnel requires an economic justification of the feasibility of its use in the structural unit of the business entity. Therefore, when searching for an effective form of labor organization, there is a need to identify a number of factors that increase the productivity of the company's personnel. In modern conditions, the process of forming an effective system of labor organization of the personnel of motor transport enterprises as a set of measures aimed at increasing the organizational effect of the joint activities of the performers is gaining special relevance. Due to the complexity, dynamism and stochastic of the production processes of maintenance (MN) and ongoing repair (OR) of cars, analytical research methods are unacceptable for solving the problems of determining the rational forms of labor organization of performers, which correspond to the forms of organization of production of repair and preventive works. In this regard, there is a need to develop new approaches to the study of complex, dynamic systems. One of these techniques is the method of simulation modeling, which is the only way to obtain information about the state of the maintenance and repair systems of cars with various forms of organization of production and work without conducting an expensive experiment on a real object.

The purpose of the study is to develop models of the organization of the labor process of repair workers of a motor vehicle enterprise, which allow to quantitatively measure the degree of effectiveness of each form of organization of the work of repairmen when using one or another form of organization of the provision of maintenance and repair services and to assess which organizational form of the labor process is in the specific conditions of the operation of the enterprise motor vehicle is the most acceptable. The data for conducting the research were obtained from the website of the YouControl company [1], as well as directly from the enterprises of motor vehicles of Ukraine. The results of the study allow employers to apply a methodical approach to choosing an effective form of work organization for the company's auto repair staff in order to obtain reserves to cover additional labor costs.

B. Alareeni et al. (Eds.): ICBT 2022, LNNS 620, pp. 252–264, 2023.
https://doi.org/10.1007/978-3-031-26953-0_25

Keywords: Form of labor organization · Personnel · Service · Motor vehicle enterprise · Simulation modeling · Organizational form of the labor process · Management · Organization · Information systems

1 Introduction

The relevance of the chosen topic is due to the fact that with the help of analytical research methods, it is possible to obtain only a certain number of finite-difference equations that take into account the specifics of the specific conditions of providing services for maintenance and repair of cars. However, these methods do not allow us to quantitatively measure the degree of efficiency of each form of organization of work of repairmen when using one or another form of organization of the production of technical influences and to assess which organizational form of the labor process is most acceptable in the specific conditions of operation of the motor vehicle enterprise. In order to make effective management decisions, it is necessary to create a modern toolkit that allows deciding an effective form of work organization of auto repair personnel of economic entities in unstable market conditions. The analysis of the development of relations between the performers of maintenance work and current repairs at motor vehicle enterprises shows that in this area in Ukraine, the interests of employees are not sufficiently taken into account and, as a result, we have low results of the work of repair zones.

The article proposes a mechanism for picking an effective form of organization of the work of the company's auto repair personnel, an adequate form of organization of the production of technical influences in order to obtain reserves to cover additional labor costs. The use of this mechanism allows, without resorting to a full-scale experiment, to get an answer to the question of how the technical service will function at motor vehicle enterprises that differ in size and operating conditions. The developed mechanism for selecting rational forms of work organization of car repair personnel contributes to the creation of a motivational environment for improving the results of the enterprise [2].

2 Literature Review

The relevance of the study of the organization of labor processes at enterprises of various branches of the economy is evidenced not only by numerous publications, but also by the practice of conducting business by leading domestic and foreign companies. Modern methodological literature offers a number of theoretical and conceptual approaches to the researched issues. In this regard, it is necessary to note the works of V.G. Vasylkov, which are devoted to the classification of labor processes; O.A. Grishnova, who developed original approaches to defining the concept of "labor process", V.M. Petyukha and V.M. Danyuk, who proposed modern mechanisms for improving labor processes. Labor organization was studied by such foreign scientists as P. Bolton, F. Bonne, G. Standing. A. Chatenier, R. Anker, D. Bescond, F. Megran, P. Baret-Reed, F. Egger, J.Ritter devoted their research to the issue of evaluating the effectiveness of the labor organization of enterprise personnel.

In the works related to the issues of the organization of the provision of services for the maintenance and repair of rolling stock of automobile transport [3, 4], the random nature of the processes of maintenance and repair of cars is indicated, which allows us to attribute the task of finding optimal forms of organization of technical influences and work to the task of theory mass service in the information system.

As shown by earlier studies [3], it is most expedient to solve such problems by the method of simulation modeling, since this method allows you to do without conducting a high-cost experiment on a real object. In addition, the simulation of the operation of maintenance and repair zones using the model allows revealing the degree of importance of each of the endogenous and exogenous variables of the system and to establish their functional relationship. And, finally, the method of simulation modeling makes it possible to use for experiments situations in which information is either completely absent or insufficient, and this, in turn, creates prerequisites for identifying production reserves and managing the technical service of enterprises.

Experiments performed by domestic researchers [5, 6] showed that in order to compare the efficiency of using different forms of organization of production of technical influences and forms of organization of work of repair workers, it is advisable to use the economic-mathematical method of modeling the work of the maintenance and repair zone of a motor vehicle enterprise, which is implemented through modernized Petri nets. Since the organizational and economic component of the motivation of performers of any work is determined by the degree of effectiveness of the form of labor organization of these workers [2], the task of determining rational forms of labor organization of performers, adequate to the forms of production organization, by methods of simulation modeling is of important scientific importance.

In the works devoted to the problems of simulation modeling of the production process of technical service using classical Petri nets [3, 4], it is proposed to determine the effectiveness of one or another system of work organization of repair and maintenance personnel from the standpoint of the quality component of preventive and repair work. At the same time, the form of organization of the process of technical and repair effects of rolling stock of road transport is not taken into account. But, as shown by previously performed studies [5–7], such tasks cannot be solved without taking into account the organization of the production of maintenance and repair of cars introduced at the enterprise. That is, the development of methodological approaches and the creation of a suitable modern mechanism for the formation of effective labor organization of the personnel of enterprises, taking into account the forms of organization of the production of maintenance and repair of rolling stock of motor vehicles, require further attention of scientists. The relevance of conducting such research is confirmed by the requirement of European standards to create a motivational environment at enterprises that is close to the norms of the Concept of Decent Work.

Thus, in the works of domestic and foreign scientists, the problems of choosing an effective form of labor organization, an adequate form of organizing the production of maintenance and repair of cars, are not sufficiently disclosed and require further elaboration.

3 Research Methodology

A set of well-known scientific methods and techniques was used to achieve the set goal of the research and to solve the relevant tasks: the method of simulation modeling, namely, the apparatus of classic Petri nets – to create a mechanism for choosing rational forms of labor organization for the personnel of motor transport enterprises; abstract-logical method – for generalization, formulation of conclusions and recommendations. The method of logical synthesis was applied for the theoretical justification of the importance of studying the problems of assessing the effectiveness of the application of forms of labor organization of performers, adequate to the forms of organization of the production of technical influences. The use of methods of analysis and synthesis made it possible to show the peculiarities of the use of modern methods of evaluating the effectiveness of modern labor organization systems in Ukraine. The method of constructing schemes and models was used for visual presentation of research results and their schematic interpretation.

4 Results

Let's consider the main forms of labor organization of personnel, which are most often found at enterprises of the motor transport industry. Studies have shown that the following forms of labor organization of repair workers can be used at road transport enterprises: individual form of labor organization; a collective form of labor organization with the organization of complex teams; a collective form of labor organization with the organization of teams specialized in the types of influences; a collective form of labor organization with the organization of teams specialized in types of aggregates [3, 4]. With the help of modernized Petri nets with programmed behavior of transitions in the information system, we will build dynamic models for the above forms of labor organization of repair workers.

A significant reserve for the growth of the efficiency of the technical service of the motor vehicle enterprise (MVE) consists in increasing the organizational and technical level of providing services for maintenance and repair of the rolling stock of road transport (RSRT) due to the modernization of the production and technical base; introduction of more advanced technology; implementation of organizational programs, the main of which are: scientific organization of work, creation of production management centers, implementation of complex maintenance and PR quality management systems, improvement of repairmen's work organization forms [3].

One of the ways to increase the efficiency of the production of maintenance and repair of cars is the improvement of the forms of work organization of repairmen. Moreover, in the recent period, when a steady trend of growth in the costs of maintaining rolling stock has been established, more and more attention is being paid to collective forms of labor organization.

The diversity of technical, technological, organizational and social conditions in different sectors of the economy and insufficient unification of the organizational structures of enterprises of the same industry serve as the reason for the emergence of many varieties of teams. The existence of many types of teams is an acceptable type of collective form of labor organization of repair and maintenance personnel in the given conditions of the enterprise's operation.

In the literature [3, 8] there are two methodological approaches to the issue of choosing rational forms of organizing the work of repairmen. The first of them is based on the statement that, regardless of the form of production organization, the brigade form of labor organization is considered the most effective, and as a result, a methodology for implementing a collective form of labor organization is being developed [3]. The second approach is characterized by the fact that the "accepted level of labor organization" is the criterion for the effectiveness of one or another form of labor organization [8]. But none of the indicators of this criterion can be considered effective (see Table 1).

Table 1. Characterization of indicators of the level of labor organization of personnel of motor transport enterprises

№	Indicators of the level of labor organization	Disadvantages
1	An integral indicator of the assessment of the level of labor organization, production and management	1. Heterogeneity in the content of individual indicators for each of the 3 groups 2. There is no single basis for comparison
2	System of indicators: labor cooperation, division of labor, organization of workplaces, working conditions, labor safety, labor regulation, use of working time	Indicators characterizing the organization of workplaces, occupational safety and the qualifications of employees are accepted conditionally
3	Indicator of the degree of use of working time at each workplace	1. The indicator is not integral 2. It does not take into account some peculiarities of the production of maintenance and repair of cars

Source: formed by the authors on the basis [3, 8]

To the unresolved part of the problem and the shortcoming of both approaches, it is necessary to attribute the fact that the forms of labor organization are considered in isolation from the form of organization of production of maintenance and repair of rolling stock adopted in MVE: the power of MVE, the degree of equipment of his garage equipment, the condition of rolling stock, conditions and nature of operation of cars on the line; there is no substantiation and selection of types of production crews that correspond to the specifics of the provision of car maintenance and repair services [3, 4].

In order to implement the task of determining the effective form of work organization of car repair personnel, it is necessary to model the labor process of repair workers and determine under which conditions of maintenance and repair work it would be appropriate to use certain forms of work organization of performers. The most appropriate mathematical apparatus for achieving the set goal is a modernized classical Petri net.

Two methods of using Petri nets are known – for designing and analyzing the system of technical influences and current repairs in MVE. The first consists in the modeling and analysis of the system, after which the deficiencies identified in the design process are eliminated [9]. The cycle is repeated until the results of the analysis satisfy the needs of the system designers. The second, more radical method reduces the entire modeling process to analytical calculations [10]. But this method requires a large array of experimental data, so in the absence of information about the behavior of the system, it is better to use the first method.

Consider the structure of the Petri net. It consists of four elements (sets): positions *P*, transitions *T*, input functions *I*, and output functions *O*. If we take into account the above conventions, then the state of the system can be described by the dependence $C = (P, T, I, O)$. To visualize the data, let's use a graphic representation of Petri nets, which are a bipartite-oriented multigraph built using the following elements: positions (marked by *O*); transitions (they are marked | in the network) and oriented arcs that connect them (displayed in the network with a ↑ sign). Given the above notation, we can write that $G = (V, A)$; $V = PUT$; $P \cap T = \acute{\emptyset}$; where *A* is a set of arcs.

Next, we will mark the *P* positions by assigning them μ chips. A Petri net is characterized by the number and distribution of chips in the network. A transition in the system occurs when each of the input positions has a number of chips equal to the number of arcs. In general, everything described above reflects an event or action that takes place in the system and characterizes a transition and a condition – a predicate or a logical condition for describing the state of the system.

The modernization of the classic Petri net is as follows. Each chip in the network is assigned some named set of labels {φ}. During the transition, when output functions *O* or input functions *I* are triggered, the value of the labels changes according to the function algorithm. When analyzing the condition (predicate) before making the transition, the value of all chip marks are analyzed, and a conclusion is made about the possibility of transition for this chip. The initial value of the chip marks is set before starting the simulation.

Let's move on to the description of the functioning of the repair service from the point of view of the Petri net. In Fig. 1 shows a Petri net model that describes the operational post form of the organization of repair crews. Each car in such a network is a chip in positions *P1-P3*, a free repair team is a chip in positions *P7, P10, P12, P14, P16*. Accordingly, the busy repair team is a chip in positions *P6, P9, P11, P13, P15*. The chip in positions *P4* and *P5* is a car in the service queue, and the chip in positions *P17* and *P18* is a car that has already been serviced.

Transitions *T4-T1-T8-T9* describe maintenance No. 1 (MN-1), which is performed independently of maintenance No. 2 (MN-2) and ongoing repairs with this form of work organization of repair crews. Transitions *T5* and *T6* describe the formation of a queue for servicing MN-2 and OR, respectively. Transition *T8* describes the exit of the car from the queue and the start of service. During maintenance (transitions *T10…T17*), the analysis of the marks of the chip {φ} is performed. If the car needs only current repairs (the value of its token [type] = OR), then maintenance is carried out at one post *k*. If its chip takes the value [type] = MN-2, then the car is serviced by all four posts *P9-P16* in turn. Service begins with a free post (denote its number $k = \overline{1...4}$).

When performing function *O* as a result of servicing the car, the chip is assigned the corresponding mark $\{\varphi\}\{\varphi_K\}$, which automatically blocks entry for this car to the post numbered *k*, but does not block its movement to other posts. After maintenance, transition *T18* is activated only when the chip has all the marks φ_k, $\forall k = \overline{1...4}$. In any other case, it is the transition T19 that is triggered, and the car gets back to the service queue.

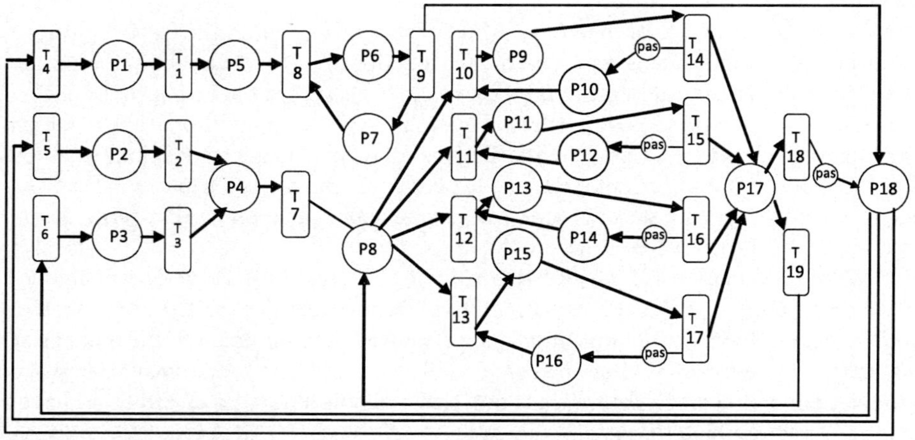

Fig. 1. The model of the Petra network, which reflects the operational post form of the organization of repair crews. Source: author's development

To facilitate the perception of the service process, the label conversion functions is highlighted in separate blocks (marked ".pas") on the diagram. Label conversion occurs in any chip passing through this transition. It should be noted that execution of transition T is possible only if the predicate condition is true, which is analyzed before execution of the transition (the analysis algorithm is embedded in the transition).

During maintenance, the "repair crews" and "cars" chips are in the same position and are displayed as a single unit – the occupied brigade. At this time, their marks overlap, forming a union of two sets. After the service is performed, the chips are separated again, but they already have the same marks. Redundant marks of the "repair crews" chip are destroyed when performing function O, shown in Fig. 1 in the form of a ".pas" object, while the "cars" chips store in their marks' information about which crew performed the repair of this car. The above-mentioned modernization of the classic Petri net greatly facilitates the creation of a discrete model of repair production of MVE, as it allows to flexibly take into account many conditions during modeling, which cannot be reflected when using classical Petri nets.

The specified method is used to model any organizational forms of the labor process of the personnel of motor transport enterprises, which can be presented as a system of events and conditions (preconditions and postconditions). According to the simulation results, data on the number of transitions and positions, duration of transitions of one step, limitation of the number of steps, warning about the number of possible conflicts are obtained. All this information makes it possible to simulate the work of the repair service of the motor vehicle enterprise in real time, taking into account the risks that arise in the process of providing services from technical influences and current repairs of rolling stock of motor transport.

In Fig. 2 shows a Petri net that reflects a discrete system for the aggregate-zonal organization of production with the work organization of repair and maintenance personnel specialized in terms of types of influence. With this form of labor organization of repair workers, serviceable cars that do not require maintenance are in position *P11*. After waiting, displayed in transitions *T4 (T5, T6)*, cars get into the queue for servicing MN-1 (MN-2, OR) and move to positions *P1 (P2, P3)*, respectively. Waiting time *T4*, *T5* and *T6* depends on the make of the car (model parameter – service mileage, average time for failure after current repair).

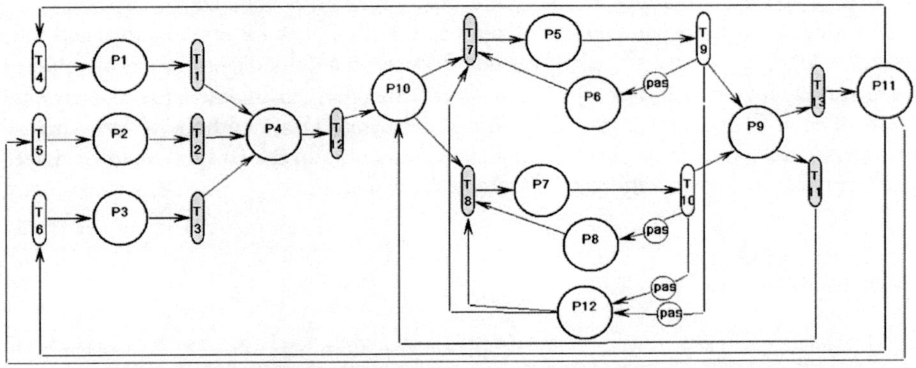

Fig. 2. The model of the organization of the work of car repair personnel in a brigade specialized in the types of impact with an aggregate-zonal form of organization of repair production. Source: author's development

We assume that the flow of applications for OR is subject to the Poisson distribution law, and the flow of applications for MN-1 and MN-2 has a normal distribution law and is characterized by a small variance, which reflects the influence of random factors on the deviation of the maintenance execution time from the planned one. Then, transitions *T1-T3* are immediately activated, which reproduce the movement of cars to the general queue for servicing *P4*. From the general queue *P4*, the cars get to the distribution point *P10*: there, service of the application is carried out in two steps - *P5* and *P7*, moreover, the jobs with the shortest service time are performed first.

Events *T7* and *T8* characterize the expectation of release of resource *P6* (repair station free) and *P12* (specialized repair team free). The events are triggered only if the repair team, which is in position *P12*, has the necessary specialization (by type of impact): MN-1, MN-2 or OR to service the current request. The initial marking *P12* contains information about the number of repair crews at the enterprise and their specialization, *P6* and *P12* reflect the number of maintenance stations. The service time for the first half of the work (transition *T9*) and the service time for the second half of the work (transition *T10*) depends on the labor intensity of the work performed and is subject to the Poisson distribution law. After the end of the service, the time of which is displayed using transitions *T9 (T10)*, the repair crew returns to position *P12*, the post is vacated, and the application moves to position *P6 (P12)*. After the service is completed, the car is moved to position *P9* (applications that have been completed). If its service is partially completed, then event *T11* is triggered and the car takes the internal queue *P10* to finish service. In the event that the car is fully serviced, transition *T13* is triggered, and the car moves to position *P11* (cars on the line). The following statistical data are used in the modeling process: waiting time for the car (the time of the application from the moment of receipt to *P4* to the moment of receipt of *P11*); car service time (the total time of the application from the moment of receipt to *P10* to the moment of receipt of *P9*); the car's operating time on the line (the time spent at position *P11*); working hours of service posts (time of stay and number of markers at positions *P5* and *P7*); idle time of service posts (time spent and number of markers at positions *P6* and *P12*); downtime of repair crews (time spent and number of markers at position *P12*).

In order to carry out an analysis of the peculiarities of work organization and the motivation of repair workers, the study used information obtained from 19 motor vehicle enterprises, which, in addition to freight transportation, perform maintenance and current repairs of cars. In the Table 2 provides a list of these enterprises with the address, organizational and legal form of business, registration number in the Unified State Register of Enterprises and Organizations of Ukraine (USREOU).

Table 2. Enterprises that provided information on the organization of the labor process of car repair personnel

№	Name of Company	USREOU code	Location	OLF*
1	Joint-stock company "Kyiv" Production company "Rapid"	05475156	Kyiv	JSC
2	Public joint-stock company "Kharkiv motor vehicle enterprise 16363"	01332106	Kharkiv	JSC
3	Private joint-stock company "Autotransport enterprise 13555"	05465755	Kropyvnytskyi	PJSC
4	Private joint-stock company "Transport and forwarding combine "Zakhidukrtrans"	13825481	Drogobich	PJSC
5	Private joint-stock company "ATP 11263"	03116157	Dnipro	PJSC
6	Public joint-stock company "ATP 13058"	05475147	Kyiv	JSC
7	Private joint-stock company "ATP-1"	03746384	Kyiv	JSC
8	Public joint-stock company "Kyivske ATP 13061"	05440979	Kyiv	JSC

(*continued*)

Table 2. (*continued*)

№	Name of Company	USREOU code	Location	OLF*
9	Limited Liability Company "Truck Service Lviv"	31417137	Lviv region	LLC
10	Limited Liability Company "112 Ukraine"	20856507	Lviv	LLC
11	Company with additional liability "Autotransport enterprise 11262 Vazhavtotrans"	03116163	Dnipro	CAL
12	Company with additional responsibility "Kharkivske ATP 16363"	01332106	Kharkiv	CAL
13	Zahid BP Group Limited Liability Company	40037324	Dubliani	LLC
14	Limited Liability Company – "Divitrax"	32089226	Rivne region	LLC
15	Private enterprise "Intertraffic"	33312872	Odessa	PE
16	Limited Liability Company "Ukrainian Logistics Systems"	30784014	Kyiv	LLC
17	ROMDI Ukraine Limited Liability Company	36834332	Lutsk	LLC
18	Raben Ukraine Limited Liability Company	32306522	Brovary	LLC
19	Private joint-stock company "Ivano-Frankivsk-auto"	05495466	Ivano-Frankivsk region	JSC

* Note: OLF is an organizational and legal form; JSC – public joint-stock company; PJSC – a private joint-stock company; CAL – company with additional liability; PE – private enterprise; LLC – a limited liability company. Source: author's development.

The enterprises that provided information about the organization of the labor process have different organizational and legal forms of business, different terms of existence, are located in different regions of Ukraine, that is, it can be considered that the sample is representative and reflects the situation that has developed at the enterprises of the motor transport industry regarding the application of forms organization of production of MN and OR of cars and forms of organization of work of car repair personnel.

The implementation of the simulation model of the functioning of the technical service of the motor vehicle enterprise was carried out with the help of a program developed to solve the set tasks, written in the Delphi programming language. The created program for simulating production processes of maintenance and repair, as well as labor processes that take place at motor vehicle enterprises under different operating conditions of the enterprise, made it possible to obtain the results presented in the Table. 3.

Table 3. Recommendations regarding the choice of the form of labor organization of the company's personnel, which allows employers to keep reserves to cover additional labor costs

The size of the motor vehicle enterprise, the listed number of cars	The form of work organization of car repair personnel	Peculiarities of influence of forms of labor organization on labor productivity of car repair personnel	Expected socio-economic effect
1–50 units	Individual	Obtaining additional work results due to the improvement of professional skill, culture and quality of customer service	Improving the quality characteristics of the personnel structure of employees, improving the image and reputational positions of the enterprise, the possibility of applying flexible work schedules
50–100 units	Collective with the organization of complex tef1ams	Increase of work results due to reduction of time for service execution due to parallel execution of works	Attracting more customers, reducing the time customers wait for orders to be fulfilled
100–400 units	Collective with the organization of teams specialized in the types of influences	Increasing work results due to the expansion of the volume of services provided through the performance of specialized operations	Increasing the volume of services provided, improving the quality of their performance, more effective use of production facilities

Source: author's development

Listed in the Table 3 recommendations allow employers to determine a rational form of work organization of repair workers performing maintenance and repair of cars, taking into account such an important factor as the size of the motor vehicle enterprise.

Thus, as a result of the formalization of the task of choosing rational forms of work organization of repair and maintenance personnel and the construction of a simulation model of the operation of the technical service of a motor vehicle enterprise with the help of the mathematical apparatus of modernized Petri nets, a rational form of work organization of car repair personnel for the specific working conditions of the enterprise was determined (Table 3), which is the main element of an effective motivational strategy for organizing the work of car repair personnel.

5 Conclusion

As a result of the research, the effectiveness of the mathematical apparatus of modernized Petri nets was proven, which made it possible to build dynamic models of labor organization of work performers and to determine effective forms of labor process organization of repair and maintenance personnel in the specific working conditions of the enterprise. Thus, a tool was obtained for determining the form of work organization that is effective for the specific conditions of work of a motor transport enterprise, which satisfies the requirements for the formation of an effective motivational strategy for the organization of the work of the personnel of motor transport enterprises. At the same time, the initial data are the statistical reporting of economic entities, as well as the performance indicators of the technical service of motor vehicle enterprises. Since the motivational strategy for organizing the work of the company's auto repair personnel proposed in the study involves increasing labor costs due to the rational organization of the labor process, the recommendations developed during the simulation regarding the choice of the form of labor organization of the company's personnel, which allow employers to obtain reserves to cover additional labor costs.

The prospect of further research in this direction may be the improvement of organizational forms of production of technical influences at motor vehicle enterprises, on the basis of which rational forms of work organization of car repair personnel are selected and effective systems of material stimulation of performers are developed. The obtained research results can be used in the process of finding and implementing an effective salary system for the personnel of motor vehicle enterprises in accordance with the requirements of corporate social responsibility.

References

1. The YouControl system is an online company verification service [website]. https://youcontrol.com.ua/ru/. Last accessed 25 Sep 2022
2. Antonenko, N., Bazyliuk, A., Ilchenko, V., Nadiia, R.: Methodological tools of verbal evaluation of efficiency enterprise's personnel payment systems in the context of corporate social responsibility. In: Alareeni, B., Hamdan, A. (eds.) Financial Technology (FinTech), Entrepreneurship, and Business Development. ICBT 2021. Lecture Notes in Networks and Systems, vol. 486. Springer, Cham. (2022). https://doi.org/10.1007/978-3-031-08087-6_36
3. Bednyak, M.N.: Modeling of Car Maintenance and Repair Processes. Kyiv (1983)
4. Bidnyak, M.N.: Organization of Management: Education. Manual. Kyiv (2003)
5. Nesterenko, B.B.: Modeling Parallel Processes: From Petri nets to Neural Networks: Monograph. NAS of Ukraine, Kyiv (2004)
6. Kryvoruchko, O.M.: Quality Management at Road Transport Enterprises: Theory, Methodology and Practice: Monograph. Khnadu, Kharkiv (2006)
7. Bakalinsky, O., Lozhachevska, O., Ilchenko, V., Kovalenko, N.: Analysis of trajectory of client's attitude formation in managerial decisions for improving the customer service value. In: Alareeni, B., Hamdan, A., Elgedawy, I. (eds.) ICBT 2020. LNNS, vol. 194, pp. 1947–1957. Springer, Cham (2021). https://doi.org/10.1007/978-3-030-69221-6_140
8. Organization of work: training and part-time. manual Babenka, A.G. (eds.) Dnipropetrovsk (2014)

9. Peterson, J.L.: Petri Net Theory and the Modeling of Systems. Prentice Hall, Englewood Cliffs (1981)
10. Barad, M.: Petri nets—a versatile modeling structure. Appl. Math. **07**(09), 829–839 (2016). https://doi.org/10.4236/am.2016.79074
11. Nadiia, R., Innola, N., Olesya, L., Oleksandr, N., Yulia, K., Yulia, H.: Socio-economic processes functioning and innovation education development. In: Alareeni, B., Hamdan, A. (eds.) Innovation of Businesses, and Digitalization during Covid-19 Pandemic: Proceedings of The International Conference on Business and Technology (ICBT 2021), pp. 749–763. Springer International Publishing, Cham (2023). https://doi.org/10.1007/978-3-031-08090-6_47
12. Martyn, A., et al.: Gender equality in access to the profession of land surveyor and geodesist & land appraiser in Ukraine: national and regional assessment. Int. Trans. J. Eng., Manag., Appl. Sci. Technol. **13**(2), 13A2S: 1–8 (2022)
13. Mykhaylichenko, M.V., Hridin, O.V., Vitkovskyi, Y., Hryschenko, N.V., Reznik, N.P.: Improvement of the personnel management system in the process of employment as a factor of increasing the competitiveness of enterprises. In: AIP Conference Proceedings 2413, 040004 (2022). https://doi.org/10.1063/5.0091674
14. Yuliia, M., et al.: Financial opportunities management of ensuring enterprise investment costs. Int. Trans. J. Eng. Manag., Appl. Sci. Technol. **13**(2), 13A2I: 1–10 (2022)
15. Nadiia, P.R., Oleksandr, V.H., Ivanna, V.C., Oleksiy, O.K., Mykola, V.M.: Mechanisms and tools of personnel management in institutional economic. AIP Conf. Proc. **2413**, 040012 (2022). https://doi.org/10.1063/5.0089330
16. Kalna-Dubinyuk, T., Zbarska, A., Ovadenko, V., Barylovych, O., Heraimovych, V., Reznik, N.: The role of information technologies in access to rural tourism education. In: Alareeni, B., Hamdan, A. (eds.) Innovation of Businesses, and Digitalization during Covid-19 Pandemic: Proceedings of The International Conference on Business and Technology (ICBT 2021), pp. 739–747. Springer International Publishing, Cham (2023). https://doi.org/10.1007/978-3-031-08090-6_46

Impacts of PR and AI on the Reputation Management: A Case Study of Banking Sector Customers in UAE

Riadh Jeljeli(✉), Faycal Farhi, and Alaaldin Zahra

College of Communication and Media, Al Ain University, Abu Dhabi, UAE
{Riadh.jeljeli,faycal.farhi,alaaldin.zahra}@aau.ac.ae

Abstract. The banking sector organizations focus on improving their reputation through Public Relations and communication. However, today, when Emotional Intelligence accompanies technology, providing clients with intelligent PR communication is significantly helping in reputation management. Emotional Intelligence and Artificial Intelligence also focus this study on the PR practices widely accompanied by Reputation Management Purposes. The conceptual model is supported by a symmetric communication model that is further tested using Structural Equation Modelling. Results revealed that Public Relations significantly affect Artificial Intelligence, indicating Emirati banks have widely integrated Artificial Intelligent systems in their clients' support systems ($p > .006$). Besides, the effect of Emotional Intelligence on Artificial also remained significant, indicating the integrated PR system as having Artificial Intelligence ($p > .082$). The effect of Artificial Intelligence on Reputation Management also remained significant ($p > .010$), indicating the role of AI in improving organizational reputation. Finally, the mediating effect of Symmetric Communication also remained significant ($p > .000$), affirming its role in indirectly affecting Reputation Management. The current study concluded that Artificial Intelligence is integrated into Public Relations Practices and Emotional Intelligence. As a result, these factors play a constructive role in Reputation Management. While Symmetric Communication is further adding more to the Reputation Management process as an indirect variable, also highlighting the importance of communication for an organization.

Keywords: Public relations · Emotional intelligence · Artificial intelligence · United Arab Emirates · Reputation management

1 Introduction

The financial sector globally pays special consideration to improve and upgrade its services. Particularly, banking organizations remain comparatively more concerned about their internal and external relationship development and management [1]. According to Dacko-Pikiewicz [2], the goal of relationship management is not only to stay in touch both externally and internally, as these organizations equally value communication to build and sustain their reputation in the business arena. Especially banks remain concerned about building relationships with their relationships to meet their organizational

B. Alareeni et al. (Eds.): ICBT 2022, LNNS 620, pp. 265–277, 2023.
https://doi.org/10.1007/978-3-031-26953-0_26

goals. Notably, a healthy client-organizational relationship ensures loyal customers, positive Word of Mouth, and attaining desired goals [3]. Talking particularly about relationship management, Kazankova [4] cited an example of Public Relations professionals that ensure communication among the organization and their clients. These PR professionals value the importance of communication, provide on-time support, and adopt strong problem-solving behavior.

Consequently, the improved reputation of an organization becomes inevitable. As noted by Szwajca [5], organizations, specifically the banking sector, aim to position themselves in the spotlight using different communication strategies. The relevant communication involves details about services and promotional packages and provides a customer-support system. Despite, many stakeholders considering advertising as a significant factor, the communication by Public Relations professionals affirms the strategic attainment of organizational goals including reputation management [6].

Similarly, Kivayilu and Wanjira [7] consider Public Relations practices as widely accompanied by improved communication practices, particularly in the United Arab Emirates-based banking sector organizations. Their Public Relations professionals are well trained and recruited according to their intelligence capabilities and communication skills. These PR professionals are considered to have strong empathy and the ability to recognize, manage, and use communication skills that add value to their expertise [8]. Altogether, Naeini and their colleagues [9] attributed these capabilities to Emotional Intelligence, which enhances goal-oriented behavior among PR professionals. However, today, when technology has integrated into our lives, human force is also advanced and sometimes replaced [10], by certain technological inventions, indicating a major revolution in PR practices. For example, using Artificial Intelligence in online PR practices can be observed in many cases. According to Arief and Saputra [11], this AI-enabled online system provides clients with on-time response and service delivery. Mainly known as "AI Bots," the relevant technology provides the clients with twenty-four-hour services, with just a single click, where communication becomes a fundamental part of the problem-solving and support system. Notably, the AI-enabled systems contain improved yet complex systems to ensure effective human-computer interaction. These systems have enhanced the conventional patterns of Public Relations practices, where the availability of the human force to communicate was the only way to access the services [12]. As a result, when consumers are provided with an instant support system and efficient service delivery and their problems are solved, they feel motivated to consider the same banking service again. Besides, sharing their experiences with others also results in increased client turnover and improved organizational management [13, 14].

Thus, this chapter also systematically and empirically tests the role of Public Relations and Emotional Intelligence in Reputation Management in Emirati banking sector organizations. However, both exogenous variables are proposed as resorted to Artificial Intelligence to meet the desired goals. Notably, the role of symmetric communication is also tested to highlight the primary characteristics, distinguishing the role of Artificial Intelligence (AI Bots) in Reputation Management. The researchers have divided this chapter into five consistent sections, providing support to assess and make the conclusions accordingly.

2 Literature Review

2.1 Artificial Intelligence in Public Relations

According to Biswal [15], Artificial Intelligence is observed in almost every field today. Cision and Google Analytics are the two most common and observed AI-enabled tools that witness their importance. Besides, personal assistants such as Siri by Apple, Alexa by Google, and others are some prominent examples of AI in our daily lives. However, talking specifically about the role of Artificial Intelligence in communication and Public Relation practices, Szwajca [5] considers Artificial Intelligence as providing Public Relations professionals ta flexible environment where they can brainstorm and focus on creative prices such as crafting potentially compelling messages and also planning for a strategic communication outreach. According to Award [16], enhanced predictive analysis, Chatbots, sentiment analysis, Natural Language Generation (NLG), and others strongly affect organizations. The more these organizations upgrade their PR systems, the more they enjoy clients' attention, loyalty, and improved reputation among their rivals [17].

H1: Public Relations has a significant effect on Artificial Intelligence.

2.2 Emotional Intelligence in Artificial Intelligence

Emotional Intelligence in Public Relations is one of the key elements that help the professional effectively persuade clients. When the PR professional understand the client's need, answer their queries, and provide them with the best suitable services they want, the clients will feel valued [18]. Consequently, increased attention and loyalty will become an inevitable phenomenon for the organization [19]. Notably, Public Relations professionals confront several emotions from the clients, i.e., impatience, anger, excitement, and others. According to Becker and Lee [1], the PR professional should be emotionally intelligent to cope and settle t down with the relevant matters. In this regard, the Robotics and software developers are designing Chatbots and AI-enabled service representatives enriched with Emotional Intelligence. The relevant system aims to improve clients' sales and purchases, support, and technical solutions. The developers are focused on stepping up their approaches, as the aim is to provide strong support to eh clients [20].

H2: Emotional Intelligence has a significant effect on Artificial Intelligence.

2.3 Artificial Intelligence in Reputation Management

According to Liew [21], clients do not worry about the representative as a human or a chatbot as they are mainly concerned about solving their problems and answering their questions. Clients expect the service provider to provide them with a quick response regardless of the complexity and nature of their inquiry. Consequently, today organizations are integrating and using Chatbots as one of the most preferred marketing and Public Relations tools to increase their reputation and attain their goals [22]. Shami and Ashfaq [23] further argued that Chatbots are a strong example of Artificial Intelligence contributing to organizational reputation management. These bots are getting smarter as

Emotional Intelligence is a critical component. Today, organizations not only chat with clients but also gratify their needs with voice chatbots [24]. AI developers are training these bots to understand the client's needs, intentions, data, and context leading to an increased and ever-improved emotional intelligence among them [25].

H3: Artificial Intelligence has a significant effect on Reputation Management.

2.4 Symmetric Communication for Reputation Management Purposes

Symmetric communication is widely discussed and highlighted due to its applicability and perceived benefits [26]. In general terms, Moschella and Pinto [27] consider symmetric communication as providing an equal chance for both parties to communicate and increasing the sense of understanding and mutual agreement. Especially when the service providers practice symmetric communication, they provide a more democratic and equal chance to their clients to share their opinion and feedback about the relevant services [28]. Especially for an organization, symmetric communication is one of the crucial factors to success [29]. As noted by Sizaro [30], symmetric communication allows an organization to know its clients, acknowledge their needs, and solve their concerns in the best possible way. As a result, when today twenty-four presence and communication are available through different platforms, reputation management is not a questionable phenomenon. It also demands that organizations not only understand but also ensure quick service and support to eh clients [31]. As noted by Olaniran [32], only when the clients are satisfied with an improved reputation and desired financial goals are attainable.

H4: Symmetric Communication has a significant indirect effect on Reputation Management.

3 Theoretical Framework

The Two-way symmetric model of Public Relations theoretically supports this research. As noted by Bourne [22], the two-way symmetric model is one of the most ethical and sophisticated practices of Public Relations [33]. As noted by Biswal [15], categorical components of fairness, equality, and justice further help to guide ethical decisions according to the two-way communication model. Two-way communication comes under the universal laws of ethics proposed by Immanuel Kant, which he thinks should be met by the people. Two-way communication helps organizations to ensure mutually beneficial relationships [34]. For example, organizations offering banking services to clients rely on the proposition of "utilitarianism," which emphasizes the actions ensuring the common good for the clients and the organization. As a result, attaining Emotional Intelligence and further integrating Artificial Intelligence is based upon the clients' getting their queries answered effectively, enjoying good quality services that are just a single click away [35]. Under the relevant argument, Tarasov [36] considers organizations subservient to social needs as Public Relations practices seek the greatest good for all [37] (Fig. 1).

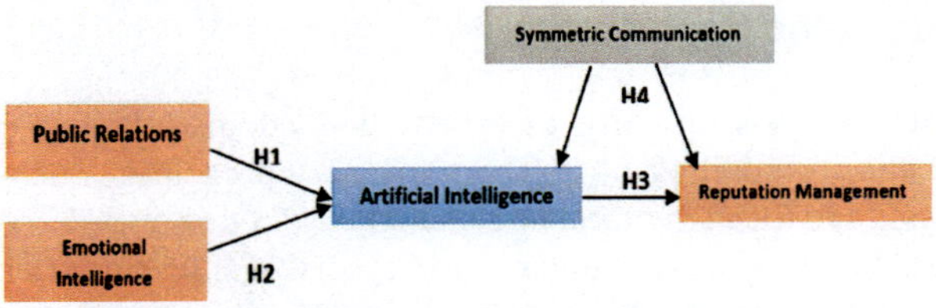

Fig. 1. Conceptual model of current research

4 Methodology

This study is based on an experimental approach. The experimental design was the most suitable for the current research based on the problem, objectives, and questions. According to [38], experimental research is based on scientific constructs with cause and effect relationships [39]. Notably, any changes in the dependent variables are primarily caused by the independent variables. Exploring these changes is the main objective of an experimental approach. Further, the research used self-proposed survey questionnaires designed on a Five-point Likert scale [40]. The questionnaires were sent through email and personal visits to the required organizations. The data gathering was done from June 2022 to August 2022. After the data gathering, the researchers performed data analysis, specifically structural equation modeling, by using the Amos Ver 23.

4.1 Sampling Approach

The study population is comprised of all the clients of the Emirati banking sector. Further, the researchers narrowed down the sample size and selected a sample of $n = 400$ clients from Abu Dhabi and Al Ain city. Selecting the sample of n = 400 individuals was important as the study involves Structural Equation Modeling. According to Kelcey [41], the studies containing Structural Equation Modeling should contain a minimum sample of $n = 200$ respondents to ensure the reliability of the results. Consequently, the sample size of $n = 400$ individuals was ideal [42]. Further, the researcher randomly selected the respondents, regardless of their affiliation with any specific bank. Instead, the only criterion was their experience with the online customer support system through the official websites/apps of the relevant banks. Thus, after the data gathering, the researchers removed some questionnaires as they were incompletely filled. Also, 3 of the questionnaires were missing. Thus, the total response rate remained at 97.2% ($n = 389$).

4.2 Research Ethics

The researchers ensured respondents that their data would be kept confidential and their details would not be used for commercial purposes. Further, informed consent was also considered an important research ethic [43]. Finally, the respondents were informed that they could quit filling out the questionnaires whenever they wanted without further obligations and questioning.

5 Analysis and Study Findings

To conduct the data analysis formally, the researchers adopted the two-stage approach of Structural Equation Modelling [44]. The relevant approach first involves measurement model analysis and then structural model analysis. In this regard, the measurement model analysis first involves the convergent validity analysis. As summarized in Table 1, the researchers calculated the Factor Loading, Lambda, Expulsion, Average Variance Extracted, Composite Reliability, and Cronbach Alpha values. It was found that a majority of Factor Loading values were greater than the threshold value of 0.5. Besides, the Average Variance Extracted value also remained greater than the threshold value of 0.5 (.733–.856). Further, regarding the Campsite Reliability, all the relevant values (.724 to .833) are greater than the threshold value of 0.7 [44]. Additionally, the Cronbach Alpha values (.734–.833) were greater than the designated value of .07, indicating that the convergent validity of the measurement model is affirmed and the items contain internal consistency [45].

Table 1. Convergent validity of Measu2ewzrement model

Constructs	Survey Items	Loadings	LAM	APRIL	AVE	CR	CA
Public relations	PR1	.705	0.724	0.276	.829	.833	.827
	PR2	.851	0.651	0.349			
	PR3	.807	0.602	0.398			
Emotional intelligence	EI1	.776	0.781	0.219	.830	.821	.802
	EI2	−.172	0.788	0.212			
	EI3	.884	0.712	0.288			
Artificial intelligence	AI1	.888	0.687	0.313	.856	.799	.799
	AI2	.648	0.405	0.595			
	AI3	.844	0.58	0.42			
Symmetric communication	SCM1	.829	0.654	0.346	.733	.734	.762
	SCM2	.537	0.724	0.276			
	SCM3	.637	0.651	0.349			
Reputation management	RMG1	−.087	0.602	0.398	.785	.817	.812
	RMG2	.762	0.781	0.219			
	RMG3	.809	0.788	0.212			

After assessing the convergent validity, the researchers examined the discriminant validity as the second important step in measurement model analysis. Notably, for discriminant validity, the research uses two criterion-based approaches, as suggested by Baggozi and Yi [46]. First, examining the Fornell-Larcker criterion revealed all the squares of Average Variance Extracted values are greater than the correlation values in Table 2, indicating a significant difference among all the mentioned values. Further, the

Heterotrait-Monotrait Ratio scale revealed the HTMT value at .391, smaller than the threshold value of .90, as suggested by Tahedoost [47]. Thus, the measurement model is affirmed as having discriminant validity (Table 3).

Table 2. Fornell-Larcker criterion

	PR	HI	AI	SCM	RMG
PR	**.687**				
HI	−.116	**.688**			
AI	.107	.045	**.732**		
SCM	.431	.008	.239	**.537**	
RMG	.076	.003	.050	.001	**.616**

Table 3. Heterotrait-Monotrait ratio

	PR	EI	AI	SCM	RMG
PR					
HI	.134				
AI	−.006	−.045			
SCM	−.425	−.055	−.214		
RMG	−.085	.012	−.050	.046	

The goodness of fit is an important component of the measurement model analysis. It examines the extent to which the observed data fits well with the expected results [48]. Thus, the goodness of fit in this research revealed the chi-square value at $x^2 = .201(18)$, with the probability level at .003. Besides, the Non-Fit Indices value remained at .516, and the Standardized Root Mean Square value at .399. Thus, the calculations validated the goodness of fit (Fig. 2).

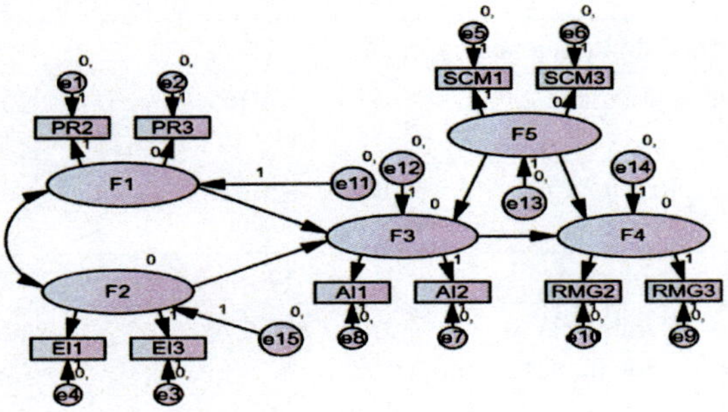

Fig. 2. Goodness of Fit

Coefficients of Determination R^2 involves determining the predictive power of the Independent Variable. Pasha and their colleagues [49] further define coefficients of Determination R2 as assessing the extent to which the Independent Variable is causing variance in the dependent variable(s). Thus, the relevant analysis in current research revealed a 50.0% variance in AI and 47.0% variance in Reputation Management, and 56.7% in Symmetric Communication (See Table). Thus, we found a strong predictive power of the independent variables in the current research (Table 4).

Table 4. Coefficients of determination *R2*

	R2	Strength
Artificial intelligence	.500	Strong
Symmetric communication	.567	Strong
Reputation management	.470	Moderate

Finally, the researchers examined the relationships between study variables proposed in the conceptual model. For the relevant purposes, the researcher conducted path analysis, including regression weights, *t*-values, path values, and significance values [50]. First, the proposed effect of Public Relations on Artificial Intelligence remained significant, with the path value at .905 and the significance value at $p > .006$. These findings indicated consistency with the arguments given by Liew [21]. As noted, the integration of Artificial Intelligence in Public Relations practices remained significant in many ways. Especially when it is about clients' queries requiring quick support, AI-enabled Public Relation systems are an ideal consideration for most organizations. In the second hypothesis, the research proposed a significant effect of Emotional Intelligence on Artificial Intelligence, as also witnessed by Suciati and their colleagues [18] in their study. Thus, the findings remained online with the study by Suciati and their colleagues, with the path value at .533 and the significance value at $p > .082$ (Table 5).

Table 5. Path analysis & regression weights

Hypotheses	*B*	*t*	*Sign*	*Decision*
Public relations→ Artificial intelligence	.905	2.272	.006	Accepted
Emotional intelligence→ Artificial intelligence	.533	6.159	.082	Accepted
Artificial intelligence→ Reputation management	1.741	5.927	.010	Accepted
Hypotheses	***B***	***Indirect Effects***	***Sign***	***Decision***
Artificial intelligence→ Symmetric communication→ Reputation management	1.040	7.270	.000	Accepted

Additionally, the third hypothesis assumed a significant effect of Artificial Intelligence on Reputation Management. The relevant proposition remained significant ($p > .010$) and compatible with the argumentation made by Alward and their colleagues [16]. As noted, Chatbots answer the clients' queries using Artificial Intelligence and Machine Learning approaches. The organizations focusing on providing intelligent support to their clients are enjoying comparatively stronger clients'-organization relationships and improved reputations among them. Finally, the fourth hypothesis was based on the significant, indirect effect of Symmetric Communication on Reputation Management. The relevant assumption was based on the argumentation given by Ardila [17]. As noted, communication and reputation management are two indispensable phenomena. Through communication, it is to ensure the clients that their needs are fulfilled. Further, their presence is fully acknowledged, leading to an increased trust in the organization and positive perceptions about it. Thus, the mediation of Symmetric Communication also remained significant, with the value of the indirect effect at 7.270 and the significance value at $p > .000$.

6 Discussion and Conclusion

According to Kuteynikov and their colleagues [51], symmetric communication provides transparency and disclosure, providing several benefits to an organization. When there is direct communication with an equal chance of interaction, it builds trust between the organization and its clients. Public Relations experts build a relationship with clients by considering the importance of symmetric communication. However, today, the role of technology, particularly Artificial Intelligence, is further accelerating Public Relations practices with the goal of reputation management [52]. As noted earlier, AI-enabled chatbots are always available for the clients' support in the banking sector services. These bots are enriched with empathy, awareness, and problem-solving behavior. According to Szwajca [5], the contributions of Artificial Intelligence indicate its obvious presence in the clients' support and service. From a communication point of view, it is practical and revolutionary to ensure client satisfaction and improved organizational reputation [53]. This research also focused on the relevant argumentation with n intent to provide empirical evidence. More specially, the effect of Public Relations practices was significant as indicating a wider inclination of Emirati baking organizations on technology incorporation and integration in PR-based activities. The study respondents agreed that their organizations had integrated Artificial Intelligence into their communication systems leading to improved support services and an opportunity for the experts to brainstorm more creative tactics regarding clients' persuasion [54].

Further, the respondents also agreed that the Artificial Intelligence-based system contains Emotional Intelligence to understand, acknowledge, suggest and gratify the client's needs in the best possible manner [36]. Additionally, the respondents indicated that Artificial Intelligence significantly contributes to organizational Reputation Management. The role of Artificial Intelligence in reputation management was highlighted because it provides quick support and services to the clients, provides them with better suggestions, and helps them in the decision-making process. Consequently, clients not only rely on the services but also build trust in the organization, leading to improved reputation and positive opinion among the clients. Finally, the mediating effect of Symmetric

Communication on Reputation Management also remained significant as the respondents revealed communication as one of the crucial factors in their experiences with the organization [11]. Finally, it is concluded that Artificial Intelligence is integrated into Public Relations Practices and Emotional Intelligence. As a result, these factors play a constructive role in Reputation Management. While Symmetric Communication is further adding more to the Reputation Management process as an indirect variable, highlighting the importance of communication for an organization.

6.1 Limitations and Contributions

This nature contains some primary limitations. First, the researchers focused only on the banking sector, whereas other industries consider and practice public relations. Second, the researchers only adopted Chatbots as a part of Artificial Intelligence. However, other aspects, i.e., recommendation systems, are also helping the clients and organizations. Finally, the third limitation involves the geographical generalizability of the findings. As this research is conducted in the United Arab Emirates, its results cannot be generalized to other regions. However, more studies can further dig out the generalizable results, especially the same conceptual framework in other regions. Besides, using a symmetrical model of communication can additionally help to highlight the significance of Artificial Intelligence in general.

References

1. Becker, K., Lee, J.W.: Organizational usage of social media for corporate reputation management. J. Asian Finance Econ. Bus. **6**(1), 231–240 (2019). https://doi.org/10.13106/JAFEB.2019.VOL6.NO1.231
2. Dacko-Pikiewicz, Z.: Reputation management and family. Business (2021). https://doi.org/10.4324/9781003226215
3. Munyoro, G., Magada, E.: The significance of corporate reputation management in the public sector : a case study of national social security authority (NASSA). Res. J. Manag. **4**(4), 1–20 (2016)
4. Kazankova, T.: Artificial intelligence as an object of legal protection. In: Ashmarina, S.I., Mantulenko, V.V. (eds.) Current Achievements, Challenges and Digital Chances of Knowledge Based Economy. LNNS, vol. 133, pp. 371–376. Springer, Cham (2021). https://doi.org/10.1007/978-3-030-47458-4_44
5. Szwajca, D.: The role of social media in corporate reputation management-the results of the polish enterprises. Found. Manage. **9**(1), 161–174 (2017). https://doi.org/10.1515/fman-2017-0013
6. Choi, S.B., Min, H., Joo, H.Y.: Examining the inter-relationship among competitive market environments, green supply chain practices, and firm performance. Int. J. Logist. Manage. **29**(3), 1025–1048 (2018). https://doi.org/10.1108/IJLM-02-2017-0050
7. Kivayilu, D.K., Wanjira, J.: Influence of education program as a corporate social responsibility strategy on competitive advantage of soft drink manufacturers in Nairobi City County, Kenya. J. Soc. Sci. Educ. **2**(2), 88–97 (2020)
8. Adaurhere, E., Musonda, I., Okoro, C.S.: Construction Contingency Determination: A Review of Processes and Techniques. In: Ahmed, S.M., Hampton, P., Azhar, S., D. Saul, A. (eds.) Collaboration and Integration in Construction, Engineering, Management and Technology. Advances in Science, Technology & Innovation, pp. 271–277. Springer, Cham (2021). https://doi.org/10.1007/978-3-030-48465-1_45

9. Naeini, A., Mosayebi, A., Mohajerani, N.: A hybrid model of competitive advantage based on Bourdieu capital theory and competitive intelligence using fuzzy Delphi and ISM-Gray DEMATEL (study of Iranian food industry). Int. Rev. **439**(1–2), 21–35 (2019). https://doi.org/10.5937/intrev1901021n
10. Yu, Z., Liang, Z., Wu, P.: How data shape actor relations in artificial intelligence innovation systems: an empirical observation from China. Ind. Corp. Change **30**(1), 251–267 (2021). https://doi.org/10.1093/icc/dtaa063
11. Arief, N.N., Saputra, M.A.A.: Kompetensi baru public relations (PR) pada era artificial intelligence: case study Praktisi PR di Indonesia. J. Sist. Cerdas **2**(1), 1–12 (2019)
12. Türksoy, N.: The future of public relations, advertising and journalism: how artificial intelligence may transform the communication profession and why society should care. Türkiye İletişim Araştırmaları Dergisi **26306220**, 394–410 (2022). https://doi.org/10.17829/turcom.1050491
13. Gouda, N.K., Biswal, S.K., Parveen, B.: Application of artificial intelligence in advertising & public relations and emerging ethical issues in the ecosystem. Int. J. Adv. Sci. Technol. **29**(06), 7561–7570 (2020)
14. Jeljeli, R., Farhi, F., Setoutah, S., Laghouag, A.A.: Microsoft teams' acceptance for the e-learning purposes during Covid-19 outbreak: a case study of UAE. Int. J. Data Netw. Sci. **6**(3), 629–640 (2022). https://doi.org/10.5267/j.ijdns.2022.4.010
15. Biswal, S.K.: The space of artificial intelligence in public relations: the way forward. In: Kulkarni, A.J., Satapathy, S.C. (eds.) Optimization in Machine Learning and Applications. AIS, pp. 169–176. Springer, Singapore (2020). https://doi.org/10.1007/978-981-15-0994-0_11
16. Alawaad, D.H.A.: The role of artificial intelligence (AI) in public relations and product marketing in modern organizations. Turk. J. Comput. Math. Educ. (TURCOMAT) **12**(14), 3180–3187 (2021)
17. Ardila, M.M.: The rise of intelligent machines: how artificial intelligence is transforming the public relations industry, p. 79 2020
18. Suciati, P., Maulidiyanti, M., Wiwesa, N.R.: The public relations acceptance towards press release application with artificial intelligence. Commun. J. Commun. Stud. **8**(1), 20 (2021). https://doi.org/10.37535/101008120212
19. Galloway, C., Swiatek, L.: Public relations and artificial intelligence: It's not (just) about robots. Public Relat. Rev. **44**(5), 734–740 (2018). https://doi.org/10.1016/j.pubrev.2018.10.008
20. van Heerden, J., Rensburg, R.: Investigating social media conversations: towards implementing an online reputation management framework for NPOs. COMMUNITAS **23**(1), 53–70 (2018). https://doi.org/10.18820/24150525/Comm.v23.4
21. Liew, F.E.E.: Artificial intelligence disruption in public relations: a blessing or a challenge? J. Digit. Mark. Commun. **1**(1), 24–28 (2021). https://doi.org/10.53623/jdmc.v1i1.45
22. Bourne, C.: AI cheerleaders: public relations, neoliberalism and artificial intelligence. Public Relat. Inq. **8**(2), 109–125 (2019). https://doi.org/10.1177/2046147X19835250
23. Shami, S., Ashfaq, A.: 1Strategic political communication, public relations, reputation management & relationship cultivation through social media. Public Relat. **55**(2), 16 (2018)
24. Mazouz, A., Alnaji, L., Jeljeli, R., Al-Shdaifat, F.: Innovation and entrepreneurship framework within the middle east and North Africa region. Afr. J. Sci. Technol. Innov. Dev. **11**(6), 699–710 (2019). https://doi.org/10.1080/20421338.2019.1573959
25. Habes, M., Ali, S., Elareshi, M., Tahat, K.M., Ziani, A.: Understanding users' social TV content using artificial intelligence approach: a survey. In: *2021* International Conference of Modern Trends in Information and Communication Technology Industry (MTICTI), Sana'a, Yemen, pp. 1–7 (2021). https://doi.org/10.1109/MTICTI53925.2021.9664762

26. Wardhani, T.E.: Human resource management for increasing employees' satisfaction and enhancing competitive strengths. Int. J. Econ. Bus. Adm. **VIII**(3), 543–553 (2020). https://doi.org/10.35808/ijeba/552
27. Moschella, M., Pinto, L.: Central banks' communication as reputation management: how the fed talks under uncertainty. Public Adm. **97**(3), 513–529 (2019). https://doi.org/10.1111/padm.12543
28. Cheng, Y., Jiang, H.: Customer–brand relationship in the era of artificial intelligence: understanding the role of chatbot marketing efforts. J. Prod. Brand Manag. **31**(2), 252–264 (2022). https://doi.org/10.1108/JPBM-05-2020-2907
29. Cismaru, D.M., Gazzola, P., Ciochina, R.S., Leovaridis, C.: The rise of digital intelligence: challenges for public relations education and practices. Kybernetes **47**(10), 1924–1940 (2018). https://doi.org/10.1108/K-03-2018-0145
30. Sizaro, N.: A review in the framework of reputation management and consumer relations: online reputation components model proposal. Doğuş Üniversitesi Derg. **23**(1), 219–242 (2022). https://doi.org/10.31671/doujournal.957602
31. Hall, K.: Public penitence: Facebook and the performance of apology. *Soc. Media Soc.* 6(2) (2020). https://doi.org/10.1177/2056305120907945
32. Olaniran, B.A.: Social media as communication channel in emerging economies: a closer look at cultural implications. J. Adv. Manag. Res. **15**(2), 130–145 (2018). https://doi.org/10.1108/JAMR-04-2017-0050
33. Osawa, H., Miyamoto, D., Hase, S., Saijo, R., Fukuchi, K., Miyake, Y.: Visions of artificial intelligence and robots in science fiction: a computational analysis. Int. J. Soc. Robot. (2022). https://doi.org/10.1007/s12369-022-00876-z
34. Bokovnya, A.Y., Begishev, I.R., Khisamova, Z.I., Bikeev, I.I., Sidorenko, E.L., Bersei, D.D.: Pressing issues of unlawful application of artificial intelligence. Int. J. Criminol. Sociol. **9**, 1054–1057 (2020). https://doi.org/10.6000/1929-4409.2020.09.119
35. Rizqiyah, R.N., Kriyantono, R., Sujoko, A.: Literacy studies media relations studies on digitalization and artificial intelligence on company reputation in the digital era in 2008–2019. Adv. Educ. Philos. **5**, 236–248 (2021)
36. Tarasov, I.N.: Problems of legislation as exemplified by the concept of artificial intelligence. SHS Web Conf. **134**, 00111 (2022). https://doi.org/10.1051/shsconf/202213400111
37. Jeljeli, R., Farhi, F., Hamdi, M.E.: The effect of consumer review on the perceived trustworthiness of online retailers: Item response theory perspective. Int. J. Data Netw. Sci. **6**(4), 1341–1350 (2022). https://doi.org/10.5267/j.ijdns.2022.6.001
38. Lee, S.H., Goddard, M.E., Wray, N.R., Visscher, P.M.: A better coefficient of determination for genetic profile analysis: a better coefficient of determination. Genet. Epidemiol. **36**(3), 214–224 (2012). https://doi.org/10.1002/gepi.21614
39. Jeljeli, R., Alnaji, L.A., Khazam, K.: A Comparison between moodle, Facebook, and paper-based assessment tools: students' perception of preference and effect on performance. Int. J. Emerg. Technol. Learn. IJET **13**(05), 86 (2018). https://doi.org/10.3991/ijet.v13i05.8091
40. Piepho, H.: A coefficient of determination (R 2) for generalized linear mixed models. Biomed. J. **61**, 860–872 (2019). https://doi.org/10.1002/bimj.201800270
41. Kelcey, B., Cox, K., Dong, N.: Croon's bias-corrected factor score path analysis for small- to moderate-sample multilevel structural equation models. Organ. Res. Methods **24**(1), 55–77 (2021). https://doi.org/10.1177/1094428119879758
42. Gentles, S.: Sampling in qualitative research: Insights from an overview of the methods literature. Qual. Rep. **20**(11), 25 (2019)
43. Henry, G.: Practical Sampling. SAGE Publications, Thousand Oaks (1990). https://doi.org/10.4135/9781412985451

44. Soltanzadeh, A.: Path analysis of occupational injuries based on the structural equation modeling approach: a retrospective study in the construction industry. Iran Occup. Health **16**, 11 (2015)
45. Ali, S., Habes, M., Youssef, E., Alodwan, M.: A Cross-Sectional Analysis of Digital Library Acceptance, and Dependency during Covid-19. Int. J. Comput. Digit. Syst. **10**(1), 1415–1425 (2021). https://doi.org/10.12785/ijcds/1001125
46. Bagozzi, R.P., Yi, Y.: Specification, evaluation, and interpretation of structural equation models. J. Acad. Mark. Sci. **40**(1), 8–34 (2012). https://doi.org/10.1007/s11747-011-0278-x
47. Taherdoost, H.: Sampling methods in research methodology; how to choose a sampling technique for research. SSRN Electron. J. (2016). https://doi.org/10.2139/ssrn.3205035
48. Barrett, P.: Structural equation modelling: adjudging model fit. Personal. Individ. Differ. **42**(5), 815–824 (2007). https://doi.org/10.1016/j.paid.2006.09.018
49. Pasha, S.A., Youssef, E., Sharif, H.: *Role of virtual reality in improving students' lms experiences: structural equation modelling based study,* p. 7 (2021). https://doi.org/10.1109/MTICTI53925.2021.9664769
50. Fowler, D., et al.: Negative cognition, depressed mood, and paranoia: a longitudinal pathway analysis using structural equation modeling. Schizophr. Bull. **38**(5), 1063–1073 (2012). https://doi.org/10.1093/schbul/sbr019
51. Kuteynikov, D., Izhaev, O., Lebedev, V., Zenin, S.: Legal regulation of artificial intelligence and robotic systems: review of key approaches. Cuest. Políticas **40**(72), 690–703 (2022). https://doi.org/10.46398/cuestpol.4072.40
52. Vodák, J., Cakanova, L., Pekar, M., Novysedlak, M.: Influencer Marketing as a Modern Phenomenon in Reputation Management. Manag. Glob. Transit. **17**(3), 211–220 (2019). https://doi.org/10.26493/1854-6935.17.211-220
53. Yau, K.-L.A., et al.: Augmented intelligence: surveys of literature and expert opinion to understand relations between human intelligence and artificial intelligence. IEEE Access **9**, 136744–136761 (2021). https://doi.org/10.1109/ACCESS.2021.3115494
54. Berg, K.T.: The ethics of artificial intelligence: superintelligence, life 3.0 and robot rights. J. Media Ethics **33**(3), 151–153 (2018). https://doi.org/10.1080/23736992.2018.1482722

Production and Institutional Contribution to the Competitiveness of MSMEs: The Mediation Role of MSME Performance Based on Green Economy

Dian Retnaningdiah and Muafi Muafi(✉)

Management Department, Universitas Islam Indonesia, Sleman, Daerah Istimewa Yogyakarta 55283, Indonesia
muafi@uii.ac.id

Abstract. Most MSMEs in their activities are only concern about how to get the highest possible profit without paying attention to the negative impacts. The purpose of this study is how to increase competitiveness by financial performance through strengthening production and institutions in MSMEs based on a green economy. The methodology used in this study is observations and questionnaires for MSMEs, tabulated questionnaires, and analyzed using the structure equation model (SEM) and AMOS. Based on statistical analysis, the results obtained that financial performance does not have full capabilities as an intervening variable and power in MSMEs based on a green economy.

Keywords: Production · Institutional performance · Competitiveness · Green economy

1 Introduction

Small and Medium Enterprises (MSMEs) are small business units that are able to contribute and function as safety valves both in providing alternative productive business activities, alternative lending, as well as in terms of employment [1]. Industrial development must pay attention to backward linkage with the agricultural sector or the primary sector while forward linkage must pay attention to increase added value and good marketing so that the products produced are not in vain.

When the guidelines for the installation and management of standards for small enterprises have not been created by the authorities, the rise in small-scale industries appears to be one of the major causes of air pollution in the environment [2]. The Green Economy concept, a new economic paradigm and sustainable development plan that prioritizes a balance between economic, social, and environmental goals, is encouraged by this occurrence. This model can address the shortcoming of the previous development plan, which simply prioritizes growth.

B. Alareeni et al. (Eds.): ICBT 2022, LNNS 620, pp. 278–288, 2023.
https://doi.org/10.1007/978-3-031-26953-0_27

The involvement of small-scale businesses in green industries development is more urgent because small-scale businesses not only place greater pressure on the environment, but also have large number of actors. Ignoring the involvement of small-scale businesses in environmental conservation programs carries a greater risk of environmental degradation [3].

Preliminary studies about the implementation of the Green Economy on MSMEs development have been conducted [4] in Sidoarjo Regency. The foundation of the green economy is the understanding of how crucial it is for an ecosystem to maintain a balance between economic actors' actions and resource availability. The green economy strategy also aims to combine the three fundamental values of profit, people, and the environment. According to this perspective, economic actors are encouraged to contribute to society and protect the environment in addition to maximizing profits [4].

The objectives of green economy include achieving harmony between the economy and the environment, converting environmental protection technology, adopting clean industrial methods, and implementing sustainable economic growth. As more people become conscious of the need to conserve the environment, the term "green" is currently used broadly across many spheres of society, including green economies, green consumption, green tourism, green marketing, and green agriculture.

In globalized condition, MSMEs must upgrade their capabilities by innovating and adopting advanced technology and communication in order to increase the ability of entrepreneurs to improvise management without compromising their ability to meet the needs of society in the future. In other words, development operations must be able to provide future generations with the same level of welfare as we obtained from past generations, including science and technology, environmental assets, and natural resources [5].

The globalization era provides opportunities as well as challenges for Indonesian entrepreneurs, including MSMEs because in this era the competitiveness of products is very high, the product life cycle is relatively short according to market tastes, and the ability to innovate products is relatively fast. Small and Medium Enterprises are an important part of the economy of a country or region, including in Indonesia [6].

MSMEs must continue to innovate and be creative to make special strategies in order to strengthen production and institutions. Based on those conditions, it shows that MSMEs are economic pillars that should be maintained and developed, while environmental sustainability must be a concern so that environmental ecosystems are maintained. The initial findings that are expected from this study are being able to identify which MSMEs have and have not conducted green economy-oriented business activities and how they affect the company's performance. After that, the next finding is how to strengthen the model for companies that have done and companies that have not conducted business activities that are oriented towards green economy. The end result will be a strengthening model for companies that have and have not implemented activities based on the green economy and what are the implications for company performance.

This paper makes various novel contributions in an effort to address the research gap, including:

1) Prior study has shown that large enterprises, as opposed to MSMEs, are more likely to apply the concepts of production, competitiveness, and the green economy. This

is because MSMEs' adoption of green economy in their business activities is still constrained by their weak innovative capabilities and simplistic organizational structures. MSMEs can, however, make a contribution to the green economy through their commercial activities.

2) The author answers to the call from previous studies by investigating the relationship between the production, competitiveness, and performance in a larger scale rather than individually, using the context of MSMEs.

This study has the main objective to investigate and analyze the impact of MSME production, institution, and performance on MSME competitiveness, while also considers the mediating role of MSME performance. This study has three sub-main part, namely:

- Introduction, which describes the theoretical problem and business phenomenon that is related to the production, competitiveness, and performance in MSMEs context.
- Literature Review and Methods, which includes theories of MSME performance, production, institution, competitiveness, and green economy.
- Results, discussion, and research limitations.

2 Literature Review and Hypotheses

2.1 MSME Performance

Performance is a multifaceted notion, and the correlation between entrepreneurial orientation and performance depends on the indicators used to assess performance [7]. A number of prior research report differences in performance indicators [8]. Generally, the differences are between financial performance and non-financial performance measures. Non-financial performance evaluation also assesses how well owners or managers are doing in achieving business objectives like customer satisfaction and success on a global scale. Financial performance evaluation evaluates elements like ROI and sales growth [9].

Regarding financial performance, there is often a low convergence between different indicators[10]. At the conceptual level, one can distinguish between measurement of growth and profitability. Although these concepts are empirically and theoretically related, there are also important differences between the two [8].

Businesses with a strong entrepreneurial orientation can target the premium market sector, set a high selling price, and have a competitive advantage, which will, of course, result in higher earnings and a faster pace of expansion [11]. However, data collected by entrepreneurs themselves can provide a great opportunity to examine multiple dimensions of performance, such as comparisons with competitors [12]. Such measures can be subject to bias due to social appropriateness, memory impairment, and/or variations in methods commonly used.

2.2 Production

Production is defined as the use or utilization of resources that transform one commodity into another that is completely different, both in terms of what, where or when these commodities are allocated, as well as in terms of what consumers can do with the commodity. Besides the commodities such as raw materials and machinery, capital is also important as well as human resources and standard operating procedures [13].

2.3 MSME Institutions

Institutions are often considered as a serious obstacle in determining the success of rural community development, especially in the field of agro-complex which involves rural communities with various forms of small businesses. Not only that, the increasing development of MSMEs in terms of quantity has not been matched by an increase in the quality of MSMEs. These factors can be caused by internal factors and external factors from the MSME institution itself, namely internal factors in the form of labor, mastery of technology, financial management, access to financing, and weak entrepreneurship of MSME actors. Meanwhile, the external problems faced by MSMEs include the acquisition of formal legality which until now has become a fundamental problem for MSMEs.

One of the various obstacles experienced by MSME actors is strengthening MSME institutions. It is essential to improve the quality of MSME so that they can compete in both regional and international markets. In institutional strengthening, it is necessary to pay attention to various aspects that need to be considered and improved, namely legalization, capacity building, financial management, access to finance[14].

2.4 Competitiveness

Efforts to increase competitiveness need to be conducted by strengthening the integrated institutional system. The development of an integrated institutional system can streamline the supply chain which will reduce price margins so that product prices can be cheaper and more competitive. In addition, increasing competitiveness is conducted by implementing the right strategy through analysis of internal and external factors [15].

2.5 Green Economy

Green Economy is one of the activities that result in increased human well-being and social equity, significantly reducing environmental risks and ecological scarcity. The Green Economy is built on knowledge and technology and strives to reduce the effects of human economic activity on climate change and global warming while recognizing the interrelationships between human resources and natural ecosystems.

The green economy is a paradigm of economic growth that has gained significant traction in recent years as a new economic development ideology. The United Nations Environmental Program (UNEP) mentions as a new global agreement on how the government can support economic transformation towards a greener economy. Green economy academics criticize and argue that the global power of capitalism is clear evidence and has consequences for the destruction of existing environmental conditions [16].

2.6 Relationship Between Variables

2.6.1 Production to Performance

Since production is closely tied to output/product, it plays a significant role in enhancing performance. Performance may be impacted internally by the element of production [17]. Efficient production by using sophisticated machines will be able to increase production which will ultimately improve performance. In addition, internal factors (production) are more influential than external factors [18].

2.6.2 Production to Competitiveness

Since the New Order era until now, the government has made many efforts to help the development of MSMEs and cooperatives in various programs, ranging from providing cheap credit to technical assistance. Technical assistance is provided through technical guidance and training in production and production management. Through changes in technology, machinery, product design, and production efficiency, it can improve quality and ultimately increase competitiveness [13,19] .

2.6.3 Institutional to Performance

Governments in several developing countries are more interested in supporting large industries than MSMEs [20],because MSMEs in the future are able to face the challenges of a completely open free market in competition with outside economic actors. Institutional strengthening is an important factor for improving the quality of MSMEs. Temtime and Pansiri [21] find that human resource development, organizational development, organizational structure, are important components that affect the performance of small and medium enterprises (MSMEs).

2.6.4 Institutional to Competitiveness

Institutional is an indicator that measures how far the social, political, legal and security aspects are able to positively affect economic activity. The effect of institutional factors on competitiveness is based on several basic principles, such as unprocessed physical strength or unskilled labor, as well as advanced sources of knowledge and research obtained from scientific institutions [22].

2.6.5 Performance to Competitiveness

The company's performance is very necessary to support the survival of the company, but in fact if the company does have a good performance, it would not only have an impact on the survival of the company but also the competitiveness of the company. According to, performance improvement is needed to be able to strengthen competitive advantage for an industry.

H1. MSME production has significant positive effect on MSME competitiveness
H2. MSME institution has significant positive effect on MSME competitiveness
H3. MSME production has significant positive effect on MSME Performance
H4. MSME institution has significant positive effect on MSME performance
H5. MSME performance has significant positive effect on MSME competitiveness
H6. MSME performance mediates production and institutional relations to competitiveness

3 Methodology

The methodology in this study is observations and questionnaires for MSMEs, then the questionnaires are tabulated and analyzed using the structure equation model (SEM) and then analyzed using AMOS. Respondents answer questionnaires related to production, institutions, performance, and competitiveness. The population in this study is the entire MSME actors in Sleman Regency. The target sample is 300 MSMEs. It turned out that the number of those who returned the questionnaire is 257 MSMEs. After further inspection, it turned out that there are 162 MSMEs whose questionnaires had been declared complete and eligible for further processing. The sampling technique is purposive sampling method. The results of testing the validity and reliability are valid and reliable. The results of the normality test state that all data are normal and no outliers. Statistical technique using AMOS 24.0.

4 Results

Respondent Description

MSMEs apply green economy principles and have been operating for more than 2 years (100%), female (100%), 40–59 years old (70%), and have fashion and craft businesses (55%).

Model Fit Test

The research structural equation model is as shown in Fig. 2.

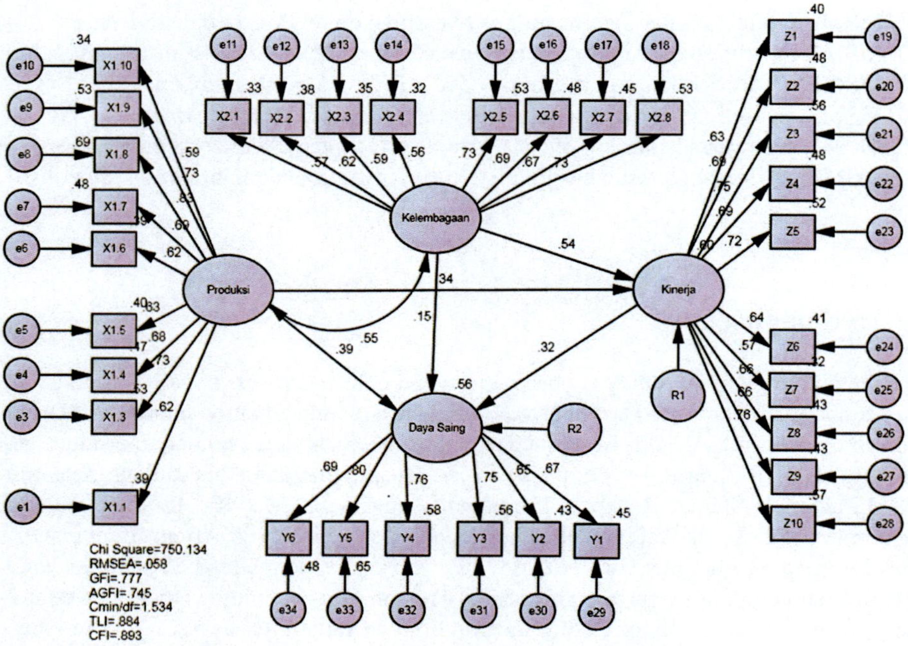

Fig. 2. Production and institutional contribution model to MSME competitiveness: the mediation role of MSME performance based on green economy

The results of the estimation of the structural model can be summarized by the value of the suitability test as shown in Table 1. It can be concluded that the GoF results can still be used because some criteria are still fit and marginal.

Table 1. Goodness of fit index modified SEM model

Goodness of fit index	Model 2		
	Cut off value	Model result	Description
Chi Square	525,756	604,317	Poor
Probability Chi-Square	≥0,05	0,000	Poor
RMSEA	<0,08	0,042	Fit
GFI	≥0,90	0,817	Marginal
AGFI	≥0,90	0,783	Poor
CMIN/DF	≤2,0	1,275	Fit
TLI	≥0,95	0,940	Marginal
CFI	≥0,95	0,946	Marginal

Hypothesis Testing

Here are the Regression Weight and Standardized Regression Weight structural equation models:

Table 2. Model I structural equation hypothesis testing

Causality relationship	Std. Estimate	Critical ratio	P-Value
H1. Production → Competitiveness	0.514	4.415	0.000
H2. Institutional → Competitiveness	0.296	2.923	0.003

In Table 2. The causal relationship of the production variable to competitiveness shows that the production variable has significant effect on competitiveness with a p value of 0.000 so that **H1 is accepted**. Likewise, the causality relationship of institutional variables on competitiveness shows that institutional variables have significant effect on competitiveness with a p value of 0.003 so that **H2 is accepted.**

Table 3. Model II structural equation hypothesis testing

Causality relationship	Std. Estimate	Critical Ratio	P-Value
H3. Production → Performance	0.324	3.420	0.000
H4. Institutional → Performance	0.555	4.573	0.000
H5. Performance → Competitiveness	0.316	2.383	0.017

In Table 3. The causal relationship of the production variable to the performance of MSMEs shows that the production variable has significant effect on the performance of MSMEs with a p value of 0.000 so that **H3 is accepted.**

The causal relationship of institutional variables on the performance of MSMEs shows that the institutional variables have significant effect on the performance of MSMEs with a p value of 0.000 so that **H4 is accepted.**

The causal relationship of the MSME performance variable to competitiveness shows that the MSME performance variable has significant effect on MSME competitiveness with a p value of 0.017 so that **H5 is accepted.**

Table 4. Indirect effect

Relationship	Indirect effect
Production → Performance → Competitiveness	0.102
Institutional → Performance → Competitiveness	0.175

The results in Table 4 show that in accordance with the provisions previously mentioned, the indirect effect between production on competitiveness through the performance variable is 0.12. Meanwhile, the value of the direct effect of production on competitiveness after including the performance variable as a mediation is also still significant, so it can be concluded that there is an indirect effect of production on competitiveness through performance.

The performance variable in this relationship is concluded to be partial mediation because the direct effect between production on competitiveness is still significant. Meanwhile, the indirect effect between institutions on competitiveness through the performance variable is 0.175. The value of the direct effect of institutions on competitiveness after including the performance variable as a mediation is concluded to be insignificant, so it can be concluded that there is an indirect effect between production on competitiveness through performance. The performance variable in this relationship is concluded to be full mediation because the direct effect between institutions on competitiveness is not significant.

Discussion

Based on the results of statistical tests, it is stated that institutional strengthening and production have significant effect on performance. These results prove that strengthening MSMEs based on green economy performance can be done through strengthening production and performance [12, 23]. Based on the final results, it can be seen that the results obtained are poor, this indicates the model that is prepared is not suitable, this shows the role of financial performance as a moderating variable is not optimal, because from 2 (two) model equations, partial mediation and full mediation are obtained.

As is well known, the business strategy conducts in the production department has a big responsibility for good performance [24], besides when the production of innovation is done well it can create a clean environment [25]. A clean environment and good performance are the impact of the production process, with the right process, maximum performance is obtained. In addition, management encouragement also has a big role in this implementation, a proactive environment, and good environmental performance will result in good environmental performance. Environmental assessment is very influential on the value of the company, if the environmental assessment is bad, it would automatically reduce the value of the company itself [26]. To do this, it is necessary to carry out targeted strategies [27]. In order to achieve this condition, a comprehensive assessment is conducted so that environmental damage can be avoided.

Institutional arrangements also have an important role, good governance, and clear work standards produce maximum performance. Good performance and produce a green economy orientation will have an advantage to compete [28]. In the future, the demands for the creation of a green economy are even greater, so that the first step that must be taken is to start implementing green management and prepare a green management blueprint. The government must prepare supporting policies in order to produce green economy activities. So that the growth of MSMEs will result in green growth-oriented growth [29]. To obtain this, the environmental assessment must be conducted properly, one of which is to monitor that there is no uncertainty in environmental conditions that often change. The role of Financial Performance as an intervening variable cannot be a full mediation in every model structure that is formed but can also be a partial moderation.

5 Conclusion

5.1 Managerial Contribution

The findings of this study suggested that all hypotheses proposed are accepted. The production, institution, and performance of MSMEs affects its competitiveness. Furthermore, the MSME production and institution positively influences the performance of MSMEs. This study provides implications for MSMEs owners or managers to encourage an increase in the production and institutional arrangements of MSMEs, as both will lead to higher performance and competitiveness, thus enabling the MSMEs to compete and sustain in green economy. In addition, this study confirms the mediating role of MSMEs performance. The owners or managers of MSME should realize that when they can produce high performance in their business through their production and institutional arrangements, this will eventually lead to higher ability to compete within the industry.

5.2 Limitation and Recommendation for Future Research

This study has several limitations, among others:

1) This study has not yet grouped the MSMEs based on their type of business, such as culinary, fashion, or creative-digital. It only focuses on MSMEs which have applied green economy principles. Future studies are recommended to distinguishes the MSMEs based on their type of business or other characteristics, so that there will be a more comprehensive knowledge in this issue.
2) Based on the data acquired, it is evident that all of these MSMEs' owners are female. Future study can broaden the research population by working with MSMEs that are owned or managed by males, allowing it to be determined whether a company's values are influenced by gender.

References

1. Ardiana, I., Brahmayanti, I.A., Subaedi, S.: Kompetensi SDM UKM dan Pengaruhnya Terhadap Kinerja UKM di Surabaya. J. Manaj. dan Kewirausahaan **12**(1), 42–55 (2010)
2. Bhanarkar, A.D., Gajghate, D.G., Hasan, M.Z.: Assessmentof air pollution from small scale industry. Environ. Monit. Assess. **103**(3), 239–248 (2003)
3. Engle, K.K.: Problem-based learning: an approach to medical education. Am. J. Occup. Ther. **35**(8), 539 (1981)
4. Sriyono, S.: Implementation of green economy on the development of SMEs in Sidoarjo district, Manag. Penang, Malaysia, 6–7, no. November, pp. 6–7 (2014)
5. Nagayya, D., Rao, T.V.: Enabling small and medium enterprises to target globalization. IUP J. Manag. Econ. **9**(4), 15–32 (2011)
6. Mariana, K.: Peran Strategis Usaha Kecil Menengah (UKM) dalam Pembangunan Nasional. Informatika **3**(I Jarnuari), 15 (2012)
7. Lumpkin, G.T., Dess, G.G.: The entrepreneurial clarifying it construct and linking orientation. Acad. Manag. Rev. **21**(1), 135–172 (1996)

8. Combs, J.G., Crook, T.R., Shook, C.L.: The dimensionality of organizational performance and its implications for strategic management research, vol. 2, no. 05, pp. 259–286 (2005)
9. Sriyono: Kemampuan Kinerja Keuangan Untuk Meningkatkan Daya Saing UKM. J. Competency Bus. **2**(Ii), 84–105 (2018)
10. Murphy, G., Trailer, J., Hill, R.: Measuring research performance in entrepreneurship. J. Bus. Res. **36**(1), 15–23 (1996)
11. Zahra, S.A., Covin, J.G.: Contextual influences on the corporate entrepreneurship-performance relationship: a longitudinal analysis. J. Bus. Ventur. **10**(1), 43–58 (1995)
12. Wiklund, J., Shepherd, D.: Entrepreneurial orientation and small business performance: a configurational approach. J. Bus. Ventur. **20**(1), 71–91 (2005)
13. Guide Jr, V.D.R.: Production planning and control for remanufacturing: industry practice and research needs. J. Oper. Manag. **18**(4), 467–483 (2000)
14. Brown, L., LaFond, A., Macintyre, K.: Measuring Capacity Building, no. March, pp. 919–966 (2001)
15. Anantanyu, S.: Kelembagaan Petani: Peran Dan Strategi Pengembangan Kapasitasnya, vol. 7, no. 2, pp. 102–109 (2011)
16. Purwanto: Penerapan Produksi Bersih Di Kawasan Industri, Semin. Penerapan Progr. Produksi Bersih dalam Mendorong Kaw. Eco-indstri di Indones., pp. 1–10 (2005)
17. Ali, M.A.: Industrial Clusters and Social Networks and their Impact on the Performance of Micro- and Small-Scale Enterprises Evidence from the Handloom Sector in Ethiopia.
18. Purwaningsih, R., Kusuma Damar, P.: Analisis faktor- faktor yang mempengaruhi kinerja usaha kecil dan menengah (ukm) dengan metode structural equation modelling (studi kasus UKM berbasis industri kreatif Kota Semarang). Pros. SNST Fak. Tek. Univ. Wahid Hasyim Semarang **6**(1), 7–12 (2015)
19. Yue, D., Brychko, A.: Competitiveness of production as a determining factor of effective management of the enterprise in market conditions. Ekon. ta Upr. APK **2**(151), 67–75 (2019)
20. Nguyen, T.V., Le, N.T.B., Bryant, S.E.: Sub-national institutions, firm strategies, and firm performance: a multilevel study of private manufacturing firms in Vietnam. J. World Bus. **48**(1), 68–76 (2013)
21. Temtime, Z.T., Pansiri, J.: Small business critical success/failure factors in developing economies: some evidences from Botswana. Am. J. Appl. Sci. **1**(1), 18–25 (2004)
22. Campbell, J.: Institutional competitiveness: a broad perspective on the Danish case. Economy, no. January 1991 (1991)
23. Triono, T.A., Kirana, K.C., Fadhilah, M.: Improved performance of business owners with social capital, human capital, and entrepreneurial competence. Bus. Account. Res. **5**(4), 224–253 (2021)
24. Bey, C.: Changing economies of scale - Synergies between implementation of an ecological tax reform and development of industrial ecosystems. Bus. Strateg. Environ. **10**(6), 383–393 (2001)
25. Kuswanto, F., Rosli, M.M., Kader, R.A.: Innovation in distribution channel, cost efficiency & firm performance: the case of Indonesian small & medium enterprise scales. Int. J. Bus. Humanit. Technol. **2**(4), 23–38 (2012)
26. Noble, B.F.: Auditing strategic environmental assessment practice in Canada. J. Environ. Assess. Policy Manag. **5**(2), 127–147 (2003)
27. Noble, B.F.: Integrating strategic environmental assessment with industry planning: a case study of the Pasquai-Porcupine forest management plan, Saskatchewan, Canada. Environ. Manag. **33**(3), 401–411 (2004)
28. Kraja, Y., Osmani, D.: Competitive advantage and its impact in small and medium enterprises (SMEs) (Case of Albania). Eur. Sci. J. **9**(16), 1857–7881 (2013)
29. Wang, H., Wu, C.: Green growth as the best choice for Chinese small and medium enterprises in sustainable development. Asian Soc. Sci. **7**(5), 81–84 (2011)

Does Financial Literacy or Digital Literacy Determine a Consumer Use of FinTech?

Malik Taufiq, Tin Fah Chung(✉), and Ayu Chrisniyanti

HELP University, 50490 Kuala Lumpur, Malaysia
tinfah@yahoo.com

Abstract. Although there are numerous options for Financial Technology (FinTech) to assist in enhancing financial well-being, financial literacy among Malaysians appears to be low. This leads to the question that this study seeks to answer: Does financial literacy impact the adoption of FinTech? This study seeks to examine the factors influencing consumers' desire to adopt FinTech in the Klang Valley with a particular focus on the impact of consumers' financial and digital literacy on the willingness to adopt FinTech using the Technology Acceptance Model (TAM). We employ non-probability sampling to gather 302 data from individuals who have engaged in any form of FinTech. The findings from the Structural Equation Modeling (SEM) indicate that perceived usefulness and ease of use, mediated by attitude, are determinants of consumers' willingness to adopt FinTech in Malaysia. We discovered a strong correlation between individuals' intention to use FinTech and their level of financial literacy. The findings of this study would assist financial services marketers, FinTech companies, and policymakers to encourage FinTech adoption in Malaysia.

Keywords: Fintech · Financial literacy · Digital literacy · TAM · Structural equation modeling

1 Introduction

Over the past few decades, information technology has become a driver of innovation in the global financial sector, including Financial Technology (FinTech), which enables financial services across the globe to advance in the development of innovative and distinctive digital solutions. Numerous opportunities are offered to consumers, such as greater transparency in financial transactions, eliminate intermediaries, reduced transaction fees, and easy access to financial transformation [23]. As a result, FinTech is seen as a sector that has produced immense progress in the digital economy with significant adoption rates, notably in a number of nations like China, India, Brazil, the United Kingdom, and Australia [13].

In Malaysia, the Gross Domestic Product (GDP) is benefited from the digital economy to the tune of 18% in 2019 [5]. In terms of FinTech, the adoption rate has accelerated since 2019 with a surge in online banking transaction volume of RM2.5 billion in 2020

B. Alareeni et al. (Eds.): ICBT 2022, LNNS 620, pp. 289–298, 2023.
https://doi.org/10.1007/978-3-031-26953-0_28

compared to RM1.7 billion in 2019 as it has become the preferred method to make digital payments, supported by an electronic wallet (e-wallet) transaction volume of RM0.6 billion in 2020 [6].

Furthermore, the COVID-19 epidemic and nationwide movement control order (MCO) in 2020 had a devastating impact on the entire world, including Malaysia. Malaysians from all socioeconomic groups were forced to adopt FinTech to conduct everyday transactions, which resulted in an additional three million consumers of e-wallet and over 400.000 registered companies in Malaysia to accept payment via QR codes, leading to an increase of 164% in 2020 [6]. In order to avoid physical contact and hence slow the spread of the COVID-19 epidemic, FinTech can be beneficial for both consumers and businesses.

However, some people seem to be skeptical about using FinTech for several reasons, including financial risk, legal risk, and operational risk which negatively impact their intention. This demonstrated how financial and digital literacy are linked to the adoption of FinTech, and both components remain limited in the Malaysian context, where 70% of Malaysians require financial literacy and financial education since the early age [14]. Therefore, we aim to examine the factors influence consumers' readiness to adopt FinTech in the Klang Valley, with an emphasis on the impact of consumers' financial literacy and the moderating effect of digital literacy on the willingness to adopt technology using the Technology Acceptance Model (TAM).

2 Literature Review

2.1 Theoretical Framework

It is universally recognized that the Technology Acceptance Model (TAM) to be the most convincing and critical theory of technology acceptance in which can be viewed as a theory portraying how consumers reach, accept, and adjust to certain technology [3]. It aims to assess the impact of external factors on beliefs, attitudes, and intentions by investigating essential variables that affect technology adoption and examining their relationship [4]. The TAM contends that consumers' intentions—defined as "how much trust is invested and how much effort is exerted" by an individual [10]—determine whether utilizing latest technology is useful or not [15].

Additionally, consumers who have a more optimistic response toward latest technology seem to be more likely to utilize it behaviorally [3], where a good attitude of consumers leads to trust, confidence, and a readiness to engage with a technology [21]. However, when consumers are introduced to a new sort of system, two essential constructs—perceived usefulness (PU) and perceived ease of use (PEOU)—affect their intention and acceptance of technology and have an impact on their decision [19]. One of the most powerful predictors of consumers' intentions to use a particular system is perceived usefulness, which was proven to have a significant impact on behavioral intentions [20]. Consumers are more likely to adopt new technology if it lives up to their perceived expectations of performance of improved efficiency and satisfaction standards, which are simply established through their interactions with technology [2]. On the other side, perceived ease of use signifies to "the degree of belief that using a particular technology would be effortless" [3, 9], where consumers are more likely to adopt a system if it is simple and easy to use [16]. Perceived ease of use has been demonstrated to have a significant impact on consumers' attitudes and intentions toward technology [16].

It is critical to have a good level of financial literacy these days, as it is basic comprehension of financial concepts and behavioral capability in applying financial knowledge to make educated, astute financial decisions to improve personal financial well-being [1]. Financial literacy levels have a considerable influence on perceived ease of use by those who have invested knowledge in the decision-making of adopting new technology [22]. However, there is yet a standard measurement for financial literacy or financial knowledge, where both terms are used interchangeably [12]. It further mentioned that time value of money, inflation, and risk diversification are the three most popular methods used to gauge financial literacy [12].

Not only financial literacy, but digital literacy is equally critical, where it signifies to the degree of individuals ability to access, manage, understand, integrate, communicate, evaluate, and create information safely and appropriately through digital devices, and networked technologies for participation in economic and social life [11]. Further, digital skills are a requirement for professions, disciplines, and personal lives—it is vital to inculcate them in society to mitigate the risks of digitalization [17].

2.2 Hypotheses

In order to investigate the factors influencing consumers' motivation to accept and use FinTech in the Klang Valley, Malaysia, we have developed the following five hypotheses:

H_1: There is a positive relationship between attitude and intention to adopt financial technology (FinTech)
H_2: There is a positive relationship between perceived usefulness and intention
H_3: There is a positive relationship between perceived ease of use and intention
H_4: There may be a significant relationship between financial literacy and consumers' intention
H_5: The positive relationship between attitude and intention to adopt FinTech will be stronger when digital literacy is high.

A conceptual framework has been established and developed based on the Technology Acceptance Model (TAM) in order to get a better understanding (see Fig. 1).

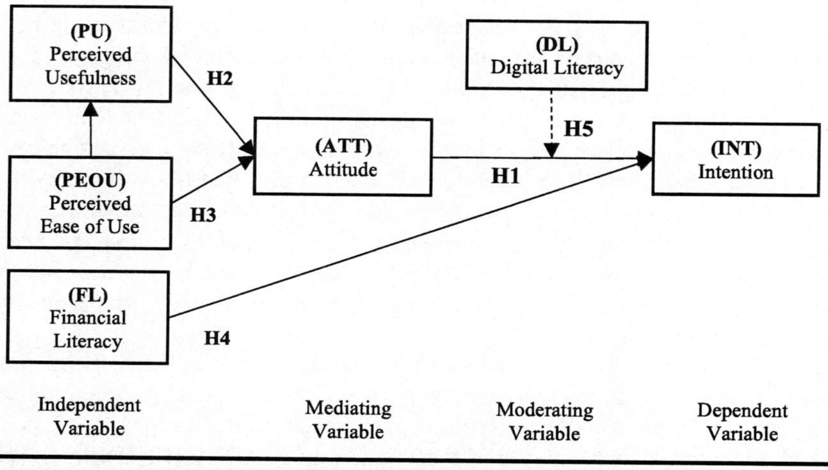

Fig. 1. Proposed conceptual framework

3 Research Methodology

In this study, we aim to investigate the factors affecting a behavioral intention to adopt and utilize FinTech among the male and female population of the Klang Valley specifically those in Kuala Lumpur, Gombak, Klang, Petaling, and Hulu Langat, aged 18 to 51 years old and above and have engaged in any kind of FinTech. Due to its socioeconomic status, in which the people seem to have a high household income and strong purchasing power, this demographic has been considered. Additionally, we discovered that they are educated and tech-savvy.

Additionally, a non-probability sampling is employed to collect samples faster and easy. However, owing the COVID-19 epidemic and nationwide social restriction in Malaysia, the entire data was collected online over the period of ten days, from August 5 to August 13, 2021. A self-administered (SAQ) was created using Google Form and a total of 302 data are collected. Further, we used a Four-Point Likert scale ranges from 1 ("strongly disagree") to 4 ("strongly agree") and no neutral option scale to remove personal bias, prevent the respondents from misusing the midpoint, and get specific results.

4 Data Analysis

A total of 302 responses were gathered and met the requirements of the questionnaire, where we used the Statistical Package for Social Science (SPSS) v.26.0 to carry out a statistical descriptive analysis of the demographic. In this study, we also used the SmartPLS v.3.2.9 to perform further investigation of the statistical association between all the objective measurements of the constructs (see Table 1).

Table 1. Demographic statistics

Demographic	Characteristics	Percentage (%)
Gender	Male	51.7
	Female	48.3
Age	18–24	2.3
	25–30	7.0
	31–35	14.9
	36–40	22.2
	41–45	30.5
	46–50	10.9
	51 and above	12.3
Education level	SPM	5.3
	Diploma	15.9
	Advanced Diploma	4.0
	Bachelor's degree	52.9
	Postgraduate	21.9
Occupation	Employed	51.7
	Self-employed	36.4
	Retiree	4.0
	Homemaker	3.6
	Student	0.7
	Unemployed	3.6
Monthly Income	Below RM5,000	29.8
	RM5,0001 to RM10,000	39.1
	Above RM10,001	31.1
TOTAL		**100.0**

Source: Authors' survey

4.1 Measurement Model Assessment

Using the composite reliability (CR) and the average variance extracted (AVE), we have assessed the measurement model of this study. Several measures were excluded from the model in order to get the square root of the AVE greater than the cut-off of 0.50 and outer loadings greater than 0.70 [7]. Furthermore, the composite reliability and Cronbach alpha of the constructs are demonstrated in Table 2 below, and it was deemed to qualified since all values are greater than 0.70. Additionally, the variance inflation factor (VIF) values were lower than five, implying absence of collinearity issues in the data.

Table 2. Measurement model assessment

Construct	Items	Loading	VIF	CR	α	AVE	Rho_A
Perceived usefulness	PU1	0.786	1.832	0.888	0.843	0.614	0.844
	PU2	0.777	1.897				
	PU3	0.818	2.022				
	PU4	0.812	1.833				
	PU5	0.721	1.503				
Perceived ease of use	PEOU1	0.711	1.533	0.908	0.872	0.664	0.876
	PEOU2	0.854	2.508				
	PEOU3	0.869	2.483				
	PEOU4	0.845	2.500				
	PEOU5	0.786	1.819				
Attitude	ATT1	0.879	1.604	0.892	0.761	0.806	0.777
	ATT2	0.916	1.604				
Financial literacy	FL8	0.818	1.694	0.852	0.767	0.590	0.768
	FL9	0.706	1.294				
	FL10	0.766	1.518				
	FL11	0.779	1.623				
Digital literacy	DL1	0.828	1.456	0.864	0.764	0.678	0.770
	DL3	0.836	1.656				
	DL4	0.807	1.593				
Intention	INT1	0.896	1.692	0.901	0.780	0.819	0.785
	INT2	0.914	1.692				

Notes: All measurement items with outer loadings lower than 0.70 were removed from the model, including FL1, FL2, FL3, FL4, FL5, FL6, FL7, DL2 and DL5
Source: Authors' survey

Table 3 demonstrates the Fornel-Larcker criterion of this study, which shows that the square root of the AVEs of each item has the highest value on its own and is bigger than all of its inter-constructs. On the other hand, in order to attain discriminant validity, the HTMT value should fall within the confidence interval of −1 and 1 [8], where discriminant validity was established in this study as all values were substantially different from 1 and greater than −1 (see Table 4).

4.2 Structural Model Assessment

The proposed hypotheses were evaluated, where Table 5 displays the results of the relationships among the constructs. The results show that four hypotheses were accepted and discovered to have significant effect toward the intention of using FinTech with the

Table 3. Fornel-Larcker criterion

	ATT	DL	FL	INT	PEOU	PU
ATT	**0.898**					
DL	0.399	**0.824**				
FL	0.414	0.362	**0.768**			
INT	0.688	0.373	0.444	**0.905**		
PEOU	0.713	0.434	0.299	0.596	**0.815**	
PU	0.563	0.577	0.351	0.597	0.700	**0.784**

Notes: The gray highlight on the Fornel-Larcker Criterion table indicates that the square root of the AVE has the highest value on itself and is low on other constructs.

Source: Authors' survey

Table 4. Heterotrait-Monotrait ratio

	ATT	DL	FL	INT	PEOU	PU
ATT						
DL	0.518					
FL	0.538	0.474				
INT	0.877	0.479	0.574			
PEOU	0.874	0.526	0.364	0.722		
PU	0.698	0.720	0.439	0.730	0.785	

Source: Authors' survey

t-values were greater than 1.96 and *p*-values are higher than 0.05. The moderating effect (digital literacy), however, was rejected.

Table 5. Hypotheses results

	Descriptions	β	*t*-values	*p*-value	Decision
H1	ATT → INT	0.583	12.227	0.000***	Supported
H2	PU → ATT PU → ATT → INT	0.127 0.074	2.067 1.970	0.039** 0.049**	Supported
H3	PEOU → ATT PEOU → ATT → INT	0.624 0.363	11.088 8.226	0.000*** 0.000***	Supported
H4	FL → INT	0.173	3.693	0.000***	Supported
H5	ATT → DL → INT	−0.068	0.656	0.512	Not Supported

Notes: ** $p < 0.05$; *** $p < 0.001$
Source: Authors' survey

The coefficient of determination (R^2) of ATT, INT, and PU were 0.516, 0.516, and 0.489 respectively, where it is clear that predictive capability has been formed, with 51.6% of changes in intention being attributed to shifts in consumers' attitudes toward adoption of FinTech. Further, Q^2 values were examined using the blindfolding on Smart-PLS, and the results indicate that the model has sufficient predictive relevance since the Q^2 values are greater than zero (see Table 6).

Table 6. Predictive relevance

Variables	R^2	Q^2
ATT	0.516	0.409
INT	0.516	0.404
PU	0.489	0.292

Source: Authors' survey

5 Conclusion and Discussion

In order to determine whether financial literacy has a significant relationship with intention, and ultimately to investigate the moderating influence of digital literacy, this study focuses on exploring and assessing the intention of consumers who reside in the Klang Valley to adopt and use FinTech. Four of the proposed hypotheses were found to be supported or accepted, whereas one hypothesis was not. The results of this study specifically showed that attitude, perceived usefulness, perceived ease of use, and financial literacy all had a favorable effect on consumers' desire to adopt and utilize FinTech. On the other hand, we found that digital literacy has partial significant effect towards intention, thus, digital literacy was not accepted.

5.1 Implication of the Study

The implication of the study reinforces the TAM framework to understand the behavioral intention towards the latest technology. These results emphasize the importance of financial literacy, where the positive significance of financial literacy implies that financial knowledge has bearing on a consumer's propensity to adopt and use FinTech in the Klang Valley, Malaysia. By improving the consumers' financial literacy skills, FinTech services and products besides e-wallet and e-payment may be used more frequently. Therefore, financial literacy training should be made available in the nation's educational system, given by financial services to their current and future clients, and encouraged through policy to increase the financial literacy levels of the general public. Financial and Digital Literacy should be streamlined in the Malaysia Digital economy blueprint.

5.2 Limitations and Recommendations for Further Research

This study was conducted successfully. As the authors when through, they discovered a number of limitations in this study. The first drawback is the use of convenience sampling that is a non-probability which limits generalization and representativeness of the findings [18]. The selection of this method is attributed to the limited resources and time available to reach the respondents as a result of the COVID-19 epidemic and nationwide social restriction in Malaysia. Consequently, using probability in future research is suggested to improve statistical inference; Second drawback is that that the questionnaire was only available in English, while it might have been translated into Bahasa Malaysia (Melayu); Third, the ethnicity of the respondents was not taken into account in the survey, which might be added in future research to collect demographic data; Finally, this study was carried out in Klang Valley, Malaysia, with respondents chosen from Malaysian consumers who have experience in adopting and using FinTech products or services, and those without experience were e disqualified from the study. This implies that the results of this study may only be applicable in the Malaysian market. Based on the results and the drawbacks of this study, it would be interesting for future research to take into account these limitations including a sample of consumers from other countries and comparing the results, where this would help FinTech companies and policymakers to have better understanding on how to promote the adoption of FinTech among society in Malaysia.

References

1. Atkinson, A., Messy, F.A.: Measuring financial literacy: results of the oecd infe pilot study. OECD Working Papers on Finance, Insurance and Private Pensions, 15, 1–73 (2012)
2. Chuang, L.M., Liu, C.C., Kao, H.K.: The Adoption of Fintech Service: TAM perspective. Int. J. Manag. Adm. Sci. (IJMAS) **3**(7), 1–15 (2016)
3. Davis, F.D.: Perceived usefulness, perceived ease of use, and user acceptance of information technology. MIS Q. **13**(3), 319–325 (1989)
4. Davis, F.D., Bagozzi, R.P., Warshaw, P.R.: User acceptance of computer technology: a comparison of two theoretical models. Manage. Sci. **35**(8), 982–1003 (1989)
5. DOSM.: Information technology and communication satellite account (2019) [Press release]. https://www.dosm.gov.my/v1/index.php?r=column/pdfPrev&id=cHBrejFuMlVMdFZzc2VrNjlIZEZQUT09. Accessed 30 June 2021
6. Fintech News Malaysia.: Fintech Malaysia report 2021 - fintech reaches an inflection point in Malaysia (2021). https://fintechnews.my/27070/malaysia/fintech-malaysia-report-2021/. Accessed 08 Aug 2021
7. Hair, J.F., Hult, G.T.M., Ringle, C., Sarstedt, M.: A Primer on Partial Least Squares Structural Equation Modeling (PLS-SEM). SAGE Publications, New York (2014)
8. Henseler, J., Ringle, C.M., Sarstedt, M.: A new criterion for assessing discriminant validity in variance-based structural equation modeling. J. Acad. Mark. Sci. **43**(1), 115–135 (2014). https://doi.org/10.1007/s11747-014-0403-8
9. Jahangir, N., Begum, N.: The role of perceived usefulness, perceived ease of use, security and privacy, and customer attitude to engender customer adaptation in the context of electronic banking. Afr. J. Bus. Manage. **2**(1), 32–40 (2008)
10. Jayawardhena, C., Foley, P.: Changes in the banking sector – the case of internet banking in the UK. Internet Res. **10**(1), 19–31 (2000)
11. Law, N., Woo, D., Wong, G.: A Global Framework of Reference on Digital Literacy Skills for Indicator 4.4.2 [Information Paper No.51]. UNESCO, Paris (2018)
12. Lusardi, A., Mitchell, O.S., Curto, V.: Financial literacy and financial sophistication in the older population. J. Pension Econ. Finance **13**(4), 347–366 (2014)

13. Milian, E.Z., Spinola, M.D.M., Carvalho, M.M.D.: Fintechs: a literature review and research agenda. Electron. Commer. Res. Appl. **34**, 10–27 (2019)
14. MohdAziz, N.I., Kassim, S.: Does financial literacy really matter for Malaysians? a review. Adv. Int. J. Banking Account. Finance **2**(2), 13–20 (2020)
15. Putranta, Y.E., Alamsyah, R., Tan, L., Tamara, D.: The determinant factors of mobile payment adoption. Eurasian J. Soc. Sci. **8**(3), 134–147 (2020)
16. Radner, R., Rothschild, M.: On the allocation of effort. J. Econ. Theory **10**(3), 358–376 (1975)
17. Rodríguez-de-Dios, I., Igartua, J.J., González-Vázquez, A.: Development and validation of a digital literacy scale for teenagers. In: Proceedings of the Fourth International Conference on Technological Ecosystems for Enhancing Multiculturality, vol. 02, no. 04, pp. 1067–1072 (2016)
18. Taherdoost, H.: Sampling methods in research methodology: How to choose a sampling technique for research. Int. J. Acad. Res. Manag. (IJARM) **5**, 1–15 (2016)
19. Taylor, S., Todd, P.A.: Understanding information technology usage: a test of competing models. Inf. Syst. Res. **6**(2), 144–176 (1995)
20. Venkatesh, V., Davis, F.D.: A model of the antecedents of perceived ease of use: development and test. Decis. Sci. **27**(3), 451–481 (1996)
21. Wenan, T., Shrestha, D., Shrestha, D., Jeong, S. R.: Attitude of international tourist towards ICT and digital services in tourism industry of nepal. In: Proceedings of the International Conference on Advanced Information Science and System (2019)
22. Yoshino, N.: Financial literacy and fintech adoption in Japan. Asian Development Bank (2020). https://www.adb.org/publications/financial-literacy-fintech-adoption-japan. Accessed 11 July 2021
23. Zavolokina, L., Dolata, M., Schwabe, G.: The FinTech phenomenon: antecedents of financial innovation perceived by the popular press. Financial Innov. **2**(1), 1–16 (2016)

A Conceptual Model for Servant Leadership and Organizational Citizenship Behavior

Jin Lu[1], Phaik Kin Cheah[1](✉), and Mohammad Falahat[2]

[1] Faculty of Arts and Social Science, Universiti Tunku Abdul Rahman (UTAR), Bandar Barat, 31900 Kampar, Perak, Malaysia
cheahpk@utar.edu.my

[2] Faculty of Accountancy and Management, Universiti Tunku Abdul Rahman (UTAR), Bandar Sungai Long, 43000 Kajang, Selangor, Malaysia

Abstract. In response to a request for further research on servant leadership, this study proposes an integrative conceptual framework for servant leadership, organizational citizenship behavior, relational identification, perceived organizational support, and workplace loneliness. A proposed conceptual model postulates that servant leadership can lead to organizational citizenship behavior. Furthermore, the model posits that relational identification and perceived organizational support are mediators, and workplace loneliness is a moderator. This research fills a knowledge gap and presents propositions to develop a conceptual framework for an improved comprehension of the impact between servant leadership and organizational citizenship behavior.

Keywords: Servant leadership · Organizational citizenship behavior · Relational identification · Perceived organizational support · Workplace loneliness

1 Introduction

Due to the importance of successful leadership in achieving goals in business, politics, and organizations, there is a lot of interest in leadership research. Many leadership theories developed as a result related to efficacy and ethics, including servant leadership (Brownell 2010). Servant leadership is regarded as a crucial topic of research as it can promote positive effects on organizations (Harrison 2018). Robert K. Greenleaf proposed the word "servant leadership" first in 1970. Greenleaf's concept of servant leadership emphasized strongly on meeting followers' needs and encouraging their growth and development (Greenleaf 2002). Servant leadership is a crucial component to developing organizations, which in turn motivate followers to provide their best efforts to the organization to achieve success.

Some researchers argued servant leadership was unrealistic because it overlooked the fact that individuals had varying levels of competence (Lee and Zemke 1993). Stone et al. (2003) thought servant leadership "systematically undefined and lacking in empirical support". Conceptualizing servant leadership is not easy due to the lack of a generally accepted definition, attributed, and characteristics (Lacroix et al. 2017), but it cannot stop

B. Alareeni et al. (Eds.): ICBT 2022, LNNS 620, pp. 299–308, 2023.
https://doi.org/10.1007/978-3-031-26953-0_29

the empirical study (Henning 2016). Over the past 10 years, increasing empirical research has emphasized the significance and value of servant leadership as a management strategy that enables organizations to create and preserve a competitive edge (Newman et al. 2017).

Servant leaders might help their organizations compete more successfully. This may depend on the ability of servant leaders to produce positive mindset and actions (Eva et al. 2019). Followers feel compelled to give back when they find a servant leader prioritizes their needs (Madison and Eva 2019). Employee organizational citizenship behavior is one significant byproduct of servant leadership (Newman et al. 2017). Organ introduced the phrase "organizational citizenship behavior" for the first time in 1988. Organizational citizenship behavior, described as personal behavior not included by the formal incentive system, is what any organization depends on to run smoothly (Organ 1988) and is necessary to improve interpersonal relationships and productivity at the individual level, team level, and organization level (Podsakoff et al. 2014).

Although previous studies have made an effort to explain the correlation between servant leadership and organizational citizenship behavior, its mechanism is still not very clear (Walumbwa et al. 2010). Previous research was lacking in examining individual differences in followers' responses to servant leadership, limited research tested whether the influence of servant leadership may be aggravated or weakened through followers' personalities (Newman et al. 2017).

The overall structure of the paper takes the form of six chapters, including introduction, review of the literature, conceptual framework, conclusion, recommendations and limitations, future research suggestions.

2 Review of the Literature

2.1 Servant Leadership

Servant leadership is a particular type of leadership that strong motivation to help others is a trait of servant leaders, and servant leaders will inspire others through their enthusiasm and confidence, and ultimately realize the identity of a leader by serving others (Greenleaf 2002).

Servant leaders consider followers as their responsibilities, focusing on followers first, then focusing on the organization, and finally on themselves (Patterson 2003). Scholars have explored different dimensions of servant leadership; there is no unified dimension in academia, but most scholars will use the dimension and scale developed by Liden et al. (2015) in their research. Researchers have examined the antecedents of servant leadership, including personal motivation, personal characteristics, values, organizational characteristics, and social culture (Eva et al. 2019). Researchers have also discovered results of servant leadership, including psychological atmosphere, interpersonal trust, employee performance, organizational performance, job satisfaction, loyalty, turnover rate, and organizational social responsibility (Langhof and Güldenberg 2020).

2.2 Organizational Citizenship Behavior

Organizational citizenship behavior is a voluntary individual behavior that is not included in job responsibilities and the organization's formal reward and punishment system

(Organ 1988). For example, employees take the initiative to help colleagues solve the difficulties encountered in the work process and take the initiative to make suitable suggestions for the development of the organization. Despite the fact that it is outside the scope of employee regular job duties, organizational citizenship behavior is conducive to organization general performance (Podsakoff et al. 2014).

2.3 Relational Identification

Sluss and Ashforth (2008) define relational identification as "the (partial) self-definition of an individual based on a specific and important role relationship". A person will assess the relationship status of the other party after entering a relationship with a role. If the individual positively evaluates the two identities of the other party, it indicates that the one identifies with their relationship with the other, that is, relationship identification (Sluss and Ashforth 2008).

Existing literature found that relationship identification has a positive correlation with self-efficacy, team identification, organizational identification, creativity, satisfaction with group members, organizational citizenship behavior, and reduce intrateam competition (Yuan et al. 2017; Walumbwa et al. 2010). Relational identification refers to the followers' identifying the relationship between the leader and themselves in this paper.

2.4 Perceived Organizational Support

Perceived organizational support is related to how employees view the organization's treatment of their contributions and their attention to their interests (Eisenberger et al. 1986). Organizational support also included instrumental support, such as work environment, resource information, tools, equipment, and training (McMillin 1997; Liu 2018). When employees' needs are met, they will have good attitudes and positive behaviors toward the organization (Rhoades and Eisenberger 2002).

When employees feel appreciation and recognition from the organization, they will have a favorable view of the organization, think they belong there, then rely on it psychologically (Rhoades and Eisenberger 2002).

2.5 Workplace Loneliness

According to Wright (2005), there are many different types of complicated interpersonal connections in the workplace, which causes a significant gap between what people expect from interpersonal relationships and what actually occurs. If they cannot make up for this gap, they will feel lonely, this is workplace loneliness. Lam and Lau (2012) pointed out that the rise of virtual teams and the popularity of network technologies have diminished possibilities for face-to-face contact among employees, who increasingly transfer information over the Internet at work. In addition, intensified competition in the workplace makes the organization members not be sincere, then the problem of workplace loneliness has become increasingly. In this context, the term loneliness in the workplace has increasingly attracted the attention of scholars.

3 Conceptual Framework

3.1 Servant Leadership and Organizational Citizenship Behavior

According to previous researches, servant leadership significantly improves organizational citizenship behavior (Neubert et al. 2016; Newman et al. 2017). The "service" and "altruism" are considered as core of servant leadership (Greenleaf 2002). It makes employees feel valued by their leader and the organization (Newman et al. 2017; Zou et al. 2020), which leads to a positive emotion to deal with their work and motivates them to do more behaviors that go beyond work responsibilities and benefit the organization (Anand et al. 2018). Servant leadership is employee-focused, focus on their growth and development, provide assistance, and solve difficulties (Liden et al. 2015). Depending on Social Exchange Theory, employees are grateful for the support and assistance provided by servant leaders, and employees are more willing to perform behaviors beyond work responsibilities to return the care and support of their leaders and organizations (Anand et al. 2018; Zou et al. 2020). So, our proposition is as below:

P1: Servant leadership has a positive impact on organizational citizenship behavior.

3.2 The Mediating Effect of Relational Identification

Employees can form self-concepts and change behavior through their relationships with the organization or with another member of the organization (Zhou et al. 2016). But there has not been much focus on individual identification at the interpersonal level (Guo 2020a, b). In practice, good interaction between leaders and employees facilitates the formation of stable role relationships (Sluss and Ashforth 2008), promotes positive perceptions of leadership, and further promotes employees' relational identification with their leaders (Ma and Liu 2020).

It has been shown that employees will appreciate, respect, and trust servant leaders, then build strong relationships, enhance the relational identification between employees and servant leaders (Yoshida et al. 2014; Zou et al. 2020). According to Social Identity Theory, employees will pay greater attention to the growth of their relationship with the leader when they identify with the leader strongly. Which will motivate employees to take the leader's benefit as their own (Guo and Xiao 2016), and be more willing to show the behavior expected by the leader. Therefore, employees will complete their own work and be likely to engage in positive behavior outside of job responsibilities that is beneficial to the leader and organization; it is organizational citizenship behavior. To get a more complete understanding of servant leadership, future research frameworks may benefit from taking relational identification as an additional social resource (Christensen-Salem et al. 2021). Therefore, our proposition is as below:

P2: Relational identification plays a mediating variable between servant leadership and organizational citizenship behavior.

3.3 The Mediating Effect of Perceived Organizational Support

Perceived organizational support is related to how employees view the organization's treatment of their contributions and their attention to their interests (Eisenberger et al. 1986). Simply put, it is the extent to which employees feel the organization cares about them, and one of the important factors influencing perceived organizational support is leadership support (Rhoades and Eisenberger 2002).

Research has shown that employees with higher level of perceived organizational support are more motivated to put in extra effort to support their leader and organization and gain better recognition (Chen and Ge 2020), prompting employees to care more about the organization and to engage in more beneficial behaviors (Edwards 2009). Organ et al. (2005) argues that there is a positive reciprocal process between perceived organizational support and organizational citizenship behavior. When employees perceive support from the leader, they perceive it as organization support; according to Social Exchange Theory, employees will develop reciprocal beliefs, they will voluntarily contribute to the organization, and are willing to put in extra work to pay back the leader and the organization, generating organizational citizenship behavior (Thompson et al. 2020). Thus, we hypothesize:

P3: Perceived organizational support plays a mediator role between servant leadership and organizational citizenship behavior.

3.4 The Chain Mediating Effect of Relational Identification and Perceived Organizational Support

Research on the relationship between relational identification and perceived organizational support is rare, but there may be a connection between relational identification and perceived organizational support. Steffens et al. (2014) demonstrate that relational identity positively affects employees' perceptions of leadership. The level of relational identification has also been shown to affect employees' perceptions of social support (White et al. 2021). A good relationship between employees and leaders affects the perceived organizational support of the employee (Kim et al. 2021). It has also been found that perceptions of leadership support led to perceived organizational support (Eisenberger et al. 2002), but the relationship between perceived leadership support and perceived organizational support is based on employees' identification with their leaders (Rhoades et al. 2001).

Based on the logic of the above study, it can be concluded that the level of relational identification affects employees' perceptions of the leader as well as their perceptions of the organization support.

When servant leaders show service behaviors such as caring, compassionate, and healing toward employees, employees consider service behaviors as recognition and support from the leader and consider them as a higher quality relationship with the leader, creating a strong identification with the leader (Liden et al. 2014; Newman et al. 2017). Leaders are frequently viewed as representatives of the company (Eisenberger et al. 2002), and when employees have a stronger identification with the relationship between themselves and the leader, the more likely they are to feel valued by the organization, the more likely they are to feel supported by the organization.

According to social exchange theory, the more employees feel supported by the organization, the more likely they are to develop mutual beliefs that lead to positive organizational citizenship behaviors (Farh et al. 2007; Zou et al. 2020). Based on this, we proposed the following:

P4: Relational identification and perceived organizational support play a chain-mediating role between servant leadership and organizational citizenship behavior.

3.5 The Moderating Effect of Workplace Loneliness

Wright's (2005) study confirmed that workplace loneliness can cause employees to lack identification of organizational membership. Highly lonely individuals show more social difficulties (Qu et al. 2010). These can have a variety of negative consequences. For example, employee performance may suffer as a result of workplace loneliness, and extra-role behaviors may also diminish (Lam and Lau 2012).

For employees with a high workplace loneliness, it is difficult to initiate quality relationships with their leaders due to negative self-perceptions, lack of identification with their identity (Wright 2005), and poor recognition of their relationships with their leaders; it is difficult to take the initiative to establish a high-quality relationship with their leaders (Ozcelik and Barsade 2011). Employees with high levels of loneliness believe that they do not meet the expectations of their leaders and are reluctant to actively communicate with the leader (Zhu 2017), and increase interpersonal mistrust and doubts about relationships with leaders (Dai 2021).

In spite of servant leadership concern and help employees, employees with high workplace loneliness are not willing to accept services and support and are skeptical of their leaders' altruistic behaviors. Therefore, they cannot develop a good perception of their leaders and organizations and are not willing to face the risks associated with social communication with their leaders and organizations (Lam and Lau 2012; Li and Ye 2015), which in turn diminishes the impact of servant leaders on employee relational identification. This ultimately leads to a reduction in altruistic behavior in the organization. On this basis, we propose:

P5: Workplace loneliness reversely regulates the relationship between servant leadership and employee's relational identification. When the employee has high workplace loneliness, the impact of servant leadership on employee's relational identification is weakened; On the contrary, when the employee has low workplace loneliness, the influence of servant leadership on employee's relational identification of the employee increases.

The above mediating proposition proposes that relational identification and perceived organizational support play a chain mediating role between servant leadership and organizational citizenship behavior, and the moderating proposition proposes that the direct interaction between workplace loneliness and servant leadership affects employees' relational identification and finally affects organizational citizenship behavior. Based on these hypotheses, it is reasonable to assume that workplace loneliness has a moderating effect on the mediating effect. Synthesizing the above analysis, this study proposes the following hypotheses:

The theoretical framework of this study is constructed as follows (see Fig. 1):

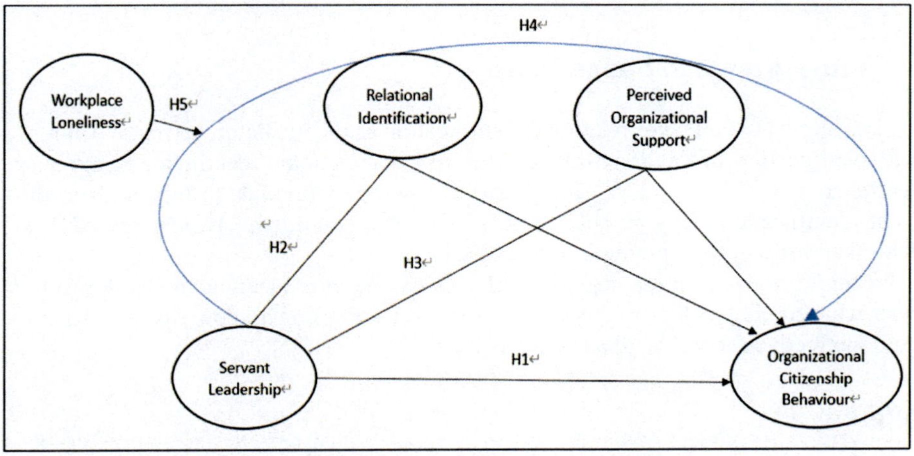

Fig. 1. Conceptual framework

4 Conclusion

This paper proposes a conceptual model for servant leadership. It proposes the mediating and moderating effect in the relationship between servant leadership and organizational citizenship behavior, which can help scholars better understand the effectiveness of servant leadership. It will expand the research on the intermediary mechanism between servant leadership and organizational citizenship behavior. Furthermore, this model introduces personal negative emotion to explore when the effectiveness of servant leadership will be diminished. It makes up for the shortcoming that many studies focus on the positive features and find out the reasons for the negative effects, avoiding invalid behaviors as much as possible when implementing servant leadership in the future.

5 Recommendations and Limitations

5.1 Recommendations

Servant leadership takes "service" and "altruism" as the core concepts, which can improve employees' perceived organizational support and relational identification with their leader, and then motivate employees to show more organizational citizenship behaviors, and finally improve organizational performance. Therefore, servant leadership should be introduced to more enterprises.

However, the characteristics of each person are different, employees with negative emotion may not accept servant leadership. Therefore, attention should be paid to the personal characteristics of employees when promoting servant leadership.

5.2 Limitations

This is only a conceptual model, especially the chain mediation effect between perceived organizational support and relational identification has not been supported by many empirical studies.

6 Future Research Suggestions

The conceptual model takes relational identification as the mediator, there are still multiple identification in the organization, such as organizational identification and team identification at collective level, leader identification at individual level, and these different identifications may be able to transform into each other. The synergy of these identifications will be explored in future research.

In future research, more negative variables can be introduced as moderator to test under what conditions the effectiveness of servant leadership will be weakened, so as to avoid ineffective servant leadership behaviors.

References

Anand, S., Vidyarthi, P., Rolnicki, S.: Leader-member exchange and organizational citizenship behaviors: contextual effects of leader power distance and group task interdependence. Leadersh. Q. **29**(4), 489–500 (2018)

Brownell, J.: Leadership in the service of hospitality. Cornell Hospitality Q. **51**(3), 363–378 (2010)

Chen, X., Ge, Y.: Influence of family—work conflict on career growth of female employees: a regulated mediation model. Chin. Pers. Sci. **3**, 31–41 (2020)

Christensen-Salem, A., Zanini, M.T.F., Walumbwa, F.O., Parente, R., Peat, D.M., Perrmann-Graham, J.: Communal solidarity in extreme environments: the role of servant leadership and social resources in building serving culture and service performance. J. Bus. Res. **135**, 829–839 (2021)

Dai, Y.: The effect of workplace loneliness on employees' emotional labor: the moderating effect of job embeddedness. Master's thesis, Beijing Jiaotong University (2021)

Eisenberger, R., Huntington, R., Hutchison, S., Sowa, D.: Perceived organizational support. J. Appl. Psychol. **71**(3), 500 (1986)

Eisenberger, R., Stinglhamber, F., Vandenberghe, C., Sucharski, I.L., Rhoades, L.: Perceived supervisor support: contributions to perceived organizational support and employee retention. J. Appl. Psychol. **87**(3), 565 (2002)

Edwards, M.R.: HR, perceived organisational support and organisational identification: an analysis after organisational formation. Hum. Resour. Manag. J. **19**(1), 91–115 (2009)

Eva, N., Robin, M., Sendjaya, S., Van Dierendonck, D., Liden, R.C.: Servant leadership: a systematic review and call for future research. Leadersh. Q. **30**(1), 111–132 (2019)

Farh, J.L., Hackett, R.D., Liang, J.: Individual-level cultural values as moderators of perceived organizational support–employee outcome relationships in China: comparing the effects of power distance and traditionality. Acad. Manag. J. **50**(3), 715–729 (2007)

Greenleaf, R.K.: Servant Leadership: A Journey into the Nature of Legitimate Power and Greatness. Paulist Press (2002)

Guo, S., Xiao, M.: Collectivism-oriented human resource management and employees' positive and negative reciprocity: the differently mediating role of organizational and relational identifications. J. Bus. Econ. **12** (2016)

Guo, S.: How does self-sacrificing leadership bring advice? The mediating of relationship identity and the adjustment of error management culture. Predictions (2020a)

Guo, S.: The influence of self-sacrificing leadership on employees' voice behavior: the mediation of relational identity and the moderation of error management culture. Forecasting (2020b)

Harrison, C.: Leadership Theory and Research: A Critical Approach to New and Existing Paradigms. Springer, Heidelberg (2018)

Henning, P.: The relationship between servant leadership and employee job satisfaction in a Colorado nonprofit organization (Doctoral dissertation, Colorado Technical University) (2016)

Kim, T.W., You, Y.Y., Hong, J.W.: A study on the relationship among Servant Leadership, Authentic Leadership, Perceived Organizational Support (POS), and Agile Culture using PLS-SEM: mediating effect of POS. Ilkogretim Online **20**(3) (2021)

Lacroix, M., Pircher Verdorfer, A.: Can servant leaders fuel the leadership fire? The relationship between servant leadership and followers' leadership avoidance. Adm. Sci. **7**(1), 6 (2017)

Lam, L.W., Lau, D.C.: Feeling lonely at work: investigating the consequences of unsatisfactory workplace relationships. Int. J. Hum. Resour. Manag. **23**(20), 4265–4282 (2012)

Lee, C., Zemke, R.: The search for spirit in the workplace. Train.-N. Y. Then Minneap. **30**, 21 (1993)

Liden, R.C., Wayne, S.J., Liao, C., Meuser, J.D.: Servant leadership and serving culture: influence on individual and unit performance. Acad. Manag. J. **57**(5), 1434–1452 (2014)

Li, Ye: Review on workplace loneliness. Hum. Resour. Dev. China (11), 29–37 (2015)

Liden, R.C., Wayne, S.J., Meuser, J.D., Hu, J., Wu, J., Liao, C.: Servant leadership: validation of a short form of the SL-28. Leadersh. Q. **26**(2), 254–269 (2015)

Liu, L.: A review of perceived organizational support. DEStech Trans. Econ. Bus. Manag. (ICSSED) (2018)

Langhof, J.G., Güldenberg, S.: Servant leadership: a systematic literature review—toward a model of antecedents and outcomes. German J. Hum. Resour. Manag. **34**(1), 32–68 (2020)

McMillin, R.: Customer satisfaction and organizational support for service providers. Doctoral dissertation, University of Florida, Florida (1997)

Madison, K., Eva, N.: Social exchange or social learning: a theoretical fork in road for servant leadership researchers. In: Sendjaya, S. (ed.) Leading for High Performance in Asia: Contemporary Research and Evidence-Based Practices, pp. 133–158. Springer, Singapore (2019). https://doi.org/10.1007/978-981-13-6074-9_7

Ma, J., Liu, L.: Study on the relationships among inclusive leadership, relational identification, and followership: the moderation of power distance. Hebei J. Ind. Sci. Technol. **37**(2), 59–68 (2020)

Neubert, M.J., Hunter, E.M., Tolentino, R.C.: A servant leader and their stakeholders: when does organizational structure enhance a leader's influence? Leadersh. Q. **27**(6), 896–910 (2016)

Newman, A., Schwarz, G., Cooper, B., Sendjaya, S.: How servant leadership influences organizational citizenship behavior: the roles of LMX, empowerment, and proactive personality. J. Bus. Ethics **145**(1), 49–62 (2017)

Organ, D.W., Podsakoff, P.M., MacKenzie, S.B.: Organizational Citizenship Behavior: Its Nature, Antecedents, and Consequences. Sage Publications (2005)

Organ, D.W.: Organizational Citizenship Behavior: The Good Soldier Syndrome. Lexington books/DC heath and com (1988)

Ozcelik, H., Barsade, S.: Work loneliness and employee performance. In: Academy of Management Proceedings, vol. 2011, no. 1, pp. 1–6. Academy of Management, Briarcliff Manor (2011)

Patterson, K.A.: Servant Leadership: A Theoretical Model. Regent University (2003)

Podsakoff, N.P., Podsakoff, P.M., MacKenzie, S.B., Maynes, T.D., Spoelma, T.M.: Consequences of unit-level organizational citizenship behaviors: a review and recommendations for future research. J. Organ. Behav. **35**(S1), S87–S119 (2014)

Qu, K., Zou, H., Yu, Y.: The relationship between loneliness and personalities in adolescent students: the mediating effect of interpersonal competence. Psychol. Explor. (6) (2010)
Rhoades, L., Eisenberger, R., Armeli, S.: Affective commitment to the organization: the contribution of perceived organizational support. J. Appl. Psychol. **86**(5), 825 (2001)
Rhoades, L., Eisenberger, R.: Perceived organizational support: a review of the literature. J. Appl. Psychol. **87**(4), 698 (2002)
Sluss, D.M., Ashforth, B.E.: How relational and organizational identification converge: processes and conditions. Organ. Sci. **19**(6), 807–823 (2008)
Steffens, N.K., Haslam, S.A., Reicher, S.D.: Up close and personal: evidence that shared social identity is a basis for the 'special' relationship that binds followers to leaders. Leadersh. Q. **25**(2), 296–313 (2014)
Stone, A.G., Russell, R.F., Patterson, K.: Transformational versus servant leadership: a difference in leader focus. Leadersh. Organ. Dev. J. **25**(4), 349–361 (2003)
Thompson, P.S., Bergeron, D.M., Bolino, M.C.: No obligation? How gender influences the relationship between perceived organizational support and organizational citizenship behavior. J. Appl. Psychol. **105**(11), 1338 (2020)
Wright, S.L.: Organizational climate, social support and loneliness in the workplace. In: The Effect of Affect in Organizational Settings. Emerald Group Publishing Limited (2005)
Walumbwa, F.O., Hartnell, C.A., Oke, A.: Servant leadership, procedural justice climate, service climate, employee attitudes, and organizational citizenship behavior: a cross-level investigation. J. Appl. Psychol. **95**(3), 517 (2010)
White, C.A., Slater, M.J., Turner, M.J., Barker, J.B.: More positive group memberships are associated with greater resilience in Royal Air Force (RAF) personnel. Br. J. Soc. Psychol. **60**(2), 400–428 (2021)
Yoshida, D.T., Sendjaya, S., Hirst, G., Cooper, B.: Does servant leadership foster creativity and innovation? A multi-level mediation study of identification and prototypicality. J. Bus. Res. **67**(7), 1395–1404 (2014)
Yuan, Q., Niu, W., Chen, L.: A literature review of relational identification in organizations. Soft Sci. **31**(8), 47–50 (2017)
Zhou, R., Long, L., He, W.: Self-sacrificing leadership and employee counterproductive behavior: the role of leadership identity and psychological rights. Forecasting **35**(3), 1–7 (2016)
Zou, W., Hu, D., Tang, L.: The impact mechanism and effectiveness of the servant leadership on work-family facilitation. J. Guangdong Univ. Financ. Econ. **35**(5), 69–79+ (2020)
Zhu, D.: The relationship between perceived insider status and workplace loneliness: a multiple mediation model through role stress and burnout. Master's thesis, Jiangxi Normal University (2017)

The Effect of Supply and Demand Attributes Towards of Talent Shortage: A Mixed Method Approach

Mohd Ikhwan Aziz[1](✉), Satishwaran Uthamaputhran[1], Hasannuddin Hassan[1], Marlisa Rahim[2], Md Zaki Muhammad Hasan[1], Mohd Rafi Yaacob[1], Azwan Abdullah[1], and Noor Raihani Binti Zainol[1]

[1] Faculty of Entrepreneurship and Business, Universiti Malaysia Kelantan, Pengkalan Chepa, Kota Bharu, Kelantan, Malaysia
{ikhwan.a,sathiswaran,hasann,zaki.mh,rafi,azwan.a,raihani}@umk.edu.my

[2] Faculty of Tourism, Hospitality, and Wellness, Universiti Malaysia Kelantan, Pengkalan Chepa, Kota Bharu, Kelantan, Malaysia
marlisa@umk.edu.my

Abstract. Over the past years, Malaysian Islamic banking has shown promising rapid growth concurrently with global legality, infrastructure, talent, and product development. Islamic financial sector has anchored on the real economy, as Islamic banking services should ultimately improve lives and livelihoods – assisting individuals or businesses in growing their wealth, making payments, owning a home, growing their enterprise, managing financial risks and adverse events, creating high-quality jobs, and other economic outcomes. As a consequence of this, the widespread good development of IBIs has resulted in a talent conflict among the industry players. Each of these institutions needs skilled talent to be occupied within their organisation, and the talent conflict has resulted in a competition for talent. To date, the talent shortage has become more critical compared to previous years, and despite an increase in Islamic banking education and knowledge service providers, talent shortage issues remain unsolved. This paper focused on distinguishing the talent shortage in Malaysian IBIs from the talent demand and supply with the impact of government interventions. A mixed method approach was used to generate a result from 371 respondents and a group focus group discussion from three participants. Both generated results are later triangulated to acquire the ultimate. The result showed that all the variables positively affect talent shortage, with government intervention mediating the result. Finally, the contribution of the data will clear up the root cause of talent shortage among graduates at entry level in Malaysia IBIs.

Keywords: Talent shortage · Supply and demand · Islamic banking · Government intervention

B. Alareeni et al. (Eds.): ICBT 2022, LNNS 620, pp. 309–323, 2023.
https://doi.org/10.1007/978-3-031-26953-0_30

1 Introduction

The shortage has affected the employers as they struggle to fill their talent vacancies because the previous staff have moved to other banks or baby boomers' talent has reached their retirement age. As a result, employers have to hire graduates who do not have an Islamic banking qualification background. Apart from that, [1] Malaysian talent shortage has increased to 82% in IBIs compared with the average industrial and conventional financial services standards at 76% and 78%, respectively. Despite the tremendous growth in talent shortage, there is an increasing concern over the talent shortage in IBIs in Malaysia. From that particular matter, BNM has taken proactive action by initiating the human capital development plan through its education, knowledge and agencies service providers. As an outcome, various education and knowledge service providers offered numerous programmes related to Islamic baking.

This study will discuss three fundamental problems of the talent shortage in Malaysian IBIs. First and foremost, talent shortage occurred in IBIs due lack of collaboration between industries and higher education institutions [2]. The study indicated that this issue remains the talent development problem, as discussed in past studies [3]. Despite numerous collaborative programmes with the industries the students have to participate in, they further need to undergo the internship programme. Nevertheless, the issue remains unsettled and persists unchanged [3]. The findings do not mean that the curriculum does not reflect the industry's needs. Hence, to confront the rapid growth in the Islamic banking industries, the curriculum must be improvised.

The second issue is the talent shortage, which occurred in IBIs because they could not attract young talent into their organisation. IBIs could not attract young talent due to outdated employees proposition such as unattractive employee benefits, flexible working hours and basic salaries, which the benefits did not reflect young talent preferences. As an outcome, most IBIs leaders experienced difficulties engaging suitable and qualified talent for various banking positions [4, 5].

Last but not least, according to Asian Islamic Finance (AIF), the major issue of talent shortage in the IBIs is the mismatch of skills provided between IBIs and Islamic banking education institutions (IBEIs). Furthermore, AIF highlighted that the essential skills in Islamic banking are English proficiency and leadership skills. However, on the other hand, the graduates believe they have equipped sufficient skills to fit into IBIs [6, 7].

Since academia and industry have not collaborated to develop a comprehensive curriculum in Islamic finance, a report by the Financial Accreditation Agency of Malaysia states that graduates do not possess the skills required by the business. This is due to the fact that academia and industry have not worked together to create this curriculum. Because they have been educated in a system that places a greater emphasis on theory than on practise, graduates do not enter the workforce with the necessary level of preparation. This would ensure a steady stream of personnel who are both competent and adaptable, which is necessary to support the rising innovation and dynamism within the Islamic financial business. The Islamic Financial Services Industry (IFSI) will be able to become more competitive and efficient in the delivery of value, which will ultimately result in the restoration of confidence and trust if new technologies and systems are utilised as the foundation. The business of providing Islamic financial services is a specialised industry; therefore, it is essential to acquire the appropriate skills and skill sets in order to ensure its long-term viability and further growth [4, 8, 9].

Education and training in Islamic finance have received a substantial amount of work, yet there are still many problems that need to be handled. According to the findings of [1], 68% of the businesses surveyed felt that inadequate talent management programme execution is the primary reason for the talent shortage. As a result, it has turned into a question mark as to whether or not the educational system in Malaysia is capable of producing well-equipped talent that can satisfy both the demand for talent and the supply of talent in companies and universities. To add to the facts that have already been stated, the method for talent management has become a dubious issue as a result of the current practise in IBIs. The lack of available talent may have an effect on the business conducted by the IBI across the entire organisation.

In general, the purpose of this study is to determine the underlying factor that contributes to the talent gap in IBIs, as well as to determine the degree to which the talent gap exists and which factors have the greatest influence on whether the talent gap is caused by a lack of supply or demand. In a broader sense, the purpose of this research is to investigate the factors that contribute to the scarcity of talent in IBIs in Malaysia.

2 Literature Review

2.1 Talent Shortage

Several scholars have presented the definition of a talent shortage. The authors of [10] used harsh language to portray executive talent deficit as the death of executive talent. The definition provided by Dytchwald et al. was contrasted with the word "workforce crisis" [11]. Hewlett et al. (2005), however, defined the term "talent shortfall" in terms of the "hidden brain drain.". Conveyed the opposite meaning, Miner (1974) defined *talent shortage* as a lack of talent equivalent to a human constraint within an organisation or institution. Therefore, regardless of the definition established by academics in the past, it is evident that it has contributed to a talent, human capital, or labour deficit in both demand and supply institutions. The talent shortage will be covered in greater detail later on.

A holistic talent management approach should be used as an effective strategy in each specific organisation or industry. Improving management methods is now the best strategy to manage a talent shortage. Additionally, the protracted negative impact of talent scarcity could restrict industry expansion, sparking a fierce competition for talent worldwide [14], increasing high mobility turnover, and impairing business performance. The market for IBIs is experiencing a growth in talent need, which has caused the demand for talent to rise quickly. Since then, IBIs have sparked a talent war within the Islamic banking industry. Because of these factors, the skills shortage at IBIs is at an acute level. Employers had trouble locating and hiring the ideal candidates, which led to a talent deficit in their organisation.

A comprehensive value of talent development would be provided by combining fundamental resources in higher education institutions, including talent, university infrastructures, and curriculum [2]. Higher education institutions frequently emphasise a conflict in how talent develops. However, various talent development programmes have been put into practice. There appears to be a mismatch between the skill set supplied and what is needed in the current and past plans [15].

As a result, a talent shortage in a particular organisation could have disastrous effects on the organisation's development in the future. The ideal method to address the talent shortage challenges is a comprehensive, integrated approach to talent management between talent supply and demand of each party, including government, industrial organisations, and higher education institutions. It seems to strike a balance between talent availability and resource.

2.2 Supply and Demand of Talent

According to the theory of supply and demand, if there is a need for a particular item, there will also be a supply available to fulfil that demand. The researcher will elaborate on the idea of talent supply in the following part of the report. [16] defined the term "talent supply" as higher education institutions that have developed talent for any reputable industry. This definition can help you better understand the concept of "talent supply." In the year 1988, [17] defined a different concept of talent supply, which included the following four categories: (1) new entrants into the system; (2) former talents who re-enter the system; (3) holdover talents from the previous year; and (4) immigrant talents who have been transferred into another system. [18], expanding on the prior idea of talent supply, suggested that talent supply may be generated and grown within businesses or organisations. [18] stated that this is possible. In situations like these, talent supply is the location to transform talent to the next level; thereafter, it can be globally outsourced based on two criteria: the level of expertise and capabilities of the firm.

The production of talents that are a good fit for the available jobs is the objective of talent supply. A considerable demand-supply gap has been established as a result of the growing number of retirements among baby boomers in the workforce [19]. This is in addition to the previously mentioned factors. On the other hand, the long-term viability of talent production within the talent supply is currently being called into doubt. It is essential to keep responding to talent needs [20], since this is a factor that can be considered sustainable. As a result of the inelastic nature of the talent supply, skilled workers may find themselves unemployed as the market moves toward decentralisation. Due to the fact that the supply of talent is unable to decide the quantity of talent that will be retained, the talent will have to compete for the position. In such a scenario, the individual on the losing side would be forced to seek new employment or relocate to a different place. The elastic condition of talent supply happens when talents are hired in a state of equilibrium and revenue sharing lowers in proportion to the amount of talent hired. This is because the elastic condition of talent supply allows revenue sharing to drop as more talents are hired.

The investigation focuses on three key factors: the demand for talent, the supply of talent, and the shortage of talent. As a consequence of this, the researcher began the conversation by focusing on the demand for talented individuals before moving on to other variables. According to [21], the demand for talent is a reflection of the availability of talent, and it may be forecast so that the organisation may plan for its future. However, during the past ten years, various research have attempted to explain the phenomenon [23,

24], and [22] have emphasised that talent demand refers to enterprises or organisations that engage in a talent war.

On the other hand, the term "talent demand" refers to those individuals who participate in talent management activities within an organisation. At a minimum, it encompasses the following three fundamental aspects of people management: recruiting, retention, and development [25]. At the same time, a number of academics had a variety of perspectives regarding the desire for talent. According to [26], this is the portion in which they highlight the traits that are associated with the talent function in their own organisations. In addition, this is the location where it occurred that a job function was offered for each work scope. This has been the case. When they have recruited or employed a talented individual into their organisation, one researcher offers a different perspective on the demand for talents. After that, they provide the talent with the appropriate compensation, employee benefits, and incentives [27].

In addition to this, the policy of the government has begun to reflect the demand for talent, with such organisations having their own corporate policy, rules, and regulations to which they must conform. However, this must still be done in a manner that is consistent with the official policy. In that case, the particular organisation in question will be obstructing the goal of the government for the development of the industrial sector and will be susceptible to financial penalties. The policy recommendations are not only for a single organisation or firm, but also for particular industries, and they are based on the vision and purpose of the current government [28–30].

2.3 Talent Management

Research conducted by Lewis and Heckman [31] suggests that the talent management paradigm ought to contain a least of three pillars. These pillars are recruiting, leadership development, and succession planning. During the initial phase of the hiring process, one of the primary emphasis is on sourcing qualified applicants from existing talent pools in order to fill available roles. At this point in the procedure, the recruiter for the organisation will search the talent pool for a candidate who possesses the abilities necessary to be successful in the organisation. The second pillar of the approach is primarily concerned with promoting members of the organization's existing talent pool into positions of leadership responsibility. The talent will now take part in a range of different leadership-related courses and programmes in order to develop and improve upon the leadership talents they already possess. In order to enable the establishment of an additional internal talent pool, the department of human resources developed a succession planning strategy. This strategy lays an emphasis on the third pillar, which was formed in order to facilitate the creation of the talent pool. When it came time to talent mobility either within or across the organisation, the time gap for the open post would be minimised as a result of the fact that the probable applicants for that position had been identified in advance. This would be the same regardless of whether the mobility of talent occurred internally or externally in the organisation.

A recent theory that was created by Cappelli [21] is predicated on the concept that talent management should be related to the operational management practises that are actually carried out in organisations. This theory is built on the pillars of four fundamental concepts. The first part of two more principles is used to define the risk that is involved in calculating demand, while the second part of two more principles is used to define the uncertainty that is associated with supply. Both of these parts are used in conjunction with two more principles. In order to reduce risk and make up for any potential talent shortages, the organisation should, in accordance with the first principle, actively seek out potential applicants for employment from outside the company. To maximise the return on investment, the second concept is to make improvements to the employee training programmes that are currently in existence (ROI). It is essential to maintain excellent connections with current as well as former employees in order to increase the likelihood that they will volunteer their skills or make a financial contribution in the future toward the cost of training a new employee. These connections can be maintained via email, phone calls, and in-person meetings.

The third guiding principle is to respond to the evolving demand for talent rather than adopting the traditional approach for management trainees, which requires three years of training. This is in contrast to the typical manner for management trainees, which requires three years of training. It is essential to differentiate between these two things. Before sending the selected employees back for additional specialised training, the author suggested sending them through a management programme that lasted for 18 months rather than sending them back to their original functions. This would be done instead of sending them back to their original functions. As a consequence of this, the organisation is in a position to foster the development of a wider pool of talent.

It is crucial to establish a middle ground between the competing priorities of employers and workers in order to protect the investment and adhere to the fourth guiding principle. In order to do so, it is necessary to create a compromise. If an employee is dissatisfied with their current role and there are better chances available elsewhere, they will eventually look for work elsewhere. Talent is a scarce resource in any organisation. The company needs to come to a choice regarding future promotions that strikes a healthy balance between the opposing goals of the employer and the employee in order for it to be able to protect its investment asset. This will allow the company to protect its investment asset. The company needs to come to a choice regarding future promotions that strikes a healthy balance between the opposing goals of the employer and the employee in order for it to be able to protect its investment asset. This will allow the company to protect its investment asset.

On the other hand, it is not very normal to hear sources from the talent supply to demand talk about talent management. The majority of the study that is being taken into consideration focuses on talent management within businesses or talent management on the premises of the corporation. According to Wu et al. [32], in order for talent management programmes to become a concept at talent sustainability for industry, they should be put into practise from the very beginning of the learning process in institutions of higher education. This is necessary in order for talent management programmes to become a concept at talent sustainability for industry. [Further citation is required] Birch and Ristevska [33] used the collaborative approach that was provided by the mentorship programmes as additional confirmation of the successful talent management practises across both the supply and demand. This was accomplished by utilising the collaborative approach that was provided by the mentorship programmes.

On the other hand, Sharma [34] discussed the ways in which higher education institutions have the potential to assist in the development of talent through the application of competency mapping and the talent management model. Specifically, he focused on the ways in which these institutions have the ability to contribute. In the meantime, study carried out by Barron [35] voiced concerns over the talent management gaps that exist between the hotel industry's current employability needs and the staff that is now available.

The data that have been presented make it abundantly clear that the implementation of best practises for talent management will help alleviate problems that are caused by a shortage of personnel. In spite of this, it appears as though the integration strategy between supply and demand parties would benefit from their combined efforts in promoting and improving the development of talent in order to make it more robust and all-encompassing. This conclusion is based on what has been said, and it seems as though these efforts would benefit the integration strategy. It is common knowledge that the employer will only request the services of qualified employees at the times when those individuals are actually needed. In addition, the employer will necessitate that such services be provided at a price that is within the scope of what the business is able to pay given its resources. People who are talented almost often place a high importance on the possibility of advancement in their careers as well as having some say in how their professions progress. It is essential that talent management be involved in the implementation process right from the start of the learning phase. This will allow the talent to get a deeper comprehension of the give and take that exists between both sides at an earlier stage.

3 The Mixed Method

The scope is limited to the Malaysian Islamic banking sector. It considers the public universities in Malaysia. Thus, the study aims to understand the best practice of talent management currently used in managing talent in Malaysian higher education and examine the effect of supply and demand derived from the objective of this paper.

3.1 Research Design

A substantial strategy that includes methods and procedures for data collecting and analysis is called a research design [36]. In order to fill the void in research regarding the shortage of talent in the IBIs market, this study utilised a combination of research approaches. Only twenty percent of research projects use a mixed-method design, despite the fact that mixed-method research has a long history in administrative sciences [37]. On the other hand, practical studies on talent shortage are either quantitative or qualitative, and only twenty percent of research projects use a mixed-method design [38]. Additionally, the use of qualitative and quantitative methodologies in the same research endeavour improves the reliability and validity of the findings by capitalising on the benefits offered by both approaches [39, 40].

As a result, a sequential exploratory strategy as a particular mixed-methods design has been adopted for the purpose of this study. The first step, which consists of the gathering and analysis of qualitative data, is then followed by the second stage, which consists of the collecting and analysis of quantitative data. Because it enables a researcher, during the qualitative phase of the research, to better comprehend the current conditions of a phenomenon and, as a result, to generate suggestions that can be tested later, during the more extensive second phase of the research, this strategy is suitable for the research objectives (quantitative). Taking into consideration the different study approaches that were produced by the researchers.

This study incorporates the qualitative multi-method approaches that are typically utilised in empirical qualitative research as well as talks held in focus groups. A concise explanation of the session's overarching aim, in the form of a rundown, was offered. The participants in the focus group discussion were then asked to introduce themselves prior to the beginning of the formal conversation. This includes the head of the department as well as the chief executive officer of the company. This took roughly 5 min of my time. On the day that was planned, one question was intended to collect information for approximately forty to sixty minutes. It was important to provide an answer to this question so that the findings from the brainstorming session could be validated and the scope of the individual interviews could be defined while simultaneously working towards the first research objective. This procedure was helpful in discovering the themes that were associated with the shortage of talent.

Approximately forty to sixty minutes was spent with the participants exchanging ideas and information. The moderator of the focus group session inquired of the participants, just before to the conclusion of the meeting, regarding any concluding thoughts or further observations. At the very end, the moderator and the facilitator concluded the meeting by expressing their gratitude to the attendees for their participation. The extensive debate made it abundantly clear that the participants experienced feelings of involvement and motivation in relation to the research issue. Following the conclusion of this meeting, the researcher analysed the data and created a narrative summary of the proceedings before beginning the transcribing processes. The audio of the focus group session was captured in MP3 format, and then it was transcribed. The spontaneous nature of the participants' comments was preserved throughout the process.

After that, the researchers put together a document that had the interview notes as well as a full transcription of the conversation. Before beginning the process of transcription, the researchers conducted follow-up interviews, analysed the information gleaned from those interviews, and drafted synopses of the events. After each interview was transcribed, the spontaneous nature of the speakers' responses was preserved by recording them verbatim in MP3 format and playing back the files.

After then, the quantitative method was applied as a supplement to the data gathering that had been done for the paper. The plan for the research design calls for the application of a quantitative technique that features descriptive, explanatory, and confirmatory components [41]. The strategy of selecting respondents by a straightforward random process was utilised for the study of university students [42]. Because of its favourable probability profile and compatibility with structural equation modelling, the sampling method is favoured for data collection (SEM). For the unit analyses of the study, a total of 371 participants were chosen as respondents. The research instrument consisted of a questionnaire and a structured interview, both of which were administered to respondents in order to collect data that was then distributed among university students. The questionnaire is broken up into four sections: A (Demographic), B1 (Talent Demand), B2 (Talent Supply), C (Government Policy Support), and D (Talent Shortage), and each section uses a 10-point Likert scale [43]. This scale is utilised for data-sensitive input from survey takers in addition to avoiding the answer that comes naturally.

The proposed analyses will make use of three distinct kinds of tools: first, a qualitative tool called Atlas.Ti; second, a quantitative tool called IBM-SPSS; and third, an IBM-AMOS tool (quantitative). These two tools are designed to assist in doing three distinct sorts of analyses: thematic, descriptive, and inferential. In order to discover the overarching theme of this investigation, a thematic analysis was carried out. The recurring themes It places a strong emphasis on recognising, analysing, and interpreting patterns of meaning that can be found within qualitative data. In factor analysis, the variability of a group of observed and correlated variables is characterised in terms of a potentially smaller number of unobserved variables that are referred to as factors.

In addition, confirmatory analysis, also known as CFA, is utilised in order to validate the construct's measurement model and test the hypothesis that is expressed in the path model. The procedure that was followed in order to validate the path model of SEM. The ordinary least squares regression has some drawbacks, and this method addresses those constraints by concentrating on how to deal with latent constructs in the model.

4 Finding

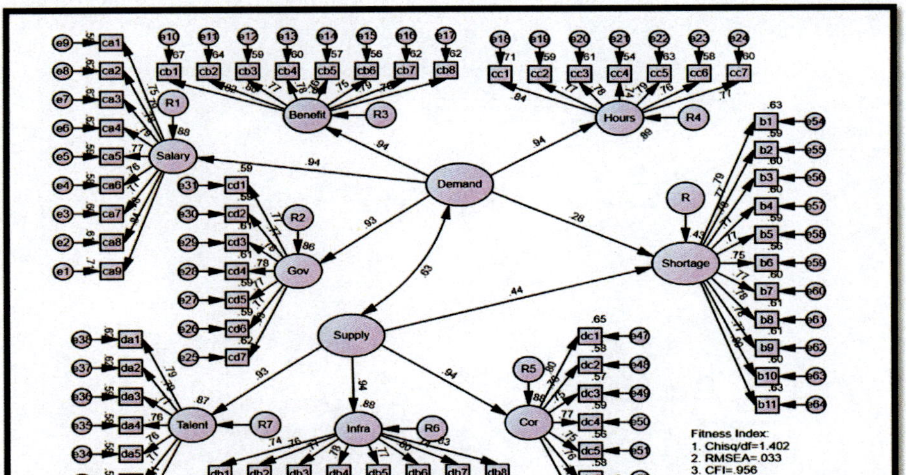

Fig. 1. Structural model

Table 1. The regression weight for supply and demand in predicting shortage

			Estimate	S.E	C.R	P	Result
Shortage	< ---	Supply	0.435	0.062	6.967	***	Significant
Shortage	< ---	Demand	0.273	0.058	4.714	***	Significant

Both Hypotheses 1 and 2 investigated the consequences of the relationship between the demand for skill and the supply of talent. In light of the findings, it was discovered that both the supply and demand of talent were statistically significant contributors to the talent gap (=.435, CR = 6.967; = .273, CR = 4.714). The researcher will utilise the crucial ratio (CR) as a guideline to highlight the study hypotheses while composing the statistical significance. In order to establish a significant likelihood of either accepting or rejecting the null hypothesis, the CR needs to be greater than the rule of thumb value of 1.96.

In addition to this, it was discovered that each and every path estimate had a favourable effect on the talent deficit. When the demand for talent increases by one unit, the talent shortfall increases by 0.273 units; conversely, when the supply of talent increases by one unit, the talent shortage increases by 0.435 units. As shown in Fig. 1, it was discovered that the proposed model described an acceptable significant percentage (43%) of the variance in talent shortage. [44] said that an R2 value that is greater than 26% is regarded to affect the causal model significantly. As a result, the structural model that measures the amount of talent scarcity is deemed to be both attainable and relevant. This is due to the fact that it makes a significant contribution to the research on talent shortage in Malaysian IBIs. In addition, the investigation offered support for hypothesis 1 and 2, as shown by the results of the SEM analysis in Table 1.

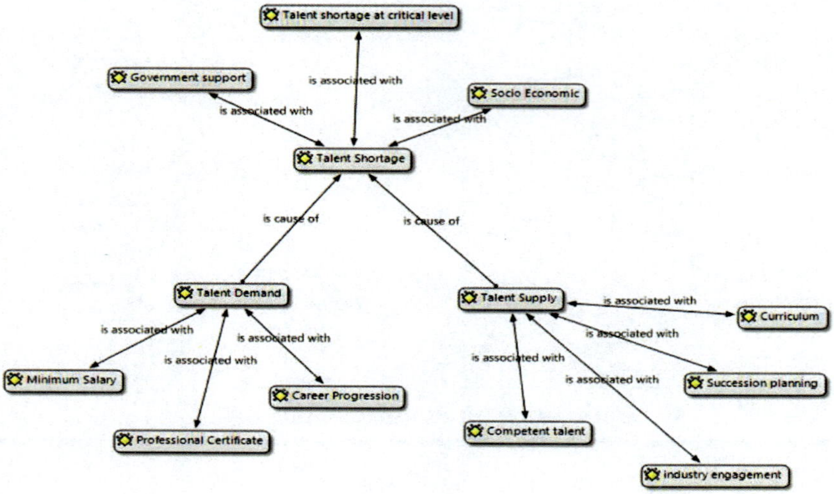

Fig. 2. Network diagram for talent shortage

Figure 2 network diagram for talent shortage has displayed a collective finding. The Talent Demand variable has three associate dimensions: Minimum Salary, Professional Certificate, and Career Progression. Those three associated dimensions of Talent Demand have caused the talent shortage in Malaysian IBIs. Meanwhile, Talent Supply was found to have four associate dimensions: Competent Talent, Industry Engagement, Succession Planning, and Curriculum. Those three associated dimensions of Talent Supply have caused Malaysia's talent shortage. In extent of explanation for Talent Shortage, it has been found that three associated dimensions are reflected (Government Support, Talent Shortage at Critical Level, and Socio Economic).

5 Conclusion

The research was conducted based on the three main problems amongst Malaysian UA graduates. Currently, there is a large talent pool in the job market. However, the unemployment rate rises to 33.8% due to tertiary education graduates. Nevertheless, this issue is not new, especially when it has always been a highlight each year at the national level. To illustrate this issue, many graduates are found to be unemployed, even though many industries or employers have reported that a large amount of talent is required by their organizations. Even so, no candidate with the right talent for the organization could be found. Consequently, many organizations are facing an acute talent shortage. This issue also happens to IBIs.

Moreover, developing graduates' skills is necessary to minimize the unemployment rate. The government has drafted and implemented various planning policies to build the talented skills required by the industry. Those policies, such as MEB, 11th Malaysia Plan, and FSB, are developed for a better balance in talent supply and demand. Based on this situation, which party is truly responsible for this development, be it the government, IBIs, IBEIs, talents, or all of them, becomes a question. Therefore, this research will point out and highlight the responsibilities played by the parties in propelling the IBIs' talent development for a better future.

Overall, the scenarios mentioned above have grabbed the attention of researchers to study those issues extensively. Apart from that, they had also become keen to discover the extent of talent shortage in Malaysia's IBIs and the levels of talent shortage. Besides that, finding the current employability skills that are always desired by IBIs and the factor of talent supply and demand, which gives the most impact on talent shortage, had become the focus of the researchers. The same case went for the proposal of modeling the talent development model for talent shortage in Malaysian IBIs. Three main objectives and seven research questions were derived based on what the researchers sought.

6 Discussion

Then, to answer all the research objectives and questions, this research was divided into two main phases. The first phase of the research was the fundamental phase, which aimed to identify and analyse the factors of talent shortage in IBIs. Under this phase were two sub-phases divided through the preliminary factor identification and confirmation of the factors. The preliminary phases consisted of fieldwork, where the researcher went to interview the Islamic banking practitioner. This interview aimed to obtain the general idea and information regarding the talent shortage issue in IBIs. In the second sub-phase, a literature review from numerous and various sources was made. The literature review's purpose was to develop a conceptual model for talent shortage in IBIs. The first research objective was fulfilled by doing so, and two research questions were answered. The first research objective was to identify the factor affecting the talent shortage in Malaysia's IBIs.

Meanwhile, to what extent of talent shortage affects the IBIs in Malaysia, and what are the factors of talent shortage in Malaysia's IBIs were the questions of this research. To answer these questions, the findings stated that Malaysia's IBIs faced an 82% talent shortage rate, where most affected areas were between the middle and specialist levels. In addition, the identified factors of the talent shortage were basic salary, employee benefits, FWH, and government policy for talent supply. Additionally, talent demand consists of talent, university infrastructure, and curriculum. Through this identification of the factors, a conceptual model was built.

Next, the data was gathered from Malaysia's Islamic banking and finance student, who was in their final year at UAs. The research objective was to analyse the effects of talent supply and demand on talent shortage in Malaysia's IBIs. The research findings showed that talent demand and talent supply significantly affect talent shortage at $p < 0.005$. In the research, the extension of talent and talent supply affected the talent shortage rate at 43%, and both research variables showed a positive relationship. To illustrate this point, between talent supply and talent demand, talent supply had the most impact on talent shortage, with a β value of 0.435. The last part focuses on the third research objective and the answer to the seventh research question. Moreover, the model has achieved all the fitness index levels: Chisq/df at 1.402, RMSEA at 0.033, CFI at 0.956, IFI at 0.956, and TLI at 0.955.

Therefore, the structural model that has been developed is valid since it contributes a high impact to the research on talent shortage in Malaysia's IBIs. At the same time, the research supported H1 and H2. In addition, all the data were analysed using the SEM approach to generate the results for quantitative data. The second phase of the research was the complementary part of the research. This phase encompassed a qualitative approach involving a group of experts from IBIs. The data was obtained from them through an in-depth interview. This phase aimed to discover the reality of talent shortage in IBIs.

Nevertheless, the second research phase was meant to be the complementary part. Somehow, certain parts of the research objective and question were reaffirmed, and a value was added to them. The first objective of the research was to identify the factors of talent shortage from qualitative findings. To fulfil this objective, it was suggested by the expert that a new factor was to be included in talent demand and supply measurement. Succession planning and industry engagement were the new factors of talent supply, while competent talent was the factor of the value provided to the research objective and questions. Then, two factors of talent demand, namely professional certificates and career progression, were proposed.

However, considerable new factors should be included in the talent shortage variable. Out of all the new factors, two of them were recommended, which were socio-economic and government agency support. Therefore, the second research question was answered through the indirect introduction of these new factors, and the first research objective was accomplished. Moreover, the first research question was answered in the second phase of the research, where all the interviewed experts agreed that talent shortage was the current challenge faced by IBIs.

On top of that, to achieve the second objective of the research, the feedback given by the experts in determining the factors gave the most impact on talent shortage. The findings showed that talent supply is critical in impacting the talent shortage. Furthermore, to accomplish the third objective of the research, experts were asked about the functionality and practicability of the model. They verified that the model developed in this study was valuable and operational. Besides that, they also extended the duration of the application of the model's operation. The model can be generalised to all of Malaysia's IBIs, including the IBIs from foreign countries that practice Islamic banking operations.

In conclusion, all the research objectives and questions have been achieved and answered, and this study has supported all hypotheses. Based on the results displayed above, all dependent variables provided a significant value towards talent shortage. To illustrate this point, the factors listed in the dependent variables were the critical factors which reflected the talent shortage in Malaysia's IBIs. Meanwhile, the factor newly added to the list of all the variable factors verified the significant value of the factor to strengthen the model for talent shortage in Malaysian IBIs. Moreover, this result substantiated the talent supply provided by the parties responsible for developing better talent for IBIs. In order to develop a better talent for IBIs, the curriculum content needs to be improvised so that the talent could complement the employability needed by the employer.

Acknowledgement. This paper supported by Ministry of Higher Education under the grant named Fundamental Grants Research Scheme (FRGS) with code grant: FRGS/1/2020/SS02/UMK/02/2.

References

1. Aziz, M., Adnan, A., Afthanorhan, A., Foziah, H., Ishak, S., Rashid, N.: The influence of employer value proposition in talent demand towards talent shortage in the Malaysian Islamic banking institutions: a SEM approach. Manag. Sci. Lett. **9**, 843–850 (2019)
2. Aziz, M.I., Afthanorhan, A., Awang, Z.: Talent development model for a career in Islamic banking institutions: a SEM approach. Cogent Bus. Manag. **3**, 1186259 (2016)
3. Aziz, M.I., Hassan, H., Uthamaputran, S., Merican, R., Rahim, M.: Modeling talent shortage for entrepreneurship among student with government policy support as mediating effect: SEM approach. In: Alareeni, B., Hamdan, A., Elgedawy, I. (eds.) ICBT 2020. LNNS, vol. 194, pp. 2011–2028. Springer, Cham (2021). https://doi.org/10.1007/978-3-030-69221-6_145
4. Ali, N.A.M., Kasim, N.: Talent management for Shariah auditors: case study evidence from the practitioners. Int. J. Financ. Res. **10**, 252–266 (2019)
5. Bindu, P.H.: Engaging the diverse workforce–an empirical study of generation X and Y. Int. J. Res. Econ. Soc. Sci. IJRESS. 7 (2017)
6. Damayanti, L., Gafur, A.: English proficiency of students at politeknik negeri balikpapan based on TOEIC. Teknosastik. **18**, 50–58 (2020)
7. Franco, A., Roach, S.S.: An assessment of the English proficiency of the Thai workforce and its implication for the ASEAN economic community: an empirical inquiry. Open J. Bus. Manag. **6**, 658–677 (2018)

8. Aasheim, C.L., Li, L., Williams, S.: Knowledge and skill requirements for entry-level information technology workers: a comparison of industry and academia. J. Inf. Syst. Educ. **20**, 10 (2019)
9. Arntz, M.: What attracts human capital? understanding the skill composition of interregional job matches in Germany. Reg. Stud. **44**, 423–441 (2010)
10. Crainer, S., Dearlove, D.: Death of executive talent. Manage. Rev. **88**, 16–23 (1999)
11. Dychtwald, K., Erickson, T.J., Morison, R.: Workforce Crisis: How to Beat the Coming Shortage of Skills and Talent. Harvard Business Press, Cambridge (2006)
12. Hewlett, S.A., Luce, C.B., Shiller, P., Southwell, S.: The Hidden Brain Drain: Off-Ramps and On-ramps In Women's Careers. Harvard Business School Press, Cambridge (2005)
13. Miner, J.B.: The Human Constraint; The Coming Shortage of Managerial Talent (1974)
14. Beechler, S., Woodward, I.C.: The global "war for talent." J. Int. Manag. **15**, 273–285 (2009)
15. Rü\s smann, M., et al.: Industry 4.0: the future of productivity and growth in manufacturing industries. Boston Consult. Group. **9**(1), 54–89 (2015)
16. Romer, P.M.: Should the government subsidize supply or demand in the market for scientists and engineers? Innov. Policy Econ. **1**, 221–252 (2000)
17. Haggstrom, G.W.: Assessing Teacher Supply and Demand. ERIC (1988)
18. Stephan, M., Silvia, M., Arie, Y.L.: A dynamic perspective on next-generation offshoring: The global sourcing of science and engineering talent. Acad. Manag. Perspect. **22**, 35–54 (2008)
19. Irshad, M.: Talent production in the face of retirement of baby boomers in developed countries. Abasyn J. Soc. Sci. **9**, 141 (2016)
20. Dubey, R., Gunasekaran, A.: Supply chain talent: the missing link in supply chain strategy. Ind. Commer. Train. **47**(5), 257–264 (2015)
21. Cappelli, P.: Talent management for the twenty-first century. Harv. Bus. Rev. **86**, 74 (2008)
22. Michaels, E., Handfield, H., Axelrod, B.: The War for Talent. Harvard Business Press, Cambridge (2001)
23. Chambers, E.G., Foulon, M., Handfield, H., Hankin, S.M., Michaels, E.G.: The war for talent. McKinsey Q. **3**, 44–57 (1998)
24. McCracken, M., Currie, D., Harrison, J.: Understanding graduate recruitment, development and retention for the enhancement of talent management: sharpening 'the edge'of graduate talent. Int. J. Hum. Resour. Manag. **27**, 2727–2752 (2016)
25. Iles, P., Chuai, X., Preece, D.: Talent management and HRM in multinational companies in Beijing: definitions, differences and drivers. J. World Bus. **45**, 179–189 (2010)
26. Brown, P., Tannock, S.: Education, meritocracy and the global war for talent. J. Educ. Policy. **24**, 377–392 (2009)
27. Lawler III, E.E.: Motivation in work organizations (1973)
28. Burn, J., Robins, G.: Moving towards e-government: a case study of organisational change processes. Logist. Inf. Manag. **16**(1), 25–35 (2003)
29. Cram, L.: The European commission as a multi-organization: Social policy and IT policy in the EU. J. Eur. Public Policy. **1**, 195–217 (1994)
30. Marsh, J.C., D'Aunno, T.A., Smith, B.D.: Increasing access and providing social services to improve drug abuse treatment for women with children. Addiction **95**, 1237–1247 (2000)
31. Lewis, R.E., Heckman, R.J.: Talent management: a critical review. Hum. Resour. Manag. Rev. **16**, 139–154 (2006)
32. Wu, M.-C., Nurhadi, D., Zahro, S.: Integrating the talent management program as a new concept to develop a sustainable human resource at higher educational institutions. Int. J. Organ. Innov. **8** (2016)
33. Birch, L., Ristevska, D.: A collaborative approach to developing successful mentoring programs: the case of William Angliss institute. In: Cauthe 2016: The Changing Landscape Of

Tourism And Hospitality: The Impact Of Emerging Markets And Emerging Destinations, pp. 800–808. Blue Mountains International Hotel Management School Sydney (2016)
34. Sharma, P., Khriney, P.D., Pradhan, P., Chettri, P., Dahal, P., Bhatt, S.: Knowledge, attitude, and practices of dental students regarding medical emergency preparedness in paonta sahib. Himachal Pradesh. Int. J. Community Dent. **7**, 11 (2019)
35. Barron, P.: Education and talent management: implications for the hospitality industry. Int. J. Contemp. Hosp. Manag (2008)
36. Creswell, J.W., Clark, V.L.P.: Designing and Conducting Mixed Methods Research. Sage publications, New York (2017)
37. Bulsara, C.: Using a mixed methods approach to enhance and validate your research. Bright. Group Res. Cent. **16**, 1–82 (2015)
38. Gallardo-Gallardo, E., Thunnissen, M.: Standing on the shoulders of giants? a critical review of empirical talent management research. Empl. Relat. (2016)
39. Connelly, L.M.: Mixed methods studies. Medsurg Nurs. **18**, 31–33 (2009)
40. Creswell, J.W., Creswell, J.D.: Research Design: Qualitative, Quantitative, and Mixed Methods Approaches. Sage publications, New York (2017)
41. Sekaran, U., Bougie, R.: Research Methods for Business: A Skill Building Approach. Wiley & Sons, New York (2016)
42. Hair, J.F.: Research methods for business (2007)
43. Hair, J.F., Gabriel, M., Patel, V.: AMOS covariance-based structural equation modeling (CB-SEM): Guidelines on its application as a marketing research tool (2014)
44. Cohen, J.: Set correlation and contingency tables. Appl. Psychol. Meas. **12**, 425–434 (1988)

Financial and Economic Basis of Ensuring the Competitive Potential of the Enterprise

Olha Bielienkova[1(✉)], Mykola Stetsko[2(✉)], Lesya Sorokina[1], Tetiana Tsyfra[1], Viktoriya Tytok[1], and Davyd Kalashnikov[1]

[1] Kyiv National University of Construction and Architecture, Kyiv, Ukraine
{bielienkova.oiu,sorokina.lv,tsyfra.tiu,tytok.vv, kalashnikov_dp}@knuba.edu.ua

[2] Department of Financial Technologies and Banking Business, West Ukrainian National University, Ternopil, Ukraine
nsteco6691@gmail.com

Abstract. The article examines the structure of assets and liabilities of enterprises with different levels of competitive potential, which was measured by the following three indicators: increase or decrease in assets, increase or decrease in the ratio of income from sales of products, works, services to cost, increase or decrease market share. The sample of twenty-five development companies was divided by cluster analysis into three groups - enterprises with low, medium and high levels of competitive potential. The analysis of the structure of assets revealed that developers with a high level of competitiveness are characterized by a slightly higher share of current assets than other groups of enterprises, and this cluster is characterized by a higher percentage of equity in the structure of liabilities. The construction companies are recommended to increase their level of financial stability, which is generally lower, to direct profits to the development fund and pursue a moderate policy of attracting long-term sources of financing for production activities. The practical use of the results of the study will allow companies to establish a system of asset management and the formation of attracted and equity and, as a consequence, create conditions for increasing competitive potential.

Keywords: Structure of assets · Structure of liabilities · Financial stability · Financial management · Competitive potential

1 Introduction

The structure of capital and assets are important strategic parameters for any company, which characterize its potential for operational, investment and financial activities. The optimal structure of assets allows to achieve long-term goals of the company's development - increasing market share, generating stable cash flow, etc., and the capital structure provides a financial basis for achieving these goals. The capital structure characterizes the ratio of sources of equity and equated to equity and borrowed funding and characterizes the financial stability of the company in the strategic perspective. Proper condition

B. Alareeni et al. (Eds.): ICBT 2022, LNNS 620, pp. 324–332, 2023.
https://doi.org/10.1007/978-3-031-26953-0_31

and structure of assets is the key to successful operations in modern markets and the implementation of the company's market strategy.

Both the structure of sources of financing and the structure of assets of companies may differ depending on the sector in which the company operates, and the maximum efficiency and feasibility - to differ for each activity. That is why for the management staff of enterprises asset management and the formation of a rational capital structure is one of the priorities.

2 Literature Review and Methodology

The structure of capital, its importance for the development and financial stability of enterprises studied in the works of (Izmailova and Bolila 2018; Izmailova and Parkhomenko 1997; Singh 2016; Tian 2016). The efficiency of investment activity of companies and the structure of assets were studied (Sorokina and Hoyko 2017).

Financial factors that affect competitiveness and determine the success or failure of the company in the long or short term have identified in their work many scientists (Alfan and Zacharia 2013; Bolila 2008; Chen 2009; Chiu and Wang 2011; Halim et al. 2014; Hoiko et al. 2017; Lai et al. 2014; Omopariola et al. 2021; Shpakov et al. 2021; Tucker et al. 2015; Vakhovych and Molodid 2013; Lehenchuk et al. 2020; Fedun et al. 2020; Kozyk and Zahoretska et al. 2020).

In order to group enterprises by their competitive potential, which can be determined by a set of indicators (Bielienkova 2020): market share index, change in production efficiency and change in capital value index of the development company, cluster analysis was performed by k-means method, which divided enterprises into three groups according to their competitive potential? All calculations were performed in the program Statistica 5.0.

Since the structure of capital and assets of enterprises in different sectors of the economy differs, the analysis chose the development of the construction sector in 2000–2020 and a sample of 25 construction companies that are developers in the housing market in Kiev in 2004–2020 (93 observations), all analyzed enterprises belong to group F (41.20 - construction of residential and non-residential buildings and 41.10 - organization of construction of buildings).

Also, all surveyed enterprises operate and build facilities in Kyiv and Kyiv region, among them: PJSC HC "Kyivmiskbud", PJSC "Miskbud-4", PJSC "Trust Kyivmiskbud-2", PJSC" Trust "Kyivmiskbud -6", PJSC "Trust Kyivmiskbud-3", PJSC Trust "Kyivmiskbud-1 named after M.P. Zagorodny", PJSC HC "Eco-Dim", State Public Joint Stock Company "Construction Enterprise UkrBud", PJSC "ISA Prime Developments", PJSC "House-Building Plant №3", PJSC "Foundation", PJSC "KP", PJSC "KP Zhytloinvest-UKB", PJSC "Ukrrestavratsiya", PJSC "Domobudivnyi Kombinat №4", PJSC "Trust Kyivmiskbud-1", LLC "KAN", LLC "TMM", PJSC "Energopol-Ukraine", LLC "Liko-Holding", SE "Intergal Bud", LLC "BC Miskbudinvest", LLC "BC Granit", PJSC "Poznyaky-Zhyl-Bud", KP "Spetshitlofond".

A comparative analysis of twenty-five development companies was conducted, as a result of which they were divided into groups according to "high", "medium", "low" level of competitiveness. It was found that enterprises with a high level of competitiveness have a higher share of current assets in the capital structure, as well as a higher level of equity than the average in the construction sector (autonomy ratio 0.29), compared to other groups and construction averages (0.13 for construction). It is expedient to continue the conducted research in the direction of increasing the research sample and conducting correlation-regression analysis between the indicators of the structure of capital and assets and the level of competitiveness of the enterprise. It is also advisable to identify sectoral patterns between indicators of capital and asset structure and the potential for competitiveness of enterprises.

The practical use of the results of the study will allow companies to establish a system of asset management and the formation of attracted and equity and, as a consequence, create conditions for increasing competitive potential.

3 Research Results

It should be borne in mind that currently the activities of all enterprises without exception are influenced by the process of digitalization and transition to virtual reality of many business processes, which also affects the structure of assets and liabilities of enterprises and their financial results (Zeltser et al. 2019).

However, the study of capital structure and asset structure of enterprises with different competitive potential requires additional research, as they will determine additional parameters to ensure the competitiveness of the enterprise.

An analysis of the development of development companies and the selection of factors that can be the basis for the formation of their competitiveness in the strategic perspective. The initial data on the activities of companies were selected for four - six-year periods during 2010–2020.

It is proposed to choose as indicators of competitive potential for developers asset growth indices (A), market share (CR) and efficiency of operating activities - the ratio of sales revenue to cost of sales (E), which are calculated according to the financial statements of the analyzed companies.

The first group (high competitive potential) included enterprises that increased their market share in the analyzed period, the efficiency of their activities decreased slightly, and assets increased (Fig. 1, Table 1).

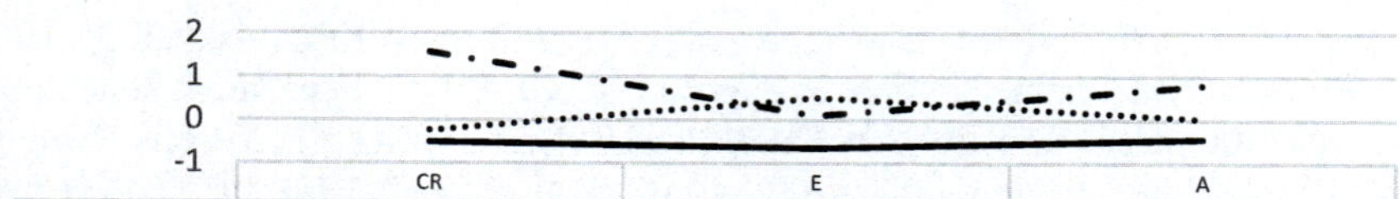

	CR	E	A
High	1.564822894	0.100029421	0.8015291
Medium	-0.235646635	0.51232624	0.0235647
Low	-0.51220552	-0.652674364	-0.4401109

Fig. 1. The average values of clusters 1–3

Table 1. Distances to the center of the cluster of enterprises of the groups

Enterprise	Distance
Group "high competitive potential" - 5 enterprises	
C_1	1,184691
C_2	0,512057
C_3	1,669065
C_4	0,831635
C_6	1,549287
Group "medium competitive potential" - 12 enterprises	
C_5	0,676625
C_7	0,585365
C_9	0,363953
C_10	0,315658
C_11	0,301783
C_13	0,259892
C_14	0,141872
C_19	0,565353
C_20	0,395946
C_21	1,060723
C_22	0,859648
C_23	0,585126
Group "low competitive potential" - 8 enterprises	
C_8	0,462429
C_12	0,482598
C_15	0,144644
C_16	0,200080
C_17	1,238254
C_18	0,345167
C_24	0,511350
C_25	0,657282

The first group included PJSC "Miskbud-4", PJSC "Fundament", PJSC "BC Capital", PJSC "UkrBud" and KP "Zhitloinvest-UKB". The enterprises of this group are quite different in size. They have asset values ranging from UAH 1,686 million in PJSC BC Stolitsa to UAH 3,753.8 million in Zhytloinvest-UKB. The average value of assets in the group for the study period amounted to UAH 805.683 million.

The structure of assets in the group as a whole differs slightly from the industry average in the direction of increasing the weight of non-current assets (Table 2), namely: 40% are non-current assets and 60% - current. The largest weight of non-current assets in the capital structure of the analyzed companies in PJSC "Miskbud-4" - 77%. Accordingly, only 23% of the company's capital is current assets. PJSC BC Capital, on the other hand, is characterized by an almost complete lack of fixed assets on the balance sheet - the company's non-current assets amounted to 5 to 20% of capital in different years. Accordingly, from 80 to 95% of the company's capital is in circulation. That is, successful companies use completely different asset formation strategies, while achieving success in the market.

Table 2. Asset structure of enterprises with low, medium and high competitive potential and the construction sector

Indicator	High	Medium	Low	Construction sector
Non-current assets, %	40	40,4	48	36
Current assets, %	60	59,6	52	64
Total	100	100	100	100

The structure of assets of enterprises "construction" compared to previous periods has shifted towards an increase in working capital. However, such values may be due to the imperfection of the asset valuation system, as both current and non-current assets are usually reflected in the balance sheet at their value at the time of acquisition and non-current assets at the current time, in the absence of revaluation.

Equity of enterprises of the first group averaged 29% of the amount of funding sources, and attracted - 71% (Table 3). The share of long-term funding sources was 10%. The share of own sources of capital in successful enterprises was 14 points more than the industry average, and the share of long-term sources in liabilities - 14 points less. That is, most of the sample companies are trying to pursue a more prudent fiscal policy than most of the construction sector companies.

Table 3. Structure of sources of financing of enterprises with low, medium and high competitive potential and the construction sector

Indicator	High	Medium	Low	Construction sector
Equity, %	29	1	8	13
Short-term liabilities, %	61	41	59	63
Long-term liabilities, %	10	58	33	24
Total	100	100	100	100

The second group includes 12 companies with an average level of competitive potential, namely: LLC "KAN", LLC "TMM", PJSC HC "Kyivmiskbud", PJSC

"Energopol-Ukraine", LLC "Liko-Holding", PJSC "House-Building Plant №4", PJSC "Kyivmiskbud 2", SE "Intergal Bud", LLC BC "Miskbudinvest", LLC BC "Granit", PJSC "Poznyaky-Zhyl-Bud", PJSC "Kyivmiskbud 6".

The structure of assets of enterprises in the study period almost coincides with the structure of capital of the first group, respectively 40.4% non-current assets, 59.6% - current assets. The average share of equity of enterprises in this group was about 1%, while borrowed capital - 99%. The weight of long-term funds in the structure of sources of financing is high - 58%. That is, enterprises of this group finance their own activities mainly through long-term borrowed funds.

At the same time, the share of equity in PJSC "House-Building Plant №4" is quite high - up to 0.39. Whereas in PJSC "Kyivmiskbud-2" due to the long-term loss of the company, equity is negative, i.e. the company operates entirely on borrowed funds, which is quite a risky financial policy. Among the enterprises of the group, four companies experience an acute shortage of own funds, which is 30% (5 - 41.6) of analyzed sample. In addition to PJSC "Kyivmiskbud - 2", it is LLC "KAN", PJSC "ENERGOPOL-Ukraine", LLC BC "Miskbudinvest", PJSC "Poznyaky-Zhyl-Bud".

The peculiarity of the enterprises of the analyzed group is that they actively attract long-term sources of financing. Thus, only two companies (PJSC Domobudivny Kombinat G4 and PJSC ENERGOPOL-Ukraine) out of 12 do not have long-term loans, which is 16.7% of the analyzed population.

Comparison of the structure of assets and liabilities of the analyzed three groups of enterprises is shown in Fig. 2 and 3.

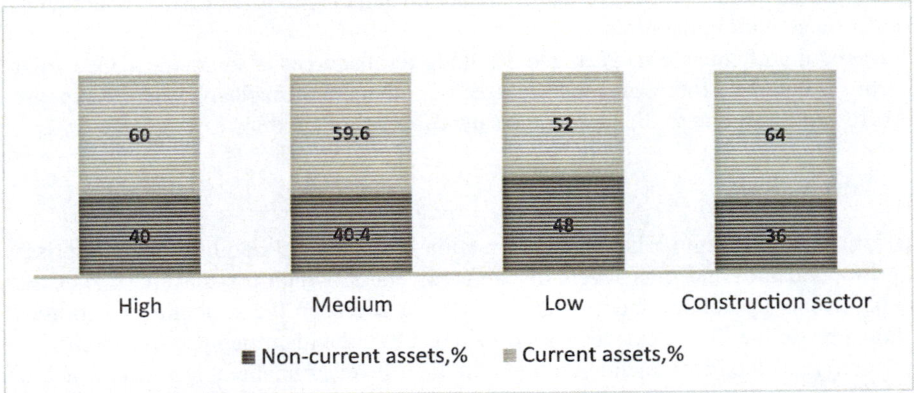

Fig. 2. The structure of assets of enterprises with low, medium and high competitive potential and the construction sector

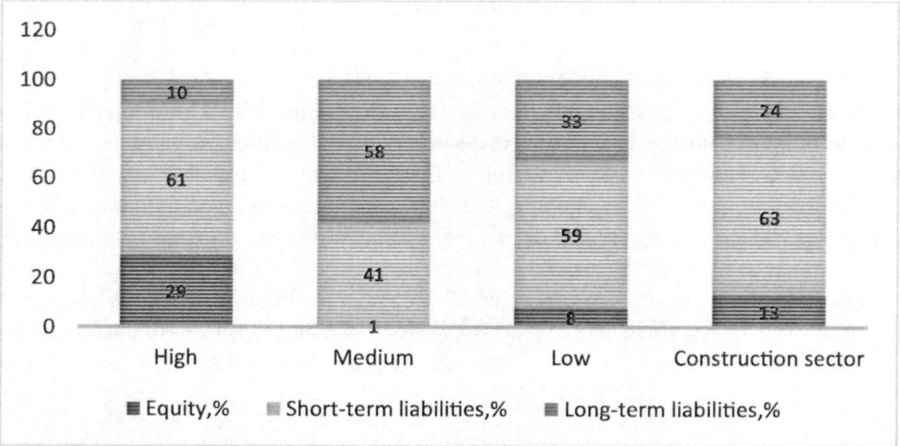

Fig. 3. Structure of sources of financing of enterprises with low, medium and high competitive potential in comparison with indicators of the construction sector

The third group - enterprises with a low level of competitive potential (PJSC "Ukr-restavratsiya", PJSC "Domobudivnyi Kombinat № 3", PJSC "ISA Prime Developments" PJSC "Trust" Kyivmiskbud-1 "PJSC named after MP, PJSC "Kyivmiskbud-3" have a slightly different capital structure compared to the first two groups.Thus, non-current assets of the companies in this sample average 48%, respectively, non-current - 52%, which is much more than the average construction and may indicate on the reduction of production activity in the group.

The equity of enterprises averaged 8%, long-term sources of financing - 33%. Four enterprises out of six (66,7%) in the analyzed period operated exclusively at the expense of borrowed funds due to the lack of their own sources of funding.

4 Conclusions

The article reveals regularities in the formation of assets and liabilities of enterprises with low, medium and high levels of competitiveness, which are that the higher the competitiveness potential, the greater the share of equity in the structure of liabilities and the greater the size of current assets. As a result of the identified patterns, construction companies are recommended to increase their level of financial stability, which is generally lower, to direct profits to the development fund and pursue a moderate policy of attracting long-term sources of financing for production activities. The practical consequence of the study is the recommendations in attracting new financing schemes to the implementation of their own construction projects, increasing attention not only to profitability, but also to ensuring financial stability in the long run.

References

Alfan, E., Zacharia, Z.: Review of financial performance and distress: a case of Malaysian construction companies. Br. J. Arts Soc. Sci. **12**, 1–15 (2013)

Bielienkova, O.Iu.: Strategy and mechanisms for ensuring the competitiveness of construction enterprises based on the model of sustainable development: monograph. Lira-K, Kyiv (2020)
Bielienkova, O., Moholivets, A., Loktionova, Y.: Measurement of the level of economic security of enterprises by the indicator of the attractiveness for hostile takeovers. Baltic J. Econ. Stud. **7**(5), 29–40 (2021). https://doi.org/10.30525/2256-0742/2021-7-5-29-40
Bolila, N.V.: Ways to prevent the emergence of a crisis state of the enterprise. Ways to increase the efficiency of construction in the conditions of the formation of market relations, vol. 19, pp. 67–69 (2008)
Chen, H.L.: Model for predicting financial performance of development and construction corporations. J. Constr. Eng. Manag. **135**, 1190–2000 (2009)
Ching-Yi, C., Wang, M.-W.: An integrated DEA based model to measuring financial performance of construction companies. WSEAS Trans. Bus. Econ. **1**, 1–15 (2011)
Halim, M., Jusoh, M., Osman, A., Amlus, M.: Determining the financial performance factors among Bumiputera entrepreneurs in Malaysian construction industry. Aust. J. Basic Appl. Sci. **8**, 18–24 (2014)
Hoiko, A.F., Sorokina, L.V., Skakun, V.A.: SVP is a financial security mechanism construction enterprises in unstable economic condition. Ways to increase the efficiency of construction in the conditions of the formation of market relations, vol. 35, pp. 13–36 (2017)
Izmailova, E., Bolila, N.: Models of forecasting the loss of financial strength as a component of the system of economic security of contracting enterprises. Ways to increase the efficiency of construction in the conditions of the formation of market relations, vol. 38, pp. 225–233 (2018)
Izmailova, K.V., Parkhomenko, V.V.: Simulation modeling of financial indicators of investment activity of the enterprise. Ways to increase the efficiency of construction in the conditions of the formation of market relations, vol. 22, pp. 73–76 (1997)
Ying Lai, H., Rashid Abdul Aziz, A., Khuan Chan, T.: Effect of the global financial crisis on the financial performance of public listed construction companies in Malaysia. J. Financ. Manag. Property Constr. **19**(3), 246–263 (2014). https://doi.org/10.1108/JFMPC-02-2014-0002
Omopariola, E.D., Windapo, A., Edwards, D.J., El-Gohary, H.: Level of financial performance of selected construction companies in South Africa. J. Risk Financ. Manag. **14**, 518 (2021). https://doi.org/10.3390/jrfm14110518
Shpakov, A., Stetsenko, S., Shpakova, H., Sorokina, L., Akselrod, R.: Assessment of the influence of adaptability factors on the effectiveness of managing changes in enterprises by fuzzy logic. Sci. Horiz. **24**(10), 72–82 (2021)
Singh, D.: A panel data analysis of capital structure determinants: an empirical study of non-financial firms in Oman. Int. J. Econ. Financ. Issues **6**(4), 1650–1656 (2016)
Sorokina, L.V., Hoyko, A.F. (eds.): Econometric toolkit for financial security management of construction companies. Kyiv National University of Construction and Architecture, Kyiv (2017). [in Ukrainian]
Stetsenko, S.P.: Economic security: security and structure. Invest. Pract. Addit. **24**, 104–106 (2013)
Tian, Y.: Optimal capital structure and investment decisions under time-inconsistent preferences. J. Econ. Dyn. Control **65**, 83–104 (2016)
Tucker, G.C., Windapo, A., Cattel, K.: Exploring the use of financial capacity as a predictor of construction company corporate performance: Evidence from South Africa. J. Eng. Design Technol. **13**, 1–18 (2015)
Vakhovych, I.V., Molodid, O.O.: Ensuring the economic stability of small construction enterprises is the basis of their economic security. Ways to increase the efficiency of construction in the conditions of the formation of market relations, vol. 30, pp. 249–255 (2013)
Zeltser, R.Ya., Bielienkova, O.Yu., Novak, Ye., Dubinin, D.V.: Digital Transformation of resource logistics and organizational and structural support of construction. Sci. Innov. **15**(5), 38–51 (2019)

Lehenchuk, S.F., Valinkevych, N.V., Vyhivska, I.M., Khomenko, H.Y.: The significant principles of development of accounting support for innovative enterprise financing. Int. J. Adv. Sci. Technol. **29**, 2282–2289 (2020)

Fedun, I.L., Novikova, I.V., Igor, L.V., Marina, M.K., Marta, O.G.: Concept of innovative marketing in management of enterprise. J. Adv. Res. Dyn. Control Syst. **12**(7 Special Issue), 352–358 (2020)

Vorobec, S., Kozyk, V., Zahoretska, O., Masuk, V.: Simulation model of planning financial and economic indicators of an enterprise on the basis of business model formalization. In: Kryvinska, N., Greguš, M. (eds.) Data-Centric Business and Applications. LNDECT, vol. 30, pp. 299–318. Springer, Cham (2020). https://doi.org/10.1007/978-3-030-19069-9_12

The Impact of Bahrain's Adaptive Sports on Quality of Life

Noor S. J. I. Ahmed[1], Ali Moosa[2], Allam Hamdan[1](✉), and Siraj Zahran[3]

[1] Ahlia University, Manama, Bahrain
ahamdan@ahlia.edu.bh
[2] Arabian Gulf University, Manama, Bahrain
[3] Jeddah College of Engineering, University of Business and Technology, Jeddah 21448, Kingdom of Saudi Arabia

Abstract. The importance of research and review of the advantages of physical exercise for humans and the importance of sports and the extent of its impact on the lives of individuals in various aspects. Bahrain Economic Vision 2030 consider sport is one of the most important factors to increase immunity in the human body, and that sport is important to confront COVID-19 and all diseases, so there are positives to the need to adhere to sports practices. Literature Review shows Public sector support for sport and physical activity unleash promote wellbeing and health, pursue positive social goals and promoting genders equality (Giulianotti et al. 2019) and how Physical activities enhance quality of life as exercise are very great for the general health and physical and psychological health of those who exercise regularly. (Shen et al. 2020).The method of the research is systematically literature review the articles related how sport enhancing quality of life, global initiatives regarding sport as social innovation and Bahrain overview about physical activity. In the Conclusion, Sport is one of the main things in a healthy person's life, which maintains public health and prevents diseases. Also, Bahrain 2030 vision improve health system by promoting and encouraging a healthy lifestyle by offering more attractive public spaces and modern facilities to boost sports activities. Finally, For future research academic institutions need to conduct researches about how sport can result outcomes beyond the playing field (Sanders et al. 2017; Camp 2020), what is the role of innovation in sport for development and peace to develop solutions for social challenges (Svensson and Cohen 2020).

Keywords: Life quality · Sport · Development · Physical activity

1 Introduction

Active people's social networks have the potential to favorably affect their physical activity by offering possibilities for social support, social influence, social engagement, access to resources, and personal interaction. The importance of research and review of the advantages and advantages of physical exercise for humans and the importance of sports and the extent of its impact on the lives of individuals in various aspects. A physical exercise has helped many people with genetic diseases and diabetes to improve

B. Alareeni et al. (Eds.): ICBT 2022, LNNS 620, pp. 333–338, 2023.
https://doi.org/10.1007/978-3-031-26953-0_32

treatment and develop a better psychological state due to physical exercise. In the current circumstances, there are many sports motivators and e-sports programs that make people continue to care about public health and exercise in various circumstances because technology has helped people communicate even through social networking sites and attend various sports courses.

The Kingdom of Bahrain was distinguished by its interest in developing the technology sector, as the Kingdom of Bahrain considered it one of the foundations for development and leadership in the Kingdom of Bahrain and to support the economic aspect and the sustainable development process, leading to the realization of Bahrain Economic Vision 2030, by focusing on technology that contributed greatly to building a better future, and providing All business needs, even entrepreneurship, to build an integrated economy. In the time of the COVID-19, it was confirmed by specialist consultants for infectious diseases in the Kingdom of Bahrain that sport is one of the most important factors to increase immunity in the human body, and that sport is important to confront COVID-19 and all diseases, so there are positives to the need to adhere to sports practices. Most of the health and guiding instructions in the Kingdom of Bahrain were rational and balanced to allow people to walk outside without requiring them to wear a medical mask. Physical activity will help in recovery from COVID-19.

2 Literature Review

2.1 Public Sector

Public sector support for sport and physical activity unleash promote wellbeing and health, pursue positive social goals and promoting genders equality (Giulianotti et al. 2019). Sport by focusing on social networks can build entrepreneurship into ecosystem and by evaluate, contribute and facilitate attractive ecosystem will be created (Ratten 2019). Invest in sport carry benefits for physical and mental health access to financing rise the level of sport activity as crowd funding in sport emotional investment created as support for favorite sport club, players or activity (Krupa et al. 2020). Also, as social innovation Sport can make positive changes in the society (Dadelo 2020) by Technological innovations develop new strategies to impact society, (Ratten 2020) such as sport for disabled people can increase self-esteem (Sorrentini et al. 2019).

(Nałęcz et al. 2020) Sport, depending on its level of professionalism or discipline, can fulfill social objectives such as pro-health, volitional development, socialization, education, and integration. In Bahrain, there is a link between sports participation, social capital, and creativity. (Escamilla-Fajardo et al. 2020) When it comes to surviving crisis circumstances, sports entrepreneurship has been seen as an important component of sports organizations in Bahrain. There is an influence of the crisis obtained from COVID-19 on sports entrepreneurship and whether there are disparities in entrepreneurship prediction on service quality in non-profit sports organizations.

2.2 Physical Activities and Quality of Life

A systematic review on the positive physical health-related effects of traditional physical activity experiences (Mays and Brevetti 2020) on young people in various countries of the world, specifically from the age of eight to the age of twenty-five.

In various systematic studies, (Ahmed et al. 2021) the study proved that here are many benefits of traditional methods, not only on health, but also on the social and psychological aspect of sports practitioners directly, better than the modern party, because playing together makes young people more powerful and free to express their opinion, manage their team, and know their points (Akbar et al. 2020) strength and weakness.

In addition, (Akbar et al. 2020) the benefits of exercise are very great for the general health and physical and psychological health of those who exercise regularly. (Shen et al. 2020) The various diversity of sports experiences were clarified for the activities of children, coaches and youth, and to motivate them to develop and develop a healthy culture for humans (Grimaldi-Puyana et al. 2018), and to pay more attention to sports through social networking sites or recorded or live videos in advanced technology around the world. In different circumstances, provided the descriptive information about the risk that faced them by two differences of mobility outcomes (Baia do Nascimento et al. 2013).

The percentage of motivation for physical activity is more than the traditional health culture that is taught in educational institutions because physical activity is one of the most activities that everyone wants to try and practice in various types of sports activities (Gill et al. 2017). Some types of medical treatments and physical therapy require the injured or patient to exercise and some physical activities to recover better and faster to ensure the development of the health level and the recovery of the injury case (Xu et al. 2021).

(Gately et al. 2013) The teachers' point of view has been added to the existing literature to support the interrelationship between the studies and the teachers' views to reach a coherent and real conclusion. Instead, the school evaluates students on an academic level without focusing on physical health. However, (Warden et al. 2017) the present research shows that physical exercise should be fostered at a young age and sustained throughout early adulthood to reap the most bone advantages. The observations shown them the advantages of doing a physical sport and practices that will increase their health level and improve their basic results when they compare them with who they are not doing these kinds of practices (Hämäläinen 2013; Zander et al. 2015). Moreover, physical activity throughout the early stages of puberty has been shown to have increased skeletal advantages.

3 Research Methodology

The method of the research is systematically literature review the articles related how sport enhancing quality of life, global initiatives regarding sport as social innovation and Bahrain overview about physical activity. The problem statement that has been found it as the finding of the research or it as a result. In 2010 more than 2,430,000 documents found by Google when you research about phrase "sustainable development through sports" now it has increased more than 5 times (Alvina et al. 2020; Farmasi and Dan 2017) and the researches remain not sufficient as it is new and growing field of study need more research (González-Serrano et al. 2020; Panwala et al. 2017; Chawansky et al. 2018).

4 Conclusion

Many of the positive aspects of sports have been discussed and previous studies have shown that coaches have succeeded in achieving innovation to communicate between them and the members of the sports team who follow a sports system for a specific goal. Moreover, encouraging cooperation and communication between coaches and athletes in the field of sports technology and promoting common interest in the future of sports technology. Sport is one of the main things in a healthy person's life, which maintains public health and prevents diseases.

Technology has allowed many coaches and specialists in the field of sports to go and teach others with ease from all countries of the world through social networking sites and educational, motivational, and educational courses for sports and the provision of electronic and printed courses and videos recorded through the most popular YouTube and Instagram channels in the last years. Creativity is essential to the success of various sports program development efforts. Creative innovation requires sports teams to integrate multiple team members to diversify their sports and cultural programmers. The style of team communication is an important determinant of team creativity and change from routine programs in physical sports.

Bahrain 2030 vision stands on sustainability, competitiveness and fairness to enhance living standards by change economic and social life. As part of Bahrain 2030 vision improve health system by promoting and encouraging a healthy lifestyle by offering more attractive public spaces and modern facilities to boost sports activities. For example, Supreme Council for Youth and Sports policies are ensure physical activities facilities for all and encourage sports for a healthy lifestyle. Also, Ministry of Youth and Sport Affair translate government vision through a new website to provide information, display ministry plans and interact with the community.

For future research Sport innovation is a new field of research within the sport sciences (Tjønndal 2016) as academic institutions need to conduct researches about how sport can result outcomes beyond the playing field (Sanders et al. 2017; Camp 2020), how sport can contribute to the world from development perspective (Author et al. 2018), what is the role of innovation in sport for development and peace to develop solutions for social challenges (Svensson and Cohen 2020). In addition, a partnership between the government and stakeholders relate to sport and health sector highly required. (Deemua 2019; Jamil 2018; Piercy et al. 2018). Finally, Resulting covid-19 physical activity is required to inspire people return to a good lifestyle routine (Mattioli et al. 2020).

References

Ahmed, F., Zuk, A.M., Tsuji, L.J.S.: The impact of land-based physical activity interventions on self-reported health and well-being of indigenous adults: a systematic review. Int. J. Environ. Res. Public Health **18**(13), 7099 (2021). https://doi.org/10.3390/ijerph18137099

Akbar, L., Zuk, A.M., Tsuji, L.J.S.: Health and wellness impacts of traditional physical activity experiences on indigenous youth: a systematic review. Int. J. Env. Res. Public Health **17**(21), 8275 (2020). https://doi.org/10.3390/ijerph17218275

Author, T., Date, I., Version, T., Doi, U.R.L.: Osaka University Knowledge Archive : OUKA キャリア発達と組織コミットメントの関係性に関する考察内炭広志・羽豆智之・小合祐輔・. Report I (2018)

Baia do Nascimento, M.I., da Silva, G.L.: Information and its technologies: a descriptive and correlative analysis of the disciplines of information technology (it) of undergraduate programs in executive secretariat area in public institutions in Brazil. Revista De Gestão e Secretariado **4**(3), 01–29 (2013). https://www.proquest.com/scholarly-journals/information-technologies-descriptive-correlative/docview/1505071509/se-2?accountid=44753

Chawansky, M., et al.: Digital Commons @ Otterbein Health and Sport Sciences A systematic review of sport for development interventions across six global cities (2018)

Dadelo, S.: The analysis of sports and their communication in the context of creative industries. Creativity Stud. **13**(2), 246–256 (2020). https://doi.org/10.3846/cs.2020.12206

Escamilla-Fajardo, P., Núñez-Pomar, J.M., Calabuig-Moreno, F., Gómez-Tafalla, A.M.: Effects of the COVID-19 pandemic on sports entrepreneurship. Sustainability **12**(20), 8493 (2020). https://doi.org/10.3390/su12208493

Farmasi, J., Dan, S.: No 主観的健康感を中心とした在宅高齢者における健康関連指標に関する共分散構造分析 Title. **14**(1), 55–64 (2017)

Gill, T.M., et al.: The effect of intervening hospitalizations on the benefit of structured physical activity in promoting independent mobility among community-living older persons: secondary analysis of a randomized controlled trial. BMC Med. **15**(1), 65 (2017). https://doi.org/10.1186/s12916-017-0824-6

Giulianotti, R., Itkonen, H., Nevala, A., Salmikangas, A.K.: Sport and civil society in the Nordic region. Sport Soc. **22**(4), 540–554 (2019). https://doi.org/10.1080/17430437.2017.1390906

González-Serrano, M.H., Sanz, V.A., González-García, R.J.: Sustainable sport entrepreneurship and innovation: a bibliometric analysis of this emerging field of research. Sustainability **12**(12), 5209 (2020). https://doi.org/10.3390/su12125209

Grimaldi-Puyana, M., Alcaráz-Rodríguez, V., Sánchezo-Oliver, A.J.: Job profile and motivation of workers in sports programs with intellectual functional diversity. J. Phys. Educ. Sport **18**, 1386–1390 (2018). https://doi.org/10.7752/jpes.2018.s3205

Hämäläinen, M.: A situational theory of advantages in sport. Phys. Culture Sport **60**(1), 5–13 (2013). https://doi.org/10.2478/pcssr-2013-0028

Jamil, A.T.: Physical activity is a priority in public health. **32**, 25–38, 2017–2018 (2018). https://doi.org/10.1249/MSS.0000000000001943.2018

Krupa, D., Walczak, D., Żołądkiewicz-kuzioła, A.: Original Article Crowd funding in financing sport activities. J. Phys. Educ. Sport **20**(5), 2959–2966 (2020). https://doi.org/10.7752/jpes.2020.s5402

Mays, B.J., Brevetti, M.A.: Lessons on ways to develop self-empowerment: a phenomenological study of the lived experiences of higher education learners with physical disabilities. J. Multicultural Educ. **14**(2), 61–84 (2020). https://doi.org/10.1108/JME-06-2019-0047

Nałęcz, H., Skrok, Ł., Majcherek, D., Biernat, E.: Through sport to innovation: sustainable socio-economic development in European countries. Sustainability **12**(24), 10489 (2020). https://doi.org/10.3390/su122410489

Panwala, T., et al.: No 主観的健康感を中心とした在宅高齢者における健康関連指標に関する共分散構造分析. PLoS Neglected Trop. Dis. **9**(5), 1–14 (2017)

Ratten, V.: Sport entrepreneurial ecosystems and knowledge spillovers. Knowl. Manag. Res. Pract. **19**(1), 43–52 (2019). https://doi.org/10.1080/14778238.2019.1691473

Ratten, V.: Coronavirus disease (COVID-19) and sport entrepreneurship. Int. J. Entrepreneurial Behav. Res. **26**(6), 1379–1388 (2020). https://doi.org/10.1108/IJEBR-06-2020-0387

Sanders, B.E.N., Coning, C.D.E., Keim, M.: Going to Scale ? A critique of the role of the public sector in sport for development and peace in South Africa. Afr. J. Phys. Act. Health Sci. **23**(December), 514–532 (2017)

Shen, J., Cheng, J., Huang, W., Zeng, F.: An exploration of spatial and social inequalities of urban sports facilities in Nanning city, China. Sustainability **12**(11), 4353 (2020). https://doi.org/10.3390/su12114353

Sorrentini, A., Palma, D.D., Madonna, G.: Sport e Disability Management per l ' Inclusione scolastica Sport and Disability Management for the School Inclusion, pp. 69–74 (2019)

Svensson, P.G., Cohen, A.: Innovation in sport for development and peace. Managing Sport Leisure **25**(3), 138–145 (2020). https://doi.org/10.1080/23750472.2020.1728068

Tjønndal, A.: BBR-Brazilian business review sport, innovation and strategic management: a systematic literature review sport, innovation and strategic management: a systematic literature review. BBR-Brazilian Business Review Bus. Rev. (Engl. Ed., Online) **13**(13), 38–56 (2016). http://www.redalyc.org/articulo.oa?id=123052598003%0A. https://doi.org/10.15728/edicaoesp.2016.3

Warden, S.J., Weatherholt, A.M., Gudeman, A.S., Mitchell, D.C., Thompson, W.R., Fuchs, R.K.: Progressive skeletal benefits of physical activity when young as assessed at the midshaft humerus in male baseball players. Osteoporos. Int. **28**(7), 2155–2165 (2017). https://doi.org/10.1007/s00198-017-4029-9

Xu, J., Yang, S., Lin, Y., Yang, R.: An evaluation of coupling coordination between sports industry and health service industry in china. PLOS ONE **16**(8), e0256333 (2021). https://doi.org/10.1371/journal.pone.0256333

Zander, A., Rissel, C., Rogers, K., Bauman, A.: Active travel to work in NSW: trends over time and the effect of social advantage. Health Promot. J. Austr. **25**(3), 167–173 (2015). https://doi.org/10.1071/HE14004

Justification of Directions of Agricultural Waste Usage as Biomass

Bugaychuk Vita[1], Valinkevych Nataliia[1](✉), Grabchuk Inna[1], Opalov Oleksandr[2], Khodakyvskyy Volodymyr[1], and Tymchak Vira[3]

[1] Department of Economics, Entrepreneurship and Tourism, Polissia National University, Zhytomyr, Ukraine
{natali1573,xxxx-vova}@ukr.net

[2] Department of International Economic Relations and European Integration, Polissia National University, Zhytomyr, Ukraine
opsan@ukr.net

[3] FG (Farm Enterprise) «Agroinnovatsiia», Chernivtsi, Ukraine

Abstract. The article is devoted to the research of agricultural waste usage as biomass in order to obtain new products and energy, as well as the implementation of the principles for sustainable development of the state. This approach is aimed at greening production, protecting the environment and improving the efficiency of enterprises.

The methodological basis of the study is a system of certain methods and techniques used to assess the generation and use of agricultural waste. In particular, the article conducted a correlation-regression analysis to establish cause-and-effect relationships in the field of waste management.

It is proved that at each stage of waste generation the directions and methods of their repeated recycling should be planned in advance. Agricultural waste forms biomass, from which in the process of recycling there is an opportunity to return the useful values of waste in the life cycle. Forecasting of agricultural waste generation in Ukraine for the period up to 2023 was carried out, the results of which confirmed the need to implement a comprehensive waste management policy, taking into account the positive experience of the European Union.

The result of scientific research was the formulation and justification of the algorithm for the reuse of biomass, which includes the most important control units for the generation and use of agricultural waste. Under these conditions, they become biomass and become a potential factor in additional resources, reducing production costs and increasing the profitability of economic entities.

The practical application of the above algorithm will increase the production of innovative products, reduce waste into the environment and save costs on their treatment. The main guideline for solving the problem of generation and accumulation of agricultural waste should be the concept of «zero waste».

Keywords: Agriculture · Waste · Biomass · Raw materials · Recycling · Ecology · Efficiency

B. Alareeni et al. (Eds.): ICBT 2022, LNNS 620, pp. 339–355, 2023.
https://doi.org/10.1007/978-3-031-26953-0_33

1 Introduction

As part of the implementation of the European Commission's action plan (since December 2015) for the transition of the economy to a closed production cycle (until 2019) [19], Ukraine has approved the National Waste Management Strategy, which provides (until 2030) the introduction of the latest models that form the basis for the sustainable development of the national economy. The primary goals of sustainable development of Ukraine for the period until 2030 are overcoming poverty, achieving food security, improving nutrition and promoting sustainable agricultural development through the introduction of a systematic approach to waste management at the state and regional levels, reducing waste generation and increasing the volume of recycling and reuse [10, 14]. However, the implementation of the National Strategy is constrained by inadequate level of compliance with legal requirements; low awareness and knowledge of the agricultural sector about the benefits of agricultural waste treatment or reuse and opportunities for the manufacturing of new products; lack of an effective mechanism for attracting private investment into the development of agricultural sector facilities; low efficiency of strategic waste management facilities [10, 11].

Agricultural waste management should start from the moment of planning the use of natural resources in the process of basic production as raw materials for agricultural products and continue during the design of its entire life cycle, regardless of its generation in the crop or livestock sectors. Most importantly, at each stage of waste generation, directions and methods of its multiple recycling should be planned in advance. Agricultural waste produces biomass, which in the process of recycling can return the beneficial aspects of waste to the life cycle [12].

Due to its relevance, this research requires resolving the problem of the current state of agricultural waste generation and its recycling as biomass.

In terms of production capacity and economic advantages, the scientific hypothesis of the research is to justify the algorithm of waste recycling as biomass, the practical application of which will increase the production of innovative products, reduce the flow of waste into the environment and save capital costs for its management.

2 Methodology

The aim of the research is justification of methodical provisions and practical recommendations for enhancing the manufacturers' understanding and perception of waste as biomass for the production of biotechnology-based products.

Research tasks to analyze the processes of waste generation in agricultural production and determine their distribution by type and explore the tools for their management; to identify the tools of waste recycling and its usage in agricultural production; to explore the interconnection between gross output and reuse of agricultural waste; to connect the volumes of waste generation with the costs of its management; to determine the impact of costs for waste management and the amount of waste flowing into the environment; to justify the algorithm of waste recycling as biomass based on biotechnology.

A set of general and special methods was used to conduct the research and achieve its objectives: analytical method (to analyze the theoretical and methodological basis for the generation and use of waste as biomass); methods of analysis and synthesis (to test the hypothesis in the work); abstraction method (during the process of identifying tools for the generation and reuse of waste); methods of induction and deduction (to determine the impact of waste generation, utilization and use of funds for waste management on the volumes of waste generation and distribution in territorial areas); graphical method (for interpretation of the research results regarding the generation and feasibility of waste recycling as biomass); multiple regression analysis (to establish the following connections: firstly, between the growth of gross output and the recycling of agricultural waste; secondly, between the volumes of waste generation, its inflow in the environment and the costs of its management; methods of systematization, grouping and logical synthesis (for information processing, drawing conclusions and academic offerings).

3 Literature Review

The originality of the research lies in establishing the following correlative-regression connections: firstly, between the growth of gross output and recycling of agricultural waste; secondly, between the volumes of waste inflow into the environment and its utilization and combustion to produce energy; thirdly, between the volumes of waste generation and the costs of waste management; fourthly, between the capital costs of waste management, its utilization and combustion to produce energy.

The scientific novelty lies in the development of an algorithm for the recycling of waste as biomass, which includes practical tools for the application of waste-free and biological technologies in traditional agricultural production systems, taking into account inhibitors (legal nihilism on the part of state authorities; excessive sanctions for non-fulfillment of the EU Directive in the field of energy efficiency and renewable energy) and catalysts (encouragement of the use of energy produced from renewable energy sources; regulation and support of bioenergy technology development; encouragement of bioenergy development, etc.)

The practical value of the research lies in the justification of the main reference point of solving the problem of agricultural waste generation and accumulation, which has to become «Zero waste» concept. The proposed algorithm for the use of waste in the production processes of enterprises as a secondary raw material will enable its transformation into a valuable product with its subsequent extensive use in various fields of agriculture and processing industry.

Since the second half of the twentieth century, the problem of accumulation and recycling of waste as additional material resources has gradually taken the main position in the world science and has become practically significant.

As Karl Marx pointed out that as a result of social labor, «waste is generated in such significant volumes that it becomes the subject of trade again, and therefore new elements of production» [7, p. 320]. The author drew attention to the feasibility of waste recycling as the main reserve of production resources as well as to the growth of economic efficiency of their use. At the same time, the problem of waste accumulation and its storage in the environment remained undisclosed in that research.

The founder of the environmental movement was Rachel Carson, who emphasized in her book «Silent Spring» published in 1962 that «those who value profits and technological progress above all else unconsciously believe that man's appearance on the stage of history abolishes the problem of balance in nature» [18]. In other words, the balance in nature is based on the internal connections of the living world and its connections with the environment. This does not mean that human beings should not try to tip the balance in their favour, but under any circumstances they should remember what to do and foresee the consequences of their steps». The author started the ecological movement in the society and raised the issue of connections between economic growth of production of social products and environmental degradation.

In the words of Kenneth Boulding, it was Rachel Carson's publication «Silent Spring» that inspired him to write an influential essay «The Economics of the Coming Spaceship Earth» (1966). In this essay, the author outlined the need for an economic approach to the ecological system with its limited resources [23]. He admitted the existence of material and energy limitations in the economy and suggested a transition from the «cowboy economy» to the «spaceman economy». According to the «cowboy economy», success is determined by the number and speed of production and consumption. In the «spaceman economy», on the contrary, «we are primarily concerned with maintaining life support, so that any technological change that leads to the ability to maintain a given general level of life support with a decreased flow of resources (i.e. with less production and consumption) is obviously a benefit». Therefore, the necessity of developing a model of economy that would take into account the interests of society and the consequences of economic activity for the environment was proclaimed.

Another challenge to economic science is posed by Herman Daly, who proves that economics is represented mainly by neoclassical postulates and does not take into account the capacity of economic processes in the biosphere, which develop on the verge of the natural limit of resources. Therefore, it is impossible to adequately assess their efficiency, both at the local and planetary levels. The scientist draws a disappointing conclusion that the desire to constantly expand the scale of the economy pushes humanity to the resource crisis and ecological collapse in general. Thus, in order to change this threatening trajectory, it is necessary to move to a new paradigm of management - sustainable development, which scientists define as a harmonious, balanced, conflict-free progress of the entire civilization, which simultaneously solves a set of issues related to environmental protection, poverty eradication and discrimination of each individual and entire peoples or groups of people [3, p. 6]. In his research, H. Daly identified the vectors of further recycling of waste as an additional resource. However, there was a need to develop practical tools that could be applied at the level of individual economic structures for waste-free production and to achieve results from such activities.

The response to this challenge, in 1997, was provided by scientists E. Weizsäcker, E. Lovins, L. Lovins in the work «Factor Four: Doubling Wealth, Halving Resource Use». Their work is based on the «resource productivity» concept, which the authors define as the ability to obtain twice as much efficiency and at the same time spend twice as few resources, which gives grounds to conclude that it is possible to quadruple productivity (hence the title of the book) [29]. «Factor Four» suggests a new approach to management, setting the main task - to increase the productivity of resources (electricity, water, fuel, materials, soils, etc.) used in economic processes and in everyday life, without additional costs and even with greater benefits.

In further studies, scientists claim that most technical solutions for the waste recycling already exist, but they remain unrealized opportunities for economic and environmental development. However, they should and can be used [28]. The authors present dozens of examples for optimizing the use of resources - the construction of hypercars (which allows to reduce the cost of car production and save a significant amount of resources during their use), the development of electronic communication (in particular, e-mail, which allows to save huge forest areas that will not be used for paper production), video conferencing (an idea that once seemed to be naive to critics of the concept, but today has become an integral part of our lives), innovative bio and agricultural technologies (which allow to increase the productivity of agricultural production, reducing the use of limited natural resources and saving financial resources), eco-friendly and ergonomic models of household appliances (which reduce costs and environmental impact).

Walter Stahel's research «The Performance Economy» (2010) presents convincing evidence of increased material production in industrial waste-free systems by recycling waste as biomass [30]. The living world operates in a circular cycle, where the by-product of one biological species easily provides the raw material for another. However, there arises a need for producers to be responsible for the results of their activities, and first of all - at the legislative level.

The problems of waste accumulation and its use as biomass in the national economy are regulated by the execution of laws and implementation of scientific developments into practice.

Nowadays, waste management in Ukraine is regulated by the Law of Ukraine «On Waste», the State Waste Classifier, State Standards, etc. Particularly, the Law of Ukraine «On Waste» focuses on the legal aspect of waste management that is the intentions or obligations of the waste producer to perform certain actions with waste. However, the environmental and economic aspect of the waste use as biomass is not reflected in these legal documents [9].

DSTU 2195-99 (National Standard of Ukraine) states that waste in the environment is considered firstly as a pollution that occupies a certain space in it and causes negative impact on other living and non-living objects and substances, and secondly, as secondary material and energy resources for possible utilization immediately after generation or after appropriate processing [4]. At the same time, attention is drawn to the use of waste as an alternative source of material goods and energy production.

The research of H. Heletukha, T. Zhelyezna, M. Zhovmir, Yu. Matvyeyev, O. Drozdova [2] resulted in the assessment of the potential of waste as biomass available for energy production in Ukraine, in particular its components such as primary and secondary agricultural waste, wood biomass. It has been analyzed the changes in the potential in previous years and made a forecast for the next period. The research considers approaches and methods for calculating the energy potential of biomass.

H. Kaletnyk in the paper «The use of biomass as an energy source in the process of selecting the location of the enterprise» points out the necessity of considering the possibilities of using biomass as an energy source when selecting the location of the enterprise. The author describes the benefits of using and long-term prospects of biomass in the Ukrainian market. Taking biomass supply logistics as an example, he analyzes and recommends three ecological models for practical application: a model of maximum distances of energy-efficient transportation for different types of biomass, an economic model, and a model of heuristic selection of the enterprise location by decreasing the radius of biomass collection [6].

Ukrainian scientists G. Golub, S. Kukharets, V. Kukharets, Y. Yarosh in the study «Integrated use of bioenergy conversion technologies in agroecosystems» indicate that the issue of providing humanity with energy resources is one of the priorities, considering the growth of population on the globe. Moreover, the significant cost of purchased energy resources increases the cost of food products obtained in agricultural production. These problems can be largely solved by using biomass of agricultural origin as biofuels. In this case, agricultural production will solve a triple task, namely food production, energy production and environmental protection by ensuring a positive balance of humus in the crop rotation soils. In turn, a comprehensive solution of this problem is impossible without increasing the energy and environmental efficiency of technical and technological processes of energy production from biomass [21].

However, the use of biomass as a biofuel raises reasonable concerns about the reduction of food production and environmental damage, expressed through soil dehumidification, which causes the scientific problem of reconciling food and energy production and the preservation or restoration of the ecological state of agroecosystems.

The results of scientific research of I. Grabchuk, I. Petrunenko, T. Vlasenko, E. Petrova, L. Strikha «Ensuring food security of EU countries in the context of sustainable development» (2021), indicate that the policy of the European Union is characterized by inconsistency of sustainable development goals in the context of food security, in particular in such areas as: migration policy, which increases inequality and problems of access to food; limited internal resources of EU countries, in particular, land areas for the development of organic production to ensure food quality and safety (Ireland, Denmark, Germany and Finland); lack of effect of public funding for research and development of the agricultural sector in case of excessive allocation of funds to the sector; excessive regulation of technology development to ensure production growth through public bias against potential risks. This means that the production approach to food security cannot be effective due to numerous contradictions between environmental and food safety issues. Therefore, it is reasonable to transform the EU food security policy in the context of sustainable development by defining goals for technological sustainability of the agricultural sector [27].

In the research conducted by O. Ivanenko, V. Bugaychuk, S. Belei, N. Grynchuk, T. Kulinich «Financial equalization of territorial development of eastern European countries and its impact on quality of life» (2021) identified that one of the main aspects of the financial policy of EU Member States is budget allocations for economic, environmental, technical-technological and social purposes. The costs of environmental protection and environmental protection are significant compared to those in Ukraine. This contributes to the development of innovative technologies for the reuse of waste and, in particular, the development of biotechnologies in agricultural production, the implementation of which ensures the efficiency of their usage, reflected in the increase in agricultural production per unit of resource consumption and profitability [26].

Thus, the research «Efficiency of the Innovative Use of Waste from Fruit-and-Vegetable Production» proved the effectiveness of innovations in the use of waste from fruit and vegetable production. In particular, experiments on the preservation of apple pomace with a natural sorbent as bentonite and the introduction of canned waste into the daily diets of pigs grown in the farm «Agroinnovation» of Zastavna district, Chernivtsi region, convincingly prove that feed consumption is reduced, self-cost is reduced, and profit is increased with a simultaneous increase in pork production [22].

As a result, modern civilization has reached a level of development at which production growth in virtually all sectors of the economy can occur without the involvement of additional resources and energy, creating the necessary conditions and opportunities for the recycling of accumulated waste.

4 Results

Intensification of agricultural production and expansion of agricultural land use have led to an increase in the amount of waste and its impact on the environment. The percentage of agricultural waste generation is a relatively small share of the total amount of generated waste, that is, no more than 4% of all waste from I to IV classes generated during the period of statistical observations.

There are several potential factors of negative impact on the environment: organic waste from crop production; organic waste from livestock and poultry farming; bio-waste; residual amounts of fertilizers, chemical and biological plant protection products, veterinary drugs. Traditionally, agricultural waste in Ukraine is divided into livestock waste, crop waste and agrochemical waste.

Over the 2005–2020 period and according to the logarithmic trend forecast until 2023, waste generation in agricultural production tends to increase (Fig. 1).

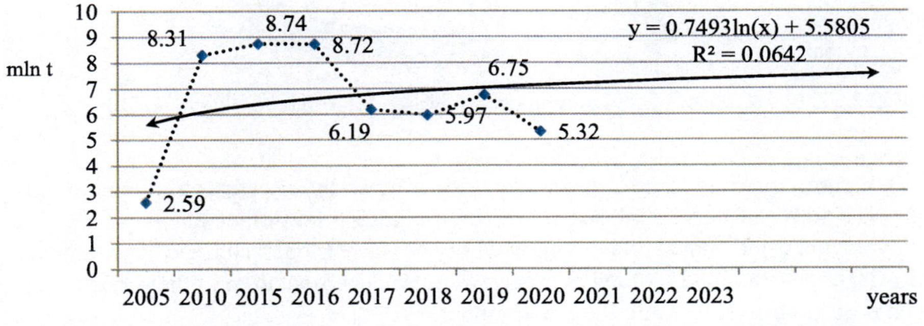

Fig. 1. Forecast of waste generation in the agricultural sector of Ukraine until 2023. Reference: based on [15, 25]

According to the Fig. 1, the amount of waste generated in 2020 increased by 2.73 million tons compared to 2005, and decreased by 3.42 million tons compared to 2015. At the same time, the analysis of production volumes shows that in 2005–2020 there was an increase in almost all types of products, except for sugar beet and milk. The decline in sugar beet production by 6317.6 thousand tons is explained by the reduction of sown areas, and the decline in milk production by 4450.8 thousand tons - by the decline in the number of dairy herds. During the research period, the decline in the number of animals is observed in all categories. However, the largest reduction was in the number of cattle - 3640.1 thousand heads and pigs - 1176.2 thousand heads. It can therefore be concluded that the reduction of agricultural waste in Ukraine in 2020 is caused by a decline in livestock production resulting from a reduction in the number of animals.

There is a big gap in waste generation, in particular, in 2005 it was 2.59 million tons, in 2015 - 8.74 million tons, and in 2020 - more than 5.32 million tons. It is worth noting that this situation is explained not only by the growth or decline in agricultural production, but also by the accounting of waste generation in agricultural enterprises and its display in statistical reporting. The reliability of waste generation in agricultural production in Ukraine is one of the issues connected with its formation as biomass, which should be involved in recycling in order to increase production. At the same time, according to the forecast until 2023, the growth of waste is inevitable.

According to the statistical observations, on average for 2005–2020, out of more than 6.7 million tons of agricultural waste generated, the most massive flows are: plant origin waste - 57%; animal excrement, urea and manure - 37%; animal waste and mixed food waste - 6% (Fig. 2).

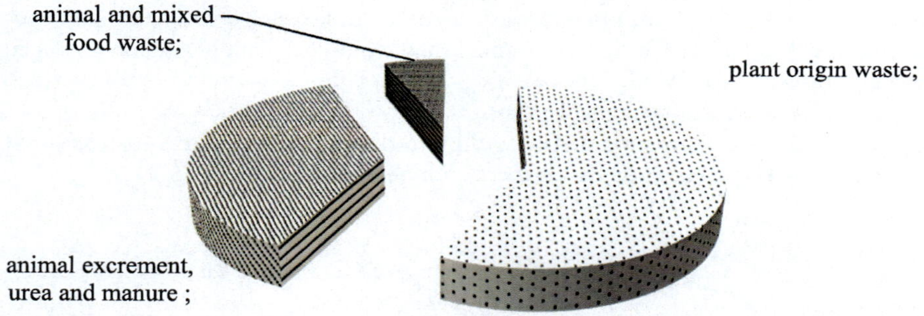

Fig. 2. The structure of agricultural production waste, %. Reference: based on [15, 24].

The largest share is plant origin waste, including straw of cereals and leguminous crops, dry stalks of corn, sunflower, soybeans, rapeseed and other plants, as well as forage perennial and annual crops. This category of waste is characterized by significant territorial congestion and biological processes of its decomposition in these territories, which poses a serious threat to the environment.

In addition to the negative impact caused by improper waste management, the agricultural sector in Ukraine itself contributes to environmental degradation: agricultural waste is often burned in fields, causing general inconvenience, air pollution and danger to the rural population. Inappropriate use of humus causes water pollution, and inappropriate storage and unreasonably high rates of mineral fertilizers and pesticides per 1 ha of agricultural arable land are still one of the main environmental threats.

At the level of 50 million tons of gross grain harvest, there may be produced up to 25 million tons of straw. This number may increase to 40 million tons if the gross harvest increases to 80 million tons. According to the National Waste Management

Strategy of Ukraine until 2030, livestock and poultry farming constitute another source of agricultural waste generation, where manure is the main type of agricultural waste. There are about 13 million heads of livestock in Ukraine. On average, one animal produces 20 kg of manure per day, which means that up to 100 million tons of manure can be produced per year. The number of one and a half million chickens produces more than 100 tons per day. In general, about 500 thousand tons of poultry manure is produced annually in Ukraine. Additionally, there is waste from slaughter and mortality of poultry. According to the standards, mortality can be 3.5% of the total amount per year, which is at an average bird weight of five hundred grams equal to 3,570 tons of bio-waste, excluding bio-waste generated during slaughter [15, 16].

However, waste recycling as biomass is expected to contribute not only to production growth, but also to providing the basis for an efficient, cost-effective and safe agricultural waste management system in Ukraine aimed at stabilizing and ultimately eliminating the negative impact of agricultural waste production on the environment and human health. The Law of Ukraine «On Waste» defines that one of the main principles of waste management in Ukraine as well as the future task in the field of waste management is to promote the maximum possible waste utilization in the form of recycling or alternative use of valuable waste resources as biomass through biotechnology.

Biological mass (biomass) belongs to renewable energy sources. Potential opportunities for obtaining biological raw materials are significant. In agricultural production, particularly in crop production, the main source of biogas is manure and by-products of crop production such as straw of cereals, bread, corn stalks, sugar beet tops, potato tops. More than 90% of this organic matter is straw. Biomass utilization produces biogas, which is an unconventional energy source. In addition, there remains undecomposed organic mass (sludge) and liquid, which is called supernatant. The sludge can be used as fertilizer.

The design or selection of a biogas plant project requires data on the chemical composition of biomass, its volume and dynamics of supply. Thus, manure is supplied from livestock complexes evenly, straw - seasonally, although its reserves can be significant and it can be used throughout the year. The basis of biogas production is methanogenesis which is a process of fermentation of biomass, including manure, with the help of natural methanogenic microflora. The process lasts 26–30 days (sometimes longer). Depending on the quality, about 0.2–0.7 m^3 of biogas is produced from 1 kg of dry manure. The amount of biogas obtained from cattle manure is 0.2–0.4, from pig manure - 0.3–0.7, from chicken manure - 0.8–1.2 m^3 [16].

Animal waste recycling in Ukraine is carried out by specialized enterprises (units) for the recycling of animal waste and cannot be carried out by enterprises that produce animal products intended for human consumption, except for cases of carrying out such activities by specialized units for the utilization of animal waste. All animal waste produced on the territory of Ukraine is transferred by its producers to the enterprises for its recycling [9].

Waste recycling, in our opinion, includes a set of technological operations and methods of biotechnology to neutralize the impact of waste on the environment and its use as a raw material in the form of biomass for the production of alternative energy sources of biogas, dry organic fertilizers and other products.

Table 1 illustrates the structure of agricultural waste recycling in 2015–2020.

Table 1. Structure of agricultural waste recycling

Type of waste	Specific weight of recycled waste, %				Specific weight of incinerated waste for energy production, %				2020 to 2015 by specific weight of recycled waste (+, –)
	2015	2018	2019	2020	2015	2018	2019	2020	
Animal and mixed food waste	24,53	48,61	52,15	52,66	0,22	1,55	2,65	1,17	+28,13
Plant origin waste	34,54	33,70	29,26	27,28	5,97	5,35	5,72	8,93	–7,26
Animal excrement, urea and manure	65,44	71,14	66,62	70,79	0,00	0,00	0,00	0,00	+5,35
Total	45,12	44,85	41,23	44,09	3,42	3,67	3,90	5,37	–1,03
Total of all types of waste	29,61	29,42	24,47	22,65	0,36	0,29	0,24	0,23	–6,96

Reference: based on [15].

It should be mentioned that in compliance with the explanations of the State Statistics Committee of Ukraine, the accounting of agricultural waste recycling by its types, specifically animal and plant origin and animal excrement, urine and manure, started in 2010. The actual data of waste recycling for 2010–2014 are insignificant. The largest amount of waste was recycled in 2015 - 29.61%, and in 2018 - 29.42%. Among all types of waste, animal excrement, urine and manure are most often subject to disposal - in 2020 - 70.79%, which is 5.35% more than in 2015.

The share of recycled waste of animal and mixed food origin increased significantly in 2020 compared to 2015 - by 28.13%. It should be mentioned that this trend is positive. At the same time, there is a slowdown in the recycling of waste of plant origin by 7.26%. This is explained by a significant decrease in the number of animals, changes, violations and non-compliance with the structure of sown areas. Moreover, the export-oriented market of cereals and legumes also contributes to the reduction of waste. The grain received from the fields undergoes only the primary cleaning at elevators and then goes for sale, further cleaning is carried out at the place of its use in production processes. The practice of Ukrainian farmers demonstrates that straw is used for feeding animals and bedding, as well as for soil fertilization, and its possible residue is 20–40%. Straw is successfully used for the production of energy or building materials, for mushroom growing, and it is traditionally used for making hats such as bryls, women's hats, etc. The share of incinerated crop waste for energy production increased by 2.96% within the analyzed period (in 2020 it was 8.93%).

Long-term research and economic analysis have shown that despite the use of straw in Ukraine for fertilizers and fuel, the potential of this natural resource in agriculture is still underutilized. After all, the presence of the straw residues, which are often burned in the fields (the volumes of carbon dioxide emissions increase, which leads to an increase in the «greenhouse effect», and in this case the normal biological functioning of the soil is restored only after 2–3 months), indicates an imbalance between livestock and agriculture. According to the average data of analyses, 5.2 kg of nitrogen, 2.3 kg of phosphorus, 10.3 kg of potassium, 3.2 kg of calcium, 1 kg of magnesium and other trace elements are transferred to the soil with 1 ton of straw, depending on the conditions of crop cultivation [17].

Overall, the specific weight of incinerated waste used for energy production is insignificant and in 2020 is 0.23% of the total amount of waste generated in agricultural production. The problem lies in the lack of financial resources of agricultural enterprises for the purchase of biological plants and technologies.

In this regard, the EU experience is valuable, where the costs of environmental protection and development are significant. Budget allocations contribute to the development of innovative biotechnologies that intensify the processes of waste recycling [26].

Table 2 illustrates the waste management costs for 2015–2020.

Table 2. Waste management costs in Ukraine for 2015–2020, mln UAH

Type of costs	2015	2018	2019	2020	2020 to 2015, (+, –)
Total	7,5	10,0	16,0	14,1	+6,6
Capital investments, total	0,7	1,2	5,7	2,9	+2,2
Of which for capital repairs	0,04	0,06	0,07	0,2	+0,16
Current costs	6,8	8,8	10,2	11,2	+4,4

Reference: based on [15].

According to Table 2, in 2015–2020 waste management costs tend to increase. However, even this increase in 2020 by UAH 6.6 million is insignificant. According to the Law On 2022 State Budget of Ukraine, it is planned to spend approximately 0.2% of Ukraine's GDP on nature protection, while EU member states spent almost 2% of GDP on environmental protection in 2021.

The main reasons for the low level of waste recycling in agriculture are not only insignificant budget expenditures, but also the fact that their structure does not change from year to year. Expenditures are used for the following items: protection and use of natural resources, elimination of pollution, preservation of the reserve fund, waste recycling, scientific research. Another factor of low efficiency of using waste management expenditures is inflation.

World trends currently demonstrate the production of various products, material goods and energy through the use of renewable organic raw materials based on the extensive use of biotechnology. In the context of an enterprise, bioeconomy is an integral part of the economy, which provides an opportunity for increasing energy efficiency and level of waste recycling, developing renewable energy based on biomass, greening production, increasing its sustainability, producing new products, etc. In other words, bioeconomy creates bioeconomic advantages in the interaction between the economy, environment, technology and society through the use of scientific and technological innovations that serve as drivers [1].

As a result of research, it has been determined that agricultural waste is biomass, which in the process of recycling in production processes can be converted into products. Thus, biomass has its contribution to the final outcomes of agricultural enterprise activity, in particular to the volume of gross output.

Based on the content of the resource approach, which gives the opportunity to determine the state and value of the totality of production resources in the process of production using a linear multivariate regression model, it has been determined the impact of recycling waste as biomass on the volume of gross agricultural production.

The linear regression equation takes the following form

$$y = -612121{,}5 + 385{,}5x_1 + 5376{,}4x_2 + 3261{,}9x_3 \quad (1)$$

where: y - gross agricultural output (at constant prices of 2016), mln. UAH; x_1 – received animal and mixed food waste, mln. tons; x_2 – received plant origin waste, mln. tons; x_3 – received animal excrement, urine and manure, mln. tons.

The obtained results indicate the existence of a close correlation between the effective feature and the selected research factors. It means that the recycling of a larger amount of animal and mixed food waste, plant origin waste, animal excrement, urine and manure contributes to the growth of gross agricultural production. The correlation coefficient value of 0.84 indicates that the obtained econometric model adequately describes the researched economic dependence which means that 99% of the variation in gross output depends on the selected factors and only 1% depends on other random factors.

Thus, from the perspective of the economy, the recycling of waste indicates an increase in gross output and there arises the possibility of establishing environmental priorities, namely, to establish a correlation-regressive dependence between the volume of waste flow into the environment and its recycling and combustion for energy.

The equation of the correlation-regression model takes the following form

$$y = -5{,}32 + 203x_1 + 1466{,}8x_2 + 2309{,}1x_3 + 4{,}5x_4 + 480{,}3x_5 + 0x_6 \quad (2)$$

where: y – the amount of waste flow into the environment, mln. tons; x_1 – the amount of recycled animal and mixed food waste, mln. tons; x_2 – the amount of recycled plant origin waste, mln. tons; x_3 – the amount of recycled animal excrement, urine and manure, mln. tons x_4 – animal and mixed food waste incinerated for energy, mln. tons; x_5 – plant origin waste incinerated for energy, mln. tons; x_6 – animal excrement, urine and manure incinerated for energy, mln. tons.

As a result of the factors formation, it was found that according to statistical data, animal excrement, urine and manure wastes were not incinerated for energy (x_6).

The calculated correlation coefficients indicate that the flow of agricultural waste is closely related to all factors. The overall correlation coefficient is $R^2 = 0.8373$. The coefficients of determination indicate that the distribution of waste on the adjacent territories of agricultural enterprises is more dependent on their recycling.

Due to the fact that the waste recycling as biomass requires the involvement of financial resources, the application of a multi-factor linear regression modelhas made it possible to establish the connection between the volumes of waste generation and the costs of waste management.

Consequently, the multiple regression equation that characterizes the dependence on the factors takes the following form:

$$y = -5{,}32 + 2{,}9x_1 + 0{,}2x_2 + 11{,}2x_3 \quad (3)$$

where: y – volume of waste generated, mln UAH; x_1 – capital investment costs, mln UAH; x_2 – capital repair costs, mln UAH; x_3 – current costs, mln UAH.

The economic interpretation of the model demonstrates the possibility of increasing the parameters, specifically, an increment of x_1 by 1 unit leads to an increase in y by 2.9 units and respectively x_2 - by 0.2, x_3 - by 11.2. On the basis of the maximum coefficient $\beta 4 = 14.843$, we conclude that the factor x_1 - capital investment costs has the greatest impact on the outcome y.

At the same time, financial capital costs for waste management in agricultural production are insignificant. Therefore, there is a need to justify the effectiveness of their use for the purchase of innovative technologies for the agricultural waste recycling. Rational use of its mass can produce biogas, bacterial protein and ecologically clean fertilizer for environmentally friendly technologies of growing field crops. Therefore, along with the traditional use of manure and other biomass residues, especially straw, it is important to recycle them to produce biogas which is a valuable fuel and bacterial protein. These methods of using agricultural by-products are not only economically viable, but also environmentally friendly.

Therefore, establishing connections between capital costs of waste management, its recycling and incineration for energy production will allow to ensure the feasibility of waste recycling and cost savings for its management.

The regression equation takes the following form:

$$y = -2{,}9 - 203x_1 - 1466{,}8x_2 - 2309{,}1x_3 - 4{,}5x_4 - 480{,}3x_5 + 0x_6 \quad (4)$$

where: y – capital costs of waste management; x_1 – recycled animal and mixed food waste, mln. tons; x_2 – recycled plant origin waste, mln. tons; x_3 – recycled animal excrement, urine and manure, mln. tons x_4 – animal and mixed food waste incinerated for energy, mln. tons; x_5 – plant waste incinerated for energy, mln. tons; x_6 – animal excrement, urine and manure incinerated for energy, mln tons.

As mentioned above, according to the statistical data, animal excrement, urine and manure were not incinerated for the purpose of producing energy mln (x_6).

The results of the correlation objective indicate an inverse connection of the effective feature with the selected factors. Thus, the more waste is recycled as biomass and incinerated for energy production, the less money has to be spent on waste management.

Analyzing the current situation in the field of agricultural waste management, a number of major problems can be defined. The main problem faced by the agricultural sector is the improper implementation and compliance with the legislative requirements. Moreover, there is a lack of awareness and knowledge of the agricultural sector about the opportunities and benefits of agricultural waste treatment or reuse, as well as a lack of awareness of potential investors about the potential for the new production of products out of agricultural waste.

5 Conclusions

The reference point for solving the problem of formation and accumulation of agricultural waste has to be the «Zero waste» concept, which is based on 2 cycles: biological – for organic materials and substances that after their useful life can be safely returned to the natural environment, where they can be involved in biological cycles as a result of the vital activity of individual microorganisms; technological – for materials that can and should be reused in the economic system. Thus, considering absolutely all waste as a source of resources, this concept serves to justify the mandatory waste distribution: first into organic and inorganic, and then by the type of waste and the degree of recirculation [21]. All this determines the development of an algorithm for ensuring the process of biomass recycling (Fig. 3).

The algorithm of waste recycling as biomass in agricultural production includes practical tools for the application of waste-free as well as biotechnologies in traditional

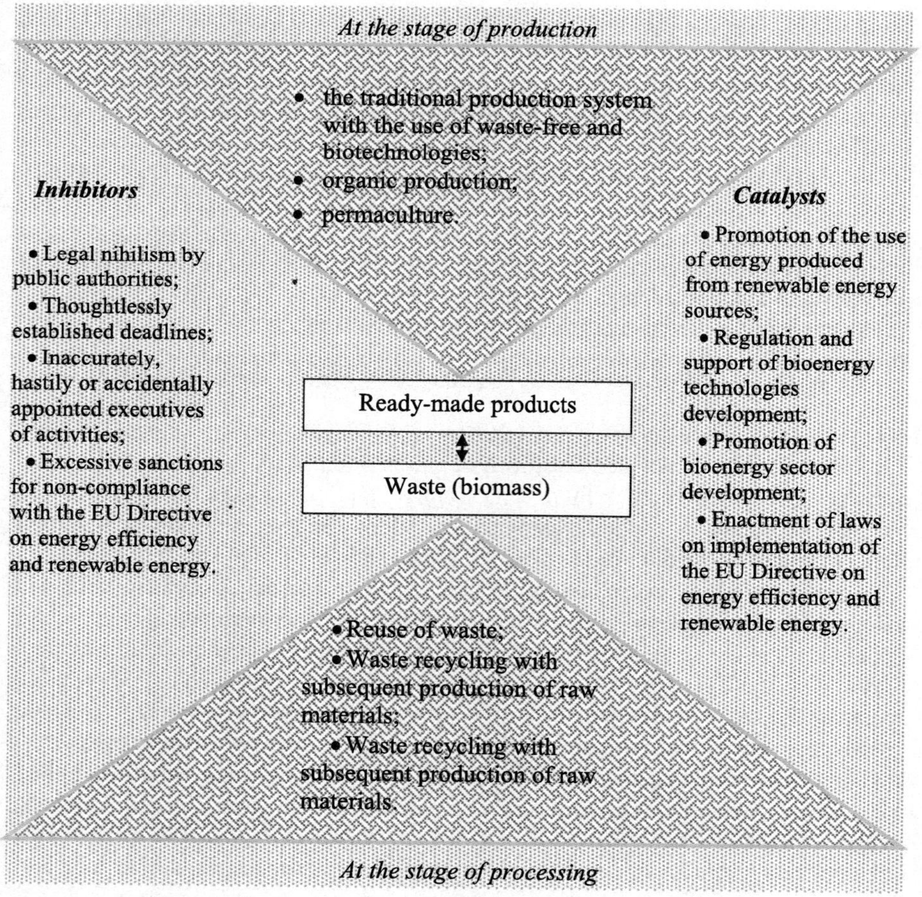

Fig. 3. Algorithm for providing the process of biomass recycling. Reference: own researches

agricultural production systems, development of organic production and permaculture for waste reuse, waste processing with further production of raw materials and energy.

The use of innovative production systems in agriculture and systems for processing raw materials and waste (biomass) requires the implementation of a recycling process algorithm, which is an economically, environmentally and socially efficient strategic way to generate and reuse waste in Ukraine. This algorithm aims to solve the urgent issues of energy and resource independence of the state, as well as saving natural, material and energy resources.

References

1. Bugaychuk, V., Opalov, O., Grabchuk, I., Razumna, K: Bioeconomic advantages of developing enterprises operating in dairy products industry. Sci. Horiz. **08**(93), 72–78 (2020). https://doi.org/10.33249/2663-2144-2020-93-8-72-78
2. Heletukha, H.H., Zhelyezna, T.A., Zhovmir, M.M., Matvyeyev, Yu.B., Drozdova, O.I.: Otsinka enerhetychnoho potentsialu biomasy v Ukrayini [Assessment of biomass energy potential in Ukraine]. Promyshlennaya Teplotekhnika – Ind. Heat Eng. **32**(6), 58–65 (2010)
3. Daley, G.: Poza zrostanniam. [Beyond rising]. Ekonomichna teoriia staloho rozvytku. Intelsfera, Kyiv, 298 s (2002)
4. DSTU 2195-99. Okhorona pryrody. [Conservation.]. Povodzhennia z vidkhodamy. Tekhnichnyi pasport vidkhodu. Sklad, vmist, vyklad i pravyla vnesennia zmin (62119). https://dnaop.com/html/62119/doc-%D0%94%D0%A1%D0%A2%D0%A3_2195-99
5. Zakharchenko, O.V.: Utylizatsiia vidkhodiv pryrodnoho pokhodzhennia ahrarnykh formuvan: problemnyi aspekt. [Utilization of waste of natural origin of agrarian formations: problematic aspect.]. Visnyk KhNAU im. V.V. Dokuchaieva. № 1. Seriia «Ekonomichni nauky» (2017)
6. Kaletnik, H.M., Honcharuk, T.V.: Innovatsiyne zabezpechennya rozvytku biopalyvnoyi haluzi: svitovyy ta vitchyznyanyy dosvid [Innovative support for the development of the biofuel industry: world and domestic experience]. Biznes Inform. **9**, 155–160 (2013). http://nbuv.gov.ua/UJRN/binf_2013_9_26
7. Marks, K.: Kapytal. Krytyka polytycheskoi эkonomyy. [Capital. Critique of political economy.]. Per. Y. Y. Stepanova-Skvortsova, proverennы y yspr. M.: Hospolytyzdat, T. 1, Kn. 1: Protsess proyzvodstva kapytala, 794 s (1950)
8. Miurrei, R.: Tsel – Zero Waste. [The goal – Zero Waste.]. M.: OMNNO «Sovet Hrynpys», 232 s (2004)
9. Pro vidkhody. [About waste]. Zakon Ukrainy № 187/98-VR vid 05.03.1998. Vidomosti Verkhovnoi Rady Ukrainy, № 36–37. St. 242 (1998). https://zakon.rada.gov.ua/laws/main/187/98-%D0%B2%D1%80
10. Pro skhvalennia Natsionalnoi stratehii upravlinnia vidkhodamy v Ukraini do 2030 r. [On approval of the National Strategy for Waste Management in Ukraine until 2030]. Rozporiadzhennia Kabinetu Ministriv Ukrainy vid 8 lystopada 2017 r. № 820-r. https://zakon.rada.gov.ua/laws/show/820-2017-%D1%80#Text
11. Pro Stratehiyu staloho rozvytku «Ukrayina – 2020». [On the Strategy for sustainable development «Ukraine – 2020»]. The Decree of the President of Ukraine on 12 January 2015. http://zakon5.rada.gov.ua/laws/show/5/2015
12. Stratehiia staloho rozvytku Ukrainy – 2030. [Strategy of sustainable development of Ukraine - 2030]. Vyp. 21. 89 s (2021). http://nbuviap.gov.ua/images/praktuka_susp_peretvoren/2021/21.pdf

13. Pro rishennya Rady natsional'noyi bezpeky i oborony Ukrayiny vid 6 travnya 2015 roku «Pro Stratehiyu natsional'noyi bezpeky Ukrayiny» [«The decision of the Council of national security and defense of Ukraine from may 6, 2015 «About Strategy of national security of Ukraine»]. The Decree of the President of Ukraine on 06 May 2015. http://zakon.rada.gov.ua/laws/show/287/2015
14. Uriadovyi portal. [Government portal.]. Yedynyi veb-portal orhaniv vykonavchoi vlady Ukrainy. Tsili staloho rozvytku ta Ukraina (2017). https://www.kmu.gov.ua/diyalnist/cili-stalogo-rozvitku-ta-ukrayina
15. Utvorennia ta povodzhennia z vidkhodamy. [Waste generation and management] (2022). http://ukrstat.gov.ua
16. Utylizatsiia vidkhodiv roslynnytstva za dopomohoiu metodiv biotekhnolohii [Utilization of crop waste using biotechnology methods] (2022). https://bio.ukr.bio/ua/articles/397/
17. Shuvar, I.A., Sendetskyi, V.M., Tymofiichuk, O.B.: Soloma dopomozhe rodiuchosti gruntu. [Straw will help soil fertility]. Ahrobiznes sohodni **17**(312), 40–43 (2015)
18. Rachel, C.: Silent spring, London: Houghton mufflin (1962). https://www.paypervids.com/silent-spring-rachel-carson/
19. Directive 2009/28/EC of the European Parliament and of the Council of 23 April 2009 on the promotion of the use of energy from renewable sources and amending and subsequently repealing Directives 2001/77/EC and 2003/30/EC. http://eurlex.europa.eu/legal-content/EN/TXT/?uri=CELEX:02009L0028-20151005
20. Environmental Services Association. Going for growth, a practical route to circular economy (2013). http://www.esauk.org/esa_reports/Circular_Economy_Report_FINAL_High_Res_For_Release.pdf
21. Golub, G.A., Kukharets, S.M., Yarosh, Y.D., Kukharets, V.V.: Integrated use of bioenergy conversion technologies in agroecosystems. INMATEH – Agric. Eng. **51**(1), 93–100 (2017)
22. Grabchuk, I., Bugaychuk, V., Tymchak, V., Orlykovskyi, M.: Efficiency of the innovative use of waste from fruit and vegetable production. Manag. Theory Stud. Rural Bus. Infrastruct. Dev. Res. Pap. **45**(2), 119–128 (2019). https://doi.org/10.15544/mts.2019.16
23. Sabine, H.: 'Carrying capacity'–the moral economy of the 'coming spaceship earth'. Atenea **26**(1), 59+ (2006). Gale Academic OneFile, https://link.gale.com/apps/doc/A170372836/AONE?u=anon~839c256d&sid=googleScholar&xid=0d658e8f
24. IEA. Statistics. Total. (2012). http://www.iea.org/topics/renewables
25. IEA. Statistics. Ukraine (2012). http://www.iea.org/statistics/statisticssearch/report/?country=UKRAINE&product=balances&year=2012
26. Ivanenko, O., Bugaychuk, V., Belei, S., Grynchuk, N., Kulinich, T.: Financial equalization of territorial development east european countries and its impact on quality of life. Int. J. Qual. Res. **15**(4), 1301–1316 (2021). https://doi.org/10.24874/IJQR15.04-18
27. Petrunenko, I., Grabchuk, I., Vlasenko, T., Petrova, E., Strikha, L.: Ensuring food security of eu countries in the context of sustainable development. J. Manag. Inf. Decis. Sci. **24**(3) (2021). https://www.abacademies.org/articles/ensuring-food-security-of-eu-countriesin-the-context-of-sustainable-development-10266.html
28. Von Weizsacker, E., Hargroves, K., Smith, M., Desha, C., Stasinopoulos, P.: Factor Five: Transforming the Global Economy through 80% Increase in Resource Productivity. Germany: Earthscan (2009). https://doi.org/10.4324/9781849774475
29. Von Weizsacker, E., Lovins, A.B., Lovins, L.H.: Factor Four: Doubling Wealth, Halving Resource Use, p. 320. Earthscan Publications Ltd., London (1997)
30. Stahel, W.R.: The Performance Economy (2010). https://www.globe-eu.org/wp-content/uploads/THE_PERFORMANCE_ECONOMY1.pdf
31. World Economic Forum Annual Meeting (2014). https://www.weforum.org/events/world-economic-forum-annual-meeting-2014/programme

32. Olena, A., Reznik, N., Dmytro, S.: Organizational and economic mechanism of the development of foreign economic activity of industrial production. In: Alareeni, B., Hamdan, A., Elgedawy, I. (eds.) ICBT 2020. LNNS, vol. 194, pp. 1011–1022. Springer, Cham (2021). https://doi.org/10.1007/978-3-030-69221-6_77
33. Alekseieva, K.A., Melnyk, L.V., Ostapchuk, A.D., Horská, E., Reznik, N.P.: Theoretical and methodological approaches to development of the concept of digital economy. Test Eng. Manag. (2020). http://www.scopus.com/inward/record.url?eid=2-s2.0-85083002183&partnerID=MN8TOARS
34. Podtserkovnyi, O., Reznik, N., Zghama, A., Tsurkanu, S., Kovalevych, I.: Problems of cryptocurrency transactions: foreign experience and development prospects in Ukraine. In: Alareeni, B., Hamdan, A., Elgedawy, I. (eds.) ICBT 2020. LNNS, vol. 194, pp. 847–857. Springer, Cham (2021). https://doi.org/10.1007/978-3-030-69221-6_65

Insurance Instruments in Covering Foreign Economic Risks of an Enterprise: Ukrainian Experience

P. Reznik Nadiia[1](✉), Slatvinskyi Maksym[2], Chvertko Liudmyla[2], Kyryliuk Iryna[2], Demchenko Tetyana[2], and Kosmidailo Inna[3]

[1] National University of Life and Environmental Sciences of Ukraine, Kyiv, Ukraine
nadya-reznik@ukr.net
[2] Pavlo Tychyna Uman State Pedagogical University, Uman, Ukraine
ms@udpu.edu.ua, {luda_um,irina_kurul}@ukr.net
[3] European University, Kyiv, Ukraine
kosm_inna@i.ua

Abstract. The article deals with implementing economic and mathematical analysis for the use of managing system foreign economic risks of an enterprise. It also denotes the possibility to form effective tendencies of their improvement while using insurance instruments. The realized analysis proves that covering negative foreign economic risks influences on economic activity realized by a business entity needs heavy expenses. It also reasons the search for stimuli to realize the risks including the instruments for foreign economic operating. Here we imply the use of insurance instruments as one of effective mechanisms for positive impact of foreign market trade, formation of sources to cover foreign economic risks and minimization of their influence on commercial activity. The process is caused by the fact that foreign economic contracts ignoring or underestimating may bring weakening of the positions in the corresponding branch of business relations. Defiance may also cause less expected profits. The used economic and mathematical modelling helped reveal the dependence between the amount of foreign trade operations and the insurance amount of these operations. The authors recommend a number of managing decisions that may rise the effective use of insurance instruments to cover foreign economic risks of an enterprise.

Keywords: Foreign trade operations · Foreign economic risks · Insurance · Stimulation · Risk management

1 Introduction

The desire to develop new foreign markets and spread the presence on the current markets of other countries is accompanied by the influence of immense risks, the possibility of their realization usually exceed the risks that are typical for inner economic activity in Ukraine. Determinants that cause the high level of risks for foreign economic operations include the factors of time, distance, determination in legal country foreign economic

B. Alareeni et al. (Eds.): ICBT 2022, LNNS 620, pp. 356–367, 2023.
https://doi.org/10.1007/978-3-031-26953-0_34

activity, as well as of intergovernmental unions and economic zones. The necessity to cross the border, cultural specific features, etc. are among the determinants.

Under current global conditions, when it is the time for strengthening protectionisms tension and active international trade wars for developing countries the support of foreign trade operations, including at the expense of insurance is of specific urgency. It determines the practical interest for revealing interdependent activity to conduct the operations and the number of insuring services that are consumed by exporters and importers. Here the factor of forces to influence the activity is of great importance.

Financial services, especially insuring ones, are significant and effective instruments of economic increase in economic policy of the countries with market economy. In accordance with scientific research data, separate channels of risky insuring influence on economic increase insert trade development, savings mobilization, risks minimization and avoiding, supply of effective capital distribution.

Insuring services have been closely connected with foreign trade operations for long time. They are an important reason for their conducting; they influence trade balance indirectly that cause contribute to economic growth of the country. Cargo and transport insurance, civil liability insurance, etc. are among the most spread insuring services today.

2 Literature Review

The scientific works of the researchers are devoted to the study of the use of insurance tools to cover foreign economic risks and minimize their impact on the company's activities: Brainard, L. [10], Prokopieva, Y. [5], David, L. [6], Zhukova, N. [3], N. V. Mayer [4], Outreville, J. F. [9], Slatvinskyi, M. [8], Sviderska, A. [1], Sysiuk, S. V. [2], Skipper, H. D. [14], Semchuk, J. V. [7] and others. The researchers focused on defining the content, principles, and functions of the use of insurance tools to cover foreign economic risks.

According to A. Sviderska, foreign economic risks are a specific risks category that are to be interpreted as possibilities for positive and negative deviations from predicted desired result of the decisions made that are connected with the integration of native business entities into the world economy, as well as realizing foreign trade agreements [1].

N. Maier examines risks managing that appear while organizing and full filling foreign trade contracts. He describes the following approaches to choose the protecting strategies from risks that are connected with the activity: decision making on the necessity of specific measures on risks insurance; distinguishing the part of a contract or a credit agreement, open currency transaction that is to be insured; choosing a definite way and method of risk insurance [4].

Insurance of foreign trade enterprises activity is international economic relationships with the sense of effective protection of property interests of business entities (enterprises) from different risks during the fixed (specific) period. There take place fixed events (insurance cases) at the expense of property funds that are being formed of insurance premiums paid by them (enterprises) [3].

Prokopieva, Y., Tankov, Y., Shibaeva, T. and Perekhozheva, E. presented their researches of risks systematization on different levels of economic system, generalization of motivation factors of insurance consumers, organization of descriptive, economic and mathematical behavioral models of business entities in risky situations. The received results include a behavioral model of the interaction between insurers of the market subjects. It determines the possibilities of effective and mutually beneficial coordination of their activity that draws considerable attention as there was used economic and mathematical modelling in order to define possible interaction ways between the given business entities [5].

Studies of David L. Eckles, Robert E. Hoyt, Steve M. Miller underline that an important aspect is the statement formation describing the fact that the use of an enterprise's risks managing reduces its expenses on risks diminishing, particularly when insurance events occur. The authors specify the system accepting the enterprise's risks management as a radical view change from the traditional method of individual risks managing to collective risks managing. It lets companies that have received the system of the enterprise's risks managing, show better identification of natural hedging, determine the hedging priorities on the risks that contribute to the general enterprise's risk for the best, and optimize the evaluation and choice of the accessible hedging instruments. The authors prove that these advantages give the opportunity to the companies that implement the system of the enterprise's risks managing, to provide more risk reduction on every spent dollar, as well as to feel the stimuli for the profit maximization to the risk diminishing [6].

In their work J. V. Semchuk, O. S. Protseviat, R. M. Skrynkovskyi state that under present economic conditions of Ukrainian economy, when an enterprise has often financial and economic crises, stagnation periods, it should be sure for his activity to have no crisis tomorrow, as well as it will not occur its liquidation as the worst scenario. In order to bypass the barriers, it is the insuring that is to help an enterprise. As a participant of the world business, an enterprise should insure directly its commercial activity from some unpredictable risks in the direction [7].

Insurance in foreign economic activity that is studied by the authors performs the functions: organization of specific insuring money funds, compensation for damage and personal money support of insurance participants; prevention of an insurance event and losses minimization in the situation when it has already occurred. Here it is obliged for the complete measures complex to be formed if the actions would not bring arising an insurance event or minimizing its negative results. Considering it, the authors emphasize that if to use the proper insuring system at the enterprise, it will fully protect its responsibility, leave the results of its work untouchable by risks, or it will need to reduce the negative results in the worst. [7].

The works of the above mentioned scientists are sapid and substantially discover the solution of scientific problems. Simultaneously we cannot find an exhaustive answer to the question of effectiveness and necessity to stimulate insuring foreign trade operations in the present global conditions.

3 Purpose of the Study

The purpose of the article include the study of specific features of using insurance instruments to cover foreign trade risks and determine the ways of activating foreign trade operations by means of stimulating the development of insurance market in the developing countries, organize practical recommendations.

4 Methodology

The theoretical and methodological basis of our study is the scientific provisions of the use of insurance tools to cover foreign economic risks and determination of the ways of activating foreign trade operations by means of stimulating the development of insurance market in the developing countries. Both general scientific and specific methods were used to solve the research tasks set in the work, in particular, statistical method, graphical method, methods of regression analysis, method of simultaneous equations solution.

5 Findings and Discussion

Overcoming negative influences of the risks of foreign economic activity on the results of economic activity of the enterprises of Ukraine requires immense expenses and causes the necessity to search the stimuli for their performance, as well as the instruments of risks management that accompany foreign trade activity. The use of insurance instruments is explained as one of the effective mechanisms to maximize the positive effect when appearing on foreign markets, form the sources to cover foreign economic risks and minimizing their influence on the enterprise's commercial activity.

Most scientists consider risks of foreign trade operations for business entities as a danger of the losses that may influence the activity results on foreign market and threaten the participants' interests due to the fault of the counterparty or of the change of political, economic or other situation in the partner's country, etc. (see Fig. 1).

Ignoring or underestimation of the risks of foreign trade contracts may bring weakening of the settled positions in the place of business relations and receiving less income [10].

In economic realities of Ukraine, the problem of involving insurance sector to the development of foreign trade is of great urgency. In accordance with present specific features of foreign trade, there recently have appeared international character of separate insurance services as transit and transport insurance, export credit insurance [8]. Simultaneously, specific studies prove the key role of insurance in the development of foreign trade and investments [11], as it is the risks insurance that is enough for covering foreign trade risks.

The study is based on the data analysis on 55 developing countries (according to World Bank classification), including Ukraine, for the period 1996–2020 – 32 countries with above average income (58,2%), 23 countries with above average income (41,8%). The period and the countries for the study have been determined of the necessary data availability. Database of the World Bank was their source [11, 12].

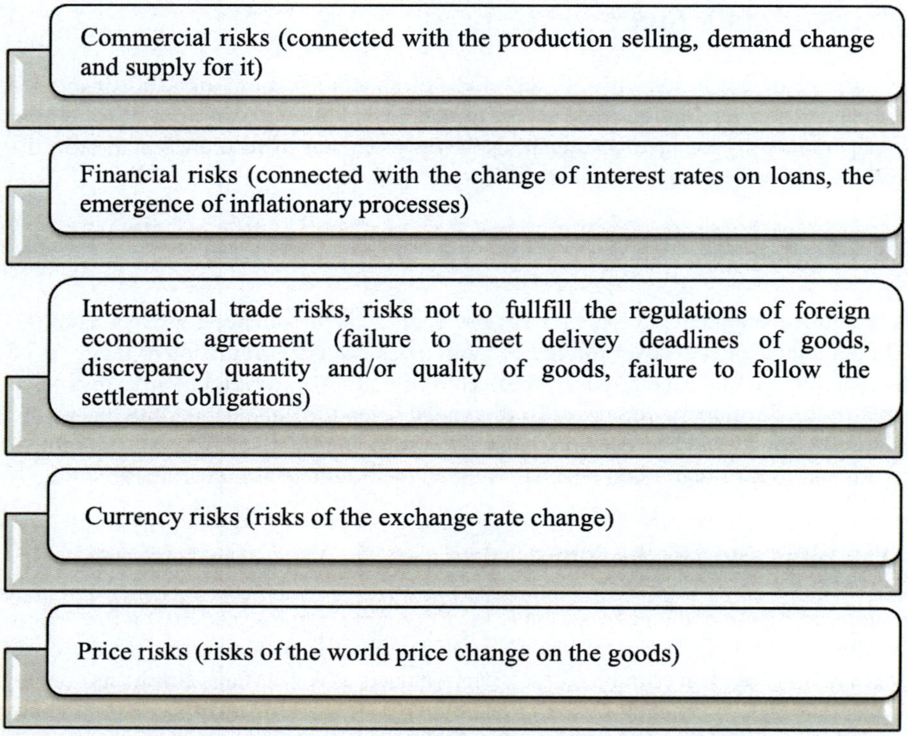

Fig. 1. Types of the risks of foreign economic activity. Source: Compiled by the authors on the basis of [9].

Volumes of foreign trade operations as a resulting indicator (a part of foreign trade operations in relation to GDP that is defined as the sum of exports parts of export and import to GDP (in %)) are of main interest for us. Influencing factors on the operations volumes are explained according to the study purpose, part of risks insurance premiums (in % to GDP) and the results of previously conducted studies [9, 10, 13, 14], parts of net incomes of direct foreign investments, inner crediting of the economy private sector and the usage charges of natural resources (in % to GDP). At the same time, the important index that influences the volumes of foreign trade operations is exchange rate fluctuations. However, in accordance to the present list of countries, data are not complete that prevented us from including it in calculations.

Descriptive variable data statistics that have been used to estimate the dependence between the volumes of foreign trade operations and influencing factors on them, including the volumes of risks insurance, are shown in Table 1.

Table 1. Descriptive variable model statistics

Index	Average value	Standard deviation	Minimal value	Maximum value
Export and import operations, % to GDP	76,808	33,48	15,636	220,407
Risks insurance premiums, % to GDP	0,982	0,84	0,00002	14,723
Direct foreign investments, net incomes, % to GDP	3,78	4,263	−6,37	55,07
Inner crediting of the economy private sector, % to GDP	44,452	31,881	3,476	165,39
Aggregate usage charges of natural resources, % to GDP	5,544	7,889	0,001	55,852

Source: Calculated by the authors

Average value of the volumes of foreign trade operations (in % to GDP) for the period of 1996–2020 is 76,808%, combining immense spread of values (from 15,636% in Brazil in 1996 to 220,407% in Malaysia in 2000). During the same period the volume of risks insurance premiums (in % to GDP) is 0,84%, (from 0,00002% in Peru in 2001 to 14,723% in Ukraine in 1996). The same immense spread values are observed of the other indexes.

Figure 2 and Fig. 3 graphically show the changes within average values of the above mentioned indexes. Both on the general graphic (Fig. 2) and on the graphic of the countries with below average income (Fig. 3a), we may see the dependence with time lag in four years between the indexes of volumes of foreign trade operations and risks insurance premiums. Here we may prove co-integrating relations between the indexes. Thus, on the graphic of the countries with above average income (Fig. 3b) we confirm no clear dependence.

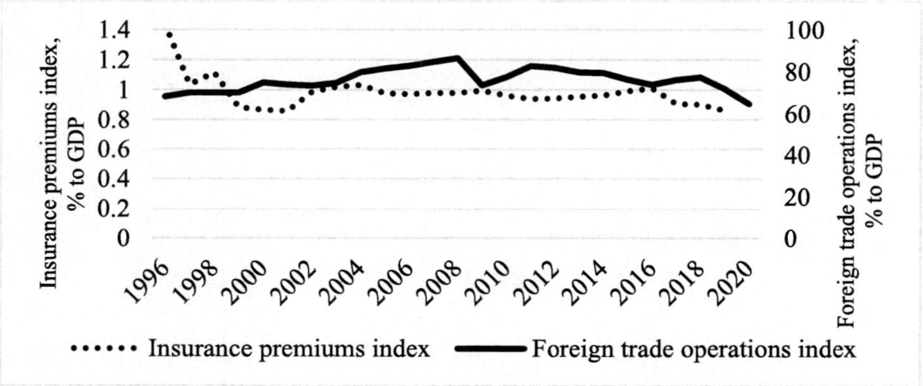

Fig. 2. Average values of indexes of foreign trade operations and risk insurance premiums for developing countries. Source: Calculated by the authors

a) Countries with incomes below-average

b) Countries with incomes above-average

Fig. 3. Average values of indexes of foreign trade operations and risk insurance premiums by groups of developing countries with incomes below- and above-average. Source: Calculated by the authors

The nonlinear relationship between the studied data series can be expressed in the form of a quasi-linear function. In this case, the linear regression equation takes the following form:

$$y = \beta_0 + \beta_1 \times x_1 + \beta_2 \times x_2 + \beta_3 \times x_3 + \beta_4 \times x_4 + \varepsilon \quad (1)$$

where y – resulting variable (share of foreign trade operations);

x_1, x_2, x_3, x_4 – explanatory variables (shares of risk insurance premiums, net inflows of direct foreign investments, internal crediting of the private sector of the economy, payments for the use of natural resources, respectively);

β_0 – free regression term;

$\beta_1, \beta_2, \beta_3, \beta_4$ – coefficients for explanatory variables;

ε is the random error of the regression.

The results of the regression analysis, the significance of which is separately confirmed by establishing a linear relationship between the resulting and explanatory variables, are shown in Table 2.

World practice evidences insurance to be widely used by institutions that implement the state policy promoting exports to minimize obstacles while entering foreign markets for goods and services. In particular, export credit agencies and export promotion organizations ensure the promotion of enterprise products to international markets and support of investments abroad through the export credit insurance mechanism.

Table 2. Results of regression analysis

Variable	Coefficient of determination	Regression significance (F-criterion)	Signicance of F-criterion (α)	t-criterion of the corresponding coefficient	Conclusions concerning the coefficients regression
Regression equation	$y = 45,13 + 4,29x_1 + 2,04x_2 + 0,39x_3 + 0,33x_4$				
x_1	0,24	102,6	$6,9 * 10^{-76}$	4,1	Significant
x_2				10,9	Significant
x_3				14,2	Significant
x_4				3,1	Significant

Source: Calculated by the authors

In addition, non-created products of export credit insurance by the portfolio method and factoring insurance have been operating for over 100 years and have proven to be an effective tool for stimulating exports, increasing the competitiveness of goods and services on the world market. In particular, the Export Credit Guarantee Department (ECGD) has been operating in Great Britain since 1915, the Euler Hermes agency since 1917, Atradius (Netherlands) since 1925, Eximbank (USA) since 1934. The countries of Eastern Europe, namely, Poland, Hungary, the Czech Republic, Slovakia and others created EKA back in the 1990s [15]. In Ukraine, in 2018, PJSC «Export Credit Agency» was established to provide credit and insurance support to exporters, but the activity of this institution cannot be called full-fledged due to existing legislative and institutional problems.

At the same time, it should be noted that in order to stimulate the large-scale expansion of Ukrainian export – works and services, goods by the processing industry, the recapitalization of the export credit agency to UAH 2 billion was completed in 2021 [16]. According to the experience of the agencies of neighboring states, in the first full year of operation, the export credit agency is able to cover 0.3–0.5% of the export of goods and services of Ukrainian origin and, according to the Government's estimates, contribute to an increase in the volume of total exports of Ukraine by 2–3% [17]. Currently, the agency is not active. According to official information [17], only 7 agreements were concluded (for a total amount in terms of currencies: 12.1 million hryvnas, 396.6

thousand US dollars and 6.2 thousand euros) (Table 3), in addition, there is no export credit insurance products using the portfolio method and factoring insurance.

According to the rules of activity of the private JSC «Export Credit Agency» [18], if the insured amount for a separate subject of the insurance contract exceeds 10% of the amount of the paid-up authorized capital and the formed free reserves and insurance reserves, the EKA is obliged to conclude a reinsurance contract to ensure the fulfillment of debt obligations under such an insurance contract. At the same time, the export credit agency must purchase reinsurance services from export credit agencies of other states or non-resident reinsurers according to the rules of the Law of Ukraine «On Public Procurement» [19].

Table 3. Agreements of Privat joint-stock company "Export Credit Agency" for the period of 2021–22.

№	Name of the agreement	Date of the insurance contract	Counterparty	Amount of the deal	Currency of the deal
1	Insurance contract for foreign economic agreements (contracts)	21.10.2020	Public joint-stock company «Kryukiv Carriage Building Plan»	6 210,06	euro
2	Export credit insurance contract	21.12.2020	Public joint-stock company «State Export-Import Bank of Ukraine»	2 213 437,50	hrn
3	Export credit insurance contract	10.02.2021	Public joint-stock company «State Export-Import Bank of Ukraine»	2 162 737,50	hrn
4	General Agreement on Insurance of Foreign Economic Agreements (Contracts)	05.04.2021	«PATON INTERNATIONAL» LIMITED LIABILITY COMPANY	219 300,00	US dollar
5	Export credit insurance contract	26.03.2021	Public joint-stock company «State Export-Import Bank of Ukraine»	3 360 562,00	hrn

(*continued*)

Table 3. (*continued*)

№	Name of the agreement	Date of the insurance contract	Counterparty	Amount of the deal	Currency of the deal
6	General Agreement on Insurance of Foreign Economic Agreements (Contracts) Insurance certificate №1	08.07.2021	Joint-stock Company «Zhytomyr Maslozavod»	177 277,78	US dollar
7	Export credit insurance contract	28.07.2021	Joint-stock Company 2State Export-Import Bank of Ukraine»	4 395 000,00	hrn
8	Export credit insurance contract	17.12.2021	PUBLIC JOINT-STOCK COMPANY JOINT-STOCK BANK «UKRGAZBANK»	1 105 000,00	hrn
9	Export credit insurance contract	25.01.2022	Joint-stock Company «State Export-Import Bank of Ukraine»	2 211 250,00	hrn

6 Findings and Discussion

Based on the economic and mathematical calculations, the following conclusions can be drawn:

- the hypothesis of the connection between the volume of foreign trade operations and the volume of insurance of these operations is substantiated, which can be confirmed with a degree of reliability of 95%;
- changes in the volume of risk insurance premiums, net receipts of foreign direct investments, internal crediting of the private sector of the economy, payments for the use of natural resources provide only 24% of changes in the volume of foreign trade operations;
- the variables listed above, only insurance has the greatest impact on the development of foreign trade operations (4.29%).

The performed calculations, reflecting the presence of interdependencies between the selected indicators, require further development to identify all the resulting factors of influence on the volume of foreign trade operations. At the same time, the obtained calculation results are sufficient to consider the insurance of foreign trade operations as a stimulating factor for the activation of these operations. In addition, it is necessary to take into account the importance of insurance to prevent the risks of losses during the international movement of goods and services.

So, as the results of the research showed, there is a positive relationship between insurance and the volume of foreign trade operations. The ways to find opportunities for the development of insurance presuppose the analysis of the level of risks in foreign trade and the means of accelerating the formation of the export insurance system, which will contribute to ensuring the efficiency of export activities of Ukrainian enterprises.

The basis of further research in this field is the development of further foreign trade relations and mechanisms to permit an intensified use of insurance opportunities for a growing number of export operations.

References

1. Sviderska, A.V.: Risk management of foreign economic activity of the enterprise: thesis Cand. of Econom. Sciences : 08.00.04/Khmelnytskyi national university. Khmelnytskyi. http://www.khnu.km.ua/root/res/2-21-65-10.pdf (2016). Last accessed 19 Aug 2022
2. Sysiuk, S.V.: Otsinka ta strakhuvannia ryzykiv u zovnishnyoekonomichnii diialnosti: oblikovyi aspect (Evaluation and insurance of risks in foreign trade activities: the accounting aspect). Visnyk 3. Zhytomyrskoho derzhavnoho tekhnolohichnoho universytetu. Seriia: Ekonomichni nauky **2**(72), 86–91 (2015)
3. Reznik, N.P.: Organizational and economic mechanism of involvement investments in the agro-industrial complex of Ukraine. Economy of agro-industrial complex (7), 70–77 (2013)
4. Majer, N.V.: Risk management in the currency transactions management system. Issues Econ. Law **3**, 267–270 (2011). https://law-journal.ru/files/pdf/201103/201103_267.pdf. Last accessed 9 Aug 2022
5. Zhukova, N.: «Teoretychni zasady strakhuvannia u sferi zovnishnyoekonomichnoi diialnosti» (The theoretical basis of insurance in the sphere of foreign economic activities). Visnyk Kyivskoho natsionalnoho universytetu imeni Tarasa Shevchenka **106**, 21–22 (2008)
6. Prokopjeva, E., Tankov, E., Shibaeva, T., Perekhozheva, E.: Behavioral models in insurance risk management. Inv. Manag. Financ. Innovations **18**(4), 80–94 (2021). https://doi.org/10.21511/imfi.18(4).2021.08
7. Eckles, D.L., Hoyt, R.E., Miller, S.M.: The impact of enterprise risk management on the marginal cost of reducing risk: evidence from the insurance industry. J. Bank. Finance **43**, 247–261 (2014). https://doi.org/10.1016/j.jbankfin.2014.02.007
8. Semchuk, J.V., Protseviat, O.S., Skrynkovskyy, R.M.: «Osoblyvosti strakhuvannya zovnishn'oekonomichnoyi diyal'nosti pidpryyemstv» (Features of insurance international business). Effective Econ. 10 (2013). http://www.economy.nayka.com.ua/?op=1&z=2382. Last accessed 27 Aug 2022
9. Slatvinskyi, M., Demchenko, T., Kosmidailo, I., Reznik, N.: Asset management performance: evidence from Ukraine. In: Alareeni, B., Hamdan, A. (eds.) Sustainable Finance, Digitalization and the Role of Technology: Proceedings of The International Conference on Business and Technology (ICBT 2021), pp. 51–62. Springer International Publishing, Cham (2023). https://doi.org/10.1007/978-3-031-08084-5_6
10. Reznik, N.P.: Foreign experience of investing in agro-industry complex. Culture of the peoples of the Black Sea region (147), 77–80 (2008)
11. Outreville, J.F.: The relationship between insurance and economic development: empirical papers for a review of the literature. Risk Manage. Insur. Rev. **16**(1), 71–122 (2013)

12. Brainard, L.: What Is the Role of an Entrepreneur in Economic Development? Zurich Government and Industry Affairs Thought Leadership Series, pp. 1–11. Zurich, Zurich Insurance (2008)
13. Global Financial Development Database: The World Bank, https://www.worldbank.org/en/publication/gfdr/data/global-financial-development-database. Last accessed 7 Dec 2022
14. World Bank's World Development Indicators database: The World Bank, https://databank.worldbank.org/source/world-development-indicators. Last accessed 1 Aug 2022
15. Reznik, N.P.: Peculiarities of leasing investment in agriculture. Herald Agrarian Sci. (11), 72–73 (2011)
16. Sawadogo, R.: Insurance development and international trade in developing countries. The Int. Trade J. **33**(3), 239–254 (2019)
17. Skipper, H.D.: Foreign insurers in emerging markets: issues and concerns. International Insurance Foundation, Washington, DC (1997)
18. What obstacles prevent the creation of an export and credit agency? Business. https://business.ua/finances/4582-yaki-pereshkody-ne-daiut-stvoryty-eksportnokredytne-ahentstvo. Last accessed 4 Aug 2022
19. Government portal: https://www.kmu.gov.ua/news/minekonomiki-zavershuye-dokapitalizaciyu-eksportno-kreditnogo-agentstva-na-sumu-2-mlrd-griven. Last accessed 10 Jul 2022
20. Government portal: The Ministry of Economic Development, Trade and Agriculture of Ukraine, the Government increased the authorized capital of «The Export and Credit Agency» to UAH 2 billion. https://www.kmu.gov.ua/news/uryad-zbilshiv-statutnij-kapital-eksportno-kreditnogo-agentstva-do-2-mlrd-grn-igor-petrashko. Last accessed 01 Jul 2022
21. ECA. Insurance, reinsurance, guarantee, interest rate compensation agreements for export credits have been concluded. https://www.eca.gov.ua/ukladeni-eka-ugody/. Last accessed 1 Aug 2022
22. CMU: Resolution dated August 21, 2019. N. 772, https://zakon.rada.gov.ua/laws/show/772-2019-%D0%BF#Text. Last accessed 7 Aug 2022

The Role of Strategic Leadership in Improving Business Performance (The Implementation Case of Metaverse Oriented Health Safety Environment/HSE)

Muafi Muafi[1] and Retno Purwani Setyaningrum[2](✉)

[1] Department of Management, Business and Economics Faculty, Universitas Islam Indonesia, Yogyakarta, Indonesia
[2] Universitas Pelita Bangsa, Cikarang, Indonesia
retno.purwani.setyaningrum@pelitabangsa.ac.id

Abstract. The purpose of this study is to analyze the role of strategic leadership in encouraging business performance improvements in oil companies that utilize the Metaverse implementation. Businesses with metaverse experiences are now here to create the ability to teleport from one business process experience to another. Companies must find ways to connect their online platforms to each other in order to operate more efficiently and effectively. This study is descriptive qualitative research whose data collection was carried out through interviews and independent interviews with participants. The companies studied are companies that had implemented metaverse, especially for the HSE sector in Jakarta. Researchers visit to the company's location and conducted open interviews with unstructured and non-standard questions. Interviews were conducted more than once with a certain schedule that was strictly followed. The results of the study explain that metaverse is needed in the HSE simulation in oil companies to reduce the accident rate so that it can have an impact on improving the company's business performance. All of them require a visionary strategic leadership role.

Keywords: Strategic leadership · Business performance · Metaverse-oriented health safety environment/HSE

1 Introduction

Leadership is a subject that has long attracted the attention of many people. This is a term that connotes the image of a strong and dynamic individual who has successfully led the military, led a company that was at the peak of its glory, or even led a country. This term is often used to describe courage and leadership in various legends and myths. For the most part, the historical picture is the story of military, political, religious and social leaders who are praised or blamed for an important historical event, although sometimes people don't really know how that event happened or how much influence the leadership had. Leadership has such appeal perhaps because the process is mysterious

B. Alareeni et al. (Eds.): ICBT 2022, LNNS 620, pp. 368–378, 2023.
https://doi.org/10.1007/978-3-031-26953-0_35

and it touches the lives of many. Various questions about leadership have long been the subject of speculation, but scientific research only began after the twentieth century. The focus of attention from researchers is more on issues of leadership effectiveness, inherent traits, behavior, ability to lead or other aspects that determine how leaders can influence their members. Some progress has been made in research on leadership. The leadership literature in the 1950–1980s period has a lot to do with supervisors and middle managers in organizations. In recent years the attention of many leadership theorists has shifted to executives and the top management team. To become effective strategic leaders in the twenty-first century, managers must play an active role in changing terms of reference to anticipate changes that occur so quickly. One of the strategic issues that is currently rife is the world of metaverse.

Snow Crash has covered the metaverse world which was published in 1992. Neal Stephenson envisions human beings meeting in realistic 3D buildings and other virtual reality environments. The term metaverse refers to a 3D virtual world inhabited by avatars of real people, where it imagines lifelike avatars meeting in realistic 3D buildings and other virtual reality environments. Zuckerberg as the founder of Facebook has described the metaverse as a virtual environment that can be entered by just looking at the screen and playing its role on the screen. This makes the metaverse an endless world of interconnected virtual communities [1, 2]. Metaverse players are actually not only those from Facebook but are also starting to engage other companies through Nvidia's Omniverse platform, the video game company behind the popular video game Fortnite, Epic Games, the gaming platform Roblox, as a place where people can come together to have 3D experience for work, play, socialize, learn and create. [2]. Brands such as Italian fashion house Gucci are also seen collaborating with Roblox to sell a collection of digital-only accessories. In addition, Coca-Cola and Clinique are also selling digital tokens as a stepping stone to the metaverse [2].

Metaverse development has also begun to be used by oil companies studied by researchers, namely companies in the HSE field. These companies have an innovative metaverse platform dedicated to training, animation and organization in the Health, Safety and Environment (HSE) field. The digitization and platform created in metaverse aims to demonstrate access to online training for all users of PC, mobile, virtual reality headsets, Windows tablets and Mac versions which will be available soon [3]. Training and practice in a safe yet challenging environment is necessary to properly prepare employees to perform well in life-threatening situations and emergencies [4]. With the HSE metaverse, the fatal accident rate has fallen by 83%, and overall there has been a significant reduction in injury rates across all major sectors [5].

This research aims to analyze the role of strategic leadership in the implementation of the metaverse, especially in the field of HSE in oil companies in Indonesia in improving business performance.

2 Literature Review

2.1 Strategic Leadership: Contribution to the Metaverse Business World

Strategic leadership is the ability to anticipate, have vision, maintain flexibility, and empower others to create the strategic changes needed. In similar vein, strategic leadership is seen as multifunctional, primarily involving managing through other people and helping organizations to cope with changes that appear to be growing exponentially in today's global environment. It demands the ability to accommodate and integrate external and internal conditions and the ability to manage ambiguity and engage in complex information processing. Through effective strategic leadership, organizations are able to utilize the strategic management process successfully [6]. In strategic leadership, top-level managers are important assets for the company so that the most important component for the company is to have a top management team that has better managerial abilities. [6] identified that there are several components that top management should have in relation to strategic leadership, including; vision, business knowledge, contacts, ability to integrate top management team, ability to make changes, special skills, leadership skills, relationships with key stakeholders and general managerial skills.

If it is associated with the decision-making hierarchy in a company, it usually consists of three levels. At the top of the corporate (company) hierarchy, it consists primarily of the board of directors, chief executive, and administrative officers. They are responsible for the company's financial performance and for achieving non-financial performance. In a multi-business company, corporate-level executives determine the areas of business that the company will serve. They also set goals and formulate strategies that cover the areas of activity and functional areas of the business. While corporate and business level managers focus their attention on 'doing the right things, functional level managers focus their attention on 'doing things right' [7–9]. It can be concluded that decisions at the corporate level tend to be more value oriented, more conceptual and less concrete than decisions at the business and functional levels. Decisions at the corporate level are often characterized by the potential for greater risk, cost and profit and longer time. Meanwhile, the functional level decisions implement the overall strategy formulated at the corporate and business levels. These decisions concern operational issues that are action-oriented, exist in a relatively short term, are low risk, and require moderate costs because they only depend on available resources. Functional level decisions are relatively more concrete and quantifiable. Meanwhile, business-level decisions help bridge decisions at the corporate and functional levels. Elaborated further by[6]. There are several important actions in strategic leadership including: (1) determining strategic direction, (2) utilizing and maintaining core competencies, (3) developing human capital, (4) maintaining an effective corporate culture, (5) emphasizing ethical practices and (5) develop strategic control.

In connection with the Metaverse, the term was first introduced by Neal Stephenson. His novel Snow Crash, which deals with the metaverse world, was published in 1992. Neal Stephenson envisions lifelike avatars meeting in realistic 3D buildings and other virtual reality environments. The term metaverse refers to a 3D virtual world populated by avatars of real people. Neal envisions lifelike avatars meeting in realistic 3D buildings and other virtual reality environments. Etymologically, metaverse itself comes from the word 'meta' which means 'beyond' and 'verse' which means 'universe'. So the metaverse can be interpreted as a space filled with matter that transcends all things seen in this world [10]. Metaverse is an online digital world, combining smart technologies such as video, audio and virtual reality, creating a place for humans to interact virtually with each other and with the digital environment of digital things from anywhere, anytime [11].

Persistent: Metaverse will continue to run without ever resetting, pausing, and never ending.

1. Real-time: Although the events in the metaverse space are pre-designed, users can experience each experience in real-time
2. Economic Functions: As in the real world, the economy in the metaverse space is fully functional. Various buying and selling activities, product and service transactions are carried out with blockchain-based native currency.

[12] conducted research by offering a novel concept in the service context. It gives virtual world circumstances by creating the connection of installed beacons which are installed in real spaces, namely exhibition halls, to head-mounted displays (HMDs). In addition, the service integrates narrative features for various user experience specifications through conducting presentation regarding the attribute and narrative about museum artifacts. The existence of service design enables both virtual and direct experience in museum become more significant. As for the content of the Metaverse exhibition, it is created from a successful collaboration of augmented reality and the virtual world. Furthermore, the results of the research recommend that the data utilized, or logs built within the metaverse can be used as updated content that can be made available via mobile phones or the web [12].

A research by [13] researched about the UXBDI of the Metaverse platform as distinguished among the channel types of the Metaverse platform. The Metaverse platform stand for a collective term of services which comprises of augmented reality, lifelogging, mirror worlds, and cyberspace. User Experience-Based Design Innovation (UXBDI) is distinguished by the unconventionality of its product and service design that satisfies the user experience. The author examines the influence of the Metaverse platform's UXBDI on the user-Metaverse platform. The results confirm that the UXBDI sub-dimension's attractiveness and interaction scale can improve the identification and commitment of the Metaverse-user platform. Likewise, the identity of Metaverse platform can improve user's identification of platform, whether it is Metaverse in the virtual world and mirrors world. The implication for marketing managers is a finding that suggests the urge to establish durable strategy using UXBDI management to increase the loyalty towards certain Metaverse platform. Various firms are now preferring to have a strategy of before, during, and after interaction for increasing the involvement with the Metaverse platform, which comprise of all kinds of emotional, cognitive, and physical reactions. This approach proposes that the Metaverse platform is distinguished by functional, symbolic, and aesthetic aspects which together shape user response. Marketers can increase the prominence of UXBDI's dynamic effects over time by arranging and executing motivating force with promotions in Metaverse platform users to increase their commitment.

[14, 15] examined innovation and the effect of imitation on the adoption of Metaverse services. "Metaverse Services" is a collective term for services such as Augmented

reality, Life logging, Mirror world, and Virtual world. Twitter, Google, iPhone, and Secondlife (T.G.I.S) are today's services/products. For measuring that this product or service is acquired, the most commonly used are IP traffic and iPhone sales. For this reason, we measure adoption by assessing changes in the volume of IP traffic of Twitter.com, Maps.Google.com, Secondlife.com, as well as the sales for iPhones over a 2-year period (from the first quarter of 2008 to the fourth quarter of 2009). In analyzing the data series to disclose the effects of innovation and imitation, the researchers used the Bass model. The results showed that each of these services resulted in different innovations and the value of the coefficient of imitation. The imitation effect for all Metaverse services is greater compared to the innovation effect, and Secondlife's innovation effect is greater than the others. Also, iPhone sales, as a measure of information and communication technology (ICT) products, indicated a greater innovation effect compared to other services. Suggestions are drawn to clarify this difference, such as, the Google-map imitation effect is based on network externalities, while the imitation effect is brought up by individual communication; iPhone sales innovation effect is elucidated by measurement time [16] discussed the potential of recent developments in cyberspace to offer new value propositions for the digital change of society and trade. Specifically, considering the function of Metaverse, it is conceived as a globally accessible 3D virtual space and computing infrastructure—and currently still a conceptual vision—as a mediator between trending innovation, technology, as well as social and commerce application. It is necessary to outline the role of current technology patterns that can be connected to high value added application scenarios through Metaverse as a space mediation design. There are two effects, namely the push effect, which is a new technology that drives radical changes in society and business, and the pull effect, namely radical ideas that encourage technological development. However, leveraging both impacts to form high value-added applications requires integration, mediating the design space, as it can possibly be gained through the advancement of cyberspace towards the Metaverse.

2.2 Metaverse the Real World of Digital Energy

Many companies are in a good position to participate in their own or others' metaverses and prepare for the impending land rush. With more than two years of a global pandemic, climate change, and pressure from the public, the oil and gas industry faces unparalleled challenges. Companies are constantly adapting to find creative solutions to new barriers, from collection regulations, lockdown restrictions and emission targets.

The oil and gas industry takes this opportunity to transform the industry for the better. The secure and sustainable operation of oil and gas facilities generally depends on manpower. This is when the industry embarks on a long-awaited recovery including the use of metaverse technology in the HSE field. There are two main challenges within the field of worker health and security, namely: [17]

1. Manage operations with a less experienced workforce. It has been shown that younger employees are twice as likely to be harmed on the job than experienced employees. The digital metaverse technology together with the extended reality technology can offer a virtual platform to train field experts or employees for unfamiliar tasks in

new equipment or hazardous environments. Such virtual training can help reduce the number of accidents that may occur when new workers operate physical assets.
2. The behavior or response of employees at oil and gas companies in learning how to move forward work safety by implementing the metaverse concept towards human capital.

3 Research Method

This study chose the type of qualitative descriptive research by preferring to use data collection methods through interviews and in-depth interviews with participants. The participants who were interviewed were three managers who had worked at least 7 years in the company. The researchers visited the company's location and conducted open interviews with unstructured and non-standard questions. The interview consisted of a keyword, agenda or list of topics, namely the role of strategic leadership in driving the metaverse business in the company. The researchers conducted interviews more than once and tried to avoid the dross rate, i.e. avoiding the amount of information and material that was not useful. The case study was chosen in this research because the researchers conducted research at an oil and gas company in Jakarta. The companies studied are companies that had used metaverse for their business processes, especially in the HSE field. The researchers seek to emphasize the reasons why they used metaverse and how the implementation of metaverse in the business world was. This was once suggested by [18, 19].

This oil & gas company was chosen because it used the metaverse facility. This study involved managers in companies to be the research participants because they implemented metaverse in their business operations. Validity and reliability tests are needed in order to achieve the research objectives as expected and carried out by triangulation [18, 20]. The researchers cross-checked the data in the field by conducting interviews with other managers, users and leaders in the company where they worked.

4 Research Result and Discussion

This study discusses the role of strategic leadership in driving business performance in the case of metaverse implementation in the HSE field. Three key participants concluded the need for strategic leadership within their companies, including in the metaverse-facilitated HSE management. Leadership and strategic thinking skills in the oil & gas industry are complex and rapidly evolving. The Oil & Gas industry is undergoing significant changes with new business models, especially in the current metaverse era. Therefore, it is imperative for organizational leaders to be able to identify opportunities and successfully navigate the complex and evolving risk landscape to become key players of the future. The current challenges faced by the oil & gas industry, however, require companies to integrate sustainability, innovation and technology strategy management, including within the metaverse. Key components of a Technology Management strategy include R&D strategy, innovation, organizational efficiency and operational performance [21].

It is no exaggeration to say that safety in the oil and gas sector is very important. Employees in this industry are exposed to a variety of potential hazards at various stages of production—whether they are on power lines, using heavy equipment, or working around flammable gases. For this reason, companies are more likely to experience a stagnant or gradual increase in safety. Organizations that approach safety as a cultural value and commitment to people and their work can drive meaningful change that can and should be a part of everyone throughout the organization [22]. Metaverse technology is considered a panacea for QHSE-related risks. This not only does the 'what-if' scenario but the company has to think about how to increase production, and also reduce downtime and HSE risk. This is because volatile crude oil prices can significantly affect the profitability of crude oil companies [23, 24]. Important methods of strategy development can be incorporated into the decision-making framework in three steps [25]:

1. Input stage, the basic information needed to formulate a strategy and identify it.
2. Comparison stage, taking into account possible strategies, a kind of balance is established between the main internal and external factors of the organization using SWOT analysis.
3. Decision stage, only one unique method is used, namely the quantitative strategic planning matrix

There are interviews with key participants regarding strategic leadership and QSHE as well as the use of metaverse in oil & gas companies. In more detail, the results of the interview can be written as follows:

Participant 1:

> "The company has been using a strategic leadership style in the management system. The company has taken advantage of strong management, and access to the right information in the field. Companies can make effective decisions at the right time in, so the company is able to move into the future. This is marked by digital oil fields that are presented by new technologies and new media. This is a visionary strategic direction that we have taken."

Participant 1:

> "The oil and gas industry is undergoing a period of significant change. The prospect of fluctuating oil and gas prices is difficult to predict, and this is under pressure from alternative energy sources. At the same time growth in global demand is coming from emerging industrial markets as well as recovery in more mature economic zones. This has major implications for the company's sustainability. Therefore, the oil and gas business will require a different strategic approach in the future. Management must maintain high performance in a rapidly changing market with a strategic leadership style that must be able to design and manage new approaches with new technologies. Therefore, there is a need for having strategic control in the company's business operations and managing the best human capital for our company"

Participant 2.

In high-risk industries, employee safety is paramount. Companies require thorough product testing before bringing any technology solution to site. This testing takes time and money and can sometimes be a drawn-out process but is necessary to ensure proper safety standards. HSE is the key to the company's sustainability. The best approach for HSE is "prevention" to help map the preparedness pathway with mitigation, evacuation, and containment plans. Metaverse is used as an important parameter in oil and gas operations, which can be useful in evaluating the performance of various previous processes. Through virtual reality it is possible for companies to create metaverses by replicating the performance of oil fields, refineries or other operational units on virtual platforms. We continue to adhere to ethical business practices and in accordance with work safety norms".

"HSE at our place is also implemented by holding virtual training in order to help reduce the number of accidents that may occur when new workers operate physical assets. This metaverse technology can also assist in conducting emergency evacuation training without increasing facility downtime or creating unnecessary risks for personnel on board. This case is for offshore/or oil fields in the sea. It should be noted that during the training, equipment may be required"

"Metaverse at our HSE oil company uses 3D for risk assessment by providing groundbreaking capabilities, including smoke dispersion prediction, incident investigation, passive fire specification and protection and placement of safety equipment".

Participant 3.

"Metaverse-oriented HSE in oil and gas companies encourages business sustainability. With this metaverse technology, a new dimension of sustainability opens up significant challenges to global climate goals and new opportunities to achieve them so that business performance can improve. As important as environmental sustainability, social sustainability is also very important. To build an accessible and inclusive metaverse, it's time to build sustainability into the front end of the virtual world rather than trying to fix it after exponential growth. The Metaverse concept that uses virtual reality technology is a highly developed technology. The application of the metaverse concept is used for safety campaigns for employees".

"HSE through metaverse can implement an integrated management system for all employees, customers and contractors that are guaranteed to remain safe regarding work systems, equipment and operations. With metaverse at HSE we are able to improve systems to ensure regular and effective hazard and risk assessments and implementation of effective control measures. Training and supervision on the basis of HSE and the regular flow of related information ensures that the organization is able to maintain the highest standards of safety for its employees, and this can have an impact on business sustainability".

From the results of interviews with managers, and literature reviews of previous studies, the researchers conclude several advantages:

1. Companies that use Metaverse as the next step can become more secure and productive organizations.
2. The foundation of metaverse applications lies in digital twins (DTs) technology, the power of tools such as digital twins, mixed reality, and autonomous systems.
3. Metaverse can be turned on and synchronized with the physical world via a two-way internet of things (IoT) connection. When these spaces come together, new worlds, opportunities and transformative solutions can be created. Modern software techniques such as analytics, simulation, autonomous control, and interaction is able to be implemented within digital models to explore their untapped potential.
4. The adoption of metaverse technology is increasing in several verticals, especially in the oil and gas (O&G) sector. This can happen because it is a highly regulated industry due to the inherent health, safety and environmental (HSE) risks related to exploration and production, particularly in drilling activities.
5. Metaverse can reduce HSE risks such as being able to control production safely so that employees can work safely and also can protect the environment. Metaverse can be used as a control parameter that can only be applied to physical assets.
6. With the metaverse there will be new breakthroughs, including smoke dispersion prediction, incident investigation, passive fire specifications and protection and placement of safety equipment.
7. Digital twin technology can put individuals at the center of catastrophe of risk management so core competencies are needed.

5 Conclusion and Implication

Oil and gas companies need to have strategic leadership that is able to maintain the company's revenue and profitability. Metaverse technology has been used in oil and gas companies, especially in improving the quality of HSE.

It is important for companies, both oil and gas and manufacturing companies, to have strategic leadership in seizing opportunities and avoiding threats due to the rapidly developing information technology. The implementation of metaverse technology in the field of HSE is needed to reduce workhazards and improve employee safety and health. Strategic leadership greatly influences employee engagement and motivation, especially in a challenging business environment. This study also reveals the importance of trust in the employee-leader relationship in implementing the strategic steps that have been written by [25]. Meanwhile, metaverse technology in the field of HSE is very important for oil companies and manufactures, because it has been proven to be able to maintain and improve the quality of HSE. Companies, therefore, will gain the trust of both their own employees and colleagues so that they can encourage business performance, environmental performance and social performance.

6 Limitation and Future Research Directions

This study is carried out using case study in certain companies, thus it is not yet able to generalize the case in oil companies in Indonesia. Furthermore, the focus of this study is on metaverse application on certain department of the firm, namely HSE, thus future studies is required to explore other departments. In addition, this study is highly dependent to the interpretation of the authors. Since the authors attempt to understand the meanings in the interview process with the existing question, it is feared that there would be a bias in this study. It is suggested for future studies to use triangulation method through utilizing research method and processes, thus the results can be more valid and reliable.

References

1. "https://inet.detik.com/cyberlife/d-5857985/apa-itu-metaverse-dan-koin-kripto-yang-masuk-di-dalamnya. Retrieved on 28 Aug 2022," *detik.com* (2022)
2. Dewi, I.R.: Mengenal Apa Itu Metaverse dan Bagaimana Cara Kerjanya, *Cnbcindonesia.Com*. https://www.cnbcindonesia.com/tech/20211216163806-37-299867/mengenal-apa-itu-metaverse-dan-bagaimana-cara-kerjanya (2021)
3. Fajar, M., et al.: The metaverse to enhance safety campaign toward shipping industry (case study: the development and implementation of metaverse in government-owned corporation). Int. J. Emerg. Technol. Adv. Eng. **12**(08), 1–11 (2022). https://doi.org/10.46338/ijetae0822
4. Cullum, A., Matland, T., Berge, T., Kvamme, T.B.: Scenarios of virtual world functionality. Soc. Pet. Eng. Intell. Energy Conf. Exhib. Intell. Energy **2**(2), 952–957 (2008). https://doi.org/10.2118/112230-ms.
5. Granger, S., Joyce, J.: Evolution of health and safety regulation, management, and a profession in the UK. In: ASSE Professional Development Conference and Exposition 2011 (2011)
6. Hitt, M.A., Ireland, D.R., Hoskisson, R.E.: Strategic management : competitiveness and globalization: Concepts and Case. In: Canada: Cengage Learning,. Cengage Learning (2017)
7. Robinson, R., Pearce, J.: Strategic Management. 13e. McGraw-Hill Education (1997)
8. Pearce, J.A., Robinson, R.B.: Strategic Management: Strategy Formulation and Implementation, p. 988. University of California, Irwin (1988). https://books.google.com/books?id=CFq1AAAAIAAJ&pgis=1
9. Robinson, R.B., Pearce, J.A.: Strategic Management Formulation, Implementation and Control -12/E, McGraw-Hill. McGraw-Hill (2011)
10. Detik.com, "'apa-itu-metaverse-dan-koin-kripto-yang-masuk-di-dalamnya," *Detik.com* (2022)
11. Cann, G.: The MetaVerse and Digital Energy - Geoffrey Cann - Energy News for the Canadian Oil & Gas Industry _ EnergyNow. EnergyNow Media (2021)
12. Choi, H.-S., Kim, S.-H.: A content service deployment plan for metaverse museum exhibitions—Centering on the combination of beacons and HMDs. Int. J. Inform. Manag. **37**(1), 1519–1527 (2017). https://doi.org/10.1016/j.ijinfomgt.2016.04.017
13. Jeon, J.E.: The effects of user experience-based design innovativeness on user– metaverse platform channel relationships in South Korea. J. Distrib. Sci. **19**(11), 81–90 (2021). https://doi.org/10.15722/jds.19.11.202111.81
14. Lee, S.-G., Trimi, S., Byun, W.K., Kang, M.: Innovation and imitation effects in Metaverse service adoption. Serv. Bus. **5**(2), 155–172 (2011). https://doi.org/10.1007/s11628-011-0108-8
15. Metaverse Roadmap_ Pathways to the 3D Web. https://www.metaverseroadmap.org/ (2010)
16. Rehm, S.-V., Goel, L., Crespi, M.: The metaverse as mediator between technology, trends, and the digital transformation of society and business. J. Virtual Worlds Res. **8**(2), 1–6 (2015). https://doi.org/10.4101/jvwr.v8i2.7149
17. Mohamed, Y.: Harnessing metaverse technology for QHSE Excellence _ Egypt Oil & Gas. Egypt Oil & Gas (2021)

18. Yin, R.K.: Case Study Research: Design and Methods, 4th edn. Sage Publications, Thousand Oaks, CA (2009)
19. Stake, R.E.: The Art of Case Study Research. Sage Publications, Thousand Oaks, California (1995)
20. Guba Egon, G., Lincoln Yvonna, S.: Competing Paradigms in Qualitative Research, pp. 105–117. Handbook of qualitative research (1994)
21. Singh, A.: Strategies for oil and gas companies to remain competitive in the coming decades of energy challenges. Chem. Eng. (2010)
22. Spinoy, B.: How oil and gas company Total transformed its safety culture _ McKinsey. Mc Kinsey & Company (2019)
23. Hasan, S., Mahbobi, M.: The increasing influence of oil prices 0on the canadian stock market. The Int. J. Bus. Financ. Res. **7**(3), 27–39 (2013). http://search.proquest.com/docview/1313208447?accountid=48385
24. Jiménez-Rodríguez, R.: Oil price shocks and stock markets: testing for non-linearity. Empirical Economics **48**(3), 1079–1102 (2015). https://doi.org/10.1007/s00181-014-0832-8
25. Eslami, D., Rezvani, S.: Strategic analysis of oil and gas industry to provide appropriate strategies. In: 5th Internataional Conference on Applied Researchs in Science and Engineering Institution of Engineering and Technology of London, November 2020 University of Amsterdam (2020) https://www.researchgate.net/publication/344888231

Methods of Calculating the Integrated Indicator for Assessing the Socio-Economic Development of the Territory: A Marketing Approach

Oklander Mykhailo[1], Valinkevych Nataliia[2], Oklander Tatyana[3], Pandas Anastasiia[3], Radkevych Larysa[1], and P. Nadiia Reznik[4(✉)]

[1] Department of Marketing, Odessa Polytechnic National University of Ukraine, Odesa, Ukraine
imt@te.net.ua

[2] Department of Economics, Entrepreneurship and Tourism, Polissia National University, Zhytomyr, Ukraine
natali1573@ukr.net

[3] Department of Economics and Entrepreneurship, Odessa State Academy of Civil Engineering and Architecture of Ukraine, Odesa, Ukraine
oklander@ogasa.org.ua

[4] Department of Management, National University of Life and Environmental Sciences of Ukraine, Kyiv, Ukraine
nadya-reznik@ukr.net

Abstract. The modern strategic development of any territory is being actively discussed today within a reasonable combination of market instruments to increase the attractiveness of the territory and improve the quality of life. To assess the effectiveness of territorial marketing activities, the integrated indicator or the system of the indicators is needed, which could be used to monitor how the needs for the formation of long-term competitive advantages of legal entities and improving the life quality are met. It should be noted that the development of approaches to the assessing the socio-economic development of the territory has become a necessary major trend in their attractiveness.

The aim of the work is to develop scientific and methodological tools for calculating the integrated indicator for assessing the socio-economic development of the territory on the basis of a marketing approach. In the article was formed the author's methodology of estimation of socio-economic policy of territories. A set of indicators characterising the quality of territorial population life has been formed; the assessment methodology of development priority of territorial clusters is offered; the integrated indicator of the efficiency of the territorial marketing system has been developed. Specified that the review of the territorial complex is based on the indicators of the quality of life of the population, the map of the quality of life is developed, which takes into account the key subjective and objective components. The direction of further research is based on development of territorial marketing activities tools, for solution of socio-economic imbalances of territories.

The practical value of the article lies in the possibility of using theoretical and methodological developments for assessing the socio-economic development of the territory based on the marketing approach of state authorities in developing policy documents in socio-economic development of regions and settlements.

B. Alareeni et al. (Eds.): ICBT 2022, LNNS 620, pp. 379–391, 2023.
https://doi.org/10.1007/978-3-031-26953-0_36

Keywords: Territorial marketing · Marketing activities · Quality of life · Competitiveness

1 Introduction

In modern conditions, the most important task of management remains stable, harmonious and systematic social and economic development, based on research of territorial processes, analysis of problem moments, revealing of disproportions for possibility of further stable development of business activity and improvement of population's quality of life.

Competitive economy should be considered the territorial economy, which provides: supporting for the development of living standards and maintaining such positions when the goods of the territorial enterprises are successfully sold in outer markets.

A territorial approach is an important component of the creation strategy economic system development. Development of territories is a push for production, business, migration processes at different territorial levels, to ensure achievement of strategic goals of socio-economic development of regions and the country in general. Territorial development component as economic systems is an effective marketing activity, which should be based on an integrated indicator or system of indicators, which allows to form long-term competitive advantages for business and to increase the indicator of population's quality of life. The essence of managing the process of territorial transformation is the purposeful development of a set of economic and marketing mechanisms, regulators, which allow to choose the directions of sustainable development for territories.

Territorial attractiveness includes such key elements as competitive advantages and capacity to attract people, investments, resources, etc. that makes it interesting for different target audiences [1].

The development of the market relations in Ukraine has led to the intensification of the activities of economic entities aimed at achieving the highest possible commercial results. As a result, there are "external externalities" (secondary effects), which often have a negative impact on humans. Accordingly, on the one hand, there is a necessity to ensure ecological, social and food safety [2]. And, on the other hand, the objective proposals were made to take into account and assess the impact of these factors on vital activity. A new approach has been developed that offers to research not only the level but also the quality of life. Quality of life is a complex concept that is interpreted and defined in a number of ways within and between various disciplines [3].

Until the end of the 1980s, the concept of living standards prevailed in the world, which was based on the cost quantitative indicators calculated as the value of gross domestic product (GDP) or national income (NI) per capita [4]. According to the UN definition (1961), the standard of living includes the following concepts and indicators: health, food consumption, lifestyle and activities, living conditions, social security, clothing, leisure, human rights.

Nowadays the concept of life quality dominates in the world, which has included the concept of living standards and significantly added it. In 1990, the UN introduced the integrated indicator – the Human Development Index (HDI), which consists of three components: life expectancy, education and living standards. The quality of life of the individual meets with the quality of life of the community or the societal quality within it [5]. The concept of "quality of life" reflects a set of conditions that determines physical, mental and social well-being. These are both external factors that ensure human well-being (food, housing, employment, services, education, etc.) and the subjective understanding of well-being (health, family relationship, work, financial situation, educational opportunities, self-esteem etc.), pleasure, happiness. According to this interpretation, the quality of life can be identified with the help of the system of objective and subjective assessments [2]. The first group is the set of certain objective indicators that compare the actual values with the normative ones. The second group contains subjective results of marketing research in the form of population surveys [6]. Of course, this technique can be used periodically and "point by point" for strategic assessment of life quality, but requires significant time, labor and financial costs [7].

The specific set of the indicators can be expanded or narrowed. But, in any case, such a map will give an idea of the real situation and allow you to model an adequate territorial development policy. The important indicator of competitiveness in the world market, which demonstrates the strength of economic positions, is leadership in a particular branch [8]. However, in general, the structure of the Ukrainian economy is extremely distorted both at the national level and at the level of individual regions. This problem is one of the key ones. Thus, in 2020, 42.8% of the products produced in Ukraine were in Kyiv and six of the 24 regions (Dnipropetrovsk, Donetsk, Zaporizhia, Kyiv, Odesa, Kharkiv) had 31.4%. In the other 18 regions only 25.8% of the total industrial output was produced [9]. Taking into account the dynamic understanding of the term "territory", territorial development can be described as oriented changes that by stimulating driving forces, such as: entrepreneurship, balancing endo-and exogenous factors as well as competitiveness and cooperation [10, 11].

The important component of territorial development is the implementation of balanced structural policy, as well as the formation of an effective management system that would be able to determine the priorities of economic development and thus contribute to the formation of the optimal economic structure [12]. This problem is multifaceted and requires a fundamentally new approach to their functioning and improvement. The management mechanism is needed, which would contain a set of economic methods, levers, regulators to choose the priority areas of territorial economic development for forming a rational structure of the territorial production complex. The cluster policy has a large potential in solving this problem at the regional level, the instruments of which are territorial clusters that improve the region's competitiveness, as well as integration and diversification of the regional economy [13].

The aim of the paper is to substantiate the theoretical and methodological principles and develop scientific and methodological tools for calculating the integrated indicator for assessing the socio-economic development of the territory on the basis of a marketing approach. It is planned to form a set of indicators that characterize the quality of life of the population of the territories, based on the quality of life map, to develop a methodology for assessing the priority of territorial clusters and to develop an integrated indicator of territorial marketing.

2 Literature Review

The economic aspect of sustainable development is connected with the transition from the current economy of resource use to the economy of their systemic reproduction, while subjects of sustainable development are called to become the territories of different spatial groupings – cities, agglomerations, regions. In modern conditions, considerable attention is paid to the development of territories on the basis of marketing of territories, which is considered as one of the areas of socio-economic policy. The essence of the marketing approach is that the territorial body must take into account the harmonization of its interests with the interests of community members and businesses. For example, among the presented studies it is necessary to note the results obtained by Kulikova E. [14], Rehailia H. and Saadi I. [15], Oklander M. Chukurna O., Oklander T. and Yashkina O. [16].

The necessity to achieve stable, dynamic and sustainable economic growth and on this basis to improve the life quality of the population, the maximum approximation of the welfare to the European standard necessitates the development and implementation of territorial economic policy. The aim of the regional public administration should be to create favourable conditions for the formation of long-term competitive advantages of economic entities and on this basis to improve the life quality of the local population. These statements can be traced in the papers of the following scientists (Haraldstad K., Wahl A. and Andenæs R. [3], Murgas F. and Klobučník M. [5], Pandas A. [2]). The goals of territorial economic policy and the goals of the concept of socio-ethical marketing are essentially the same, and, therefore, the application of the concept of socio-ethical marketing, methodology and technology of marketing at the territorial level will increase the economic and social level of the territory.

Working on the issue of determining the essence of territorial marketing, it is first necessary to determine what is meant by the territory to which marketing efforts are applied [14, 15]. Traditionally, it is customary to understand the territory of the economy as a historically established geographic place, a locus characterised by individual characteristics and certain resources (available and potential) to meet the needs and demands of internal and external consumers, viewed as partners, with which long-term mutually beneficial relations are built, according to an ideal scenario [17].

Socio-economic development of the territory largely depends on the size and efficiency of its potential. That is why, with all complexity and multiaspect of territorial problems, one of the most important modern directions of analysis is development of scientific and methodical approaches to assessing the effectiveness of its use. The dynamism of economic systems necessitates the formation of a mechanism for the effective use of territories and should combines the system of economic and marketing instruments that are able to ensure effective use of investment, innovation and labor potential [10, 13].

3 Materials and Methods

For the formation of author's tools for the development and evaluation of socio-economic policy of the territories. The paper proposes the following stages that characterize the course of the study:

- the method of constructing a graphic multi-vector model of the "competitiveness polygon" was adapted to the specifics of territorial economic systems based on the assessment of the priority of territorial clusters through the formation of a list of 12 indicators;
- the scientific and methodical approach to calculation of an integrated efficiency indicator of functioning of territorial social and economic system is offered. This approach consists of the following stages:

a) selection, ranking and determination of the level of local indicators to meet the interests of: consumers (increasing the competitiveness of economic entities and quality of life), the executive branch and society as a whole;
b) division into two groups of selected indicators depending on whether the growth rate is positive or negative;
c) determining indices for each local indicator separately for each group;
d) assignment to each local indicator of a weighting factor;
e) definition of the system of integrated local indicators: satisfaction of consumer interests. Namely, increasing the competitiveness of economic entities and the quality of the population life, meeting the interests of the executive branch, meeting the interests of society.
f) determination of the integrated indicator of the efficiency of the territory system. This indicator characterizes the formation of long-term competitive advantages of legal entities and improving the quality of life, increasing budget revenues, neutralizing the negative effects of economic activity.

For an effective evaluation of the effectiveness of the territorial complex, it is advisable to use only a set of objective indicators of population quality (Fig. 1).

The following list of the assessing indicators of the priority of territorial clusters that correspond to the radius vectors of the graphical multi-vector model can be proposed:

1. total production;
2. level of labor productivity;
3. availability of material and technical base;
4. profitability (loss) of products;
5. relative share of profitable enterprises;
6. cash flow;
7. number of employees;
8. contribution to GDP (GVA);
9. the share of exports in sales of the territory;
10. trained personnel;
11. implementation of fixed funds;
12. production index.

The assessment of the potential possibilities of each cluster is carried out depending on the corresponding area of the polygon. The larger this area, the higher the competitiveness rating of the cluster. Thus, the area of such a polygon becomes a kind of competitive

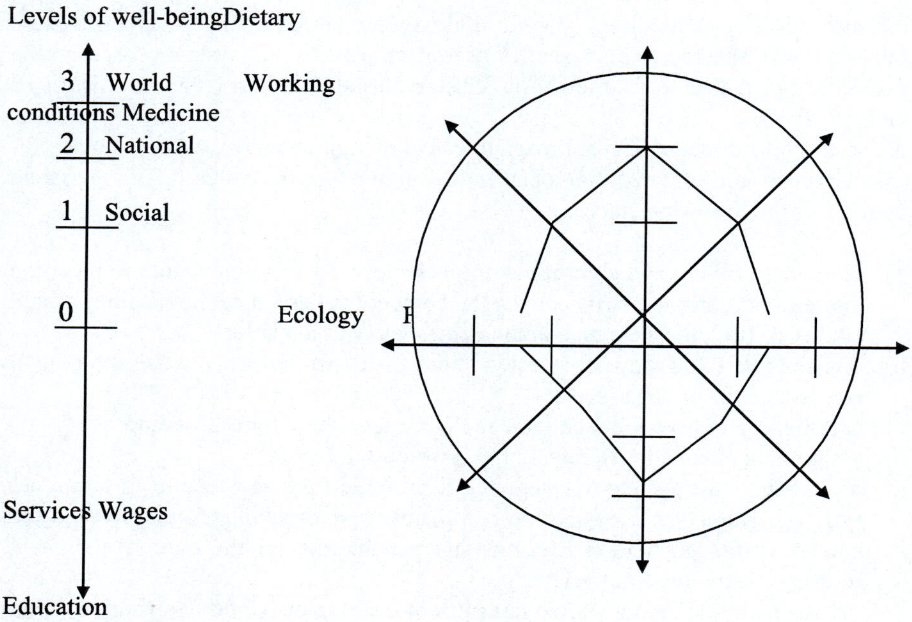

Fig. 1. Map of life quality (model of forming an integrated indicator of life quality–HDI) source: developed by the authors.

criterion of the cluster. The criterion is anestimation way of the choice validity of priority areas of the activity in the territorial economic system (country, region, city) [16].

The construction of a graphical multi-vector model is carried out according to the following algorithm:

1. The angles between the radius-vectors are the same and are determined by the formula:

$$\alpha = \frac{360}{m} \tag{1}$$

where α–the angles between the radius vectors;

m– the number of radius-vectors used in the model.

2. Length normalisation of i radius-vector for the indicators whose growth rate is desirable and reflects increasing of the competitive level of the cluster is carried out by the formulas:

$$R_i = \frac{X_{ACT_i}}{X_{MAX_i}} \tag{2}$$

where R_i–normalised length of i radius-vector, $i = \overline{1, m}$;

X_{ACT_i}–the actual value of the indicator of i radius-vector (the values of the selected indicators, which cannot be found due to the lack of the methods or relevant data, are defined by the expert assessment.);

X_{MAX_i}– the maximum value of i radius-vector.

$$R_{MAX_i} = 1 = X_{MAX_i}$$

where R_{MAX_i} –the maximum normalised length of i radius-vector is equal to one at the maximum value of the corresponding indicator.s

3. The length normalisation of i radius-vector for the indicators whose growth rate is not desirable and reflects decreasingof the competitive level of the cluster is carried out by the formulas:

$$R_i = \frac{X_{MIN_i}}{X_{ACT_i}} \tag{3}$$

where X_{MIN_i} –the minimum value of i radius-vector.

$$R_{MAX_i} = 1 = X_{MIN_i},$$

where R_{MAX_i} –the maximum normalised length of i radius-vector is equal to one at the minimum value of the corresponding indicator.

4. The area of "competitiveness polygon" can be determined by the formula:

$$S_{CL_j} = 0,5 \times \sin\alpha \times (X_1 \times X_2 + X_2 \times X_3 + X_3 \times X_4 + \ldots + X_{m-1} \times X_m) \tag{4}$$

where S_{CL_j}–the area of the competitiveness polygon of$j$ cluster, $j = \overline{1, n}$;
n –the number of the clusters in the structure of the territorial economic complex.

Thus, the area of the competitiveness polygon is a function of the length of the radius-vectors:s

$$S_{CL_j} = f(X_1, X_2, X_3, X_4, \ldots X_{m-1}, X_m).$$

The difference in the areas of the competitiveness polygons of the clusters will reflect their relative priority. Therefore, the condition of competitiveness, for example, cluster 1 over cluster 2 has the form:

$$S_{CL_1} - S_{CL_2} > 0.$$

Since the activities of territorial authorities are carried out in order to achieve indicators above the normative or average, the following criterion is proposed for the effectiveness of the territorial system: the system is effective if its indicators are not less than the normative or average in a given area.

4 Results and Discussion

Assessing and analysing the effectiveness of marketing activity within macro- and meso-economic systems, it is impossible to limit only the life quality indicator, because in this case the structure of the indicator is oversimplified. In addition, achieving a certain life quality is long. An integrated indicator of the effectiveness of the territorial marketing system is proposed, with the help of which it is possible to control how the needs for:

- formation of long-term competitive advantages of legal entities and improvement of life quality of the population;
- increasing budget revenues;
- neutralisation of negative consequences of economic activity.

These needs are different and therefore the relevant local indicators have different dimensions. Therefore, to determine the integrated index, it is necessary to use an index form, in which the influence of dimensionality is eliminated and there is an opportunity for mathematical calculations. The algorithm of the proposed calculating method of the effectiveness integrated indicator of the territorial marketing system consists of six stages.

The first stage is the selection, ranking and determination of the level of the local indicators to meet the consumers' interests (competitiveness increasing of businesses and life quality), the executive branch and society as a whole.

1. The indicators of customers' satisfaction.

   1.1 The indicators of competitiveness increasing of the economic entities:

   - GDP or GVA per capita;
   - value of investment per capita;
   - innovation spending per capita;
   - average educational level of employees;
   - profitability of economic activity;
   - net exports.

   1.2 The indicators of life quality of the population:

   - life expectancy;
   - education level of the population;
   - employment rate;
   - inflation rate;
   - level of average income per capita;
   - savings in banks per capita;
   - poverty level (minimum consumer budget);
   - housing.

2. The satisfaction indicators of executive bodies:

   - the capacity of tax revenues and fees;
   - degree of budget balance.

3. The indicators of public interest (in this group of the indicators, all the coefficients are calculated per 100,000 population):

   - natural increase (+), decrease (−) of the population;

- crime rate;
- crime rate for especially serious crimes;
- the incidence rate of particularly dangerous infectious diseases;
- AIDS incidence rate;
- drinking water quality of conformance to world standards;
- atmospheric air quality of conformance to world standards.

The second stage is the division of the selected indicators into two groups, the growth rates of which are:

1) positive value;
2) negative value.

The following Table 1 shows two groups of indicators for which growth is desirable, group 1 and undesirable, group 2. The indicators include the same key components, namely: consumer satisfaction indicators, which are divided into indicators of increasing the competitiveness of economic entities and quality of life, then indicators of satisfaction of the interests of executive bodies and indicators of public interest. Table data provide a detailed view of the components of key indicators for which growth is a positive and a negative dimension.

The third stage is to determine the indices for each local indicator separately for groups 1 and 2.

For group 1, the index is calculated by the formula:

$$I_i = \frac{IND_{ACT_i}}{IND_{NORM_i}} \tag{5}$$

where I_i–the index$i$of local indicator, $i = \overline{1, n}$;

n–the number of indicators;

IND_{ACT_i}– theactual level i of local indicator;

IND_{NORM_i}– the normative or average national level i of local indicator.

For group 2, the index is calculated by the formula:

$$I_i = \frac{IND_{NORM_i}}{IND_{ACT_i}} \tag{6}$$

The fourth stage is the assignment each local indicator of weighing factor (φ_i), which can take two forms:

a) to achieve the validity and reliability of the results, it is recommended to use the method of expert evaluation. At the same time, it is necessary to ensure the accuracy and understand ability of the questions, to involve a wide range of experts, to achieve full independence of inner-directedness;
b) taking into account the place of the indicator (m), which it occupies in each list.

The most significant indicators are in the first place, then in descending order of importance.

Table 1. Distribution of key local indicators of positive and negative response to growth

Indicators for which growth is desirable (group 1)	Indicators for which growth is undesirable (group 2)
1. Indicators of consumer satisfaction 1.1. Indicators of competitiveness increasing of economic entities: - GDP or GVA per capita; - the value of investment per capita; - innovation costs per capita; - median educational level of employees; - profitability of economic activity; - net exports 1.2. Indicators of life quality: - lifetime; - education level of the population; - employment rate; - income level per capita; - the amount of savings in banks per capita; - provision of housing	1. Indicators of consumers' satisfaction 1.1. Indicators of competitiveness increasing of economic entities: - no 1.2. Indicators of life quality: - inflation rate; - poverty rate (minimum consumer budget)
2. Indicators of satisfaction of executive bodies' interests: - the amount of tax revenues and fees	2. Indicators of satisfaction of executive bodies' interests: - the degree of budget balance
3. Satisfaction indicators of public interests: - natural population increase (+); - drinking water quality of conformance to world standards; - atmospheric air quality of conformance to world standards	3. Satisfaction indicators of public interests: - natural decrease (–) of the population; - crime rate; - crime rate for especially serious crimes; - the incidence rate of particularly dangerous infectious diseases; - AIDS incidence rate

Source: developed by the authors

Weighing coefficient:

$$\varphi_i = \frac{m}{2^m} \tag{7}$$

The fifth stage is the definition of the system of integrated local indicators: managing customers' needs (increasing the competitiveness of economic entities E_1 and quality of life E_2; satisfaction of executive interests E_3; satisfaction of public interests E_4. Each of the four integrated local indicators is determined by the formula:

$$E = \sum\nolimits_{i=1}^{n} \varphi_i I_i \tag{8}$$

The sixth stage is to determine the integrated indicator of the efficiency of the territory system.

$$E = \frac{E_1 + E_2 + E_3 + E_4}{4} \tag{9}$$

Considering the developed scientific and methodological tools for calculating an integral indicator for assessing the socio-economic development of a territory based on a marketing approach, we can conclude that the problematic point is the complexity of its measurement.

However, this technique is extremely effective, because through the quality of life of society as a whole and its various social groups, it is possible to carry out an integral assessment of the effectiveness of territory management.

A significant advantage of the proposed integral indicator is the ability to consider long-term competitive advantages for business and improve the quality of life of the population, as well as show the direction of development of the territory and evaluate its effectiveness. In addition, the methodology makes it possible to analyze individual components of partial indices, which makes it possible to solve problems in certain areas of the life of the population of a particular territory.

The integral indicator presented in the paper is of a complex nature, in turn, the existing studies on this topic have a narrower focus. For example, the methodology of Ayvazyan S. [18] is aimed only at assessing the quality of life of the population of the territory and includes several indicators within itself: 1. Quality of the population; 2. Welfare; 3. Social security; 4. Environmental quality; 5. Natural and climatic conditions.

Research by Tarasov P., Smirnova D. [19] considers the indicators adjusted taking into account the marketing strategy of the analyzed territory and the systems of indicators considered according to the method of Ayvazyan S. based on the multifactorial modified Fishbein formula [20].

The marketing approach regarding the socio-economic development of territories allows to take advantage of all the opportunities of a particular territory to improve the quality of life of the population and the efficiency of the activities of management and business entities.

5 Conclusions

The aim of the study is achieved. A method for calculating an integrated indicator for assessing the socio-economic development of the territory is developed. This approach allows harmonizing the interests of citizens, businesses and public and territorial authorities.The development of a scientific and methodological approach to the calculation of the integrated indicator of the effectiveness of the territorial system allows to developing strategies and programs for socio-economic development of territories. This process contributes to the formation of long-term competitive advantages of legal entities, improving the quality of life, increasing budget revenues, neutralizing the negative effects of economic activity.

The results of the study can be used in the work of regional authorities to create favorable conditions for the formation of long-term competitive advantages of economic entities and, on this basis, to improve the life quality of local people.

References

1. Bondarenko, A., Zakharkina, L., Syhyda, L., Saher, L.: The economic and marketing attractiveness of countries: measurement and positioning in terms of economic security. Int. J. Sustain. Dev. Plan. **15**, 439–449 (2020). https://doi.org/10.18280/ijsdp.150404
2. Pandas, A.: Suburbanisation as a factor in increasing life expectancy. Ekonomia–Wroclaw Econ. Rev. **26**(3), 65–74 (2020). https://doi.org/10.19195/2658-1310.26.3.5
3. Haraldstad, K., et al.: A systematic review of quality of life research in medicine and health sciences. Qual. Life Res. **28**(10), 2641–2650 (2019). https://doi.org/10.1007/s11136-019-02214-9
4. Zinchuk, T., Kutsmus, N., Prokopchuk, O., Kovalchuk, O., Usiuk, T.: Disproportionality of Ukraine's participation in international business: rethinking under conditions of new stage of global development. Stud. Appl. Econ. **39**(6) (2021). http://ojs.ual.es/ojs/index.php/eea/article/view/5312
5. Murgas, F., Klobučník, M.: Quality of life in the city, quality of urban life or well-being in the city: conceptualization and case study. Ekológia (Bratislava) **37**, 183–200 (2018). https://doi.org/10.2478/eko-2018-0016
6. Babenko, V., Baksalova, O., Prokhorova, V., Dykan, V., Ovchynnikova, V., Chobitok V.: Information and consulting service using in the organization of personnel management. Stud. Appl. Econ. **3**(1) (2021). https://doi.org/10.25115/eea.v38i4.3999
7. Oklander, M., Oklander, T., Pedko, I., Yashkina, O.: Development of the subsystem of forecasting for the system of marketing information management at an industrial enterprise. Eastern-Eur. J. Eenterp. Technol. **5/3**(89), 39–51 (2017). https://doi.org/10.15587/1729-4061.2017.111547
8. Zinchuk, T., Kovalchuk, O., Kutsmus, N., Charutska O.: Challenges of sustainable development of rural economy. Manag. Theory Stud. Rural Bus. Infrastruct. Dev. **40**(4), 609–619 (2018). https://ejournals.vdu.lt/index.php/mtsrbid/article/view/204
9. Official site of the State Statistics Service of Ukraine. (n.d.). https://ukrstat.org/uk/operativ/operativ2012/fin/roz_p.htm
10. Skydan, O., Sheludchenko, B., Kukharets, S., Medvedskyi, O., Yarosh, Y.: Analytical study of multifractal invariant attributes of traffic flows. Eastern-Eur. J. Enterp. Technol. **3**(3–99), 22–29 (2019)
11. Kozlovskyi, S., Khadzhynov, I., Lavrov, R., Skydan, O., Ivanyuta, N., Varshavska, N.: Economic-mathematical modelling and forecasting of competitiveness level of agricultural sector of Ukraine by means of theory of fuzzy sets under conditions of integration into European market. Int. J. Recent Technol. Eng. **8**(4), 5316–5323 (2019)
12. Tulchinskiy, R., Chobitok, V., Dergaliuk, M., Semenchuk, T., Tarnovska, I.: Strategic guidelines for the intensification of regional development under the impact of potential-forming determinants in the conditions of digitalization. Int. J. Comput. Sci. Netw. Secur. **21**(8), 97–104 (2021)
13. Aparina, N., Mediantseva, S., Kavkayeva, N., Surtseva, A.: The effects of territorial clusters formation. In: Advances in Economics, Business and Management Research, 2nd International Scientific Conference on «Competitive, Sustainable and Safe Development of the Regional Economy», pp. 258–263. Atlantis Press, Amsterdam (2019)
14. Kulikova, E.S.: Territory marketing strategy: processes digitization and investment development. In: Advances in economics, business and management research: 1st International Scientific Conference «Modern Management Trends and the Digital Economy: from Regional Development to Global Economic Growth», pp. 518–521. Atlantis Press, Amsterdam (2019)
15. Rehailia, H., Saadi, I.: Territorial marketing: a tool for developing the attractiveness of territories. A/Z: ITU J. Faculty Archit. **15**, 61–69 (2018). https://doi.org/10.5505/itujfa.2018.14622

16. Oklander, M., Chukurna, O., Oklander, T., Yashkina, O.: Methodical approach to calculating information value in pricing policy in supply chains. Estudios de Economía Aplicada. **38**(4) (2020).https://doi.org/10.25115/eea.v38i4.4009
17. Bondarenko, V.A., Kalieva, O.M., Pisareva, E.V.: Marketing management of the territory in the aspect of the regional brand formation. Eur. Res. Stud. J. **21**(2), 72–78 (2018)
18. Aivazian, S.A.: Methods of Econometrics. INFRA-M Publ, Moscow (2010)
19. Tarasov, P.S., Smirnova, D.S.: Integral indicators for evaluating the effectiveness of territorial marketing. Bulletin of the Astrakhan State Technical University. Series: Economics, no. 2, pp.152–155 (2012)
20. Naumov V.N.: Models of behavior of consumers in marketing systems. Izd-vo SPbGUEF, Saint Petersburg (2009)
21. Reznik N.P.: Osoblyvosti lizynhovoho investuvannia sil's'koho hospodarstva. [Features of leasing investment in agriculture]. Visnyk ahrarnoi nauky, no 11, pp. 72-73 (2011)
22. Reznik, N.P., Gupta, S.K., Sakovska, O.M., Ostapchuk, A.D., Demyan, Y.Y.: A research of state regulation of stock exchange in Ukraine: significance and growth for economic development, Int. J. Eng. Adv. Technol. **8**(6), 3851–3857 (2019). ISSN 2249-6958
23. Reznik, N.P.: Innovative activity as a factor increasing production efficiency. Economics: problems theories and practices: Collection of science works. Dnipropetrovsk, no.198, pp. 979–985 (2004)

Exchange Rate Volatility and Its Impact on FDI Inflows in India Using Maki Cointegration Approach

Erum Fatima[1], Mohammad Asif[1](✉), Raj Bahadur Sharma[2], and Anjali Chaudhary[3]

[1] Department of Economics, Aligarh Muslim University, Aligarh, India
asifalig71@gmail.com, m.asif.ec@amu.ac.in
[2] College of Business Administration, University of Bahrain, Zallaq, Kingdom of Bahrain
[3] College of Business and Administration, Princess Nourah Bint Abdulrahman University, Riyadh, Kingdom of Saudi Arabia
archaudhary@pnu.edu.sa

Abstract. This paper examines the impact of exchange rate volatility and FDI inflows in India for the period 2000 to 2019 on a monthly basis. The role of exchange rate and its fluctuations possess a vital role in influencing key macroeconomic variables in any economy. The present study has been significantly important to assess the exchange risk which arises when exchange rate of any country has been too volatile which impacts the international trade and investment decisions. It has utilized the technique of Generalized Autoregressive Heteroscedasticity as a proxy variable in order to capture the volatility in exchange rate of India. To investigate the cointegration between FDI inflows and exchange rate volatility, the approach of Maki cointegration which provides information up to five unknown structural breaks has been employed. Further, a comparative analysis of simple ARDL and ARDL with structural breaks has been analyzed. Firstly, the result of cointegration by Maki approach has been consistent with ARDL bound test. The model with structural breaks stood superior as its ARDL bound test was far higher than simple model and it also removed the problem of heteroscedasticity which was observed in model 1 (simple ARDL). The findings suggest that exchange rate volatility has been negatively associated with FDI inflows in both the models signifying that volatility in exchange rate deter FDI inflows due to uncertainty which rises among foreign investors regarding the returns, though it was negative but statistically insignificant in our study.

Keywords: Foreign direct investment (FDI) · Exchange rate volatility · Financial management · Maki cointegration

JEL Classification: F31 · F32

1 Introduction

Exchange rates have been vital and have influential implications for the nation and its economy. Its dynamic role in shaping the competing strength of the economy is indubitable. However, after adoption of flexible exchange rate regime by many countries,

B. Alareeni et al. (Eds.): ICBT 2022, LNNS 620, pp. 392–406, 2023.
https://doi.org/10.1007/978-3-031-26953-0_37

exchange rates have been extremely sensitive to small changes observed domestically or at global levels. Particularly in short run, exchange rates are prone to overshoot its long run equilibrium as investors rearrange their financial assets to attain a newly balanced portfolio as a result of changes in wealth, interest rate expectations etc. "There is a growing agreement in the literature that prolonged and substantial exchange rate misalignment can create severe macroeconomic disequilibria" (Aliyu and Rao 2010). On the other hand, few decades have noticed a significant upsurge in foreign direct investment (FDI) that has outshined both trade and output (UNCTAD 2017). The main motive to encourage more FDI is based on the multidimensional benefits prevailing such as technical know-how, growth in productivity as well as in production and labor empowerment (Sadder, 1999). Moreover, FDI is regarded as a non-debt producing and non-volatile resource for economic development in developing countries. Hence it becomes important to investigate the relationship between exchange rate volatility and foreign direct investment. Exchange rates have been susceptible to be volatile in many emerging economies. These random and frequent changes in exchange rate raises ambiguity among foreign investors regarding interpretation of these changes. Therefore, investors in decisive mode, may delay the investment which would deter foreign direct investment. Traditional theories states that exchange rate volatility influences FDI through two main channels i.e. production flexibility and risk aversion approach. The former theory states multinational firms are capable to invest abroad as volatility in exchange rate rises in host country. To be precise, producers are committed to foreign and domestic capacity ex ante and have the ability to adjust its variable factors like labor, capital ex post, see (Aizenman 1992). If the investing firm can opt to function in foreign market through exports or FDI then any significant volatility in exchange rate might lead the firm to substitute FDI for export as FDI decreases exposure of its profits to exchange rate risk. Similarly, (Cushman 1988)and (Goldberg and Kolstad 1995) demonstrated the prominence of allowing for post FDI changes in the exposure of multinational's profit to exchange rate risk. On the other hand, theory of risk aversion states that firms decide to operate abroad only when the expected returns are equal to the combined value of cost and payment for the risk arise due to volatility in exchange rate. Therefore, FDI declines with increase in exchange rate volatility because high variation in exchange rate lowers the certainty which is equivalent to expected exchange rate. Moreover, since investments are assumed to be irreversible in nature (Dixit and Pindyk 1994), adjustment cost in investments are asymmetric. As there is risk of over accumulation of capital if event turns unfavorable. An investment decision is made when expected profits are higher than capital cost (Serven 2003). However, (Bernanke 1983) suggests that even if uncertainty may increase the profitability, their relative ranking would still remain unclear. Thus, it is clear that the relationship between exchange rate volatility and FDI is found to be mixed. The present study aims to find the connection between exchange rate volatility and FDI inflows in India. Since India has seen a significant rise in capital flows after the economic reforms of 1991.With liberalization, Indian Rupee has been witnessing to frequent changes. Moreover, empirical studies on volatility in exchange rate and FDI in India are less corpus and traditional whatever studies have been found in investigating the relationship between exchange rate volatility and FDI are old-fashioned and indifferent of advance econometrics techniques except few.

2 Review of Literature

The study has made an attempt to review various related works that has been based on several countries. Most of the relevantly associated work regarding the concerned paper has been discussed in this section.

2.1 Effects of Exchange Rate on FDI

Since exchange rate is an important tool to determine various macroeconomic dynamics. Therefore, quite a few studies have been conducted to define the role of value of currency. For instance, (Vernon R. 1996) stated that depreciation in exchange rate attracts FDI through cost reduction in international investment and increasing returns to foreign investment than exports. Similarly, studies like (Froot et al. 1991), (Klien and Rosengren 1994) (Kiyota and Urata 2004)predicted that depreciation in exchange rate of host country increases relative wealth of foreigners and thereby attract FDI. On the other hand, (Lipsey and Feliciano 2016)have analyzed data from 1988 to 2006 to enquire that acquisition and setting up of new multinationals tends to occur at periods of high US growth. Moreover, they found that Host depreciation increases foreign acquisitions but is insignificant for new firms. (Kogut and Chang 1996) found that as home currency appreciates, FDI outflows (OFDI) decreases. Some of the studies also found that relationship between exchange rate levels and FDI were insignificant see, (Marchant et al. 1999) (Yang et al. 2000) and (Trevino et al. 2002).

2.2 Effects of Exchange Rate Volatility on FDI

Since exchange rate volatility has been a major concern among economists. Various empirical studies have been done to investigate about linkage between exchange rate volatility and FDI. In some studies, relationship has been positive, negative and even insignificant based on distinct countries and numerous macro-economic factors. (Muhammad et al. 2018) conducted a study in Nigeria for period from 1970 to 2014 by using Auto Regressive Distributed Lag (ARDL) model found that effect of exchange rate and volatility is higher in short run while devaluation increases FDI but volatility can deter FDI as investors are suspicious with higher uncertainty. This study suggested a democratic regime to be more stable for currency behavior then fixed regime. Likewise see, (Gopinath et al. 1998), (Kyereboah-Coleman and Agyire- Tettey 2008), (Ullah et al. 2012) and have found that exchange rate volatility deters FDI because of costs involved in volatility risk. (Beanco and Loan 2017)validated the option value and theory of hysteresis. (Khan et al. 2017) studied that among exchange rate volatility, Current account balance,GDP, and trade openness; FDI is negatively influenced by volatility and current account balance in short as well as long run. On the contrary, (Cushman 1988) found that greater exchange rate risk is positively correlated with FDI. (Goldberg and Kolstad 1995) and (Ménil 1999) also depicted a positive relationship between exchange rate variability and FDI. (Lin et al. 2010) examined the effect of exchange rate volatility on FDI based on motives. Volatility might delay FDI of market seeking firm but stimulate FDI inflow of export substituting firm. The rationale behind it is that market seeking firm will increase exchange rate risk while export substituting firm will reduce it. In several studies, effect of exchange rate volatility on FDI has been either insignificant see, (Iannizzotto and Miller 2005) or different impact on different countries (Crowley and Lee 2003).

2.3 Measures of Volatility

Volatility is generally defined as a rate at which security price rises or falls at a particular set of returns. It measures the risk involved in a security. Since volatility has been often witnessed in financial time series. Therefore, this area of study has become a great deal of discussion among researchers in order to capture the volatility with most exact measure. There are various measures to capture volatility in exchange rate. Many studies have measured by calculating the standard deviation where volatility in exchange rate is measured based on the degree of fluctuations in exchange rate in relation to its mean overtime (Gadanecz and Mehrotra 2013). Studies which have used the standard deviation of the annualized/monthly returns over a given period of time, see (Arratibel et al. 2011), (Esquivel and Larraín B. 2002), (Al-abri and Baghestani 2015) but this measure has been considered as unconditional (Cheung and Sengupta 2013). (Brooks 2008) explained volatility clustering as a tendency of large changes in asset prices to follow large changes and vice versa. More robust research papers have been inclined towards model like Auto Regressive Conditional Heteroscedasticity (ARCH) and Generalized Autoregressive Conditional Heteroscedasticity(GARCH) which are used to forecast and model volatility that allows the behavior of the series to follow diverse process at different point of time. Some of the studies haves used this measure of volatility, see (Muhammad et al. 2018) (Ullah et al. 2012), and (Muhammad et al. 2018).

3 Data Collection and Econometric Modelling

Total foreign direct investment inflows into India have been extracted from the Reserve Bank of India's Handbook of Indian Economy, nominal exchange rate of Indian Rupee against US dollar and Gross Domestic Product of India has been used as a proxy of market size which have been taken from official website of Federal Reserve of St. Louis. As the study has been focused on capturing volatility by conditional measure like GARCH, nominal exchange rate of Indian Rupee against US dollar has been analyzed. Since GARCH could not captured any volatility in effective nominal exchange rate of India (Durairaj and Nirmala 2012). The data has been analyzed on monthly basis for time period from 2000 to 2019.

3.1 The Model

The concerned study aims to investigate the influence of exchange rate volatility on Volume of FDI inflows in India. Since volatility in exchange rate is invisible component, it has been captured by the proxy measure, using, GARCH. Several studies have inquired into the relationship with different models and qualifications. Similarly, the present study aims to model the following specifications.

$$lfdi = \beta_0 + \beta_1 lex + \beta_2 lvol + \beta_3 lGDP \quad (1)$$

where in Eq. (1), lfdi is the log of total volume of FDI inflows in India, Lex is log of exchange rate of India against US dollar, lvol is log of volatility captured in exchange rate and lGDP is the log of GDP as a proxy variable of market size in India. The variables have been transformed into natural logarithm in order to create uniformity in variance of the series.

3.2 Measuring Exchange Rate Volatility

Fluctuations in exchange rate are undetectable directly. Therefore, it is the urgent requirement of the model to estimate its proxy variable by skilled and effective measurement. As discussed above, the most traditional and common method to estimate volatility is by the standard deviation of moving average of logarithm of exchange rate. This method is simple and easy to work out but is unsuitable and unconditional. Moreover, it has also been seen that prices of assets are not only defined by large changes but also linked with volatility clustering. Generally, volatility clustering is defined as when inclination of the series of large changes to follow large changes and small changes to follow small changes. As a result, (Engle 1982)propounded one of the most popular method, namely ARCH (Auto Regressive Conditional Heteroscedasticity). It is conditioned on the past behavior of the series. Due to drawback of the ARCH that it looks more like a moving average specification than auto regression (Engle 1982). Therefore, in the proposed study, a more generalized version of this model is used. The new idea was developed which aimed at including lagged conditional variance terms in autoregressive order (Asteriou and Hall 2006). Hence, (Bollerslev 1986) developed this new version acknowledged as Generalized Autoregressive Conditional heteroscedasticity (GARCH) which is used in this paper to detect volatility in exchange rate.

Specification of the GARCH model is as follows.

$$DLex_t = \alpha_\circ + be_{t-1} + \eta_t \tag{2}$$

$$\begin{aligned} &\eta_t \Omega \sim \mathrm{N(0, h_t)} \\ &h_t = \gamma_\circ + ce_{t-1}^2 + dh_{t-1} \end{aligned} \tag{3}$$

Here, $DLex_t$ is logarithmic difference of exchange rate from period t to t-1 and h_t is the variance of the error term η_t. The GARCH model empower us to examine variance as dependent on time. This is against to the usual assumption depicting error term as possessing constant variance in moving average (MA) process. Therefore, GARCH model provides the independence to study the patterns observed in volatility of asset price changes. Most importantly, stationarity of the variables must be checked at foremost.

3.3 Maki (MBk Approach)

The prevailing literature has often revealed various cointegration techniques with structural break such as (Gregory and Hansen 1996)and (Hatemi-J 2008)). In spite of this, these approaches have been ineffective in performance than Maki (MBk) Cointegration approach in case of multiple breaks. The MBk method is simple to compute and considers up to five unidentified structural breaks stemming from the data at the time of examining the cointegrating relationship. Hence, this paper aims to analyze the relationship between exchange rate volatility and FDI inflows in India with the help of MBk method. Moreover, regime shift has been applied which would allow structural breaks in level, regressor and also regime shift model with regressors, levels and trends with structural break. The null hypothesis states no cointegrating relationship while alternative suggest presence of cointegrating relationship.

Level Shift

$$y_t = \mu + \sum_{t=i}^{k} \mu_1 D_{it} + \beta' x_t + \mu_t, \tag{4}$$

Regime shift

$$y_t = \mu + \sum_{i=1}^{k} \mu_1 D_{it} + \beta' x_t + \sum_{i=1}^{k} \beta' x_t D_{i.t} + \mu_t, \tag{5}$$

Level Shift with Trend

$$y_t = \mu + \sum_{i=1}^{k} \mu_1 D_{it} + \gamma t + \beta' x_t + \sum_{i=1}^{k} \beta' x_t D_{i.t} + \mu_t \tag{6}$$

Regime shift with trend

$$y_t = \mu + \sum_{i=1}^{k} \mu_1 D_{it} + \gamma t + \sum_{i=1}^{k} \gamma_i t D_{i.t} + \beta' x_t + \sum_{i=1}^{k} \beta' x_t D_{i.t} + \mu_t, \tag{7}$$

In the above equations, t depicts time period such as t = 1,..T; y_t is the dependent variable and x_t is a set of regressors. The value of $D_{i.t}$ is 1 if t > $T_{Bi}(i = 1 \ldots K)$ and = 0 if t < T_{Bi}. Here, T_{Bi} depicts different periods of structural breaks and K is maximum number of lags. The advantage of MBk model is that it provides cointegrating relationship as well as structural breaks.

3.4 The ARDL Bound Testing Approach to Cointegration

In order to examine the long run and short run relationship among the variables, we employ ARDL bound testing. Further, we also include dummies which are obtained with help of MBk test. ARDL model is advantageous in the sense that it is sufficiently flexible to take into account different orders of integration of the variables simultaneously. It is also suitable for small samples as it gives unbiased results, overcome the concern of endogeneity which is based on selecting the optimum lags and deals with autocorrelation.

The specification for ARDL bound approach to cointegrations is as following.

$$F_t = f(\text{lexch}_t + \text{lgdp}_t + \text{lvol}_t + D_1 2002M07 + D_2 2006M02 + D_3 2007M06 + D_4 2008M10 + D_5 2011M04)$$

The general specification of the ARDL model with structural breaks is as follows:
Model 1

$$\Delta lfdi_t = \alpha_0 + \gamma_1 \Delta lfdi_{t-i} + \gamma_2 lex_{t-i} + \gamma_3 lvol_{t-i} + \gamma_4 lgdp_{t-i} + \theta_1 D2002M07_{t-i} + \theta_2 D2006M02_{t-i} + \theta_3 D2007M06_{t-i} + \theta_4 D2008M10_{t-i} + \theta_5 D2011M04_{t-i} + \sum_{b=1}^{a} {}_i \Delta fdi_{t-b} + \sum_{i=1}^{s} \pi_i \Delta Flex_{t-i} + \sum_{j=1}^{u} \rho_j \Delta lvol_{t-j} + \sum_{v=1}^{k} \sigma_v \Delta lgdp_{t-v} + \sum_{m=1}^{b} \Omega_m \Delta D2002M07_{t-} + \sum_{w=1}^{y} \delta_w \Delta D2006M02_{t-w} + \sum_{q=1}^{l} \emptyset_q \Delta D2007M06_{t-q} + \sum_{c=1}^{h} \tau_c \Delta D2008M10_{t-c} + \sum_{k=1}^{b} \gamma_k D2011M04_{t-k} + \varepsilon_t \quad (8)$$

Above equations represent an ARDL model based on the structural breaks found in regime shift with trend., α_0 is the intercept and $\gamma_1, \gamma_2, \gamma_3, \gamma_4$ are the long run parameters. The concerned null hypothesis suggests no correlation between FDI and its regressors including the dummies.

F test is employed in order to investigate the existence of a cointegrating relationship among variables. Therefore, the null hypothesis in Eq. (8) is $H_0 = \gamma_1 = \gamma_2 = \gamma_3 = \gamma_4 = 0$ against

$$H_1 \neq \gamma_1 \neq \gamma_2 \neq \gamma_3 \neq \gamma_4$$

Moreover, if the ARDL bound test confirms cointegration among variables then the estimated long run equation is as follows:

$$lfdi_t = \beta' + \sum_{k}^{s} \Omega' lfdi_{t-k} + \sum_{i=1}^{q} \pi_i' lex_{t-i} + \sum_{j=1}^{p} \rho_j' lvol_{t-j} \sum_{v=1}^{k} \rho_v' lgdp_{t-v} + \sum_{w=1}^{y} \psi_s' D2002M07_{t-w} + \sum_{m=1}^{b} \emptyset_v' \Delta 2006M02_{t-v} + \sum_{s=1}^{f} \varphi_q' D2007M06_{t-q} + \sum_{c=1}^{h} \tau_c' D2008M10_{t-c} + \sum_{z=1}^{y} \partial D2011M04_{t-z} + \varepsilon_t$$

where $\Omega' \pi_i', \rho_j', \rho_v', \psi_s', \emptyset_v'., \varphi_q', \varphi_q', \tau_c' \varphi_q', \tau_c'$ and ∂ are the long run coefficients. Therefore, once there is presence of long run relationship confirmed, then the concern model is estimated to analyze the short run dynamics with the help of error correction method (ECM).

$$\Delta lfdi = \beta^* + \sum_{i=1}^{q} \pi_i^* \Delta lfdi_{t-i} + \sum_{j=1}^{p} \rho_j^* \Delta lex_{t-j} + \sum_{k=1}^{o} \rho_v^* \Delta vol_{t-v} + \sum_{l=1}^{n} \rho_x^* \Delta lgdp_{t-x} + \sum_{o=1}^{l} \psi_s^* \Delta D2002M07_{t-s} + \sum_{p=1}^{k} \emptyset_w^* \Delta D2006M02_{t-w} + \sum_{n=1}^{m} \Omega_z^* \Delta D2007M06_{t-z} + \sum_{q=1}^{j} \varpi_\mu^* \Delta D2008M10_{t-\mu} + \sum_{p=1}^{k} \eta_p^* \Delta D2011M04_{t-p} + \lambda ECM_{t-1} + \varepsilon_t$$

In the above equation, short run coefficients are $\pi_i^*, \rho_j^*, \rho_v^*, \rho_x^*, \Omega_z^*, \varpi_\mu^*, \eta_p^*$ and the coefficient of ECM_{t-1} represnts the speed at which adjustment takes place (Pearsen and Shin 1999).

4 Results

Before proceeding further, it is utmost important to get familiarize with data. Therefore, unit root has been conducted to check the stationarity of the data. All the variables like

Table 1. Result of unit root tests

Variables	ADF Test		DF GLS		PP Test	
	Level	First difference	Level	First difference	Level	First difference
lfdi	−1.574	−14.037***	0.099	−0.416***	−2.578	−52.65***
lex	−0.476	−11.025***	−0.418	−11.03***	−0.229	−10.98***
lgdp	−0.784	−2.620*	−0.112	−0.802	−0.115	−21.30***

Note: i) All variables are stationary at first difference. ii)lfdi, lex, lgdp represent natural log of fdi, exchange rate and gross domestic product respectively. Iiii) Authors' calculation

FDI, exchange rate and GDP have been stationary at first difference. The results are given in Table 1.

It is important to note that the Schwarz criterion (SC) and Akaike Information criterion (AIC)have been used as a tool for model standard as heteroscedasticity might possess an autoregressive form, so the ARCH/GARCH can be used to model the volatility in the given data. Outcomes of the GARCH (1,1) model are presented in Table 2 which is based on the optimum lag criterion as stated above and with no ARCH effect left after GARCH estimation.

Table 2. Estimation of the GARCH type model for nominal exchange rate of India against US$

	Coefficients	p value
ARCH 1(α)	0.091***	0.0013
GARCH1(β)	0.907***	0.0000
C	0.03E*	0.0388
ARCH LM test		0.3197

Note: i). Null hypothesis for ARCH-LM test: "No ARCH effect". ii) author's calculation

ARCH term (α) which captures the volatility clustering and GARCH term(β) represents the persistence in conditional volatility. In the table 2, all the three terms are significant signaling presence of volatility in exchange rate of Indian rupee. The estimated GARCH model is well specified as sum of $\alpha + \beta < 1$ and there is no ARCH effect left after estimation.

4.1 Maki Cointegration Approach

Since the empirical findings of (Gregory and Hansen 1996) and (Hatemi-J 2008) are not suitable if the series comprises of more than two breaks. Hence in such a case, a comparatively better approach has been opted to look for the presence of more than two breaks i.e. MBk.

The present study aims to compare the results of simple ARDL and ARDL technique with breaks in order to bring out the efficiency of the results. Therefore, the study has applied the Maki cointegration technique to examine the long run relationship between FDI and exchange rate volatility and GDP. The results of MBk approach have been presented in Table 3, suggesting the cointegration with structural breaks among the variables in all the four models i. e level shifts, level shifts with trend, regime shifts and regime shifts with trend. Therefore, the empirical evidence has been able to confirms the presence of relationship among variables when structural breaks have been incorporated in the model.

Hence, in order to include the effect of structural breaks, five dummies have been incorporated 2002M07,2006M02,2007M06,2008M10 and 2011M04 from the regime shift with trend model in order to examine its impact on FDI which may be negative or positive. Therefore, to fulfill this purpose, the study would be applying ARDL model.

Table 3. Maki cointegration analysis with structural breaks

Regime	Test statistic	CV(5%)	CV(1%)	Break year
Level Shifts	− 13.01	− 6.038	− 6.555	2004M08-2006M03-2007M10-2001M11-2014M12
Level Shifts with trend	− 14.27	− 6.25	− 6.784	2002M07-2006M03-2007M10-2009M09-2017M02
Regime Shift	− 15.19	− 8.11	− 8.673	2002M07-2006M03-2007M10-2009M09-2011M04
Regime shift with trend	− 16.85	− 8.88	− 9.428	2002M0-2006M03-2007M06-2008M10-2011M04

Note: *,** is significance level at 5% and 1%

Firstly, the study would analyze the cointegrating relation between variables by applying simple ARDL model and then it would trace the long run relationship by using ARDL model with structural breaks. This would help in establishing a clear picture regarding the efficacy of results with and without the structural breaks.

The outcome of ARDL bound test with and without breaks have been shown in Table 4 and 5. Before applying ARDL bound test, it is important to select an optimum lag length for the model. Hence, for selecting optimum lag length, SBC criterion has been used in the present study. In both the cases, the value of the ARDL F statistics has been higher than the upper critical value at 1 per cent level of significance. This confirms the strong evidence of cointegration among FDI and its determinants.

Table 4. Results ARDL co-integration bound test without break

ARDL function	F statistics	K
	5.6615	3
	Critical bound F - value	
Significance level	Lower	Upper
1%	2.72	3.77
2.5%	3.32	4.35
5%	3.69	4.89
10%	4.29	5.61

Source: Author's calculation

In fact, the F statics is quite higher than upper critical bound in case of structural breaks than in simple ARDL model as reported in Table 5 below.

Table 5. Results ARDL co-integration bound test with breaks

ARDL function	F statistics	K
	18.929	8
	Critical bound F - value	
Significance level	Lower	Upper
1%	1.95	3.06
2.5%	2.22	3.39
5%	2.48	3.7
10%	2.79	4.1

Source: Author's calculation

Hence, it can be considered that the presence of cointegration which is determined by ARDL bound test also validated the empirical results of Mbk approach. Therefore, this depicts the sturdiness of the empirical outcomes. As the long run relationship among the variables has been established, the results of cointegration analysis of both the models (ARDL without breaks and with breaks) has been depicted in Table 6.

The empirical results of long run relationship have been shown in Table no 6. On one hand exchange rate and FDI establishes a negative relationship suggesting that as the value of the Indian Rupee depreciates, FDI inflows are encouraged and vice versa. It is worth noting that negative association between exchange rate FDI is significant in model 1. However, it is negative but insignificant in model 2 with structural breaks. Further, exchange rate volatility and FDI have a negative but insignificant relationship in both the cases. The negative association implies that volatility in exchange rate would create uncertainty among foreign investors and thus deter FDI inflows. Further, FDI inflows and GDP possess a positive relationship in both the models. This suggest that market size has

Table 6. Long run analysis

	Model 1 (without breaks)	Model 2 (with breaks)
Variables	Coefficients (p values)	Coefficients (p values)
LEX	– 2.72*** (0.0036)	– 0.180 (0.7384)
VOL	– 0.0304 (0.7483)	– 0.023 (0.454)
LGDP	4.1478*** (0.000)	1.728*** (0.000)
2002M07	–	– 0.42*** (0.0019)
2006M02	–	1.406*** (0.000)
2007M06	–	0.624*** (0.000)
2008M10	–	– 0.62*** (0.0001)
20011M04	–	0.2376 0.1073
Constant	– 45.9*** (0.000)	– 19.4*** (0.000)
$\in_t$	– 0.31*** (0.000)	– 0.85*** (0.000)
Heteroscedasticity (p value)	0.005	0.135
Serial Correlation (p value)	0.112	0.328

Note: *,**,*** denotes 10%, 5% and 1% level of significance and where Model 1 denotes simple ARDL model, Model 2 denotes ARDL model with dummies representing structural break
Source: Author's Calculation

a huge impact on appealing FDI inflows in India. Since model 2 comprises of structural breaks which are represented in the form of several dummies in the model. The economic reforms of India in 1991, helped its economy to strengthen and stabilize and also to gain momentum in receiving capital inflows. Further, dummy 2002M07, comprises a negative and significant relationship between FDI inflows in India. This could be the result of various reasons like an attack on parliament of India, such disturbances slumped down FDI inflows during this period. Moreover, dummies 2006M02and 2007M06 also depict a significantly positive relationship with FDI inflows in India. In 2006–07, FDI increased to nearly about 80% (Mahanti 2007). Moreover, India stood at rank second to attract the highest FDI inflows after China in 2007. On the other hand, all the developed economies faced a major setback due to global financial Crisis. Emerging countries like India also faced a decline in growth of FDI inflows due to shaky confidence of foreign investors. As a result, the dummy 2008M07 is negatively associated with FDI inflows in India. Lastly the dummy 2011 M0 is positive but insignificant with FDI inflows in India.

5 Conclusion

The present study aimed at examining the impact of exchange rate volatility and FDI in India for the period from 2000 to 2019 on monthly basis. The variables like exchange rate, volatility in exchange rate and gross domestic product as a proxy variable of market size has been analyzed as determinants of FDI.Further, five dummies are also incorporated which have been obtained with the help of Maki Cointegration approach. Moreover, Mbk cointegration analysis has been utilized in order to obtain the existence of long run association in the presence of multiple breaks. The Mbk approach has found five structural breaks in regime shift with trend represented with dummies. Further, the same has been confirmed with ARDL bound test. This study had compared the results with and without breaks using ARDL model. The results reveal that exchange rate though negative but becomes insignificant in model 2. Exchange rate volatility has been negative and insignificant in both the models and market size is positive signifying that FDI inflows are market seeking in India. Moreover, as stated above dummy 2002M07 has been decline in FDI inflows due to mis-happenings like parliamentary attacks in 2000 and other external factors like attack on World Trade Centre in 2001. Further 2006M02 and 2007M06 has been seen India as one of the top destinations of FDI inflows and lastly 2008M10 is negative and significant for which reason of global crisis cannot be ruled out. The error correction model has been negative and significant for both the models. Further, model1 comes out to be heteroscedastic while this correction has been addressed and improved by model 2 which has been homoscedastic. Also the value of ARDL bound test is much higher in model 2 than in model 1. Hence this provides an upper hand to the efficiency of model with structural breaks. Therefore, the paper signifies that volatility has no impact on FDI. Due to presence of negative sign, it may indicate that volatility would have deterred FDI because fluctuations in exchange rate creates uncertainty regarding returns among foreign investors. Hence, they prefer to avoid entering in such type of foreign market. Though the volatility is negative but insignificant in the case of India.

Based on the above findings, the study suggests that a model with structural breaks is more well defined and efficient then without breaks as exchange rate becomes insignificant after adding breaks into the model. This explains the importance regarding efficiency of the results when structural breaks are present in the model. Moreover, the model also becomes homoscedastic, once the breaks are incorporated. Since, the volatility is negatively insignificant, this implies that the model could add more variables to check the significance. Even though, the negative sign indicates that volatility might hinder FDI inflows for which government should strategize while making policy to ensure steadiness of Indian rupee intact. It is important for the Reserve Bank of India to timely intervene in order to keep a regular check regarding stability of exchange rate. Since volatility in exchange rate impacts several macro-economic factors like exports, capital flows, inflation etc. as it increases the exchange rate risk.

Lastly the study is limited to few variables which give the scope to make analysis on wider scale also based on different time periods and data frequency to verify the consistency of the results and suggests advancements. Also the study is open for comparison between pre pandemic and post pandemic after the availability of sample data.

References

Aizenman, J.: Exchange Rate Flexibility, Volatility, and the Patterns of Domestic and Foreign Direct Investment. Massachusetts, U.S.A: Working Paper No. 3953, National Bureau (1992)

Lipsey, R.E., Feliciano, Z.: Foreign entry into U.S. Manufacturing by Takeovers and the Creation of New Firms. Eastern Econ, J. Palgrave Macmillan;Eastern Econ. Assoc. **43**(1), 1–16 (2016)

Al-abri, A., Baghestani, H.: Foreign investment and real exchange rate volatility in emerging Asian economies. J. Asian Econ. **37**, 34–47 (2015)

Aliyu, R.S.U.: Exchange rate volatility and export trade in nigeria: an empirical analysis. Appl. Financ. Econ. **20**(13), 1071–1084 (2010)

Arratibel, O., Furceri, D., Martin, R., Zdzienicka, A.: The effect of nominal exchange rate volatility on real macroeconomic performace of CEE countries. Econ. Syst. **35**, 261–277 (2011)

Asteriou, D., Hall, S.G.: Applied Econometrics. Palgrave Macmillan, London (2006)

Beanco, S.D., Loan, N.C.: FDI inflows, price, exchnage rate volatility: new empirical evidence from Latin America. Int. J. Financ. Stud. **5**, 6 (2017)

Bernanke, B.S.: Irreversibility, uncertainty, and cyclical investment. Quart. J. Econ. **97**(1), 85–106 (1983)

Bollerslev, T.: Generalised autoregressive conditional heteroskedasticity'. J. Econometr. **31**, 307–321 (1986)

Brooks, C.: Introductory Economterics for Finance. Cambridge University Press, New York (2008)

Cheung, W.Y., Sengupta, R.: Impact of exchange rate movements on exports: an analysis of indian non financial sector firms. J. Int. Money Finance **39**, 231–245 (2013)

Crowley, P., Lee, J.: Exchange rate volatility and foreign investment: international evidence. Int. Trade J. **17**(3), 227–252 (2003)

Cushman, D.O.: Exchange-rate uncertainty and foreign direct investment in the United States. Weltwirtschaftliches Archiv **124**, 322–336 (1988)

Dixit, A., Pindyk, R.: Investment under Uncertainity. Princeton University Press (1994)

Durairaj, K., Nirmala, V.: Do exchange rates and its volatility deter foreign direct investment (FDI) to India? Indian Econ. J. **60**(1), 130–144 (2012)

Engle, R.F.: Autoregressive conditional heteroskedasticity with estimates of the variance of United Kingdom inflation. Econometrica (1982)

Esquivel, G., Larraín B.F.: The Impact of G-3 Exchange Rate Volatility on Developing Countries. CID Working Paper no. 86 (2002)

Froot, K.A., Stein, J.C.: Exchange rates and foreign direct investment: an imperfect capital markets approach. Quart. J. Econ. **106**(4), 1191–1217 (1991)

Gadanecz, B., Mehrotra, A.: The exchange rate, real economy and financial markets. BIS papers, no 73 (2013)

Goldberg, L., Kolstad, C.: Foreign direct investment, exchange rate variability and demand uncertainity. Int. Econ. Rev. **36**, 855–73 (1995)

Gopinath, M., Pick, D., Vasavada, U.: Exchange rate effects on the relationship between FDI and trade in the U.S. Food Process. Indus. Am. J. Agric. Econ. **80**(5), 1073–1079 (1998)

Gregory, A.W., Hansen, B.E.: Residual-based tests for cointegration in models with regime shifts. J. Econometr. **70**(1), 99–126 (1996)

Harthman, R.: The effects of price and cost uncertainity on investment. J. Econ. Theory **5**, 258–266 (1972)

Hatemi-J, A.: Tests for cointegration with two unknown regime shifts with an application to financial market integration. Empirical Economics **35**, 497–505 (2008)

Lannizzotto, M., Miller, N.J.: The effect of exchange-rate uncertainty on foreign direct investment in the United Kingdom. In: Graham, E.M. (eds) Multinationals and Foreign Investment in Economic Development. International Economic Association Series, pp. 163–178. Palgrave Macmillan, London (2005)

Khan, U.U., Sultan, F., Rehman, Z.U.: An analysis of volatility and FDI inflow in Pakistan using ARDL bound testing technique. Int. J. Appl. Econ. Stud. **5** (2017)

Kiyota, K., Urata, S.: Exchange rate, exchange rate volatility and foreign direct investment. World Econ. Wiley Blackwell **27**(10), 1501–1536 (2004)

Klien, M.W., Rosengren, E.: The real exchange rate and foreign direct investment in the United States: relative wealth vs. relative wage effects. J. Int. Econ. **36**, 373–389 (1994)

Kogut, B., Chang, S.J.: Platform investments and volatile exchange rates: direct investment in the U.S. by Japanese electronic companies. Rev. Econ. Stat. **78**(2), 221–231 (1996)

Kooper, P., Kohlhagen, S.: The effect of exchange rate uncertainity on prices and volume of international trade. J. Int. Econ., 483–511 (1978)

Kyereboah-Coleman, A., Agyir, K.: Effect of real exchange rate volatility on FDI in sub Saharan Africa: the case of Ghana. J. Risk Finance **9**, 52–70 (2008)

Kyereboah-Coleman, A., Agyire-Tettey, K.: Effect of exchange-rate volatility on foreign direct investment in Sub Saharan Africa the case of Ghana. J. Risk Finance **9**(1), 52–70 (2008)

Lin, C.-C., Chen, K.-M., Rau, H.-H.: Exchange rate volatility and the timing of foreign direct investment: market-seeking versus export-substituting. Rev. Develop. Econ. (2010)

Mahanti, T.K.: The Economic TimesNews (2007). economictimes.indiatimes.com: https://economictimes.indiatimes.com/news/economy/indicators/fdi-increases-80-in-2006-07/articleshow/2423632.cms

Marchant, M.A., Saghaian, S.H., SVickner, S.S.: Trade and foreign direct investment management strategies for U.S. processed food firms in China. Int. Food Agribus. Manag. Rev. **2**(2), 131–143 (1999)

Ménil, G., d.: Real capital market integration in the EU: How far has it gone? What will the effect of the euro be? Econ. Policy **14**(28), 166–201 (1999)

Muhammad, S.D., Azu, N.P., Oko, N.F.: Influence of real exchange rate and volatility on FDI inflow in Nigeria. Int. Bus. Res. **11**(6) (2018)

Pearsen, H.M., Shin, Y.: An Autoregressive Distributed Lag Modelling Approach to Cointegration Analysis. Econometrics and Economic Theory in the 20th century: The Ragnar Frish Centennial Symposium .Cambridge University Press (1999)

Pozo, S.: Conditional exchange-rate volatility and the volume of international trade: evidence from early 90s. Rev. Econ. Stat. **74**(2), 325–329 (1992)

Sadder, F.: Attracting Foreign Direct investment into Infrastructure.Why is it so Difficult? The World Bank (1999)

Serven, L.: Real exchange rate uncertainty and private investment in developing countries. Rev. Econ. Stat. **85**, 212–218 (2003)

Trevino, L.J., Daniels, J.D., Arbelae, H.: Market reform and FDI in Latin America: an empirical investigation. Trans. Corporations **11**(1), 29 (2002)

Ullah, S., Haider, S.Z., Azim, P.: Impact of exchange rate volatility on FDI : a case study of Pakistan. Pak. Econ. Soc. Rev. **50**, 121–138 (2012)

UNCTAD. Investment and Digital Economy. World Investment Report (2017)
Vernon, R.: International investment and international trade in the product cycle. Q. J. Econ. 190–207 (1996)
Yang, J., Groenewold, N., Tcha, M.: The determinants of foreign direct investment in Australia. Aust. Econ. Rec. **76**, 45–54 (2000)

Growth of Farm Mechanization in Karnataka: A Longitudinal Study

Roopa Adarsh[1] and K. Sivasubramanian[2(✉)]

[1] Department of Economics, Mount Carmel College (Autonomous), Bangalore, India
[2] Department of Economics, Kristu Jayanti College, Bangalore, India
sivasubramanian@kristujayanti.com

Abstract. Agriculture is a valuable substance in India's economic development as it offers employment and income generation to the mass. A massive subcontinental size like India, with a manifest of resource endowment and population density, regional diversities in an agro-climatic environment, is likely to be characterised by irregular economic and agricultural development among various regions. Agriculture plays a vital role in ornamental inclusive enlargement in rural India. As a decisive factor for the development of an economy, Farm Mechanisation has marked and paved the way bright for ruralisation in India. Inclusive growth is considered in the reduction of poverty and terms of employment opportunities. Reduction in economic inequalities, achievement of sustainability and inclusive development are correlated to farm mechanization in Karnataka. Farm mechanization is a significant component of the modernization of agriculture. Farm Productivity is positively correlated with the accessibility of farm power attached to efficient farm implements and their cautious utilization. As a result of diverse programmes implemented by the Government of India over the years and equal participation from the Private Sector, the level of mechanization has been escalating steadily over the years. This is evident from the sale of tractors and power tillers, taken as an indicator of the adoption of the mechanized means of farming. The Department of Agriculture and Cooperation is following a multi-pronged strategy for promoting Farm Mechanization. This paper seeks to explore the intensification of Farm Mechanisation in Karnataka and its impact on sustainable and comprehensive growth and development.

Keywords: Farm mechanisation · Machinery · Inclusive growth

1 Introduction

There has been a significant advancement in agriculture mechanization over the years. A momentous transformation shifts of ideology towards moving from using animate sources to mechanical equipment has empowered the agricultural activities of farmers across the country. Mechanicals are generally being used by the farming neighbourhood are equipped with various mechanical equipment like farm operating tillage, sowing, irrigation, plant protection and threshing, etc., As a result of escalating farm mechanization trends, the agricultural equipment market has witnessed strong intensification in the past

B. Alareeni et al. (Eds.): ICBT 2022, LNNS 620, pp. 407–416, 2023.
https://doi.org/10.1007/978-3-031-26953-0_38

few years. This market is currently being ambitious by several factors such as easy availability of credit, government incentives, the emergence of contract farming, increasing rural incomes and rising agricultural efficiency. Farm mechanization incorporates the use of tractors, tube wells, and plant protection measures. So, the application of machinery is greater than the labour force in farm fields. Agricultural mechanization plays an important role in sustaining and improving agricultural productivity, enabling the farming operations to be more efficient, improving the timeliness of operation, increasing cropping intensity, and minimising rigid labour in the fields. On the other hand, conventional corn farming practices are inefficient, laborious, and high-priced in terms of production cost.

In the context of farm economics, farm mechanization plays a vital role in increasing the effectiveness of agricultural operations, sinking the cost of production. It also comes in handy in reducing the drudgery of farm work when the farm labour is becoming gradually scarcer. A Centrally Sponsored Scheme for Farm Mechanisation was introduced in the year 2001–02 as a recognition advantage, 25% subsidy was provided under this scheme. In the year 2002–03, the State Government hiked the subsidy to 50% by contributing 25% as its share. Of late, the government provided an 80% subsidy on machinery and equipment for farmers to maintain the stubble.

2 Literature Review

Systemized farming is the method of applying agricultural machinery to mechanize the work of farm to hike the agricultural output and productivity (Shoba et al. 2018). Currently, Indian farmers are applying and adopting farm mechanization at very faster rate as compare with the recent past. The Indian farm sector is assorted and capable of producing many food and commercial crops (Tiwari et al. 2019).

Due to the use of machineries in agricultural sector, the sale of tractors and power tillers have been consistently increased by 6 percent over the last two decades (NABARD (2018). The growth of formal and non-agricultural informal employment attracting more labour force from rural areas. It leads to increase the wage rate for agricultural workers. The increase in wages causes a substitute of workers for machine power (Rajkhowa and Kubik 2021). The mechanization on agriculture has increased the productivity of farm sector (Afridi et al. 2020). In developing economies, small-holding farmers augment their agricultural labor requirement with family workers (Daum and Birner 2020). The review reveals that there is a fast growth of machine power in the country. However, the growth is not uniform as it is concentrated in a few states only. Therefore, there is necessary to spread the same in all the country's regions. Hence, in this study, an effort will be made to understand the growth of machine power in the state of Karnataka. The above studies reveal that farm mechanization led to an increase in production and productivity. However, unfortunately, they indicate that animal and human labour are being displaced.

3 Objectives of the Study

To study the growth of farm machinery in Karnataka between 1996–2017 based on Census Reports.

3.1 Scope of Study

The results of the present study would be useful in finding out the facts about the existing situations in the selected regions regarding mechanization and its impact on the growth of farm machinery across four divisions of Karnataka State. It would help to save a farmer's time, and labour charges and increase productivity. Also, it helps the planners and policymakers identify the problems in the mechanization of farms and find possible remedies for the same.

3.2 Methodology of Study

This study is based on the secondary sources of data from statistics pages of Government published sources such as Input Survey – inputsurvey.dacnet.nic.in, Other Secondary data was collected from books, journals, manuals, and articles on websites. The secondary data were used to trace the growth of farm machinery and implements for the years 1996–2017 (According to Census Report Published). The data collected includes 4 decades. For the study, the districts of Karnataka are divided into four zones/divisions namely Bangalore Division, Belgaum Division, Gulbarga Division, and Mysore Division. The following table gives the details about the divisions (Table 1).

Table 1. Administrative division/ regions of Karnataka

Bangalore division	Belgaum division	Gulbarga division	Mysore division
Bangalore Rural	Bagalkote	Bellary	Chamraj Nagara
Bangalore Urban	Belgaum	Bidar	Chikkamagaluru
Chikkabalapur	Bijapur	Gulbarga	Dakshina Karnataka
Chitradurga	Dharwad	Koppal	Hassan
Davangere	Gadag	Raichur	Kodagu
Kolar	Haveri	Yadgir	Mandya
Ramnagara	Uttara Karnataka		Mysore
Shimoga			Udupi
Tumakuru			

Source: www.wikipedia.org

Analysis and Interpretation of Study

The first part of the study analyses the distribution and the growth of farm machinery in the state of Karnataka:

Table 2. Percentage distribution and growth of major farm machinery and growth in the state of Karnataka

Year	Power operated sprayers and dusters	Diesel engine pump sets	Electric pump sets	Power tillers	Tractors for agricultural purpose
1996–1997	54313	180433	676384	42212	203768
2001–2002	62943	151396	1014052	127558	555399
2006–2007	106129	249721	1108284	205710	1759586
2011–2012	209765	263317	1292881	310292	1671060
2016–2017	344938	334216	1619411	227762	431511
AAGPA	27%	4%	7%	22%	6%

Source: input survey (http://inputsurvey.dacnet.nic.in).

The above Table 2, analyses the percentage distribution of the major machinery taken for this study in Karnataka state. Power-operated dusters and sprayers showed the maximum numbers with 27% of the Average Annual Growth rate per annum (AAG P.A) from 1997 to 2017 on an average of 2 decades on the count. The second place was occupied by the power tillers with a 33% hike in their usage, third stood by Electric pump sets with 7% & followed by with 6% of Tractors for agricultural purposes & in the last place by diesel engine pump sets with 3% growth indicating the oil consumption expensiveness towards machinery. The table reveals the expansion and extensive use of mechanisation in the state of Karnataka.

Status of Distribution of Major Farm Machineries Across the Divisions of Karnataka State

The second part of the study sees the farm machinery's growth across the divisions/regions of Karnataka state. The below tables explain the availability of Farm Machineries across the Divisions of Karnataka (Tables 3, 4, 5, 6, 7 and 8).

Table 3. Power-Operated Sprayers and Dusters ('000 Numbers)

Bangalore	1996–1997	2001–2002	2006–2007	2011–2012	2016–2017
Bangalore rural	2	2	8	2	2
Bangalore urban	1	1	1	3	2
Chikabalabura	0	0	0	2	2
Chitradurga	1	1	6	3	7
Davangere	0	3	3	4	8
Kolar	1	2	14	3	3

(continued)

Table 3. (*continued*)

Bangalore	1996–1997	2001–2002	2006–2007	2011–2012	2016–2017
Ramnagara	0	0	0	4	1
Shimoga	4	9	2	2	1
Tumkur	0	4	5	8	0
Total	9	23	39	33	25

Source: input survey (http://inputsurvey.dacnet.nic.in).

Table 4. Status of distribution and growth of major farm machineries in the state of Karnataka, Gulbarga division

Gulbarga division	1996–1997	2001–2002	2006–2007	2011–2012	2016–2017	Total
Bellary	5	0	2	1	28	36
Bidar	1	1	2	10	5	18
Gulbarga	11	22	123	1	21	177
Koppal	0	1	1	8	20	29
Raichur	1	1	1	8	23	34
Yadgir	0	0	0	70	1	71
Total	17	25	128	97	97	365

Source: input survey (http://inputsurvey.dacnet.nic.in).

Table 5. Status of distribution and growth of major farm machineries in Belgaum division, Karnataka

Belgaum division	1996–1997	2001–2002	2006–2007	2011–2012	2016–2017	Total
Bagalkote	0	2	0	1	22	25
Belgaum	3	4	8	12	43	71
Bijapur	7	0	2	10	21	40
Dharwad	1	2	1	0	20	24
Gadag	0	1	0	2	28	31
Haveri	0	0	3	1	13	16
Uttar Kannada	2	1	2	3	8	15
Total	13	9	16	29	154	222

Source: input survey (http://inputsurvey.dacnet.nic.in).

Table 6. Status of distribution and growth of major farm machineries in Mysore division, Karnataka

Mysore division	1996–1997	2001–2002	200–2007	201–2012	2016–2017	Total
Chamraj Nagara	0	0	0	5	3	8
Chikmagalur	2	1	7	15	25	50
Dakshin Kannada	7	4	1	6	5	23
Hassan	2	5	4	24	32	67
Kodagu	0	2	0	6	15	23
Mandiya	0	0	2	0	16	18
Mysore	5	8	2	16	22	53
Udupi	0	2	0	2	1	5
Total	16	22	16	74	119	245

Source: input survey (http://inputsurvey.dacnet.nic.in).

Table 7. Status of distribution of major farm machineries across the divisions of Karnataka state: power operated sprayers and dusters (in thousand)

Division	Power operated sprayers and dusters	AAP
Bangalore	102892	11%
Belgaum	221697	24%
Gulbarga	364963	30%
Mysore	247417	26%

Source: Input Survey (http://inputsurvey.dacnet.nic.in).

Above tables explains the growth and numbers of machine distribution of major farm machinery, Power-operated sprayers & dusters across the division of Karnataka. It is very much interesting to observe Gulbarga division stood first in the application of mechanization with 39% of (AAG P.A) for overall growth rate for the entire state, Mysore division occupied the second place but with 26%, followed by Belgaum division with 24% and Bangalore division stood at the last with 11% in terms of usage of Power operated sprayers & dusters in their fields across Karnataka.

Table 8. Power tillers ('000 Numbers)

Division	Power tillers	AAP
Bangalore	117406	13%
Belgaum	300947	33%
Gulbarga	127427	14%
Mysore	365142	40%

Source: Input Survey (http://inputsurvey.dacnet.nic.in).

The above table explains the growth and numbers of machine distribution of major farm machinery, Power Tillers across the division of Karnataka. It is very much remarkable to observe Mysore division stood first in the application of mechanization with 40% of (AAG P.A) compared to the previous consecutive years for overall growth rate for the entire state, this is a remarkable trend and further, this must be promoted. Belgaum division occupied second place with 33%, followed by the Gulbarga division with 14% and the Bangalore division stood at the last with 13% in terms of usage of Power Tillers in their fields across Karnataka (Table 9).

Table 9. Diesel engine pump sets ('000 Numbers)

Division	Diesel engine pump sets	AAP
Bangalore	151162	12%
Belgaum	327223	26%
Gulbarga	150631	12%
Mysore	606340	50%

Source: Input Survey (http://inputsurvey.dacnet.nic.in).

The above table explains the growth and numbers of machine distribution of major farm machinery, Diesel Engine pump sets across the division of Karnataka. This is a welcome trend and further this must be promoted. Diesel Engine Pump Sets with a record accounted 50% of the share consumption is set foot and royally retained by Mysore division alone, followed by with 26% by Belgaum division, and the third position occupied by both Bangalore division and Gulbarga division with 12% respectively (Table 10)

Table 10. Electric engine pump sets ('000 Numbers)

Division	Electric pump sets	AAP
Bangalore	1195564	28%
Belgaum	1586510	37%
Gulbarga	481427	11%
Mysore	990579	24%

Source: Input Survey (http://inputsurvey.dacnet.nic.in).

The above table, explains the growth and numbers of machine distribution of major farm machinery, Electric Engine pump sets across the division of Karnataka. This is a relaxed trend and further, it requires to be encouraged. In the availability of Electric Engine Pump Sets in their farms, Belgaum division stood first by 37% (AAG P.A), Bangalore division with 28% stood second followed by 24% by Mysore division, and finally in the last position was Gulbarga division with 11% of availability of Electric Engine pump sets respectively (Table 11).

Table 11. Tractors for other agricultural purposes ('000 Numbers)

Division	Tractors for agricultural purpose	AAP
Bangalore	1395120	40%
Belgaum	327223	10%
Gulbarga	515543	14%
Mysore	1249837	36%

Source: Input Survey (http://inputsurvey.dacnet.nic.in).

The above table explains the growth and numbers of machine distribution of major farm machinery, Tractors for other Agricultural purposes across the division of Karnataka. This seems to be the favourable trend for the state of Karnataka as tractor usage has been accepted by the farm operators in terms of tractor availability and there is a bright future which requires to be stimulated. In the availability of Tractors for other Agricultural purposes on their farms Bangalore division ranked first with 40% (AAG P.A), the second position was occupied by the Mysore division with 36%, in the third position was the Gulbarga division with 14%, and finally in the last position Belgaum with 10% of availability of Tractors for other Agricultural purposes respectively.

4 Summary of Findings, Conclusion and Suggestions

4.1 Findings of the Study

The review of the literature exposes that though there is a lot of scope for tractorisation in Indian agriculture, the improvement achieved is rather at a very low pace. Indian farmers have faced a lot of uncertainty at different stages. Altering assertiveness among Indian farmers and aggressive structures has undoubtedly fascinated the farmers to embrace advanced technology viz., HYVs, fertilizers, pesticides, and farm implements. It analyses the percentage distribution of the major machinery taken for this study in Karnataka state. It is very much interesting to observe that Power operated dusters and sprayers showed the maximum numbers with 27% of the Average Annual Growth rate per annum (AAG P.A) from 1997 to 2017 on an average of 2 decades on the count for the entire state in the application of mechanization. The second place was occupied by the power tillers with a 33% hike in their usage. Third stood by Electric pump sets with 7% & followed by with 6% of Tractors for agricultural purposes & in the last place by diesel engine pump sets with 3% growth indicating the oil consumption expensiveness towards machinery. The table reveals the expansion and extensive use of mechanisation in the state of Karnataka.

It explains the growth and numbers of machine distribution of major farm machinery, Power-operated sprayers & dusters across the division of Karnataka. It is very much interesting to observe Gulbarga division stood first in the application of mechanization with 39% of (AAG P.A) for the overall growth rate for the entire state. Table 2 explains the growth and numbers of machine distribution of major farm machinery, Power Tillers across the division of Karnataka. It is very much remarkable to observe Mysore division stood first in the application of mechanization with 40% of (AAG P.A) compared to the previous consecutive years for overall growth rate for the entire state, this is a remarkable trend and further, this must be promoted. Belgaum division occupied second place with 33%, followed by the Gulbarga division with 14% and the Bangalore division stood at the last with 13% in terms of usage of Power Tillers in their fields across Karnataka. It explains the growth and numbers of machine distribution of major farm machinery, Diesel Engine pump sets across the division of Karnataka. This is a welcome trend and further this must be promoted. Diesel Engine Pump Sets with a record accounted 50% of the share consumption is set foot and royally retained by Mysore division alone, followed by with 26% by Belgaum division, and the third position occupied by both Bangalore division and Gulbarga division with 12% respectively. It explains the growth and numbers of machine distribution of major farm machinery, Electric Engine pump sets across the division of Karnataka. This is a relaxed trend and further, it requires to be encouraged. In the availability of Electric Engine Pump Sets in their farms Belgaum division stood first by 37% (AAG P.A), Bangalore division with 28% stood second followed by 24% by Mysore division, and finally in the last position was Gulbarga division with 11% of availability of Electric Engine pump sets respectively. It explains the growth and numbers of machine distribution of major farm machinery, Tractors for other Agricultural purposes across the division of Karnataka. This seems to be the favourable trend for the state of Karnataka as factorization has been accepted by the farm operators in terms of tractor availability and there is a bright future which requires to be stimulated. In the availability of Tractors for other Agricultural purposes on their farms Bangalore division ranked first by 40% (AAG P.A).

4.2 Conclusion

Production and productivity have enhanced enormously due to mechanization. They indicate that animal and human labour are being replaced. In the state of Karnataka, machine power has advanced. However, there is no uniformity in the growth rate as it is only intense in a few divisions. Therefore, there is a requisite to spread across the regions of the country. Hence, in this study, an effort is made to comprehend the growth of machine power in the state of Karnataka. One of the interesting observations is that in all the divisions there is an optimistic growth rate in terms of possessing agricultural machinery, there happen to be regional dissimilarities in terms of the Mechanisation of agriculture and therefore, there is a need for the government to intervene and handle the balance to set right a kind of constant growth for inclusive development of the state's agriculture sector. As an overall observation from the analysis, there is an improvement of agricultural machinery over a period. There is a growing demand for agricultural equipment in the state. This is not only a welcome inclination but it must be valued because the state is empowered with farmers' possession of their equipment. However, one must estimate the impact of the farm machinery on the increase of production and productivity of the crops in the state.

4.3 Suggestions

- To elaborate a national policy and an apex body to implement farm mechanization in terms of assisting industries sales, servicing of equipment through their capacities in means of motivating farmers to adopt mechanization.
- To begin adequate training centres for exposing the mechanics, engineers, and technicians to farm power and machinery in terms of proper selection, operation repairs and maintenance of machines.
- To start a tractor testing station on the lines of international testing stations. Giving access to industrial policy about improving better quality of implements and machines. Post-harvest technology deserves special attention.
- Rural area needs special attention in means of proving Custom hiring system.

References

Afridi, F., Bishnu, M., Mahajan, K.: Gendering technological change: evidence from agricultural mechanism. In: IZA Discussion Paper, no. 13712 (2020)

Daum, T., Birner, R.: Agricultural mechanism in Africa: myths, realities and an emerging research agenda. Glob. Food Sec. **26**, 100393 (2020)

NABARD, Sectoral Paper. Farm Mechanism. Farm Sector Policy Department Mumbai (2018)

Rajkhowa, P., Kubik, Z.: Revisiting the relationship between farm mechanization and labour requirement in India. Indian Econ. Rev. **56**, 487–513 (2021)

Shoba, H., Rajeshwari, N., Yogeeshappa, H.: A study on farm mechanization level of farmers in North Karnataka, India. Int. J. Curr. Microbiol. App. Sci. **7**(2), 652–657 (2018)

Tiwari, P.S., Singh, K.K., Sahni, R.K., Kumar, V.: Farm mechanism – trends and policy for its promotion in India. Indian J. Agric. Sci. **89**(10), 1555–1562 (2019)

Understanding the Use of Artificial Intelligence (AI) for Human Resources in the Dubai Government

Amal Almesafri[1] and Mohammad Habes[2(✉)]

[1] Faculty of Management and Economics, Universiti Pendidikan Sultan Idris, Tanjong Malim, Malaysia

[2] Faculty of Mass Communication, Radio and TV Department, Yarmouk University, Irbid, Jordan

mohammad.habes@yu.edu.jo

Abstract. The use of technology to enhance the quality of life and facilitate everyday activities is growing. Similarly, organizations are also enjoying technology implementation and its perceived benefits. This analysis also shows that AI can help address a wide range of regional economic and social challenges. The researcher particularly focused on the human resources departments to apply artificial intelligence in government institutions in the United Arab Emirates (Dubai government). To achieve this, the study followed the quantitative approach by interviewing $n = 27$ employees of the Dubai government at the different administrative levels, upper, middle, and executive, in several institutions in the UAE. The data was gathered specifically about the impact of artificial intelligence applications on human resources departments in Dubai. Important results and recommendations were reached, the most important of which was the establishment of public policies and ethical laws for using artificial intelligence applications in human resources in Dubai. Also. The obligation for all employees of government institutions to take training and development courses provided to them through distancing learning. Finally, deepening the leading role of the UAE in leading the applications of artificial intelligence in all fields will bring positive, constructive changes in the relevant institutions leading to greater national progress and growth.

Keywords: Artificial intelligence · Dubai · Human resources · UAE · Machine learning · Technology

1 Introduction

Artificial intelligence (AI) is considered one of the most important outcomes of the Fourth Industrial Revolution due to its multiple uses in various fields [1]. As with the tremendous and accelerating technological development in different areas, i.e., management, human and technical resources, medical, educational and service applications, etc., all are witnessing greater social revolutions after the Fourth Industrial Revolution. As a result, technology such as Artificial Intelligence is accelerating progress, growth,

B. Alareeni et al. (Eds.): ICBT 2022, LNNS 620, pp. 417–428, 2023.
https://doi.org/10.1007/978-3-031-26953-0_39

and prosperity during the next few years, also leading to several innovations to establish a new and even transformative world that was earlier merely a concept for many of us [2, 3]. According to [4], it is also expected that Artificial Intelligence will make a real breakthrough in business management and fundamentally affect how employees' work patterns. Big data is related to various fields, including business management, to the extent that some people imagine that systems based on Artificial Intelligence and machine learning will take over management positions in the future [5]. Through its ability to understand and analyze a large amount of data and reach informed decisions, the world will witness a gigantic transformation in business and organizational management within the next few years [6, 7].

Thus, greater economic opportunities are provided by Artificial Intelligence to many sectors in the country. Its ability to achieve huge profits while applying its uses and relying on the accurate information and advice it provides, as well as its positive effects in reducing dependence on the human element and employment, indicate the prospects offered by AI in even macro-level sectors [8]. For example, in the United Arab Emirates, the state has adopted many mechanisms to promote the development and accelerate the activation of Artificial Intelligence applications at all government and private levels [9] The aim is to not only improve project performance but also to reduce the number of expatriate workers and modify the imbalance in the labor market structure and demographics [10]. According to a recent study /conducted by the consulting company Accenture [11] on the uses of artificial intelligence in the UAE, it is found that governments in the Middle Eastern region have responded to several interrelated strategies. The most important of which is economic diversification. Notably, this economic diversification aims to develop non-oil sectors to provide sustainable employment and less reliance on public sector jobs, with significant improvements in education and training to prepare the next generation, simplification, and modernization of regulation and governance [11]. All these strategies have an ambitious, comprehensive, and well-founded consideration of Artificial Intelligence. According to [12], AI applications can potentially raise economic growth rates in the region, adding $215 billion and $182 billion in annual gross value added to the economies of Saudi Arabia and the UAE, respectively, by 2035. Here comes the role of human resource management in developing a strategy that may create a balance between the needs of the organization and the individual, determined mainly by the internal and external factors of the relevant institution [13]. With the increasing applicability of Artificial Intelligence and the role of the workforce in the organization, an effective and key role of human resources management, applying artificial intelligence in various fields is expected to introduce and implement employee-friendly policies leading to greater facility and useful outcomes [14]. Artificial Intelligence is expected to perform different functions, i.e., job analysis, planning, appointments, designing wages and incentives, ensuring employee services and benefits, and evaluating performance. Additionally, it may also help in conducting training and developing programs and planning each employee's career path [12]. Hence, despite the multiplicity of studies showing the importance of Artificial Intelligence in many fields, there is a gap that needs more studies to understand its role in the UAE's public sector organizations. In this regard, the main idea of this study is to understand the role of artificial intelligence in the management of human resources in Emirati institutions and to identify the challenges that human resources departments in government institutions may face in the future. Also, there is a need to know the role of technology, especially artificial intelligence, in Human resource management. By keeping in view its prospects even for the future generations, employing Artificial Intelligence will help fulfill their needs and cope with the complex organizational challenges [13, 15]. For the relevant purpose, there are a set of questions that this study seeks to answer through how to activate AI-based training programs, initiatives, and workshops in the human resources sector? What are the most important challenges that artificial intelligence may face in government institutions? How can we invest energy in the available human and material resources and capabilities innovatively when applying artificial intelligence? How can we improve government performance, accelerate achievement and create innovative work environments using artificial intelligence in human resources?

2 Literature Review

2.1 Future of Artificial Intelligence in HR Government

Advances in Artificial Intelligence have emerged over the past few years as, i.e., robotics and deep learning, neural networks, and others can enhance the manual work performed by humans [16]. They usually work together as Artificial Intelligence instructs the robot to do what it requires. One of the most relevant examples of this is Google's self-driving cars. Robots have transcended their traditional roles in the logistics and manufacturing sectors and have witnessed greater developments. It empowers everything, from the personal advisor "Siri" by "Apple" to the "Watson" platform by IBM. "Artificial intelligence is based on computer science called "machine learning", which teaches algorithms themselves how to perform tasks by analyzing large amounts of data [17, 18]. This development was reinforced by the great progress in the processing power of computers, the great spread of data, and the growth of open-source software [19]. Today, AI can answer legal questions, write recipes, and even automate news writing [20–22].On the other hand, Governments (e-government/smart cities) also have to keep abreast of the latest scientific and technological developments. To organize such developments and take advantage of their services, policymakers still lack sufficient knowledge and consideration towards applying AI in the relevant areas [23]. On the other hand, man times technology implementation without prior knowledge and skills has shown a failure, leading to greater concerns regarding technology usage and the relevant execution skills [24].

Likewise, the development of Artificial Intelligence has made great leaps in recent years, and "deep learning" technology is one of the most prominent manifestations. In this sense, there is an increased need to use Artificial Intelligence to draw and complete the development parameters in smart cities fully. The goals should be the highest level of qualitative development, aiming to develop technical capabilities "deep learning", solving the concerns related to Artificial Intelligence, focusing on the seven sectors that benefit most from Artificial intelligence in the future and how it is applied in the institutions [11]. At the same time, there are concerns about the rapid mechanization of jobs, as some express fears regarding artificial intelligence exceeding what the human mind can understand. However, still, the prospects weigh more than the concerns [25]. As a result, governments must find the right balance between supporting the development of robotics and AI and also address their negative effects [26]. Where the replacement of machines for humans may lead to many problems, the question is, what alleviates these problems? [27] suggest creating new job opportunities and skills that keep pace with the changes due to smart applications. Especially in the technology and technical industry, i.e., robotics and artificial intelligence applications are replacing the human force to perform many jobs [28]. Similarly, management scholars have been interested in setting updated principles and improved foundations that may help make the most of each individual in the organization through human resource management and development [26]. Human resources management has recently gained the attention of scholars, researchers, and scientists. It is no longer the traditional management that includes routine tasks such as recruitment, training, and motivation. Rather it has added technological dimensions that are intertwined with the other fields of science, administrative knowledge, and social behavior [29]. Thus, the success and effectiveness of contemporary organizations depend on the human resource and their improved policies providing the basis for creating the value represented in appropriate outputs that may enhance the reputation and reputation of the organization [25, 30].

2.2 Artificial Intelligence in the UAE

According to [11], AI can help address many regional economic and social challenges, ranging from oil price volatility to rapid urbanization, water scarcity, and food security. However, UAE has played a leading role in developing research on Artificial Intelligence. His Highness Sheikh Mohammed bin Rashid emphasized the focus on using artificial intelligence applications in all government services. He further provided an idea about imposing AI development in the nine sectors: transportation, health, space, renewable energy, water, information technology, education, environment, and the traffic sector. He also indicated that the primary goal is to provide improved services through artificial intelligence. In addition to achieving a comprehensive integration of Artificial Intelligence with medical and security services, the continuation of the smart government to achieve tangible progress in all fields of government work and improve performance both horizontally and vertically. Further, a leading consulting company, Sketchure, surveyed all executives in the UAE. Figure 1 below shows the percentage of investment in new technologies in the field of artificial intelligence by executives in the UAE [11]. According to the respondents, the application of areas Artificial intelligence in various fields is providing fruitful results. The economic growth rate is expected to increase in the UAE at a rate of 1.6 percent, reaching a gross domestic revenue of $182 billion by 2035. [31]. It is clear from the above the pioneering role that the UAE intends to achieve in being the incubator of the applications of artificial intelligence and its scientific research in the world, which will support its economy and make it a leader in this field. The researcher also sees the prominent role of the insightful future vision of the rulers of the Emirates in developing government work and its services through Artificial Intelligence by setting strategies, goals, values, and the mechanism for working on them, whether in the short or long term; Which in turn will lead to rising government efficiency and productivity as Fig. 1.

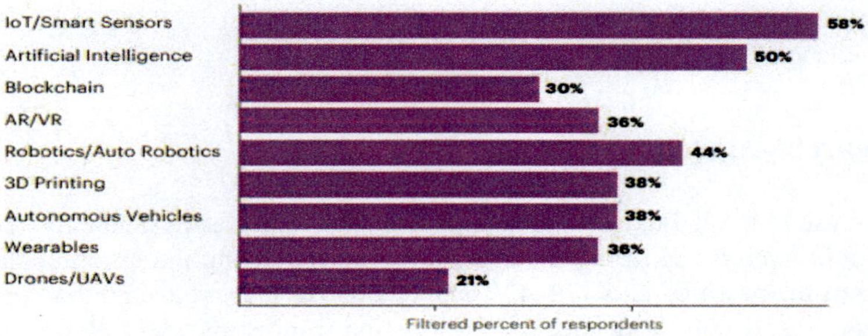

Fig. 1. The percentage of investments in (AI) in UAE [11]

It is shown from the above record that UAE intends to achieve maximum benefits by applying Artificial Intelligence as this will support the economy and bring other social changes [32]. The researcher also sees the prominent role of the insightful future vision of the rulers of the Emirates in improving government work and its services through Artificial Intelligence by setting strategies, goals, values, and the mechanism for working on them leading to raise government efficiency and productivity [33].

2.3 Use of (AI) Applications in UAE Human Resource

The UAE's strategy for Artificial Intelligence is the first of its kind in the region regarding the sectors it covers. The scope of services it includes and the complementarity of the future vision it foresees are also unique in the given contexts. The major goals are facing rapid changes and achieving qualitative development at all levels by building a complete and connected smart digital system that addresses challenges [34]. The UAE Strategy for Artificial Intelligence aims to make the UAE government the first in the world to invest in artificial intelligence in its various vital sectors, create a new promising market in the region with high economic value, support private sector initiatives, and increase productivity, Additionally, building a strong base in the field of research and development, that reliance on artificial intelligence in services and data analysis at a rate of 100% by the year 2031, which seeks to make the UAE the best in the year in all fields is also under consideration [35]. A strategy based on Artificial Intelligence that links all vital sectors also aims to achieve the UAE Centennial's goals and improve services and data analysis by 100% by 2031. According to UAE Centennial, AI will create new jobs that keep pace with changes and involve the private sector in the development of vital sectors in the country, in addition to focusing on overall public sector development [35]. These technologies are used to find innovative solutions that will adopt transparency in analyzing data to standardize it and work on sharing and making it open to all [36]. Additionally, the state-linked these expectations with its outputs. It worked to develop the education strategy to keep pace with the labor market changes, as the state's education policy was changed and linked to international standards. Hence, the UAE has issued its strategy for Artificial Intelligence, which is the next wave of smart transformation of the government in the human resources sector. (K. O'Sullivan, 2015). As Artificial Intelligence plays in many institutions, it is considered an important factor in developing new policies and procedures that regulate the process of oversight and performance to reach excellence [3].

3 Research Approach

The study used the qualitative approach, mainly the interpretive qualitative analysis, to gather information and further interpret it by understanding the meaning and interpretations of the gathered data [38–40]. The qualitative interpretive method facilitates flexibly and openly collecting data and information from employees, allowing them to answer and speak accordingly to their experiences [22, 41, 42]. It is also one of the accurate means, as there is no repetition of the answer and direct recording of answers spontaneously and spontaneously. The study sample consisted of a sample of $n = 27$ respondents represented by (Director of Department, Deputy Director, Head of Department, and Head of Division) for each of (4) government departments, namely (Dubai Police, Dubai Ambulance, Dubai Economic Department, Dubai Electricity and Water Authority "DEWA"). "), through questions prepared within personal interviews. Group interviews were directed to directors of the selected departments.

4 Analysis and Results

The interview questions presented to the respondents were concerned with the study dealing with everything that revolves around the uses and applications of artificial intelligence, its role in the future of human resources departments, and the preparation of the institutions of the United Arab Emirates. The questions were organized as follows:

1. How can the uses of Artificial Intelligence improve government performance, accelerate achievement, and create creative and innovative work environments with high productivity?

Participants responded that Artificial Intelligence is needed to increase efficiency and productivity and the need to move the human element to more creative and innovative jobs. However, the plans still depend on strategic implementation. Also, selection and recruitment of employees are still done by traditional methods, with special consideration towards employees' working skills that further help them gain new experiences for future jobs and thus work to raise their capabilities through learning on future skills. The appropriate skills provide the appropriate career path for employees and develop their capabilities to help them excel in their current positions and enhance their ambition for higher promotions. In the traditional methods, financial matters are in the hands of the decision-makers, and dependence is on the approved budget in recruitment by using online recruitment resources. Some of the applications referred to in the second chapter of the literature study can be used, such as independent advisors, independent external suppliers, smart personal assistants, and smart investment funds (Sharij, 2018), which further raise the level of government work, speed up achievement, and gain competitive advantage.

2. What is the contribution of Artificial Intelligence in preserving the accumulated human experiences?

Artificial Intelligence, specifically expert systems, can save the accumulated human experiences in an organized manner to make decisions in the future automatically. Expert systems are trained to enter data for the machines to be employed, as selection and appointment are traditionally done. The available machines are used to train new employees/customer service (automated systems to respond to customer inquiries from IBM [32, 33], and career path planning depends on the employee's experiences and qualifications. As the vision of this AI-enabled management aims to plan a career path in coordination with the Human Resources Department, smart machines will relieve employees from job stress enabling them to focus on improved performance. Artificial Intelligence techniques help recruit and assign tasks correctly, which helps to select the new employees and benefit from the human expertise.

According to [43], specializations, skills, experience, and strategic planning are all basic pillars of integrated institutional work. To ensure the sustainability of projects and services, enhance competitiveness, and achieve the vision that there is a need for a sustainable, innovative institution on a global level. As a result, Artificial Intelligence in HR is an initiative to get benefit from the skilled employees and link it to the selection, arrange appointments, and internal transfers so that the skills and experience of current employees are preserved. Even if the employees do not meet the job requirements, the external search is directed to take the necessary action. Thus, human expertise is transferred to intelligent machines through human language instead of a long traditional human recruitment process. In this way, applying Artificial Intelligence guarantees the employee to get the jobs suitable to their skills and expertise that may positively affect their performance, as job satisfaction guarantees work in a distinctive way, which positively affects the overall employee's performance.

3. What procedures and policies are followed in preparing qualified cadres capable of dealing with artificial intelligence applications?

Study participants emphasized that public sector employees are qualified to deal with modern systems where they use computers daily. In this regard, employees will not need special training to use Artificial Intelligence. For example, whoever does not find it difficult to use the smartphone will not find it difficult to deal with Siri (a system equipped with artificial intelligence techniques) [28]. As we are almost certain that using it is easier than using the phone, providing special training in the general administration of Artificial Intelligence was opened in 2017 based on the state's directions in this field. There is a plan to link it in the future with all other departments. Consequently, the systems are easy to deal with as we are familiar with them and are also based on a user-friendly approach.

4. How can the efficiency and effectiveness of the human resource be raised and the optimal investment for it at the level of government departments?

Study participants stressed that the jobs that will be completely regulated by Artificial Intelligence (such as the driver or secretary job) should be considered an example of integrating AI further to perform other organizational tasks. Notably, the greatest contribution of Artificial Intelligence is replacing some of the work related to existing jobs (Job Automation [26]. The change will not be necessary here, but the workforce efforts in these jobs will be reduced. For example, the emergence of the automated teller machine did not eliminate the presence of cashiers but reduced their numbers, which happened in human resource planning. Applying Artificial Intelligence tends to affect the selection of suitable approaches that can deal with the side of the machine by the side, and the plans for the completion of work will raise efficiency and effectiveness. Human employees tend to perform tasks and invest time and effort in work that the machine cannot work on, leading to an increased need for training sessions for the employees to fulfill the criteria regarding jobs produced by the Fourth Industrial Revolution in the future.

Artificial Intelligence has a significant impact on developing performance and understanding technology's role in helping organizations raise their efficiency and productivity. AI will enable the human cadre to focus better on the tasks that the robot cannot perform, making it achieve them more efficiently and effectively. Thus, there are opportunities for improvement in human resources through proving technology development and acceptance programs, especially in Artificial Intelligence and its deployment.

5. How can artificial intelligence assist senior government leaders in artificial intelligence applications?

The participants emphasized that using artificial intelligence does not require special capabilities, but it should be noted that senior leaders must be aware of the capabilities of Artificial Intelligence. These senior employees and policymakers should read future foresight reports to understand the expected impact on their business and develop strategies that help them keep pace with the development. Drafting clear tools, methodologies, and criteria to measure the extent to which employees obtain their basic rights regarding salaries and compensation can be further by using AI assistance. Further, wage scale approval, providing health insurance services, taking into account the exceptional circumstances about leaves and absences of emergency staff, developing the annual performance appraisal system, achieving financial sustainability, etc., all can be performed by Artificial Intelligence.

5 Discussion on Results

The study shows that the government departments did not address the applications of Artificial Intelligence in human resources, and their preparation to integrate the digital government still needs greater consideration. Although Artificial Intelligence can be used in human resources, as discussed by the literature, UAE needs a strong consideration of AI in human resource management. Also, the policymakers and organizational stakeholders should address the upcoming trends in their current jobs and train employees for skills for future jobs.

On the other hand, there are many challenges for the government institutions, including the lack of knowledge among the senior leadership in the organization regarding the applications of artificial intelligence and the lack of confidence in these applications. Further, senior leadership also overlooks other important capabilities of Artificial Intelligence, i.e., administrative decision-making, as they raise concerns from delegating authority to intelligent algorithms so that they can make independent decisions. The lack of control over information security and open data also limits the institution from giving accurate and clear data to AI-enabled software. Hence, if Artificial Intelligence is applied to give detailed reports and recommendations, it will not be accurate, leading to wrong recommendations and decisions that lead the institution to fail. The human also fears that the machine usage will be obstructive due to their resistance to change and its inclusion in their jobs and duties.

Machines should also be trained and developed by experts specialized in the field of forecasting to further facilitate the overall organization and its systems in line with the goals and objectives of collaborative trust and development, as also witnessed by the literature as indicating an optimistic stance on the provision and training of robots [11]. Also, provide a gateway to applying artificial crowd management during crises and disasters. Besides, the current employees specialized in information technology and network engineers are well qualified and should be preferred to occupy jobs related to artificial intelligence. On the other hand, the government should also introduce specialized courses and diplomas in this field. So far, we have also seen that the internal policies have been linked to the state's strategy for Artificial Intelligence. Yet, the selection and appointment according to traditional methods are one of the most important challenges faced by government institutions in motivating the future use and applicability of AI, as we cannot focus on reforming the infrastructure if we do not make primary changes in technology deployment and uses.

In addition to applying artificial intelligence to improve and protect electronic security and privacy, updating legislation, laws, and policies to employ technology is also required. The new strategic plan also seeks to quickly and accurately respond to queries in their various languages, analyze evidence with augmented reality through virtual applications, and raise the efficiency of employees' Artificial Intelligence usage and methods to achieve rational leadership directions and support the UAE's strategy for technological adoption and development as Artificial Intelligence represents device simulation for jobs. On the technical side, it also supports fraud detection, automation of knowledge documents, and others. Besides, a system of rewards and incentives is further needed to increase the efficiency and productivity of the employee so that it may highlight talents and creativity, indicating an important role of Artificial intelligence in the future.

The interviews also showed that the existing jobs are considered, and the workforce is analyzed to identify the competencies and the degree of preparation. This workforce is compared based on what sectors need from these forces and with the expansion of the use of artificial intelligence applications. However, there is a decrease in the volume of employment and a reduction in structures, and there is the replacement of non-workers facilitated by employing Artificial Intelligence. Thus, Artificial Intelligence will replace the non-suitable ones with skilled employees. The middle management category will disappear as new jobs have been created, and everyone is now working in teams. Besides, the project managers can choose their team, as the internal policies are developed and linked to the state's AI policy and strategy [1, 28]. Artificial Intelligence also accelerates human resources information system that relies on a database containing employee data. These data files include information related to the name of the workers, their health insurance numbers, and the job category they practice. With the organized availability of employee data, Artificial Intelligence makes it easy to set and get the employee information quickly when needed.

5.1 Conclusion and Future Research

Artificial intelligence contributes mainly to analyzing a large amount of data and then providing correct and accurate instructions through the electronic application about the procedures that must be followed., Especially regarding the importance of training and development and informing national cadres of the latest technical developments to benefit from national energies at work. Addressing the challenges related to artificial intelligence in the Dubai government and ethical considerations is the need of the day. Highlight robots' abilities and information that they may not cause harm; rather, protection and obeying human orders should be considered when designing Artificial Intelligence-based systems. Therefore, the researcher proposes issuing a clear legislative policy emphasizing ethical considerations regarding Artificial Intelligence applications usage. With the demand for new jobs related to Artificial Intelligence, creativity can be exported to an area similar to our culture and the Arabization of applications. Calling for establishing an independent research laboratory for Artificial Intelligence with scientific management, sufficient, sustainable resources, attention to education, research, and training, and the integration of human and material resources, to prepare the future generation for the complex technological systems. All improvement can be done by building a complete, connected, smart digital system that addresses challenges and provides practical, fast, quality, and efficient solutions. It is also recommended to support scientific research that includes algorithms related to learning machines in teaching and writing software, especially in the human resources sector in the United Arab Emirates.

References

1. Habes, M., Ali, S., Elareshi, M., Tahat, K.M., Ziani, A.: Understanding users' social TV content using artificial intelligence approach: a survey. In: 2021 International Conference of Modern Trends in Information and Communication Technology Industry (MTICTI), pp. 1–7 (2021)
2. Alhammad, K.L., Habes, M., Al Olaimat, F., Haddad, I.: Attitudes of students of the faculty of mass communication at Yarmouk University towards using video platforms in distance education. Rev. Int. Geogr. Educ. Online **11**(5), 1041–1052 (2021)
3. Salloum, S.A., AlAhbabi, N.M.N., Habes, M., Aburayya, A., Akour, I.: Predicting the intention to use social media sites: a hybrid SEM - machine learning approach. In: Hassanien, A.-E., Chang, K.-C., Mincong, T. (eds.) AMLTA 2021. AISC, vol. 1339, pp. 324–334. Springer, Cham (2021). https://doi.org/10.1007/978-3-030-69717-4_32
4. Ahmed, A.O., Idris, A.A.: Examining the relationship between soft total quality management (TQM) aspects and employees' job satisfaction in 'ISO 9001' Sudanese oil companies. TQM J. **33**(1), 95–124 (2021). https://doi.org/10.1108/TQM-05-2019-0147
5. Al-Sarayrah, W., Al-Aiad, A., Habes, M., Elareshi, M., Salloum, S.A.: Improving the deaf and hard of hearing internet accessibility: JSL, text-into-sign language translator for Arabic. In: Hassanien, A.-E., Chang, K.-C., Mincong, T. (eds.) AMLTA 2021. AISC, vol. 1339, pp. 456–468. Springer, Cham (2021). https://doi.org/10.1007/978-3-030-69717-4_43
6. Johennesse, L.-A., Chou, T.-K.: Employee perceptions of talent management effectiveness on retention. Glob. Bus. Manag. Res. An Int. J. **9**(3), 46–58 (2017)
7. Jeljeli, R., Alnaji, L., Khazam, K.: A comparison between moodle, Facebook, and paper-based assessment tools: students' perception of preference and effect on performance. Int. J. Emerg. Technol. Learn. **13**(5), 86–98 (2018). https://doi.org/10.3991/ijet.v13i05.8091

8. UNESCO: Education in a post-COVID world: Nine ideas for public action International Commission on the Futures of Education (2020)
9. Habes, M., Ali, S., Khalid, A., Elareshi, M.: E-learning acceptance during the Covid-19 outbreak: a cross-sectional study E-learning acceptance during the Covid-19 outbreak: a cross-sectional study, no. May (2021). https://doi.org/10.1007/978-3-030-77246-8
10. Mazouz, A., Alnaji, L., Jeljeli, R., Al-Shdaifat, F.: Innovation and entrepreneurship framework within the Middle East and North Africa region. Afr. J. Sci. Technol. Innov. Dev. **11**(6), 699–710 (2019). https://doi.org/10.1080/20421338.2019.1573959
11. Elsaadani, E.H.A., Purdy, M., Hakutangwi, E.: How artificial intelligence can drive diversification in the middle east. Accenture (2018)
12. Dhanabalan, T., Sathish, A.: Transforming Indian industries through artificial intelligence and robotics in industry 4.0. Int. J. Mech. Eng. Technol. **9**(10), 835–845 (2018)
13. Alhashmi, S.F.S., Salloum, S.A., Abdallah, S.: Critical success factors for implementing artificial intelligence (AI) projects in Dubai government United Arab Emirates (UAE) health sector: applying the extended technology acceptance model (TAM). In: Hassanien, A.E., Shaalan, K., Tolba, M.F. (eds.) AISI 2019. AISC, vol. 1058, pp. 393–405. Springer, Cham (2020). https://doi.org/10.1007/978-3-030-31129-2_36
14. Zhongming, Z., Linong, L., Xiaona, Y., Wangqiang, Z., Wei, L.: US-UAE joint statement on artificial intelligence cooperation (2019)
15. O'Sullivan, S., et al.: Legal, regulatory, and ethical frameworks for development of standards in artificial intelligence (AI) and autonomous robotic surgery. Int. J. Med. Robot. Comput. Assist. Surg. **15**(1), e1968 (2019)
16. Leicht-Deobald, U., et al.: the challenges of algorithm-based HR decision-making for personal integrity. J. Bus. Ethics **160**(2), 377–392 (2019). https://doi.org/10.1007/s10551-019-04204-w
17. Mom, T.J.M., Chang, Y.-Y., Cholakova, M., Jansen, J.J.P.: A multilevel integrated framework of firm HR practices, individual ambidexterity, and organizational ambidexterity. J. Manage. **45**(7), 3009–3034 (2019)
18. Elareshi, M., Habes, M., Al-Tahat, K., Ziani, A., Salloum, S.A.: Factors affecting social TV acceptance among Generation Z in Jordan. Acta Psychol. (Amst) **230**, 103730 (2022)
19. Khatri, P., Gupta, S., Gulati, K., Chauhan, S.: Talent Management in HR. J. Manag. Strateg. **1**(1), 39–46 (2010). https://doi.org/10.5430/jms.v1n1p39
20. Griffin, D.K., Mitchell, D., Thompson, S.J.: Podcasting by synchronising powerpoint and voice: what are the pedagogical benefits? Comput. Educ. **53**(2), 532–539 (2009)
21. Salloum, S.A., Al-Emran, M., Khalaf, R., Habes, M., Shaalan, K.: An innovative study of e-payment systems adoption in higher education: theoretical constructs and empirical analysis. Int. J. Interact. Mob. Technol. **13**(6) (2019)
22. Habes, M., Ali, S., Qamar, A., Elareshi, M., Ziani, A., Alsridi, H.: Public service advertisements and healthcare attitudinal changes in developing countries: pakistanis' perspectives. In: Alareeni, B., Hamdan, A. (eds.) Explore Business, Technology Opportunities and Challenges After the Covid-19 Pandemic. ICBT 2022. Lecture Notes in Networks and Systems, vol. 495, pp. 433–442. Springer, Cham (2023). https://doi.org/10.1007/978-3-031-08954-1_40
23. Shabbir, J., Anwer, T.: Artificial intelligence and its role in near future, vol. 14, no. 8, pp. 1–11 (2018)
24. IEEE UAE Cyber: Dubai Electronic Security Center (DESC) (2014). https://desc.dubai.ae/ar/strategy-ar/
25. Sharma, A., Tyagi, R., Verma, A., Paul, A.: Review on digitalisation and artificial intelligence in human resource function of energy sector. Water Energy Int. **65**(2), 38–46 (2022)
26. Tambe, P., Cappelli, P., Yakubovich, V.: Artificial intelligence in human resources management: challenges and a path forward. Calif. Manage. Rev. **61**(4), 15–42 (2019)

27. Peersman, C.: Using AI for supporting cybercrime investigations (2017)
28. Hmoud, B., Laszlo, V.: Will artificial intelligence take over human resources recruitment and selection. Netw. Intell. Stud. **7**(13), 21–30 (2019)
29. Elareshi, M., Habes, M., Ali, S., Ziani, A.: Using online platforms for political communication in bahrain election campaigns. Soc. Sci. Humanit. **29**(3), 2013–2031 (2021)
30. Oosthuizen, K., Botha, E., Robertson, J., Montecchi, M.: Artificial intelligence in retail: the AI-enabled value chain. Australas. Mark. J. **29**(3), 264–273 (2021). https://doi.org/10.1016/j.ausmj.2020.07.007
31. Qasim, A., El Refae, G., Eletter, S.: Embracing emerging technologies and artificial intelligence into the undergraduate accounting curriculum: reflections from the UAE. J. Emerg. Technol. Account. **19**, 155–169 (2021)
32. Salloum, S., Gaber, T., Vadera, S., Sharan, K.: A systematic literature review on phishing email detection using natural language processing techniques. IEEE Access (2022)
33. Jatoba, M., Gutierriz, I.E., Fernandes, P.O., Teixeira, J.P., Moscon, D. : Artificial intelligence in the recrutment & selection: innovation and impacts for the human resources management. In: 43rd International Scientific Conference on Economics and Social Development, pp. 96–104 (2019)
34. Al Darayseh, A.: The impact of COVID-19 pandemic on modes of teaching science in UAE schools. J. Educ. Pract. **11**, 110 (2020)
35. Government UAE: UAE Strategy for Artificial Intelligence. UAE (2020). https://government.ae/en/about-the-uae/strategies-initiatives-and-awards/federal-governments-strategies-and-plans/uae-strategy-for-artificial-intelligence
36. Martin, J.M.: Technology, education and Arab youth in the 21st century: a study of the UAE. no. July (2013). https://doi.org/10.13140/RG.2.1.4418.8002
37. O'Sullivan, K.: Challenges and constraints in meeting international standards in UAE education: external objectives versus local realities. Near Middle East. J. Res. Educ. **2015**(1), 4 (2015). https://doi.org/10.5339/nmejre.2015.4
38. Elareshi, M., Habes, M., Youssef, E., Salloum, S.A., Alfaisal, R., Ziani, A.: SEM-ANN-based approach to understanding students' academic-performance adoption of YouTube for learning during Covid. Heliyon **8**(4), e09236 (2022)
39. Habes, M., Salous, M.H., Al Jwaniat, M.I.: Applying the uses and gratifications theory to college major choice using social networks online video. In: Hassanien, A.E., Rizk, R.Y., Snášel, V., Abdel-Kader, R.F. (eds.) AMLTA 2022. LNDECT, vol. 113, pp. 388–400. Springer, Cham (2022). https://doi.org/10.1007/978-3-031-03918-8_33
40. Aoun, C.G., Lagadec, L., Habes, M.: An extended modeling approach for marine/deep-sea observatory. In: Hassanien, A.E., Rizk, R.Y., Snášel, V., Abdel-Kader, R.F. (eds.) AMLTA 2022. LNDECT, vol. 113, pp. 502–514. Springer, Cham (2022). https://doi.org/10.1007/978-3-031-03918-8_33
41. Habes, M., Ali, S., Pasha, S.A.: Statistical package for social sciences acceptance in quantitative research: from the technology acceptance model's perspective. FWU J. Soc. Sci. **15**(4), 34–46 (2021). https://doi.org/10.51709/19951272/Winter-2021/3
42. Tahat, K.M., Al-Sarayrah, W., Salloum, S.A., Habes, M., Ali, S.: The influence of YouTube videos on the learning experience of disabled people during the COVID-19 outbreak. In: Hassanien, A.-E., Elghamrawy, S.M., Zelinka, I. (eds.) Advances in Data Science and Intelligent Data Communication Technologies for COVID-19. SSDC, vol. 378, pp. 239–252. Springer, Cham (2022). https://doi.org/10.1007/978-3-030-77302-1_13
43. Tewari, I., Pant, M.: Artificial intelligence reshaping human resource management: a review. In: 2020 IEEE International Conference on Advent Trends in Multidisciplinary Research and Innovation (ICATMRI), pp. 1–4 (2020)

The Impact of Fatigue on Workers at Dubai Airport: Experimental Study

Amna Mohammed Humaid(✉), Norafidah Binti Ismail, and Mohammed R. A. Siam

University Utara Malaysia, Changlun, Malaysia
amnahumaid86@gmail.com

Abstract. Airport security personnel are faced with a high risk of fatigue, among other things. Employees under the said department are critically responsible for providing protection to both people and facilities in the airport. The 24-h operation of airports subject's employees to irregular shifts which result in stress, fatigue, exhaustion and many other signs of job burnout. An interview was performed with ten employees of Dubai International Airport in which nine are from the airport security department and one is from administrative security. Interview findings indicated that the most common fatigue risk factors identified by participants include work shifts, workload, and sleep problems, among others the implications of fatigue are mostly negative, as it leads to increased errors, reduced alertness, and longer response time. All these are an indicator of declined productivity and performance among airport security personnel in executing their responsibilities of providing safety and security of people and of airport facilities. The results of the research can be used as additional information to existing studies regarding the existence of employee fatigue in the aviation industry.

Keywords: Fatigue · Airport security · Productivity

1 Introduction

One of the primary stakeholders of the organization is the employees, from whom organizational success can be attributed to. The role of employees in an organization's pursuit of growth and success is recognized, thus the need to pay attention on their best interest, safety and wellbeing [1]. This is in line with the fact that the overall productivity of an organization depends largely on the workforce's welfare [2–4]. As determined in past researches, there are various factors that impact employees' wellbeing and these said factors are associated with either the demands that come with the job such as the physical and social aspects of the job that call for constant physical or mental exertions or the job resources which include material insufficiency that can affect the completion of tasks [5, 6]. All these factors can directly impact employees' health, considering that high job demands can lead to stress, burnout, and physical and/or psychological fatigue [7, 8]. Specifically, in the aviation sector, the high job demands among employees including airline pilots, cabin crew and airport security personnel have contributed to high stress and work-related fatigue [9]. With the different external threats (e.g., terrorism)

B. Alareeni et al. (Eds.): ICBT 2022, LNNS 620, pp. 429–440, 2023.
https://doi.org/10.1007/978-3-031-26953-0_40

that global airports are exposed to as a result of increased passenger volume and traffic, airport security employees have a vital responsibility in ensuring the security and safety of the people in the airport [10, 11]. As indicated by Rosenbloom et al., (2016), it is the primary duty of airport security employees to perform direct observations and to conduct patrols to identify any risk or threat to the security and safety of people and the facilities in the airport. With that responsibility comes the need for them to stay alert, reliable, confident and highly capable to adapt with pressure. Furthermore, there are several factors identified that influence the work performance of airport security personnel such as the fact that airports are risky environments, given that there are stressors present as brought about by their regular operations including noise, vibrations, etc. [12, 13]. As entailed by Adey, (2009), irregular shifts are experienced by airport personnel because airports run a 24-h operation that leads to loss of sleep, thereby causing fatigue and declined productivity. In that regard, it is anticipated that stress, fatigue, and many other signs of job burnout can possibly be felt by airport security personnel [15].

Poor performance and decreased productivity among airport security personnel is caused by fatigue and this is because of they take heavier workload and work for longer hours which result in loss of sleep and inadequate sleeping patterns. Taking that into account, there is a need to effectively address the risks brought by fatigue. This dissertation has the following objectives: To investigate fatigue-related risks experience among airport security personnel. To identify the factors that lead to physical and psychological fatigue among airport security personnel. To analyze the factors implications of fatigue on airport security employees at Dubai International Airport, specifically on their productivity and performance. The study also helps in expanding the findings of Butlewski et al., (2015) regarding the threats posed by fatigue on the performance of airport security personnel, considering that it increases risk of human errors which can result in substantial material and human injury. As well as The Dubai Airport is considered one of the busiest airports in the world in terms of passenger and cargo traffic, which requires more efforts by its employees, not to mention the tiring levels of security in it, which requires more effort from employees in implementing of adhering to these standards [11, 17]. Since all airline staff are under extreme stress at the moment, and these people work with tremendous levels of stress, for example if the check-in clerk is overworked, can it trigger a chain of events that is difficult to stop? [18] All it takes is for one person to make a mistake without thinking about the implications.

In relation, this research has established assumptions. One of which was the assumption that the participants have the basic knowledge about fatigue. Another assumption was that the study was not perceived by participants as a form of a performance assessment especially when their performance and productivity were included. Most importantly, the study assumed the honesty of the participants in responding to the interview. On the other hand, there are specific limitations in this study which include budget and time constraints that have affected the sample size. The research additionally does not assess the current strategies and initiatives implemented for fatigue management in Dubai International Airport.

2 Literature Review

There is a wide existing literature regarding human fatigue; however, it has no standard definition because of its complexity. As what [19] indicated, fatigue is regarded as a phenomenon that can be easily detected through direct observations, albeit the difficulty of making a comprehensive definition for it. This difficulty is due to the fact that fatigue is a multifaceted health issue that is caused by various factors (e.g. psychological, physical and environmental, among others). Some researches provided definitions of fatigue in relevance to its focus area. One example is using diabetes or multiple sclerosis as reference for conceptualizing fatigue [20, 21]. Even though there is still a need to develop a standard definition for fatigue, specifically occupational fatigue, past studies provided theoretical or conceptual explanations that have given the health issue a multifaceted construct. For one, [22], placed an emphasis on the different aspects that fatigue encompasses, which interact with one another thus affecting a person's experience of fatigue. As cited in [23] stated that there are two types of phenomena that occurred in human fatigue: (1) "the diminution of the muscular force"; and (2) "fatigue as a sensation" (p. 154). There are several existing literatures that have explored the definition of fatigue. One of which is the study of [24] which noted that fatigue is a psychophysiological condition associated with tiredness and sleepiness, negatively affecting human functioning which results in performance decline and negative emotions. Within the context of the aviation industry, ICAO (2018) indicated that fatigue is a multifaceted concept which involves a decline of mental and/or physical performance capacity. Also, there are risk factors to fatigue among aviation personnel which are loss of sleep and heavy workload. As there are several past researches that explored human fatigue as a health issue, its risk factors are also consequently identified which contribute to the feelings of exhaustion and fatigue of an individual. It is noted in past studies that the major risk factor of fatigue is loss of sleep or extended wakefulness. With sleep being one of basic human needs as indicated in the Hierarchy of Needs by Abraham Maslow, it is therefore considered a very important factor in human health. When an individual is experiencing sleep loss, which is the condition of a person being deprived of the needed number of hours of night sleep for maximum level of attentiveness and performance during wakefulness [25] his/her physiological functions are negatively affected which then causes fatigue. In the study of [26] it was revealed that declined cognitive performance is experienced by people who have limited number of sleeping hours ($\geq$ 6 h of sleep). It is also emphasized that even when if sleep is moderately limited, waking neurobehavioral functions can still be seriously impaired, even among healthy adults, [27] According to [25], the physical health and overall functioning of an individual can be adversely affected by loss of sleep. He identified that among the most common effects of sleep deprivation include higher risk of heart disease and Type 2 diabetes, impaired immune system, cognitive impairment, hallucinations, etc.

In the study conducted by [28], the causes and impacts of fatigue among all employees were investigated and findings from their review of relevant literature revealed that fatigue is experienced because of extended work shifts, sleep deprivation, high demand levels of job, and disruption of circadian rhythm. It is further highlighted that fatigue is one of the significant issues in the organization because it can gravely affect the health and safety of the employees. [28], also entailed that as fatigue adversely impacts the performance of employees at work, it thereby leads to increased errors and accidents and reduced productivity. [29], also pointed out that the negative implications of fatigue on the cognitive functions of the employees as well as on their mood, motivation, physiological functions and relationships with fellow employees contribute to their poor performance and productivity at work.

In line with these findings, the study [30], noted that risk factors of fatigue such as shift work and overtime can cause errors and decreased performance. This study which had involved health care personnel revealed that if any of these risk factors are present, patient safety is compromised due to increased likelihood of human errors. This can be linked with other studies such as that by [31], which examined the impacts of fatigue among transport operation personnel including airline pilots. According to the results of said study, alertness of employees is affected by fatigue which increases human errors. Awareness and processing of information in the environment can be difficult due to decreased cognitive functions, thereby leading to reduced performance and productivity. Generally, for aviation personnel, fatigue reduces attentiveness and awareness and at the same time, creates difficulties in communication, concentration and comprehension [32] For example, when aviation employees had to suddenly change their work schedule, they are more likely to have lower productivity and poorer job performance.

As indicated in some past researches, the implications of fatigue on productivity and performance are similar for employees working in the airport security department. According to Baeriswyl et al., (2016) who delved into the impacts of emotional fatigue and job burnout on airport security personnel's level of productivity and satisfaction, decrements in attentiveness and job efficiency are experienced by airport security officers as a consequence of emotional exhaustion which is brought about by heavy workload and high level of job demand. This consequently leads to job dissatisfaction. In addition, work-family conflict can have a mediating implication on the correlation between feelings of emotional exhaustion, job satisfaction and employee performance and/or productivity [9]. In relation, the linkage between fatigue and productivity and performance was examined in the studies of [9]. In both studies, it was highlighted that fatigue directly affects performance and productivity of employees and that it further leads to more problems such as reduced organizational efficiency. These studies also indicated that by lowering fatigue-related risk factors such as workload and workplace pressure, employee performance and productivity improves alongside employee satisfaction and retention.

The various adverse effects of fatigue and its risk factors prompt the need for effective fatigue management strategies and frameworks. According to the Australian Civil Aviation Safety Authority [32], Fatigue Risk Management System and the Fatigue Management Framework are two of the most common frameworks which can be used in the aviation industry to manage fatigue and reduce its negative implications on employees. US FAA (2015), on the other hand, had identified specific strategies to manage employee fatigue in the aviation industry. These include scheduled rest breaks and napping, proper shift schedules, and increased exposure to environmental stimuli. All of these initiatives can help reduce aviation personnel's sleepiness when on the job, thus improving both their productivity and performance.

Implementation of effective and adequate frameworks, strategies and systems relevant to managing fatigue and its risk factors can be performed by organizations and institutions in the aviation industry in order to effectively facilitate fatigue risk management. For instance, Air New Zealand, an airline company, adopts its own fatigue risk management framework which covers different activities and approaches that ensure constant monitoring and maintenance of risk factors associated with fatigue. It is indicated in past researches those effective frameworks, systems and strategies for managing fatigue-related risks are very important [33, 34]. As what indicated, the importance of these fatigue management systems and frameworks is reflected on how they enable protection of employees against any fatigue-related risk, denoting that it helps identify and measure the actual risk present and develop custom-made controls and strategies to reduce or even completely remove the risk. Furthermore, said systems, strategies and frameworks can be proactive and reactive since conventional approaches in managing fatigue-related risks are reactively implemented when fatigue-related accidents or events happens. On the other hand, they are also proactive because the developed controls in the system or framework are intended to resolve the sources of fatigue before said incidents happen [34, 35].

3 Methodology

The qualitative, descriptive research is used by this dissertation as its research design [36–38]. The qualitative research method covers gathering and interpretation of data which allows the researcher to develop a deeper exploration and understanding of the topic area which is fatigue and its implications on the performance of employees working in Dubai International Airport's airport security department [39–42]. According to [43, 44], qualitative research enables a more in-depth exploration of the topic, thus allowing the researcher to look into the experience of fatigue among airport security employees while also identifying different fatigue-related risk factors and the effects which fatigue has on productivity and performance [45, 46]. For this study, the importance of conducting qualitative research is reflected on its potential of producing results relevant in developing approaches and strategies essential in decreasing the negative impacts of work-related fatigue on the productivity and performance of airport security employees. More so, it is reiterated that using a qualitative, descriptive single case study design is appropriate for the research, considering that rich information will be obtained about the experiences and perspectives of the participants about fatigue, its associated risk factors, and its effects on their performance and productivity at work.

Airport security employees of Dubai International Airport are the study population for this dissertation, who are further classified into administrative security personnel or operations security personnel. Administrative security personnel are employees whose responsibility is to monitor the routine security operational processes, whereas operations security personnel refer to employees who directly carry out airport security measures and protocols. It is underpinned in past studies that fatigue and other health issues including stress and job burnout are experienced by airport and airline operations employees (e.g., cabin crew, airline pilots, ground personnel, etc.) [47–49]. Nevertheless, [9], mentioned that there were only a few studies that have fully investigated on fatigue and its impacts on airport security personnel's performance [50].

With that said, this dissertation includes the use of non-probability sampling in determining the ten airport security employees in Dubai International Airport which was initially composed of eight operations security personnel and two from administrative security [51]. However, having only one employee from administrative security showing willingness to be involved in the study, the researcher was prompted to add one more employee from administrative security, thereby making the sample participants be comprised of nine operations security personnel and one employee from administrative security. In particular, convenience sampling method was used in selecting the interview participants. Said sampling procedure involves gathering of data from conveniently available samples of the target population who are willing to be a part of the study's data collection. When it comes to determining the sample size, no pre-established rules were used and the researchers mainly used their own judgment based on the research purpose and research questions [50]. The decision to have ten participants for this study was motivated by the judgment that this specific sample size is adequate and that rich information relevant to the topic of the research can be obtained from the participants.

A semi-structured interview was conducted to ten airport security employees of Dubai International Airport. Interview responses are classified as this study's primary data. The questions used for the interview were carefully drafted in order to ensure that they can obtain the needed data to meet the study's objectives and answer its research questions. The interview questions used in this dissertation aim to gather relevant information from the participants regarding their personal experiences and insights on the existence of fatigue and its associated risk factors. In addition, participants were also asked about their perceived implications of fatigue on their work productivity and performance as well as the strategies they adopted to effectively manage said health issue. Interview questions also include the different approaches implemented by Dubai International Airport's security department to reduce fatigue and its related risks. Since convenience sampling was used in selecting the interview participants, the researcher personally approached the airport security employees to ask for their willing involvement in the study. Willing participants were given the information sheets and consent letters. At the same time, they were also invited to have a brief interview with the researcher, given their availability. Once settled, both the researcher and the participant set a scheduled time for the interview.

In conducting each session of interview which lasted 20 to 30 min, the researcher was taking notes using a reflexive journal where the responses of the participants and the perspectives of the researcher were written. Aside from the reflexive journal, a recorded audio was also made. To analyses the data gathered, thematic content analysis was used which involved identifying themes and/or data patterns from the responses of the participants. These themes were relevant in answering all the research questions of the study. Secondary data used in the literature review was also incorporated into the presentation of primary data results through triangulation to ensure the reliability and validity of the data.

Ethics plays an important role in this research and the researcher has adopted various ethical practices. One of which was providing consent letters and research information sheets to participants before the interview. The details of the research were outlined in the hand-outs and it also included an agreement emphasizing the willingness of the participants to be involved in the research. Privacy of the participants is honored all throughout the research, ensuring that the identity or any of the participants' personal information were not disclosed. Access to the data was also limited to the researcher only, guaranteeing the confidentiality of data.

4 Result

From the data collected, it is found that the respondents have worked more than two years at Dubai International Airport. It is additionally revealed by the findings that majority of the respondents enjoy their work. They mentioned that working at the airport allowed them to meet and know more people and juggle different tasks. The salary as well as the work environment also contributed to the positive experience of the respondents as airport personnel. On the other hand, they also identified certain aspects about their job which they find challenging. The workload, the shifting schedule, the overcrowding due to the high number of passengers, and the need to stay alert all throughout their shift were some of the challenges that the respondents had experienced. The lack of stability of their job due to the current global health crisis and the work routine were also identified as challenges among interview respondents.

4.1 Causes and Effects of Fatigue

In line with the challenges that the interview respondents had encountered at work, majority of them also revealed that they experienced fatigue although there were others who said the contrary. From the interview data obtained, it was found that airport personnel had felt tired and had even gotten ill because of their job. Several factors were identified as causes of the physical and psychological fatigue experienced by the respondents. These include the long working hours, the high number of employees, lack of sleep, overcrowding and the pressure to handle operational disruptions. These relate to the study of [50] with regards to the lack or loss of sleep being a risk factor of fatigue. On the other hand, for the long working hours being a factor of fatigue among the interview respondents, such finding can be aligned with what [52] had elucidated. The study noted that when an individual works for longer periods or has a shifting work schedule, sleep opportunities are reduced and the body's circadian rhythm is disrupted. As a result, the individual experiences fatigue. In the findings, it was found that respondents had to work 12 h a day, with some of them in shift work. They reported feeling tired, most especially at times when there were too many passengers and the airport was too crowded. Such finding was aligned with what [53] had mentioned about shift work and overtime being significant risk factors of fatigue because of how they alter sleep patterns, leading to loss of sleep or sleep deprivation.

Alongside the fatigue experienced by the interview respondents, majority of them also said that fatigue only led to negative effects. The findings suggested that fatigue could lead to stress, wrong decisions, lack of concentration, lack of sleep, and high physical demand among others. Similar to what [31, 54] had indicated in their study, fatigue could affective an individual's cognitive functions which makes concentrating and decision-making difficult. Alertness and response time were also adversely affected by fatigue, as both studies suggested. Furthermore, one of the respondents even mentioned that fatigue had affected their mental health. A similar notion was also stated in existing studies. For example, [34] highlighted that fatigue can cause health problems as the individual is more at risk of getting sick. Aside from the cardiovascular and metabolic diseases that individuals with fatigue can possibly have, their productivity and work performance are also adversely affected. This was also noted by [28], entailing that fatigue can make employees more prone to committing errors, further lowering their productivity levels and overall job performance. However, in the findings, majority of the respondents had reported that fatigue did not affect their performance nor their productivity. Despite experiencing fatigue, most of them did not let the stress or the negative implications of fatigue on their physical and psychological health affect their work. In this particular aspect, the study's findings did not agree with the findings of [31], which noted that employees with fatigue have latent performance due to their reduced ability to process information in their environment and their affected cognitive functions.

4.2 Strategies to Reduce Fatigue

Given the effects of fatigue on airport personnel, the importance of strategies and/or initiatives to address the issue has become more magnified. From the interview findings, it was revealed that Dubai International Airport was aware of the experiences of their employees, thus the reason why it had implemented certain measures to provide support and reduce the risk of fatigue among its workers. According to one of the respondents, surveys were oftentimes performed to assess the employees. At the same time, the workers are provided an email and a phone number which they can contact to seek support. Another initiative that was mentioned by the interview respondents was the provision of short leaves, holiday leaves, vacation leaves and sick leaves to the employees. In addition, the organization had also set up a wellbeing unit with the HR department and communicated with the employees to further understand the reason for fatigue and consequently determine what can be done to eliminate it. These strategies were also aligned with what [55] had identified. The provision of leaves to employees is related to allowing employees have their breaks during their shift. This gives them time to rest and relax themselves from work for a brief period, thus reducing fatigue in return.

Because of the efforts exerted by Dubai International Airport in reducing employee fatigue, interview findings revealed that airport personnel had perceived the strategies to be very effective and that they were highly satisfied with these initiatives. Majority of them acknowledged the support that the organization provides, especially in making sure that they are comfortable and happy at work. They also believed that these strategies had reduced the risk of fatigue and had positively affected their work productivity. Based on these responses, it could be generally satisfied with the current fatigue management framework and guidelines implemented in Dubai International Airport. This additionally means that majority of the participants had a positive perception of the effectiveness of these initiatives in reducing fatigue and in addressing fatigue-related risk factor. Despite that, there were also others who wanted to make a change in the current implementations if they had a chance. Most of them had mentioned reducing the number of working hours and to employ more people to lessen the workload. These suggested changes were all valid considering that both workload and working hours have a direct influence on employee fatigue. As [50, 51, 56] indicated, overtime, shifting work schedules and the nature of the job and the different tasks it involves can cause fatigue, hence increasing likelihood of reduced productivity and performance.

5 Conclusions

The Fatigue is a physiological condition that affects the mental and physical health and wellbeing of an individual. Airport personnel are among these individuals affected by fatigue, given the nature of their job and the number of hours they had to work. Through the qualitative research utilized for this study, it was revealed that airport security personnel of Dubai International Airport experience fatigue brought by their shifting schedule, long working hours, the high number of passengers they had to handle, and the pressure associated to the nature of their job. These fatigue risk factors had consequently led to physical exhaustion and although performance and productivity were not negatively affected, their personal life was as it became harder for them to spend time with their family. Amid the challenges encountered by the employees, the support they receive from their organization had been a significant help. The strategies implemented by Dubai International Airport did not only reduce the risk of fatigue, but had also provided comfort and happiness to the personnel. This all relates to the reason why fatigue management framework and strategies are an important component in any organization.

Considering that majority of the respondents had mentioned long working hours and shifting schedule being the main factors that caused fatigue, the recommendations then center on improving these two factors. In line with what FAA (2015) [51] had mentioned, proper shifting schedules must be established to ensure that employees get the proper amount of sleep that they need. This also helps ensure that their level of attentiveness is not affected so as their overall job performance. As for the working hours, there is a need to reassess as 12 h is too long and the employees' personal life and rest time are affected. It would be good for the airports management to educate workers about the dangers of fatigue and exhaustion and its consequences, and work to identify the sources of fatigue for workers, and work on developing specific strategies for managing fatigue and fatigue to maintain the health and safety of workers at airports and workplaces in general, and an understanding of the biological clock constitutes The daily hours of workers, and the hours of lack of sleep and rest for workers is an important factor in the ability to effectively manage fatigue and tiredness in the workplace. To expand the findings of this study for future research, a larger sample size can be used and an empirical approach can be undertaken for data collection and analysis. The findings can also be used to further investigate the linkage between fatigue, productivity and performance among airport security personnel in the UAE, not only in Dubai International Airport.

References

1. Al-Shibly, M.S., Alghizzawi, M., Habes, M., Salloum, S.A.: The impact of de-marketing in reducing Jordanian youth consumption of energy drinks. In: Hassanien, A.E., Shaalan, K., Tolba, M.F. (eds.) AISI 2019. AISC, vol. 1058, pp. 427–437. Springer, Cham (2020). https://doi.org/10.1007/978-3-030-31129-2_39

2. Shuck, B., Reio, T.G., Jr.: Employee engagement and well-being: a moderation model and implications for practice. J. Leadersh. Organ. Stud. **21**(1), 43–58 (2014)
3. Prochaska, J.O., et al.: The well-being assessment for productivity: a well-being approach to presenteeism. J. Occup. Environ. Med. 735–742 (2011)
4. Haddon, J.: The impact of employees' well-being on performance in the workplace. Strateg. HR Rev. (2018)
5. Brauchli, R., Schaufeli, W.B., Jenny, G.J., Füllemann, D., Bauer, G.F.: Disentangling stability and change in job resources, job demands, and employee well-being—a three-wave study on the job-demands resources model. J. Vocat. Behav. **83**(2), 117–129 (2013)
6. Gauche, C., De Beer, L.T., Brink, L.: Managing employee well-being: a qualitative study exploring job and personal resources of at-risk employees. SA J. Hum. Resour. Manag. **15**, 13 (2017)
7. Topcic, M., Baum, M., Kabst, R.: Are high-performance work practices related to individually perceived stress? A job demands-resources perspective. Int. J. Hum. Resour. Manag. **27**(1), 45–66 (2016)
8. Porto, R.B., Costa, R.D.R., Watanabe, E.A.D.M.: The multilevel effect of marketing activities on sales, revenue and profitability in a micro-enterprise. Rev. Bus. Manag. (2017). https://doi.org/10.7819/rbgn.v19i65.2911
9. Baeriswyl, S., Krause, A., Schwaninger, A.: Emotional exhaustion and job satisfaction in airport security officers–work–family conflict as mediator in the job demands–resources model. Front. Psychol. **7**, 663 (2016)
10. Kim, K.: Study on improving airport security to prevent air terrorism. J. Korean Soc. Aviat. Aeronaut. **21**(4), 90–105 (2013)
11. Habes, M., Ali, S., Qamar, A., Elareshi, M., Ziani, A., Alsridi, H.: Public service advertisements and healthcare attitudinal changes in developing countries: Pakistanis' perspectives. In: Alareeni, B., Hamdan, A. (eds.) Explore Business, Technology Opportunities and Challenges After the Covid-19 Pandemic. ICBT 2022. LNNS, vol. 495, pp. 433–442. Springer, Cham (2023). https://doi.org/10.1007/978-3-031-08954-1_40
12. Rosenbloom, T., Malka, Y., Israel, S.: Job burnout of security guards of aviation company. Pers. Rev. (2016)
13. Rizk, S.A.M., Sharaf, N.E., Mahdy-Abdallah, H., ElGelil, K.S.A.: Some health effects of aircraft noise with special reference to shift work. Toxicol. Ind. Health **32**(6), 961–967 (2016)
14. Adey, P.: Facing airport security: affect, biopolitics, and the preemptive securitisation of the mobile body. Environ. Plan. D Soc. Sp. **27**(2), 274–295 (2009)
15. Aoun, C.G., Lagadec, L., Habes, M.: An extended modeling approach for marine/deep-sea observatory. In: Hassanien, A.E., Rizk, R.Y., Snášel, V., Abdel-Kader, R.F. (eds.) The 8th International Conference on Advanced Machine Learning and Technologies and Applications (AMLTA2022). AMLTA 2022. LNDECT, vol. 113, pp. 502–514. Springer, Cham (2022). https://doi.org/10.1007/978-3-031-03918-8_42
16. Butlewski, M., Dahlke, G., Drzewiecka, M., Pacholski, L.: Fatigue of miners as a key factor in the work safety system. Procedia Manuf. **3**, 4732–4739 (2015)
17. Seo, E.J., Park, J.W.: A study on the effects of social media marketing activities on brand equity and customer response in the airline industry. J. Air Transp. Manag. **66**(September 2017), 36–41 (2018). https://doi.org/10.1016/j.jairtraman.2017.09.014
18. CE, E., Chika, C.: Implication of Aircraft Noise on Workers in Port Harcourt International Airport Omuagwa Rivers State. (2018)
19. Maddux, J.E.: Positive psychology and the illness ideology: toward a positive clinical psychology. Appl. Psychol. **57**, 54–70 (2008)
20. Turpin, M., Kerr, G., Gullo, H., Bennett, S., Asano, M., Finlayson, M.: Understanding and living with multiple sclerosis fatigue. Br. J. Occup. Ther. **81**(2), 82–89 (2018)

21. Chu, S.-C., Lien, C.-H., Cao, Y.: Electronic word-of-mouth (eWOM) on WeChat: examining the influence of sense of belonging, need for self-enhancement, and consumer engagement on Chinese travellers' eWOM. Int. J. Advert. 1–24 (2018)
22. Mosso, A.: Fatigue, vol. 13. GP Putnam's Sons, New York (1904)
23. Noakes, T.D.: Fatigue is a brain-derived emotion that regulates the exercise behavior to ensure the protection of whole body homeostasis. Front. Physiol. **3**, 82 (2012)
24. Craig, A., Tran, Y., Wijesuriya, N.: Psychophysiological characteristics of driver fatigue. Sleep, sleepiness traffic Safety (2011)
25. Roelen, A.L.C., Stuut, R.: Association of sleep deprivation with speech volume and pitch. Ergon. Hum. Factors 19–21 (2016)
26. Jackson, M.L., et al.: Deconstructing and reconstructing cognitive performance in sleep deprivation. Sleep Med. Rev. **17**(3), 215–225 (2013)
27. Habes, M., Salous, M.H., Al Jwaniat, M.I.: Applying the uses and gratifications theory to college major choice using social networks online video. In: Hassanien, A.E., Rizk, R.Y., Snášel, V., Abdel-Kader, R.F. (eds.) The 8th International Conference on Advanced Machine Learning and Technologies and Applications (AMLTA2022). AMLTA 2022. LNDECT, vol. 113, pp. 388–400. Springer, Cham (2022). https://doi.org/10.1007/978-3-031-03918-8_33
28. Sadeghniiat-Haghighi, K., Yazdi, Z.: Fatigue management in the workplace. Ind. Psychiatry J. **24**(1), 12 (2015)
29. Drago, R., Miller, K.: Sick at work: infected employees in the workplace during the H1N1 pandemic. Inst. Women's Policy Res. **B264**, 1–14 (2010)
30. Keller, S.M., Berryman, P., Lukes, E.: Effects of extended work shifts and shift work on patient safety, productivity, and employee health. Aaohn J. **57**(12), 497–504 (2009)
31. Phillips, R.O.: What is fatigue and how does it affect the safety performance of human transport operators. Fatigue Transp. Rep. I **1351** (2014)
32. A. C. A. S. Authority, Fatigue management strategies for aviation workers: a training & development workbook, May 2012
33. Habes, M., Alghizzawi, M., Elareshi, M., Ziani, A., Qudah, M., Al Hammadi, M.M.: E-Marketing and customers' bank loyalty enhancement: Jordanians' perspectives. In: Hamdan, A., Shoaib, H.M., Alareeni, B., Hamdan, R. (eds.) The Implementation of Smart Technologies for Business Success and Sustainability. Studies in Systems, Decision and Control, vol. 216, pp. 37–47. Springer, Cham (2023). https://doi.org/10.1007/978-3-031-10212-7_4
34. Lin, W.: Atmospheric conditioning: airport automation, labour and the COVID-19 pandemic. Trans. Inst. Br. Geogr. **47**(1), 214–228 (2022)
35. Al-Shakhanbeh, Z.M., Habes, M.: The relationship between the government's official Facebook pages and healthcare awareness during Covid-19 in Jordan. In: Hassanien, A.-E., Elghamrawy, S.M., Zelinka, I. (eds.) Advances in Data Science and Intelligent Data Communication Technologies for COVID-19. SSDC, vol. 378, pp. 221–238. Springer, Cham (2022). https://doi.org/10.1007/978-3-030-77302-1_12
36. Habes, M., Ali, S., Pasha, S.A.: Statistical package for social sciences acceptance in quantitative research: from the technology acceptance model's perspective. FWU J. Soc. Sci. **15**(4), 34–46 (2021). https://doi.org/10.51709/19951272/Winter-2021/3
37. Habes, M., Ali, S., Elareshi, M., Tahat, K.M., Ziani, A.: Understanding users' social TV content using artificial intelligence approach: a survey, no. January, pp. 1–7 (2022). https://doi.org/10.1109/mticti53925.2021.9664762
38. Al-Maroof, R.S., et al.: Students' perception towards behavioral intention of audio and video teaching styles: an acceptance study. Int. J. Data Netw. Sci. **6**(2), 603 (2022)
39. Ali, S., Habes, M., Youssef, E., Alodwan, M.: A cross-sectional analysis of digital library acceptance, & dependency during Covid-19. Int. J. Comput. Digit. Syst. (2021)
40. Elareshi, M., Habes, M., Ali, S., Ziani, A.: Using online platforms for political communication in Bahrain election campaigns. Soc. Sci. Humanit. **29**(3), 2013–2031 (2021)

41. Al-Skaf, S., Youssef, E., Habes, M., Alhumaid, K., Salloum, S.A.: The acceptance of social media sites: an empirical study using PLS-SEM and ML approaches. In: Hassanien, A.-E., Chang, K.-C., Mincong, T. (eds.) AMLTA 2021. AISC, vol. 1339, pp. 548–558. Springer, Cham (2021). https://doi.org/10.1007/978-3-030-69717-4_52
42. Habes, M., Ali, S., Salloum, S.A., Elareshi, M., Ziani, A.K.: Digital media and students' AP improvement: an empirical investigation of social TV, no. December, pp. 1–6 (2021). https://doi.org/10.1109/3ict51146.2020.9311941
43. Habes, M., Elareshi, M., Ziani, A., Almansoori, A., Alsridi, H.: Smart interaction and social TV used by Jordanian university students. Technol. Soc. 102110 (2022)
44. Creswell, J.W., Plano Clark, V.L., Gutmann, M.L., Hanson, W.E.: Advanced mixed methods research designs. Handb. Mix. Methods Soc. Behav. Res. **209**, 240 (2003)
45. Elareshi, M., Habes, M., Al-Tahat, K., Ziani, A., Salloum, S.A.: Factors affecting social TV acceptance among Generation Z in Jordan. Acta Psychol. (Amst). **230**, 103730 (2022)
46. Habes, M.: The influence of YouTube videos on ELA during the COVID-19 Outbreaks in Jordan. In: 2020 Sixth International Conference on e-Learning (econf), pp. 133–138. IEEE, 2020., 2020, no. December
47. Gander, P.H., et al.: Effects of sleep/wake history and circadian phase on proposed pilot fatigue safety performance indicators. J. Sleep Res. **24**(1), 110–119 (2015)
48. Gander, P.H., et al.: Pilot fatigue: relationships with departure and arrival times, flight duration, and direction. Aviat. Space. Environ. Med. **85**(8), 833–840 (2014)
49. Castro, M., Carvalhais, J., Teles, J.: Irregular working hours and fatigue of cabin crew. Work **51**(3), 505–511 (2015)
50. Arokkiaraj, H., Rajan, S.I.: Trafficking to gulf countries: the lived experiences of Indian female domestic workers. Asian Women **37**(1), 119–141 (2021)
51. Tabares, D.A.: An airport operations proposal for a pandemic-free air travel. J. Air Transp. Manag. **90**, 101943 (2021)
52. Parkes, K.R.: Work environment, overtime and sleep among offshore personnel. Accid. Anal. Prev. **99**, 383–388 (2017)
53. Parkes, K.R.: Shift rotation, overtime, age, and anxiety as predictors of offshore sleep patterns. J. Occup. Health Psychol. **20**(1), 27 (2015)
54. Du, J., et al.: The role of emotions and relationship quality in service failure and recovery. J. Serv. Res. (2010)
55. Gonczy, S.T.: Federal aviation administration (FAA) airworthiness certification for ceramic matrix composite components in civil aircraft systems. In: MATEC Web of Conferences, vol. 29, p. 2 (2015)
56. Seok, J.-M., Cho, J.-H., Jeon, W.-J., Ahn, J.-O.: Risk factors for fatigue and stress among Korean police officers. J. Phys. Ther. Sci. **27**(5), 1401–1405 (2015)

Challenges for Supply Chain Management (Logistics Management) in Petroleum Industry

Naser Hamad Obaid Zohari(✉)

Universiti Teknikal Malaysia, Melaka, Malaysia
Naser.zohari@adnocdistribtion.ae

Abstract. Supply chain management in the government sector is a fundamental aspect of economic stability and development. Notably, supply chains have often been considered the vertical sequential flow of interdependent transactions that eventually add value for the final consumer. By keeping in view the importance of supply chain management in the petroleum sector, this article addresses the challenges confronted by the Emirati petroleum industry. The researcher selected a sample of $n = 205$ participants and applied structural equation modelling to examine the proposed impacts of the challenges on the supply chain management in the petroleum sector. Findings showed that all the challenges (Risk management, digital transformation, shipping cost, and supply chain volatility) have a significant impact on the supply chain management in the Emirati petroleum sector. The respondents widely agreed that these factors are creating several challenges as the company lacks a skilled system to cope with them. Thus, it is concluded that there are many challenges in the Emirati supply chain management of the petroleum industry despite they have good supply chain management. These challenges need to be solved by the petroleum industry with the use of proper approaches and customizing their supply chain management system to cope with the challenges raised during the contemporary era.

Keywords: Supply chain · Logistic · Distribution process · Management · United Arab Emirates · Petroleum industry

1 Introduction

The supply chain involves all the parties' direct or indirect involvement to fulfil the customers' requests. It includes the vendors and the manufacturers, suppliers, customers, and even themselves. As a result, the supply chain involves all the functions of operations, marketing, product development, distribution, customer service, finance, distribution, operation, and others [1]. Today, the supply chain management process has significant implications in various industries. -For instance, in the automobile sector, it is perceived that the suppliers and vendors have been tiered more often like three-tiers. It is perceived that the first-tier vendors have been responsible for a total Sub-assembly [2]. On the other hand, two-tier suppliers mainly offer Seat finished elements, which the vendor of Tier-one has assembled. The tier-two suppliers are said to be supplied by the Tier three suppliers, who can be fabric manufacturers or fasteners [3].

B. Alareeni et al. (Eds.): ICBT 2022, LNNS 620, pp. 441–452, 2023.
https://doi.org/10.1007/978-3-031-26953-0_41

For example, supply chain management in the government sector is a fundamental aspect of economic stability and development. Notably, supply chains have often been considered the vertical sequential flow of interdependent transactions that eventually add value for the final consumer. However, following the findings of several researchers, previous studies do not emphasize the potential challenges and barriers of implementing a supply chain management system on an organizational basis, especially in the UAE petroleum industry [4–6]. However, the petroleum-based global value chain today is more complex and needs utmost consideration. Significantly, the petroleum-based global value chain has upstream value change that holds the crude oil resources. This involves exploring, drilling, extracting, and finally, managing the logistics to move the crude oil to the refineries. Currently, the term "supply networks" is suggested as a strong network nature of supply chains [7]. The main issue identified in the supply chain management in different organizations in UAE mainly include uncertainties in integrating technology, lack of risk management capabilities, volatility in supply chain management and others. For example [8] argued that the volatility in logistics is primarily aligned with the rising fuel price for transporting goods via air, sea, or land. This issue is one of the most significant challenges for many organizations in the UAE in terms of the supply chain. Furthermore, high costs of labour from manufacturers as well as suppliers are also a significant challenge that creates particular uncertainties over the distribution of the supply chain [8, 9].

Thus, by keeping in view the looming challenges for the Emirati challenges the supply chain management in the petroleum industry [2, 3, 10], this article focuses on identifying the potential challenges and barriers to the implementation of supply chain management system in the UAE petroleum industry. The selected platform for research is that of the United Arab Emirates. The focus of primary data will be on the ADNOC logistics and holding that is a wing under the Abu Dhabi National Oil Company for handling the supply chain management. Thus the idea of researching the challenges and barriers to supply chain management is an important aspect that needs to be taken care of with utmost priority. There lies the problem area, which is the focus of the research [11, 12].

2 Literature Review

2.1 Supply Chain Management System in Petroleum Industry

With the advancement of time and livelihood of the global population, demand for petroleum and its derivatives like petrochemical products and diesel is also increasing steadily. Such growing market demand has made the petroleum industry, and its companies enable them to extend their business reach and market share value and profitability [8]. Notably, any organization's supply chain management system depends on some key drivers. According to [13], drivers can be defined by the motivators that lead to the contribution of sustainability into the supply chain management practice originating from external and internal influences.

Similarly, cutting down the cost of products and services in the market is also an internal motivator for an organization to adopt a sustainable supply chain management system. Incorporating an effective supply chain within the business process can make an organization capable enough to shrink its budget to a considerable extent. Thus the business profit can be achieved and maintained over a long period [14]. Quality enhancement and managing the standard of products and services is considered one of the most important aspects of any organization, particularly in the petroleum industry. By using a sustainable supply chain management approach, an organization within the petroleum industry can quickly improve the quality of crude oil and refinery products, lessening waste products and emission of industrial pollutants in the environment [15].

2.2 Uses of Inventory and IT in SSCM in the UAE Petroleum Industry

Compared to the global petroleum and crude oil production rate, the United Arab Emirates holds significant energy reservoirs. The United Arab Emirates is considered the world's seventh-largest crude oil producer and the fourth-largest producer country of petroleum oil across the globe. The economic development of the entire country depends on the production and supply of petroleum and natural gases worldwide [16, 17] Notably, the United Arab Emirates produces 3,096.34 thousand barrels per day on crude oil 2,687.67 thousand barrels per day basis. Controlling consumer and distributor partnerships is a crucial component of managing supply chains. In certain instances, the principle of mutual collaboration has become the foundation of operations management. According to recent research by [18], in the business scenario, a direct link can be evident between the green practice of the supply chain management system and the progression of economic performance. According to the core approach of the triple bottom line, it can be said that integrated sustainable supply chain management in collaboration with the social and environmental supply chain management can impact positively the corporate financial performance of the industry, which in turn can be measured based on the return of assets and return of equity at long term basis [19].

2.3 Barriers to Effective Supply Chain Management

Supply chains include producers, customers, and suppliers who are the main partners and have different interests in the various components of the supply chain. Managing the diverse interests and components of the supply chain, multiple challenges and barriers arise. Proper identification of the obstacles and barriers is required to ensure the effectiveness of the supply chains [20]. The different barriers to supply chain management are - culture, organizational structure, data availability, and supply and purchase policies [7]. Likewise, the structure implemented by an organization dramatically affects the flow of the supply chain functions as the organizational structure directly impacts the services, products, and information flow within the business environment. When the partners implement similar organizational structures, it becomes easy to arrange the operations and processes within the supply chain [21]. The challenges can occur in the organizations or the surroundings of the business. Supply chain managers admit that external monitoring and internal planning failures are two barriers that significantly hinder supply chain practices. The organizational barriers impacting the supply chain generally fall under two categories, namely, organizational complexities and inter-firm rivalries. Inter-firm rivalry is represented by barriers like behavioural misalignment, lack of proper trust, weak collaborations, and turf protection [22].

2.4 Obstacles in Case of Sustainable Supply Chain

The obstacles to implementing sustainable supply chain management can be classified as internal and external. The internal barriers can be summarized as costs, lack of knowledge, lack of training, proper integration of IT systems, and poor organizational structures [23]. Customers always demand goods and services at low prices. For this, the costs incurred have to be reduced. However, it is observed that integrating sustainability within supply chain practices is significantly costly. Therefore, incurred costs are turning out to be an obstacle in applying sustainable supply chain management [22].

One of the main challenges that significantly impact the implementation of supply chain management in the petroleum industry is the logistic challenges. The Logistics network within the petroleum industry is not that flexible. The logistics network is represented by the capability of suppliers supplying crude oil, limitations regarding mode of transportation, and long lead times associated with vehicles. Thus, it is seen that every component of the logistics network is a challenge. The organizations within the petroleum industry operate on a global level. The locations are often continents apart, and therefore commodities have to be transferred over long distances. Long distances and slow and limited modes of transportation increase transportation costs [21].

2.5 Theories in SCM

Traditional Supply Chain Management: A typical SCM is characterized as an automated manufacturing method in which the manufacturers provide raw material to produce and convert into another. Results are then shipped to the distributors and eventually distributed to the consumers. The correct architecture of its supply chain reflects mainly on the desires of all consumers, and the process works in crucial steps. Modern supply chain management: New supply chains are moving even faster. The company's industrial supply chain covers all activities, including the safe and reliable production and distribution of a service or product from its inception until its demise and removal [24]. It stretches the overall supply chain and aids back to their distributor's network, and so forth. The effect of globalization, the constant and inevitable growth of science and skills, and the geographic strengths of labour and skills have generated the need for more interconnected trade in different areas of industry. Uncertainty of fuel costs has shifted the balance between the expense of stocks and buying the product. Modern supply chain management has a considerable impact and will aid in enhancing operations resulting in increasing efficiency [25]. Hence in the light of the cited literature, this research is based on the following hypotheses that are graphically illustrated in Fig. 1:

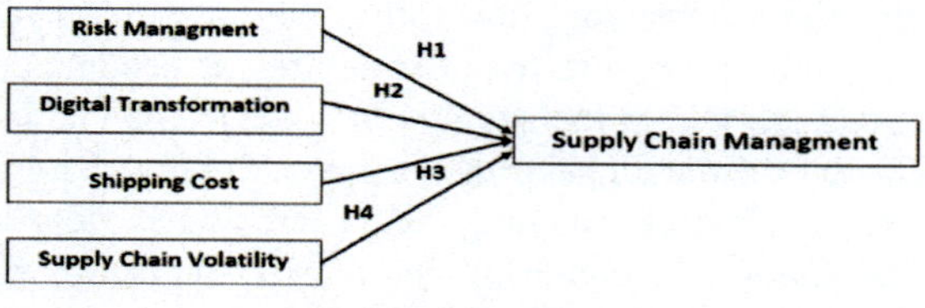

Fig. 1. Conceptual model

H1: Risk management has a significant impact on supply chain management
H2: Digital transformation has a significant impact on supply chain management
H3: Shipping cost has a significant impact on supply chain management
H4: Supply chain volatility has a significant impact on supply chain management

3 Material and Methods

For this particular research project, the researcher applied the quantitative case study technique as suggested by [26–28]. The current research consists of the collection of primary. Moreover, the most important outline is that all the associated tasks would be based on the case study of the chosen enterprise, ADNOC. Notably, the research design executed in the current research is known as the research onion. While selecting the other methods, the structure would be followed from outwards to inwards, as mentioned by [29, 30]. By maintaining this process, the research activities are conducted effectively and efficiently without any issues and errors.

3.1 Research Approach

It is an effective procedure with the help which it assists the research conductor in efficiently executing the research task, especially in the case of the collection of data and information [31]. There are three types of research philosophies within the group of research methodologies. Those are positivism, realism, pragmatism, and interpretivism. As per the aims and the target objectives of the research work, this particular research project is based on quantitative research in which the primary data are collected for analysis [32]. Furthermore, the current research involves the deductive approach as the proposed hypotheses are assessed. Further, regarding the data collection, the primary data is gathered by using both surveys as suggested by [33–35] which is later assessed by using the Structural Equation Modelling (SEM). According to [36], survey methods help to gather data directly from the individuals having direct experience of a phenomenon. They are based on a shorter time and provide generalizable results. Besides, the students of petroleum sciences are selected from the UAE for the survey purposes. Moreover, n = 205 individuals are selected by using convenience sampling. As noted by [51], despite convenience sampling having some limitations, it is one of the most preferred approaches in the business and management sciences.

Similarly, the survey Google forms automatically generated the percentage of responses regarding primary data. Around $n = 20$ questions were there in the survey questionnaires [39, 40]. As observed, there are three demographic questions where details about age, gender, and experience are gathered. Based on the demographic configuration, the report is evaluated.

4 Analysis and Result

As the study contains Structural Equation Modelling, the researchers applied two primary steps the measurement model and structural model analyses. First, the researchers examined the internal consistency of the measurement model. According to [41], examining the convergent validity of the measurement model is an important step in structural equation modelling as it helps to examine the internal consistency among the survey constructs. As summarized in Table 1, the researcher first calculated the Factor Loading and Average Variance Extracted (AVE) values. Results showed most of the Factor Loading values as greater than the threshold value of 0.7. Besides, the Average Variance Extracted (AVE) values ranged from .797 to .915. Moreover, regarding the construct reliability of the measurement model, the Cronbach Alpha values are ranging from .762 to .880 and Composite Reliability values are ranging from .790 to .926 which are greater than the threshold value of 0.7. Thus, it is found that the convergent validity of the measurement model is affirmed [16, 36].

Table 1. Summary of convergent validity

Constructs	Items	FL	AVE	CA	CR
Risk management	RMT1	.820	.848	.880	.793
	RMT2	.789			
	RMT3	.875			
Digital transformation	DTN1	.686	.797	.809	.790
	DTN2	.908			
	DTN3	−.530			
Shipping capacity (Insufficient)	SHP1	.888	.871	.799	.926
	SHP2	.855			
	SHP3	.016			
Supply chain volatility	SCV1	.894	.915	.792	.811
	SCV2	.649			
	SCV3	.936			
Supply chain management	SCM1	.012	.859	.760	.873
	SCM2	.924			
	SCM3	.794			

According to [42], two criterion-based approaches are important to determine the discriminant validity of the measurement model including the Fornell-Larcker criterion and the Heterotrait-Monotrait Ratio. Hence, regarding the Fornell-Larcker criterion, square calculations of all the Average Variance Extracted values are higher than the correlation values as mentioned in Table 2a. Further, the calculation of the Heterotrait-Monotrait Ratio (See Table 2b) revealed the HTMT value at .482 which is lower than the threshold value of .85, indicating that the discriminant validity of the measurement model is established.

Table 2. **(a)** Fornell-Larcker Criterion. **(b)** Heterotrait-Monotrait Ratio

	RMT	DTN	SHP	SCV	SCM		RMT	DTN	SHP	SCV	SCM
RMT	**.719**					RMT					
DTN	.492	**.635**				DTN	.657				
SHP	.492	.060	**.758**			SHP	.657	.945			
SCV	.220	.176	.176	**.837**		SCV	.297	.155	.105		
SCM	.225	.s119	.119	.092	**.737**	SCM	.264	.102	.100	.838	

According to [43], the goodness of fit is an important part of measurement model analysis. It determines the extent to which the observed data fits the expected data. Besides, it also examines whether the sample follows the normal distribution. Thus, Goodness of Fit in this revealed the chi-square value at .394 (18) and the probability level at .001. Further, the Standardized Root Mean Square (RMSEA) value remained at .611, which is smaller than the threshold value of .90, indicating that the observed data is well-aligned with the expected data. Figure 2 illustrates the model validating the Goodness of fit:

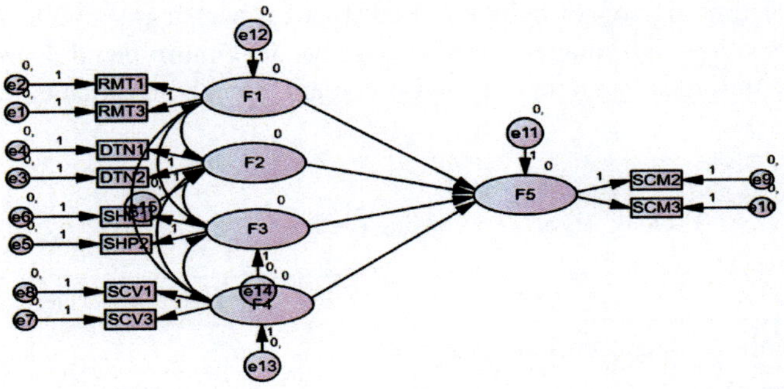

Fig. 2. Goodness of fit

Coefficients of Determination R^2 helps to determine the extent to which the exogenous variable is causing variance in the endogenous variables. Also, it assesses the predictive power of the latent variables. As shown in Table 3, the R^2 values of the latent variables are ranging from .527 to .729, indicating a fundamentally strong predictive power of the latent variables.

Finally, the researcher tested the study hypotheses by using path analysis. According to [6], despite linear regression analysis being widely used to examine the hypotheses, path analysis provides in-depth details about the structural relationships between the

Table 3. Coefficients of determination R^2

Variables	R^2	Strength
Risk management	.638	
Digital transformation	.527	
Shipping cost	.582	
Supply chain volatility	.729	

research variables. Thus, Table 4 summarizes the results of path analysis, indicating that all the proposed relationships are validated. As H1 of the current research proposes a significant impact of lacking Risk Management on the Supply Chain Management in the Emirati petroleum industry. The responses revealed that risk management in the supply chain management is required for sustainable business in the Emirati petroleum industry, but it has recently faced significant challenges. Especially, during a crisis like the Covid-19 pandemic, which negatively affects the economy and also the social situation in the country. The relevant proposition remained significant with the path value at .067 and p-value at $p > .047$. These results remained consistent with the arguments proposed by Meyer and their colleagues highlighting most government institutions lack risk management capabilities leading to unprecedented challenges for the petroleum industry across the globe [44]. Proper supply chain management can reduce production costs. According to most participants, risk management is the most critical internal driver of supply chain management, whereas, according to many, quality enhancement is also significant. The majority of participants think that proper supply chain management can give a competitive advantage to the companies. Maximum people agreed that the Emirati petroleum industry maintains environmental, social, and financial integrity in its procedures.

Table 4. Hypotheses testing (Path and Regression)

Relationships	Path	*t*	*P*	Decision
Risk management ---> Supply chain management	.067	1.998	.046	Accept
Digital transformation ---> Chain management	−.307	−7.760	.000	Accept
Shipping cost ---> Supply chain management	.268	6.777	.000	Accept
Supply chain volatility ---> Supply chain management	.696	13.538	.000	Accept

Furthermore, H2 of the current research proposed a significant impact of Digital Transformation on Supply Chain Management. As noted by [45], technology acceptance and integration have been challenging for many organizations. Especially, unfamiliarity with the digitalized methods of managing evaluation everyday responsibilities is considered a big challenge for untrained employees. The analysis also validated the relevant hypothesis with the path value at −.307 and significance value at .000. Notably, many respondents also think the most critical component of supply chain management is controlling production and distribution costs through technology integration. However, some respondents also believe that it is reducing lead time. According to the respondents, though globalization has made supply chain management more challenging, proper use of Information Technology can significantly contribute to the relevant industry. Also, according to many participants, the complexity of the new generation of IT systems is making supplies and management complicated. In contrast, some respondents also believe that the outdated IT system is creating problems.

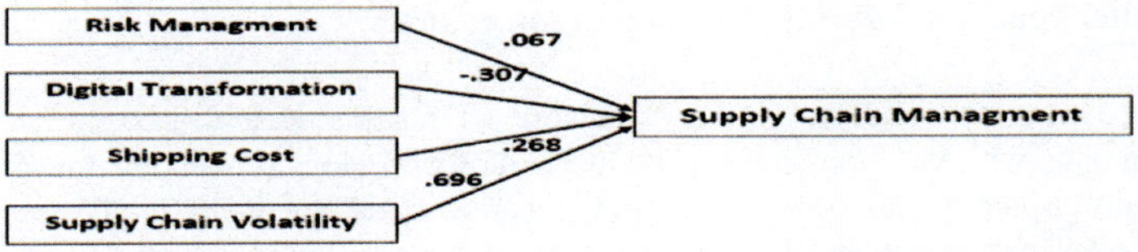

Similarly, H3 of the current research proposed a significant impact of Shipping Costs on Supply Chain Management. As per [46], shipping cost is an important challenge to determine the success or failure of the supply challenge management system. If shipping costs are high, the company will generate competitively a small revenue, leading to decreased employee performance and overall company excellence. The proposed hypotheses also remained affirmed with the path value at .268 and significance value at $p > .000$. These results also showed compatibility with the arguments given by Boneva and their colleagues [46] as respondents think that increased shipping cost is a significant challenge in the supply chain management for the Emirati petroleum industry. The survey shows that outsourcing to third-party logistics supply is a barrier in supply chain management. Study participants also think that the challenges of supply chain management in the petroleum industry are not visible. Yet, the companies should customize their supply chain management to meet production and cost allocation demands.

Finally, in the last hypothesis (H4), the researcher proposed a significant impact of Volatility on Supply Chain Management. As noted by Nitsche, managing the global supply chain is even more challenging today. Volatility in Supply Chain Management not only indicates variances in customer demands but also, a growing number of competitive companies and prices of raw petroleum products. Thus, this research also proposed a significant impact of volatility on Supply Chain Management and validated with the path value at .696 and significance value at $p > .000$ indicating consistency with the proposition given by Nitsche [47]. From the survey, it is clear that the global character of supply chain management in recent days has made it more challenging. Also, the majority of the respondents that the supplies and management of ADNOC are actually on point.

5 Conclusions

The majority of the respondents believe that the petroleum sector crisis somehow affects its logistics and supply chain management among the respondents, 50% think familiarity with the technology and risk management system are the two most critical components of supply chain management. Besides, respondents also consider shipping cost as an important factor that may affect the supply chain management either positively or negatively, depending mainly upon the company's ability to deal with the relevant challenges. Finally, most of the respondents also agreed that having a potential grip over volatility is important to overcome the challenges in the petroleum industry. Thus, it can be concluded that there are many challenges in the Emirati supply chain management of the petroleum industry despite the chosen company, ADNOC, having good supply chain management. These challenges need to be solved by the petroleum industry with the use of proper IT facilities and customizing their supply chain management system to cope with the challenges raised during the contemporary era.

Limitations. This research has two primary limitations. First, this research is based on the Emirati petroleum company while, its applicability in Gulf and other regions is questionable. Second, the researchers only selected four basic challenges, whereas there are many challenges faced by ADNOC and other petroleum companies in the United Arab Emirates that further narrow down the scope of current research.

References

1. Zigiaris, S.: Supply Chain Management. Report produced for the EC funded project, pp. 0–26 (2000)
2. Sandberg, E.: The role of top management in supply chain management practices, vol. 38, no. 1 (2007)
3. Anca, V.: Logistics and supply chain management: an overview. Stud. Bus. Econ. **14**(2), 209–215 (2019). https://doi.org/10.2478/sbe-2019-0035
4. Tahat, K.M., Al-Sarayrah, W., Salloum, S.A., Habes, M., Ali, S.: The influence of YouTube videos on the learning experience of disabled people during the COVID-19 outbreak. In: Hassanien, A.-E., Elghamrawy, S.M., Zelinka, I. (eds.) Advances in Data Science and Intelligent Data Communication Technologies for COVID-19. SSDC, vol. 378, pp. 239–252. Springer, Cham (2022). https://doi.org/10.1007/978-3-030-77302-1_13
5. Salloum, S.A., AlAhbabi, N.M.N., Habes, M., Aburayya, A., Akour, I.: Predicting the intention to use social media sites: a hybrid SEM - machine learning approach. In: Hassanien, A.-E., Chang, K.-C., Mincong, T. (eds.) AMLTA 2021. AISC, vol. 1339, pp. 324–334. Springer, Cham (2021). https://doi.org/10.1007/978-3-030-69717-4_32
6. Elareshi, M., Habes, M., Youssef, E., Salloum, S.A., Alfaisal, R., Ziani, A.: SEM-ANN-based approach to understanding students' academic-performance adoption of YouTube for learning during Covid. Heliyon **8**(4), e09236 (2022)
7. Raut, R.D., Narkhede, B., Gardas, B.B.: To identify the critical success factors of sustainable supply chain management practices in the context of oil and gas industries: ISM approach. Renew. Sustain. Energy Rev. **68**, 33–47 (2017). https://doi.org/10.1016/j.rser.2016.09.067
8. Mandira, A., Rahul, S., Lixin, M.A.: Challenges in supply chain management in upstream sector of oil and gas industry. In: Paper Conference Studies Energy, pp. 0–17 (2018)
9. Habes, M., Ali, S., Qamar, A., Elareshi, M., Ziani, A., Alsridi, H.: Public service advertisements and healthcare attitudinal changes in developing countries: Pakistanis' perspectives. In: Alareeni, B., Hamdan, A. (eds.) Explore Business, Technology Opportunities and Challenges After the Covid-19 Pandemic, pp. 433–442. Springer, Cham (2023). https://doi.org/10.1007/978-3-031-08954-1_40

10. Setyadi, A., et al.: Developing a circular-economy-based construction waste minimisation framework for Nigeria. Renew. Sustain. Energy Rev. **111**, 314–331 (2019)
11. Storey, J., Emberson, C., Godsell, J., Harrison, A.: Supply chain management: theory, practice and future challenges. Int. J. Oper. Prod. Manag. **26**(7), 754–774 (2006). https://doi.org/10.1108/01443570610672220
12. Salloum, S.A., Al-Emran, M., Khalaf, R., Habes, M., Shaalan, K.: An innovative study of e-payment systems adoption in higher education: theoretical constructs and empirical analysis. Int. J. Interact. Mob. Technol. **13**(6) (2019)
13. Wiengarten, F., Ahmed, M.U., Longoni, A., Pagell, M., Fynes, B.: Complexity and the triple bottom line: an information processing perspective (2017)
14. Tachizawa, E.M., Gimenez, C., Sierra, V.: Green supply chain management approaches: drivers and performance implications. Int. J. Oper. Prod. Manag. **35**(11), 1546–1566 (2015). https://doi.org/10.1108/IJOPM-01-2015-0023/FULL/XML
15. Felea, M., Albăstroiu, I.: Defining the concept of supply chain management and its relevance to Romanian academics and practitioners. Amfiteatru Econ. J. (2013)
16. Aoun, C.G., Lagadec, L., Habes, M.: An extended modeling approach for marine/deep-sea observatory. In: Hassanien, A.E., Rizk, R.Y., Snášel, V., Abdel-Kader, R.F. (eds.) AMLTA 2022, pp. 502–514. Springer, Cham (2022). https://doi.org/10.1007/978-3-031-03918-8_42
17. Juaidi, A., Montoya, F.G., Gázquez, J.A., Manzano-Agugliaro, F.: An overview of energy balance compared to sustainable energy in United Arab Emirates. Renew. Sustain. Energy Rev. **55**, 1195–1209 (2016). https://doi.org/10.1016/j.rser.2015.07.024
18. Rodger, J.A., George, J.A.: Accepted Manuscript (2016). https://doi.org/10.1016/j.jclepro.2016.11.089.This
19. Ahi, P., Searcy, C.: Assessing sustainability in the supply chain: a triple bottom line approach. Appl. Math. Model. **39**(10–11), 2882–2896 (2015). https://doi.org/10.1016/j.apm.2014.10.055
20. Yawar, S.A., Seuring, S.: Management of social issues in supply chains: a literature review exploring social issues, actions and performance outcomes. J. Bus. Ethics **141**(3), 621–643 (2015). https://doi.org/10.1007/s10551-015-2719-9
21. Silvestre, B.S.: Sustainable supply chain management in emerging economies: environmental turbulence, institutional voids and sustainability trajectories. Int. J. Prod. Econ. **167**, 156–169 (2015). https://doi.org/10.1016/j.ijpe.2015.05.025
22. Sajjad, A., Eweje, G., Tappin, D.: Sustainable supply chain management: motivators and barriers. Bus. Strateg. Environ. **24**(7), 643–655 (2015). https://doi.org/10.1002/bse.1898
23. Marta, B., Chacartegui, M.R., Molina-, M.: Please cite this article as: JES. Energy Build. **50**, 2331–2352 (2016)
24. Carter, C.R., Rogers, D.S., Choi, T.Y.: Towords the theory of the supply chain. J. Supply Chain Manag. **51**(2), 72 (2015)
25. Pohja, T.: Some theoretical foundations of supply chain management and supply networks: the role of social networks in selecting partners. In: 20th IMP Group Conference, Copenhagen, 2–4 September 2004, Working Paper, no. 2001, pp. 1–13 (2004)
26. Elareshi, M., Habes, M., Al-Tahat, K., Ziani, A., Salloum, S.A.: Factors affecting social TV acceptance among generation Z in Jordan. Acta Psychol. (Amst) **230**, 103730 (2022)
27. Habes, M., Elareshi, M., Ziani, A., Almansoori, A., Alsridi, H.: Smart interaction and social TV used by Jordanian University students. Technol. Soc. 102110 (2022)
28. Rahman, M.S.: The advantages and disadvantages of using qualitative and quantitative approaches and methods in language 'testing and assessment' research: a literature review. J. Educ. Learn. **6**(1), 102 (2016). https://doi.org/10.5539/jel.v6n1p102
29. Habes, M., Alghizzawi, M., Elareshi, M., Ziani, A., Qudah, M., Hammadi, M.M.A.: E-marketing and customers' bank loyalty enhancement: Jordanians' perspectives. In: Hamdan,

A., Shoaib, H.M., Alareeni, B., Hamdan, R. (eds.) The Implementation of Smart Technologies for Business Success and Sustainability: During COVID-19 Crises in Developing Countries, pp. 37–47. Springer, Cham (2023). https://doi.org/10.1007/978-3-031-10212-7_4
30. Saunders, M., Lewis, P., Thornhill, A.: Understanding research philosophies and.... 122–161 (2015)
31. Sarstedt, M., Ringle, C.M., Hair, J.F.: Handbook of Market Research (2020). https://doi.org/10.1007/978-3-319-05542-8
32. Lewis, S.: Qualitative inquiry and research design: choosing among five approaches. Health Promot. Pract. **16**(4), 473–475 (2015). https://doi.org/10.1177/1524839915580941
33. Krosnick, J.A.: Survey research. Annu. Rev. Psychol. **50**, 537–567 (1999). https://doi.org/10.1146/annurev.psych.50.1.537
34. Alhammad, K.L., Habes, M., Al Olaimat, F., Haddad, I.: Attitudes of students of the faculty of mass communication at Yarmouk University towards using video platforms in distance education. Rev. Int. Geogr. Educ. **11**(5), 1041–1052 (2021)
35. Habes, M., Ali, S., Pasha, S.A.: Statistical package for social sciences acceptance in quantitative research: from the technology acceptance model's perspective. FWU J. Soc. Sci. **15**(4), 34–46 (2021). https://doi.org/10.51709/19951272/Winter-2021/3
36. Stevens, L.M., Lynm, C., Glass, R.M.: Basic science research. J. Am. Med. Assoc. **287**(13), 1754 (2002). https://doi.org/10.1001/jama.287.13.1754
37. Habes, M., Elareshi, M., Youssef, E., Ali, S., Qudah, M.: Social impact of videos at new media platforms on the eLearning acceptance during the Covid-19. Inf. Sci. Lett. **11**(3), 913–923 (2022)
38. Tryfos, P.: Sampling methods for applied research. Text Cases, p. 440 (1996)
39. Habes, M., Ali, S., Salloum, S.A., Elareshi, M.: Digital media and students' AP improvement: an empirical investigation of social TV (2020)
40. Ali, S., Habes, M., Youssef, E., Adwan, M.N.A.: A cross-sectional analysis of digital library acceptance, and dependency during Covid-19. Int. J. Comput. Digit. Syst. **10**(1), 1415–1425 (2021). https://doi.org/10.12785/ijcds/1001125
41. Awang, Z.: Validating the measurement model CFA, Chapter 3. SEM Made Simple, pp. 54–74 (2015)
42. Golfashni, N.: Validity and reliability in social science research. Educ. Res. Perspect. **38**(1), 105–123 (2011)
43. Babbie, E.: Community-Based Research: Handbook for Native Americans, pp. 257–318 (1990)
44. Meyer, A., Walter, W., Seuring, S.: The impact of the coronavirus pandemic on supply chains and their sustainability: a text mining approach. Front. Sustain. **2**, 1–23 (2021). https://doi.org/10.3389/frsus.2021.631182
45. Boneva, M.: Challenges related to the digital transformation of business companies. In: 2018 Proceedings of the 6th International Conference, pp. 101–114 (2018)
46. Deveshwar, A., Rathee, R.: Challenges for supply chain management in today's global competitive environment. In: International Review of Business Research Papers (2021)
47. Nitsche, B.: Unravelling the complexity of supply chain volatility management. Logistics **2**(3), 14 (2018). https://doi.org/10.3390/logistics2030014

Economic and Political Challenges of Development in Ukraine Industry 4.0

Igor Fedun[1], Liudmyla Kudyrko[1], Oleksandr Shnyrkov[1](✉), Roman Bey[2], Mykhailo Yatsiuk[3], and Artem Syniuchenko[4]

[1] Department of World Economy, State University of Trade and Economics, Kyiv, Ukraine
{i.fedun,l.kudyrko,o.shnyrkov}@knute.edu.ua

[2] Sector of Scientific Bibliography and Biography Study of Institute of History of Agrarian Science, Education and Technique of National Scientific Agricultural Library of National Academy of Agrarian Sciences of Ukraine, Kyiv, Ukraine

[3] Institute of Water Problems and Land Reclamation of NAAS, Kyiv, Ukraine
mv_yatsiuk@ukr.net

[4] Department of Political Science, Taras Shevchenko National University of Kyiv, Kyiv, Ukraine

Abstract. The study found that the demand for technological transformation in Ukraine on the basis of Industry 4.0 is due to a number of factors associated with low economic and social performance of production models and the external sector that do not meet the conditions of the XXI century. Testing the results of Ukraine's international activities over the past five years according to the criteria of foreign economic security has indicated a permanent state of achieving critical levels of threats. The analysis showed that the investment resource of non-resident companies does not serve as a driver of technological innovation in Ukraine due to the low level of investment attractiveness of the country. The identification of changes that correspond to the principles of neo-industrial development has confirmed that they are not systemic in nature, but rather targeted. Modernization covers individual companies in different segments of the Ukrainian market, which belong to both traditional and new industries. Transformations related to Industry 4.0 in Ukraine are accompanied not only by economic problems, but also by the political dimension. The expected positive effects of modernization of production and its optimization on the other hand have a loss of workers in traditional industries of their jobs and income. This, in the conditions of weak financial capabilities of Ukraine as a state, leads to the deepening of the processes of precarization and impoverishment.

Keywords: Industry 4.0 · Technological structure · Migration · Ukraine · COVID-19 · Precarization · Unemployment

1 Introduction

The technological transformations that the world is currently experiencing are radically affecting the models and sources of economic growth in the world, changing the role and importance of traditional industries in ensuring employment and welfare of workers

B. Alareeni et al. (Eds.): ICBT 2022, LNNS 620, pp. 453–467, 2023.
https://doi.org/10.1007/978-3-031-26953-0_42

(Kergroach, S., 2017) [1]. This calls into question the system of social and social values that met the demands and nature of the industrial age. The driver of these processes in the last decade is the Fourth Industrial Revolution (Industry 4.0), which is based primarily on the information component (Vaidya, S., Ambad, P., & Bhosle, S., 2018) [2].The development of Internet, Internet of Things (IoT) (Lu, Y. J., and Cecil, J., 2016) [3], information and communication technologies (5G), stable communication channels, cloud technologies, the use of artificial intelligence based on large unstructured data (Big Data) and digital platforms provides the emergence of open information systems and global industrial networks that go beyond the individual enterprise and interact with each other. Such systems and networks have a transformative impact on all the sectors of the modern economy and business and bring industrial automation to a new fourth level of industrialization.

The fourth industrial revolution is no longer just a concept. Industry 4.0 standards are already being actively implemented, particularly in the real sector in the United States, the EU, some countries in the Middle East and Southeast Asia. The term "Industry 4.0" comes from the initiative of the financial and industrial complex and scientific circles of Germany as a driver in ensuring the international competitiveness of the country's industry through the use of "cyberphysical systems" Cyber Physical Systems - CPS. (Kagermann H., et al., 2016) [4]. Ukraine faces the same challenges, but in this sense there are more questions than answers.

It is important to emphasize that the transformations associated with Industry 4.0 have more than just an economic dimension. Their effects are increasingly manifesting themselves in social and political changes, beginning to be articulated by political parties and their leaders in view of the potential restructuring of the political electoral field of previous decades, the erosion of the class base of political parties, the apocalyptic consequences of losing traditional jobs E., & McAfee, A., 2014) [5].

In particular, back in 2001, Dick Morris in his article "Direct democracy and the internet" former adviser of Bill Clinton predicted that the potential changes in the Fourth Industrial Revolution through the Internet would be so dramatic that they would create the preconditions for building the potential of direct democracy and change not only the existing political system in most Western countries but also the form of government (Sparrow, J., 2017) [6].

For Ukraine, which has long been positioned on the economic map of the world as a supplier of traditional low-tech products of the third and fourth technological modes (raw materials and semi-finished products of metallurgy, chemical industries, agricultural food and agro-industrial products) it is important to overcome inefficient and consumption. Moreover, there is a critical lack of opportunities for extensive expansion of existing production, at least given the loss of production capacity (metallurgical, chemical) in eastern Ukraine due to hostilities, large-scale external labor migration, limited investment resources in traditional industries. The processes of destabilization of economic life and precarization in the modern conditions of Ukraine, on the one hand, reflect the general civilization processes and relate to technological change, the deployment of "Industry 4.0". On the other hand, they have local specifics, which only deepens the state of uncertainty with the corresponding negative consequences.

2 Source Review

Schwab, K., 2015), drew attention to the new stage of technological development of mankind [7], raising to a global level the discussion of this phenomenon in 2016 on the global WEF platform. Describing the logics of civilizational development in terms of technology, Schwab defined it as follows: the first industrial revolution involved the use of water and steam energy to mechanize production; the second is the use of electricity to create mass production; the third - electronics and information technologies are implemented to automate production; the fourth industrial revolution is based on the third, the digital revolution that has been going on since the middle of the last century. This industry is characterized by a fusion of technologies that blur the boundaries between the physical, digital and biological spheres. Thus, technology is changing the lives of present and future generations. No less promising changes have been analyzed in the book for several decades (Ross, A., 2017) [8]. We are talking about industries that will be the main drivers of economic and social change. How social, political, economic, socio-cultural life is being transformed under the influence of artificial intelligence, which is an important component of Industry 4.0 has been analyzed in a number of works (Fouad, Fekry, 2019) [9], (Saeed, Mousa, 2020) [10]. As the impact of Industry 4.0 on social development is still not unambiguously perceived, these issues are the subject of increased attention of scientists (Morrar et al., 2017) [11]. Will technological changes lead to institutional modernization, in particular in terms of macroeconomic and social policy of states? A number of European scholars are considering these topical issues (Smith J. et al., 2016; Bekbergeneva D., 2020) [12, 13].

Among Ukrainian researchers studying the development of Industry 4.0 and the digital economy, several priority areas should be identified.In the first, the sectoral aspects and prospects of implementing Industry 4.0 for industry are assessed (Chrysovatiy A.et al. 2018) [14]. The second direction involves the assessment of alternative models of implementation of the principles of Industry 4.0, variability of models of innovative development of the country, the possibility of both favorable and destructive changes associated with the formation of Industry 4.0 in Ukraine (Ilyashenko S., 2016) [15]. In other studies the paradigm of intensification of investment activity of the enterprises in the conditions of systemic crisis and falling markets the purpose of which is the developed industry with the Industry 4.0 (Buleyev I.P., Bryukhovetskaya N.Ye., 2019) [16]. There are quite a number of studies that assess the prospects, directions and mechanisms of the smart industry and the digital economy (Vishnevsky V. et al., 2018) [17]. Empirical research on the weak impact of digitalization on innovation and investment processes in Ukraine at the macro level has also been revealed (Tkalenko S. et al., 2021) [18].

Despite the fact that most Ukrainian studies on Industry 4.0 focus on the weak readiness of Ukraine's economy for technological change in the new millennium, the lack of investment support for systemic transformations of the manufacturing sector, in our opinion, no less important and urgent for modern Ukraine is the search for answers to the following questions:

- whether the existing technological structure of production has the potential for further functioning according to the criteria of international competitiveness and national security;

- which sectors and branches of the economy of Ukraine carry out modernizations that meet the principles of Industry 4.0;
- What challenges, political and economic in particular, will create the positive, at first glance, technological modernizations?

3 Purpose of Study

The aim of the article is to identify the latest challenges for Ukraine in the deployment of "Industry 4.0", identify the factors that constrain technological progress and assess the positive results of modernization processes in both traditional and new industries.

4 Methodologies

The study used the key provisions and principles of modern institutional and evolutionary methodological approach, based on both general and special methods of scientific analysis, namely: historical and logical - to systematize theoretical research on the development of Industry 4.0; method of dialectics of general, special and individual - in identifying common features and peculiarities of the development of Industry 4.0 in Ukraine in comparison with the world trends; structural factor - in disclosing the content, economic and political effects of transformational changes on the basis of neo-industrialization of the XXI century; statistical - in order to assess the quantitative characteristics of the readiness of the economy of Ukraine for technological modernization according to the criteria of economic security; expert assessments (in identifying the scale of migration flows from Ukraine, including latent, as a factor in weakening the staffing of the national economy to modernize the industry). To calculate economic security indicators as indicators of the effectiveness of the current structure of the national economy of Ukraine and its external sector a methodological approach of calculating the level of economic security according to guidelines №1277 has been used: "On approval of the guidelines for calculating the level of economic security of Ukraine" (approved by the Ministry of Economic Development and Trade Of Ukraine of October 29, 2013) [20].

5 Conclusions and Discussions

The priority of tasks related to the transformation of Ukraine's economic system in the context of Industry 4.0 is due to the fact that currently its technological basis does not meet the challenges facing the state in the XXI century. Experts of the Institute of Economic Forecasting of the National Academy of Sciences of Ukrainedetermined that about 60% of the volume of industrial production of Ukraine is the 3rd technological way, 38% - the 4th way. Higher technological systems - 5th and 6th - account for only 4%, while the 6th system, which determines the prospects for high-tech development in the future in Ukraine is virtually absent and is less than 0.1% (Mazaraki A., et al., 2021) [21].

At the time when the transition to the new sixth technological mode is beginning in the technological leaders, Ukraine has not yet overcome the initial stages of building the capacity of progressive technological modes.

The current low-tech structure of production and the external sector is characterized not only by low efficiency and socio-economic effects, but also directly creates risks for the realization of national economic interests and economic security of the state. To calculate the state of economic security, let's use the list of indicators (Table 1), which are regulated by the Guidelines for calculating the level of economic security, approved by the order of the Ministry of Economic Development of Ukraine from 29.10.2013 №1277.

We want to state that for most years, the level of economic security indicators in terms of its individual components indicates a high level of threats and corresponds to unsatisfactory and critical conditions. The exception is 2020 in terms of foreign trade security and the ratio of gross external debt to exports. We have a relatively good result

Table 1. Indicators of the state of economic security of Ukraine in terms of its foreign economic component in 2016–2020

Indicators of economic security by areas of their formation	2016	2017	2018	2019	2020
Foreign trade security					
Coefficient of coverage by export of import, times	0.86	0.81	0.83	0.85	0.97
Current account balance of Ukraine,% of GDP	-7.25	-9.71	-8.36	-7.60	-1.1
Debt security					
The ratio of gross external debt to exports,%	247	216	194	192	156
The ratio of gross external debt to GDP,%	121.7	103.9	87.7	79.2	80.8
Ratio of public and state-guaranteed debt to GDP,%	79.1	80.9	71.8	60.9	50.3
Currency security					
Gross international reserves of Ukraine, months of imports	3 , 0	3.5	3.0	3.2	3.9

The level of economic security is critical

The level of economic security is satisfactory

on the indicator of the number of potential months of import purchases, which should be at least 3.0. 2020 is especially successful in this sense.

The negative balance in foreign trade in goods and services in terms of the main items of the current account of the balance of payments is the result of complex effects of external and internal risks of destabilization for the economy focused on production of raw materials and semi-finished products with high volatility: reduction of domestic consumer and investment demand, rupture of interregional ties, reduction of public funding, narrowing of lending activity, etc. The events of the last seven years (2014–2021) in eastern Ukraine have led to the shutdown of a significant number of metallurgical, chemical, machine-building enterprises and coal mines in the region, which traditionally accounted for 20–25% of Ukrainian export. At the same time, the record harvest of grain crops and the liberalization of access of Ukrainian goods to EU markets are a stimulating factor for the economy.

In connection with the armed aggression of the Russian Federation against Ukraine and the conduct of large-scale military operations on the territories and settlements in various regions of Ukraine, the provision of conditions and the implementation of measures to adopt the accepted level of water security of the state have significant risks associated with unprecedented challenges and threats. Such a state really creates objective risks for the life of the population, threatens ecological, nuclear and hydrodynamic disasters, negative impact on the environment and problems and losses for the functioning of various economies.

Pointing to the potential of the investment sphere to bring about change in the direction of Industry 4.0, it should be noted that currently the volume of investment in the national economy to ensure the reproduction processes, according to experts (Khudoley V. Yu. et al., 2019) does not correspond to the desired level of 19–25% of GDP [24]. Figure 1 presents statistical data on the share of gross investment in nominal GDP of Ukraine. These data show that in Ukraine in recent years, investment is extremely insufficient to ensure the basic reproduction processes, even against the background of relatively small GDP, not to mention technological modernization on the basis of

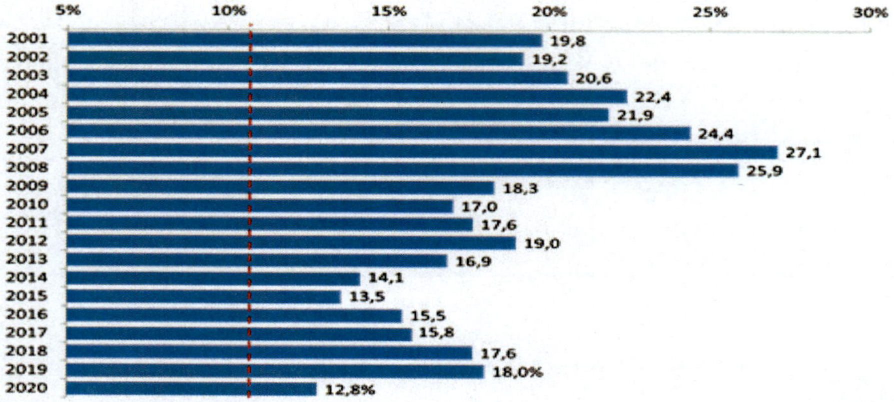

Fig. 1. Gross fixed capital formation in Ukraine in 2001–2020, % of GDP.

Industry 4.0. Trends over the past few years have shown increasing asymmetries due to cross-border flows of financial resources and capital movements.

The formation of the deficit was primarily due to the deterioration of the investment climate due to hostilities in the east of the country and the accumulation of a number of macroeconomic imbalances. This limited the private sector's involvement in both investment and credit resources to refinance current payments on previous loans. Foreign direct investment (FDI) in Ukraine, which in many recipient countries is perceived as a driver of technological innovation in the context of Industry 4.0, also has a negative trend [25]. Thus, their annual change in the balance of inflows and outflows from the country was much better in the period 2010–2013. In 2014–2015, they decreased significantly, and in the future could not recover. The outflow and inflow balance last year was minus $ 2.5 billion dollars. The total cumulative result for the end of 2020 amounted to 48.9 billion dollars, which is 30% lower than in pre-war 2013 for Ukraine and even lower than in 2014. (Fig. 2). All this indicates the lack of external investment prerequisites for the development of Industry 4.0 in Ukraine.

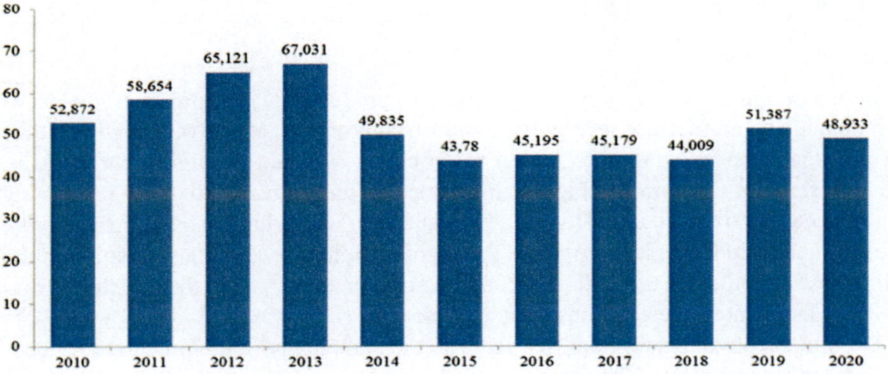

Fig. 2. Foreign direct investment in Ukraine (cumulative total), billion dollars USA

Indicative in the context of Ukraine's ability to meet the structural technological transformations of the XXI century is the international indicator - the index of industrial competitiveness UNIDO (English Competitve Industrial Performance index, abbreviated CIP), the index of readiness for the Fourth Industrial Revolution of the World Economic Forum (WEF).

Although Ukraine has made some progress on a number of components of the overall economic rankings during 2014–2019, including improving ease of doing business from 96th to 71st place and the index of economic freedoms from 162th to 147th place, these improvements have led to neither growth nor prosperity of the population. As for per capita income Ukraine remains the poorest country in Europe, and the gap in this indicator from our European neighbors is not narrowing, but growing! [26, 27] Instead, UNIDO argues that there is a close causal link between poverty and the level of development of the manufacturing industry. Therefore, it is worth looking critically at the positions of Ukraine and its neighbors on the components of **SIRI**. According to the **SIRI** report, the 2020 UNIDO report ranks Ukraine 69th out of 152 countries against 53rd place in 2010 (Table 2).

Table 2. Ratings of Ukraine and individual countries according to the UNIDO Industrial Competitiveness Index (SIRI)

Country	2020	2016	2015	2014	2013	2010
Poland	22	23	23	23	24	25
Slovakia	26	24	24	26	26	27
Hungary	27	23	26	27	28	29
Turkey	30	29	29	30	30	30
Romania	31	37	36	36	36	46
Bilorus	47	47	49	41	42	40
Ukraine	69	64	64	57	56	53
Number of countries in the ranking	152	150	150	144	144	135

In our opinion, the ability of the economy to produce finished products should be accompanied by an increase in exports of industrial products. Manufacturing industries that are unable to integrate into global value chains will not be highly competitive. The high share of medium and high-tech products in the value added of the processing industry characterizes the intensity of industrialization with a high level of productivity, innovation and technological progress. The significant loss of Ukraine's position in the UNIDO rating indicates the lack of successful structural changes in the transition from low-tech, labor-intensive activities to more high-tech.

Despite the obvious problems that indicate the complexity of the processes of technological modernization of the economy in the direction of Industry 4.0, in some sectors of the economy some positive changes can be observed. It should be emphasized that the success of individual companies is achieved not so much through state support, but in spite of institutional barriers.

Table 3 presents a list of industry leaders in Ukraine as for 2020 [28]. The list includes companies that have developed a new or improved existing product, technology or service, as well as those that use innovative approaches in business process management. Among the latter were innovations in production, logistics, management, finance and sales.

Experts evaluated companies on a 5-point scale on three parameters:

- innovation of the company's product: development and implementation of new or improved products, services, production technology.
- innovative approaches in business process management: in the organization production, logistics, personnel management, finance, sales, etc.
- the scale of the company's innovation: how widespread the innovation has become.

The analysis of the presented list indicates that it is not a question of one sector or market segment. Innovative leaders represent both industry and the sphere of finance, agro-industrial complex, telecommunications, etc. At the same time, in our opinion, two areas are significant, which belong to different technological systems, but which at the same time demonstrate truly impressive successes and results according to the criteria of Industry 4.0. We are talking about the IT sector and agro-industrial complex.

The rapid development of the IT sector is primarily due to external demand for the services of Ukrainian companies. If to treat the list of the largest Ukrainian IT companies in Ukraine, their achievements are obvious. For example, EPAM, which was founded in 1993, has offices in more than 25 countries. EPAM has a market capitalization of $ 18 billion as of 2020, headquartered in Newtown, Pennsylvania, USA. The company's annual turnover since 2019 exceeds $ 2.0 billion. Another well-known Ukrainian company SoftServe (1993) - the first known client of SoftServe was General Electric. Today, SoftServe is one of the largest software companies in Central and Eastern Europe. SoftServe has offices in the following cities: Kyiv, Kharkiv, Lviv, Dnipro, Rivne, Chernivtsi, and Ivano-Frankivsk. Abroad - Wroclaw, Warsaw, Poznan, Bialystok, Gliwice (Poland), Sofia (Bulgaria). There are business offices in the USA, Great Britain, Germany and the Netherlands. Location: head offices are located in Austin (Texas, USA) and Lviv.

Table 3. Rating of leaders of innovative companies of Ukraine for 2020

№	Name companies	Branch	The essence of innovation
1	Metinvest	Industry	Sheet rolling mill "1700" was modernized at the Ilyich MMC
2	DTEK Naftogaz»	Energy and oil and gas	DTEK Naftogaz is implementing a project to create digital field to increase efficiency gas production at great depths, as well as for development of difficult-to-recover gas reserves
3	Silpo	Retail	The company designs its stores in original concept, combining traditional retail with food court
4	AVK	FMCG	The company has changed markets and product range. It has entered markets of 60 countries over the last four years A pioneer in making snacks by extrusion method

(*continued*)

Table 3. (*continued*)

№	Name companies	Branch	The essence of innovation
5	MHP	APK	The company has implemented the largest innovation project - biogas direction of business. They have begun using chicken litter as raw material
6	Alfa-Bank Ukraine	Finances	Alfa-Bank in partnership with the open platform Innovation RE: It has launched online technology Alfa Digital platform
7	Taryan Group	Construction	Now Taryan Group is building Taryan Towers with two-storied glass bridges, a restaurant, a park, an area of innovation and entertainment on the roofs
8	Kyivstar	IT and Telecom	For online employee training via Coursera and Lynda. For middle-class managers gamified training program was introduced
9	Farmak	Pharmaceutics	Factory laboratories were transformed into intellectual clusters, where 35 candidates and doctors of science work
10	New Post	Transport and Logistics	In 2018–2019 three innovation terminals were put into operation in Khmelnytsky, Lviv and Kiev

Among other leaders of the IT industry with an international range of activities is GlobalLogic (2000), Luxoft (2000), Ciklum (2002), NIXSolutions (1994), EVOPLAY (2003), Infopulse (1991), ZONE3000 (1999).

Over the recent years there has been a trend of rising level of access of agricultural companies to advanced technologies. Thus, in their activities, small farms actively use quadcopters, robotic processes, various sensors and devices, although previously such technologies could only be used by large agricultural holdings. According to InVenture estimates, only 10% of agricultural enterprises in Ukraine use innovative technologies in their activities, and 20–30% of lands have the concept of precise farming.

One of the largest agricultural holdings in Ukraine, Ukrlandfarming, is actively implementing the achievements of Industry 4.0. The company has its own telemetry system "Stranger", which provides data collection from all the units of equipment using GPS-trackers, movement, speed, fuel use, engine load indicators. This system does not only ensure the harvest, but also prevents theft and violation of the rules of work in the fields. The economic effect of the system is about UAH 100 million. (0.3 tons per hectare of yield), and returns from detected thefts annually range from 5 to 10 million UAH [29].

Except Stranger system, the company uses remote photo-capture with the help of quadcopters or satellites, which allows assessing quickly the quality of the soil in a particular area and getting recommendations for fertilizing or treating plants from pests. Ukrlandfarming also uses technologies to control the depth of plowing, the process of preserving the crop in the elevator, loading it on the train or sending it to the port [30].

Myronivsky Hliboproduct Agricultural Holding, which owns one of the largest land banks in Ukraine, also focuses on innovative developments and actively uses them in the course of its own activities. The company successfully operates Farmmanagement systems, precise farming systems, remote sensing systems (remote sensing), GPS-monitoring systems of vehicles, agricultural production management, automation of production processes of database accumulation, structuring and analysis of information, satellite monitoring, use of unmanned aerial vehicles, geoinformation systems (for automated land bank registration).

In addition, the company uses solar energy, which provides lighting, heating and water heating for the checkpoint. In the future it is planned to install solar panels on poultry houses. The Myronivsky Plant for the Production of Cereals and Compound Feeds also has a boiler house that runs on biofuels: sunflower husks from oil press plants, which allows providing steam for technological processes and the company's own needs without the use of gas.

In 2017, Myronivsky Hliboproduct together with RadarTech and Agrohub launched the MHPA-ccelerator program, which aims to find, develop and integrate innovative projects in the agricultural sector [31].

Significant achievements in the implementation of elements "Industry 4.0" in its own activities is demonstrated by the agricultural company "Svarog West Group". The corporation consists of three companies that provide development and implementation of innovative solutions. Thus, in 2011 the corporation together with American partners created the LiveAG platform, the main product of which is precise farming technology - the system "Agro", which allows to manage agrobusiness online. The system involves the collection of data using sensors and trackers installed on the equipment, and direct transmission of the information to the owner [32].

Except the use of foreign innovative developments in their activities, Ukrainian companies and scientists are actively designing and implementing promising solutions for the agricultural sector based on elements of "Industry 4.0" (Table 4).

The technologies of domestic developers are of interest to both domestic and foreign partners, as they do not only reduce costs, use resources efficiently, improve yields, automate and control production processes in companies, but also have a lower price than global counterparts.

The steady progress of technological change has a multidisciplinary manifestation. It doesn't bring only new opportunities but also new challenges. In particular, it puts a number of questions to each nation-state as a political institution.

Yes, we are already talking about how to combat structural unemployment in the country and what will form a source of income for those segments of the population

Table 4. Digital projects in the agricultural sector of Ukraine

Name project	Characteristic
Bitrek	The company SPE "Disk Systems" manufactures equipment for GPS-monitoring and control of transport, which prevents theft grains
Jump Agro	Manufacture of devices for measuring soil moisture, air and soil temperature, speed and wind direction. The selected data are transformed into digital format and displayed in the electronic office of the agronomist in the mode of real time. The main consumers: Kernel, Ukrprominvest-Agro, and also the main importers: Australia, Canada, the Netherlands, Moldova, Germany, Poland
AgriEye	A platform based on satellites, UAVs, generates and analyzes the data from the fields and provides recommendations to the agronomist on the next steps
Kray technologies	Agrodrons for pesticides and soil fertilizers. The company performs orders for the farmers from the USA and Canada
SoilLines	Soil analyzer based on a microlaser, which provides information about the chemical soil composition and helps to determine which elements the soil shoul be fertilized with
Drone.ua	Production of unmanned aerial vehicles for agromonitoring, creation maps of fields, maps of plant litter, maps of differential fertilization
GlobalGIS	The company develops and implements geographic informational systems and technologies, data of remote sensing of the earth. The main customers of the company are: MHP, "Industrial Dairy Company", "Baryshivska Zernova company", "Green Valley", "Cousteau Agro"
AgriLab	The agroconsulting company that develops complex solutions to increase the efficiency of agricultural enterprises (technological expertise and diagnostics of lands, development of technological maps, modernization of agricultural machinery, quality control systems, technological processes in the distance)

whose professions will be replaced by digitalization and automation of production processes? In such circumstances, how to ensure political stability in the country and achieve the goals of sustainable development and social progress? It is obvious that new forms of employment, including hybrid, flexible, remote in Industry 4.0, which will replace the traditional, consistent with the realities of the twentieth century, modify the place and role of trade unions and ultimately the entire social protection system, increase uncertainty and precarization in Ukraine.

The process of precarization by its origins affects not only Ukraine but the world as a whole, the end of the twentieth century under the influence of a number of factors, both technological and migratory, cultural and so on. According to experts' and scientists' expectations, first of all, the Fourth Industrial Revolution has been a driver of precarization in the long run, due to systemic changes in production and socio-political life of modern societies. It is expected that a number of professions will lose their relevance as they will be replaced by job automation. We should also expect changes in the bureaucratic hierarchy formed in the previous decades and the collapse of the middle class. There will be workers under the contract with precarious employment; all this indicates a major change in the professional environment. At the same time, the last two years have shown increasing destabilization of public life in the world. This was largely due to negative trends in labor markets and in the system of contractual relations between employees and employers under the quarantine restrictions of COVID -19 and a significant reduction in demand for the latter in a number of professions and activities. In this sense, Ukraine faces an additional task - to fulfill its international obligations (COP-26, Glasgow 2021) and close its coal industry by 2030 as part of the decarbonisation strategy. And this is at least 35 thousand miners of state-owned enterprises, and most of those fired will relate to the private sector.

Equally important is the question of whether the regulatory political model in Ukraine is being transformed in the conditions of new technological opportunities? After all, it faces new challenges: it should preserve the principles and values of a democratic society with its inherent freedoms of economic mobility, and to some extent, counteract further risks of "intellectual" migration from Ukraine to richer countries. The latter are able, through material levers, to provide the desired international, often transcontinental redistribution of valuable highly qualified personnel. According to experts' estimations, the total numer of migrants from Ukraine working abroad after 1990 ranged from 0.8 up to 7 million persons [6], which was up to 25% of the labor market. Thus, through fully democratic instruments, in the absence of restrictions on labor migration, nation states are able to incur irreversible losses that will push their technological development back for decades?

The question also becomes relevant: how to counteract cybersecurity threats, which are quite real, in the conditions of large-scale digitalization of economic, political, cultural life of citizens and their states? After all, as the experience of other states has shown, they are able to influence directly the political choice of states both in the long run (Brexit phenomenon) and in the more predictable future (accusations of Russia in the election of US President Trump). In this context, the question of whether democracy as a product of civilization is able to preserve its nature and benefits under the influence of Industry 4.0? (Sparrow J., 2017) no longer looks inappropriate [34].

6 Conclusions

Analysis of certain aspects of the implementation of Industry 4.0 allows us to come to certain generalizations. The demand for technological transformations in Ukraine on the basis of Industry 4.0 is due to a number of factors related to low economic and social performance of production and external sector models, which were formed in the middle of the last century and do not meet the conditions of the XXI century. Testing the results of Ukraine's international activities according to the criteria of foreign economic security indicated a permanent state of achieving critical levels of threats. The international investment resource of non-resident companies does not serve as a driver of technological innovation in Ukraine due to the low level of investment attractiveness of the country. The identification of changes that meet the principles of neo-industrial development confirmed that they are not systemic, but rather targeted and cover individual companies in different segments of the Ukrainian market, relating to both traditional and new industries, Transformations related to "Industry 4.0". In Ukraine concern not only economic problems, but also have a political dimension. The expected positive effects of modernization of production and its optimization in turn have a loss of workers in traditional industries of their jobs and income. This, given the weak financial capabilities of Ukraine as a state, will lead to a deepening of the processes of precarization and impoverishment.

References

1. Kergroach, S.: Industry 4.0: new challenges and opportunities for the labor market. Foresight STI Govern. **11**, 6–8 (2017). https://doi.org/10.17323/2500-2597.2017.4.6.8
2. Vaidya, S., Ambad, P., Bhosle, S.: Industry 4.0-a glimpse. Proc. Manuf. **20**, 233–238 (2018)
3. Lu, Y.J., Cecil, J.: An IoT (IoT) - based collaborative framework for advanced manufacturing. Int. J. Adv. Manuf. Technol. **84**(5–8) (2016)
4. Kagermann H., Anderl R., Gausemeier J., Schuh G., Wahlster W.: Industrie 4.0 in a Global Context. Strategies for Cooperating with International Partners. Herbert Utz Verlag GmbH (2016). https://www.acatech.de/wp-content/uploads/2016/11/acatech_eng_STUDIE_Industrie40_global_Web.pdf
5. Brynjolfsson, E., McAfee, A.: The Second Machine Age: Work, Progress, and Prosperity in a Time of Brilliant Technologies, 1st edn. WW Norton & Company, New York (2014)
6. Sparrow, J.: Can democracy survive the fourth industrial revolution? Should it? (2017). https://www.theguardian.com/sustainable-business/2017/jan/11/can-democracy-survive-the-fourth-industrial-revolution-should-it
7. Schwab, K.: The Fourth Industrial Revolution. What It Means and How to Respond. Foreign Affairs (2015). https://www.foreignaffairs.com/articles/2015-12-12/fourth-industrial-revolution. [in English]. Accessed 12 Dec 2015
8. Ross, A.: Industry of the future. Per. with English Kyiv: Our format, 357 p. (2017)
9. Fouad, F.: The fourth industrial revolution is the AI revolution an academy prospective. Int. J. Bus. Inf. Syst. **8**, 155–167 (2019). https://doi.org/10.30534/ijiscs/2019/01852019
10. Mousa, S.: Industry 4.0: from technical to social innovation. Int. J. Adv. Sci. Technol. **29**(12), 2839–2855 (2020)
11. Morrar, R., Arman, H., Mousa, S.: The fourth industrial revolution (Industry 4.0): a social innovation perspective. Technol. Innov. Manag. Rev. **7**(11), 12–20 (2017)
12. Smith, J., Kreuztzer, S., Moeller, C., Karlberg, M.: Policy department a: economic and scientific policy – industry 4.0. European Parliament, EU. pp. 1–94 (2016)
13. Bekbergeneva, D.E.: Key directions of industry development 4.0 in modern conditions of digitalization of the economy. Econ. Sci. **185**, 61–65 (2020). https://doi.org/10.14451/1.185.61
14. Osadtsa, Yu.V.: The fourth industrial revolution: changing the direction of international investment flows: monograph/for science. p ed. Doctor of Economics, Prof. A.I. Krysovatoho and Doctor of Economics, Prof. OHM. Sokhatska. Ternopil, 478 p. (2018)
15. Ilyashenko, S.M., Ilyashenko, N.S.: Prospects and threats of the fourth industrial revolution and their consideration in the choice of strategies for innovative growth. Mark. Manag. Innov. **1**, 11–21 (2016). http://nbuv.gov.ua/UJRN/Mimi_2016_1_3

16. Buleev, I.P., Bryukhovetskaya, N.E.: Paradigm, models and mechanisms of activation of investment activity of business entities. Bus. Inform. **3**, 75–83 (2019). https://doi.org/10.32983/2222-4459-2019-3-75-8
17. Vyshnevsky, V.P., Vietskaya, O.V., Garkushenko, O.M., et al.: Smart industry in the digital economy: prospects, directions and mechanisms of development, 192 p. Institute economy prom - sti NAS of Ukraine, Kyiv (2018)
18. Tkalenko, S., Sukurova, N., Kudyrko, L., Litvin, N.: Digitization of investment-innovation development of Ukrainian economy: empirical analysis. In: Antipova, T. (ed.) ICCS 2020. LNNS, vol. 186, pp. 43–60. Springer, Cham (2021). https://doi.org/10.1007/978-3-030-66093-2_5
19. On approval of Methodical recommendations for calculating the level of economic security of Ukraine: order of the Ministry of Economic Development and Trade of Ukraine of October 29, vol. 1277 (2013). http://cct.com.ua/2013/29.10.2013_1277.htm
20. Mazaraki, A., Melnyk, T., Melnychenko, S., Kudyrko, L., Lositska, T., Pugachevska, K.: Import substitution potential in the conditions of digital transformation, vol. 164. PC TECHNOLOGY CENTER, Kharkiv (2021). https://doi.org/10.15587/978-617-7319-51-0
21. Export and import of Ukraine. https://index.minfin.com.ua/ua/economy/gdp/eximp/
22. Ukraine's foreign trade in goods and services in 2020. https://www.me.gov.ua/Documents/Detail?lang=uk-UA&id=76438c62-5e54-4931-ad98-7fe59d004a2a&title=ZovnishniaTorgivliaUkrainiTovaramiTaPoslugamiU2020-Rotsi&isSpecial=true
23. Khudoley, V.Yu., Ponomarenko, T.V., Kisilyuk, E.M.: Economic security in the conditions of the industrial revolution 4.0: case – Ukraine, 282 p. MNTU, Kyiv (2019)
24. Foreign direct investment. https://index.minfin.com.ua/ua/economy/fdi/2020/
25. Ukraine's position in international rankings in the field of industrial development. https://www.industry4ukraine.net/publications/pozycziyi-ukrayiny-v-mizhnarodnyh-rejtyngah-u-sferi-promyslovogo-rozvytku/
26. UNIDO, Competitive Industrial Performance Report, Edition 2020, Vienna 2020. https://stat.unido.org/content/publications/competitive-industrial-performance-report-2020
27. TOP-50 innovative companies of Ukraine. https://delo.ua/business/top-50-innovacionnyh-kompanij-ukrainy-361631/
28. Official site of PLC Ukrlandfarming. [Electronic resource]. https://www.ulf.com.ua/
29. Silicon fields. Ukrainian agricultural sector is becoming an IT leader. [Electronic resource]. https://economics.unian.net/agro/10251159-kremnievye-polya-ukrainskiy-agrosektor-vyhodit-v-it-lidery.html
30. Industry 4.0 in mechanical engineering: the situation in Ukraine and prospects for development. Analytical report. Association of Industrial Automation Enterprises of Ukraine. https://industry4-0-ukraine.com.ua/2018/10/18/analytical-report-industry-4-0
31. Svarog West Group: faster, cheaper, more accurate. [Electronic resource]. http://forbes.net.ua/ua/magazine/forbes/1416280-svarog-vest-grup-shvidshe-deshevshe-tochnishe
32. The Impact of Labor Migration on the Ukrainian Economy. https://ec.europa.eu/info/sites/info/files/economy-finance/dp123_en.pdf

Assessing the Service, Information, and Website Quality of the Opera Student Information System at the University of Business and Technology (UBT)

Mohammed Khouj(✉), Abdullah AlSharif, Abdulaziz AlObaid, Alaa Omar, Fekr Aazam, Majed AlGhamdi, Ziyad Durayi, and Mohammad Kanan

Jeddah College of Engineering, University of Business and Technology, Jeddah, Saudi Arabia
m.khouj@ubt.edu.sa

Abstract. This paper evaluates the service quality of the Opera website, which is a student information system that provides a variety of e-services to both students and faculty at the University of Business and Technology (UBT), Saudi Arabia. The purpose of this evaluation is to improve the service quality of the Opera website and to ensure that Opera users are satisfied with the service they receive. In this evaluation, a questionnaire with two sections (Importance and Performance) for 22 item statements based on six key dimension criteria and their associated indicators derived from the ISO 9126/2 standard was printed and distributed to 512 random samples. The reliability of the questionnaire was then evaluated using Cronbach's alpha, and the service quality of the Opera website was evaluated using SERVPERF Model and Importance – Performance Analysis. According to the results, users of the Opera website are satisfied with the overall service quality and consider the website to be user-friendly. In contrast, while using Opera, they occasionally encounter inaccurate results and technical difficulties.

Keywords: E-learning · SERVPERF · SERVQUAL · Opera

1 Introduction

In the mid-nineties, the first e-website service was initiated and spread worldwide. Thus, it started a new revolutionary wave in different sectors, such as business, finance, and economics (Sui and Rejeski 2002; UNCTAD/WTO and JEDCO 2001).

Since the beginning of the fourth industrial revolution, governments, businesses, education, banks, and many other sectors around the world have been working toward digitalizing their routine services. Along with the emergence of the COVID-19 pandemic, people around the globe realized the importance of being able to rely entirely on online services in all sectors, particularly during the full lockdown period. Furthermore, the educational sector has faced numerous challenges as it transitioned from traditional learning to full distance learning.

B. Alareeni et al. (Eds.): ICBT 2022, LNNS 620, pp. 468–479, 2023.
https://doi.org/10.1007/978-3-031-26953-0_43

Although online services have been implemented in many sectors, including the educational sector, in Saudi Arabia, the quality of the information associated with these services is poor, which negatively impacts the satisfaction of the services' beneficiaries.

Nowadays, academic websites have become one of the essentials factors that reflect on the university's reputation and rank, since it shows how advanced the university has become by showing the improvements of their own website, which provides different services for both faculty and students (users). Another indication that can be considered important is user satisfaction, which shows how successful the website has recently become (Rezaeean et al. 2012).

Opera is a student information system website at the University of Business and Technology (UBT) that accompanies UBT students during their educational journey: personal information, curriculum plans, course services, online payment services, advisories, schedules, grades, student letters, transcripts, service requests, and a variety of other online services. Faculty at UBT uses Opera to take attendance, post grades, advise students, and track their progress.

As an outcome of this paper, a website service quality assessment will be done of the Opera website utilizing ISO/IEC 9126-2 external metrics. In addition, SERVPERF (Service Performance) and Importance – Performance Analysis (IPA) will be used to obtain the service quality score and compare the importance and performance of the evaluated criteria on the Opera website using the IPA matrix. The primary goal of this paper is to evaluate the service quality of the Opera website to provide recommendations to the UBT IT department for improvement and maintenance.

2 Literature Review

In the recent past, the emphasis on assessing the quality of websites has been slightly lacking. However, the emergence of the COVID-19 pandemic in 2020 resulted in an almost complete reliance on online services in various sectors, specifically the educational sector. As a result, many researchers have published numerous papers on evaluating the quality of websites, particularly those related to education.

This section will demonstrate previous literature on website quality and used methods and models for assessing service quality.

Some researchers focused on specific criteria to evaluate the quality of university websites. Evaluating the quality of an academic information system (AIS) is vital to the development of any university, because it represents the management of all activities and operations held at the institution. One of the implementations to assess software quality is the novel model, which is used to verify if the implementation of an e-learning system will pass or fail in higher education programs (Al_Nawaiseh et al. 2020).

In Akgül (2020), all public and private Turkish universities' websites were evaluated using four different criteria: accessibility, usability, performance standards, and readability.

Moreover, IPAs and WebQual methods were used to evaluate university e-learning website. WebQual can convert user opinions into questionnaires, and the IPA method is used to manage the results of questionnaire data to generate IPA quadrants (Jundillah et al. 2019).

A usability study was conducted at twelve Saudi universities. The usability criteria included performance, load time, navigation, mobile friendliness, user satisfaction, SEO, accessibility, and security. The usability criteria were evaluated using three automated evaluation tools. The tools used are Web Page Analyzer, Qualidator, and Website Grader (Al-Omar 2017).

A similar study published by Roy et al. (2016) evaluated the usability of three academic websites from three different institutions. To determine the level of user satisfaction, the usability of these three academic websites was assessed using questionnaires. Analytic Hierarchy Process (AHP) was used to calculate the website usability based on the results of the usability attributes: attractiveness, controllability, efficiency, usefulness, and learnability.

ISO 9126 is the most globally known and implemented quality standard for specifying and evaluating software product attributes. A study was conducted to propose a quality model for evaluating e-learning software products. The study defined software product characteristics and integrated it with the ISO 9126. The characteristics are portability, maintainability, efficiency, usability, reliability, and functionality (Djouab and Bari 2016).

Another study developed a framework using ISO/IEC 9126 and the Kano model questionnaire to evaluate the quality of an academic website. Questionnaires were used to obtain students' opinions on the characteristics that can be assessed and improved on the Telkom university academic website. According to the ISO/IEC 9126 standard, the characteristics were functionality, reliability, and understandability (Suwawi et al. 2015).

In addition, Trichkova (2014) proposed a framework for evaluating the performance of an e-learning system in the field of medicine. Fifteen information technology experts were given questionnaires to give their opinions on the e-learning system, considering six target criteria including functionality, reliability, usability, efficiency, maintainability, and portability based on the ISO 9126 standard.

In a study performed by Asogwa et al. (2015), the SERVQUAL model was used to calculate the average gap between users' perceptions and their expectations. The criteria used were reliability, assurance, tangibility, empathy, and responsiveness.

Rasyida et al. (2016) combined SERVPERF and IPA in a study to assess the service quality of the service firms and identify what dimensions they must prioritize to attain customer satisfaction. The SERVPERF model was used to evaluate service quality by developing a questionnaire with two sections: importance and perception. Each section has 22 item statements. The 22 statements fall into five categories: tangibles, reliability, responsiveness, assurance, and empathy. While the IPA technique was used to prioritize service, attributes were based on importance and performance as defined by the SERVPERF model.

In 1977, the IPA matrix was presented for the first time to the marketing field by Martilla and James to assist targeting stakeholders to recognize and rate specific product or service characteristics, relying on the importance of the person who rates, and the effect on the overall performance of the organization. Using this matrix, management can gain insights into attributes that need and deserve to be improved, as opposed to those that have consumed excessive resources with little benefit to customer satisfaction (Prajogo and McDermott 2011).

Hasan (2014) investigated the relative importance of specific design criteria developed for the purpose of evaluating the usability of educational websites from the perspective of students, using a questionnaire that was distributed in nine different institutions. The criterion of the study includes navigation, architecture and organization, ease of use and communication, design, and content.

Others created frameworks with broad criteria that can be applied to all types of websites. For example, Hasan and Abuelrub (2011) aimed to create a theoretical, comprehensive, and measurable framework for assessing website quality to provide simple criteria to encourage improvements in website design and implementation. A general criterion is proposed for evaluating the quality of any website. Regardless of the service it provides. The dimensions of the criteria are content quality, design quality, organization quality, and user-friendliness.

Following a comprehensive literature review, we developed a framework for assessing the service quality for the Opera website system by implementing some of the ISO 9126/external metrics standards on the SERVPERF model and prioritizing the applicable criteria: functionality, reliability, usability, efficiency, maintainability, and user-friendliness using IPA.

3 Methodology

The service quality of the student information system "Opera" at the University of Business and Technology (UBT) was assessed. Six main dimension criteria and ten sub-dimensions derived from ISO 9126/2 standard were applied in a questionnaire and prioritized through Importance – Performance Analysis (IPA), and the service quality was calculated using the SERVPERF model. The main and sub-dimensions are provided in Table 1.

Table 1. Dimensions and Sub-dimensions

Dimensions	Sub-Dimensions
Functionality	*Suitability* *Accuracy* *Functionality Compliance*
Reliability	*Recoverability*
Usability	*Understandability* *Operability* *Attractiveness*
Efficiency	*Time Behaviour* *Efficiency Compliance*
Maintainability	*Maintainability Compliance*
User-Friendliness	

A questionnaire was printed and distributed to 512 random samples, including students and faculty, in both genders' campuses. However, 157 paper were disposed and removed due to either missing information or error occurred during inspection and evaluation, for that reason the only remaining questionnaires are 355. In addition, the questionnaire was divided into two sections: Importance and Performance. Each section consists of 22 item statements, including two extra general questions, and was provided in both English and Arabic. Students and faculty were asked to give their opinions on the importance of each item statement and evaluate the perception (performance) of the Opera website through a 1 to 5 Likert-scale rating. The 22 item statements including the two general questions are illustrated below:

1. How adequate are the functions evaluated on the Opera website?
2. How complete is the implementation of the Opera website functions in accordance with the expected specifications?
3. How correct is the implementation of the Opera website functions?
4. How stable is the functional specification of the Opera website after entering operations?
5. How acceptable do you find the differences between the actual and expected results on the Opera website?
6. How frequently do you encounter inaccurate results when using the Opera website?
7. How often do you encounter results with inadequate precision when using the Opera website?
8. How compliant is the functionality of the Opera website to your expectations?
9. How compliant are the Opera website interfaces to your expectations?
10. How do you assess the Opera website's availability for use during a specific period?
11. How capable is the Opera website of restoring itself following an abnormal event or at the request of a user?
12. How reliable is the Opera website in relation to expected standards?
13. How do you evaluate the Opera website's functions in terms of their ability to be understood correctly?
14. How do you evaluate the time needed to learn how to use the Opera website?
15. How consistent are the user interface features on the Opera website?
16. How attractive do you find the Opera website interface?
17. How do you find the time taken to complete a specified service in the Opera website?
18. How compliant is the efficiency of the Opera website to your expectations?
19. How compliant is the maintainability of the Opera website to your expectations?
20. How friendly do you find using the Opera website?
21. How Important do you think that opera website needs to be improved?
22. How frequently do you encounter technical difficulties while using Opera?

Questions 1, 2, 3, 4, 5, 6, 7, 8, and 9 fall under the functionality criteria, whereas questions 10, 11, and 12 are part of the reliability criteria. Questions 13, 14, 15, and 16 cover the usability criteria, questions 17 and 18 fall under efficiency, and the maintainability criteria is only found in Question 19. Question 20 is the only example for inquiring about the user-friendliness criteria. Lastly, Questions 21 and 22 are general questions.

Based on the SERVPERF model, by multiplying the calculated weight of each item using the perception score, the service quality (SQ) for each item is obtained. The weight for each item is calculated by subtracting the minimum importance score from the importance score of each item and dividing over the subtraction of the minimum importance score from the maximum importance score as illustrated below in Eq. (1) and (2):

$$SQi = \sum_{j=1}^{k} Wij \cdot Pij \tag{1}$$

$$Wij = \frac{Iij - Min}{Max - Min} \tag{2}$$

in which SQi is the service quality of an individual i, Wij is the weighting factor of item statement j to an individual i, and Pij is the individual i's perception of item statement j's performance.

Using the 22 item statements, the importance and performance of the criteria evaluated on the Opera website were assessed and compared in the IPA matrix using the IBM SPSS statistical software V25. The IPA matrix was originally presented as a two-dimensional matrix, with the x-axis representing "performance" (then defined as "customer satisfaction") and the y-axis representing "importance." In the IPA two-dimensional matrix, there are four quadrants.

The first quadrant indicates high performance and high importance and is called "Keep up the good work." Moving to the second quadrant, this consists of low performance but high importance and is named "Area of improvement." The third quadrant has low importance and low performance and is labelled "Low priority." The fourth quadrant consists of low importance and high performance and is called "Possible overkill" (Fig. 1).

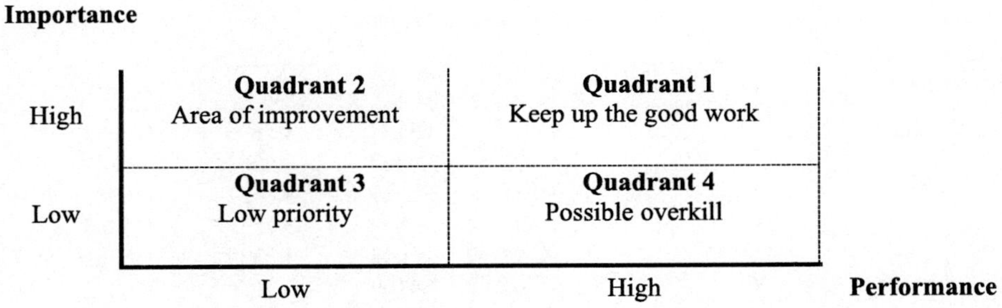

Fig. 1. IPA matrix

4 Results

Except for maintainability and user-friendliness, which have only one item statement each, Cronbach's alpha was calculated for the importance and performance of each criterion as well as the entire questionnaire using Eq. (3) to determine the questionnaire's reliability.

$$\alpha = \frac{K}{K-1}\left[1 - \frac{\sum S^2 j}{S^2 i}\right] \quad (3)$$

where K is the number of tested items, $\sum S^2 j$ is the sum of the item's variances, and $S^2 i$ is the variance of the total score.

As a result, Cronbach's alpha for all dimensions and items is above 0.70 except for the importance of usability dimension of 0.68, indicating that the questionnaire is reliable. The results for the reliability analysis are illustrated in Table 2.

Table 2. Reliability analysis

Dimensions	Cronbach's Alpha for importance	Cronbach's Alpha for performance
Functionality	0.73	0.79
Reliability	0.77	0.82
Usability	0.68	0.75
Efficiency	0.88	0.78
All Items	**0.79**	**0.89**

The average scores for the service quality, importance, performance, and weight for each item statement of all seven dimensions, as well as the extra questions for the 355 questionnaires, are shown in Table 3. Using Eqs. (1) and (2), the average scores for service quality (SQj) and weight (Wj) were calculated.

Table 3. Average scores per items

Dimensions	Questions	SQj	Ij	Pj	W
Functionality	1	2.3451	3.5127	3.7606	0.6282
	2	2.2549	3.4563	3.6563	0.6141
	3	2.4493	3.5549	3.8423	0.6387
	4	2.2979	3.3859	3.7831	0.5965
	5	2.3479	3.5042	3.7239	0.6261
	6	1.4296	3.4169	2.3634	0.6042
	7	1.5352	3.5746	2.3944	0.6437
	8	2.1958	3.4254	3.5972	0.6063
	9	2.2951	3.4366	3.7183	0.6092

(continued)

Table 3. *(continued)*

Dimensions	Questions	SQj	Ij	Pj	W
Reliability	10	2.3373	3.5775	3.6310	0.6444
	11	1.9620	3.4732	3.1408	0.6183
	12	2.1986	3.4338	3.6113	0.6085
Usability	13	2.4620	3.5577	3.8479	0.6394
	14	2.2866	3.3155	3.8986	0.5789
	15	2.4000	3.5493	3.7718	0.6373
	16	2.0338	3.3549	3.3831	0.5887
Efficiency	17	2.2725	3.4620	3.6817	0.6155
	18	2.1887	3.5070	3.4732	0.6268
Maintainability	19	1.7887	3.3493	3.0197	0.5873
User-Friendliness	20	2.5324	3.5296	3.9859	0.6324
Extra Questions	21	2.8042	3.6817	4.1606	0.6704
	22	1.8310	3.5887	2.8592	0.6472

Where the average scores for the service quality (SQ), importance, and performance for the seven dimensions including the extra questions are summarized below in Table 4.

Table 4. Average scores per dimensions

Dimensions	Average SQ	Average importance	Average performance
Functionality	2.1279	3.4742	3.4266
Reliability	2.1660	3.4948	3.4610
Usability	2.2956	3.4444	3.7254
Efficiency	2.2306	3.4845	3.5775
Maintainability	1.7887	3.3493	3.0197
User-Friendliness	2.5324	3.5296	3.9859
Extra Questions	2.3176	3.6352	3.5099

As per Fig. 2, the results of the 22 item statements are plotted on the previously mentioned quadrants. In the first quadrant "Keep up the good work," there are eight questions that fall under it: Q1, Q3, Q5, Q10, Q13, Q15, Q20, and Q21. Q7, Q18, and Q22 were in the second quadrant, which is known as "area of improvement." Moving on to the third quadrant, which is referred to as "Low priority," there are four questions: Q6, Q11, Q16, and Q19. Finally, Q2, Q4, Q8, Q9, Q12, Q14, and Q17 are located on the fourth quadrant, which stands for "Possible overkill."

This indicates that eight items' statements have high importance and performance, while three items' statements have high importance but poor performance.

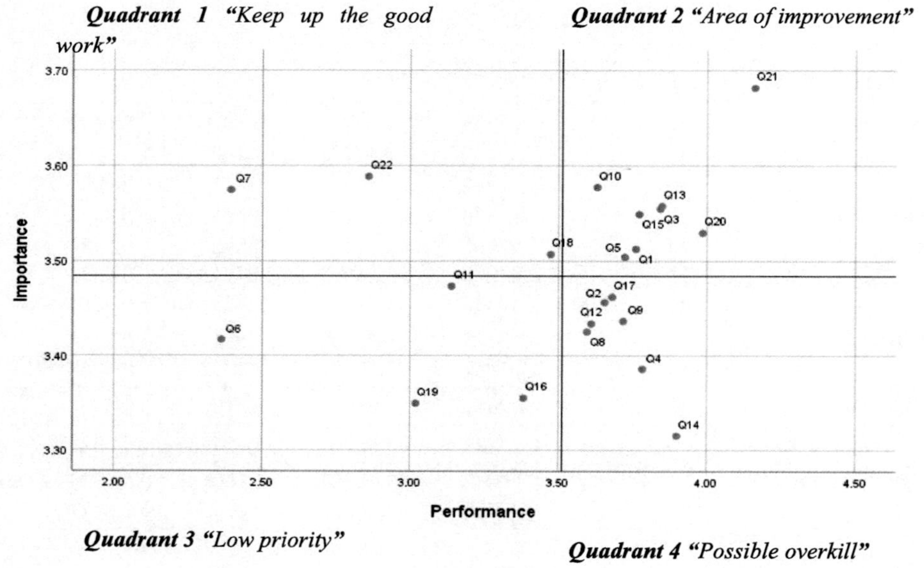

Fig. 2. IPA results

Four statements have low importance and performance, and seven statements have low importance and high performance.

5 Discussion

This section will discuss the highest and the lowest scores for the perception (performance) and importance amongst each dimension, as well as the maximum for the highest and the minimum for the lowest average scores.

Q3 of functionality, Q10 of reliability, Q14 of usability, Q17 of efficiency, and Q21 of the general extra questions have the highest average performance scores. Q21 with a value of 4.16 is identified as the maximum among them, which means that it is critical to continually update and improve the Opera website.

The questions that have the lowest average performance scores are Q6 of functionality, Q11 of reliability, Q16 of usability, Q18 of efficiency, and Q22 of the general extra questions. In addition, with an average score value of 2.36, Q6 of functionality scored the minimum value among the lowest average scores, indicating that Opera users rarely face inaccurate results while using the website.

Moreover, Q7 of functionality, Q10 of reliability, Q13 of usability, Q18 of efficiency, and Q21 of general extra questions have the highest average importance scores. Q21 has the maximum score with a value of 3.68, which means that Opera website users think that it is important to improve the website.

In contrast, the lowest average scores for the importance of each dimension are Q4 of functionality, Q12 of reliability, Q14 of usability, Q17 of efficiency, and Q22 of the general extra questions. However, with a value of 3.32, Q14 of usability is considered to be the minimum score among the mentioned dimensions. This means that Opera users find the Opera website is easy to use, and it is not important to evaluate the time needed to learn how to use the website.

Q19 of maintainability has an average score of 3.35 for importance and 3.35 for performance. In addition, the average Importance Score is 3.53, and the average performance score is 3.99 for Q20 of the user-friendly dimension. Q19 and Q20 were excluded from the previous comparison discussion since they each contained only a single question. Using the average score of the importance (Ij) and performance (Pj) for each item statement, the 22 item statements were plotted on the IPA matrix. As shown in Fig. 2, the IPA matrix's x-axis represents the performance "perception" of the Opera website for each item statement, while the y-axis implies the importance of each item statement to the Opera website's users.

Starting with the first quadrant, this contains "Keep up the good work," of questions Q1, Q3, and Q5 of functionality, Q10 of reliability, Q13 and Q15 of usability, Q20 of user-friendliness, and Q21 of the general extra questions. All questions are critical in terms of importance and performance, indicating that they should keep up high service quality in order to maintain the satisfaction of the Opera website's users.

Moving on to the second quadrant "Area of improvement," this is highly important and low in performance. As a result, the answers in this quadrant show a need for improvement to achieve high performance "perception" and therefore service quality. Q7 of functionality, Q18 of efficiency, and Q22 of the extra general questions are all questions that are plotted on this quadrant.

The low priority, which is the third quadrant, demonstrates the items that are not important to the Opera website's users and perform low. This indicates that the dimensions and the items that were included in this quadrant are not critical and do not need to be improved. The questions are: Q6 of functionality, Q11 of reliability, Q16 of usability, and Q19 of maintainability.

Finally, for the fourth quadrant, which is identified as "Possible overkill," the items located in this quadrant seem to perform high but are considered not important by the Opera users. The questions are: Q2, Q4, Q8, and Q9 of functionality, Q12 of reliability, Q14 of usability, and Q17 of efficiency. These items seem to be not important enough to be improved.

6 Conclusion

Continuous improvement and constant assessment of the service quality of university websites indicates how a university is capable of providing advanced e-services. On a daily basis, the Opera website provides a variety of e-services for both students and faculty. Hence, ensuring that Opera users are satisfied and receive high service quality is an integral part of the success of UBT e-services. This paper is an evaluation of the service quality of the student information system website Opera at the University of Business and Technology (UBT) using six key dimension criteria and their associated indicators, which are derived from the ISO 9126/2 standard as shown in Table 1. Based on these dimensions, a questionnaire was created including two sections (Importance and Performance) for 22 item statements and further distributed to 512 random samples.

The questionnaire is reliable, according to a reliability analysis that was performed by calculating the Cronbach's alpha for the importance and performance for each dimension, as well as the entire questionnaire. Further, the 22 item statements were prioritized using Importance – Performance Analysis and plotted on a scatter plot consisting of four quadrants that show how high or low the importance and performance for each item statement is. As per Fig. 2, the four-quadrant scatter plot demonstrates the items that provide high service quality and satisfy Opera users, items that need to be improved, low priority items, and items that perform high but are not important to be improved based on Opera user opinions.

As a result, Opera users believe that its features are adequate, accurate, and correctly implemented. Opera can be used whenever they choose. The features of the Opera website's user interface are consistent, and it is simple to understand how each feature works. Users of Opera perceive it to be user friendly. Contrarily, Opera users sometimes encounter inaccurate precision and technical issues when using the website, and regard Opera inefficient. Therefore, it is crucial for Opera users that the website is continually improved to address such issues.

Recommendations

It is recommended to the IT team at the University of Business and Technology to focus on improving the efficiency of the Opera website constantly, and particularly in the registration periods, to ensure that Opera users do not face any problems and receive inaccurate results. Hence, to maintain high service quality and user satisfaction.

Acknowledgement. There are not enough words to express our gratitude to our dear professors, Dr. Mohammad Khouj, dean of the College of Engineering, and Dr. Mohammad Kanan, vice dean of Scientific Research, for instructing us, extending college facilities and academic experience to the successful pursuit of our senior project thus far, as well as their endless support. Also, a special thanks to all who participated in the survey including students and faculty at UBT.

References

Akgül, Y.: Accessibility, usability, quality performance, and readability evaluation of university websites of Turkey: a comparative study of state and private universities. Univ. Access Inf. Soc. **20**(1), 157–170 (2020). https://doi.org/10.1007/s10209-020-00715-w

Al_Nawaiseh, A., Helmy, Y., Khalil, E.: New software quality model for academic information systems "case study e-learning system. Int. J. Sci. Technol. Res. **9**(01), 271 (2020). https://www.researchgate.net/publication/338987738_A_New_Software_Quality_Model_For_Academic_Information_Systems_Case_Study_E. Accessed 6 July 2022

Al-Omar, K.: Evaluating the internal and external usability attributes of e-learning websites in Saudi Arabia. Adv. Comput. Int. J. **8**(3/4), 1–12 (2017)

Anusha, R.: A study on website quality models. Int. J. Sci. Res. Publ. **4**(12), 1–5 (2014)

Asogwa, B., Ugwu, C., Ugwuanyi, F., Asadu, B., Ezema, J.: Evaluation of electronic service infrastructures and quality of e-services in Nigerian academic libraries. Electron. Libr. **33**(6), 1133–1149 (2015)

Debei, M.: The quality and acceptance of websites: an empirical investigation in the context of higher education. Int. J. Bus. Inf. Syst. **15**(2), 170 (2014)

Djouab, R., Bari, M.: An ISO 9126 based quality model for the e-learning systems. Int. J. Inf. Educ. Technol. **6**(5), 370–375 (2016)

Hasan, L., Abuelrub, E.: Assessing the quality of web sites. Appl. Comput. Inform. **9**(1), 11–29 (2011)

Hasan, L.: Evaluating the usability of educational websites based on students' preferences of design characteristics. Int. Arab J. Inf. Technol. **3**(3), 179–193 (2014)

Jundillah, M., Suseno, J., Surarso, B.: Evaluation of e-learning websites using the webqual method and importance performance analysis. In: E3S Web of Conferences, vol. 125, p. 24001 (2019). https://doi.org/10.1051/e3sconf/201912524001

Prajogo, D., McDermott, P.: Examining competitive priorities and competitive advantage in service organisations using importance-performance analysis matrix. Manag. Serv. Qual. Int. J. **21**(5), 465–483 (2011)

Rasyida, D., Ulkhaq, M.M., Priska, R., Setyorini, N.: Assessing service quality: a combination of SERVPERF and importance-performance analysis. In: MATEC Web of Conferences, vol. 68, p. 06003 (2016). https://doi.org/10.1051/matecconf/20166806003

Roy, S., Pattnaik, P.K., Mall, R.: Quality assurance of academic websites using usability testing: an experimental study with AHP. Int. J. Syst. Assur. Eng. Manag. **8**(1), 1–11 (2016). https://doi.org/10.1007/s13198-016-0436-0

SCC. Software engineering – Product quality – Part 2: External metrics (2022). https://www.scc.ca/en/standards/notices-of-intent/csa/software-engineering-product-quality-part-2-external-metrics. Accessed 6 July 2022

Suwawi, D.D.J., Darwiyanto, E., Rochmani, M.: Evaluation of academic website using ISO/IEC 9126. In: 3rd International Conference on Information and Communication Technology (ICoICT), pp. 222–227 (2015). https://doi.org/10.1109/ICoICT.2015.7231426

Trichkova, E.: ISO 9126 based quality assessment approach for e-learning system. Inf. Technol. Control **12**(1), 21–29 (2014)

Role of Remittance in Trade Deficit and Poverty Reduction - A Recent Account of an Asia Pacific Story – Bangladesh, India, Pakistan and Philippines

Hafizur Rahman[1,2](✉)

[1] Abu Dharr Gifari College, Dhaka, Bangladesh
hafi2ur@aol.com
[2] Rahman Foundation Inc, Silver Spring, MD, USA

Abstract. The concept of remittances got so much importance in the annals of modern history that United Nations General Assembly adopted June 16 as the International Day of Family Remittances (IDFR). It is estimated that, in 2014, around 80% of all global remittances went to developing countries – $436 billion out of a total of $583 billion – around double the amount of global development aid. The objectives of this paper is to examine the relationships of remittances among others, on GDP growth and more importantly on deficit (trade) decline and poverty reduction of India, Bangladesh, Pakistan and Philippines an Asia Pacific hub where around twenty four percent global population live.

Keywords: Human resource · Foreign aid · Savings · Development aid · Economic growth · Deficit financing · Poverty reduction

1 Introduction

"About one in nine people globally are supported by funds sent home by migrant workers. On average, migrant workers send between $200 and $300 home every one or two months" 3. The question among others, arises – Is remittance contributing enough to the GDP of the countries under study? Is remittance helping enough in the poverty reduction?. The other important question – did remittances play significant role in reducing balance of payment deficit for the counties under study? The research investigated responses as it quantified the cumulative effect of poverty reduction, the gap between poverty reduction and severity of poverty reduction and the dollar value of remittances in trade deficit reduction in the four countries. The study is expected to highlight policy insights and formulate achievable goals deliverables.

Executive Juris Doctor (EJD) (Concord Law School at Purdue University (Global)).
Alumnus Franklin Fellows – US Department of State.
President, Rahman Foundation Inc.

B. Alareeni et al. (Eds.): ICBT 2022, LNNS 620, pp. 480–494, 2023.
https://doi.org/10.1007/978-3-031-26953-0_44

2 Reviews of the Related Literature and Proposed Relationships in the Conceptual Model

Dean Naoyuki Youshino (a John Hopkins PhD whose thesis supervisor was Sir Alan Walters, economic adviser to former British Prime Minister Margarat Thacher) of Asian Development Bank Institute (ADBI) and coauthors Farhad Taghizadeh-Hesary, Miyu Otsuka in ADBI working paper series no 759 July 2017 made an in-depth scholarly analysis of remittances and the multiple variables. One of the important findings of their research is that a "1% increase in the international remittance flows as a percentage of the GDP can lead to a decrease in the poverty gap ratio of 22.6% and a decrease in the poverty severity ratio of 16.0%" which is extraordinary and an invaluable guide for the future researchers to conduct further study in this area.

In her article Shirin Akter employing time series data over the period 1980–2015 mentioned that there had positive effect of remittances on gross savings of Bangladesh and Philippines although there existed an insignificant negative impact for India while foreign aid had significant negative long run effects for all the three countries.

In their research paper Professors Mohammad Salahuddin and Jeff Cow concluded that there was a highly significant long-run positive relationship between remittance and economic growth (1977–2012) in Bangladesh, India, Pakistan and the Philippines. However, there was an insignificant positive association in the short term but overall positive results for the economic development of these countries.

In his article from 216 household's data Mr. Bezon Kumar asserted in his own words "I found that the level of poverty among remittance recipient households is notably lower than households that are not receiving remittances. Similarly, the probability of a household being poor is alleviated by 28.07% if the household receives remittance. It can be suggested that nursing international remittances can be useful for poverty alleviation in Bangladesh." Emigration is encouraged in the Muslim Holy Book Al-Quran vividly which goes "They (Angels) say, was not the earth of Allah wide so that you (could) have emigrated in it?" – An-Nisa – verse 97.

3 Methodology, Design of the Study and Data Collection Procedure

The study is based on secondary data primarily of international agencies (World Bank, International Monetary Fund, Asian Development Bank), country official sites as well as reputable websites The large body of data has been taken in order to verify data integrity for comparison and underlying analysis. A p-value is used using T test/Anova test whether or not there are enough evidence to reject the null hypothesis. It is used to see how likely different sets of data of India, Bangladesh, Pakistan and Philippines would have occurred under the null hypothesis ("there is no difference between certain characteristics of a population") of the statistical test or rejects the null hypothesis that relationships are significant (p value being less than 0.05). Excel Software was employed to calculate P-value & other statistical results throughout this paper.

4 Results and Data Analysis

4.1 A Glimpse Over Immigration

According to the United Nations, among ten countries that have the highest number of emigrants who born in that country and living abroad are these four countries included and ranked as follows:

Ranking	Country	# of Emigrants live abroad
1st	India	15.9 million
5th	Bangladesh	7.2 million
7th	Pakistan	5.9 million
9th	Philippines	5.4 million

Immigrants leave their home country for several reasons prominent among them are searching for economic prosperity, job opportunities, family reunification, retirement and better access to resources.

Source: Immigration By Country 2021 - https://worldpopulationreview.com/country-rankings/immigration-by-country.

4.2 Personal Remittances

Table 1. Personal remittances received (current US$ - in Million)

Year	Bangladesh	India	Pakistan	Philippines
1980	338.67	2757	2048	626
1981	381.01	2301	2067	800
1982	526.46	2618	2588	1049
1983	642.4	2660	2940	1124
1984	500.75	2295	2581	718
1985	502.5	Year	2537	806
1986	576.3	2240	2446	861
1987	747.8	2665	2181	1020
1988	763.6	2315	1872	1262
1989	758	2614	2017	1362
1990	778.9	2387	2006	1465
1991	769.40	3289	1549	1850

(*continued*)

Table 1. *(continued)*

Year	Bangladesh	India	Pakistan	Philippines
1992	911.80	2897	1574	2538
1993	1007	3523	1446	2587
1994	1161	5857	1749	3452
1995	1202	6223	1712	5360
1996	1345	8766	1284	4875
1997	1526	10331	1707	6799
1998	1606	9479	1172	5130
1999	1807	11124	996	6693
2000	1968	12883	1075	6924
2001	2105	14273	2001	8760
2002	2858	15736	2002	9735
2003	3192	20999	2003	10239
2004	3584	18750	3945	11468
2005	4315	22125	4280	13733
2006	5428	28334	5121	15496
2007	6562	37217	5998	16437
2008	8941	49977	7039	18851
2009	10521	49204	8717	19960
2010	10850	53480	9690	21557
2011	12071	62499	12263	23054
2012	14120	68821	14007	24610
2013	13867	69970	14629	26717
2014	14988	70389	17244	28691
2015	15296	68910	19306	29799
2016	13574	62744	19819	31142
2017	13502	68967	19856	32810
2018	15566	78790	21193	33809
2019	18364	83332	22252	35167
2020	21750	83149	26108	34913

Source: World Bank

Analysis: In this analysis "personal remittances will be defined as current and capital transfers in cash or in kind between resident households and non-resident households, and "take-home" compensation of employees earned by persons working in economies where they are not resident" as used by World Bank. Personal remittances received by

Bangladesh, India, Pakistan and Philippines from 1980 to 2010 are presented in Table 1 in detail. The data show growth of remittances received in Bangladesh was spectacular from $338.7 million in 1980 to $21.8 billion in 2020 (64 times) followed by Philippines from $626 million in 1980 to $34.9 billion in 2020 (56 times). During this period Indian remittances posted growth from $2.8 billion to $83.1 billion in 2020 (30 times). Pakistan posted an impressive remittance inflow of $26.1 billion in 2020 from $2.0 billion in 1980 (13 times). It is interesting to note that in 1980 the difference of remittances between Pakistan and India was only $709 million a ratio of 1:1.3 but in 2020 that ratio widened to 1:3.1. On the other hand in 1980 India received over eight times (8:1) higher remittances than Bangladesh. In 2020 that ratio declined to 4:1. Philippines' performance was also impressive. In 1980 ratio between Philippines & India was 1:4.4 which decreased to 1:2.38 in 2020. The analysis suggests that both Bangladesh & Philippines performed well relative to India while Pakistan lost its competitive edge compared to India.

4.3 Personal Remittances - Projected

Table 2. Personal remittances projected (current US$ - in Million)

Year	Bangladesh	India	Pakistan	Philippines
2021	19564	84740	24999	36224
2022	20528	87574	27008	37836
2023	21541	90504	29290	39578
2024	22603	93531	31892	41463
2025	23718	96660	34871	43505
2026	24888	99893	38297	45721
2027	26115	103234	42254	48130
2028	27403	106688	46846	50753
2029	28755	110256	52203	53615
2030	30173	113945	58484	56741
% change 2020–2030	39%	37%	124%	63%

Analysis: Other things remaining equal based on the data 2010–2020, projections are made to reflect tentative remittance trend in the above four countries. Projections reveal an interesting insight that in 2030 Pakistan will see the highest growth (124%) followed by Philippines (63%) in the next decade. Although India will remain as the highest remittance earner at $113.9 billion in 2030 it will attain a moderate growth of 37% over a period of 2020–2030. (Table 2 above). Bangladesh is expected to witness a growth of 39% during the same period and needs and its overall immigration worker program evaluated critically in order it to be a more competitive player as a prominent remittance earner.

4.4 Personal Remittances to GDP – Actual

Table 3. % of personal remittances to GDP – Actual

Year	Bangladesh	India	Pakistan	Philippines
2010	9%	3%	5%	10%
2011	9%	3%	6%	10%
2012	11%	4%	6%	9%
2013	9%	4%	6%	9%
2014	9%	3%	7%	10%
2015	8%	3%	7%	10%
2016	6%	3%	7%	10%
2017	5%	3%	7%	10%
2018	6%	3%	7%	10%
2019	6%	3%	8%	9%
2020	7%	3%	10%	10%
Average 2010–2020	8%	3%	7%	10%

Chart 1- % of Remittances to GDP

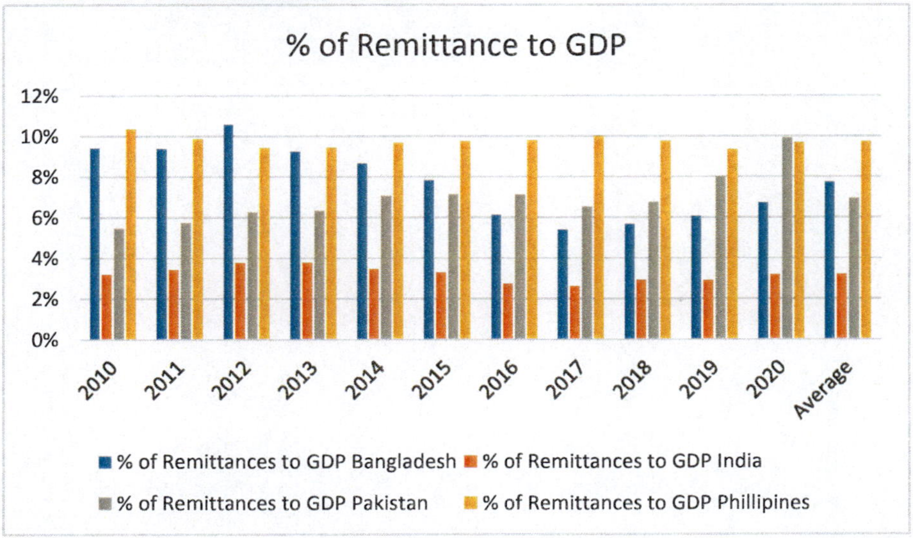

Analysis: Average contribution of remittances to GDP for the period 2010–2020 was highest for Philippines (10%) followed by Bangladesh 8%, Pakistan 7%. India's share to GDP was 3%. A recent trend shows (2020) that share of remittances to GDP was 10% for both Pakistan and Philippines, Bangladesh was trailing behind them at 7% (Table 3 & Chart 1 above). India expanded its economic vitality in other rewarding sectors like IT, manufacturing and service sectors to its advantages that led to a relatively lower share of remittance to GDP.

4.5 %Personal Remittances to GDP – Projected

Table 4. %Personal remittances to GDP – Projected

Year	Bangladesh	India	Pakistan	Philippines
2021	5%	3%	9%	10%
2022	5%	3%	9%	10%
2023	5%	4%	10%	9%
2024	4%	4%	10%	9%
2025	4%	3%	10%	9%
2026	4%	2%	11%	9%
2027	4%	2%	11%	9%
2028	3%	2%	12%	9%
2029	3%	2%	12%	9%
2030	3%	2%	13%	9%
Average 2021–2030	4%	3%	11%	9%

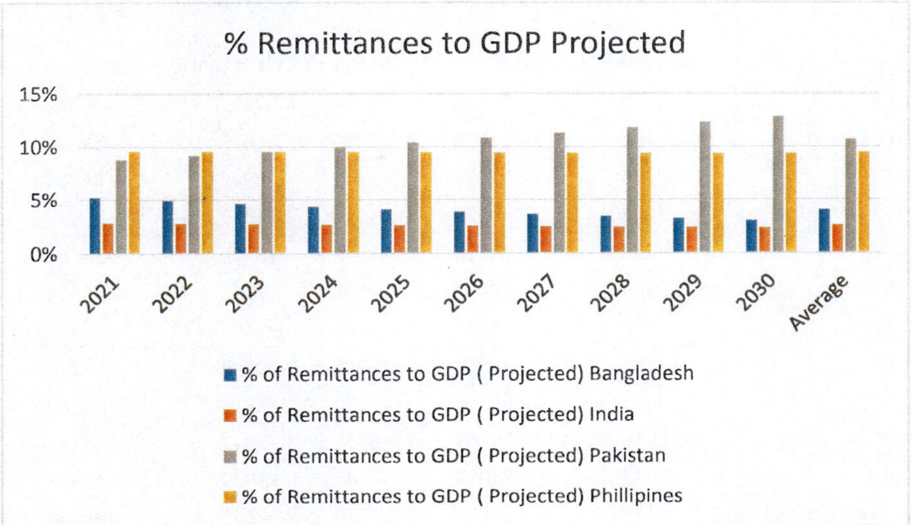

Chart 2

Analysis: Average contribution of remittances to GDP for the period 2021–2030 will be highest for Pakistan (11%) followed by Philippines 9%, Bangladesh 4%. India's share to GDP will be 3%. (Table 4 and Chart 2 above). % Remittances to GDP will see a declining trend in Bangladesh & India starting from 2024 attributable due to higher contributions of other sectors like industry, manufacturing and other service sectors. On the other hand both Pakistan & Philippines will have continued resilience of remittances for their share to the GDP.

5 Current Account Balance (Bop, Current US Million $) and Personal Remittances Received

Analysis: In Fig. 1 Anova Single factor test has been used. This tool performs a simple analysis of variance on data for two or more samples (here being four). The analysis provides a test of the hypothesis that each sample is drawn from the same underlying probability distribution against the alternative hypothesis that underlying probability distributions are not the same for all samples. Figure 1 also suggests that difference between the means of more than two groups is very significant (p value being 7.53E−10). The above tables & charts (Table 5 and 6, Chart 2) reveal that actual state of net balance of payments of four Asian countries Bangladesh, Pakistan, India and Philippines by year. Table 5 suggests that all the four countries witnessed deficits in their balance of payments position in each of the years 2017–2019 even though personal remittances received were included in the balance. The average (2010–2020) Bangladesh's deficit was lowest at $373 million, significantly lower than Pakistan ($6.2 Billion Dollar). India's average deficit exceeded $38.1 Billion dollars (India is 8 times larger than Bangladesh). Only the country in this analysis was Philippines that had an average surplus balance of $4.2

Table 5. Current account balance – in Million $ (actual)

Year	Bangladesh	India	Pakistan	Philippines	Total
2010	2109	−54516	−1354	7179	−46582
2011	−162	−62518	−2207	5643	−59244
2012	2576	−91471	−2342	6949	−84288
2013	2058	−49123	−4416	11384	−40097
2014	756	−27314	−3658	10756	−19460
2015	2580	−22457	−2803	7266	−15414
2016	931	−12114	−7191	−1199	−19573
2017	−5985	−38168	−16180	−2143	−62476
2018	−7095	−65599	−19959	−8877	−101530
2019	−2949	−29763	−8558	−3047	−44317
2020	1082	33007	245	12979	47313
Average 2010–2020	−373	−38185	−6220	4263	−445668

Source: World Bank

Anova: Single Factor

SUMMARY

Groups	*Count*	*Sum*	*Average*	*Variance*
Column 1	10	-5181	-518.1	12845726
Column 2	10	-453043	-45304.3	575410356
Column 3	10	-68668	-6866.8	40722496
Column 4	10	33911	3391.1	45430182

ANOVA

Source of Variation	*SS*	*df*	*MS*	*F*	*P-value*	*F crit*
Between Groups	15038248616	3	5.013E+09	29.731224	7.53E-10	2.866265551
Within Groups	6069678834	36	168602190			
Total	21107927450	39				

Fig. 1. Anova: Single Factor for analysis of variance on data for two or more samples (Bangladesh, India, Pakistan and Philippines)

billion. During the Pandemic year 2020 all the four countries registered surpluses in their balance of payments position (Bop). During this period the imports did decline significantly contributing to these favorable balances. When excluded personal remittances received from balance of payments position as shown under Table 6 in no year for any country under analysis was the balance a surplus. The combined cumulative deficit was $ 1.9 Trillion (2010–2020).When included remittances, the deficit reduced to a balance of $445.6 billion. The net effect o was $1.5 Trillion – that was funded from personal remittances: $ 164 billion in Bangladesh, $771 billion in India, $196 billion in Pakistan

Table 6. Current account balance – in Million $ (actual) – excluding personal remittances received

Year	Bangladesh	India	Pakistan	Philippines	Total
2010	−8741	−107996	−11044	−14378	−142159
2011	−12233	−125017	−14470	−17411	−169131
2012	−11544	−160292	−16349	−17661	−205846
2013	−11809	−119093	−19045	−15333	−165280
2014	−14232	−97703	−20902	−17935	−150772
2015	−12716	−91367	−22109	−22533	−148725
2016	−12643	−74858	−27010	−32341	−146852
2017	−19487	−107135	−36036	−34953	−197611
2018	−22661	−144389	−41125	−42686	−250888
2019	−21313	−113095	−30810	−38214	−203432
2020	−20668	−50142	−25863	−21934	−118607
Average 2010–2020	−15277	−108281	−24072	−25034	−1899303

Source: World Bank

and $322 billion in Philippines. The analysis suggests that role of remittances toward balance of payments in these Asia Pacific countries was very significant and in fact acted as the life blood for about one fourth of the world economy in terms population.

6 Examining Remittances Impact on Poverty Reduction

ADBI Working Paper Series No 759 July 2017 reveals a research result that "a 1% increase in the international remittance flows as a percentage of the GDP can lead to a decrease in the poverty gap ratio of 22.6% and a decrease in the poverty severity ratio of 16.0%." If we apply the above assumptions, then each of Pakistan, Bangladesh, & Philippines would have reduced the poverty gap completely (cumulative basis 2011–2020) and India 30% as seen in Table 7. Similarly as shown under Table 8 and Chart 3 poverty severity reduction would have been for Pakistan completely, Bangladesh 81%, Philippines 72% and India 21%.

Table 7. Poverty gap reduction (cumulative) based on ADBI assumption

Year	Bangladesh	India	Pakistan	Philippines
2011	21%	11%	27.2%	14.4%
2012	35%	7.8%	17.6%	13.4%
2013	−3.8%	1.4%	6.1%	16.8%
2014	14.7%	.5%	24.2%	15.0%
2015	3.6%	−1.6%	17.2%	8.2%
2016	−17.6%	−6.1%	4.2%	9.5%
2017	−0.7%	5.3%	0.3%	11.5%
2018	17%	8.2%	9.6%	6.5%
2019	21%	3.6%	8.6%	8.1%
2020	24%	−0.2%	33.0%	−1.6%
Cumulative (2011–2020)	114%	30%	148%	102%

Table 8. Poverty severity reduction (cumulative) based on ADBI assumption

Year	Bangladesh	India	Pakistan	Philippines
2011	15.19%	7.92%	19.27%	10%
2012	24.58%	5.53%	12.44%	10.0%
2013	−2.7%	0.99%	4.30%	12%
2014	10.37%	.33%	17.12%	11.0%
2015	2.53%	−1.12%	12.19%	6%
2016	−12.44%	−4.30%	2.95%	7%
2017	−0.46%	3.76%	0.19%	8%
2018	12.05%	5.82%	6.8%	5%
2019	14.80%	2.53%	6.09%	6%
2020	16.71%	−0.11%	23.40%	−1%
Cumulative (2011–2020)	81%	21%	105%	72%

Chart 3 – Poverty Severity Reduction (Cumulative) Based on ADBI Assumption

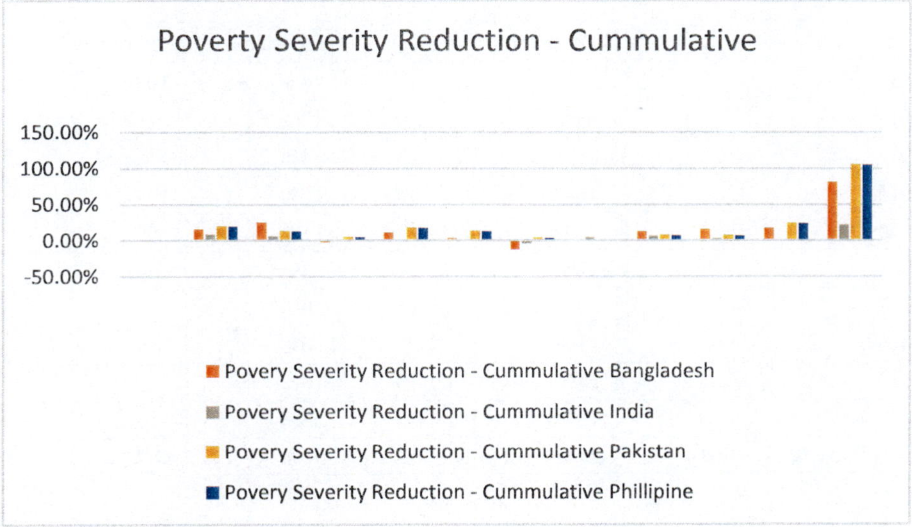

6.1 Examining Remittances Impact on Poverty Reduction Using a Moderate Assumption

Using a moderate assumption that "a 1% increase in the international remittance flows as a percentage of the GDP can lead to a decrease in the poverty gap ratio of 15% and a decrease in the poverty severity ratio of 10.0%. Following are the findings: Cumulative effects (2011–2020 Tables 9 and 10 and Chart 4 below) - poverty gave reduction - Bangladesh (76%), Pakistan (98%), Philippines (68%)" and India (20)% In the indicator poverty severity reductions the calculations show Bangladesh 50%, Pakistan (65%), Philippines 45% and India 13%. Given the reality of the economic conditions, it appears that the moderate assumptions generated a more acceptable result than ADB model.

Table 9. Poverty gap reduction (cumulative) based on the assumption - 1% increase in remittance to GDP = 15% decrease in poverty gap ratio

Year	Bangladesh	India	Pakistan	Philippines
2011	14%	7%	18%	10%
2012	23%	5%	12%	9%
2013	−3%	1%	4%	11%
2014	10%	0%	16%	10%
2015	2%	−1%	11%	5%
2016	−12%	−4%	3%	6%
2017	0%	4%	0%	8%
2018	11%	5%	6%	4%
2019	14%	2%	6%	5%
2020	16%	0%	22%	−1%
Cumulative (2011–2020)	76%	20%	98%	68%

Table 10. Poverty severity reduction (cumulative) based on the assumption - 1% increase in remittance to GDP = 10% decrease in poverty severity ratio

Year	Bangladesh	India	Pakistan	Philippines
2011	9%	5%	12%	6%
2012	15%	3%	8%	6%
2013	−2%	1%	3%	7%
2014	6%	0%	11%	7%
2015	2%	−1%	8%	4%
2016	−8%	−3%	2%	4%
2017	0%	2%	0%	5%
2018	8%	4%	4%	3%
2019	9%	2%	4%	4%
2020	10%	0%	15%	−1%
Cumulative (2011–2020)	50%	13%	65%	45%

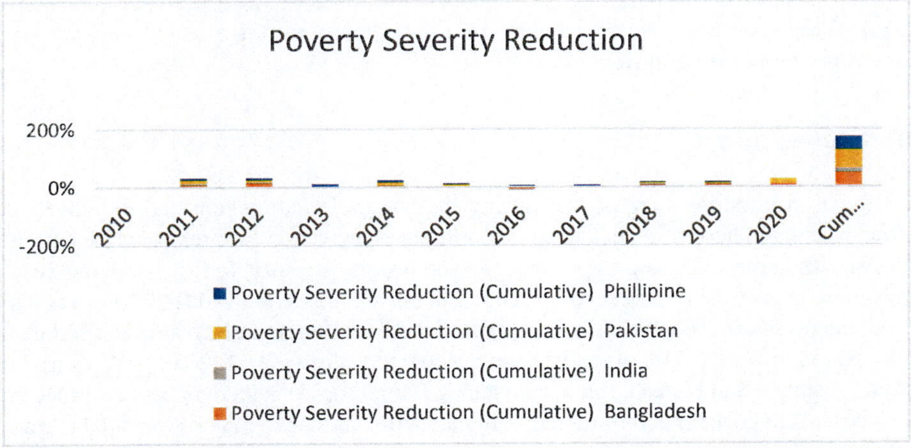

Chart 4

7 Interesting Facts

On the occasion of second International Day of Family Remittances, UN Reports observed that "between 2025 and 2030 $8.5 trillion is expected to be transferred by migrants to their communities of origin in developing countries. Of that amount more than $2 trillion – a quarter will either be saved or invested." UN reports also mentioned that around half of global remittances go the village areas where 75% of the world's poor and "food insecure live." The pandemic year 2020 witnessed a new innovation – accelerated adoption of digital technology by the migrant workers and their families. Among other things local currency depreciation took place in the recipient countries and governments in host countries increased their support.

8 Limitation

This study has been conducted based on the secondary data source. No survey was made and therefore no findings were based on the sample survey results. For projections purpose Excel software was used to generate data reported in this paper. No other programming language or modeling techniques was written in Excel to reflect the output that could be generated by using advanced Algorithms in order to make the data possible error free.

9 Recommendations

Although national governments across the world are taking necessary and specific measures to boost up their remittances and meet the underlying challenges, they should make regular and reliable disclosures on impact of remittances on each of savings, investment, foreign currency reserve, balance of payments and on poverty reduction so that concerted global policy actions can be coordinated more effectively for the greater benefits both to the host and home countries.

10 Conclusion

The above discussions suggest that during the last half century remittances played a pivotal role in all these four Asia Pacific countries, especially in the areas of trade deficit reduction and both reduction of poverty gap and poverty severity. In fact, applying even a moderate measure, poverty severity reduction during the period 2011–2020 showed a promising result.. In Pakistan it is estimated to be 65%, in Bangladesh 50%, in Philippines 45% and in India 13%. The other data above show that during 2011–2020, 79% of trade deficit together - Bangladesh, India, Pakistan & Philippines was financed by remittances received. It is believed that remittance resilience will continue to play its formidable role in the coming decade too, because the entire world is increasingly becoming more of a global village than a national cottage, Asia being in the lead. Technological advances and utilization of human resources are two giant forces will continue to play their respective dominant roles despite newer ramifications from other players like environment or business boost. Still then, impacts of remittances will remain significant.

References

Azad, A.K.: Importance of migrants' remittance for Bangladesh economy. Presented at the International Conference on Migrant Remittances: Development Impact and Future Prospects, London, 9–10 October 2003 (2003). https://web.worldbank.org/archive/website01040/WEB/IMAGES/O_A_AZAD.PDF

https://data.worldbank.org/indicator/BX.TRF.PWKR.CD.DT?locations=BD

Saritoprak, Z.: The Qur'anic Perspective on Immigrants: Prophet Muhammad's Migration and Its Implications in our Modern Society. https://jsr.shanti.virginia.edu/back-issues/vol-10-no-1-august-2011-people-and-places/the-quranic-perspective-on-immigrants/

Akter, S.: Do remittances and foreign aid augment the gross savings: Bangladesh, India and Philippines perspective? Int. Rev. Econ. **65**(4), 449–463 (2018). https://doi.org/10.1007/s12232-018-0305-z

Salahuddin, M., Gow, J.: The relationship between economic growth and remittances in the presence of cross-sectional dependence. J. Dev. Areas **49**(1) (2015). https://www.jstor.org/stable/24241287

Hassan, Z., Sirajal-Ud-Doulah, Md., Sathi, S.N.: https://www.researchgate.net/publication/338392504_Forecasting_the_Remittance_inflow_Based_on_Time_Series_Models_in_Bangladesh

https://www.un.org/development/desa/en/news/population/remittances-matter.html

https://www.migrationdataportal.org/themes/labour-migration

Kumar, B.: The Impact of International Remittances on Poverty Alleviation in Bangladesh. https://www.ceeol.com/search/article-detail?id=841090

https://d1wqtxts1xzle7.cloudfront.net/30940345/UNDP_-_BANGLADESH_MIGRATION_AND_REMITTANCES_080120-with-cover-page-v2.pdf?Expires=1636658102&Signature=ey1cQsn~ePfxJo5EBbT7U5Z7wKzMA-2YvLsRhxpcwn6fyYehcLV

Research Advances on Financial Technology: A Bibliometric Analysis

Zouaghi Adel[1](✉), Aznan Bin Hasan[1], Anwar Hasan Abdullah Othman[1], and Lammar Redhouane[2]

[1] Institute of Islamic Banking and Finance, IIU-Malaysia, Kuala Lumpur, Malaysia
zouaghi.adel@live.iium.edu.my, {haznan,anwarhasan}@iium.edu.my
[2] University of Tipaza, Tipaza, Algeria
lammar.redhouane@cu-tipaza.dz

Abstract. The term Fintech is a new term that is being used frequently now in the business and banking sector, which translates to financial technology. These technologies are being utilized and applied in the financial services sector and their intervention in mobile payments which includes money transfers, loans, fundraising, asset, and property management.

Several papers have been published to report the latest accomplishments and noted challenges faced in the financial technology field from different perspectives to address this need. Hence, a bibliometric study would be required to conduct a detailed examination of the current body of knowledge in financial technology research. To investigate the influence of financial technology on the financial industry, this paper searched a list of 764 research papers that were published between the period of 2015 and 2022. The Web Science (WoS), literature reviews, conducted systematic scientific, providing research for future research of the analysis results of co-citation and co-cited sources, disciplines, and keywords indicated a noted increase in the field publishing industry which has developed rapidly in recent years in numerous countries and interdisciplinary research.

Additionally, institutions in the United States, China, and British are adept at hosting such multidisciplinary work. Moreover, different keyword kinds show significant interactions in the visualization: (a) fintech, (b) blockchain and innovation, (c) performance impact, (d) information, and (e) financial inclusion.

Keywords: FinTech · Bibliometric analysis · Citation structure · Visualization networks

1 Introduction

The tech world started facing challenges after the last financial crisis when a growing number of start-up companies as well as some well-established tech names started developing new products and services regardless of their little or no background in the financial industry.

These new entrants were called fintech. Their rise was so dramatic that in 2008 fintech achieved about $1.4 in investments, and by 2013 it was doubled to $4 billion.

B. Alareeni et al. (Eds.): ICBT 2022, LNNS 620, pp. 495–508, 2023.
https://doi.org/10.1007/978-3-031-26953-0_45

In 2014, the amount had doubled again to nearly $12 billion. It was also estimated that investors in 2017 will pour approximately $20 billion into these companies (Bussmann 2017).

Another major factor that helped achieve the process of digital transformation in a major way is the penetration of mobile devices, as the first cash exchange operations date back to the end of the nineties when it became possible to mobilize call balances via mobile phones.

Fintech term refers to every company that occurs in this field; to propose to its clients creative or innovative technological solutions are up-start firms that attempt to capture market share at the expense of traditional market players in the financial services sector (Menad 2019).

Banks and other key players are attempting to invest to resist the competition of these new entrants. In other words, those who do not belong to the banking and financial sector (Hardie et al. 2016).

The concept of financial technology is not clearly defined, but it is closely related to information and communication technology. The term first appeared in the field of business to describe the challenge facing the financial sector due to the provision of faster and cheaper financial services. The term became a buzzword among private investors and institutions who invested more than $50 billion in the sector between 2010 and 2015 (Tan et al. 2019). China‘s forthcoming issuance and use of digital currency in the year 2020 by the China central bank may result in a revolution in the banking sector, and its impacts will extend to the rest of the world. In order to partly replace the usage of cash, the Chinese government has created a digital Yuan that is anticipated to be a blockchain offered by the People's Bank of China via a banking system. It is also very expected that it will spread and grow rapidly as the records of the distributed blockchain will make correspondent banks redundant.

This opens up new possibilities for faster and less costly payments to companies that annually remit about $124 trillion worldwide. With central banks in command centers, digital commissions can become a substitute for bank reserves, as well as banknotes, given that any holder of digital currency can have a deposit in the central bank, which may turn the state into a monopoly supplier of money for retail customers. Also, official digital currencies will allow central banks to facilitate monetary policies efficiently. As companies like Facebook move forward with research into the development of massive digital currencies, central banks around the world are being urged to look more seriously at what it means to issue their digital currency (Baker and Werbach 2019).

Many experiences in developing countries indicate that the use of financial technologies expresses a great deal of importance and effectiveness in providing financial and banking services for a group of countries that have less stringent banking systems, low-income countries, or low levels of financial competition. For example, Kenya's PESA-M, a popular digital remittance platform based on blockchain technology, allows payments, and provides mobile retail financial services to Kenyans, especially to the underserved in rural areas. The platform has achieved rapid and resounding success since its inception.

This success will inspire many developing countries to find new sources of growth and jobs. The leading Kenyan mobile phone company, Safaricom, has revolutionized the way Kenyans spend their money by operating the PESA-M platform (Mader 2016).

Bibliometrics, as defined by Wikipedia, is the statistical analysis of books, papers, and other publications, particularly in relation to their scientific content. Furthermore, the two subjects frequently overlap since they are so closely tied to scientometrics, a procedure that involves the examination of scientific metrics and indicators.

(Kamran et al. 2020) as the number of publications in the field of Fintech grows, therefore, by carefully collecting, categorizing, and assessing research articles connected to Fintech, it is necessary to present an overview of the research published on this subject. The research study investigates academic research on the issue of financial technology using bibliometric approaches to give insights to active academics and practitioners in the Fintech area. The web of science core collection database is used for bibliometric analysis in this research. Moreover, This paper aims to address challenges faced in the financial technology field by examining the current body of knowledge in financial technology research. It will look into a list of 764 research papers that were published between the period of 2015 and 2022.

The following parts of this article are structured as follows: Section Research Method and Questions discusses the research method and research questions. Section Methods and Materials deals with the methods and materials. Section Results and Discussion lays out our results and discussion. Section Conclusion and Limitation presents the conclusion and suggestions.

1.1 Research Questions

The major purpose of this study, as previously stated, is to perform a bibliometric analysis of papers indexed by the WoS core collection for researchers and practitioners. We set out to answer the following study questions to attain this goal:

1. How does the distribution of both financial technology publications and citations appear over recent years?
2. What have been the most influential Keywords, used in financial technology?
3. What is the authorship pattern of articles for a period of study?
4. Based on the number of citations, which are the most influential papers in financial technology?
5. What are the most frequent and popular publication venues and Countries for financial technology papers?

2 Materials and Methods

The research team relied on data from WoS (Falagas et al. 2008), a database from Thomson Reuters Corporation that is among the most popular in the academic community. In this work, we utilized WoS's search function to gather information by using the following search parameters: Database = Web of Science TM Core Collection database; Topic search = FinTech or "Financial technology", Timespan = 2015–2022.

Documents from the inception of WoS were combed through. In this way, 746 research articles were culled and exported as plain text files for further bibliometric study. All elements of the resultant papers are representative, including the title, abstract, keywords, citations, and references. The bibliometric research was performed using VOSviewer software.

Figure 1 shows the methodology's logical sequencing, as well as the search criteria used to find research papers.

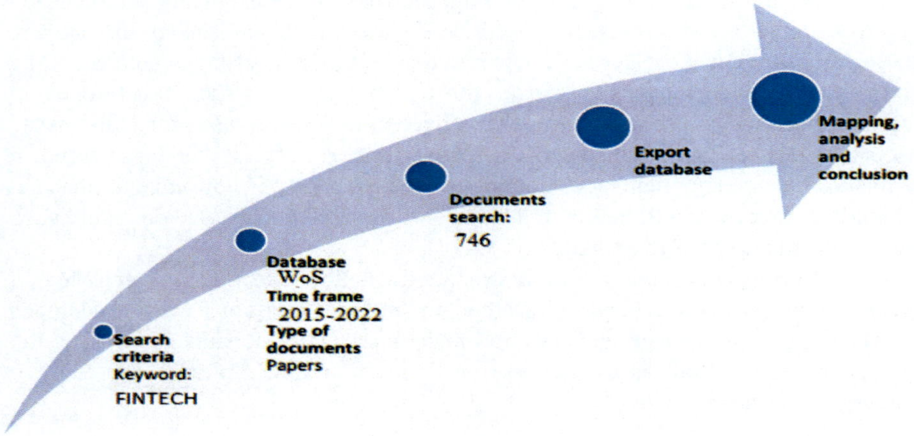

Fig. 1. Methodology

3 Results and Discussion

3.1 Scientific Output Evolution

Figure 2. This demonstrates that most of the study has been completed in the last few years.: 126 in 2019, 2021 in 202, and 297 in 2021. Why does the curve appear to be rising? It highlights also the distinctiveness of financial technology as a field of study within the financial management discipline. Web of Science database has been used for this research. A database may include a wide variety of articles, journals, conference papers, and book chapters. Articles and conference papers dominated the sorts of publications devoted to the study of financial technology, while notes, conference reviews, and letters accounted for a far smaller proportion. Based on the findings of the search, articles from 2015–2022 were found to be the most common kind of publishing. The trend has been on the upswing since 2015, and an increasing number of individuals are starting to take notice.

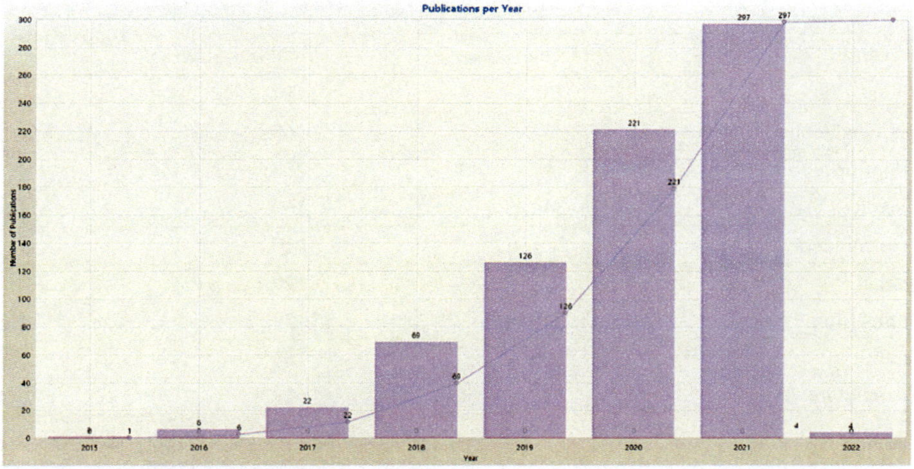

Fig. 2. Evolution of the number of articles during the study period

3.2 Keywords Analysis

The following table illustrates the 746 research articles that were published during the period between2015 and 2022 in the field of financial technology, they include 1354 keywords. Table 1 includes the 30 top frequently selected keywords, indicating the subjects in this study field that the authors believe are most important.

Table 1. Keywords on financial technology.

keyword	Occurrences	Total link strength
Fintech	343	1395
Blockchain	65	259
Innovation	65	366
Performance	55	282
Impact	51	250
Information	49	254
Model	48	173
Financial inclusion	45	199
Risk	43	216
Technology	41	273
Trust	41	272

(*continued*)

Table 1. *(continued)*

keyword	Occurrences	Total link strength
China	36	184
Financial technology	35	168
Adoption	34	268
Bitcoin	34	160
Determinants	34	210
Market	30	129
Competition	29	159
Credit	29	153
Artificial intelligence	28	113

Table 1 shows that the following keywords were used in most investigations, as shown in Table 1 Fintech, blockchain, innovation, performance, Impact, Information, etc.; This means that past research has concentrated on fintech financial technology.

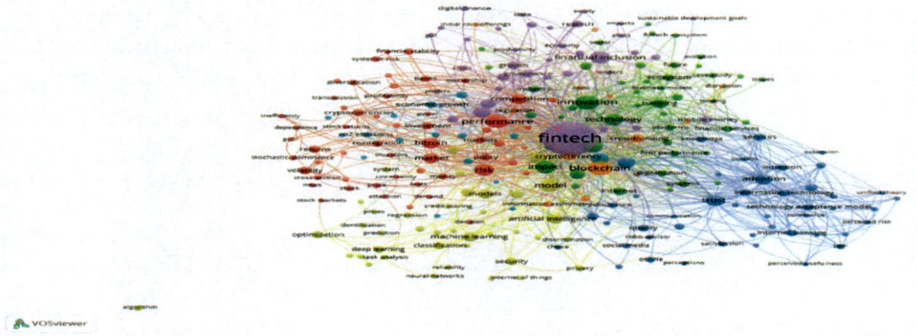

- Density

Fig. 3. Network of principal keywords, by co-occurrence

Figure 3 shows that there are 7 clusters in the network that the researcher can take in the field of financial technology as research thematic namely Fintech and its related clusters, blockchain and innovation, Performance, Impact and information, Adoption and information technology, Financial inclusion, artificial intelligence, and machine learning.

3.3 Network of Authors

- Network

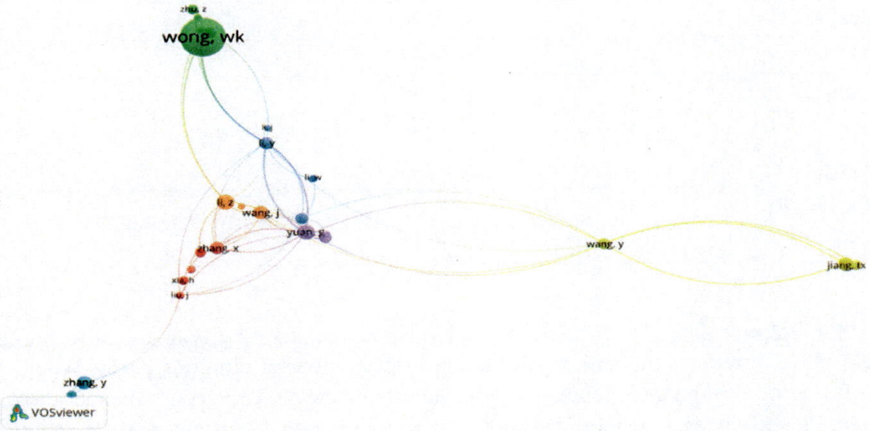

- Density

Fig. 4. Most important authors

Figure 4 shows the most productive researcher in the field of financial technology and the author with the highest number of citations.

Table 2. Most important authors

Author	Documents	Citations	Total link strength
Wong, Wk	46	246	56
Yuan, G	14	172	81
Yang, W	6	161	72
Han, J	10	81	21
Li, Y	11	77	45
Wang, X	5	69	0
Jiang, Tx	11	64	75
Zhao, Xl	11	64	75
Liu, D	5	50	8
Huang, Tz	8	47	62

(continued)

Table 2. (*continued*)

Author	Documents	Citations	Total link strength
Zhang, X	11	43	15
Chen, Z	5	40	0
Guo, X	5	40	17
He, F	7	40	7
Choi, D	7	39	0
Li, X	10	36	6
Wang, S	7	35	0
Buckley, Rp	5	31	6
Wang, J	12	31	20

Table 2 reveals a list of the most productive researchers in the financial technology field. Also, it reveals the author with the highest number of citations (246), Wong, Wk Professor of Computer Science, Oregon State University, who accomplished a total of 46 published research articles, followed by Yuan, G, with 14 articles and 172 citations. The very first appearance in this ranking is that of Yang, with 161 citations and 6 articles (Fig. 5).

3.4 Documents

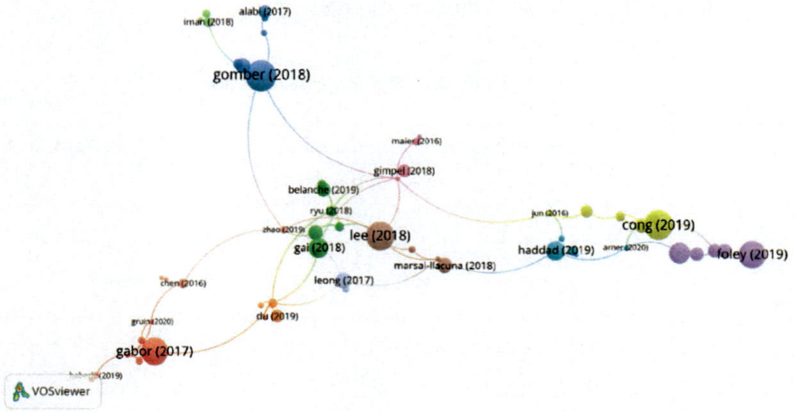

- Network
- Density

Fig. 5. Most important documents

Table 3: A list of WoS-indexed and ten most-cited financial technologies is provided. Furthermore, these publications are categorized depending on the average frequency of

Table 3. Most important documents

Document	Citations	Links
On the Fintech Revolution: Interpreting the Forces of Innovation, Disruption, and Transformation in Financial Services (Gomber et al. 2018)	142	5
Blockchain Disruption and Smart Contracts (Cong and He 2019)	131	4
Industrial Artificial Intelligence 4.0-based manufacturing systems (Lee et al. 2018)	130	5
Why do businesses go to crypto? An empirical analysis of initial coin offerings (Adhami et al. 2018)	125	0
financial inclusion digital revolution: The international development in the fintech era. (Gabor and Brooks 2017)	123	4
To FinTech and Beyond (Goldstein et al. 2019)	122	2
Challenges for Islamic Finance and banking in the post-COVID era and the role of Fintech (Hassan et al. 2020)	120	0
Financial Inclusion and Fintech during COVID-19 Crisis: Policy Solutions (Ozili 2020)	96	0
Blockchain Technology: Transforming LibertarianCryptocurrency Dreams to Finance and Banking Realities, (Eyal 2017)	91	0
Fintech, regulatory arbitrage, and the rise of shadow banks, (Buchak et al. 2018)	88	3
A survey on FinTech, (Gai et al. 2018)	83	4
The emergence of the global fintech market: economic and technological determinants, (Haddad and Hornuf 2019)	80	3
Fintech, Credit Market Competition, and Bank Risk-Taking, (Tseng and Guo 2018)	69	0
Nurturing a FinTech ecosystem: The case of a youth microloan startup in China, (Leong et al. 2017)	64	4
Future living framework: Is blockchain the next enabling network. (Marsal-Llacuna and Change 2018)	61	1

citations they get each year (as shown in the rightest column of Table 2). "On the Fintech Revolution: Interpreting the Forces of Innovation, Disruption, and Transformation in Financial Services", authored by Peter Gomber, Robert J. Kauffman, Chris Parker, and Bruce W. It is worth noting here that Weber's work has received the most citations, at 142 (to the date of conducting this research). In 2018, it was published in the Journal of Management Information Systems. Furthermore, this work has received the highest citations on average and is regarded as one of the most referenced studies done in China. Based on the average amount of citations, Financial Studies Review has published the most highly referenced publications.

3.5 Institutions and Countries' Productivity

This section investigates the outcomes taken for institutions and countries, by using indicators such as their productivity, rates of cooperation, collaborative networks based on co-authorship, and international collaborations.

Table 4 lists the twenty most productive institutions. As in the case of individual researchers, there is a predominance of institutions located in Asia, followed by the USA and Europe.

- Network

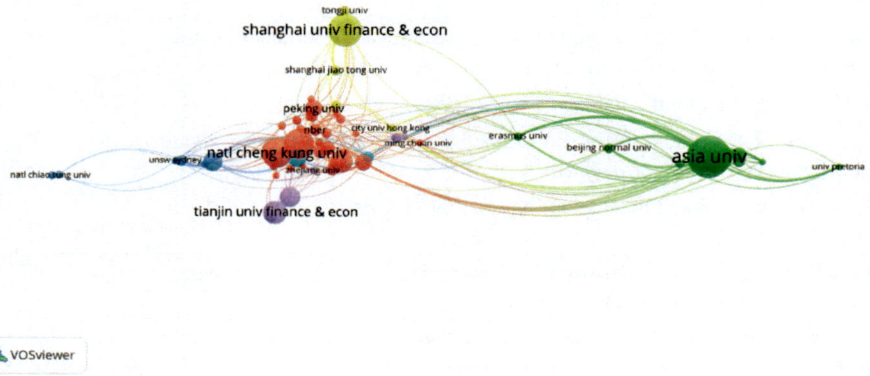

- Density

Fig. 6. Most productive institutions.

Table 4. Ranking of the most-twenty productive institutions.

Organization	Documents	Citations	Total link strength
The National Bureau of Economic Research, USA	9	386	29
The University of Sydney	15	282	97
Asia University, Australia	48	260	368
Shanghai University of Finance and Economics, China	34	214	66
Singapore Management University, Singapore	11	185	16
China med univ Hosp	35	184	315

(*continued*)

Table 4. (*continued*)

Organization	Documents	Citations	Total link strength
The University of Shanghai for Science and Technology, China	16	173	56
Columbia University, USA	6	167	12
Cornell University, USA	6	166	12
Southwestern University of Finance and Economics, CHINA	30	156	110
UNSW Sydney, Australia	6	137	24
Shenzhen University, China	21	136	11
Peking University, China	15	129	22
Sungkyunkwan University, South Korea	6	128	15
The Hang Seng University of Hong Kong, Hong Kong	34	118	299
The Hang Seng University of Hong Kong	12	113	123
Lingnan University, Hong Kong	11	113	123
Shanghai Jiao Tong University, China	8	110	20
University of Essex Colchester Campus, England	5	108	12
City University of Hong Kong	6	101	26

3.6 Countries

Figure 7 shows the collaboration map between the most important countries, The colors show the international networks, while the size of the bubbles reflects the number of articles published. In total, four international cooperation networks are identified (Table 5).

Figure 6 and Fig. 7 reveal the visual information of the bibliographical coupling network. Among them, CHINA (a total of 330 articles) made the largest contribution, and major organizations including Shanghai University of Finance and Economics, China Southwestern University of Finance and Economics, China, China med univ Hosp; Shenzhen University, China, then, the U.S.A comes second (141 articles in total). The main agencies are Columbia University, and Cornell University, USA. National Bureau of Economic Research, USA, the third in the United Kingdom (113 articles in total), with major agencies including the University of Essex Colchester Campus, England.

- Network

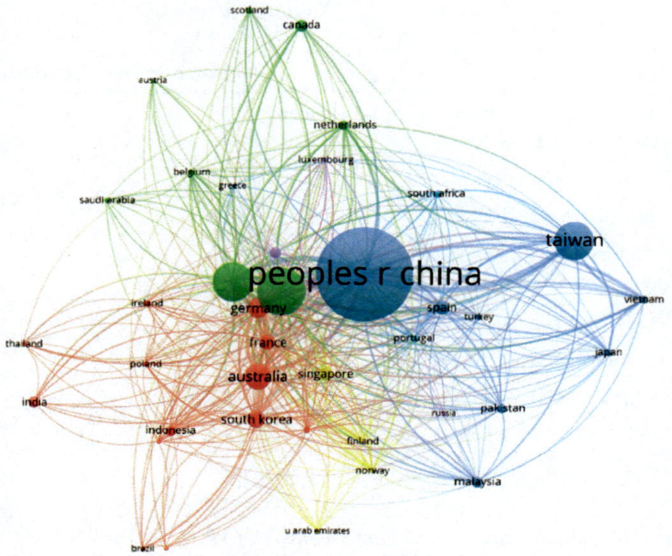

Density

Fig. 7. Most important countries

Table 5. Most important countries

Country	Documents	Citations	Total link strength
USA	141	2050	743
China	330	1861	1041
England	113	1106	622
Australia	67	620	449
Germany	49	525	375
Taiwan	109	513	424
France	35	505	371
South Korea	41	461	226
Singapore	30	232	167
Sweden	12	227	89
Canada	26	212	64

(*continued*)

Table 5. (*continued*)

Country	Documents	Citations	Total link strength
Spain	30	211	185
Denmark	6	210	58
Italy	24	203	116
Switzerland	21	187	113
Netherlands	20	149	129
Vietnam	16	98	165
Indonesia	21	88	69
Turkey	12	86	51
Japan	15	77	97

4 Conclusion and Future Work

According to this bibliometric study, researchers' interest in the field of financial technology is growing. The number of citations to relevant research papers published has been substantially increasing in recent years.

The most influential paper is "On the Fintech Revolution: Interpreting the Forces of Innovation, Disruption, and Transformation in Financial Services" written by Peter Gomber, Robert J. Kauffman, Chris Parker & Bruce W. Weber. Was published in 2018 in the Journal of Management Information Systems. This paper has obtained the highest citation average as well per year. Review of financial studies, journal of management information systems, and journal of economics and business are the most popular venues, based on the highest number of publications.

5 Limitations of Research

In general, the findings in this work are important for the next step and motivate researchers to continue their research. However, given the information in this paper is limited to the Web of Science, which limited research in articles; other databases, such as ScienceDirect or Scopus, may be suggested in future bibliometric analyses. And the search keywords are all linked to FinTech, it will need to be expanded in the future. We will devote more time and resources to FinTech innovation research and development.

Acknowledgements. The authors would like to express their appreciation to the reviewers and editor for their valuable suggestions and comments.

Conflict of Interest. They have no conflict of interest.

References

Adhami, S., Giudici, G., Martinazzi, S.: Why do businesses go crypto? An empirical analysis of initial coin offerings. J. Econ. Bus. **100**, 64–75 (2018)

Buchak, G., Matvos, G., Piskorski, T., Seru, A.: Fintech, regulatory arbitrage, and the rise of shadow banks. J. Financ. Econ. **130**(3), 453–483 (2018)

Bussmann, O.: The future of finance: fintech, tech disruption, and orchestrating innovation. In: Francioni, R., Schwartz, R.A. (eds.) Equity Markets in Transition, pp. 473–486. Springer, Cham (2017). https://doi.org/10.1007/978-3-319-45848-9_19

Cong, L.W., He, Z.: Blockchain disruption and smart contracts. Rev. Financ. Stud. **32**(5), 1754–1797 (2019)

Eyal, I.: Blockchain technology: transforming libertarian cryptocurrency dreams to finance and banking realities. Computer **50**(9), 38–49 (2017)

Falagas, M.E., Pitsouni, E.I., Malietzis, G.A., Pappas, G.: Comparison of PubMed, scopus, web of science, and Google scholar: strengths and weaknesses. FASEB J. **22**(2), 338–342 (2008)

Gabor, D., Brooks, S.: The digital revolution in financial inclusion: international development in the fintech era. New Polit. Econ. **22**(4), 423–436 (2017)

Gai, K., Qiu, M., Sun, X.: A survey on FinTech. J. Netw. Comput. Appl. **103**, 262–273 (2018)

Goldstein, I., Jiang, W., Karolyi, G.A.: To FinTech and beyond. Rev. Financ. Stud. **32**(5), 1647–1661 (2019)

Gomber, P., Kauffman, R.J., Parker, C., Weber, B.W.: On the fintech revolution: interpreting the forces of innovation, disruption, and transformation in financial services. J. Manag. Inf. Syst. **35**(1), 220–265 (2018)

Haddad, C., Hornuf, L.: The emergence of the global fintech market: economic and technological determinants. Small Bus. Econ. **53**(1), 81–105 (2019)

Hardie, S., Wood, J., Gee, D.: Mapping the fintech bridge in the open source era–fintech disruptors report. MagnaCarta Commun. **28**, 2017 (2016)

Hassan, M.K., Rabbani, M.R., Ali, M.A.M.: Challenges for the Islamic Finance and banking in post COVID era and the role of Fintech. J. Econ. Coop. Dev. **41**(3), 93–116 (2020)

Lee, J., Davari, H., Singh, J., Pandhare, V.: Industrial artificial intelligence for industry 4.0-based manufacturing systems. Manuf. Lett. **18**, 20–23 (2018)

Leong, C., Tan, B., Xiao, X., Tan, F.T.C., Sun, Y.: Nurturing a FinTech ecosystem: the case of a youth microloan startup in China. Int. J. Inf. Manag. **37**(2), 92–97 (2017)

Mader, P.: Questioning three fundamental assumptions in financial inclusion (2016)

Marsal-Llacuna, M.-L.: Future living framework: is blockchain the next enabling network? Technol. Forecast. Soc. Change **128**, 226–234 (2018)

Menad, R.: L'impact des Fintechs sur le secteur bancaire Cas pratique du la Trust banque et la fintech Kepler technologie. Université Mouloud Mammeri (2019)

Ozili, P.K.: Financial inclusion and Fintech during COVID-19 crisis: policy solutions (2020)

Tseng, P.-L., Guo, W.-C.: Fintech, Credit market competition, and bank risk-taking (2018)

The Impact of Artificial Intelligence on Enhancing Human Resource Management Functionality

Maryam Al-Jawder[1], Allam Hamdan[1](✉), and Amjad Roboey[2]

[1] Ahlia University, Manama, Bahrain
allamh3@hotmail.com
[2] College of Business Administration, University of Business and Technology, Jeddah 21448, Kingdom of Saudi Arabia

Abstract. Artificial Intelligence (AI) field of research started in the 1950s, for the purpose of understanding the nature in human intelligence (Jatobá et al. 2019). Artificial Intelligence accuracy and efficiency highly overreach the traditional management ability, AI highly beats human capabilities in accuracy and in processing data and storage, as human judgments in some cases are inaccurate and part from real situation; AI judgments are accurate and will help employees in making the right decisions and ongoing advancements can be reached (Wang and Li 2019). Human resource management (HRM) strategy is integrated with the organizations business strategy, HRM represents the organizations' high level of decision making, HRM strategy focuses on employment policies and practices which consists of recruitment, selection, evaluation, development and retention of employees, and consultation and negotiation with individuals. Artificial intelligence is extremely important to be integrated in human resource management functions to support and run human resources functions efficiently (Jatobá et al. 2019). The research method will be systematic review aiming to find out the impact of artificial intelligence on enhancing human resource management functionality. It has been found that integrating artificial intelligence to the complex and various human resources tasks will reduce the large amount of time and efforts spent on performing those tasks, leading to efficiency and quality gains for the HRM functions (Arena et al. 2018).

Keywords: Artificial intelligence · Human resources management

1 Introduction

Officially the word Artificial Intelligence has been introduced in 1956 in an eight week-long workshop Dartmouth Summer Research Project on Artificial Intelligence (DSRPAI) at Dartmouth College in New Hampshire in the United States of America. Artificial Intelligence is defined as "a systems' ability to interpret external data correctly, to learn from such data, and to use those learnings to achieve specific goals and tasks through flexible adaptation" (Haenlein and Kaplan 2019). Artificial Intelligence is being involved

B. Alareeni et al. (Eds.): ICBT 2022, LNNS 620, pp. 509–515, 2023.
https://doi.org/10.1007/978-3-031-26953-0_46

in every aspect in our society, and will become an essential part of daily life, the same way the internet has become in the past (Haenlein and Kaplan 2019). Artificial Intelligence effects will not be limited to our personal lives, AI recently joined the business environment and public conversation, it will structurally transform how organizations take decisions and deal with their external shareholders (Haenlein and Kaplan 2019). Artificial intelligence has become intelligent as human in thinking and analytical tasks; as machine learning and data analytical are the major analytical AI applications, and the best results will be between human and machines working together (Huang and Rust 2018).

The increase in business pressures in business organizations which began by the late 1970s such as globalization, deregulation and rapid technological change resulted in the development of human resource management (Ahammad 2017). Human resource departments play additional roles to the traditional HR service which are roles related to the organizational performance as a whole and roles related to the organizational strategies (Farndale et al. 2010). Artificial intelligence is changing the way organizations work and communicate, technologies and digital transformation enables employees to work anytime and anywhere. Changes in organizations caused by artificial intelligence and digital transformation will cause changes in human skills required in organizations as well. Combining human skills with machine learning and automation software is worth investing for organizations in order to increase its productivity and efficiency (Abdeldayem and Aldulaimi 2020). AI enables quicker and more accurate and precise adjusting to environmental changes; it is more relevant to use artificial intelligence technologies for decision making. AI technologies enable employees to analyze data without the need to have special data analysis skills (Buzko et al. 2016). AI will enhance all HRM functions, it will fasten evaluating the organization candidates, matching their skills and knowledge with the suitable position and measure their added value to the organization. It will enhance performance indicators to enable the organization measure how productive and efficient its staff is in order to compensate their valuable staff and avoid losing one of their best employees to one of their competitors. These are the major challenges and difficulties human resources face and artificial intelligence will solve it (Berhil et al. 2020)!

2 Literature Review

2.1 Artificial Intelligence

Artificial Intelligences' (AI) birth is approximately back to the 1940s, when an American writer published a story about a robot that has been developed by engineers (Haenlein and Kaplan 2019). Artificial Intelligence field of research started in the 1950s, for the purpose of understanding the nature in human intelligence (Jatobá et al. 2019). Officially the word Artificial Intelligence has been introduced in 1956 in an eight weeklong workshop Dartmouth Summer Research Project on Artificial Intelligence (DSRPAI) at Dartmouth College in New Hampshire in the United States of America. Artificial Intelligence is defined as "a systems' ability to interpret external data correctly, to learn from such data, and to use those learnings to achieve specific goals and tasks through flexible adaptation" (Haenlein and Kaplan 2019). Artificial Intelligence is part of computer science and it focus on creating machines that work and response like human beings. AI is described as machine learning and thought of in terms of problem solving, speech recognition and planning in ways which human brain will take a long time to be useful and might not be accurate (Weber 2019). Artificial Intelligence is being examined as an extraordinary revolutionary technology with a probability to transform humanity due to technological advances (Brock and Wangenheim 2019).

Artificial Intelligence effects will not be limited to our personal lives, AI recently joined the business environment and public conversation, it will structurally transform how organizations take decisions and deal with their external shareholders (Haenlein and Kaplan 2019). Artificial Intelligence accuracy and efficiency highly overreach the traditional management ability, AI highly beats human capabilities in accuracy and in processing data and storage, as human judgments in some cases are inaccurate and part from real situation; AI judgments are accurate and will help employees in making the right decisions and ongoing advancements can be reached (Wang and Li 2019). Arguments regarding firm goals that might be to replace workers with robots and intelligence tools or use robots and intelligence tools to support workers capabilities rather than replacement. Policy and politics play a vital role in this transformation, each country has its own governance rules (Zysman and Kenney 2018). Employment of robotics and artificial intelligence is being widely discussed, it is said that we are facing a "second machine age" that is threating jobs except jobs that require creativity and social intelligence are relatively safe (Lloyd and Payne 2019). Although this process might be imperfect and slow with lots of costs if we would evidence from the past globalization and digitalization world experience (Martens and Tolan 2018). World Economic forum reported predictions that as robotics and artificial intelligence systems increase in the workplace, jobs will be reduced, although there is insufficient analysis and predictions that robotics and artificial intelligence will replace human jobs (Upchurch 2018).

2.2 Integrating Artificial Intelligence into Human Resources Management Functions

Artificial intelligence technologies are based on replication of fundamentals of human intelligence functioning, the fast changes in business environments require fast responses, AI technologies meet these requirements, unlike the traditional information systems. AI technologies are a vital element of modern management (Buzko et al. 2016). Artificial intelligence is changing the way organizations work and communicate, technologies and digital transformation enables employees to work anytime and anywhere. Changes in organizations caused by artificial intelligence and digital transformation will cause changes in human skills required in organizations as well. Combining human skills with machine learning and automation software is worth investing for organizations in order to increase its productivity and efficiency (Abdeldayem and Aldulaimi 2020). AI enables quicker and more accurate and precise adjusting to environmental changes; it is more relevant to use artificial intelligence technologies for decision making. AI technologies enable employees to analyze data without the need to have special data analysis skills (Buzko et al. 2016).

Human resource management strategy is integrated with the organizations business strategy, HRM represents the organizations' high level of decision making, HRM strategy focuses on employment policies and practices which consists of recruitment, selection, evaluation, development and retention of employees, and consultation and negotiation with individuals. Artificial intelligence is extremely important to be integrated in human resource management functions to support and run HR functions efficiently (Jatobá, et al. 2019). Integrating artificial intelligence to the complex and various human resources tasks will reduce the large amount of time and efforts spent on performing those tasks, leading to efficiency and quality gains for the HRM functions (Arena et al. 2018). The growth of information and the huge databases result in many practical problems, integrating AI techniques in the search methods will be a good solution. Machine learning is being the trend nowadays because of the numerous possibilities of automation due to advances in artificial intelligence and HR sector is the most influenced by this trend, it is transforming from traditional way of functioning (Jatobá et al. 2019). Artificial intelligence will provide human resources management with all the necessary data based from internal data analysis and market external data. AI will enhance all HRM functions, it will fasten evaluating the organization candidates, matching their skills and knowledge with the suitable position and measure their added value to the organization. It will enhance performance indicators to enable the organization measure how productive and efficient its staff is in order to compensate their valuable staff and avoid losing one of their best employees to one of their competitors. These are the major challenges and difficulties human resources face and artificial intelligence will solve it (Berhil et al. 2020)! The efficient use of artificial intelligence in human resources management will benefit all sections in human resources such as recruitment, training, planning, performance management and predicting the labor market (Abdeldayem and Aldulaimi 2020). Integrating machines with HR employees will enhance the task efficiency, and machines might replace human employees in particular tasks. HR employees will have new roles to manage the technology and gives them space to generate value. Machines perform better than human in some HR tasks and cost less which will lead organizations to diversify their investments (Verma and Bandi 2019). Efficient searches, calculations, statistics, and analysis can be reached by implementing AI in human resources analysis (Wang and Li 2019).

The main human resources challenges are: first the loss of valuable employees due to inefficient skills management and the absence of training, second the lack of communication and concentrating on managerial objectives leads to high stress among the organization, and to deficit trust between managers and staff and demotivation of staff, next the absence of health and safety procedures which will lead to unsafe work environment, more over the absence of internal control which may lead management to treat staff with inequality, finally the high and increasing human resources costs due to lack of control, such as insufficient payroll and insurance management (Berhil et al. 2020). In addition, organizations lose a lot of costs due to bad hiring and attrition costs corresponding to rehire costs, this happens because of lack of insights to candidate profile and subjective evaluation of skills (Verma and Bandi 2019). Human resources management challenges that have been raised to information technology (IT) solutions were mainly: human resources development followed by skilled management, recruitment, turnover, and attritions. IT solutions that have been suggested were basically: artificial intelligence algorithms such as machine learning that were used mostly, next was statistics such as big data, followed by analysis such as software and websites (Berhil et al. 2020). AI and machine learning can solve HRM problems and lead to efficiency in HR functions by collecting data on candidates' CVs and job descriptions and analyzing candidate profiles. AI can also connect to existing human management systems and retrieve relevant data to perform analysis and provide the organizations with the required information accurately (Verma and Bandi 2019). Searching for IT solutions in human resource management is highly elaborating and experts have found many solutions for human resources challenges; artificial intelligence is one major solution. AI's field is continuously developing there will always be new techniques and solutions. HR is wide field that is developing as well, organizations are seeking to compete by their human capital, increasing productivity and attracting skilled workforce (Berhil et al. 2020). There are many innovative ways to apply artificial intelligence and machine learning to human resources functions effectively (Verma and Bandi 2019).

Searches for IT solutions to solve HR issues are increasingly developing and many AI solutions have been applied to solve HR problems using different methods and algorithms (Berhil et al. 2020). Artificial intelligence in human resource management is inspected in many functions such as "candidate search with knowledge-based search engine, turn over prediction with artificial neural networks, curriculum vitae data acquisition with information extraction and employee self-service with interactive voice response." (Stefan Sfrohmeier and Franca Piazza) (Buzko et al. 2016). Data and algorithm intelligence will not substitute human intelligence in the decision making, although it is necessary, but the value relies on human intelligence regarding the interpretation to take the decisions. AI does not replace HR employees, it is at their service to function the HR processes efficiently (Berhil et al. 2020). HRM transformation is trending in combining human skills with artificial intelligence, human resources functions have totally changed from the past. Now the system can be customized for each employee and help improve their work (Abdeldayem and Aldulaimi 2020). Creativity is enhanced in organizations through the collaboration between human and artificial intelligence. It has been found in the Japanese labor market that employing AI and robots support organizations efficiency and better balance between work and personal/social life for employees as it will solve the issue of long working hours (Nobuaki and Keisuke 2018).

There are major challenges in implementing AI in HRM because of the complexity of human resources, issues that may arise in response to fairness, ethical and legal constraints, and possibility of negative employee reaction to algorithm-based decisions (Tambe et al. 2019). Human resource management should consider ensuring legal compliance as HRM is responsible to keep the compliance with labor and employment law to secure the organizations' continued existence. HR should be aware of all policies, law and regulations related to employment and workers' rights and duties. Moreover, HRM should also consider equality and diversity; equality means treating staff fairly with no bias to gender, race, or religion. Whereas diversity means valuing employees' differences and creating a comprehensive culture for all staff. Equality and diversity can be achieved by lots of ways for example: treating all staff equally and ensuring equal opportunities to all staff (Ahammad 2017). Technology is developing and management transformation from big data to machine learning to artificial intelligence is rapid. Most organizations are struggling to make this transformation happen as they are not prepared for it! AI has advanced in many applications such as deep learning, language translation and pattern recognition but when it comes to management of employees this gets more complicated in decision making (Tambe et al. 2019). Successful human resource management should not only recognize the importance of artificial intelligence in human resources functions, but also it is important to recognize how changes in technology effects jobs, and how continuous learning is critical for an employee to gain new skills required for his job. In order for an organization to compete in the current global economy, AI is vital to be involved in its HR functions as it will affect the organizations decisions and help HR staff to concentrate on the organizations' strategic planning (Abdeldayem and Aldulaimi 2020).

3 Conclusion

Artificial Intelligence is defined as "a systems' ability to interpret external data correctly, to learn from such data, and to use those learnings to achieve specific goals and tasks through flexible adaptation" (Haenlein and Kaplan 2019). Artificial Intelligence is part of computer science and it focus on creating machines that work and response like human beings. AI is described as machine learning and thought of in terms of problem solving, speech recognition and planning in ways which human brain will take a long time to be useful and might not be accurate (Weber 2019). Artificial intelligence technologies help in classifying and analyzing huge data from various sources automatically, leading to efficiency and enhanced outcomes. In addition, it leads to enhancement of employees' skills and proficiency (Bolton et al. 2018).

Human resource management mainly is positioning employees and their activities within the organization to reach organizations' strategic goals and employees' recruitments and goals as well (Ahammad 2017). Human resources' role is vital in driving the organizations performance and employees' capabilities to achieve the desired outcomes and objectives. Positioning HR and organizational strategies for competitive advantage is outstanding, organizations can benefit from knowledge growth by having knowledgeable employees and deploying their skills efficiently which will lead to organizations' competitive advantage (Ramona and Anca 2013).

There are major challenges in implementing AI in HRM because of the complexity of human resources, issues that may arise in response to fairness, ethical and legal constraints, and possibility of negative employee reaction to algorithm-based decisions (Tambe et al. 2019). Human resource management should consider ensuring legal compliance as HRM is responsible to keep the compliance with labor and employment law to secure the organizations' continued existence. HR should be aware of all policies, law and regulations related to employment and workers' rights and duties (Ahammad 2017).

References

Abdeldayem, M.M., Aldulaimi, S.H.: Trends and opportunities of artificial intelligence in human resource management: aspirations for public sector in Bahrain. Int. J. Sci. Technol. Res. **9**(1), 3867–3871 (2020)

Berhil, S., Benlahmar, H., Labani, N.: A review paper on artificial intelligence at the service of human resources management. Indones. J. Electr. Eng. Comput. Sci. **18**(1), 32–40 (2020)

Verma, R., Bandi, S.: Artificial Intelligence and Human Resource Management in Indian IT Sector. SSRN-Elsevier, pp. 962–967 (2019).

Ahammad, T.: Personnel management to human resource management (HRM): how HRM functions? J. Mod. Account. Audit. **13**(9), 412–420 (2017)

Arena, D., et al.: Human resource optimisation through semantically enriched data. Int. J. Prod. Res. **56**(8), 2855–2877 (2018)

Bolton, C., Machová, V., Kovacova, M., Valaskova, K.: The power of human–machine collaboration: artificial intelligence, business automation, and the smart economy. Econ. Manage. Financ. Markets **13**(4), 51–56 (2018)

Brock, J.K.-U., Wangenheim, F.V.: Demystifying AI: what digital transformation leaders can teach you about realistic artificial intelligence. Calif. Manage. Rev. **61**(4), 110–134 (2019)

Buzko, I., Dyachenko, Y., Petrova, M., Nenkov, N., Tuleninova, D., Koeva, K.: Artificial intelligence technologies in human resource development. Comput. Model. New Technol. **20**(2), 26–29 (2016)

Farndale, E., et al.: Context-bound configurations of corporate HR functions in multinational corporations. Hum. Resour. Manage. **49**(1), 45–66 (2010)

Haenlein, M., Kaplan, A.: A brief history of artificial intelligence: on the past, present, and future of artificial intelligence. Calif. Manage. Rev. **61**(4), 5–14 (2019)

Huang, M.-H., Rust, R.T.: Artificial intelligence in service. J. Serv. Res. **21**(2), 155–172 (2018)

Jatobá, M., Santos, J., Gutierriz, I., Moscon, D., Fernandes, P.O., Teixeira, J.P.: Evolution of artificial intelligence research in human resources. Procedia Comput. Sci. **164**, 137–142 (2019). https://doi.org/10.1016/j.procs.2019.12.165

Lloyd, C., Payne, J.: Rethinking country effects: robotics, AI and work futures in Norway and the UK. New Technol. Work Employ. 208–225 (2019)

Martens, B., Tolan, S.: Will this time be different? A review of the literature on the impact of artificial intelligence on employment, incomes and growth. JRC Tech. Rep. 1–24 (2018)

Nobuaki, H., Keisuke, K.: Regional employment and artificial intelligence in Japan. Res. Inst. Econ. Trade Ind. 1–47 (2018)

Ramona, T., Anca, Ş.: Human Resource Management – From Function to Strategic Partner. Faculty of Economics, "Lucian Blaga" University of Sibiu, Romania, pp. 631–638 (2013)

Tambe, P., Cappelli, P., Yakubovich, V.: Artificial intelligence in human resources management: challenges and a path forward. Calif. Manage. Rev. **61**, 15–42 (2019)

Upchurch, M.: Robots and AI at work: the prospects for singularity. New Technol. Work Employ. 205–218 (2018)

Wang, H., Li, H.: Research on theoretical analysis of human capital of labor economics based on artificial intelligence. J. Intell. Fuzzy Syst. **37**(3), 3257–3265 (2019)

Weber, R.M.: Hey, Siri! Is artificial intelligence the ultimate oxymoron? J. Financ. Serv. Profession. **73**(4), 46–50 (2019)

Zysman, J., Kenney, M.: The next phase in the digital revolution: intelligent tools, platforms, growth, employment. Commun. ACM **61**(2), 54–63 (2018)

Check for updates

Applying Lean Six Sigma to Architectural Consultation Office Using Artificial Intelligence Technology

Rawan Althagafi and Mohammed Khouj(✉)

College of Engineering, University of Business and Technology, Jeddah, Kingdom of Saudi Arabia
m.khouj@ubt.edu.sa

Abstract. Lean and Six Sigma are famous management strategies applied in various companies in both the production and service sectors. The author works at the company used as a case study in this paper, which describes the design process, beginning with the arrival of the client. After that and based on the size of the project, each stage of the design process, which include the predesign, schematic design, design development, documentation, and building permit stages, has its own processing time. A comparison between the normal system and one modified by adding one structural engineer to the original simulation model was made using the Arena Rockwell software. After running the two alternative models, it was revealed that the modified increased the number of completed design projects from six to 10 per 360 working days. Additionally, the modified system yielded an extra giga-sized project design. Furthermore, the modified system had a lower design development waiting time of 31.6 h as opposed to 474.59 h. Finally, the modified system had a lower number of entities in the design development waiting line of 0.24 instead of 0.93.

1 Introduction

Lean and Six Sigma are famous management strategies applied in different companies, both in the production and service sectors. Recognized and applied as useful continuous improvement programs, they help to pave the way for companies to succeed and improve their competitive advantage, product quality, and customer satisfaction. Lean was created by Toyota, a globally famous car manufacturer, and was applied in the Toyota Production System (TPS) and supported by Taiichi Ohno. The enhancers of Lean management include the elimination of waste (Arnheiter and Malayeff 2005). Seven types of wastes are highlighted in particular: transportation, motion, inventory, defects, overprocessing, overproduction, and queues. Meanwhile, Six Sigma, founded by Motorola, is a systematic data-driven technique for improving process and quality, with an aim of reducing the error rate to 3.44 per 1 million opportunities. The central theme of the Six Sigma technique is the requirement for quality improvement in the manufacture of assembled products with multiple components, which increases the likelihood of final product defects (Arnheiter and Malayeff 2005). Similarly, the construction field has many activities that rely on each other.

B. Alareeni et al. (Eds.): ICBT 2022, LNNS 620, pp. 516–533, 2023.
https://doi.org/10.1007/978-3-031-26953-0_47

Lean management and Six Sigma have different applications, but they are supportive of each other in terms of achieving quality, whether in customer service, products, processes, or employee training and education (Pepper and Spedding 2010). Researchers and other studies have shown that companies that practice either Lean management or Six Sigma on their own may have lower returns. According to Snee (2010), companies that appeared to effectively implement Lean Six Sigma have achieved many benefits, with Snee demonstrating the monetary savings made using this technique. Although the combined Lean Six Sigma technique has many benefits, these two modern management theories have for the most part been implemented individually (Smith 2003).

In the area of Lean management, Bhasin (2011) describes the process of evaluating an organization's Lean status as taking a measure of its "Leanness". By implementing Lean as a philosophy, he launches an evaluation framework that helps organizations determine their stage in Lean's journey. Lean, Six Sigma, and Lean Six Sigma are essential to organizational growth, regardless of size or industry, and it must be determined how Lean Six Sigma will be implemented and how its tools and methods will be applied in the real world.

1.1 Purpose of the Study

This project aims to answer the following question: How can artificial intelligence technology help a well-known contracting company in Saudi Arabia apply the Lean Six Sigma methodology and toolset?

This study combines exploratory and descriptive methodologies, focusing on tools and methods to determine how artificial intelligence technologies can enhance the application of the Lean Six Sigma methodology application to the environment of a real company. The purpose of this study is to assess the usefulness of artificial intelligence technologies on the reduction of waste from the case contracting company employees' point of view.

2 Literature Review

2.1 Lean Manufacturing

The Lean manufacturing approach began development in the early 20th century, when Henry Ford of Ford Motor Company and Alfred Sloan of General Motors made the transition from the craft industry to manufacturing, transforming the individual worker's expertise and knowledge into a mass-production system. The TPS, also known as Lean manufacturing, was later developed by the Japanese manufacturer Toyota in the mid 20th century to enhance the production process (Womack and Jones 1996). This mass manufacturing model, which became popular in the car industry in the 1990s, was swiftly adopted by almost all North American and European automobile companies (Womack 1990).

The TPS is described by its founders as a production system based on the principle of "total elimination of all waste", exposing all parts of production in search of the most efficient methods (TOYOTA 2016). The method of production, known as both Lean manufacturing and just-in-time manufacturing, has become a well-researched phenomenon worldwide.

Following World War II, Japan's industrial base needed to recover. The industrial sector had been heavily damaged during the war, and its productivity dropped significantly behind that of the US. In particular the output of Americans was nine times that of their Japanese counterparts. Taiichi Ohno is credited with developing the TPS, which focuses on waste reduction and is believed to be the core of Lean manufacturing (Keyes 2013). Through the use of waste reduction as a strategy to close the productivity gap between Japan and the US, Toyota created many of the practices now associated with Lean manufacturing. These and other improvements, such as time study and standardized work, waste removal, and the Ford assembly line, became a part of the Lean manufacturing procedures later on. Craft manufacturing, mass production, and Fordism have all influenced the development of Lean.

The Non-Added Value Activities in Lean Production: The Seven Wastes. The concept of "waste" in this particular context must be understood to obtain a good understanding of the TPS or Lean manufacturing. Waste, in Lean language, is everything other than the minimum equipment, materials, parts, and labor time required to add value to a product.

There are seven categories of waste overall: overproduction, waiting, transporting, overprocessing, inventories, moving, and making defective parts or products.

One of the fundamentals of Lean philosophy is waste minimization. This is because actions that add value to a product represent only 5% of total effort, with the other 95% being waste.

2.2 Six Sigma

Historical Background on Six Sigma. Six Sigma was developed in the 1980s by Motorola engineer Bill Smith as a strategy for improving the work processes important to customers. Business processes can be improved by using Six Sigma concepts to reduce excessive variance in processes that lead to poor quality, shift the average, and help produce more reliable products (Shah et al. 2008). Statistically based problem-solving is the core of Six Sigma. This process uses data to drive solutions that result in significant bottom-line benefits (Snee and Hoerl 2007).

Motorola was once under significant pressure from international competitors, especially from Japan. While a specific date for the birth of Six Sigma is difficult to pinpoint, Bill Smith and his colleagues launched improvement efforts in 1987 that resembled total quality management programs in many aspects (Harry and Schroeder 1997). Mikel Harry and others eventually assisted Smith in turning this concept into a larger business endeavor focused on preserving Motorola's pager business (Eckes 2002; Pande et al. 2000). The initiative was given the name Six Sigma because the goal was to minimize the variation of specification limits for key process metrics to six standard deviations off-target (Harry and Schroeder 1997).

Six Sigma offered a "roadmap" for Motorola's problem-solving process, known as MAIC (measure, analyze, improve, control), with the individual tools being efficiently connected and integrated. As a result, employees could be taught this approach as a means to solve a wide range of problems, eliminating the need to reinvent the wheel with each new project. Six Sigma also gained strong management support, which included supporting infrastructure, such as budget line items, resources, and project selection procedures, among other things (Snee and Hoerl 2005).

In the late 1990s and early 2000s, many companies across a wide range of sectors, including DuPont, Dow Chemical, 3M, Ford, and American Express, to name a few, initiated Six Sigma programs. At the same time, the US military began making significant investments in Six Sigma. Businesses in Europe and Asia, notably including Korean companies such as Samsung, began to integrate Six Sigma to varied degrees overseas (Snee and Hoerl 2005).

2.3 Lean Architecture Engineering

Lean architecture engineering focuses on improving how we process and perceive information as well as how data moves, enabling improved communication between ourselves and service consumers.

The practice of upgrading and rethinking architectural approaches is known as Lean architecture. At the same time, the practice includes identifying what provides value and what does not.

Architectural work can be defined as organized complexity consisting of overlapping building systems and components that are logically arranged and coordinated. Planning, design, integration, and quality documentation of building systems can all be supported by well-designed procedures, which make the outputs easy to understand and follow. Because we are not producing identical appliances, architectural production should not be confused with assembly line manufacturing. However, we can learn a lot from investigating the reasoning behind successful manufacturing methods and applying the concepts and practices to the more repetitive aspects of our profession. The architectural, MEP, and structural disciplines are all considered in an architect's work to develop strategies for completing tasks quickly and accurately.

3 Perspectives

Important perspectives include approaching project development with a critical path methodology, defining what needs to happen at the correct time, integrating cross-discipline coordination efforts, and looking ahead to identify and avoid any stumbling obstacles. Preventing issues from occurring is always more effective than wasting time and effort afterward to uncover and remedy them.

Design and decision making, along with simultaneous costing and value engineering, should be completed by the conclusion of the design development stage if possible, and projects should be planned ahead of time to avoid the temptation to work as hard as one can without understanding what one wants to achieve.

3.1 Objectives

The main objective is to eliminate obstacles to job progress by evaluating the workplace and projects to see what can be done to improve the flow of work. After the necessary adjustments are made, the results and remaining issues will be observed again.

3.2 Modeling of the Case Company

The author, who works at the case company, previously described the process, which begins with the arrival of the client. The client's job is to describe their needs regarding the construction project. The project is then categorized into one of three size categories: giga, medium, or small. Determining the category is the responsibility of the design manager. Next, the pre-design and schematic design are carried out by the assigned architectural engineer. The schematic design may proceed to the design development process. If the schematic design is rejected by the client, then the design process stops for that project, and the schematic design is resumed. However, if the schematic design is accepted by the client, then the design development is carried out using the assigned resources. The assigned resources for this process include one civil engineer, one draftsman, and one architectural engineer. After that, the construction documentation phase is undertaken, for which one electrical engineer, one mechanical engineer, and one draftsman are assigned. Next, the case firm waits for the building permits to be issued. The architectural engineering design firm works finish by the end of that process. Figure 1 below describes the process.

Pre-design. This phase collects information that forms the basis of the next design phase. The main purpose at this stage is to learn as much as possible about the client's personality, lifestyle, and needs and to agree on the amount of space needed both at the current time and in the future. Additionally, it is used to determine how to use, organize, and agree on that space.

Schematic Design. In this phase, we begin to transform our program into an efficient building design. In particular, we begin exploring the design concept. It is now time to test the options and obtain an overview of what they look like.

Design Development. Design development includes a significant development of the design based on the floor plan and exterior concept approved in the previous phase.

Design Documents. At this stage, the design drawings are developed into a complete and accurate set of design documents. These drawings and specifications contain all the details, dimensions, and notes needed to convey the overall design intent to the builder.

Building Permit. This phase adds the additional information to the building document needed to obtain a building permit.

A simulation module was created in Arena Rockwell Software, and 360 days was set as the run length. The model considered that a new client arrived every 14 days. The module was decided upon prior to the predesign process. There was a 10% probability

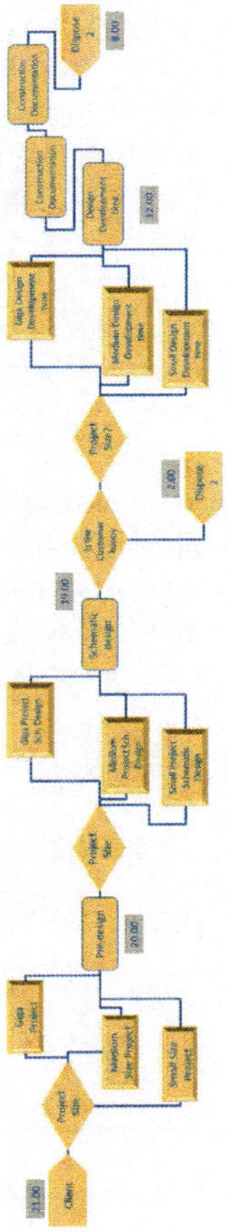

Fig. 1. The case design firm model

that each project would be giga-sized, a 20% probability that it would be medium-sized, and a 70% probability that it would be small.

A module was assigned for the giga project with an attribute assigned to identify the time necessary for the predesign process. The expected duration for the predesign process of a giga project was 28 days, while 21 days was the expected duration for the module assigned to the medium project. The module assigned for small projects had an expected duration of 14 days.

The predesign process required a single design manager, and the process time was taken from the assigned module. The module was decided on prior to the schematic design process. There were three possible modules based on the project size. The module assigned for the giga projects had an attribute that identified the time required for the schematic design process. In total, 56 days was the expected duration for the schematic design process of a giga project, while 42 days is the expected duration for the schematic design process of a medium-sized project. Finally, 28 days was the expected duration for the schematic design process of a small project. The schematic design process required a single architect. The process time was taken from the assigned module.

The module was decided on prior to the design development process. It was considered that there was a 90% probability that the client would accept the design project. The second decision module was used prior to the predesign process. There were three possible modules based on the project size.

The module assigned the giga projects had an attribute that identified the necessary time for the design development process. TRIA (84, 86, 88) days was the expected duration for the design development process of the giga project. For the medium-sized project, TRIA (70, 72, 74) days was the expected duration. For the small-sized project, TRIA (56, 58, 60) days was the expected duration.

The design development process used one draftsman, one structural engineer, and one architectural engineer. The process time was taken from the assigned module.

The documentation process used a single draftsman, one electrical engineer, and one mechanical engineer. The process time was TRIA (1, 2, 3) days.

The building permit process delays the project for a minimum of 84 days and a maximum of 168 days.

Figure 2 below shows the system resources.

Resource - Basic Process

	Name	Type	Capacity	Busy / Hour	Idle / Hour	Per Use	StateSet Name	Failures	Report Statistics
1	Design Manager	Fixed Capacity	2	0	0	0.0		0 rows	☑
2 ▶	Archtictural Engineer	Fixed Capacity	10	0	0	0.0		0 rows	☑
3	Structural Engineer	Fixed Capacity	3	0	0	0.0		0 rows	☑
4	Mechanical Engineer	Fixed Capacity	2	0	0	0.0		0 rows	☑
5	Electrical Engineer	Fixed Capacity	2	0	0	0.0		0 rows	☑
6	Draftman	Fixed Capacity	5	0.0	0.0	0.0		0 rows	☑

Fig. 2. Resources model

Figure 3 below shows the number of system outputs. It was expected that six project designs would be completed per 360-day period.

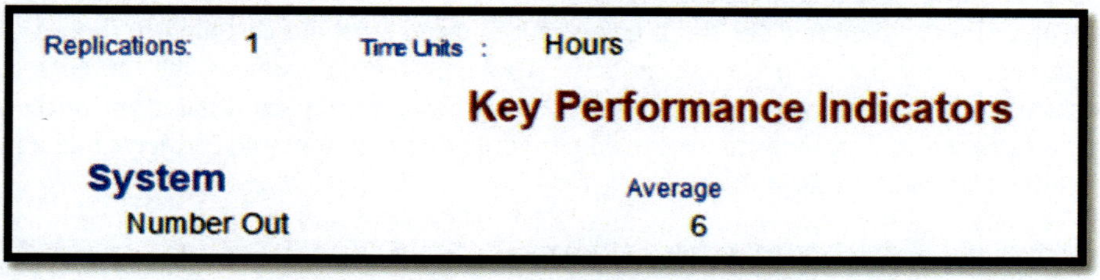

Fig. 3. The normal system output

Figure 4 below shows the value-added time for each output. It was expected that one giga project, two medium projects, and three small projects would be completed. The average total time consumed for the giga project was 8,137.06 h, that for the medium projects was 5,655.30 h, and that for the small projects was 6,180.78 h.

Total Time	Average	Half Width	Minimum Value	Maximum Value
Giga	8137.06	(Insufficient)	8137.06	8137.06
Medium	5655.30	(Insufficient)	5636.36	5674.23
Small	6180.78	(Insufficient)	5865.60	6401.34

Fig. 4. Total time for the output entities

Figure 5 below shows that among the 20 clients who visited the firm, two had giga projects, six had medium projects, and 12 had small projects. It was expected that six projects would be designed per 360 days. Additionally, it was expected there would be 1.17 giga projects, 2.76 medium projects, and 5.55 small projects waiting for processing.

Number In	Value
Entity 1	20.0000
Giga	2.0000
Medium	6.0000
Small	12.0000

Number Out	Value
Entity 1	20.0000
Giga	1.0000
Medium	2.0000
Small	3.0000

WIP	Average	Half Width	Minimum Value	Maximum Value
Entity 1	0.00	(Insufficient)	0.00	1.0000
Giga	1.1798	(Insufficient)	0.00	2.0000
Medium	2.7601	(Insufficient)	0.00	4.0000
Small	5.5515	(Insufficient)	0.00	9.0000

Fig. 5. Waiting process for the normal system

Figure 6 below shows that the highest waiting time was consumed in the design development. Thus, 474.59 h was the expected waiting time in the queue.

Queue

Time

Waiting Time	Average	Half Width	Minimum Value	Maximum Value
Construction Documentation.Queue	0.00	(Insufficient)	0.00	0.00
Design Development Time.Queue	474.59	(Insufficient)	0.00	1114.14
Predesign.Queue	56.0235	(Insufficient)	0.00	533.93
Schematic Design.Queue	0.00	(Insufficient)	0.00	0.00

Other

Number Waiting	Average	Half Width	Minimum Value	Maximum Value
Construction Documentation.Queue	0.00	(Insufficient)	0.00	0.00
Design Development Time.Queue	0.9301	(Insufficient)	0.00	3.0000
Predesign.Queue	0.1297	(Insufficient)	0.00	2.0000
Schematic Design.Queue	0.00	(Insufficient)	0.00	0.00

Fig. 6. Waiting time for the normal system

Figure 7 below shows that the highest use rate was for the structural engineer, at 86.24%.

Replications: 1 Time Units: Hours

Resource

Usage

Scheduled Utilization	Value
Archictural Engineer	0.3752
Design Manager	0.4691
Draftman	0.5339
Electrical Engineer	0.04103470
Mechanical Engineer	0.04103470
Structural Engineer	0.8624

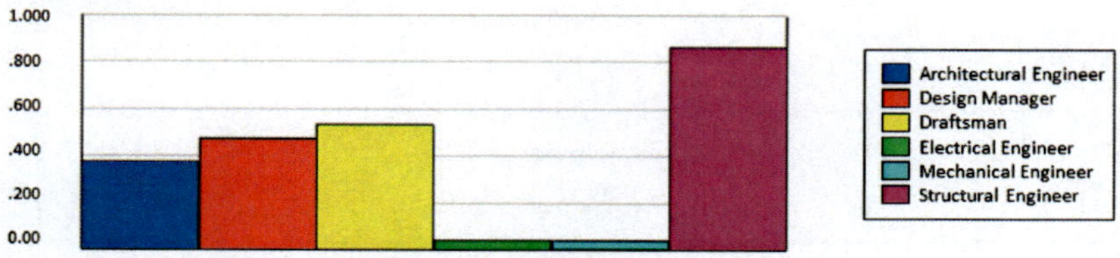

Fig. 7. The scheduled use of the normal system

Accordingly, the Lean waste could be represented as follows:

1. Waiting in process: 0.93 entities were queuing in the design development process.
2. Inventory: 20 clients entered, and 6 exited (14 in total).
3. Time: 474.59 h were consumed waiting for the design development phase.

As per the results, the author suggested raising the system capacity by highlighting the bottlenecks, with these being the highest-used resources. One structural engineer was hired to raise the system capacity. See Fig. 8 below.

Resource - Basic Process

	Name	Type	Capacity	Busy / Hour	Idle / Hour	Per Use	StateSet Name	Failures	Report Statistics
1	Design Manager	Fixed Capacity	2	0	0	0.0		0 rows	☑
2	Archtictural Engineer	Fixed Capacity	10	0	0	0.0		0 rows	☑
3 ▶	Structural Engineer	Fixed Capacity	4	0	0	0.0		0 rows	☑
4	Mechanical Engineer	Fixed Capacity	2	0	0	0.0		0 rows	☑
5	Electrical Engineer	Fixed Capacity	2	0	0	0.0		0 rows	☑
6	Draftman	Fixed Capacity	5	0.0	0.0	0.0		0 rows	☑

Fig. 8. Modification of the resource capacity (one extra structural engineer)

Figure 9 below shows the output after running the modified system. It was expected that 12 projects would be finalized in 360 days.

Replications: 1	Time Units : Hours	
	Key Performance Indicators	
System		Average
Number Out		10

Fig. 9. Output of the modified system

Figure 10 below shows the total time for each output. It was expected that two giga projects, three medium projects, and five small projects would be completed. The average time consumed on each giga project was 4,819.31 h, that for all medium projects was 3,622.7 h, and that for small projects was 5,629.61 h.

Total Time	Average	Half Width	Minimum Value	Maximum Value
Giga	4819.31	(Insufficient)	2016.00	7622.61
Medium	3622.70	(Insufficient)	504.00	5636.36
Small	5629.61	(Insufficient)	5292.75	6039.24

Fig. 10. Total time for the three entities

Figure 11 below shows that among 21 clients who visited the firm, four had giga projects, four had medium projects, and 13 had small projects. It was expected that 10 projects would be processed per 360 days. There were expected to be 1.72 giga projects, 1.45 medium projects, and 5.96 small projects waiting for processing.

Number In	Value
Entity 1	21.0000
Giga	4.0000
Medium	4.0000
Small	13.0000

Number Out	Value
Entity 1	21.0000
Giga	2.0000
Medium	3.0000
Small	5.0000

WIP	Average	Half Width	Minimum Value	Maximum Value
Entity 1	0.00	(Insufficient)	0.00	1.0000
Giga	1.7261	(Insufficient)	0.00	4.0000
Medium	1.4529	(Insufficient)	0.00	3.0000
Small	5.9694	(Insufficient)	0.00	10.0000

Fig. 11. Waiting in process parameters of the modified system

Figure 12 below shows that the highest waiting time was consumed in the design development stage. Thus, 31.6 was the expected waiting time in the queue, and 0.24 projects were queuing in the pre-design process queue.

Time

Waiting Time	Average	Half Width	Minimum Value	Maximum Value
Construction Documentation.Queue	0.00	(Insufficient)	0.00	0.00
Design Development Time.Queue	31.6096	(Insufficient)	0.00	286.92
Predesign.Queue	99.74	(Insufficient)	0.00	533.93
Schematic Design.Queue	0.00	(Insufficient)	0.00	0.00

Other

Number Waiting	Average	Half Width	Minimum Value	Maximum Value
Construction Documentation.Queue	0.00	(Insufficient)	0.00	0.00
Design Development Time.Queue	0.07693507	(Insufficient)	0.00	1.0000
Predesign.Queue	0.2424	(Insufficient)	0.00	2.0000
Schematic Design.Queue	0.00	(Insufficient)	0.00	0.00

Fig. 12. Waiting time for the modified system

Figure 13 below shows that the highest use rate was for the structural engineers, by 62.96%.

Usage

Scheduled Utilization	Value
Archtictural Engineer	0.4064
Design Manager	0.5196
Draftman	0.5196
Electrical Engineer	0.03962322
Mechanical Engineer	0.03962322
Structural Engineer	0.6296

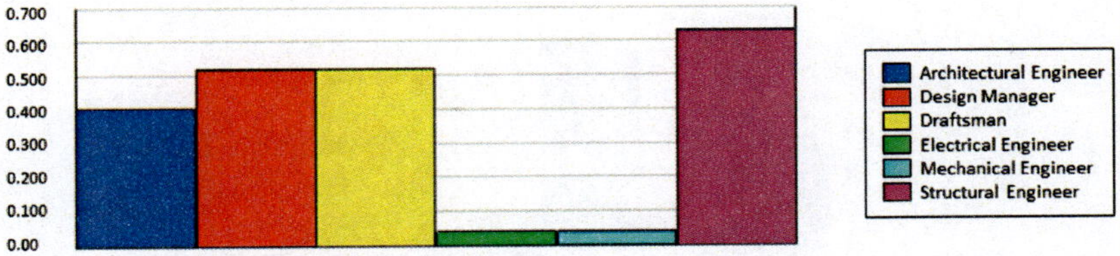

Fig. 13. Resource use of the modified system

4 Results and Discussion

4.1 Introduction

This chapter focuses on the comparison between the normal system and the modified one with one added structural engineer.

4.2 Interpretation

Figure 14 below shows that the number of completed design projects increased from six to 10 per 360 working days.

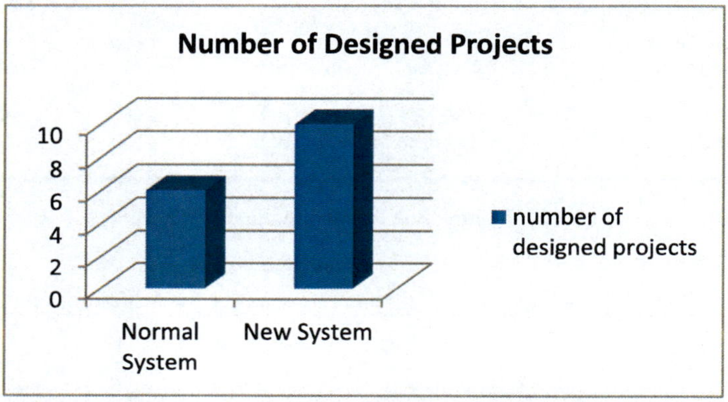

Fig. 14. Normal vs. modified system output

Figure 15 below shows that the modified system yielded an extra giga project design.

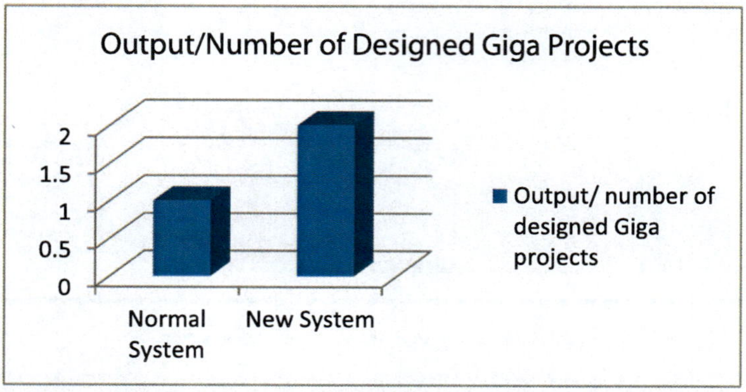

Fig. 15. Normal vs. modified system output (number of designed giga projects)

Figure 16 below shows that the modified system had a lower waiting time for giga projects under processing, namely, 1.72 giga projects instead of 1.17.

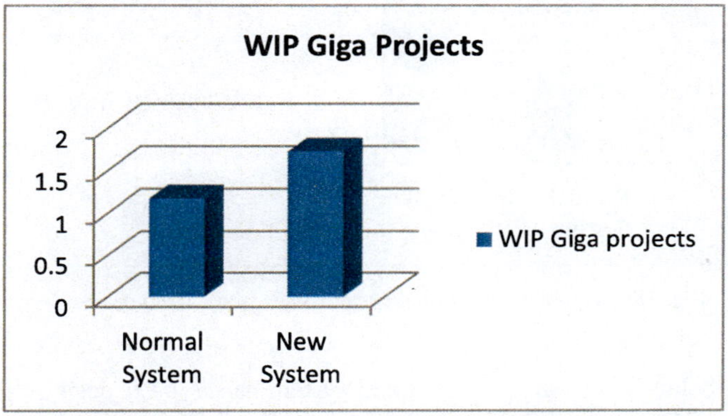

Fig. 16. Normal vs. modified system work-in-progress (WIP) giga projects

Figure 17 below shows that the modified system had a higher waiting time for medium projects under processing, namely, 1.45 medium projects instead of 2.76.

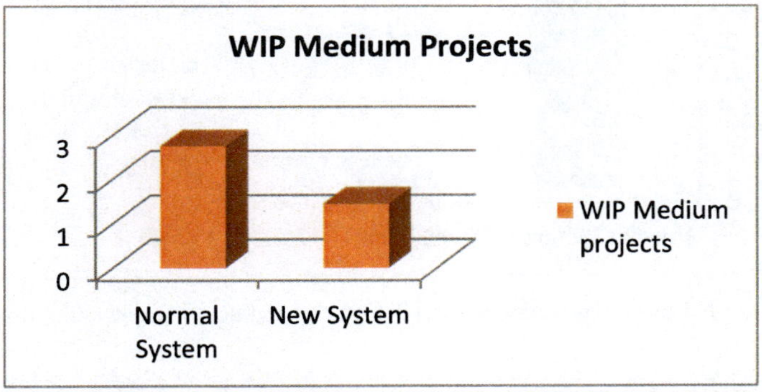

Fig. 17. Normal vs. modified system WIP medium-sized projects

Figure 18 below shows that the modified system had a lower waiting time for small projects under processing, that is, 5.96 medium projects instead of 5.55 small projects.

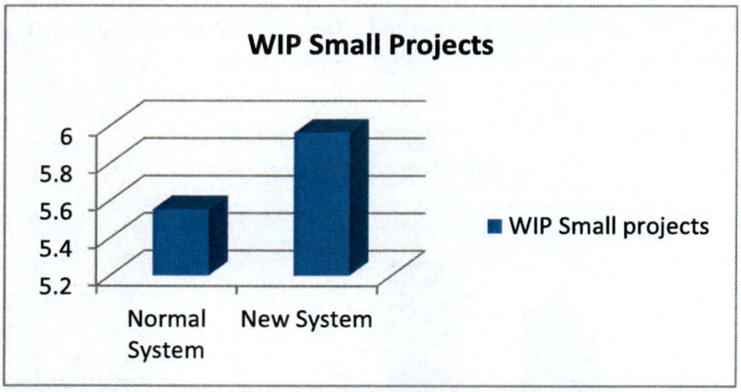

Fig. 18. Normal vs. modified system WIP small sized projects

Figure 19 below shows that the modified system had a lower design development waiting time of 31.6 h instead of 474.59.

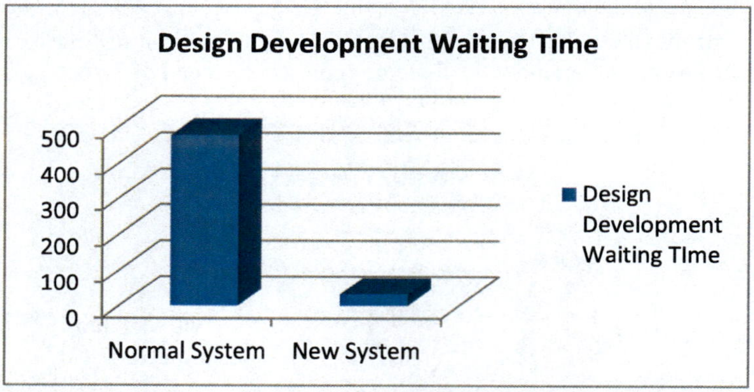

Fig. 19. Normal vs. modified system design development phase waiting time

Figure 20 below shows that the modified system had a lower number of entities in the design development waiting line of 0.24 instead of 0.93.

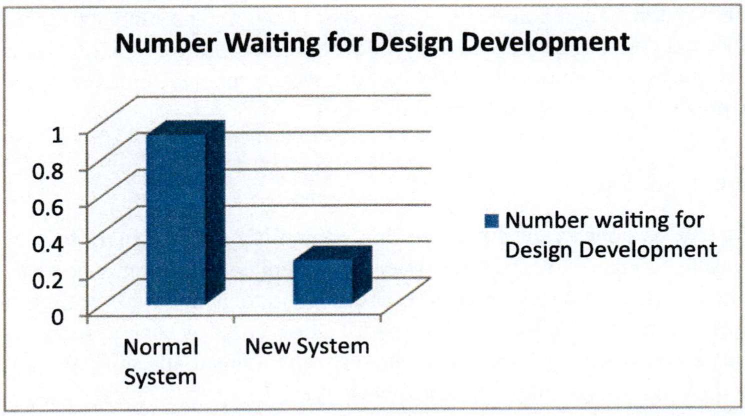

Fig. 20. Number waiting in line for design development process in normal vs. modified system design development phase

5 Conclusions and Recommendations

5.1 Conclusion

Lean and Six Sigma are famous management strategies applied in various companies both in the production and service sectors. Recognized and applied as useful continuous improvement programs, they help to pave the way for companies to succeed and improve their competitive advantage, product quality, and customer satisfaction.

The author, who works in the case company considered in this study, described the work process, which begins with the arrival of the client. The client's job is to describe their needs regarding the construction project. The project is categorized into one of three size groups: giga, medium, or small. Each category has to be predesigned by the design manager. After that, a schematic design is carried out by the assigned architectural engineer. The schematic design may proceed to the design development process. If the schematic design has been rejected by the client, then the design process stops for that project. However, if the schematic design has been accepted by the client, then the design development is carried out by the assigned resources. The assigned resources for that process include one civil engineer, one draftsman, and one architectural engineer. After that, a documentation process is made by one mechanical engineer, one electrical engineer, and one draftsman. Finally, the case firm waits for the building permits to be issued. The work of the engineering design firm finishes by the end of this process.

A comparison was made between the normal system and the modified one using the Arena Rockwell software by adding one structural engineer to the original simulation model, showing that the Lean waste in the design firm included waiting in process (0.93 entities were queuing in the design development process), inventory (20 clients entered, and six exited), and time (474.59 h were consumed waiting for the design development phase).

After running the two alternative models, in the modified model, the number of completed design projects increased from six to 10 per 360 working days. Additionally, the modified system yielded an extra giga project design. The model further showed that the modified system had a lower design development waiting time of 31.6 h instead of 474.59. Finally, the modified system had a lower number of entities in the design development waiting line of 0.24 instead of 0.93.

5.2 Recommendations

Much work can be done in the future on this system. One of the main improvements deals with system capabilities. For example, the system capability may be restricted by system bottlenecks. Bottlenecks can be reduced by increasing resources, but a future study (Allah willing) will search for the modification costs of such systems. It is also recommended that this study be benchmarked by other consultation offices. The model may be used as a generic model for firms of this type.

References

Arnheiter, E.D., Maleyeff, J.: Research and concepts: the integration of lean management and six sigma. TQM Mag. **17**(1), 5–18 (2005)

Snee, R.D.: Lean six sigma–getting better all the time. international journal of lean six sigma (2010)

Smith, J.A.: Qualitative Psychology: A Practical Guide to Research Methods. Sage Publications, Inc. (2003)

Bhasin, S.: Measuring the Leanness of an organisation. Int. J. Lean Six Sigma **2**, 55–74 (2011)

Liker, J.K.: Becoming Lean: Inside Stories of US Manufacturers. CRC Press (1997)

Womack, J.E., Cruz, J.R., Rigdon, H.K., Hoover, G.M.: Encoding techniques for multiple source point seismic data acquisition. Geophysics **55**(10), 1389–1396 (1990)

Keyes, R.J. (ed.): Optical and Infrared Detectors, vol. 19. Springer, Heidelberg (2013)

Snee, R.D., Hoerl, R.W.: Integrating lean and Six Sigma-a holistic approach. In: Six Sigma Forum Magazine, vol. 6, no. 3. ASQ (2007)

Harry, M.J., Schroeder, R.: Six sigma: the breakthrough management strategy revolutionizing the world's top corporations. A CURRENCY Book. Random House Inc. (1997)

Eckes, G.: The Six Sigma Revolution: How General Electric and Others Turned Process into Profits. Wiley, Hoboken (2002)

Pande, P.S., Neuman, R.P., Cavanagh, R.R.: The Six Sigma way: How GE, Motorola and Other Top Companies are Honing Their Performance. McGraw-Hill (2000)

Snee, R.D.: Leading business improvement: a new role for statisticians and quality professionals. Qual. Reliab. Eng. Int. **21**(3), 235–242 (2005)

Hoerl, R.W., Snee, R.D.: Statistical engineering: an idea whose time has come? Am. Stat. **71**(3), 209–219 (2017)

Goh, T.N., Xie, M.: Improving on the Six Sigma paradigm. TQM Mag. (2004)

Mitra, A.: Six Sigma education: a critical role for academia. TQM Mag. (2004)

Coleman, S.: Six Sigma: an opportunity for statistics and for statisticians. Significance **5**(2), 94–96 (2008)

Anand, G., Ward, P.T., Tatikonda, M.V.: Role of explicit and tacit knowledge in Six Sigma projects: an empirical examination of differential project success. J. Oper. Manag. **28**(4), 303–315 (2010)

George, M.L.: Leon Six Sigma. McGraw-Hill (2002)

Womack, J.P., Jones, D.T.: Lean Thinking: Banish Waste and Create Wealth in Your Corporation. Simon & Schuster, New York (1996)

Pepper, M.P.J., Spedding, T.A.: The evolution of lean Six Sigma. Int. J. Qual. Reliab. Manag. **27**(2), 138–155 (2010)

Sugimori, Y., Kusunoki, K., Cho, F., Uchikawa, S.: Toyota production system and Kanban system materialization of just-in-time and respect-for-human system. Int. J. Prod. Res. **15**(6), 553–564 (1977). https://doi.org/10.1080/00207547708943149

Deleersnyder, J.-L., Hodgson, T.J., Muller-Malek, H., O'Grady, P.: Kanban controlled pull systems: an analytic approach. Manage. Sci. **35**(9), 1079–1091 (1989). https://doi.org/10.1287/mnsc.35.9.1079

Schroeder, R.G., Linderman, K., Liedtke, C., Choo, A.S.: Six sigma: definition and underlying theory. J. Oper. Manage. **26**(4), 536–554 (2008)

Architecture Explained: The Phases of Designing & Building a Project
2020EnglishCady ChintisWC STUDIO architects

Integrating Vulnerability Assessment and Quality Function Deployment with Risk Management Process to Reduce Project Delay

Siraj Zahran(✉), Mohammad Kanan, Salem Aljazzar, and Salem Binmahfooz

Jeddah College of Engineering, University of Business and Technology, Jeddah 21448, Saudi Arabia
s.zahran@ubt.edu.sa

Abstract. This paper explores the topic of integrating vulnerability assessment and quality function deployment with a risk management process to reduce project delay risk. The main purpose of this paper is to propose an additional sub-process in the risk treatment phase to improve the results of the risk management process. The proposed sub-processes include utilizing the matrix of the quality function deployment in order to identify the related capabilities and quantify the resistance and resilience scores to be utilized as input to conduct the vulnerability assessment and decide whether the action plan requires improvement or not. Due to limitations, a medium-sized construction project for a commercial mall in Jeddah, Saudi Arabia, was selected as a test subject, and the risk register from the project was utilized to test the proposed approach. The findings of the paper exhibit an agreement with the literature review findings that proper risk management can assist in reducing the probability of project delays. The proposed approach should be tested on a larger scale in order to properly gauge its effectiveness.

1 Introduction

Project delay is a risk that many construction projects face, and the mitigation plan for this risk requires placing controls with the aim to reduce the delay. However, analyzing project risks and the vulnerabilities of the controls to manage the risks before implementing them can be a lengthy process and could potentially lead to a project delay in itself. This study develops a model that integrates vulnerability assessment and quality function deployment (QFD) concepts with the risk treatment phase to reduce project delay risk by optimizing the controls before implementing them. This paper is divided as follows: Sect. 2 contains the literature review, Sect. 3 describes the methodology, Sect. 4 analyzes a case study, Sect. 5 lists some managerial insights, and the report ends with Sect. 6, which is for conclusion and recommendations.

2 Literature Review

2.1 Project Management

In the 7th edition of the PMBOK, the Project Management Institute endorsed utilizing the system thinking approach for project management, defining it as "recognize, evaluate,

B. Alareeni et al. (Eds.): ICBT 2022, LNNS 620, pp. 534–548, 2023.
https://doi.org/10.1007/978-3-031-26953-0_48

and respond to the dynamic circumstances within and surrounding the project in a holistic way to positively affect project performance." (PMBOK, P. 64). Smit (2021) indicates that adopting a systematic approach to manage a project requires continuous evaluation of performance effectiveness, in all performance domains, through outcome-focused measures that deliver value. From a quality perspective, Juran (1986) highlights three universal steps for the quality processes (quality planning, quality control, and quality improvement) to deliver value. He states that processes run with a high level of chronic waste due to having shortfalls in the original planning, which results in the operating forces' inability to eliminate this chronic waste.

2.2 Risk Management

There are many frameworks that provide guidance in implementing risk management. This article focuses on the five steps of ISO 3100:2018 for risk management processes. Hutchins (2018, P. 204) states the following: "Risk treatment can also introduce new risks. How? Treatment may implement new controls that may surface new risks or unknown risks."

2.3 Quality 4.0 and Quality Function Deployment

The QFD, more widely known as the house of quality, is a tool developed by Hauser and Clausing (1988). It is used to pair customer wants (customer attributes or what's) with engineering characteristics (How's) in a matrix to find the links and the importance between the two.

2.4 Vulnerability Assessment

From a project perspective, Rahi et al. (2021, P. 10) have defined vulnerability as "the characteristic of a project that makes it susceptible to disruptive events." Aleksandar et al. (2017) deliberated that the assessment of vulnerability in project management follows four steps: (1) definition of project phases, (2) assessment of project phases' sensitivity to potential risks, (3) enterprise's static ability to recover its business performances (adaptive capacity), and (4) assessment of project phases' exposure to risks (susceptibility) that may occur during project delivery. They concluded that there is a lack of significant work in the area of project management, specifically in project vulnerability assessment. Thus, using the same approach to conduct the vulnerability assessment part and improve the results of the controls might be a feasible solution to improve the performance of the risk treatment controls.

3 Methodology

In this section, the authors expand the risk management process (Based on ISO 31000:2018), specifically the risk treatment step to include three additional stages as demonstrated in Fig. 1.

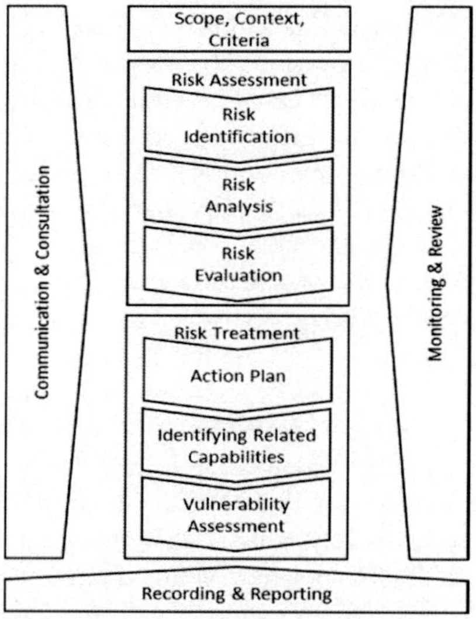

Fig. 1. The proposed risk management process

As seen in Fig. 1, the proposed model suggests adding two more stages under the risk treatment step. These steps are to identify related capabilities and vulnerability assessment, which come after formulating the action plan. The quality function deployment can be applied to identify the capabilities related to the residual risk and spot the gaps between the controls to bridge them. However, Aleksandar et al. (2017) provide four steps to identify vulnerabilities, which are (1) definition of project phases, (2) assessment of project phases' sensitivity to potential risks, (3) enterprise's static ability to recover its business performances (adaptive capacity), and (4) assessment of project phases' exposure to risks (susceptibility) that may occur during project delivery. Different tools can be used to achieve the objective of the study and support the expansion of the model, which are (1) Risk Register, (2) Quality Function Deployment (QFD), and (3) Vulnerability Assessment.

3.1 Risk Register

The risk register is the main tool that is used for the process (see Table 1).

Table 1. Proposed risk registrer

ID	Date raised	Activity	Risk description	Risk probability rating	Risk impact Without controls	Inherent Risk	Controls	Risk impact with controls	Residual score	Action plan	Action plan rating	Related capabilities	Related capabilities rating	Vulnerability rating	Improved action

The description for each column in Table 1 is shown in Table 2 below.

Table 2. Description of the columns in Table 1

Column name	Description
ID	Unique risk number
Date raised	The date the risk was documented in the risk register
Activity	The process or activity taking place
Risk description	Description of the risk event
Risk probability rating	The likelihood that an identified risk will occur
Risk impact without controls	The inherent risk impact without taking controls into consideration
Inherent risk	Risk score before controls
Controls	Measure that maintains and/or modifies risk
Risk impact with control	Risk impact rating with controls
Residual risk	The score of the risks that remain once the controls are in place, calculated by multiplying probability rating by risk impact with control rating
Action	The course of action that an organization agrees upon to address potential risks, reduce the likelihood, or lessen the impact of these risks
Action plan rating	The score of the effectiveness of the action plan
Related capabilities	The capabilities related to residual risk controls
Related capabilities rating	Rating of related capabilities on a scale from 1 to 3
Vulnerability rating	Vulnerability rating obtained by (residual risk/(action plan * related capability rating
Improved action	Any actions to be taken to improve controls

Modified Risk Register Workflow. The workflow in the proposed risk register is as follows (Fig. 2):

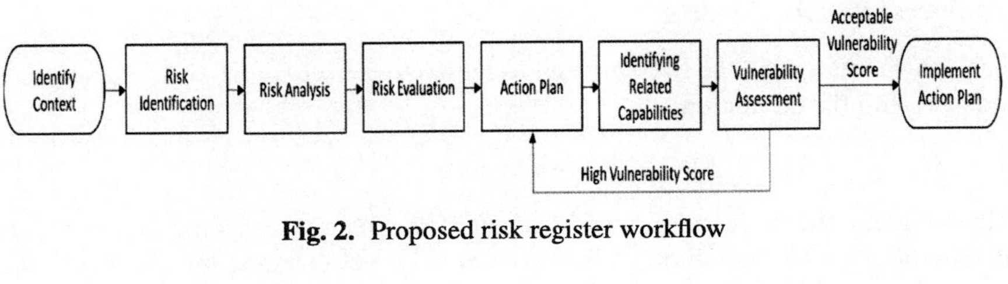

Fig. 2. Proposed risk register workflow

1. Defining the scope of the risk management activities.
2. Finding, recognizing, and describing the risks that might assist or prevent an organization from achieving its objectives.
3. Comprehending and analyzing the nature of the risk.
4. Comparing the results of the risk analysis against the identified risk criteria.
5. Writing and documenting how the chosen treatment options will be implemented.
6. Identifying and linking the capabilities required to implement the action plan.
7. Conducting the vulnerability assessment.
8. If the plan obtains an acceptable score, the action plan is implemented.
9. If the vulnerability score is high, the process is repeated from the action plan section.
10. The cycle should be repeated as the project progress or when there are changes.

3.2 Quality Function Deployment (QFD)

In this paper, the QFD is used to link the residual risks with the action plan. This is done to measure if the level of resistance provided by the action plan is properly linked to the available organization's capabilities and enables providing a score for the resilience part of the vulnerability assessment based on the criteria provided in Table 6. The modified QFD is shown in Fig. 3.

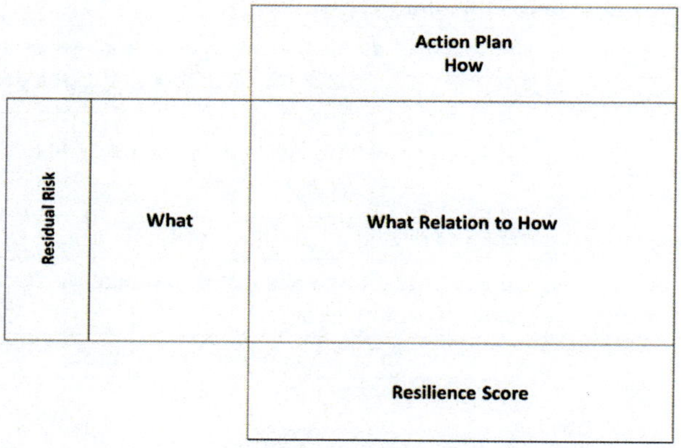

Fig. 3. Modified QFD

3.3 Vulnerability Assessment

Wisner (2004) introduced the Pressure and Release model, where risk was calculated in accordance with the equitation

$$Risk = Hazard \times Vulnerability \quad (1)$$

This indicates that vulnerability is the result of root causes, dynamic pressures, and unsafe conditions, and the authors indicate that root causes progress into dynamic pressure and then unsafe conditions as vulnerability progresses. Wisner (2004) introduces root causes as underlying causes that determine the access and distribution of power and resources, whereas dynamic pressure is introduced as all the activities and processes that transform root causes into unsafe conditions. Finally, unsafe conditions are the specific forms that vulnerability is noted through.

Westen et al. (2011) present a number of methods to calculate vulnerability. One of these methods contains the following formula:

$$Vulnerability = Exposure + Resistance + Resilience \quad (2)$$

where exposure is the at-risk property and population, resistance is the measures taken to prevent, avoid or reduce loss, and resilience is the ability to recover the prior state or achieve the desired post-disaster state. Thus, in the risk register, Eq. 2 was utilized in accordance with variables presented in Table 3.

Table 3. Vulnerability components in the risk register

Exposure	Residual Risk
Resistance	Action Plan Rating
Resilience	Related Capability Rating

Taking into consideration the elements used to obtain the variables in Eq. 2, where the exposure is obtained from the residual risk with a maximum value of 9, while the resistance and resilience will utilize the criteria identified in Tables 5 and 6, respectively. Thus, as the value of the three variables increases, the vulnerability also increases. The aim is to decrease the vulnerability score to its lowest value as much as possible. Tables 4, 5, and 6 represent the value of the rating as well as the criteria for each rate for exposure, resistance, and resilience rating, respectively, which were utilized in the risk register (Table 12).

Table 4. Exposure rating (residual risk)

Score	Rate	Criteria
7–9	High	Cost or time delay is above risk tolerance level
4–6	Medium	Cost or time delay is below tolerance level but above risk appetite level
1–3	Low	Cost or time delay is within risk appetite level

Table 5. Resistance rating

Score	Rate	Criteria
1	EAO	Residual risk is mitigated to be within risk appetite level
2	SEAP	Residual risk is mitigated to be below tolerance level but above risk appetite level
3	NEAP	Residual risk is not mitigated

Table 6. Resilience rating

Score	Rate	Criteria
1	IAA	• Real options are deployed to maximize response flexibility • Contingency and crisis management plans are in place and rehearsed regularly
2	INAA	• Some options were identified • Some contingency or crisis management plans are in place but not practiced regularly
3	NI	• Options are not identified • No contingency or crisis management plans are in place

Where IAA: Identified, available, and assigned; INAA: Identified but not available or assigned; and NI: Not identified.

4 Case Study

To test the proposed methodology, a risk register for a construction project was obtained for a mall in Jeddah, Saudi Arabia. The obtained risk register covered the construction site of the project, which presents some limitations for gauging the effectiveness of the proposed methodology. The information provided in the risk register was applied to the proposed risk register as shown in Table 12 and was completed while using the QFD as shown in Table 11, where the relationship between the controls and the residual risk and controls were completed based on the criteria provided in Tables 7 and 8, respectively.

Table 7. Correlation between controls (How)

Correlation	
Positive	+
Negative	–
No Correlation	0

Table 8. Relationship between residual risk and control (what and how)

Relationship	
Strong	9
Moderate	3
Weak	1

Furthermore, weights were assigned to each risk based on the residual risk score. Weights were also divided into three categories as demonstrated in Table 9 based on the probability of the risk with relation to the activity taking place.

Table 9. Proposed weights for identified risks

Weight	
High	3
Moderate	2
Low	1

In Table 9, the relationship between the risks and controls is established in the risk register Table 12 and identified in the action plans from a QFD matrix perspective. Each control from the action plan was placed in a column, and each risk was placed in a row. The intersection between them would represent the relationship between the risk and the control. The higher the relationship score, the more important the control would be. In this case, there were a total of 39 risks and 28 controls. Therefore, the maximum score a control could obtain is 1080, where one control affects all the identified risks. This process enables classifying the controls based on their impact and identifying the critical ones. Once that step is completed, it would be possible to identify the capabilities associated with the top-ranked controls to quantify the resilience score based on the criteria provided in Table 6. The results obtained from the QFD were then utilized in the risk register based on the criteria specified in Table 6 to evaluate the appropriateness of the response. Furthermore, the highest score for resilience that was given was 2, as in most cases the action plans were not specific and did not include controls that would provide feedback to measure the appropriateness of the applied measures. The QFD results and risk register are shown in Tables 10, 11 and 12.

Table 10. QFD matrix for residual risk with an action plan

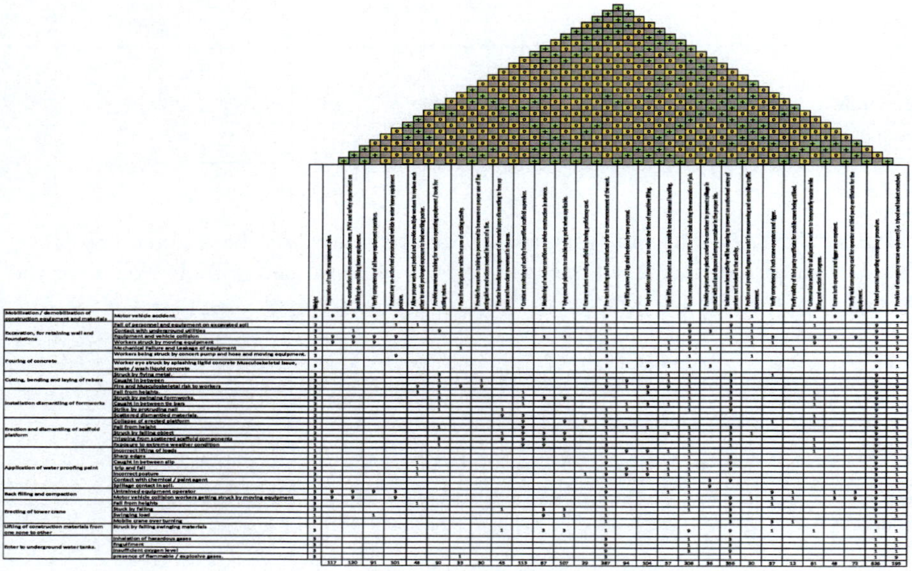

Activity	Risk	Weight
Mobilization / demobilization of construction equipment and materials	Motor vehicle accident	3
Excavation, for retaining wall and foundations	Fall of personnel and equipment on excavated soil	2
	Contact with underground utilities	3
	Equipment and vehicle collision	1
	Workers struck by moving equipment	3
	Mechanical Failure and Leakage of equipment	3
Pouring of concrete	Workers being struck by concert pump and hose and moving equipment.	3
	Worker eye struck by splashing ligiid concrete Musculoskeletal issue, waste / wash liquid concrete	3
Cutting, bending and laying of rebars	Struck by flying metal.	3
	Caught in between	2
	Fire and Musculoskeletal risk to workers	2
Installation dismantling of formworks	Fall from heights.	3
	Struck by swinging formworks.	2
	Caught in between tie bars	2
	Strike by protruding nail	2
	Scattered dismantled materials.	2
Erection and dismantling of scaffold platform	Collapse of erected platform	3
	Fall from height	3
	Struck by falling object	2
	Tripping from scattered scaffold components	2
	Exposure to extreme weather condition	1
Application of water proofing paint	Incorrect lifting of loads	1
	Sharp edges	3
	Caught in between slip	2
	trip and fall	3
	Incorrect posture	2
	Contact with chemical / paint agent	2
	Spillage contact in soil.	1
Back filling and compaction	Untrained equipment operator	3
	Motor vehicle collision workers getting struck by moving equipment	3
Erecting of tower crane	Fall from heights	3
	Stuck by falling	2
	Swinging load	1
	Mobile crane over turning	3
Lifting of construction materials from one zone to other	Struck by falling swinging materials	3
Enter to underground water tanks.	Inhalation of hazardous gases	3
	Engulfment	3
	Insufficient oxygen level	3
	presence of flammable / explosive gases.	3

After completing the steps in the QFD, the results showed that when the weights of the controls are combined based on the risks, the action plans that were developed to mitigate are combined and the risks are arranged based on that (see Table 11). Additionally, when taking into consideration the relationship between the controls, it is observed that some of the controls can contribute to other risks and were not taken into consideration in the other risk action plans to reduce the overall risk on the project. For example, the controls for risk number 11 can contribute to most other risks (see Table 11).

Contemplating the results yielded by the risk register, it is noted that two of the identified risks had a high vulnerability score, and their action plans need to be improved to lower the vulnerability score, while the rest of the risks yielded medium vulnerability scores (see Table 12). Thus, if the results were summarized, the final outcome would be that risks 1 and 4 required serious improvement to the action plans (see Table 13), while the other risk action plans would benefit from updating the procedure manuals to include controls that provide feedback on risk status, which was not apparent in the provided action plans, as the controls observed in the action plans can be categorized as corrective controls for the most part, which may contribute to decreasing the probability and impact of the risks occurring to the same degree. However, the effect may increase if preventive controls that provide feedback on the status of the risk were to be implemented to provide increased ability to monitor the situation as the project progresses.

Table 11. Risk order based on the Control Importance Score obtained from QFD

Risk #	Control	Total controls score	Ranking of controls by score
11	• Trained personnel regarding emergency procedures • Provision of emergency rescue equipment (i.e., tripod and basket stretcher)	237	1
8	• Isolate areas where the activity will be ongoing to prevent unauthorized entry of workers not involved in the activity • Position and provide a flagman to assist in maneuvering and controlling traffic movement	48	3
6	• Constant monitoring of activity from certified scaffold supervisor • Monitoring of weather conditions to advise construction in advance • Tying erected platform to suitable tying point when applicable • Ensure workers erecting scaffolding have a proficiency card	155	4
1	• Preparation of traffic management plan • Pre-coordination from the construction team, Privet Moter Vechile, and safety department on mobilizing/de-mobilizing heavy equipment	45	5
2	• Verify the competency of all heavy equipment operators • Prevent any unauthorized personnel and vehicle from entering heavy equipment posture	316	6

(*continued*)

Table 11. (*continued*)

Risk #	Control	Total controls score	Ranking of controls by score
4	• Provide awareness training for workers operating equipment/tools for cutting rebars • Place the fire extinguisher within the area of the cutting activity • Provide fire warden training to personnel for awareness of the proper use of fire extinguishers and actions needed in event of a fire	786	7
9	• Verify competency of crane operators and rigger • Verify the validity of third-party certificates for utilized mobile cranes • Communicate activity to all adjacent workers to temporarily vacate the area during the lifting and erection process	376	8
10	• Ensure operator and rigger are competent • Verify valid competency cards for the operator and third-party certificates for the equipment	130	9
3	• Allow proper work/rest periods and provide multiple workers to replace each other to avoid prolonged exposure to bad working posture	98	10
5	• Practice immediate arrangement of material upon dismantling to free up space and have clear movement in the area	819	11

Table 12. Completed test risk register

ID	Date Raised	Activity	Risk Description	Risk Probability Rating	Risk Impact Without Controls	Inherent Risk	Controls	Risk Impact with Control	Residual Score	Action Plan	Action Plan Rating	Related Capabilities	Related Capabilities Rating	Vulnerability Rating	Improved Action
1	7 June 2019	**Mobilization/Demobilization of construction equipment and materials**	• Motor vehicle accident	3	3	9	• Provision of bank man/flagman to assist with heavy equipment while mobilizing and demobilizing	2	6	• Preparation of traffic management plan • Pre-coordination from the construction team, PVM, and safety department on mobilizing and de-mobilizing heavy equipment	3	• No traffic plan	1	12	Complete traffic plan taking into consideration the site's surrounding area and all vehicle types
2	22 July 2019	**Excavation for retaining wall and foundations**	• Fall of personnel and equipment on excavated soil • Contact with underground utilities • Equipment and vehicle collision • Workers struck by moving equipment • Mechanical failure and leakage of equipment	3	3	9	• Provision of hard barricade on all excavation sites, identification of pre-existing utilities through layout and drawings • Provision of a designated spotter or flagman to assist with heavy equipment while moving at the site • Immediate removal of any leakage from equipment utilizing proper material to contain affected soil	2	6	• Verify competency of all heavy equipment operators • Prevent any unauthorized personnel and vehicle from entering heavy equipment posture	2	• Lack of updated competency records	2	10	Update procedure manual to include controls that provide feedback on risk status
3	29 July 2019	**Pouring of concrete**	• Workers being struck by concert pump and hose and moving equipment. • Worker's eye struck by splashing liquid concrete • Musculoskeletal issues, waste/wash liquid concrete	3	3	9	• Proper positioning of pump hose, provision of spotter and flagman to assist with the movement of heavy equipment • Utilize a face shield and/or safety goggles to protect eyes from splashing concrete • Provide area with washing concrete pumps with the bottom being covered by plastic to prevent soil contamination	2	6	• Allow proper work/rest periods and provide multiple workers to replace each other to avoid prolonged exposure to bad working postures	1	• Procedure manual requires improvement	2	9	Update procedure manual to include controls that provide feedback on risk status
4	29 July 2019	**Cutting, bending, and laying of rebars**	• Struck by flying metal, caught in between flying objects, fire, and musculoskeletal risk to workers	3	3	9		3	9	• Provide awareness training for workers operating equipment/tools for cutting rebars • Place the fire extinguisher within the area of the cutting activity • Provide fire warden training to personnel to raise awareness of the proper use of fire extinguishers and actions needed in event of a fire	2	• Lack of awareness materials • Emergency training for fire accidents is not adequate	1	14	Improve awareness and emergency training to include all known accident sources and how to mitigate them
5	29 July 2019	**Installation dismantling of formworks**	• Fall from heights, struck by swinging formworks, caught in between tie bars, strike by protruding nails, scattered dismantled materials	3	3	9	• Erect suitable height of the platform • Secure lifted formworks and the proper tagline attached • Coordination between each carpenter while installing and tightening the tie roads	2	6	• Practice immediate arrangement of material upon dismantling to free up space and have clear movement in the area	2	• Procedure manual requires improvement	1	9	Update procedure manual to include controls that provide feedback on risk status
6	24 August 2019	**Erection and dismantling of a scaffolding platform**	• Collapse of erected platform, fall from height, being struck by a falling object, tripping from scattered scaffolding components • Exposure to extreme weather conditions	3	3	9	• Proper leveling and compaction of ground where scaffold will be erected • Provision of fall protection to workers while erecting and dismantling; isolation of area during erection and dismantling • Securing of lifted materials with a proper tagline • Arrangement of scaffold components to present trapping incident	2	6	• Constant monitoring of activity from certified scaffold supervisor • Monitoring of weather conditions to advise construction in advance • Tying erected platform to suitable tying point when applicable • Ensure workers erecting scaffolds have proficiency cards	2	• Procedure manual requires improvement	1	9	Update procedure manual to include controls that provide feedback on risk status
7	23 November 2019	**Application of waterproof paint**	• Incorrect lifting of loads, sharp edges, caught in between slip, trip, and fall, incorrect posture, contact with chemical/paint agent, spillage/contact with soil	3	3	9	• Use an easy mode of transport, such as a trolley • Pre-task briefing to be conducted by the foreman prior to work commencement • Deployment of enough personnel for lifting depending on the type of load to be lifted • Keeping limbs away from the pinch point while placing the load • Proper personal protective equipment to be used during manual handlining of materials and loads	2	6	• Conduct pre-task briefing prior to the commencement of the work • Any lifting above 20 kg shall be done by two personnel • Deploy additional manpower to reduce the time of repetitive lifting. • Utilize lifting equipment as much as possible to avoid manual handling. • Use the required and supplied PPE for the task during the excavation of the job. • Provide polyurethane plastic under the container to prevent spillage in contact with soil and dispose of all empty containers in the proper bin	1	• Procedure manual requires improvement	1	8	Update procedure manual to include controls that provide feedback on risk status
8	12 October 2019	**Backfilling and compaction**	• Untrained equipment operators • motor vehicle collision workers getting struck by moving equipment	3	3	9	• Verify all equipment operators are authorized and certified to operate heavy equipment • Vehicles must be equipped with audible alarms, especially when backing	2	6	• Isolate activity area to prevent unauthorized entry of workers not involved in the activity • Position and provide a flagman to assist in maneuvering and controlling traffic movement	2	• Procedure manual requires improvement	2	10	Update procedure manual to include controls that provide feedback on risk status
9	9 November 2019	**Erection of tower crane**	• Fall from heights, struck by falling, swinging load, mobile crane overturning	3	3	9	• Isolation of the area where a crane is being erected • Secure approval of method statement for tower crane erection • Secure approval of the lifting plan • Properly rig and lift components of the tower crane	2	6	• Verify competency of both crane operators and rigger • Verify the validity of the third-party certificate for the utilized mobile crane • Communicate activity to all adjacent workers to temporarily vacate during the lifting and erection process	2	• Procedure manual requires improvement	2	10	Update procedure manual to include controls that provide feedback on risk status
10	15 July 2019	**Lifting of construction materials from one zone to another**	• Struck by falling swinging materials	3	3	9	• Isolation of lifting radius • Proper and secure rigging of materials	2	6	• Ensure both operator and rigger are competent • Verify valid competency cards for the operator and third-party certificates for the equipment	1	• Procedure manual requires improvement	2	9	Update procedure manual to include controls that provide feedback on risk status
11	7 December 2019	**Entering into underground water tanks.**	• Inhalation of hazardous gases, engulfment, insufficient oxygen level, presence of flammable/explosive gases	3	3	9	• Train all personnel entering the water tank • Conduct gas testing prior to entry • Provision of local exhaust ventilation to workers for proper air movement • Dedicated entry attendant and secure work permit	2	6	• Trained personnel for the emergency procedure • Provision of emergency rescue equipment (i.e., tripod and basket stretcher)	1	• Procedure manual requires improvement	2	9	Update procedure manual to include controls that provide feedback on risk status

Table 13. Risks with medium vulnerability score

ID #	Risk	Improved action plan
1	Motor vehicle accident	Complete traffic plan with taking into consideration the site's surrounding area and all vehicle types
4	Struck by flying metal, caught in between flying objects, fire, and musculoskeletal risk to workers	Improve awareness and emergency training to include all known accident sources and how to mitigate them

5 Managerial Insights

This section lists some managerial insights that help managers when making a decision.

1. Establishing the context is crucial for developing the appropriate risk management scope, architect, function design, and risk criteria. Both internal and external elements must be considered.
2. Ideally, a holistic approach should be utilized in the planning phase of a project where all the activities are considered, the risks associated with them are identified and analyzed, and an appropriate response strategy is chosen based on the company's risk appetite.
3. When identifying the risks, they must be analyzed to the appropriate level where the root causes are found to enable the treatment plan of addressing them and preventing the threat of reoccurring.
4. When developing a treatment plan, the external and internal context of the company should be taken into consideration to target what the company has control over.
5. Risk has two sides. The positive one is known as opportunities and the negative side is known as threats. Properly identifying risks and choosing the appropriate approach to addressing them can present new opportunities for the company to grow through exploring them within the respective risk appetite level.

6 Conclusion and Recommendations

The model proposed in this study showed promising signs in assisting the project management team in reducing the risk of project delay starting from the planning phase of the project if the proposed concept is implemented thoroughly. However, due to the limitation at the time of developing this paper, a finished project risk register was utilized to test the proposed methodology and serve as a proof of concept. Overall, this paper has proposed a concept to improve the efficiency of the ISO31000 framework for construction projects by adding two more steps before the implementation of action plans. The proposed two steps are identifying related capabilities and then conducting a vulnerability assessment. The score of the vulnerability assessment would serve as an indicator of whether the action plans have a good chance of succeeding or if there is room for improving the plan. When applying the concept, it is recommended to follow the timeline and order of activities as presented in the project plan in order to cover all aspects of the project. Finally, risk management should be embraced in a dynamic manner where the risk register is visited on periodic bases and when there are changes in the project, such as change in scope or timeline.

References

Acebes, F., Pajares, J., González-Varona, J.M., López-Paredes, A.: Project risk management from the bottom-up: activity risk index. CEJOR **29**(4), 1375–1396 (2020). https://doi.org/10.1007/s10100-020-00703-8

Ahmadi-Javid, A., Fateminia, S.H., Gemünden, H.G.: A method for risk response planning in project portfolio management. Proj. Manag. J. **51**(1), 77–95 (2020)

Aleksic, A., Puskaric, H., Tadic, D., Stefanovic, M.: Project management issues: vulnerability management assessment. Kybernetes **46**(7), 1171–1188 (2017)

Anysz, H., Buczkowski, B.: The association analysis for risk evaluation of significant delay occurrence in the completion date of construction project. Int. J. Environ. Sci. Technol. **16**(9), 5369–5374 (2018). https://doi.org/10.1007/s13762-018-1892-7

Bialas, A.: Vulnerability assessment of sensor systems. Sensors (Basel, Switzerland) **19**(11) (2019)

Birkmann, J.: Measuring Vulnerability to Natural Hazards: Towards Disaster Resilient Societies. United Nations University Press (2006)

Curtis, P., Carey, M.: Thought Leadership in ERM: Risk Assessment in Practice. The Committee of Sponsoring Organizations of the Treadway Commission (COSO), USA (2012)

Dinu, A.-M.: Risk treatment in projects. Knowl. Horiz./Orizonturi Ale Cunoasterii **8**(1), 160–163 (2016)

Doval, E.: Risk management process in projects. Rev. Gen. Manag. **29**(2), 97–113 (2019)

Viles, E., Rudeli, N.C., Santilli, A.: Causes of delay in construction projects: a quantitative analysis. Eng. Constr. Archit. Manag. **27**(4), 917–935 (2019)

Elsaeed, M.S.M.: Performance improvement for public construction projects using risk analysis. J. Eng. Sci. **45**(6), 878–899 (2017)

Galli, B.J.: Risk management in project environments: reflection of the standard process. J. Mod. Project Manag. 39–49 (2017)

Gunduz, M., Al-Naimi, N.H.: Construction projects delay mitigation using integrated balanced scorecard and quality function deployment. Eng. Constr. Architect. Manag. **29**, 2073–2105 (2021)

Hauser, J., Clausing, D.: The House of Quality. Harvard Business Review (1988)

Hutchins, G.: ISO 31000:2018 Enterprise Risk Management. Quality Plus Engineering (2018)

International Organization for Standardization: Risk management – Guidelines (ISO Standard No. 31000:2018) (2018)

Juran, J.M.: The quality trilogy. Qual. Prog. **19**(8), 19–24 (1986)

Kendall, K.: The increasing importance of risk management in an uncertain world. J. Qual. Particip. **40**(1), 4–8 (2017)

Keshavarzian, S., Silvius, G.: The perceived relationship between sustainability in project management and project success. J. Mod. Project Manag. **9**(3), 66–85 (2022)

Kumar, P.: Industry of quality (IoQ) – an industry 4.0 perspective. Int. J. Appl. Res. **6**, 109–114 (2020)

Kundu, A., Sarkar, D.N., Bhattacharya, A.: The effect of uncertainty on the formulation of strategies: a study of selected Indian organizations. SN Bus. Econ. **1**(1), 1–21 (2020). https://doi.org/10.1007/s43546-020-00010-z

Lindbom, H., Tehler, H., Eriksson, K., Aven, T.: The capability concept – on how to define and describe capability in relation to risk, vulnerability and resilience. Reliab. Eng. Syst. Saf. **135**, 45–54 (2015)

Murr, P., Carrera, N.: Institutional logics and risk management practices in government entities: evidence from Saudi Arabia. J. Account. Organ. Chang. **18**(1), 12–32 (2022)

Rahi, K., Bourgault, M., Preece, C.: Risk and vulnerability management, project agility and resilience: a comparative analysis. Int. J. Inf. Syst. Proj. Manag. **9**(4), 5–21 (2021)

Saeed, Y.S.: Cost and time risk management in construction projects. Tikrit J. Eng. Sci. **25**(1), 42–48 (2018)

Smit, M.J.: Will the standard for project management and a guide to the project management body of knowledge (PMBOK® Guide) - Seventh Edition recently published by the Project Management Institute (PMI, 2021) lead to more successful delivery of projects?. PM World J. **10**(10), 1–9 (2018)

Snell, M.K.: History of vulnerability assessments. Sandia National Lab. (SNL-NM), Albuquerque, NM (United States) (2015)

Sols, A.: A comprehensive approach to dynamic project risk management. Eng. Manag. J. **30**(2), 128–140 (2018)

Tworek, P.: Risk management in public sector organizations – principles, methods and tools. In: Proceedings from 8th International Conference "Managing and Modelling of Financial Risks" (PART III), Ostrava 5th–6th September 2016. VŠB – Technická Univerzita Ostrava (2016)

Walkowski, M., Oko, J., Sujecki, S.: Vulnerability management models using a common vulnerability scoring system. Appl. Sci. **11**(18), 8735 (2021). https://doi.org/10.3390/app11188735

Wisner, B.: Vulnerability as Concept, Model, Metric, and Tool. Oxford Research Encyclopedia of Natural Hazard Science (2004). https://oxfordre.com/naturalhazardscience/view/10.1093/acrefore/9780199389407.001.0001/acrefore-9780199389407-e-25. Accessed 30 Aug 2022

Westen, C.J.V., Alkema, D., Damen, M.C.J., Kerle, N., Kingma, N.C.: Multi-hazard Risk Assessment. Academia (2011). https://www.academia.edu/30308272/Multi_hazard_risk_assessment_Distance_education_course_Guide_book. Accessed 18 June 2022

Using Artificial Intelligence (AI) in the Management Process

Abdulsadek Hassan[1](✉), Mahmoud Gamal Sayed Abd Elrahman[2], Sumaya Asgher Ali[2], Nader Mohammed Sediq Abdulkhaleq[2], Mohanad Dahlan[3], and Ghassan Shaker[3]

[1] Ahlia University, Manama, Kingdom of Bahrain
dr.sadek1964@yahoo.com

[2] Faculty of Mass Communication, Radio and Television Department, Beni-Suef University, Beni Suef, Egypt

[3] University of Business and Technology (UBT), Jeddah, Kingdom of Saudi Arabia
{dahlanm,g.shaker}@ubt.edu.sa

Abstract. This study examines the use of artificial intelligence (AI) in administrative processes. The results reveal that the adoption of AI mechanisms within organizations is worth investing in, as the use of information technology benefits the management of administrative functions and relationships. Using AI enhances decision-making at all levels within an organization, and the results demonstrate that expert systems are the embodiment of AI that can serve the quality and effectiveness of administrative management. This study recommends the adoption of expert systems and AI models in various business organizations and public bodies to improve the decision-making process.

Keywords: Artificial intelligence · Management · Technology · Administrative process · Expert systems · Business organizations · Public bodies · Symbolic representation method

1 Introduction

Artificial intelligence (AI) is the next developmental step for the future of organizations, and its advantages in serving humanity at all levels cannot be overlooked. AI expands concepts and tasks, and the significant advances in research in this direction include machine self-learning with the addition of decision-making capabilities that were exclusive to human brains. Technology has advanced so much that computers can make decisions and think about the unthinkable with the help of a suite of software that can automate tasks and produce usable information [6]. Many countries and institutions are accelerating toward the future and are ready to make decisions about policies, science, and technology; for example, from the renewable energy industries to the invasion of space to the shift toward the concept of a "smart country" [14].

The latest studies also reveal that AI is still 30 years away from full self-awareness. Until then, AI can be used in projects to reduce cost, improve efficiency, and collect

B. Alareeni et al. (Eds.): ICBT 2022, LNNS 620, pp. 549–557, 2023.
https://doi.org/10.1007/978-3-031-26953-0_49

data intelligently. All these aspects will help in decision-making and forecasting future directions [5].

With the help of AI, enterprise and project management can be easier, faster, and more cost-effective, with fewer errors. There are many examples where AI technology can help improve project management, especially in automating repetitive tasks such as assembly line procedures and preparing reports [17]. AI can help prepare for future challenges in project management, as it is not just about managing and delivering basic requirements, but the machine can incorporate more trends and visualizations to come up with more consistent and confident analyses of external forces [18]. AI can reduce the risks associated with projects through data-driven forecasts. The process of entering data for every aspect of a project is a tedious task that most employees do not master [11], and a machine can be used for more comprehensive and accurate data input. Moreover, AI can be used to predict and fill in the blanks according to the available data. Data analysis is one of the tasks in which a machine can be used to unobtrusively collect metadata, which is a set of data that provides information about other data to employees. This, in turn, leads to improved coordination among team members, which affects the efficiency of project implementation [1].

Along with the tremendous progress made in AI technology, there has been a revolution brought about by other technological developments in various fields such as the internet, smartphones, robotics, the Internet of Things (IoT), self-driving cars, and digital money [7].

2 AI Applications in Business Management

Modern applications of AI in the workplace include:

1 **Recruitment and qualifications:** Today, machines scan applicants' CVs and choose the most suitable candidates for a personal interview. This is very useful in large organizations that employ hundreds of employees annually.
2 **On-the-job training:** The employee's journey with learning and growth does not end once they get the job. AI technology will play a role in the continuous training of most employees in the future as well as the transfer of skills from one generation to another [6].
3 **Enhanced workforce:** Some employees may feel uncomfortable with AI because of the belief that it will replace them at work, and, therefore, they will lose their jobs. However, the issue is more about "enhancement" than "replacement" [2].
4 **Monitoring in the workplace:** Technology can be used to monitor employees in the workplace, and this may include practices that go beyond legal surveillance to illegal "espionage" [6].
5 **Robots in offices:** It has become fashionable to see robots in factories and warehouses, but their presence in offices is relatively new. For example, delivery robots (Segway) can make their way along corridors to deliver messages and parcels from one office to another, some robots (Gamma) specialize in security monitoring, and some are responsible for parking management (ParkPlus) [6].

3 Best AI Applications in Management in 2020

AI is involved in many areas and has become a feature of civilized, technological, and economic progress, but it is probably most prominently used in management [4]. Managers in government institutions and private companies need AI to manage their organization efficiently and effectively by arranging daily, weekly, monthly, and even annual appointments and tasks [8].

Robin – AI Voice Assistant
The Robin – AI Voice Assistant is an app that obeys voice commands to help you drive safely and avoid accidents [11].

Microsoft Cortana
Microsoft Cortana is a digital assistant and one of the best AI applications that you can rely on to manage your business. It makes organizing files, including photos, videos, and important business documents, more efficient [6].

Mona
One of the best applications of AI in management, Mona is available in the iTunes app store for iPhones and iOS phones. Mona depends primarily on AI technology and provides useful information for daily life [9].

Vision – Smart Voice Assistant
One of the applications of AI in management is Vision, which is similar to the Lyra Virtual Assistant application that is used by many researchers for managing operations. For example, you can learn about weather conditions on a daily, weekly, and monthly basis to help identify the climate when you are planning a work appointment or deciding to go on a picnic [10]. You can also open smartphone files and manage photos, files, and videos using your voice [5].

4 Examples of Using AI Applications in Management

1 Independent consultant

Global companies, such as McKinsey and BCG, have turned to AI as the best strategic advisor they can rely on by employing independent algorithms [12]. A supervisory unit of workers can also determine what tasks AI solutions can perform independently and how this can be done [1].

2 Independent external supplier

Algorithms for AI applications have a different role in the field of procurement, as procurement and investment opportunities are outsourced to independent AI programs. This model is implemented within Amazon and Accenture [13], and data scientists play a pivotal role in this regard.

3 **Smart cities**

Scientists technically coordinate the required service levels and data quality standards so that applications are given autonomy in filtering opportunities [15].

4 **Freelance employees**

Smart software can be used as a 'competent colleague' who can provide the right solutions, if not the best solutions; Alibaba and Netflix have implemented smart software solutions in their organizations [7]. However, the challenge is that managers are unsure if they will get the best results when retraining people or qualifying applications while working in rapidly changing learning environments [4].

5 **Overall independence**

Investment funds and technology giants, such as Renaissance Technologies, are applying the global autonomy model of AI solutions in management, aiming to move organizations to a new stage of innovation and profitability, while taking risks at the same time [3]. It took mutual funds several years to learn which algorithms to rely on when building investment rates and forecasts [2].

5 Project Planning Methods That AI Can Help With

Make the Life of a Project Manager Easier
AI can be used in project management to automate repetitive tasks, which it already does to a large extent, especially in assembly line procedures. It can make project tasks, such as time tracking, reporting progress, etc., more efficient [15].

Better Preparation for Future Challenges
Project management is not just about managing and delivering basic requirements. The interdependence and compatibility between various tasks play a major role in exceeding expectations with the results. Project team members can estimate externalities based on intuition and past trends [17].

Data-Driven Forecasts
Entering data for every aspect of a project is a tedious task, but a machine can be used to encourage employees to provide more comprehensive and accurate data [16].

Data Analysis
Smartphones are getting smarter with each passing day and can predict what we are going to type on our phones when writing a message and when we are walking. This is an essential process for observing and predicting behavior, but it can be used in more effective ways [6].

Improve Collaboration and Coordination
Previously, a project manager was the primary control and monitoring authority on any

project. However, with the help of AI, the project manager's role becomes that of a mentor and advisor, and everyone is connected in real-time [8]. A machine can create algorithms that can determine who will complete a task in a way that is optimal for the project [6].

6 AI Features

Symbolic representation: This is the representation of information through symbols and is close to the individual's representation of the information that they possess in their daily life [6]. This is one of the first features of AI programs, as it deals with symbols that are not numerical, which is the opposite of what is familiar and acceptable in most computers of the current era that deal with numerical quantities. Nevertheless, there is nothing to prevent AI programs from performing the arithmetic operations they are used to [7].

Experimental research: AI programs tend to solve difficult problems based on certain logical steps. It follows the method of empirical research, as AI programs solve problems that do not have a known solution method, which means that the programs do not use sequential steps to find the correct solution [13]. Instead, AI programs choose a specific method while retaining the possibility of changing the method if the first option does not lead to a quick solution [17].

Embracing and representing knowledge: One of the important characteristics of AI programs is to use symbolic representation to express information and follow experimental research methods to find solutions. Therefore, AI programs must be constructed with a large base of knowledge that contains the link between cases and results. Unlike statistical programs, AI programs guarantee a method of information representation, as they use a special structure to describe the knowledge, and this structure guarantees facts and the relationships between these facts [2].

Unconfirmed or incomplete data: AI programs must be designed to provide solutions if data is uncertain or complete. This does not mean that solutions are given regardless of how correct or incorrect they are, but to perform effectively, acceptable solutions should be provided [6].

Ability to learn: The ability to learn is one of the characteristics of intelligent behavior. Whether human learning is through observation or taking advantage of previous mistakes, AI programs must be built on machine learning strategies and should have the ability to improve performance by taking previous errors into account [16].

Inferencing: This is the ability to elicit expected solutions to a specific problem from the reality of known data and previous experiences, especially for problems that cannot be solved with traditional or familiar methods [13].

7 Examples of Using AI in Business

1. **Customer service**

   Replacing humans with AI in customer service is a common example of the use of AI in business [15]. With the advent of chatbots, customers can now interact with businesses in real-time to resolve complaints, place requests, obtain information, and do just about anything they need to [7].

2 **Business intelligence**

Today's business environment is growing steadily, and finding true value or information from data has become difficult. This has resulted in the routine adoption of AI in business intelligence to obtain valuable and useful information from data [3]. This information helps companies improve marketing effectiveness, understand customers better, develop business strategies to help make corporate decisions, and leverage AI to drive business [15]. One of the common tools used in business is Microsoft Power BI, which helps companies obtain basic analytics to see which strategies or decisions have the greatest impact on business metrics [6].

3 **Target marketing**

The real key to increasing companies' revenues or sales is knowing what consumers need, what they want, and what to market to them [14]. Through data obtained from customers' online activities, businesses can now use AI to simultaneously predict and target who they should be marketing to, the effectiveness of marketing activities, and reduce the overall cost of marketing [5]. In medicine or education, AI is based on the principle of prediction and knowing what patients or students need, and it is also used in management and business [9].

4 **Product recommendations and predictive analytics**

To increase the effectiveness of marketing efforts and customer engagement with a product, companies must be able to recommend products that maintain interest and satisfy customers' desires [3]. Companies like Netflix, Spotify, and Amazon are now using AI to understand their customers' habits and behaviors to predict what products to recommend [6]. Nearly 75% of what users watch on Netflix comes from these recommendations. The company's AI recommendation also cuts expenses by about $1 billion each year [3].

5 **Natural language processing**

A common question is, "When will machines be able to read, write, and understand languages like humans?" [2]. With advances in natural language processing, companies now offer smart digital assistant products to assist users with routine tasks [16]. Companies are also using AI in business to generate automated business reports without human oversight as well as conducting sentiment analysis to understand people's brand perceptions from various online comments, tweets, and posts about the company [11].

8 What are the Benefits of AI in Business?

The benefits that companies can reap from using AI include:

- Process automation [6]
- Excellent results in sales and revenue [17]
- A better understanding of customers and improved service experience
- Cheat detection [5]
- Improved and more reliable customer service [18]

9 The Role of AI in Making Business Decisions

AI can eliminate the huge costs of a wrong decision because it can virtually eliminate human biases and errors, which, in turn, can speed up the decision-making process. We will now highlight how AI can make a difference in business and simplify decision-making [17].

1. **Marketing decision-making**
   With today's customer-driven market complexity, decision-making includes understanding customers' needs and wants and aligning products with those needs and wants. Therefore, dealing with changing customer behavior is vital to making the best marketing decisions [17].
2. **Recommendation system**
   A recommendation system (engine) is a technology that recommends products or other items to users [7]. The AI system learns what the consumer prefers based on "explicit" or "implied" feedback, as this structured information can help reduce the bounce rate and better craft targeted content or even physical products for the customer [13].
3. **Problem-solving**
   An expert system is a type of problem-solving software that attempts to replicate the knowledge and thinking styles of experts [15]. This system uses specialized thought processes to provide data, which includes assessment and recommendations for the problem [15].
4. **Opinion mining**
   AI can provide reliable insight to decision-makers; for example, using AI in business marketing can provide invaluable insights into consumers, which helps companies enhance their communication with consumers and helps retailers anticipate and respond to product demand quickly [12].
5. **Enhanced analytics**
   According to a recent Gartner press release, enhanced analytics will be the next big trend that will change the way analytics content is delivered, spent, and shared [3]. AI not only improves the performance of each team member but also improves the business's competitive advantage [10].

10 The Challenges of AI in Business Management

AI still faces several challenges, and some of the most common challenges in implementing AI include the following elements [10].

1. **Select the correct data set**
   As AI systems are powered and developed by data, obtaining the exact set of quality data for those systems must be considered first.
2. **Data and storage security**
   Most AI applications use a large amount of data to learn and make smart decisions, and the disadvantage of using large amounts of data is that it may create a storage problem for companies. Furthermore, data-driven automation in business processes may lead to problems related to data security and protection [11].

3. **Infrastructure**
   Replacing legacy infrastructure in traditional systems remains a major challenge for most organizations [4]. Most AI-based solutions have a high level of computational speed, and AI-based systems can achieve even more speed if a business has a large infrastructure and advanced processors [3].
4. **Integration of AI into existing systems**
   This may come as a surprise, but integrating AI into current business systems is the most common challenge that companies face when trying to implement AI [13].
5. **Complex algorithms and AI model training**
   The function and performance of business intelligence processes are highly dependent on AI algorithms, and companies that plan to implement AI must have a clear idea of how AI-based solutions or technologies work and can transform their outcomes [3].

11 Conclusion

There is no doubt that AI is the next developmental step for the future of companies. However, many do not realize that the future is coming sooner than expected, as concepts related to AI are expanding and research in this field is evolving to include machine self-learning and decision-making capabilities that were previously exclusive to human minds. Technology has advanced so much that even computers can make decisions, adapt, and think with the help of a set of algorithms that can automate repetitive tasks and produce usable data.

Project management in companies can include building software for logistics to financing, and each of them requires planning, management, and control.

AI is the new driver in the market, and companies can benefit by taking advantage of this powerful technology. However, many people fear that machines will replace them, and companies will make positions redundant. Although AI will reduce the number of jobs that humans are doing today, the bright side is that people will be free to perform more important functions that a machine cannot perform.

12 Study Recommendations

Business organizations and public bodies should adopt various expert systems and AI models and use them to help make decisions.

Companies should follow and keep abreast of recent developments in these fields, as the scientific arena witnesses developments and modifications every day that makes these systems more effective.

Organizations should apply AI methods and models to benefit from leading international experiences, especially regarding genetic algorithms, as they are an effective tool that helps make optimal decisions on marketing and investment options in different areas.

References

1. Al-Mushayt, O.S.: Automating e-government services with AI. IEEE Access **7**, 146821–146829 (2019)
2. Androutsopoulou, A., Karacapilidis, N., Loukis, E., Charalabidis, Y.: Transforming the communication between citizens and government through AI-guided chatbots. Govern. Inf. Q. **36**(2), 358–367 (2019)
3. Zuiderwijk, A., Chen, Y.-C., Salem, F.: Implications of the use of AI in public governance: a systematic literature review and a research agenda. Govern. Inf. Q. **38**(3), 101577 (2021)
4. Bullock, J.B.: AI, discretion, and bureaucracy. Am. Rev. Publ. Admin. **49**(7), 751–761 (2019)
5. Valtiner, D., Reidl, C.: On change management in the age of AI: a sustainable approach to overcome problems in adapting to a disruptive, technological transformation. J. Adv. Manage. Sci. **9**(3), 53–58 (2021)
6. Davenport, T.: The AI Advantage: How to Put the AI Revolution to Work. MIT Press (2018)
7. Dwivedi, Y.K., et al.: Artificial intelligence (AI): multidisciplinary perspectives on emerging challenges, opportunities, and agenda for research, practice, and policy. Int. J. Inf. Manage. **57**, 101994 (2019)
8. Jelonek, D., Mesjasz-Lech, A., Stepniak, C.J., Turek, T., Ziora, L.: The AI application in the management of contemporary organization: theoretical assumptions, current practices, and research review. In: Arai, K., Bhatia, R. (eds.) Advances in Information and Communication, pp. 319–327. Springer (2019)
9. Fleming, P.: Robots and organization studies: why robots might not want to steal your job. Org. Stud. **40**(1), 23–38 (2019)
10. Jahantigh, F., Habibi, A., Sarafrazi, A.: A conceptual framework for business intelligence critical success factors. Int. J. Bus. Inf. Syst. **30**(1), 109–123 (2019)
11. Haenlein, M., Kaplan, A., Tan, C.-W., Zhang, P.: Artificial intelligence (AI) and management analytics. J. Manage. Analyt. **6**(4), 341–343 (2019). https://doi.org/10.1080/23270012.2019.1699876
12. Kaplan, A., Haenlein, M.: Siri, Siri, in my hand: Who's the fairest in the land? On the interpretations, illustrations, and implications of AI. Bus. Horiz. **62**(1), 15–25 (2019)
13. Khatib, M., Al Jaberi, A., Al Mahri, A.: Benchmarking projects' "Lessons Learned" through knowledge management systems: Case of an oil company. iBusiness **13**(1), 1–17 (2021)
14. Kuleto, V., et al.: Exploring opportunities and challenges of AI and machine learning in higher education institutions. Sustainability **13**(8), 10424 (2021)
15. McKelvey, F., MacDonald, M.: AI policy innovations at the Canadian Federal Government. Can. J. Commun. **44**(2), 43–50 (2019)
16. Mikhaylov, S.J., Esteve, M., Campion, A.: AI for the public sector: opportunities and challenges of cross-sector collaboration. Philos. Trans. R. Soc. A **376**(2128), 20170357 (2018)
17. El Khatib, M., Al Falasi, A.: Effects of AI on decision making in project management. Am. J. Ind. Bus. Manage. **11**, 251–260 (2021)
18. Natale, S., Ballatore, A.: Imagining the thinking machine: technological myths and the rise of AI. Convergence **26**(1), 3–18 (2020)

Artificial Intelligence in the Process of Training and Developing Employees

Nawal Abd Ali Ali[1], Allam Hamdan[2](✉), Bahaaeddin Alareeni[3], and Mohanad Dahlan[4]

[1] Ministry of Labor and Social Development, Riyadh, Saudi Arabia
[2] Ahlia University, Manama, Bahrain
allamh3@hotmail.com
[3] The Middle East Technical University (METU) in Turkey – Northern Cyprus Campus, Ankara, Turkey
[4] College of Business Administration, University of Business and Technology, Jeddah 21448, Kingdom of Saudi Arabia

Abstract. Artificial intelligence is grabbing attention and increasing the interest of researchers and academics towards itself. Artificial intelligence can do a broad range of tasks that AI can perform including management of performance of employees, staffing, compensation for employees, training, and development to enhance their skills, etc. The purpose of this research study to understand the relationship between the artificial intelligence and training and development of the employees in an organization. The research questions were developed to get a clear path for the research study including (1) How does artificial intelligence improves the HR processes, specifically the training and development of employees? (2) How effective artificial intelligence is in the field of HR? (3) What is the perception of employees related to the utilization of artificial intelligence in their training and development? (4) What are the Pros and Cons of using artificial intelligence in HR practices? It is hypothesized that artificial intelligence predicts an increase in the efficiency of employees through their training and development. Data is collected with the help of questions obtained from "Employee Basic Task Performance Scale" and Quality of work life". Data was obtained from the employees working in the organizations. Analysis of the data is completed with the help of Statistical Analysis Software (SPSS). The results show a positive relationship between the artificial intelligence in the training and development of the employees. Most of the employees appear satisfied with the trainings provided in their organizations. Hence, the hypothesis developed for this research study is accepted. Further research study is required to understand the type of artificial intelligence activities used in the training and development of employees in organizations.

Keywords: Artificial intelligence · Employee performance · Training and development · Efficiency

B. Alareeni et al. (Eds.): ICBT 2022, LNNS 620, pp. 558–568, 2023.
https://doi.org/10.1007/978-3-031-26953-0_50

1 Introduction

Artificial intelligence technology has created a new generation of labor. These labor include human intelligence labor that is struggling hard to survive and transform themselves in a constantly changing and evolving technology. Artificial intelligence is strengthening its demand across the globe in every sphere of life (Ertel 2018). It is observable that artificial intelligence is grabbing attention and increasing the interest of researchers and academics towards itself after Google successfully launched Alpha Go System (Vincet 2019). It is observed that the human resource department is also focusing to optimize their operations by increasing automated operations with the help of AI. The use of technology in the training and development of employees makes the process simple and easy, giving enough time to the human resource department employees to increase creativity, efficiency, empowering their employees and improving their relationship and sincerity with the organization.

Context: Since a lot of research has been done on this topic, but it is difficult to determine which strategy used in an organizational setting for the training and development of employees through artificial intelligence is effective and efficient. This research will provide us with an opportunity to compare and contrast different strategies used in the training and development of employees and access the employee's feedback on the effectiveness of each strategy.

Therefore, we hypothesize that there is a positive relationship between artificial intelligence & the training and development of employees. This paper is based on developing information related to six dimensions of human resource management while the focus of the study will be only on one element that is training and development; and combining this process with a potential artificial intelligence application. Moreover, the pros and cons of artificial intelligence will also be highlighted in the paper.

Need for Research:
The research study will benefit those who are working in organizations. The main aim of the paper is to determine effective ways in which artificial intelligence can be incorporated in an organizational setting for the training and development of employees. With the help of this research study, organizations will be able to improve their HR practices, incorporation of AI in a more effective and efficient way, hence, decreasing or reducing the burden of employees.

Artificial Intelligence: The ability of a digitalized machine or computer-controlled robots that have an ability to understand the data, interpret that data, and perform functions based on data provided. These abilities are similar to that of human intelligence, for instance, discover meaning, reasoning, learning from past experiences and generalizing.

Training and Development: A process in which the employee's skills and knowledge are enhanced by providing them with information on the topic. Furthermore, they are taught how they can use their skills effectively and efficiently.

2 Literature Review

One of the most important assets of an organization is its human resource department. It is a crucial managerial duty that needs successful management. According to a research study by Devanna et al. (1982), there is a broad range of tasks that AI can perform including management of performance of employees, staffing, compensation for employees, training, and development to enhance their skills, etc. According to a report by IDC (Framingham and Mass 2016), artificial intelligence usage and purchase has been increased from $8 billion to $47 billion from the year 2016 till 2020. These statistics clearly show that every organization and company is evolving itself according to its needs. However, further research is required in the training and development area of employees to determine the usage of artificial intelligence in a more effective way.

2.1 Human Resource and Intersection of Artificial Intelligence

There is a wide range of tasks that are both challenging and tiring for HR departments. It not only consumes energy but also a lot of time. Beginning from finding the best suitable candidate for the right job, training, and development of new or existing employees, managing payrolls, and providing benefits to all the employees. All these tasks can be easily accomplished in today's world in a cheaper and faster way with the help of technology. The importance of data is now being accepted worldwide, which helps the hr professionals to take a decision. In this respect, artificial intelligence has helped HR professionals in analyzing huge piles of data that was quite tough before artificial intelligence. It not only predicts the trends for their practices but also helps them through its suggestion that is developed based on the data. Artificial intelligence also enables the HR department to foresee the requirements needed in the training and development of the employees. It also enables HR executives to develop and implement strategies in a more effective way. Furthermore, by using old and new data, artificial intelligence can provide details for the best practices in an organization, problems associated with and impactful decisions and solutions to overcome the issue. It is also observed that the HR decisions are biased, they favor the organizations more in the activities and strategies they implement (Maduravoyal 2018). Furthermore, employees have observed this biased nature of the HR department is present in all of their activities. Moreover, it is also observed that when an interviewee is sitting in front of the interviewer mostly perceptional and behavioral biases occur that might affect the recruitment and selection process. According to HR leader Richard Coombes of Deloitte, using artificial intelligence in the recruitment process eliminates the perceptive and behavioral bias that commonly occurs in human interaction (Christopher 2019). This major issue that has been faced by many employees in the different organization will reduce or swiped out, once Artificial intelligence is incorporated in their organizations. Artificial intelligence provides a more transparent strategy and implements these strategies without their personal biases. Moreover, the functions associated with the AI will be more employee-oriented as compared to organizational oriented.

2.2 Meta-cognitive Theory

The theory postulated by Flavell (1979), states that when an individual is aware of his strengths and weaknesses, they learn in the best way. This usually happens with the help of quizzes, pre-assessments, feedback, and tests. On the basis of this data collected from all the employees in an organization, an AI utilizes this data and predicts the weaknesses and strengths of employees. With the help of Artificial intelligence-based training, these weakest areas of every individual are targeted and help them to focus on enhancing their strengths and cope up with their weak areas. Another research study by Clark (2008), states that metacognition improves the ability of employees in task performance and provide quality work in time. The researchers focus on the development of meta-cognitions in which the employees show improvement in the self-regulated learning process and improved work performance.

2.3 Theory of Deliberate Practice

The theory holds that by knowing the weak areas of individuals, the practice can enhance and improve those weak areas. According to a research study by McEdwards (2014), the efficacy of employees is increased by using deliberate practice through asynchronous training technology. It is indicated in the research study that deliberate practice among the selected participants shows significant improvement. In another research study by Ericsson and Moxley (2012), the skills of employers who worked hard and keep on practicing improved with time. Furthermore, creativity is observed in the people who focus on deliberate practice. Due to the presence of automated intelligence, there was no need of external judgment or biases that mostly affect the performance or practice of an individual. Therefore, the data collected through artificial intelligence enables the researchers to determine the efficiency of deliberate practice.

2.4 Ways in Which AI is Reinventing Human Resource Process

There are several processes in which AI is reinventing the human resource process. Some of the following are discussed below:

Decision Making Process
In a research study by Masum et al. (2018), it is stated that decision support system (DSS) is utilized by many organizations that support them in predicting human resource needs, strategic planning, and evaluating HR practices and policies. The tasks that need high attention and hard work of the HR department, becomes easier with the help of technology. However, DSS has a certain limitation in the decision-making process that's why researchers develop a more effective output system i-HRIS model that constitutes of ten HR Module applications i.e. Strategic HR and Planning Module, Employee Relations, Training, and Development Module, Recruitment and selection module, Performance evaluation module, health, and safety module, Compensation and benefit module, talent management module, payroll interface module, employee self-service module. Hence,

a comprehensive framework is provided for the solution of HR problems in the form of the proposed i-HRIS model.

Automated Performance of Repetitive Tasks

Director of human resource operations Kate Guarino states that Artificial intelligence provides an opportunity for the human resource department to automate low value and repetitive tasks with the help of artificial intelligence. By doing so, they can shift their focus more on strategic planning and development processes (Nicastro 2018). These strategies have been implemented in the human resource department, where the repetitive tasks such as marking attendance, collecting data for the hours an employee present in the office, development of payroll and many more are becoming easier. While most of the repetitive tasks are performed with the help of artificial intelligence, the human resource departments focus more on developing and enhancing the skills of their employees. Furthermore, by focusing on the provision of a supportive environment to the employees, human resource departments are trying to provide the best, safe and secure working environment (Isa and Atim 2019).

2.5 Artificial Intelligence Recruiters and Its Impact on Labor Force

In many fields, technology is replacing humans being, hence a decrease in jobs for labor is observed worldwide. It has a dramatic effect on the income and labor of middle and lower class employment and daily wages type jobs. Hod Lipson who is an engineer at Cornell University argues that it was perceived for a long time, technology is created new jobs that are better than previous ones, however, these jobs will be fewer in number. A similar warning was issued by Ford, where he claims that the penetration of technology in the industrial and corporate sectors will provide us with a bulk of products but will ultimately lead to a decrease in the demand for human labor (Ford 2009). Moreover, the issues related to the utilization of human labor in an adequate way needs immediate attention. If the issue is not catered timely, it may lead to downward economical issues. Furthermore, there are people who do not have adequate education to grab good salary jobs, hence, the average people in our population will face the most difficulties from the automation of industries across the globe. Therefore, they will not be able to find any new job once downsizing occurs due to the incorporation of artificial intelligence in organizations.

Therefore, the above-cited literature provides us with information that enables us to determine the utilization of strategy in our research work. The literature review enables us to determine and develop a hypothesis for research purposes. Therefore, the hypothesis developed is mentioned below:

2.6 Scope of the Study

The scope of the study is that it will enable the organizations to develop a strategy that is more effective and efficient in providing training and development to their employees. Furthermore, human resource departments will be able to understand different techniques that are considered effective in training employees through artificial intelligence. Techniques that are based on theoretical information such as deliberate practice and metacognition will improve the efficiency of the employees and provide quality work on time. Students and researchers will get a direction for future researches and utilize the information provided in the research. In the next section, the methodology adopted for the completion of research article is discussed in detail.

The methodology that is used in this research paper is discussed in this section.

3 Research Methodology

3.1 Research Model

The research is aimed at analyzing the use of artificial intelligence in the training and development of employees of the organizations present in Bahrain. Furthermore, which strategy used in the form of artificial intelligence is more efficient and useful in training the employees. The study aims to critically analyze the present usage of artificial intelligence in the training and development of employees. For this reason, we have adopted primary and secondary data analysis. The primary analysis will be based on the data collected through questionnaires while the secondary data is based on the research publications present online related answering research question.

3.2 Research Methods

Methods of Data Collection

In all the researches we have observed so far, the mode of data collection is through surveys or interviews. In this research, we will use quantitative data gathered with the help of selected questionnaires to develop an understanding of the research question. Later on, this research data will be analyzed with the help of SPSS, a statistical software that will provide us insight into this topic. Different tests will be utilized in the process to understand the relationship between the variables.

Measures Used

Consent Form

A consent form is a form on which instructions related to the topic of the research, pros, and cons of research and other information related to the research are provided. Once the research participant will read this consent form, they will allow the researchers to use the information provided to them. However, the researchers will allow the participants to leave the research at any point and they will not be penalized for this. Furthermore, it will be made clear to them their confidentiality will be maintained, and the data obtained from them will only be used for the research and publication purpose. Their data will not be exposed at any time i.e., during or after the research.

Demographic Form

The demographic form utilized for the research purpose will include the information related the gender (male, female, other) age, marital status (married, single, divorce), income group, work position, education (k-12, graduate, undergraduate, diploma or other), employment status (full-time, part-time or other), the industry you work for, a primary functional area of work.

Employee Basic Task Performance Scale
Tsui, Pearce, Porter, and Tripoli (1997) were the researchers who developed this questionnaire that aimed at measuring the quality, quantity, and efficiency of the employee. It is 4 points Likert scale that is utilized in this research study.

Training and Development
The Questionnaire named "Quality of work life" is utilized in the research paper, where one component of the questionnaire training and development is considered. The questionnaire is developed by Swamy et al. (2015), where they assess a different aspect of an employee's working life. However, only the training and development part of the questionnaire is utilized in the research. They operationalized training and development as a way to increase the employee's performance. Moreover, Conditions and environment that increases the skills and potential of an employee is also the part of an operational definition. Question number 22, 21, 22 and 23 are categorized for the training and development of employees.

Data Analysis Software
Once the data is gathered, the data will be gathered in the SPSS and analyzed by using different testing and analyzing processes. The data is presented both in tabular and graphical form. SPSS is a very easy and accurate statistical data measuring tool, that provides us with an in-depth analysis of the data. Research Participants:

4 Results

Table 1. Descriptive statistics

Gender		N	Minimum	Maximum	Mean	Std. deviation
Male	Age	26	3.00	9.00	5.1154	2.28608
	Gender	26	1.00	1.00	1.0000	.00000
	Department	26	1.00	5.00	2.2692	1.58890
	Experience	26	2.00	6.00	3.8077	1.62528
	Education	26	1.00	5.00	2.6154	1.23538
	Income Group	26	2.00	4.00	3.0000	.74833
	Employment Status	26	1.00	3.00	1.1923	.56704
	Valid N (list wise)	26				

(continued)

Table 1. (*continued*)

Gender		N	Minimum	Maximum	Mean	Std. deviation
Female	Age	33	2.00	8.00	4.6061	1.47774
	Gender	33	2.00	2.00	2.0000	.00000
	Department	33	1.00	5.00	2.6667	1.57454
	Experience	33	1.00	6.00	3.3939	1.53987
	Education	33	1.00	5.00	2.6364	1.22010
	Income Group	33	2.00	4.00	2.8788	.69631
	Employment Status	33	1.00	3.00	1.1515	.44167
	Valid N (list wise)	33				

Table 2. Descriptive statistics

Gender		Mean	Std. deviation	N
Male	HRTD	15.5000	3.47851	26
	Job_Sat	10.1538	1.84808	26
Female	HRTD	15.6061	2.34440	33
	Job_Sat	9.3939	1.19738	33

Table 3. Correlations

Gender			HRTD	Job_Sat
Male	HRTD	Pearson Correlation	1	.709**
		Sig. (2-tailed)		.000
		N	26	26
	Job_Sat	Pearson Correlation	.709**	1
		Sig. (2-tailed)	.000	
		N	26	26
Female	HRTD	Pearson Correlation	1	.558**
		Sig. (2-tailed)		.001
		N	33	33
	Job_Sat	Pearson Correlation	.558**	1

(*continued*)

Table 3. *(continued)*

Gender			HRTD	Job_Sat
		Sig. (2-tailed)	.001	
		N	33	33

**Correlation is significant at the 0.01 level (2-tailed).

5 Discussion

The result statistics reveal that about 26 male while 33 female participants responded to the questionnaire. The total number of participants appeared to be (N = 58) (See Table 1). The age range of the participants was divided into seven age range. About 38.5% (N = 10) of male participants were from the age range between 24–28 years of age, 15.4 (N = 4) from the 34–38 age range, 11.5% (N = 10) from the 29–33 age range while 7.7% (N = 4) from 39–43 and 49–53 age range respectively. The female participants were 24.2%, and 24.2% from 24–28 and 29–33 (N = 16) respectively. About 21.2% (N = 7) and 15.2% (N = 5) of the female participants were from 34–38 and 39–43 age group respectively. About 34.6% of male participants have more than 1–5 years of experience in the field as compared to the female participants who have similar experience in their respective fields. 23.1% of male participants reflected to have more than 20 years' experience of their job as compared to 15.2% of the females. Most of the female (30.3%) and male (57.7) participants are working in the education department (teachers, trainers, facilitators, and admins). The education of most of the participants appears to be of a bachelor's level showing a percentage of 69.2% of male participants and 60.6% of female participants. However, 19.2% of male participants have a diploma as compared to 18.2% of the female participants. The income group of 46.2% of male participants appears to be in the range of BHD 601 to BHD 1200 as compared to 51.5% of the female participants who lie in the same income group. The demographical analysis also reflects that 87.9% of female participants and 88.5% of male participants are full-time employers. The analysis of the questionnaire utilized in the research study demonstrated a positive response from the male participants when asked about the training they receive and their satisfaction with the training provided to them. However, the female participants also appear to be satisfied with the training they receive in their respective organizations. The male participants show satisfaction and commitment towards their organization and have strongly agreed on the point where they were asked to refer their company to their friends for working purposes. Contrary to this attitude, the female participants have a difference of opinion. Some of them considered the values of the company coherent to their values, while a major portion of the female participants does not reflect the same. The overall analysis of the response reflects that participants have a positive correlation with the human resource department training and development practices with job satisfaction (Tables 2 and 3).

Most of the employees who were formally asked at the end of the questionnaire about their views on the training and the type of technology utilized in the training and development of employees. Most of the employees provided the information that they have been trained with the help of artificial intelligence that not only improves their efficiency in the work but also make it easier for them to understand complex system. Moreover, they claimed that they do not feel or faced any discrimination as they usually face in the traditional training and development method. A similar phenomenon is in accordance with the research study by the Maduravoyal (2018), where he stated that the employees have observed this biased nature of the HR department is present in all of their activities. With the help of artificial intelligence, companies and organizations can overcome the issues that employees faced from the human resource department. Therefore, our hypothesis 1 gets accepted which claims that Artificial intelligence predicts an increase in the efficiency of employees through their training and development while H2 is rejected where it claims that Artificial intelligence does not predict an increase in the efficiency of employees through their training and development.

6 Conclusion

The research study enables us to understand the employee's perception, attitude and the problems they faced in the traditional training and development process. The use of artificial intelligence in the training and development of employees improves the strengths and abilities of the employees. Hence, it is recommended that organizations utilize artificial intelligence while training their employees. The limitation of the study includes the small sample size due to which the research results cannot be generalized to a larger group. Moreover, the employees from a different perspective, every department have different training and development method; it was not certain to what extent artificial intelligence is being used in their organizations. Therefore, further research is required to understand the limitations, pros, and cons of the usage of artificial intelligence in the organizational sector for the training and development of the employees.

References

Christopher, A.: Use of Artificial Intelligence in Human Resource Management (2019). Accessed 6 Mar 2020

Clark, R.: Metacognition and human performance improvement. Perform. Improv. Q. **1**(1), 33–45 (2008). https://doi.org/10.1111/j.1937-8327.1988.tb00005.x

Devanna, M.A., Fombrun, C., Tichy, N., Warren, L.: Strategic planning and human resource management. Hum. Resour. Manag. **21**(1), 11–17 (1982)

Ericsson, K., Moxley, J.: The expert performance approach and deliberate practice. In: Handbook Of Organizational Creativity, pp. 141–167 (2012). https://doi.org/10.1016/b978-0-12-374714-3.00007-0

Ertel, W.: Introduction to Artificial Intelligence. Springer, New York (2018)

Flavell, J.H.: Metacognition and cognitive monitoring: a new area of cognitive–developmental inquiry. Am. Psychol. **34**(10), 906 (1979)

Framingham, Mass: Worldwide Cognitive Systems and Artificial Intelligence Revenues Forecast to Surge Past $47 Billion in 2020, According to New IDC Spending Guide. IDC (2016). https://www.idc.com/getdoc.jsp?containerId=prUS41878616. Accessed 18 Mar 2020

Isa, K., Atim, A.: Working environment: how important is it to make your employees happy. Int. J. Eng. Adv. Technol. **9**(1), 6505–6509 (2019). https://doi.org/10.35940/ijeat.a1269.109119

Maduravoyal, C.: Artificial intelligence in human resource management. Int. J. Pure Appl. Math. **119**(17), 1891–1895 (2018)

Ford, M.: The Lights in the Tunnel: Automation, Accelerating Technology, and the Economy of the Future. Also see his more recent book, Rise of the Robots: Technology and the Threat of a Jobless Future, Basic Books (2009)

Masum, A.K.M., Beh, L.S., Azad, M.A.K., Hoque, K.: Intelligent human resource information system (i-HRIS): a holistic decision support framework for HR excellence. Int. Arab J. Inf. Technol. **15**(1), 121–130 (2018)

McEdwards, C.E.: The efficacy of deliberate practice delivered using asynchronous training technology. Int. J. Adv. Corp. Learn. (iJAC) **7**(1), 43–46 (2014)

Nicastro, D.: 7 Ways Artificial Intelligence is Reinventing Human Resources (2018). https://www.cmswire.com/digital-workplace/7-ways-artificial-intelligence-is-reinventing-human-resources/. Accessed 6 Mar 2020

Swamy, D.R., Nanjundeswaraswamy, T.S., Rashmi, S.: Quality of work life: scale development and validation. Int. J. Caring Sci. **8**(2), 281 (2015). http://translateyar.ir/wp-content/uploads/2018/12/6_swamy.pdf

Tsui, A.S., Pearce, J.L., Porter, L.W., Tripoli, A.M.: Alternative approaches to the employee organization relationship: does investment in employees pay off? Acad. Manag. J. **40**(5), 1089–1121 (1997)

Vincet, J.: Former Go champion beaten by DeepMind retires after declaring AI invincible (2019). https://www.theverge.com/2019/11/27/20985260/ai-go-alphago-lee-se-dol-retired-deepmind-defeat. Accessed 6 Mar 2020

Artificial Intelligence Application in the Fourth Industrial Revolution

Noor Jawad Jassim Abdulla[1], Allam Hamdan[2], and Mohammad Kanan[3](✉)

[1] College of Business and Finance, Manama, Bahrain
[2] Ahlia University, Manama, Bahrain
[3] Jeddah College of Engineering, University of Business and Technology, Jeddah 21448, Kingdom of Saudi Arabia
mkanan977@gmail.com

Abstract. In this paper the Fourth Industrial revolution meaning and history is being discussed and how artificial intelligence (AI) is part of this revolution.

The fourth industrial revolution can be divided into 3 clusters digital, biological and physical clusters and each cluster it has its own applications that are connected to AI such as 3d printings, autonomous vehicle, new materials and advanced robotics implemented in this revolution AI has advanced this technology to another level where it started to affect our society in negative or positive way where our ethical value can be put in to consideration. Also, the impact of COVID-19 is discussed accordingly for on the Fourth Industrial Revolution.

Keywords: Fourth industrial revolution · Artificial intelligence · COVID-19

1 Introduction

Artificial intelligent (AI) application in the fourth industrial revolution (4IR) when this term comes through a person mind a lot of question will appear as what AI and what is 4IR and how are they connected together how do they affect the word and what are their applications?

Before we can understand 4IR lets review their industrial revolution history and meaning.

The term industrial revolution means the change that occurs relatively fast and modifies the essence of social structures or organizational practices. As the first industrial revolution (1IR) happened in Brittan in 1970's which conclude the use of mechanism approach as steam engine was invented, canals and railways network expanded and the communication ability expanded. The second industrial revolution (2IR) was on early 19th century with the invention of electricity that allowed many companies to be developed by using machine powered by electricity. In the mid-19th, the nuclear power was a developed and use of electronics was widely increased which generated the third industrial revolution (3IR). One of the most important achievements for 3IR was the Internet which open a lot of doors into technology (Kayembe and Nel 2019).

B. Alareeni et al. (Eds.): ICBT 2022, LNNS 620, pp. 569–575, 2023.
https://doi.org/10.1007/978-3-031-26953-0_51

The fourth industrial revolution (4IR) is the world where physical and virtual element cooperate together. It also interacts with digital, physical and biological domains which the previous industrial revolution could not achieve that goal (Schwab and Davis 2018).

Artificial Intelligence (AI) can be considered one of the fourth industrial revolution application as it will be explained later the meaning of AI and how it connects and affect the fourth industrial revolution.

2 Artificial Intelligence (AI)

It is important to note that AI is not a human intelligence. Some AI can simulate human intelligence relying on algorithm and data processing. This may or may not achieve the human goals. Computers can act as a human by succeeding in turning tests such as CAPTCHA or answering true and false questions. Computers can also perform human tasks with intelligence such as driving a car or observing human behavior and studying how human act to be added into their data base (Mueller and Massaron 2022).

AI can be found in many applications that is being used today and no one will take notice of its smart application such as siri, alexa and google assistant can be considered AI technology (Mueller and Massaron 2022). Smart thermostat that can control a room heating is also an AI application. AI can appear in car, work place and even outdoors in our modern society.

3 AI and Fourth Industrial Revolution

As have been known Fourth Industrial Revolution is when physical and virtual element work together and it can be categorized into three clusters: physical, digital and biological. These clusters can benefit from each other based on their progress in technology (Schwab 2017).

3.1 Physical Clusters

Physical clusters can be put int four main category which Artificial Intelligence (AI) can be implemented on them.

1. 3D Printing
2. Autonomous Vehicles
3. New Materials
4. Advanced Robotics

AI and 3D Printing

3D printing can be achieved by using a physical device that print layer over a layer of a digital 3D model and transform it to a real physical object it also can be called additive manufacturing (AM). 3D printing was not a technology that produced valuable results until NASA used it for International Space Station (ISS). As ISS have all tools needed when its leaves earth some of these tools can be missing or broken. 3D printing managed to solve this issue by creating the missing or broken tools on space (Heiden et al. 2021).

Adding AI to 3D printing can reduce the mistakes in 3D technology as AI can deduct the defective products and improve the efficiency of the work by overcoming the process of speed and cost. As now days it became an important issue for production company and supply chains to respond to the market huge requests of products and the idea of addictive manufacturing and artificial intelligence can be very effective in this field. However, it is still a new field and need to be implemented more in the future as product quality still cannot be guaranteed.

AI and Autonomous Vehicles

There are many driverless vehicles have been implemented with AI technology such as cars, trucks, drones aircrafts and boats. AI technology for autonomous vehicles is improving rapidly and after few years it will be available commercially.

Driverless Cars and Trucks (Mueller and Massaron 2022)

Driverless cars and trucks can be safer and more efficient than people on wheel. However, they are still a prototype and not on the road as Society of Automotive Engineer (SAE) International published a slandered classification for autonomous car which include the below:

1. Control should on hand of driver and the car can acts as an assistance such as controlling speeds and brakes.
2. The driver can let the car act instead of him and only be on alert when needed.
3. The car can drive by itself under specific restriction and rules such speed limits and specific places.
4. While driving the car can monitor the road changes and condition this condition is yet to be achieved and under further research
5. The car can take control of all the driving without any intervention this is a future level that will be focused on after achieving the fourth condition.

By implementing AI to the car, it determines the path of a car and by using machine learning and algorithm it can predict the future step while carrying the task assigned to prevent collision.

AI and New Materials

New materials are the basic of thee fourth industrial revolution as the ability to implement new materials can provide a lot of solutions. Society have transferred from stone age to the bronze age and to the iron age by implementing new materials which started the industrial revolutions. Semiconductors and nanotechnology are the starters of fourth industrial revolution. Implementation of AI and robotics platform together combined with new materials can create new technologies which will affect human life (Schwab 2017).

With significant improvement in revolution new material are requested to be found and this will be a one society future challenges. Beside that implementing new materials takes a long timeline and capitally intensive (Schwab and Davis 2018).

AI and Advanced Robotics

First robotics idea was introduced in 1960 and it has improved rapidly using machine learning technology which allows robot to act like human. However, they are still lacking in the human sense.

Range of service AI and robotics have been provided to us such as navigation of vehicles, medical diagnosis for certain diseases, supporting cybersecurity and smart phones technology which include Apple's Siri and google search algorithm.

AI, robots and humans combined together will produce a better work. For example, Carnegie Mellon's CoBot program it is a program that use robots into involving humans or visitors to help in some tasks. As AI systems needs people helps to set some goals if not it may lead to different or a harmful task to be done. Robots can be categorized into three famous categories.

Manipulators

They also called the robot arm. Manipulators are famous for their speed that can exceed the human work. Unfortunately, their motion is limited as they cannot be moved without human help. Manipulators can mostly be seen in industrial factories and can be seen sometimes in medical field to assist on surgical operation.

Mobile Robots

They are the second famous category and unlike the manipulators they have the ability to move using wheels, motors or legs. Mobile robots can be controlled remotely. One of their most famous applications is the flying robots.

Flying robots are also called drones. They are more advances than any other mobile robots because they can carry weight while flying. They are mostly used by the military and can be commercially available. They also can be used in aerospace by carrying a surveillance camera. Moreover, they can be used to aid rescue operation by providing medicine and food. Drones can also carry bombs and other dangerous equipment which can be an ethical issue to society as the way that we use technology can be harmless or harmful.

Mobile Manipulator

The third category can be known as a combination of both the first and second category. It can be described as a moving robot that has a robot arm. The most famous examples of mobile manipulators are the Sophia robot that can show human like expressions and the robot vacuum cleaner which can clean the floor by its self.

3.2 Digital Clusters

The main application of digital clusters is the Internet of All Things (IOT) which can be described as a relationship between product, services or places. There are many devices around us that use IOT such as computers, smartphones and tablets that connect to the Internet.

IOT consider to be a network of materials that have electronic software, sensors and actuators that links together and generate data. With the amount of data collected by IOT devices AI can process and analyze real time data to improve the public services and value in society (Kankanhalli et al. 2019).

IOT and AI can provide many services to society such as transportation by installing camera around the street IOT can collect the information that needed with a real time analysis and provide road status to drivers such as traffics, construction site and best route available. Healthcare is also be a service by using IOT and AI technology enable various medical monitor devices to be connected together such as heart rate, blood pressure and monitor devices. This will allow patient to be treat from home (Perrakis and Sixma 2021).

The major concern for IOT is the cybersecurity risks as the usage of unsecured devices that connect to internet increase the cybersecurity hazards will increase this can be one of the future issues for IOT to be revised (Schwab and Davis 2018).

3.3 Biological Clusters

It is considered to be a genetic innovation. Due to the improvement of computing power the progress for this field has been increasing rapidly as the first genetic project took more than ten years to be implemented while nowadays it only takes few ours for it to be implemented. Some major health issue has genetic component such as heart disease and cancer. By using AI in genetic innovation doctors can easily make effective decisions about patience treatments. In late 2021, AI machine AlphaFold has managed to create almost a perfect protein fold prediction which will be huge embark into AI revolution.

The ability to edit in genetic biology can apply to any cell type as genetic science is progressing rapidly and the limitation are more of legal and ethical issue than its technical. This will affect our society and when have to question our morality what is humanity, do we have the right to share our genetic information with each other and changing the structure of genetic for future generation.

4 The Impact of Fourth Industrial Revolution

With the rapid growth of AI in the fourth industrial revolution and its application it has provided fundamental change into our economic, social and political life.

4.1 Economical Change

As the rapid development of AI and robotics grows it will increase the economic growth and provide a higher productivity. However, this will affect the human workers as AI can mostly do the human work accurately and with faster speed than human which will lead to damage the human market and people will lose their jobs (Furman and Seamans 2019).

4.2 Social Change

As we know that AI and the fourth industrial revolution affect the human market with the continuous change in technology people need to be prepared for the future by implementing new skills and knowledge. Humans need to be agile and can quickly adapt to changes as many may not last without the proper skills.

With the use of smart devices such as smart phones, tablets and computer cyberbullying has become an issue in society. Harassment, threating and humiliating messages has been sent to people using social networks. Also, human privacy has been difficult to maintain. All of these has become a huge issue that is affects the young generations (Issa et al. 2016).

However, there is still a positive impact on our society as with smart devices as distance between people has been extinct. People can communicate with each other regarding of their place. As people from different country can see, track or even talk with each other's. Thus, it effected our society in many different ways as different cultures, knowledge and experiences have been introduced to the young generation (Issa et al. 2016).

4.3 Political Change

Political change is related to the government, with rapid development of AI technology it will enable the public freedom of speech and allows them to express their opinion in a way that overcome the government supervision.

Thus, government need to adapt quickly to the changes by collaboration with the business and civil society to create regulations and rules that can be safe, fair, reliable and competitive to survive the rapid technology development and not to be outpowered (Kankanhalli et al. 2019).

5 Fourth Industrial Revolution and COVID-19

When COVID-19 started and the decisions for lock down and social distance made by the government, society started to seek other method to enable interaction between people and AI and IOT were mostly used. For example, smart monitoring devices implemented from IOT is used to track and monitor infected people this reduced the workload of medical staff and allowed the work to be efficient. As, well as that information of patient are collected to prevent the same mistake to happen again. Thus, Governments started to implement mobile application to improve and develop the connection between society and the health care services such as ArogyaSetu app in India and Close Contact in China (Singh et al. 2020).

COVID-19 has a huge effect on the supply chain resilience as when the economy was locked the resources were limited and it was hard to collect or deliver assets to clients. Many companies were struggling and faced with a lot of hardship and loss. Thus, the importance of supply chain resilience was learned as AI can strength the supply chain resilience and many company sought to implement it. AI can plan and analyze the impact of sourcing quickly. It also can handle the distribution and scheduling of supply chain activities. AI has the ability to assess risks and prioritize the plan of action to be taken for strategic decisions in terms of logistic planning to meet the clients demands (Modgil et al. 2021).

6 Conclusion

In this paper we have discussed a brief introduction to the Fourth Industrial Revolution (4IR) and some of its application and how Artificial Intelligence (AI) can implemented on it. As many applications can be implemented commercially and many still are being in the research or protype phase which conclude there still more to fourth industrial revolution to come and we are still in the lookout for fifth industrial revolution and how it implemented, however with this rapid change in technology can affect the world economically, socially or politically either in a positive way or in a negative way where our ethical value and judgment can be put in to consideration. COVID-19 has a huge impact on the Fourth Industrial revolution growth as people started to depend more on smart technology with the lockdown and seek an easy and comfortable life.

References

Kayembe, C., Nel, D.: Challenges and opportunities for education in the Fourth Industrial Revolution. Afr. J. Public Affairs **11**(3), 79–94 (2019)

Schwab, K.: The fourth industrial revolution. Currency (2017)

Mueller, J.P., Massaron, L.: Artificial Intelligence for Dummies. Wiley, New York (2022)

Heiden, B., Alieksieiev, V., Volk, M., Tonino-Heiden, B.: Framing artificial intelligence (AI) additive manufacturing (AM). Procedia Comput. Sci. **186**, 387–394 (2021)

Schwab, K., Davis, N.: Shaping the Future of the Fourth Industrial Revolution: A Guide to Building a Better World. Crown Publishing Group, New York (2018)

Kankanhalli, A., Charalabidis, Y., Mellouli, S.: IoT and AI for smart government: a research agenda. Gov. Inf. Q. **36**(2), 304–309 (2019)

Perrakis, A., Sixma, T.K.: AI revolutions in biology: the joys and perils of AlphaFold. EMBO Rep. **22**(11), e54046 (2021)

Furman, J., Seamans, R.: AI and the economy. Innov. Policy Econ. **19**(1), 161–191 (2019)

Issa, T., Isaías, P., Kommers, P.: The impact of smart technology on users and society. J. Inf. Commun. Ethics Soc. **14**(4), 310–312 (2016)

Singh, R.P., Javaid, M., Haleem, A., Suman, R.: Internet of things (IoT) applications to fight against COVID-19 pandemic. Diabet. Metab. Syndr. Clin. Res. Rev. **14**(4), 521–524 (2020)

Modgil, S., Gupta, S., Stekelorum, R., Laguir, I.: AI technologies and their impact on supply chain resilience during COVID-19. Int. J. Phys. Distrib. Logist. Manag. **52**, 130–149 (2021)

Introducing Artificial Intelligence to Human Resources Management

Zahra Almaghaslah[1], Allam Hamdan[2](✉), and Weam Tunsi[3]

[1] College of Business and Finance, Manama, Bahrain
[2] Ahlia University, Manama, Bahrain
Allamh3@hotmail.com
[3] College of Business Administration, University of Business and Technology, Jeddah 21448, Kingdom of Saudi Arabia

Abstract. Artificial intelligence can be defined as a discipline of computer science whose main purpose is to provide solutions to cognitive problems related to human aptitude and beyond. AI makes machines think like a man and thus able to perform responsibilities, for instance, solving problems, reasoning, and understanding language. Artificial intelligence is based on two major technology approaches machine education and deep learning. Every single day, artificial intelligence and present technology are increasing. The fast-growing technology of Artificial intelligence is crucial since it helps solve problems in institutional management. For instance, in the human resource department of any organization, artificial intelligence has been greatly employed for efficient performance. Mainly AI aids in the automation of the HR processes through quick and accurate data analysis. Such enhancement of HR by artificial intelligence promotes successful hiring and efficient management of the company's employees. Moreover, a company with competent employees that adopts Artificial Intelligence is more likely to save on the production cost and enhance talent value among employees. In addition, the firm will improve the efficiency of the worker's team strategy. This research paper will focus on addressing how the concept of AI affects human resource management and analyzes the future of HR with the emergence of AI.

1 Introduction

In contrast to human beings and other animals, computers show artificial intelligence, which is frequently referred to as machine intelligence (Ahmed 2018). Computer study defines AI as any sophisticated equipment that senses the surrounding environment, understands such environment, and may act to enhance the environment. The technology conducts activities that aim at effectively attaining its aims of enhancing the environment. An AI computer takes external input and analyse the data to learn and apply that knowledge to accomplish a specific objective. The evolution of artificial intelligence robots, particularly how they grow and learn like a baby, has been studied extensively. To define robots that replicate human cognitive capabilities such as education and problem-solving, artificial intelligence has been frequently utilized (Ahmed 2018). 'The advancement of AI towards analysis and representation has generated a

B. Alareeni et al. (Eds.): ICBT 2022, LNNS 620, pp. 576–583, 2023.
https://doi.org/10.1007/978-3-031-26953-0_52

cognitive intelligence system that reflects human functioning in absentia and performs responsibilities via experience learning to influence future judgments' (Ahmed 2018). Machine learning has advanced to the point that it is now being utilized in a variety of ways to enhance both the environment and the result of an activity.

A company's department of human resources is always evolving. Managing human resources is a valuable department that helps the company accomplish its goals and objectives. Institutions must be modernly equipped and developed to cope with anything that emerges and to serve customers in the best manner possible to maximize profit after growing their market share in light of modern market changes and uncertainty. Modern markets demand firms to be more inventive in order to automate processes and better concentrate on all stakeholders, the author says.. Institutions must, without a doubt, look for methods to increase their competitive advantages in order to perform better and remain viable in the market. Two elements are critical when looking at today's rapidly evolving and fiercely competitive marketplaces (Jatoba et al. 2019). Implementing the latest technologies and recruiting the top personnel are among these reasons. As the brightest minds in the business expand the functioning of technology, it guarantees that things are done correctly and properly in the company. Joining cutting-edge technology with the greatest talent makes a company invincible. Managers and organizations have become more aware of the connection between AI and talent management as a result of the expansion of markets and competition.

Modern organizational practices have deeply concentrated on machine learning and its impact on institutional processes. Artificial intelligence is associated with the changing nature of work, the process, and changes in economic mechanisms and models (Jatoba et al. 2019). These changes are anticipated to change the overall organizational management significantly. The changes in business orientation and marketing plans have made artificial intelligence a significant player in the organization process. Notably, human resource management is an area that has benefited from artificial intelligence. Studies show that artificial intelligence is increasingly used in the recruitment and selection functions of human resources management in an organization (Niehueser and Boak 2020). There is significant evidence that artificial intelligence is used as a game-changer in modern organizational activities and performs fairly in various aspects.

AI is used in the employment and selection procedure of human resource management because it aids in finding special and hidden talents (Niehueser and Boak 2020). Modern markets require organizations to look at all possibilities and qualities of an individual. Singh and Shaurya (2021) inform that human resource managers do not solely focus on professional qualities but also consider deeds, words, and the candidate's appearance in the recruitment process. These are not factors that can be easily ignored because they have a significant impact on the productivity of an organization. Institutions must look at appropriately perfect employees for the organizational culture and traditional practices. These factors are depended on personality and individual qualities such as communication skills and other related aspects of the performance. Artificial intelligence makes the work easier for organizations in the selection process because the qualities they are looking for can be run through algorithms created by the machine to pick the best candidate (Singh and Shaurya 2021). AI is the best tool for the recruitment and selection of employees because it can gather and analyze candidate data more efficiently and fast than human recruiters (Niehueser and Boak 2020). Artificial intelligence has made human resource activities possible through deep machine learning that incorporates the running algorithms to create an all-new experience (Berhil et al. 2020). The activity of artificial intelligence is applicable, logical, and exhibits new and developed levels of practice. AI is best suited for human resources because it improves performance, increases efficiency, and enhances quality services. The development of artificial intelligence has taken different phases. The AI algorithms learn to run applicants' details by working with humans for some time to understand what an organization is looking for. Learning-based algorithms compare data across the different available details of applicants and analyze it to understand various needs and requirements necessary to fit in a position in an organization.

Artificial intelligence has different algorithms that apply to human resource management activities, including predictive, analysis, and forecasting. These algorithms work together to deliver unique work values as required to execute duties. More specifically, these algorithms run to forecast the future of the employment markets and the required skills needed in the market. Predictive algorithms are essential in an organization because they increase planning by making it efficient and offering required resources to the HR team (Singh and Shaurya 2021).

The use of AI in HRM is increasing. In the United Arab Emirates organizations, AI looks at the future of human resource management and the markets to explain how it impacts change and improves management in an organization (Singh and Shaurya 2021). The AI in human resource management is explored by enabling market structures and improving employees' outcomes. The last two years have been marked by working from home because of Covid-19. The research seeks to uncover how artificial intelligence helped improve HR management during covid and how such practices could be implemented to be compatible with the Abu Dhabi vision 2030 (Singh and Shaurya 2021). Currently, many organizations have started testing for conventional recruitment of employees. There are incentives to introduce AI in different organizations for hiring and recruiting employees. The UAE must investigate possibilities and the strategic mechanisms for making AI productive in human resource management (Singh and Shaurya 2021). The HR professionals focus on the strategic planning that requires running data through the AI algorithms for analysis and prediction.

2 Literature Review

2.1 Strategic HR Planning Through AI

Without strategic planning, human resource management would be impossible to operate. When strategizing, the human resources manager is supposed to conduct an in-depth assessment of the company's prior work and keep a close eye on the company's future development. The proof of AI is that human resources are necessary to do efficient planning, sensitivity, and innovation on behalf of managers (Iqbal 2018). Artificial Intelligence (AI) masters data uniquely for each individual. Due to the task's content and complexity, this has a significant impact on human resource planning. The company's characteristics, data optimization, industry trends, and future demand estimates are all matched through big data analysis. As a result, the human resources department at the current organization can quickly collect and aggregate data, provide pertinent recommendations, and conduct data analysis.

Iqbal (2018) asserts that predictive analytics may generate trustworthy forecasts about the future by analyzing historical and current data using statistical models, analysis techniques such as data extraction and machine education, and other methods. Predictive processes are critical for increasing the efficacy of human resource planning (Raub 2018). Singh and Shaurya (2021) outlined that algorithms and artificial intelligence result in unequal access to and discrimination against work opportunities without accountability. However, businesses interested in incorporating AI into their human resource operations should select a program that encourages responsible algorithms and advocates for long-term reforms in an environment devoid of gender or racial diversity.

2.2 Smooth Recruitment and Selection Process

When HR interviewers evaluate a candidate's words, behavior, and appearance, they are influenced by their professional qualities and manner, and demeanor. Lee et al. (2019) asserted that certain businesses could not afford to disregard selecting individuals who are not a good fit for their organization, even if they use an authoritative and competent quality evaluation technique. Rather than being merely integrated into human resources, AI will exhibit its dynamic aspect concerning human psychology, behavior, and so on. And so, on Candidate evaluation in a quality assessment process is feasible through the use of several algorithms and intelligent learning approaches that can be used to discover the most competent personnel (Ahmed 2018).

According to Jatobá et al. (2019), AI automates the process of scanning resumes and extracting essential data. Without automated technologies, human resources departments may find themselves unable to keep up with the demand for skilled individuals due to the increase in job applications. An applicant rating system can assist recruiters in expediting the resume evaluation process by automating operations that human resources departments traditionally undertake (Berhil et al. 2020). A candidate ranking system can function successfully and efficiently due to AI algorithms and human recruiters supplying training data. Consequently, candidates can use HireVue to evaluate and analyze their body language, facial expressions, and voice tones (Ahmed 2018). This tool compares interviewers to the company's greatest employees and then makes recommendations to recruiters. Hilton was able to save significant time and money during the recruitment and selection process for new workers through video interviews.

Chichester and Giffen (2019) suggests that chatbots as a tool can engage with candidates in a more personal and up-to-date manner via text messages, dialog boxes, or emails. Recruiters have less work to complete due to computer-assisted job-matching technologies. A good example is resume sorting software, which uses machine learning techniques and processes (Wright and Atkinson 2019). Additionally, such data may be gathered by AI-powered rating algorithms, which may be used to evaluate which candidates are most likely to succeed in a certain post. While these attributes are normally evaluated during a job interview, early information can be gathered via web searches. Instead, a linguistic examination of a blog post or a LinkedIn profile will reveal details about an individual's personality, feelings, and mood.

2.3 Planned Training and Development Process

Employee work satisfaction will increase if a company utilizes AI technology to build exercise and development chances for its personnel. Iqbal (2018) asserts that Developing career development programs tailored to employees' specific needs increases output and decreases absenteeism. Another AI-enabled breakthrough in the training and development arena is the establishment of individualized learning paths for each employee. What we refer to as "mobile coaches" assist employees in assessing their needs and developing a customized program for their personnel (Alam et al. 2020). As a result, the software is used to assess individuals' job needs and assist them in planning their future careers or team development. Employees have access to coaching and mentoring possibilities due to various mini-courses.

Due to the large costs and expenses connected with staff training, a reasonable return on investment is rare. According to Adams-Prassl (2019), HR might employ the training speaker's behavior, professional knowledge, and appearance style to assist trained personnel in generating a large amount of judgment and psychological definitions during the training procedure. It was more difficult for an intelligent machine to negatively affect since it was steadier in its immediate vicinity's psychology. Artificial intelligence can be created by employing keynote speakers or corporate training. While AI enables the investigation of numerous models of appropriate corporate training, it also enables the monitoring and balancing of training circumstances for employees throughout their participation in organizational education. AI can assist HR in teaching employees by assuming the role of the trainer and boosting the training effect. On the other hand, artificial intelligence (AI) can serve as a housekeeper and a personal trainer for each employee.

2.4 Tactical Performance Appraisal

Employee attitudes and informal organizational structures might influence performance evaluations due to emotional exchanges between co-workers. AI can precisely and empirically record information to decrease the regular performance evaluation inaccuracies caused by employees' mental suffering (Adams-Prassl 2019). On the other hand, artificial intelligence may be utilized to design a fair and objective performance evaluation system by analyzing data on corporate development, urban development, and industrial performance, to name a few. This results in a more pleasant work environment for employees, reduced waste associated with enterprise trash, and improved overall business performance.

Conventional employee associations are more concerned with regular responsibilities such as leave administration, social security payment, processing, and resolving employee contract issues (Raub 2018). Machine learning techniques and technology enable AI to perform functions other than assisting humans in their daily tasks. As a result, it assists both parties in quickly picking the optimal plan, allowing them to develop defensible Labor partnerships under fair and objective conditions (Wright and Atkinson 2019). AI's more positive recommendations have improved career planning for organizations and employees. Organizations cannot simply cut Labor costs by establishing a favorable culture of employee Labor relations.

2.5 Ease of Use and Efficient HR Practices

Organizations constantly look for new methods to enhance their decision-making, performance, and forecasting capabilities. According to the Society for Human Resource Management (SHRM), one of the most significant advancements in human resource management has increased employees' ability to make sound decisions (Raub 2018). In comparison, the service sector must understand its industry and how data and technology can help with decision-making to harness AI to enhance human resource management fully. AI possesses capabilities that facilitate three distinct commercial functions (Adams-Prassl 2019). Additionally, AI can help businesses automate their processes by embedding cognitive capabilities into the program itself. For instance, AI might accelerate the automation of decision-intensive activities such as supply chain management and loan processing and provide cognitive visions into customer purchasing behaviors.

Additionally, by providing cognitive insights, AI can assist organizations in making more informed decisions. Massive amounts of data are sorted through to uncover previously unseen patterns using algorithms and machine learning (Raso et al. 2018). Customers' purchasing habits are predicted, and artificial intelligence organizations implement targeted marketing and fraud detection (AI). To increase their campaigns' effectiveness while reducing costs, candidates have used artificial intelligence (AI) (Singh and Shaurya 2021). Additionally, Artificial Intelligence (AI) could rapidly diagnose cancer patients.

Consequently, the program automatically makes and integrates judgments with the least amount of human input feasible using AI-enabled procedure computerization. According to Raso et al. (2018), AI-based expert systems can mimic the decision-making of employee benefits specialists. Therefore, an AI-enabled human resource management system would not simply indorse benefits from which employees can choose but would instead select such benefits for employees based on the research findings.

2.6 Automation of Administrative Tasks

Like other industries and specialties, human resource operations can benefit from artificial intelligence. Routine administrative tasks can be automated to free up HR professionals' time to contribute more strategically to the organization (Adams-Prassl 2019). According to Nawaz et al. (2020), benefits administration, candidate pre-screening, and interview scheduling can all be automated with the help of smart technologies. It is important to note that, while each of these activities helps a company's overall performance, completing the responsibilities associated with such processes is frequently time-consuming, leaving HR specialists with less time to contribute more meaningfully to their employees' well-being (Adams-Prassl 2019). This burden can be lessened by automating administrative tasks with AI technologies. According to a survey conducted by Eightfold, human resource professionals who use AI software perform administrative tasks 19% more efficiently than those who do not. Human resource managers will have more time to devote to their firms' long-term planning.

2.7 Preparing for the Future of Human Resources Management

While AI will likely positively affect human resource management in the future, HR managers should be prepared for the challenges that may arise from this technological progress (Wright and Atkinson 2019). Human resource leaders are particularly concerned about making AI more accessible and safer. Indeed, security and privacy issues are the primary reasons artificial intelligence is not deployed in the workplace. According to Raub (2018), 31% of respondents indicated that they would prefer to work with a human rather than a machine. In the future, human resource professionals must be prepared to address these concerns by staying current on emerging trends and technologies.

When implementing this technology, "individuals must bear ethical and privacy concerns. In the case of human resources, artificial intelligence (AI) may require the utilization of private data to provide sensitive insights. Many people expect their companies to safeguard their personal information and obtain their consent before using such technology to collect data about them (Singh and Shaurya 2021). When securing the firm, human resource professionals must consider security measures. The best strategy to prepare for a future job in human resources management is to stay current on industry trends and build a solid foundation of HR knowledge.

3 Conclusion

Concerning problems being experienced currently with the emergence of artificial intelligence, human resources play an important role in helping leaders and organizations adopt new technologies, helping employees adapt to innovative working outlines, promoting the development and change of national rules and guidelines, and encouraging the adoption of new technologies in the workplace by companies. The concepts discussed in this report show that AI will greatly assist employees rather than replace them. However, human beings are considered exceptional in the Human resources department due to their ability to reason beyond the scope of the problem being solved. (Go-meta). This is an ability that AI does not have. They are, however, excellent at executing problems which well-defined. Artificial intelligence will guarantee a number of advantages to the human resource department. The human resource manager will effectively carry out their administrative functions. Employee quality performance is guaranteed through talent enhancement by AI. Managers can easily perform hiring and decision-making processes effectively. Even though AI's advantages to human resources, institutions need to take care of the issues that may arise with it, such as data leakage to unauthorized parties.

References

Adams-Prassl, J.: What if your boss was an algorithm? Economic incentives, legal challenges, and the rise of artificial intelligence at work. Comp. Lab. L. Poly J. **41**, 123 (2019)

Ahmed, O.: Artificial intelligence in HR. Int. J. Res. Anal. Rev. **5**(4), 971–978 (2018)

Alam, M.S., Dhar, S.S., Munira, K.S.: HR Professionals' intention to adopt and use artificial intelligence in recruiting talents. Bus. Perspect. Rev. **2**(2), 15–30 (2020)

Berhil, S., Benlahmar, H., Labani, N.: A review paper on artificial intelligence at the service of human resources management. Indones. J. Electr. Eng. Comput. Sci. **18**(1), 32–40 (2020)

Chichester, M.A., Jr., Giffen, J.R.: Recruiting in the robot age: examining potential EEO implications in optimizing recruiting through artificial intelligence. Comput. Internet Lawyer **36**(10), 1–3 (2019)

Iqbal, F.M.: Can artificial intelligence change how companies recruit, train, develop and manage human resources in the workplace? Asian J. Soc. Sci. Manag. Stud. **5**(3), 102–104 (2018)

Jatobá, M., Santos, J., Gutierriz, I., Moscon, D., Fernandes, P.O., Teixeira, J.P.: Evolution of artificial intelligence research in human resources. Procedia Comput. Sci. **164**, 137–142 (2019)

Lee, J., Suh, T., Roy, D., Baucus, M.: Emerging technology and business model innovation: the case of artificial intelligence. J. Open Innov. Technol. Mark. Complex. **5**(3), 44 (2019)

Nawaz, A., Su, X., Barkat, M.Q., Asghar, S., Asad, A., Basit, F., Iqbal, S., Zahoor, H., Raheel Shah S.A.: Epidemic spread and Its management through governance and leadership response influencing the arising challenges around COVID-19 in Pakistan-A lesson learnt for low income Countries with limited resource. Front Public Health (2020). https://doi.org/10.3389/fpubh.2020.573431. Dec 10;8:573431, PMID: 33363079; PMCID: PMC7758222

Niehueser, W., Boak, G.: Introducing artificial intelligence into a human resources function. Ind. Commer. Train. **32** (2020)

Raso, F.A., Hilligoss, H., Krishnamurthy, V., Bavitz, C., Kim, L.: Artificial Intelligence & Human Rights: Opportunities & Risks. Berkman Klein Center Research Publication, Cambridge (2018). (2018-6)

Raub, M.: Bots, bias and big data: artificial intelligence, algorithmic bias and disparate impact liability in hiring practices. Ark. L. Rev. **71**, 529 (2018)

Singh, A., Shaurya, A.: Impact of artificial intelligence on HR practices in the UAE. Humanit. Soc. Sci. Commun. **8**(1), 1–9 (2021)

Wright, J., Atkinson, D.: The impact of artificial intelligence within the recruitment industry: defining a new way of recruiting. Carmichael Fish. **10**, 1–39 (2019)

Artificial Intelligence and Human Resource Management in Public Sector of Bahrain

Mariam Juma Khamis Alfulaiti[1], Allam Hamdan[2](✉), and Rania Baashira[3]

[1] College of Business and Finance, Manama, Bahrain
[2] Ahlia University, Manama, Bahrain
allamh3@hotmail.com
[3] Jeddah College of Engineering, University of Business and Technology, Jeddah 21448, Kingdom of Saudi Arabia

Abstract. The study's major goal will be understanding the importance of artificial intelligence (AI) in the human resources management, particularly in Bahrain. Also, to look into the issues that have arisen as a result of training programs launched in Bahrain's public sector. The training and development of employees are two of the most important topics tackled. The importance of training in delivering the required knowledge and capabilities should not be overlooked. As a result, the training must be successful enough to meet these goals. This study examines a number of contextual elements that have been discovered to have an impact on vocational training in different combinations with other previously examined effective factors. It will enquire into their relationship. It also identifies the various types of relationships that exist between the efficiency of training and environmental elements. The Kirkpatrick training model is used to build the study's framework. This paradigm has four stages that measure the success of training: reaction, learning, behavior, and result. This modest study project offers a glimpse into the future of employing AI to better understand HR practitioners' attitudes and perspectives across a variety of systems. According to the study, Bahrain's public sector has been a tremendous potential in order to keep pace with the digital revolution by implementing its vision (2030 vision). As a result, the staff mix of commercial organizations has changed. It gives men and women the opportunity to compete in a variety of occupations. As a result of the requirement for gender equality, human resource management will be burdened in new ways. This also allows for a nice feminist component to be included. It is suggested that current artificial intelligence be used as a foundational method for firms that operate in a volatile environment.

Keywords: Artificial Intelligence (AI) · Human resource management · Public sector · Kingdom of Bahrain

1 Introduction

Artificial Intelligence (AI) has risen in popularity in Bahrain's government sector in recent years. As a consequence of the Kingdom's attempts to modernize services in

B. Alareeni et al. (Eds.): ICBT 2022, LNNS 620, pp. 584–593, 2023.
https://doi.org/10.1007/978-3-031-26953-0_53

order to make them more effective and dependable, artificial intelligence has become necessary for the Kingdom's numerous branches. The Kingdom of Bahrain is not alone in its transition to artificial intelligence. Artificial Intelligence is increasingly being used to replace the human aspect in modern society in favor of more effective and long-lasting machine-based systems. As noted in the referred literature below, this study covers the rising phenomena of artificial intelligence applications in many governmental sectors around the world. This presentation will examine artificial intelligence applications both local and regional prospective. The study analyzes the legal, logistical, and administrative aspects that hinder artificial intelligence from being used. It will also go through the difficulties that artificial intelligence can help tackle. As the paper will describe, artificial intelligence has been shown to boost productivity and efficiency. It can also nearly completely eliminate the margin for error by replacing error-prone humans with trustworthy machines. When artificial intelligence is implemented, it improves more than just efficiency. As the study explains, fair systems and moral are the product of a service system that was programmed very well. Taking everything into account, Artificial intelligence application prospects may act as a catalyst for economic development by providing decision makers with a special machine-based, and also non-biased viewpoint. All of this is in accordance with Bahrain's 2030 economic agenda's aims and vision. Ai technology is a true game-changer in the world of corporate governance, and it will also have a significant impact on how workers operate, particularly in terms of human resources and employment. (AI) has a number of impacts on human resource management in variety of ways. Artificial intelligence (AI) is a word that refers to technology that has been used to fulfill tasks that need intellect. To put it another way, this is a machine that has been programmed to perform jobs that a human can. When utilized properly, artificial intelligence may aid in the completion of human resource management activities such as job appraisal, HR planning, evaluation of performance, employee training needs, Job assessment, as well as anticipating the labor market and its demands and indications [1, 2]. According to a study conducted by a renowned provider of cloud-based software for certain industries, Due to the quick development of technology, we are already seeing an example of creative use of AI in a way that may give additional positive benefits to the workflow. Human resources and recruitment department employees go about their jobs. Several businesses and organizations have already demonstrated how artificial intelligence (AI) may increase treatment quality while also lowering costs [3]. Half of all employment will be outmoded or obsolete in about 20 years, and healthcare will be no exception [4]. It's just as vital to know the advantages and disadvantages of alternative tactics as it is to develop the right algorithms and data architecture. "There are many opportunities in the future to increase the quality of career possibilities, reduce the gender gap, repair the harm caused by inequality at the global level, and much more," said the International Labour Organization (ILO) in a research published in January 2019. New technologies, demographic upheavals, and climate change are all continuous economic developments that will have both detrimental and beneficial consequences on our jobs and economies, according to the study. Significant expenditures are necessary to understand the magnitude of these transitions that contribute to the development of decent work. As a result, in accordance with the 2030 Agenda for Sustainable Progress, countries should now prioritize long-term environmental expenditures that promote human growth while also conserving the environment. Communication technology advancements have boosted our capacity to track progress toward long-term development objectives. It enables us to rationalize and improve the efficacy and efficiency of all of our development efforts. According to a study conducted by the Globe Economic Forum, many employees will need to radically alter their talents in order to assist the world. Organizations must help their employees develop new skills, especially if they want to keep up with the Fourth Industrial Revolution's fast changes. Organizations that value people and capabilities, especially when data analytics is employed, will thrive. Administrators will also need to learn how to plan ahead, assess talent gaps, and reorganize their teams to address present and future challenges. Future talents is a collection of six choices for overcoming the crucial skills gap in the future world, according to statistics given by the World Summit of Governments in February 2019. In partnership with McKinsey & Company, they confirmed that digital transformation and automation will result in a wide range of job needs, which will become more vital in order to remain engaged in society. The findings point to a wide range of potential occupational changes in the coming years, all of which will have Significant ramifications on worker skills and pay. While there may be enough labor to sustain full employment until 2030 in most scenarios, the transitions will be extremely difficult, equaling or maybe surpassing the extent of earlier agricultural and manufacturing transformations. More than 800 million people (one-fifth of the world population) According to McKinsey Global Institute research conducted in 46 nations in over 800 cities, people would lose their employment and be replaced by robots until 2030[5]. According to McKinsey Global Institute research including 46 countries and over 800 persons, by 2030, more than 800 million people (one-fifth of the global workforce) would lose their employment and be replaced by robots. In a connected domain, 25 percent of today's workforce By 2020, they'll either need to find interests or substantially improve their technical and digital citizenship skills, as well as their conventional talents, bolstering their interdisciplinary capabilities, depending on the current quo. Programming, job flexibility, and adaptation are among these abilities, and primary school kids must be flexible and adaptable, as 85 percent of them will work in professions that do not yet exist by 2030. Over the next five years, however, Artificial intelligence adoption as a measure of global economic development, as well as per capita Fourth Industrial Revolution technology, will be included, Along with national income, GDP, inflation, and other economic indicators, the state's economic might is measured. To summarize, the study's goal is to gain a better knowledge of the artificial intelligence challenge in human resources, particularly in Bahrain. Which provide a forward-looking perspective on using AI to better understand HR practitioners' attitudes and viewpoints across a variety of frameworks.

2 Methodology

In order to achieve the study's purpose, we look at a range of literary critiques. Focus on an authoritative narrative review of AI applications in management and HR in particular. For a current overview, we selected electronic databases, notably Science Direct, Google Scholar, and Emerald, as the most efficient method of conducting a literature search. Because AI is a new technology, we include some material from international organization papers, magazines, journals, and newsprint. During the search, the following keywords were used: Artificial Intelligence, Human Resource Management Furthermore, we looked at English and Arabic language papers that were relevant to the study's core purpose and were published after 2012. We also examine how innovative HRM methods are being implemented in prominent firms around the world, as well as their implications for Bahrain's economic environment.

3 Literature Review

Wang et al. claim that [6] "Artificial Intelligence is the activity dedicated to making robots intelligent, and intelligence is the trait that permits an entity to perform successfully with foresight in its circumstances." Machine intelligence, or intelligence displayed by computers, is referred to as artificial intelligence (AI) in computer science, as opposed to natural intelligence, People and other living things both express this.. Computer science defines "Any technology that detects the surroundings and takes actions that optimize its chances of effectively achieving its goals," according to the research of "intelligent agents." AI is defined by Kaplan and Haenlein [7] as a system's ability to effectively absorb external information, learn from it, and use what it has learned to achieve specified objectives and tasks through flexible adaptation. Artificial intelligence is described as a machine that can do "cognitive" skills associated with human brains, such as "learning" and "problem solving."

4 Opportunities of Artificial Intelligence in Human Resources Department

In order to improve productivity and boost employee labor levels in today's world, artificial intelligence is reinventing how organizations manage their human resource planning and workforce. Employees' interests, on the other hand, must be represented not just in their talents, but also in how well they match vacancies and how inviting they are to new individuals. In today's millennium business sector, which is undergoing a revolution in work-life balance, ethics, and job content, the employment of new science-backed technologies is vital. Today's workers are mainly self-sufficient and adaptable. Technology, particularly smartphone or self-service applications, is the most effective way to engage with people. In order to recruit the best employees, service customers, and compete, institutions must undertake digital transformation in AI. Furthermore, Workplace technology should empower employees by allowing them to access their tasks at any time and from any location, in keeping with the current digital world. In February 2019, during the World Government Summit on Artificial Intelligence, the poles of the global artificial intelligence community gathered professionals, experts, government officials, and policymakers in speeches and sessions at the Global Forum on Artificial Intelligence Governance. With over 250 experts and professionals from many sectors debating AI governance and the structure of its function in research, engineering, health, and communications, we may be on the verge of creating a fascinating future. Between 2015 and 2020, it has been confirmed that the human competencies necessary in companies would change by 35%. Experts also stressed the need of governments collaborating with a wide range of partners and sectors to improve human resources. It will be able to maintain the considerable changes in the nature of vocations and career specialties that In the future, artificial intelligence will bring. And begin a community-wide, human conversation on the best methods to create new jobs and high-quality investment possibilities. Machine integration and leading automation solutions that increase productivity and cost-effectiveness while also encouraging entrepreneurship and innovation 1. As a result of rapid advances in HR technology, major trends are obviously developing, including privatization, which offers a substantial opportunity to depart from the way HR programs have been delivered in the past. Previously, there was a one-size-fits-all approach, AI technology allows us to create customized settings for each employee, motivating them to utilize business platforms while also giving them with data from their personal life. In other words, the system supports and rewards people as they attempt to improve their work experience. Continuous learning will be critical for human resource performance in the future as automation and the use of technology grow. Full and continuing support for lifelong learning will be critical for human resource success in the future. It's not only about aiding workers with human resources issues like hiring new employees or assigning new responsibilities; it's also about ensuring that employees are always considering how corporate changes and technology impact their careers. And the abilities they'll need to maintain their current level of success. To stay up with industry developments, Businesses must provide an infrastructure that encourages people to develop new skills and broaden their horizons.

4.1 Career Path

Artificial intelligence with training modules and learning management systems has been used in the HR sector for many years to develop employees' abilities and provide them with the right career path, in addition to assisting them in excelling in their existing roles and increasing their desire for greater promotions AI technologies could use increasingly sophisticated big data technology to mobilize large and diverse data sets, such as terabytes of biographies and performance reviews, as well as tons of historical data, to improve and reveal training and education models tailored to a specific experience or career level in this approach. Many companies across the globe utilize artificial intelligence to educate, train, and empower their employees, improving the attractiveness of the workplace and making it a destination for expertise and skills. Using sentiment analysis to help workers play a more active part in their own professional growth. In recent years, the Emotional analysis approach has been used to reveal employees' positive and negative attitudes and prejudices towards anything from utilizing social media such as Instagram and Twitter. Emotion analysis apps will become more widely used in the human resource sector in the next years as more entrepreneurs go into the field and apply these technologies. Employee happiness, engagement, and role are all evaluated using this method. When particular user answers are collected, the key words are entered into a lexicon and given good or negative ratings. These firms who are ready to put in the effort to deploy AI will get multiple benefits as a result of the end result. The path to successfully adopting artificial intelligence in HR operations will be long, but success will be worth the effort. The most significant challenges confronting the future workforce, according to a recent

World Economic Forum study (October 2018), are the availability of skills necessary to keep up with fast technological development. It's critical to make sure that the workforce has all of the necessary skills to keep up with changing technologies. As a result of the digital revolution, the digital divide between developed and poor nations has widened substantially. More severe, not just as a result of some cost-cutting measures, but also owing the technical and professional abilities required to develop, operate, and maintain digital infrastructure at a high level. The importance of comprehending core abilities as well as information and communication technology. Skills are seen to be helpful for reducing inequality and closing the knowledge gap between employees.

4.2 Recruitment

Artificial intelligence is used in many facilities' recruiting processes, whether for selection, assessment, or recruitment. You might also utilize chatbots or other tools. A local business recently astonished me by proving that it can properly rank hundreds of candidates by applying personal analysis algorithms to evaluate the video filmed by the candidate himself. In terms of technology, this is a huge step forward. According to Housman [8] in a service-based economy, When a company's most valuable asset is its people, utilizing decision support to make the best personnel decisions might provide an edge over its competitors (Page 9). Finding the right proof at a reduced price, in a shorter amount of time, and in a secure manner aids in developing momentum step by step, starting with the hiring process. The AI may then be seamlessly integrated into the new employee onboarding process. New employees may be unsure of where to go for opportunities to communicate with coworkers and learn more about the company. Artificial intelligence has contributed significantly to improving the efficiency of the recruiting system and attracting talent to institutions and businesses. Monitoring of employment indicators in real time. Furthermore, this criterion mitigates disparities in impact based on race and ethnicity [9]. When a database is kept up to date, managers get a clear picture of their employees' skills and experience. They find the best individual to do a task in a matter of seconds. Managers can develop a chart using predictive analytic techniques that illustrates what skills and people they'll need in the coming year and the year after that. What a simple and quick approach to arrange the workforce. What if, instead of manually looking for candidates throughout the hiring process, your technology found and communicated with the best candidate for the job? What if, before the interview, the system could answer all of the candidates' questions? HR professionals may wonder whether they know anyone in the region who might be a good fit for your position. What should they do if a group of employees wants to finish within the next year or two? Simply enter some data into the rating tool, and it will provide you with all available forecasts as if you were in Delphi, Greece. The way HR controls workforce planning will be transformed by employee travel risk models and predictive talent evaluations. To me, it's almost miraculous that you can automatically feed data into the ranking engine and obtain rich, relatively valuable results in such a short amount of time. However, because the actual world is more complicated, developing a trip risk model for a variety of circumstances demands a significant amount of human input. Employees will be able to focus on roles that provide value to the firm as AI technologies take over monotonous manual chores that need employees' competencies and knowledge. As a result, HR professionals will have more time and resources to devote to the personal care of other employees.

4.3 Talent Acquisition

The most essential priority of HR departments is the appointment of bright people. The long-term demands of the company are addressed by talent management. It selects the finest applicant and positions required for growth within a strategic framework to fill jobs necessary for the company's future ambitions. Companies having the time and resources to build a meaningful and effective talent management program will become a key element of their long-term human capital strategy. Most definitions of talent management include implementing integrated strategies or processes targeted at increasing employee recruitment and development, maintaining critical abilities and being ready to fulfill current and future organizational demands. Institutions and businesses must now adapt their conventional recruiting methods and embrace a new strategy to attract talent and skills, as well as preserve internal consistency and stay up with the newest technological and artificial intelligence advances, As a result of the rapid worldwide developments, and take use of these technologies in the construction of a system that supports work activities [14]. To attract qualified individuals, many worldwide organizations' human resources departments are becoming more reliant on AI-assisted recruiting. Talent acquisition tools, for example, can read, scan, and analyze candidates, automatically dismissing 75% of them from the hiring process. This is a significant advantage since the recruiter can devote more time to reviewing and evaluating a fewer number of qualified candidates. HR departments are dramatically improving the quality of hiring judgments in such scenarios. Robots should communicate with employees as soon as possible and answer typical inquiries. Such as the amount of an employee's remaining leave or other routine questions. As a result, the HR Analyst is focused on dealing with more complicated situations. Human resource managers can also keep a tight eye on employee morale by conducting regular, near-instant analyses. The Human Resources Department employs a variety of strategies to ensure employee commitment, including the implementation of effective incentive systems.

4.4 Training and Development

All personnel must continue to study and improve their abilities in order to keep up with the rapid progress of technology. AI could design, plan, and co-ordinate training programs for everyone in the team. The most frequent solutions in this field are online courses and virtual classrooms. According to surveys, the average employee has fewer than 25 minutes per week to train and learn; as a consequence, allocating that time to the development of employee skills through a simple acquisition technique is critical. Team managers may offer digital training opportunities for the team based on skill-gap evaluations. Using computerized information systems, workers may now pursue a professional path in a straightforward and cost-effective manner to access information about a person's skills and competencies, as well as via electronic access and from the computer to the database, how they develop his or her performance and work, and what positions and tasks he has held within the firm during his career. Working people can use several self-assessment informatics programs to help them create and evaluate their careers.; Professional Planning Center, Parys, an expert in human resources management, and Miram, an authority in this sector, are among the programs available in the market arena to aid workers in moving along their career paths [10].

4.5 Performance Analysis

Performance appraisal is one of the most crucial processes in organizations since it influences every aspect of the firm, from top management to employees. When it comes to the lowest-paid jobs, one of the administrative control tools is the minimum production plan. The actual performance is compared to the project manager's expectations. As a result, the evaluation process and the process of accomplishing the desired objectives. Given AI technology's ability to rapidly recruit and assess employees in a short amount of time, as well as link their goals to their performance [11, 12]. Managers may utilize artificial intelligence in HR to help them set appropriate goals and guarantee that all departments are working toward the same goal. It helps management to make intelligent decisions based only on factual information while conducting employee performance reviews. During the goal-setting process, individuals are given specific objectives or quotas to fulfill within a defined time range. AI can assist in real-time tracking of this progress and quick feedback based on the conditions and circumstances. Because of the impression she makes on her mind, the Human Resources Officer's motivating policy is critical; If employees agree on their objectives, the incentives may shift in type and shape. Keeping human abilities is challenging since physical factors are involved. From a moral sense, it may be the primary reason why Human resources have migrated from developing to developed countries. In her perspective, there is a lack of a defined policy for retaining these skills.

4.6 Compensations

Computers may now be used to create pay and compensation lists, and software solutions to assist speed up and enhance the review process are now available. Concentrate on areas where there are now processes in place in this business to calculate compensation and evaluate individual performance, Especially for huge or massive businesses. When it comes to providing awards, these methods have also aided in the categorizing and upholding of justice, such as a tool that assists pension funds in calculating various types of compensation.

5 Conclusion

According to an economic analysis, there is a compelling need to embrace human resources in light of the impact of artificial intelligence and automation. Because the digitization of human resources has a significant impact on facilities and their work, human resources personnel must be prepared to assist in the digital transformation of their workplaces. Human resources would otherwise be at danger of falling behind and losing track of other departments. Human resources, on the other hand, By becoming more digital and data-driven, it is possible to align with the facility and add substantial value. We hope that, in the next years, our beloved monar-capabilities chy's will allow for the digitalization of human resources. Also, notice how effectively it converts [12, 13]. The current research looks into the influence of artificial intelligence on human resources and offers a viewpoint on the subject. It also discusses how AI might aid in improving a transformation's creativity and developing a new app's user experience. The major conclusion is self-evident after analyzing available options and potential. In order to bring HR functions into the digital era, artificial intelligence (AI) will be critical. Without a question, the rising use of technology and data within enterprises has altered the sorts of businesses and abilities required for specific tasks. To adapt to the changes that have occurred on the inside with particular occupations and personnel who may be dispensed, training, development, and organization will become increasingly crucial, especially owners with basic abilities and regular labor. As a consequence, To solve the difficulty that most large businesses are facing today and increase their capacity to provide relevant information to management so that they can make educated human resource choices, HR should have basic information established based on computer services. If businesses want to stay competitive in today's global economy, they must invest in technology. They'll have to think about how to include conversational AI for HR transactions into their decision-making. In order for HR departments to become more efficient, organizations must rely on AI to do administrative tasks. HR experts will have more time to devote to the development of company strategies [14]. The public sector in Bahrain has a tremendous chance to keep up with the digital revolution thanks to the execution of Vision 2030. As a result of IT capabilities in business management, the company's employee makeup has changed. It allows women and men to compete in a variety of jobs as a result of the drive for gender equality, putting a new demand on human resource management. It also enables for a strong feminist element to be included. This necessitates top-level sponsorship (planning), and HR should be prepared to respond to the employment of a person of the opposite gender. AI will have a variety of effects on employees, therefore focusing on their requirements and potential consequences is critical. Human resource functions, on the other hand, can benefit from AI in terms of reducing the amount of time. HR professionals spend on administrative tasks, reducing the burden of shared service centers and help desks by performing HR transactions and responding to routine queries, recruiting and retention, and ROI measurement.

References

1. Aldulaimi, S.H., Qadir, O.A.: Human resources performance measurement approaches compared to measures used in master's theses in applied science university. Int. Rev. Manag. Mark. **6**(4), 958–963 (2016)
2. Abdeldayem, M.M., Aldulaimi, S.H.: How changes in leadership behaviour and management influence sustainable higher education in Bahrain. Int. J. Sci. Technol. Res. **8**(11), 1926–1934 (2019)
3. Mesko, B.: The role of artificial intelligence in precision medicine. Expert Rev. Precis. Med. Drug Dev. **2**, 1–3 (2017)

4. Frey, C.B., Osborne, M.: The future of employment: how susceptible are jobs to computerisation? (2013)
5. McKinsey & Company. Disruptive technologies: Advances that will transform life, business, and the global economy. McKinsey Global Institute, New York City (2013). https://www.mckinsey.com/quarterly/the-magazine
6. Wang, D., et al.: A problem solving oriented intelligent tutoring system to improve students' acquisition of basic computer skills. Comput. Educ. **81**, 102–112 (2015)
7. Kaplan, A., Haenlein, M.: Siri, Siri, in my hand: Who's the fairest in the land? on the interpretations, illustrations, and implications of artificial intelligence. Bus. Horiz. **62**(1), 15–25 (2019)
8. Housman, M.: Decision support comes to HR: How technology is enabling the right decisions about your workforce. Workforce Solut. Rev. **7**(5), 9–11 (2016)
9. Nicastro, D.: 7 Ways Artificial Intelligence is Reinventing Human Resources. CMS (2018). https://issuu.com/ihrimpublications/docs/wsr_sept19all
10. Schermerhorn, J.R., Hunt, S.G., Osborn, R..N.: Comportement Humain et Organisation, (Village mondiale, 2emeédition. Paris, France, imprimé au Canada,), pp. 75–80 (2012)
11. WirePiazza, L.N.: How can artificial intelligence work for HR? SHRM(2018)
12. Abdeldayem, M.M., Aldulaimi, S.H.: The economic islamicity index, between islamicity and universality: critical review and discussion. Int. Bus. Manag. **12**(1), 46–52 (2018)
13. Abdeldayem, M.M., Aldulaimi, S.H.: Privatization and financial performance in Egypt since 1991. Asian Econ. Finan. Rev. **9**(4), 461–479 (2019)
14. Alsamman, A.M., Aldulaimi, S.H., Alsharedah, M.: Training effectiveness and commitment to organizational change: Saudi Arabian ARAMCO. Manag. Adm. Sci. Rev. **5**(3), 128–142 (2016)

Artificial Intelligence in Accounting and Auditing Profession

Maryam Ali Mansoor[1], Ebtisam Moh'd Salman[1], Nayef A. Rahman Al Jasim[1], Abdulla Adel Al Mannaei[1], Allam Hamdan[2(✉)], Ayman Zerban[3], and Esmail Qasem[4]

[1] College of Business and Finance, Manama, Bahrain
[2] Ahlia University, Manama, Bahrain
allamh3@hotmail.com
[3] College of Business Administration, University of Business and Technology, Jeddah 21448, Kingdom of Saudi Arabia
[4] The Islamic University of Gaza, Gaza, Palestine

Abstract. The aim of this study is to explore the relationship and impact of the new innovation of "Artificial Intelligence" on the "accounting" as well as on the "auditing" professions including the impact on future loss of jobs within the accounting and auditing industry. In order to fulfil the aim and objectives of the study, the research paper is divided into three primary chapters starting with introduction containing the problem statement, research aim and objectives, research questions and hypotheses. The chapter two portrayed the literature review thereby setting the theme for the last chapter of conclusion and recommendations for future study. The study was conducted through a thorough review and analysis of the past literature base concerning the primary area of research that is the impact of AI on the accounting and auditing profession. Finally, it was concluded by the research study that the AI and its implementation have a positive impact on the accounting as well as auditing profession thereby facilitating a better professional decision making. It was also concluded that the AI and its implementation do not have any material impact on the future loss of jobs within the accounting and the auditing industry.

1 Introduction

1.1 Background of the Study

Artificial intelligence according to the Dick (2019), refers to the overall process of human intelligence simulation by the man-made machines and computer software. The AI which is the abbreviation for the 'Artificial Intelligence" refers to a brand-new innovation which can be applied in any established and separate field or subject of study or discipline as well as on profession which reduces the cause of human errors, reduces mistakes, misrepresentations, anomalies, irregularities along with enhancement of overall quality and substance of the work or service thereby enhancing the overall decision-making process as well. The AI is that that specific set of modern technology which can easily imitate the human intelligence and work just like the humans would have done

B. Alareeni et al. (Eds.): ICBT 2022, LNNS 620, pp. 594–603, 2023.
https://doi.org/10.1007/978-3-031-26953-0_54

under different pre-programmed situations. So, AI and its implementation enhances the overall process of the work along with reducing the errors, mistakes thereby enhancing the overall quality as well as viability and feasibly of the work performed or services rendered. It is to be noted that the implementation of AI within a business or decision making process such as the accounting or auditing profession enhances the overall process of the work along with enhancing the quality of work done and services rendered by any professional mainly because of the following reasons which initiates with "learning process" which in simple sense denotes the capability of AI empowered system and software to continuously learn new things and also develop new traits to carry-out the work with enhanced quality and craftmanship. For instance, the AI empowered systems are programmed to continuously acquire and gather new information and data and subsequently develop new rules or algorithms which in turn will convert the gathered data into real actionable information which will assist the user of such information in better and prudent decision-making or will improve the quality of the product developed or services rendered.

Secondly, the AI empowered systems have an in-built reasoning process pre-programmed which allows these systems to choose the correct and appropriate algorithm in order to provide the desired outcome as desired or needed by its user. The final eminent feature of an AI empowered system is having an in-built self-correction process where the system itself continuously works towards fine tuning its algorithms based on the data collected and knowledge gathered in due process in the light of its pre-desired outcome thereby providing the best and most accurate results every time it functions which eventually enhances over the time with the gathering of quality knowledge and subsequent best algorithm selection by the system. Therefore, from the above-explanation and introduction of AI it is quite evident what AI can do and how it does improve the overall quality of any work done and any services rendered where the noble professions of accounting and auditing and its functions are no exception. It is to be noted that the accounting as well as auditing requires a huge and mundane job of repetitive tasks which are not only monotonous but is also prone to human errors and mistakes along with the fact that there might be involvement of intentional errors or frauds which requires special attention. This is where the AI empowered systems might come into work which can resolve the above-mentioned issues of carrying out the repetitive but important work of an accountant or an auditor as a professional delivering professional service upon whom the stakeholders of a business or an organisation may depend to frame their important financial as well as economic decisions. Finally, it is also to be noted here that the AI implementation within the noble professions of accounting as well as auditing may involve loss of jobs for the people as machines may take over many important job roles of an accountant as well as an auditor and may deliver these jobs in a speedy way with proper quality of work as well. So, there exists a fear amongst the accounting as well as auditing professionals relating to the implementation of AI within the accounting and auditing profession as well. Hence, the aim of this study is to explore the relationship and impact of the new innovation of "Artificial Intelligence" on the "accounting" as well as on the "auditing" professions including the impact on future loss of jobs within the accounting and auditing industry.

1.2 Problem Statement

The problem statement in truest sense demonstrates the main issue which exists within the base of literature regarding the present topic under consideration and discussion. This is the part which frames the theme for developing the aims, objectives as well as the research questions that are to be investigated through the study thereby allowing the researcher to systematically bridge the gap in literature which have been identified within the literature review. It is to be specifically noted here that there have been quite a few studies that were undertaken on the impact of artificial intelligence on the accounting and auditing as a separate set of discipline such as the ones conducted by the Kokina and Davenport (2017), Munoko et al. (2020) and many more. However, no single researcher has considered the accounting as well as auditing as different professions within a same study in recent years. This means that no past researcher who have conducted similar studies in recent years by testing the impact of AI on the noble professions of accounting as well as auditing at the same time. Above all, no past researcher has duly studied the different impacts of AI on both the accounting and auditing industry as well as profession at the same time where the impact of AI on the future jobs of both of these professions have not been considered by any of the researcher along with all other above-mentioned factors at the same time in recent years. This is the problem which exist within the literature base and is to be duly resolved through the conduction of the present study under consideration.

2 Literature Review

2.1 Artificial Intelligence and Its Types

Artificial Intelligence (AI) can be referred to as a wide-ranging branch of computer science which is mainly concerned with the development of smart machines with the capabilities to perform task requiring human intelligence. In the most modest terms, AI is a set of systems or machines that imitate human intelligence for performing tasks and has the capability of enhancing itself on the basis of the information it collects (Makridakis 2017). There are a number of forms in which AI appears; and some of them are the use of AI in chatbots for understanding the problems of customers faster while providing more efficient answers, the use of AI in intelligent assistants for analysing critical information from large sets of datasets for enhancing scheduling, and many others.

There are two types of classifications of AI: First classification and Second classification. The first classification is done as per AI and AI-enabled machines, their resemblance to the human mind, and their capability of ‘thinking’ and even ‘feeling’ as humans. The types of AI under this classification include relative machines, limited-memory machines, theory of mind, and self-aware AI.

Relative Machines – Basic operations are performed by this AI where companies use these machines for mechanically replying to a combination of inputs. Memory-based operations are not performed by these machines and their capability is limited as they were the first AI-based system machines (Yarlagadda 2018).

Limited-Memory Machines – These are the machines that, apart from obtaining the competences to have virtuously reactive machines, have the capability to learn from historical data for informing subsequent decisions. This category of AI includes almost all present applications of AI.

Theory of Mind – It is the next level of AI mechanism which is presently considered as a concept which is the work-in-progress stage. This AI system is capable of better understanding entities with whom they interact by discriminating their requirements, opinions, feelings, and thought process (Rabinowitz et al. 2018).

Self-Aware AI – It is the final phase of AI development and is the ultimate ambition of AI research. It involves AI systems that have experienced progression to the point where they can be equalled to the human brain in that they have developed self-awareness.

The second classification is undertaken based on technology; and the types of AI under this classification are Artificial Narrow Intelligence (ANI), Artificial General Intelligence (AGN) and Artificial Super Intelligence (ASI).

ANI – All the current AI are represented by ANI where these machines' functionalities are grounded on precisely what they are programmed to do. Consequently, they obtain a narrow variety of capabilities.

AGN – It refers to an AI agent's capability of learning, understanding, observing and functioning entirely like a human being. It is concerned with solving specific programs.

ASI – The development of ASI is considered as the ultimate of AI research. If accomplished, the development of ASI will change the way of life as its main objective is the development of a machine with higher cognitive function than a human being (Hassani et al. 2020).

2.2 Importance of AI and Its Benefits

When discussing about the importance of AI, it is noteworthy to mention that AI is a key source of business value when done right. AI has large significance to the businesses because it helps them in cutting costs and bringing new levels of steadiness, rapidity and scalability to the business processes; and it results in saving a huge amount of time of the businesses (Al-Sayyed et al. 2021). Agility and competitive advantage are other two aspects which should be considered. AI is not just about streamlining laborious tasks efficiently because deep learning and machine learning enable the AI applications to learn from data and results in real time while undertaking the analysis of new information from numerous sources precisely which is irreplaceable to the higher management. In this way, businesses become able in adapting at speed with a steady stream of insight for driving innovation and obtaining competitive advantage (Siau and Wang 2018).

There are numerous advantages of AI. It increases end-to-end efficiency by eliminating resistance and enhancing resource consumption and analytics across organizations, leading to a reduction of significant amount of cost. It systematises multifaceted processes while curtailing interruption by forecasting maintenance requirements. A crucial advantage of AI is that it results in enhanced accuracy and decision-making by supplementing human intelligence with rich analytics and pattern prediction competences for enhancing the quality, effectiveness and originality of employee decisions (Wamba-Taguimdje et al. 2020). Since AI possesses the capability of thinking different from human beings, gaps and opportunities in the market can be exposed by it rapidly, which assists the companies in presenting new products, services, business models and channels with a level of speed and superiority that was not imaginable before. AI has a key role to play in empowering the employees as the ordinary activities can be tackled by AI which enables the employees in spending time on more rewarding high-value tasks (Wamba-Taguimdje et al. 2020).

2.3 AI in Accounting Sector

AI has created a significant impact in the accounting and finance sector. Actually, AI-enabled finance and accounting systems are the technique for the companies for staying as strong candidates in a more and more competitive market because they save time while providing deep insight. Industry 4.0 is being shaped by new technologies in every aspect with intelligent replies to varying prospects of the key stakeholders like suppliers, customers, vendors and partners. The use of AI enables a decrease of a huge amount of time taken beforehand by the employees to perform different and monotonous tasks manually. It reduces human error which leads to an improved quality of output (Mohammad et al. 2020).

AI has automated almost all accounting tasks, including tax, banking, payroll and audits, which is disrupting the accounting sector while generating a major change in how accounting is done. In the accounting sector, both efficiency and output quality are improved by AI which contributes to greater transparency and auditability. A wide range of opportunities is provided by AI while curtailing the outdated laborious responsibilities of the accounting team so that they can look at more venues for business growth (Maione and Leoni 2021). AI has a key role to play to forecast accurate finance statements, as the accounting and finance professionals become able in forecasting future trends on the basis of historical data.

There are many applications of AI in the accounting sector. AI enables the accounting and finance professionals to process and automate different documents for enhancing internal accounting processes like invoicing, procurement and purchasing, expense reports, purchase orders, accounts payables and receivables, and others. Accounting documentation is undertaken by AI in real-time through the use of computer vision and natural language processing for generating reports in real-time. Accountants receive insight from such reporting to make sure the activeness of the companies (Lee and Tajudeen 2020).

2.4 AI in Auditing Sector

AI is to further transform the auditing profession. For example, machine learning, a key part of AI, can be used by the auditors for automatically coding accounting entries. Fraud detection by the auditors can be enhanced by developing refined machine learning-based audit models. The use of AI in auditing helps the auditors in analysing unstructured data like social media posts, emails and conference call audio files. Contract revenue can be regarded as an example of how AI can be used in auditing. Machine learning tools permit the auditors in analysing a huge number of contracts, like leases, in a much shorter timeframe than is possible with an outmoded review that is undertaken manually (Kokina and Davenport 2017).

The use of AI in auditing is already being tested by the audit firms. The use of Argus by Deloitte is a key example. Argus is a machine learning tool having the capability of reading documents like lease contracts, derivative contracts and sales contracts. Algorithm is used for programming Argus which permit it in identifying important contract terms and trends (Ucoglu 2020). Another key example is the use of Halo by PricewaterhouseCoopers. Halo is used for the analysis of journal entries and possesses the capability of identifying possible problematic areas, like entries with questionable nature, entries made from unauthorised source, or posting a large number of journal entries just under the authorized limit (Kokina and Davenport 2017).

It implies that the use of AI is making is possible for the auditors to work in a smarter and improved manner. Consequently, it assists the auditors in optimising their time while enabling them in exercising their human judgments for analysing a wider and deeper set of documents and data. As a result, it becomes possible for the auditors to ask better questions and to cooperate more with the client's senior management teams. Thus, AI has a key role to play to ensure audits with superior quality.

2.5 Benefits of AI Within the Accounting and Auditing Industry

AI is changing the manner in which the accountants work. AI can assist the accounting professionals to become more efficient and productive. A huge reduction in the time that the accountants take to do tasks will permit them in being more attentive to provide advice to their clients. Including AI to the accounting operations also boost the service quality because of the reduction of errors. The adoption of AI makes the accounting firms more attractive as employers (Balakrishnan et al. 2019). The use of Robotic Process Automation (RPA) permits the accounting professionals in completing time-consuming and repetitive tasks easily. The accessibility of RPA ensures that the time that is used by the accountants on monotonous, time-consuming tasks is now available for advisory and more strategic tasks. Furthermore, the use of AI frequently provides the accountants with real-time status of different financial issues since it is able in processing documents through the use of computer vision and natural language processing that makes daily financial reporting possible and cheap (Rahim et al. 2018).

Alike accounting, there are many benefits of using AI in auditing. The use of AI in auditing helps in automating the manual tasks of auditors like documentation. The use of machine learning in auditing is useful in the analysis of the complete volume of both unstructured and structured data which the auditors obtain from financial records by analysing data. Most importantly, the auditors become able in identifying anomalies in the obtained clients' accounting records, like unusual activities or payments that would be possible to identify by conducting manual auditing (Martínez and Fernández 2019). The use of AI enables the auditors in predicting about future risks and events through the review and analysis of historical data on different types of transactions. The auditors can use AI for harnessing the advantages of machine learning to go beyond sampling in order to review all available information to identify the documents with high risk (Zemánková 2019).

2.6 AI and Decision-Making Process

As AI has been introduced in a more extensive manner in accounting and auditing, the role of the accountants will pivot from gathering and analysing data to the interpretation of results and supporting the decision-making process. Because AI has the capability of analysing a large amount of accounting and financial data, accounting professionals have the option of leveraging AI's insights to assist with better business decision-making on the basis of the data. Transactions and streams of income across various parts of the organizations can be compared by the accountants, which helps them in obtaining insights into the businesses in manner that would not be possible previously through the use of manual means. AI has put the accountants in the frontline of digital conversion within their own companies, because proficiency in control designs and knowledge in data biases can add value to other departments within their enterprise as they look to obtain benefits from AI's advantages (Odoh et al. 2018).

In addition, the enormous data-crunching abilities of AI will permit the accounting professionals in deriving insights from transactions, demographic and other external data in real-time basis. It will help their projections which implies that the accountants will be able in helping the clients in detecting financial challenges so that appropriate actions can be taken before these challenges become crises. Recent successes in AI is taking a very dissimilar approach to assist in the decision-making process. Instead of trying to execute a top-down model of rules, a bottom-up approach is being taken by AI whereas it learns the rules on the basis of observation of what occurred previously (Elliot et al. 2020). Patent recognition is used by it which is known as machine learning. By the combination of machine learning and other key developments of AI like reasoning and knowledge representation, accountants can use computers to assist in the decision-making process.

2.7 AI in Accounting and Auditing Sector and Loss of Jobs

A common dilemma is there that humans will be replaced by robots due to financial automation, which is not true at all.

The fact cannot be denied that the revaluation of AI in the accounting and auditing professions will not be slowed down in any way. There is not any doubt that AI technology possesses the potential of performing all routine tasks of the accountants. However, the technological changes due to AI do not have the capability of replacing humans in any circumstance because there will always be the requirement of human intelligence for performing and efficiently running technology. Firms will always need the accountants to analyse and interpret the accounting and financial data acquired by AI-powered machines. Moreover, a major role will be played by the accountants to do higher-order tasks, like financial consultation better than the machines. Instead of spending needless time in doing repetitive, time-consuming tasks, the auditor will become able in investing their time and focus on more important tasks like consulting services and data analysis. Hence, instead of replacing the accountants, AI will assist the accountants in doing their regular tasks in a more technologically advanced way (Leitner-Hanetseder et al. 2021).

In many ways, AI will provide assistance to the accountants in enhancing their services. The use of AI technology will enhance the accuracy of data entry while lessening the liability risk for the accounting professionals. Moreover, emerging technologies like AI have more competence in detecting fraud, which adds an additional layer of protection for accounting professionals and their clients (Kruskopf et al. 2020). On the overall basis, there is no need for the accountants to worry about AI will replace their jobs any time in near future, as they will always be required by the companies to analyse and interpret data.

2.8 Research Gap

The key gap of the research is the aspect that accounting and auditing have not been considered as different professions within a same study over the years by any researcher to test the impact of AI on these professions. It implies that there is a lack of recent researches on assessing the impact of AI on the accounting and auditing profession separately. Another key gap is the absence of any recent study on the impact of AI on the accounting and auditing jobs. This research takes an attempt to fill these research gaps.

3 Conclusion

3.1 Summary, Conclusion and Implications

The aim of the study was to explore the relationship and impact of the new innovation of "Artificial Intelligence" on the "accounting" as well as on the "auditing" professions including the impact on future loss of jobs within the accounting and auditing industry. The study had four primary research objectives which are as follows:

- To investigate into the impact of AI within the accounting industry.
- To investigate into the impact of AI within the auditing industry.
- To ascertain impact of AI on the decision-making process within the accounting and auditing profession.
- To derive the impact of AI on future accounting and auditing jobs.

It is to be specifically noted that the study was conducted using collection of qualitative secondary information from the peer-reviewed journals and articles, the synopsis of which is well-described within the literature review section. According to Mohammad et al. (2020), the AI have made a significant positive impact on the accounting sector where the time lag in delivery of different accounting services have decreased significantly along with reduction of human errors in accounting as well as mistakes thereby enhancing the overall accounting quality. AI have also made the role of auditors much easier where the machine learning has enabled the auditors to conduct automatic coding of accounting entries. Accounting irregularity as well as anomaly can also be detected by the AI enables systems very easily thereby immensely assisting the auditors and the auditing profession. This is how the decision-making process also enhanced within these two professions as the overall quality of accounting as well as audit immensely enhanced with the implementation of AI. Finally, it can also be said that the AI empowered machines can carry out routine tasks but cannot replace the human intelligence requirement as humans are required to interpret the accounting and auditing data and also to carry out certain inferential jobs which is not routine at all and mainly requires analytical skills. So, the AI implementation can be said to be beneficial and positively impacting the accounting and auditing professions. Finally, it can also be interpreted that the implementation of AI will not affect the jobs within the accounting as well as auditing industry that is will not cause loss of accosting and auditing jobs.

3.2 Limitations

Limitations of a study portrays the shortcomings of the specific research study that can be fulfilled by the future researches. Firstly, it should be noted that the present research study has only been conducted by analysing the past researches on the specific area of research that is on the possible impact of "Artificial Intelligence" on the accounting as well as on the "auditing" profession. So, the study did not take into consideration any primary or secondary data collection of quantitative type where all data was of secondary qualitative type only that is ascertained from the pre-existing literature base concerning the area of research. The data for this study have been collected from the peer-reviewed journals and articles only with no intervention of primary data collection and analysis that is collection of data from the primary respondents. If the perspectives and inputs of the professional accountants and auditors could have been captured using primary research may be through a survey or an interview method of data collection where the collected data could have been analysed using a statistical analysis method for quantitative primary data and thematic analysis method for the qualitative primary data, then a holistic view could have been provided on the topic and the hypothesis could have been tested very easily in the light of the primary quantitative as well as qualitative data. This is the primary limitation for the present research study. The other eminent limitation for the study is that the study is completed after a thorough review of a few peer-reviewed journals and articles that too published within a time period ranging between 2016 and 2022 where the prior published material was not considered. It is also to be noted that a huge number of peer-reviewed journal and articles was not been able to be reviewed owing to the time-constraint within which the dissertation was needed to be completed. So, these are the two primary limitations which is applicable for the present research study.

3.3 Suggestions and Recommendations for Future Research

The suggestions for the future research on similar topics would firstly include the conduction of a primary research on the present research study in order to provide a holistic view on the area of research. This means that any new researcher willing to conduct their research on similar topics must conduct the primary research along with the secondary research using a mixed method of data collection where both quantitative data can be collected using the survey method with the assistance of a close-ended semi structured questionnaire and the qualitative data can be collected using the interview method. Lastly, the quantitative data can be analysed in a statical way using an eminent statistical tool such as the "IBM SPSS" or any other well-established statistical platform and the qualitative data can be analysed using the thematic analysis technique which could provide a holistic view on the research area.

References

Al-Sayyed, S., Al-Aroud, S., Zayed, L.: The effect of artificial intelligence technologies on audit evidence. Accounting **7**(2), 281–288 (2021)

Balakrishnan, K.P., Prakash, L., Ramesh, L.: Impact of AI technology in accounting and finance. Int. J. Anal. Exp. Modal Anal. **XII**, 629–636 (2019)

Dick, S.: Artificial intelligence (2019)

Elliot, V.H., Paananen, M., Staron, M.: Artificial intelligence for decision-makers. J. Emerg. Technol. Account. **17**(1), 51–55 (2020)

Hassani, H., Silva, E.S., Unger, S., TajMazinani, M., Mac Feely, S.: Artificial intelligence (AI) or intelligence augmentation (IA): what is the future? Ai **1**(2), 8 (2020)

Kokina, J., Davenport, T.H.: The emergence of artificial intelligence: How automation is changing auditing. J. Emerg. Technol. Account. **14**(1), 115–122 (2017)

Kruskopf, S., Lobbas, C., Meinander, H., Söderling, K., Martikainen, M., Lehner, O. Digital accounting and the human factor: theory and practice. ACRN J. Financ. Risk Perspect. (2020)

Lee, C.S., Tajudeen, F.P.: Impact of artificial intelligence on accounting: evidence from Malaysian organizations. Asian J. Bus. Account. **13**(1), 15–29 (2020)

Leitner-Hanetseder, S., Lehner, O.M., Eisl, C., Forstenlechner, C.: A profession in transition: Actors, tasks and roles in AI-based accounting. J. Appl. Account. Res. (2021)

Maione, G., Leoni, G.: Artificial intelligence and the public sector: the case of accounting. In: Visvizi, A., Bodziany, M. (eds.) Artificial Intelligence and Its Contexts. ASTSA, pp. 131–143. Springer, Cham (2021). https://doi.org/10.1007/978-3-030-88972-2_9

Makridakis, S.: The forthcoming artificial intelligence (AI) revolution: Its impact on society and firms. Futures **90**, 46–60 (2017)

Martínez, M.D.C.F., Fernández, A.: AI in recruiting. multi-agent systems architecture for ethical and legal auditing. In: IJCAI, pp. 6428–6429 (2019)

Mohammad, S.J., Hamad, A.K., Borgi, H., Thu, P.A., Sial, M.S., Alhadidi, A.A.: How artificial intelligence changes the future of accounting industry. Int. J. Econ. Bus. Adm. **8**(3), 478–488 (2020)

Munoko, I., Brown-Liburd, H.L., Vasarhelyi, M.: The ethical implications of using artificial intelligence in auditing. J. Bus. Ethics **167**(2), 209–234 (2020)

Odoh, L.C., Echefu, S.C., Ugwuanyi, U.B., Chukwuani, N.V.: Effect of artificial intelligence on the performance of accounting operations among accounting firms in South East Nigeria. Asian J. Econ. Bus. Account. **7**(2), 1–11 (2018)

Rabinowitz, N., Perbet, F., Song, F., Zhang, C., Eslami, S.A., Botvinick, M.: Machine theory of mind. In: International Conference on Machine Learning, pp. 4218–4227. PMLR (2018)

Rahim, S.M., Mohamad, Z.Z., Bakar, J.A., Mohsin, F.H., Isa, N.M.: Artificial intelligence, smart contract and islamic finance. Asian Soc. Sci. **14**(2), 145 (2018)

Siau, K., Wang, W.: Building trust in artificial intelligence, machine learning, and robotics. Cutt. Bus. Technol. J. **31**(2), 47–53 (2018)

Ucoglu, D.: Current machine learning applications in accounting and auditing. Press Acad. Proc. **12**(1), 1–7 (2020)

Wamba-Taguimdje, S.L., Wamba, S.F., Kamdjoug, J.R.K., Wanko, C.E.T.: Influence of artificial intelligence (AI) on firm performance: the business value of AI-based transformation projects. Bus. Process Manag. J. **26**(7), 1893–1924 (2020)

Yarlagadda, R.T.: The RPA and AI automation. Int. J. Creat. Res. Thoughts (IJCRT) (2018). ISSN 2320-2882

Zemánková, A.: Artificial intelligence and blockchain in audit and accounting: literature review. WSEAS Trans. Bus. Econ. **16**(1), 568–581 (2019)

Artificial Intelligence for Decision Making in the Era of Big Data

Badreya Alqadhi[1], Allam Hamdan[2(✉)], and Hala Nasseif[3]

[1] College of Business and Finance, Manama, Bahrain
[2] Ahlia University, Manama, Bahrain
allamh3@hotmail.com
[3] College of Business Administration, University of Business and Technology, Jeddah 21448, Kingdom of Saudi Arabia

Abstract. This paper aims to study the impact of Artificial intelligence (AI) on decision making in the era of Big Data, and what are the main limits or challenges to this topic, and to provide a better understanding on AI in decision making. AI history consist of attracting different level of interest through spring and winter seasons ever since the 1950's. Nowadays, number of research papers and articles discussing ways into which AI tools are incorporated in decision making is increasingly rising, all sectors and fields are involved including healthcare, gaming, weather forecasting, defense and security, construction planning and business to name a few. Evidence highlighted by research papers show how decision-making tools through AI and Big data contributed to the prosperity of humans and the planet and attempts to solve numerous issues accompanied with each field in decision making, AI infused decision making versus human decision makers has shown several impacts on time saving, cost cuttings, creating enhanced products, and contributing to sustainability.

1 Introduction

The role of Artificial intelligence (AI) has grown significantly in the past few years, leaving no sector behind. (AI) can be explained as the ability of a digital computer system or computer-controlled robot to perform tasks commonly associated with human brain intelligence (Copeland 2020). The term is also applicable to the development of systems with built-in intellectual abilities or characteristic of humans, for example the ability to reason, use logic, search for meaning, generalize, or even be taught from past experiences (Copeland 2020). The terms AI and AI systems were first introduced in the 1950s. Since then, AI has experienced its ups and downs (Duan et al. 2019).

Increasing data emerges every single day and in massive quantities. One example is the data collected through social media platforms, millions of users connect on daily basis where they share and contribute, this is where Big Data comes into light. Big data science deals with huge amount of data in various formats and is handled through mediums or tools to be processed and stored (Hussain 2019). Big data in combination with (AI) technologies can process information according to prespecified systems using machine learning and predictive analytics. These technologies purposes are to significantly reduce

B. Alareeni et al. (Eds.): ICBT 2022, LNNS 620, pp. 604–612, 2023.
https://doi.org/10.1007/978-3-031-26953-0_55

the time needed to process and analyze large quantities of data to provide unbiased conclusions (Stieger 2019).

In today's era, With the rapid advancement of Big Data technologies, e.g., improved computing storage capability and super-fast speed of data processing machines, AI is being revitalized with the availability and power of Big Data (Duan et al. 2019).

Nowadays, decisions must be made in an exceeding speed to keep up with the fast-moving world. All of this while trying to eliminate as much of human errors as possible. Past experiences have shown major success in replacing elementary skills such as operator or data entry with AI systems and robots (Roberts 2019), but can it expand to more complicated skills? Are Skills usually associated with human brains such as decision making? And If using the correct inputs, can AI play a major role in decision making in the era of big data? How much has been done in this regard?

This paper aims to study the impact of Artificial intelligence (AI) on Decision making in the era of big data, and what are the main limits or challenges to this topic. The research is to provide a better understanding on AI in decision making for decision makers in both public and private sector who are interested in implementing AI to enhance work process in their fields.

The next sections will discuss first the history of AI briefly, secondly it will discuss skeptic and optimistic views in literature of AI in decision making, then it will go through past articles testing AI tools in decision making in various fields, after that a highlight of the role of Big Data in decision making, and finally conclusions will summarize the paper and limits and opportunities.

2 A Brief History of AI

Humanity long since time have dreamt of the creation of machines that would carry on the actions humans would do, or even better what humans can not do at least not in such short period of time. However, in reality, only few decades back did Humans started to advance in AI machine buildings and technologies (Buchanan 2005). Science fiction writers and philosophers inspired engineers to invest in the first AI machines (Haenlein and Kaplan 2019). AI was considered an academic discipline as of the 1950's.

AI history was stretched through what is referred to spring and winter seasons consisting of high and low periods (Gomes 2020). A major turnaround in AI was in 1950, when a British mathematician Alan Turing created what most would consider the first mechanical computer called 'The Bombe', he created the machine as a member of staff in the British government for the purpose of decoding the German Enigma code used in the World War II. Turing later published an article "Computing machinery and intelligence" in which he explained how to build an AI machine and how to test it, it is still considered to be a standard test in identifying machine intelligence (Haenlein and Kaplan 2019).

Among inspired scientists was Marvin Minsky an American computer scientist who co-founded MIT AI laboratory with his partner John McCarthy, in 1956 they hosted a famous Dartmouth Conference called Dartmouth Summer Research Project on Artificial Intelligence (DSRPAI), this workshop marked the beginning of the AI spring, where many participants were later considered the founders of AI such as the IBM creator Nathaniel Rochester (Haenlein and Kaplan 2019). In those spring season many AI projects bloomed and had plentiful success. In the 50's for example, IBM scientist Arthur Samuel created the first game playing program, this game was able to learn and play checkers and it even played against world champions with great success (Delamater 2018). Another example, in the 60's when computer program called ELIZA was created by Joseph Weizenbaum, this machine studied natural language communication and was able converse with humans, it was also considered the first chatbot existed (Natale 2018).

Up until the early 1970's the golden age of AI was coming to an end, and was followed by the winter season of AI, where funding of AI related research was wearing, and most has lost interest in the field. Others stated that AI advancement cannot possibly go far, thus for shutting many governments supports towards AI projects in both the United Kingdom and in the U.S. (Haenlein and Kaplan 2019). Reasoning behind this slow down, was suspected to the disappointment due to unmet expectations of how far AI can achieve in that period of time (Gomes 2020). Although AI was not a trendy subject anymore, some deep work was taking place that would set as a foundation for later AI algorithms creation (Gomes 2020).

Chess is game where scientists thought that it is a human mind game, where you must predict future moves in order to win. Allen Turing in 1945, predicted that in the future a computer would be able to play and win at chess games against humans, his assumption proved to be true when in 1997 a computer program called 'Deep blue' built by IBM won against a world chess champion Garry Kasparov in a six-game match (Copeland 2020). The program was able to process 200 million possible moves per second and then predicting the best possible next move (Haenlein and Kaplan 2019). Another milestone was in for AI was in 2012 when a machine also created by IBM won on the TV game show "Jeopardy" against the biggest all-time winner on the show 'Brad Rutter' (Vardi 2012). Still, AI pessimists were not convinced and accounted these achievements to be an ingenious computer program rather than a computer that can think which is the main aim of AI (Vardi 2012).

AI in the next decades had more highs and lows but has shown significant progress and more public interest since the nineties, and it is a growing movement ever since. Predictions state that it will have a fundamental role in everyday life and transform the way we live (Haenlein and Kaplan 2019).

3 AI in Decision Making, Between Skepticism and Optimism

As with everything new, relativity new, expectations to see both an opposing side and going for side must exist. AI in decision making is no exception. Phillips-Wren and Ichalkaranje (2008) Discuss in their paper to build a well-developed AI tool or system, first one needs to understand how actual humans make decisions and then mimic that to an AI which in turn should represent optimal decisions in all time. This is to be tested to acquire if this theory is realized at the present time, and how far has it come. The authors search back in time and go through past papers and research on the topic, to understand how much humans understand decision making and how much is it transferred to AI based systems. Findings on AI systems and how do they relate to decision making was categorized into three groups. The first group is AI systems that has to adapt to changing environment such as adapting to change of time and speed of time, the authors believe that this to be the least challenging. The second group AI systems adapting to similar environments and this is done by finding past settings and then generalize or adapt to the case in hand, the authors believe this more challenging than the first group. The third group and the most challenging is adapting to new unexplored situations, for these settings authors believe there is along way to achieve in AI systems.

Another paper discussing limits and challenges by Lysaght, Lim, Xafis, and Ngiam (2019), it aims to understand both stands of AI in healthcare. On one hand, AI has high potential in improving and advancing in the healthcare field, on the other hand raise the question of how much AI is to be trusted and how ethical is it regarding transparency and accountability. Research was based on the framework of how balance can be created in producing AI systems that improve the fieldwork of healthcare systems while keeping it ethical and accountable. The authors highlighted several areas where AI systems might be raising issues and its summarized in professional integrity, accountability, group harm, justice, and transparency.

On the other hand, Duan, Edwards, and Dwivedi (2019) in their paper had different views. Their aim was to determine challenges associated with using AI in decision making and to provide research proposals for AI in information systems (IS), by going through all previous literature related to AI published in the International Journal of Information Management (IJIM). The authors believed AI to have a profound impact on decision making, plus they proposed 12 research areas for (IS) in terms of theoretical progression, AI interaction with humans, and AI implementations. Another research attempt to examine the extent to which different human characteristics is correlated with their perception of AI and decision making tools, and how it differs across media, health and judicial aspects (Araujo Vrees et al. 2020), a sample of 958 were driven from a scenario-based survey, which included questions about extent of knowledge about AI and automated decision making (ADM) in addition to questions regarding perspectives about these tools. While some findings did display a concern from individuals on risks associated with ADM. However, other views demonstrated that automated decisions were more accurate than decisions of experts in their fields.

Through all AI skepticism, the number of research papers and articles discussing ways into which AI tools are incorporated in decision making is increasingly rising. All sectors and fields involved too, as seen in the next chapter, including healthcare, gaming, weather forecasting, defense and security, construction planning and business to name a few.

4 Applications of AI in Decision Making

This section will review a number of papers discussing AI in decision making in different field. All papers discussed presented a positive impact of AI in the decision making regardless of the field. For example, in the healthcare field AI based decision tools were considered improving the healthcare systems (Bennett and Hauser 2013) and (Vaishya et al. 2020). In Bennett and Hauser research (2013), the aim was to study improvement in health field in general using AI tools and the need to create a framework to explore possible obstacles through, a framework for the overall health systems in regard to policies, spending, and other decision areas. In addition to a more ambitious goal which is to create clinical AI and as the authors described "AI that thinks like a doctor". Clinical data were tested in an AI based tool called Markov Decision Process and by using decision network tools to develop strategies by testing possible sequential decisions that can be taken. As the paper stated, AI approach won over the traditional approach and produced 50% more success in decisions taken and cutting the cost to a half, thus optimizing health decision-making through challenging environment.

A more recent paper discussing AI in health field decision making in the pandemic of Covid-19 (Vaishya et al. 2020), as the authors confirmed the urgency in finding means to better perform in the health industry and as fast as possible, authors argue that AI technologies is the solution, because this pandemic is a major global concern and is characterized by fast disease spread, thus creating a necessity for a tool to help in speedy decision making in healthcare systems. The main aim of this study is to review AI technologies in analyzing and predicting in the goal to prevent pandemics such as Covid-19. The methodology used relied on literature available, where the authors with the aid of PubMed and Google Scholar databases searched and collected information of possible applications to use AI in Covid-19 pandemic decision making and analyzed all outcomes. Findings pointed to seven main fields in which AI tech can play a major role in the healthcare system: (1) Early diagnosis of the disease, (2) Monitoring effected individuals, (3) Trace the disease to stop the spread, (4) Prediction of spread area,(5) Faster development of solutions, (6) Reduction of healthcare workload, (7) Prevention. Further findings include AI tech impact in fighting the disease by predicting future spread of the diseases using analysis of available data.

Several other research papers review different AI approaches in different fields, a review from the oldest to the newest will be discussed. In the gaming world, Frutos-Pascual and Zapirain (2015) demonstrate how the gaming industry is thriving and has been expanding greatly ever since the beginning of the 21st century. Ever since diverse types of gaming emerged, one type is serious games which is the focus of the paper, where the authors aim to explore the use of certain AI tools on the creation of serious games. The authors collected all available literatures in the past decades regarding serious games to determine the use of AI in it, coming up with 129 papers that discuss the topic. Findings include a proven existence the use of AI tools and technique in building these types of games, in addition to AI tools enhancing the game creation and make it more appealing to and has better user experience.

A 2017 paper reviews AI decision making in weather forecasting (McGovern et al. 2017). Although existing models and computing powers are getting better in the weather forecasting field, the authors still believed more advancement could be made via AI, the authors deliberate on the need for better prediction to assist vastly not only in severe weather predictions and taking precaution measures, but also for support in taking more sustainable decisions. AI techniques were used in determining various type of severe weathers, such as storms, severe wind, and severe hail. Results were compared then to traditional forecasting models usually used. Results shows AI approaches in weather forecasting is helping save time and energy otherwise people would have individually gone through tedious data points to forecast. Also, AI tools that has been built on existing forecast models proved to have greater abilities into providing accurate predictions and identifying differences between various models. In addition, AI tools helped in direct decision making, where automated forecasting in the event of a weather fluctuations would generate power to create sustainable energy through wind, solar systems and many more.

Similar optimization was found in a defense field related paper (Dear 2019). The author predicts that AI will transform decision making in defense and security fields in four ways. First, AI will enable projections based on analysis and therefore contribute in early intervention. Second, to let AI systems to be the ones responsible for decision-making entirely. Third, give the right advice based on technologies used that might be hard to comprehend. Finally, aid human decision makers in the process of decision making based on AI analytical power. Although the author implies that these predictions are defense and security related, by observation it can be said that these predictions are general and may apply to various fields.

Coming next to the construction planning sector, where a study was published of AI infused decision making tool on tunnels construction (Mahmoodzadeh et al. 2020). Reasoning included that tunnel constructions are expensive and time consuming, above that most construction schedules fail to presume the cost and the time needed, and delays are often expected. The authors tested how to embed AI in the tunnel construction plans to gain more accurate estimations. Two AI tools were assumed in the research to better understand earth geology and time management, the first, Gaussian Process Regression (GPR), the second, Support Vector Regression (SVR). Available data and past observation from previous road constructions were used in the two AI tools, more data points were added as a tunnel construction was going on at that time. Comparisons were made based on the comparisons between previous and added data sets, and comparisons between the two AI tools used and findings were that both tools GPR and SVR tools had good acceptable predictions. Also, the cost, time, and geology assumptions had decreased prediction errors. As per the paper, GPR tool was more accurate than SVR. It was also found that more updates and data points in the tools the more accurate results are.

In business, Farrokhi et al. (2020) aimed to explore the role of AI in anticipating crisis in business settings in companies and firms, in addition to explore AI role in crisis-communication theory (SCCT) which stands for Situational Crisis Communication Theory. Logic used in that succeeding in managing crisis by using an AI and data driven strategy will boost organizational performance. The research paper proposed a model consisting of statistical and sentimental analytical method and used an Enron public database of 517401 data points for 150 unique users mostly senior managers for the period 1999 to 2001. Research findings implicated day-to-day data communications via emails might lead to signaling an alarm to future firm crisis, thus this model can be great detection and aid in business decision making. However, the dataset used to test the hypothesis was limited to one company (Enron), and results therefore are subject to bias to that specific company environment which may not reflect all companies.

Overall, all papers displayed a positive view of the use of AI in decision making, all agreed on AI making better decisions or aiding decision-makers in making better decisions, Some highlighted that decision-making powered by AI supported minimizing costs like in the healthcare field (Bennett and Hauser 2013) and in weather forecasting (McGovern et al. 2017) and in the construction field (Mahmoodzadeh et al. 2020). Another observed benefit was the speed in which AI can enhance decision making (McGovern et al. 2017) and (Mahmoodzadeh et al. 2020). In the gaming industry, observed better product production when implementing AI infused decision making (Frutos-Pascual and Zapirain 2015). In addition to AI tools making sustainable decisions (McGovern et al. 2017).

5 The Role of Big Data

Machine Learning in AI uses algorithms and certain methodologies in finding patterns to analyze and predict opportunities, and it depends greatly on data in order to do so (Lovis 2019). When huge volume of data exists in a dynamic form and in a big set of variety, better AI applications are possible. This type of data is defined as 'Big Data', data that is high in volume, velocity, and variety (Power 2015). The role of Big Data in AI infused decision making maybe obvious in the general sense. However, few notable implications were observed, in many research papers discussed in the previous sections, AI tools were always accompanied with data science and big data (Duan et al. 2019) and (Vaishya et al. 2020). Assumptions stated that Big Data is inseparable when using an automated decision-making tool, because AI mimic human decision-making by referring to previous experiences a.k.a. Big Data (Duan et al. 2019). Some argues that Big data advancement and availability is the reason behind AI techniques and systems flourishing in our present time, and perhaps past days of winter seasons of AI was due to lack of available big data technologies(Duan et al. 2019).

Big data and data science seem to be a trendy topic lately. However, A point to remember is that accuracy and quality of the data is crucial to obtain an accurate decision made, if decision makers acquire the time, effort, talent, and cost to capture Big Data (Power 2015). Some argues that as much as the AI tools are advanced, a low-quality data will steer the decision-making process on the less accurate path. For example, data acquired for a certain purpose is then sometimes is re-acquired for another, which may lead to bias analysis in the latter study (Lovis 2019). However, another argument state that in Big Data the acquired data is assumed to be collected with no intention to what purpose this data will be used for, thus eliminating the bias effect (Lovis 2019).

Future trends seem to support Big Data and AI roles for global prosperity. The World Economic Forum in its annual report 'The future of jobs' (2020) as with previous releases of the report state that Big data science along with AI and other fields are the future. By surveying companies throughout the world, results reveal that most companies are reporting making priority in adopting Big data by 2025, it is the second higher technology adopt predicted after cloud computing. It is also interesting to highlight that AI is making the list in the fifth place of most likely technologies to be adopted by 2025. Looking at future jobs demanded as per the report, AI and Machine Learning Specialists comes in second place and Big Data Specialists comes in third, while Data Analysts and Scientists take the first place of the most demanded job in the future.

6 Conclusion

Impact of Artificial intelligence on decision making in the age of big data were observed through literature in various ways, either by the assumption that AI tools helps human decision makers make better choices or through the creation of auto-decision-making (ADM) systems. Although AI is considered a new science, much advancement is already on the table, AI is showing a promising future just as early scientist such as Allen Turing and founders of AI predicted, considering what is already achieved today (Copeland 2020). Further optimism is widely acknowledged in the use of AI tools in decision making, although some concerns are presented through some literature such as the concern if AI infused decision making is actually taking the proper decision (Phillips-Wren et al. 2008), or concerns regarding how ethical and transparent AI tools are (Lysaght et al. 2019). Most literature reviewed identified with a positive relation in AI tools in decision making.

Evidence highlighted by research papers show how decision-making tools through AI and Big data has been contributing to the prosperity of humans and the planet, and in different sectors and fields attempting to solve numerous issues accompanied with each field in decision making. Further speculations suggest AI making covering some milestones in saving time, cutting costs, creating enhanced products and contributing to sustainability as highlighted previously (Bennett and Hauser 2013) and (McGovern et al. 2017) and (Mahmoodzadeh et al. 2020) and (Frutos-Pascual and Zapirain 2015).

Sufficient number of literatures discuss different fields in which AI may be a handful decision tool, however, a question can rise if different AI tools give the same level of success in decision making, and to what extend does each AI tool is a game changer in decision making. In a construction related paper (Mahmoodzadeh et al. 2020), it mentioned using two types of AI tools in testing, of which one (GPR) has given better results than then the other (SVR). Further research on AI and big data role tools with comparisons between various tools maybe a subject of interest to be taken by researchers, for getting the best of AI technologies and its advancements. Another aspect of interest for further research is the accuracy and quality of big data used in AI decision making tools.

References

Araujo, T., Helberger, N., Kruikemeier, S., de Vreese, C.H.: In AI we trust? Perceptions about automated decision-making by artificial intelligence. AI Soc. **35**(3), 611–623 (2020). https://doi.org/10.1007/s00146-019-00931-w

Bennett, C.C., Hauser, K.: Artificial intelligence framework for simulating clinical decision-making: a Markov decision process approach. Artif. Intell. Med. **57**(1), 9–19 (2013). https://doi.org/10.1016/j.artmed.2012.12.003

Buchanan, B.G.: A (Very) brief history of artificial intelligence. AI Mag. **26**(4), 53–60 (2005). 10.1609

Copeland, B.: Artificial intelligence, 8 Nov 2020. Encyclopædia Britannica. https://www.britannica.com/technology/artificial-intelligence. Accessed 1 Feb 2021

Dear, K.: Artificial intelligence and decision-making. RUSI J. **164**(5–6), 18–25 (2019). https://doi.org/10.1080/03071847.2019.1693801

Delamater, N.: A Brief History of Artificial Intelligence and How It's Revolutionizing Customer, 2 Aug 2018. CHATbeacon.io. https://uploads-ssl.webflow.com/5a8f3fa77e521a0001a8fd5b/5b11846536a35217d252d95b_ChatBotWhitePaper2018.pdf

Duan, Y., Edwards, J.S., Dwivedi, Y.K.: Artificial intelligence for decision making in the era of big data – evolution, challenges and research agenda. Int. J. Inf. Manag. **48**, 63–71 (2019). 10.1016

Farrokhi, A., Shirazi, F., Hajli, N., Tajvidi, M.: Using artificial intelligence to detect crisis related to events: decision making in B2B by artificial intelligence. Ind. Mark. Manag. **91**, 257–273 (2020). https://doi.org/10.1016/j.indmarman.2020.09.015

Frutos-Pascual, M., Zapirain, B.G.: Review of the use of AI techniques in serious games: decision making and machine learning. IEEE Trans. Comput. Intell. AI Games **9**(2), 133–152 (2015). https://doi.org/10.1109/TCIAIG.2015.2512592

Gomes, P.: THE SEASONS OF ARTIFICIAL INTELLIGENCE. Retrieved from critical software (2020). https://blog.criticalsoftware.com/

Haenlein, M., Kaplan, A.: A brief history of artificial intelligence: on the past, present, and future of artificial intelligence. Calif. Manag. Rev. **61**(4), 5–14 (2019). https://doi.org/10.1177/0008125619864925

Hussain, S.J.: What is Big Data & why is Big Data important in today's era. Retrieved from medium, 10 Feb 2019. https://medium.com/@syedjunaid.h47/what-is-big-data-why-is-big-data-important-in-todays-era-8dbc9314fb0a

Lovis, C.: Unlocking the power of artificial intelligence and big data in medicine. J. Med. Internet Res. **21**(11). https://doi.org/10.2196/16607

Lysaght, T., Lim, H.Y., Xafis, V., Ngiam, K.Y.: AI-assisted decision-making in healthcare. Asian Bioeth. Rev. **11**(3), 299–314 (2019). https://doi.org/10.1007/s41649-019-00096-0

Mahmoodzadeh, A., Mohammadi, M., Daraei, A., Faraj, R.H., Omer, R.D., Sherwani, A.F.: Decision-making in tunneling using artificial intelligence tools. Tunn. Undergr. Sp. Technol. **103** (2020). https://doi.org/10.1016/j.tust.2020.103514

McGovern, A., et al.: Using Artificial intelligence to improve real-time decision-making for high-impact weather. Bull. Am. Meteor. Soc. **98**(10), 2073–2090 (2017). https://doi.org/10.1175/BAMS-D-16-0123.1

Natale, S.: If software is narrative: joseph weizenbaum, artificial intelligence and the biographies of ELIZA. **21**(3), 712–728 (2018). https://doi.org/10.1177/1461444818804980

Phillips-Wren, G., Ichalkaranje, N., Jain, L.C.: Intelligent decision making: an AI-based approach. Comput. Intell. **97** (2008). https://doi.org/10.1007/978-3-540-76829-6

Power, D.J.: 'Big data' decision making use cases. In: Delibašić, B., et al. (eds.) ICDSST 2015. LNBIP, vol. 216, pp. 1–9. Springer, Cham (2015). https://doi.org/10.1007/978-3-319-18533-0_1

Problem Solving and Decision Making (Solving Problems and Making Decisions). (n.d.). http://managementhelp.org/personalproductivity/problem-solving.htm. Accessed 1 Jan 2021

Roberts, C.: FUTURE OF DATA ENTRY WITH ARTIFICIAL INTELLIGENCE. Retrieved from ingenium web, 26 June 2019. https://www.ingeniumweb.com/blog/post/future-of-data-entry-with-artificial-intelligence/452

Stieger, G.: The Role of the Market Researcher in the Era of Big Data and Artificial Intelligence, 16 May 2019. glsresearch. https://www.glsresearch.com/the-role-of-the-market-researcher-in-the-era-of-big-data-and-artificial-intelligence/#:~:text=Artificial%20intelligence%20(AI)%20can%20process,data%20to%20provide%20unbiased%20results

Vaishya, R., Javaid, M., Khan, I.H., Haleem, A.: Artificial Intelligence (AI) applications for COVID-19 pandemic. Diabetes Metab. Synd. Clin. Res. Rev. **14**(4), 337–339 (2020). https://doi.org/10.1016/j.dsx.2020.04.012

Vardi, M.Y.: Artificial intelligence: past and future. Assoc. Comput. Mach. **55**(1), 5 (2012). https://doi.org/10.1145/2063176.2063177

Artificial Intelligence (AI) in the Education of Accounting and Auditing Profession

Sara Mohammed Ali[1], Zainab Jawad Hasan[1], Allam Hamdan[2](✉), and Mohammed Al-Mekhlafi[3]

[1] College of Business and Finance, Manama, Bahrain
[2] Ahlia University, Manama, Bahrain
allamh3@hotmail.com
[3] Accounting Department, College of Business Administration, University of Business and Technology, Jeddah 21448, Saudi Arabia
m.mekhlafi@ubt.edu.sa

Abstract. The study aimed to explore the impact of artificial intelligence on the performance of counting and checking operations. This work provides a review of the literature on artificial intelligence and its use in the accounting and auditing professions. A narrative approach was used to analyze related articles and micro-research to provide a comprehensive overview of the topic. Particularly with regard to the accounting and auditing professions, artificial intelligence has lately undergone earlier breakthroughs that have caused a shift in their attention from paper to computer entries. The goal of artificial intelligence is to demonstrate how computer technology can execute activities as effectively and efficiently as humans, if not better. The future of the accounting and auditing professions depends critically on artificial intelligence technology since it gives us the means to perform our duties more effectively and efficiently. AI has significantly improved operations, reporting, and decision-making processes in accounting and auditing, among other fields.

Keywords: Artificial Intelligence (AI) · Accounting education

1 Introduction

The development in artificial intelligence has changed the way in which the world operates, which also includes the world of business. Technology is changing so much of our lives, and businesses are now using technology and Artificial Intelligence to discover and make use of new opportunities for having better business operations and increasing profitability and return while reducing costs of business operations and aiming to be more sustainable and competitive in the market.

Artificial Intelligence (AI) is quickly changing how financial institutions operate. Artificial Intelligence is now used in relation to the accounting and auditing profession, changing how financial institutions operate, switching their operations from paperwork to computer data and software entries. Artificial Intelligence can be referred to as the

B. Alareeni et al. (Eds.): ICBT 2022, LNNS 620, pp. 613–621, 2023.
https://doi.org/10.1007/978-3-031-26953-0_56

application of computational tools which once required human intelligence [1]. The accounting field has incorporated Artificial Intelligence for over 25 years, mainly in the areas of financial and auditing reports [2]. Nowadays, Artificial Intelligence is applied in all accounting and auditing operations, which has led to the diminishing of the use of human accountants in organizations, possibly eliminating their relevance in the near future [3]. Accountants are on the verge of losing their jobs with their roles being taken over by computers and machines. With the progression in technology, many jobs and professions are eliminated as others are created. The use of Artificial Intelligence now is making the accounting profession easier, more precise, and much more efficient.

Accounting is a main field in business where information technology (IT) is applied in all aspects of basic accounting systems, financial software, and analytical accounting aspects. The speed in which information technology is advancing has allowed the integration of technology in all business aspects and operations. Artificial Intelligence is now a core part of the technology industry and a main function in the technological era we all live in. It is used to conduct more sophisticated human projects and responsibilities, competing them with less risk and more efficiency. Lombardo (2015) found that the power of computer technology lies within its energy trust and not only in its complexity and intelligence [4]. Artificial Intelligence and computers are bringing in other technologies and changing the way in which businesses function and operate. All businesses, big and small, are now being supported by technology, with both the public and private sectors now using Artificial Intelligence for regulatory compliance, quality, data entries, operations, and the detection of fraud [1]. Accounting and auditing systems have moves from paper entries and ledgers to the use of computer-based data formats and system entries powered with Artificial Intelligence to reach optimal results.

Technological changes and advancements are continuously increasing [5]. The integration of more advanced Artificial Intelligence has resulted in more enhanced systems. New technologies are now leading to more changes in the business world, affecting how they operate and sustain themselves in their respective fields. Artificial Intelligence has increased flexibility and efficiency and improved data-based systems, resulting in more effective and efficient reporting, and increasing the speed of reporting and operations. The changes in technology are changing how accounting and auditing operations are conducted. The Association of Chartered Certifies Accountant (ACCA 2013) found that Artificial Intelligence is de-skilling the accounting human profession. The continuous use and reliance on Artificial Intelligence has highlighted the fact that technological changes and advancements may lead to new skills being focused on, while also rethinking the concept of "work" in the accounting world [6]. Accountants may be at risk of being replaced by artificially intelligent software and expert systems, making Artificial Intelligence one of the top five threats facing the accounting and auditing professions [6]. Murungi and Kayimba (2015) found that the failure to incorporate and make effective use of systems technologies in businesses may reflect on inaccurate financial information which is required as a means of gaining competitive advantage in the business world [7].

Through the application and adopting of the right Artificial Intelligence technologies, the entry and transmission of information and data can be significantly improved through the increase of speed, reduction of error and risk, reducing costs, and overcoming problems. The adoption of Artificial Intelligence technologies can provide cheaper and more accessible data for more effective and applicable idea generating and decision making, and to spend less time on data collection and system entries and more time of more important tasks and aspects such as the need for better critical thinking, problem solving, setting effective strategies to be applying for more actual and functioning operations, more enhanced relationship building methods, and better methods of leadership within the organization.

This paper aims at presenting how Artificial Intelligence technology has affected the accounting and auditing professions, assessing whether the professions are to remain the same or change, and to show what changes have occurred in those professions. The study contributes to existing literature as it presents relevant literature related to the use and application of Artificial Intelligence in the accounting and auditing professions while showing how Artificial Intelligence has major potential in those professions.

2 Literature Review and Theoretical Framework

Artificial Intelligence has played a crucial role in improving the roles and proficiencies of auditing and accounting professionals, with the potential of furthering the possibilities of the accounting processes [8]. Artificial Intelligence covers a broad area, though not all functions and uses of Artificial Intelligence are used within the scope of accounting and auditing. Though the technologies incorporated within Artificial Intelligence don't all fall under the scope of business, its influence and benefits have made it a core part of doing business through different business functions such as marketing and management, production and distribution, sales, research and development, and accounting practices, amongst others.

2.1 Artificial Intelligence in Accounting and Auditing

Accounting systems describe the processes of keeping track of different business operations which occur in organizations in a specific period in time for them to maintain accurate and timely records of transactions in the organization. Tarmidi et al. (2018) supported the fact that accounting systems and the use and adoption of Artificial Intelligence technologies would lead to great results in accounting and auditing processes as it would make use of the Artificial Intelligence systems to modify processes in the fields and have more accurate and efficient financial records [9]. Accounting systems are seen to be used as a means of replacing the human labor force, therefore changing the dependence of businesses on humans to computers and machines through the implementation of those computerized systems [10]. This would lead to increasing the effectiveness and efficiency of accounting and auditing functions in the workplace, which would result in more automation of internal audit control and financial statements and reporting.

Moreover, the auditing process is a method of finding misstatements in the financial statements and reports of the company. Artificial Intelligence is used to integrate technologies in the accounting and auditing process in order to enhance the competence and usefulness of business operations. Artificial Intelligence has a crucial role in predicting possibility of financial distress which may be seen through the audit process. Hansen et al. (1992) stated that computer models have great effect on the decisions made in the processes of accounting and auditing, which result in reducing the financial issues and distress which may be found in the financial reports [11].

While performing an audit, auditors seek to provide information that is correctly and fairly presented. Ivy et al. (2020) showed the importance of having practical governance on the methods used to implement Artificial Intelligence in the auditing process to be able to have ethical decision making through the achievement of correct and reliable audits in the financial reports of an organization [12]. Rezaee et al. (2002) noted that the use of electronic auditing methods allows auditors to present a higher level of confidence on the data they are processing and information they are presenting [13]. Paperless accounting systems have increased the standards of financial reporting, reduced the problems associated with accounting, and increase the value of the financial statements which are presented in a more accurate and efficient and timely manner.

A study by Longinus (2018) focused on evaluating the application of Artificial Intelligence for the purpose of record keeping in Banks in Nigeria [14]. The results of the study showed that there is a significant positive relationship between Artificial Intelligence and accurate financial reporting. Mohammad (2012) and Jessie (2019) conducted studies which showed that the automation of accounting and auditing practices don't have any influence on customer satisfaction, though they do have effect on the transparency and accuracy of reporting financial statements [15]. Also, a study by Jooman (2019) aimed at evaluating the affect of adopting Artificial Intelligence technologies on internal auditing practices in South Africa [16]. The findings of the research portrayed that the current affect of Artificial Intelligence on internal auditing is still in its beginning stages and that internal auditors don't yet really understand the vast possibilities that can be achieved through the correct and effective application of different aspects of Artificial intelligence technologies in the workplace. With the majority of accounting and auditing functions still being conducted by humans, there is less use of Artificial Intelligence as it is not being properly utilized in the accounting and auditing setting. Mhlanga (2020) aimed at assessing the influence that Artificial Intelligence has on digital financial inclusion in areas which include problems associated with the detection of risk, management, the asymmetry of information, fraud detection, and cyber security measures [17]. The researcher found that Artificial Intelligence has a significant effect on those aspects, and should be further utilized and incorporated within more aspects of the accounting and auditing functions in businesses, such as record keeping and financial reporting and decision making, in order to better utilize the resources they have while having better and quicker operations conducted in the accounting sectors. Dongre et al. (2020) noted that the use of Artificial Intelligence will enhance the value of accounting as it increases the capability and decision-making ability of the organization [18].

2.2 Benefits of AI Incorporation in Accounting and Auditing

Davenport and Ronanki (2018) conducted a study which suggested that businesses should put more emphasis on the use and adoption of Artificial Intelligence technologies for their extensive capabilities in business processes and not only for their technological uses [19]. That is, Artificial Intelligence can assist businesses in meeting their main objectives by having more automated business processes, more accurate data analysis and financial reporting, and by building better and stronger relationships with both their customer and their employees. Chukwuani and Egiyi (2020) conducted a study to evaluate the effect of Artificial Intelligence in the accounting industry by portraying the level of technological advancements existing within the accounting industry through the automation of different accounting and auditing processes and procedures [20]. The researchers highlighted the place of accountants in an automated world and how those accountants can make use of modern technologies to enhance the businesses. The application of Artificial Intelligence on accounting functions was seen to have a positive effect on businesses. The researchers also suggested that Artificial Intelligence would affect accounting by reducing the cases of fraud and enhancing the quality and value of accounting information presented by the company.

Different studies referenced by Omoteso (2012) focused on the benefits of using Artificial Intelligence and incorporating it within the functions of accounting and auditing [21]. The benefits of Artificial Intelligence were many, including more effectiveness and efficiency, more consistency and structure in auditing tasks, better thinking and decision making, enhanced communication throughout the business, better training programs for staff, and shorter decision and implementation times for various business functions. Those who accept and incorporate the implementation of Artificial Intelligence are more ready to take on risks to gain a better place in the market by holding a competitive advantage in their respective industry [22].

Artificial Intelligence is crucial in determining the future accounting success of businesses. Small businesses which don't comply with adopting such technological advancements may face the major risk of being left behind in the market they operate within. As technology has been incorporated within the accounting and auditing functions of business operations, it is now vital for businesses of all sizes and structures to keep up with the changes and trends of technology in order to remain competitive in the business world [23]. Accounting firms are now effectively using resources to target the development of Artificial Intelligence technologies and solutions to meet the requirements of their clientele. Deloitte Touch Tohmatsu Limited has made an agreement to make more use of innovation and computer learning in the workplace. Deloitte developed 'Argus', which is a tool used solely for the purpose of auditing [24]. The organization also uses a 'Guided Risk Assessment Personal Assistant (GRAPA)' to work over ten thousand cases and assess over 50 risks to help auditors compare the risks strategy adopted. Also, Ernst and Young uses machine reading (e.g. QR and barcode labels) as well as drones for inventory observation and analysis [25]. When new regulations are introduced, Ernst and Young uses NLP to gather information and a human professional to validate results instead of re-examining all contracts the firms deals with. The company also uses its own developed tool that is the EY Helix GL Anomaly Detector (GLAD) in order to detect fraud in their journal entries [24].

A study by Cheyenne and Matthew (2018) was conducted to evaluate the perception of the implementation of Artificial Intelligence and the risks associated with it [26]. The research found that the implementation of Artificial Intelligence would result in the reduction of time spent on repetitive tasks and paperwork and improve the efficiency of the work performed in the field. Though the study did find that there is not enough education related to the most effective use of Artificial Intelligence as employees need to adapt to the new reality of their professions in the accounting and auditing fields.

Artificial Intelligence can and should be used to enhance business processes and activities to have better investment decisions and value and quality of business. The implementation of Artificial Intelligence in accounting has a means of being more appealing and useful to not eliminate the accounting and auditing work force but to better their work and enhance their capabilities while taking over time consuming repetitive tasks while ensuring accuracy, reliability, and less risks associated with the work. Artificial Intelligence can be used to enhance areas of business operations while digitalizing work processes, all while decreasing costs associated with business performance and increasing the effectiveness and efficiency of the organization.

2.3 Risks of AI Incorporation in Accounting and Auditing

Omoteso (2012) also noted that there are different disadvantages to the adopting of Artificial Intelligence technologies in accounting and auditing, and they should be considered along with the many benefits that such technology provides for businesses, specifically for the accounting and auditing aspects and business functions [27].

The downsides of the adoption of Artificial Intelligence technologies in accounting and auditing include the need for more time to assess different options related to decision making, as the process of decision making and providing options is short, due to the many alternatives presented this can take time to make the best decision possible. Also, there is a large cost to be considered for building, maintaining, and continuously updating the systems incorporated within the business. There is also a risk of such tools and updates being used by competitors, and the possibility of them being used against the auditor for legal purposes for having over-relied on financial evidence using aids that are technological. Bizarro and Dorian (2017) stated that as much as Artificial Intelligence technologies are efficient and reliable for practices in accounting and auditing, it still cannot replace some human factors, which include the ability to be creative, the ability to reason and judge, the ability to express motions and show disbelief in a subject matter [28]. Doshi et al. (2020) noted that although Artificial Intelligence technologies can bring about great benefits and major aids for businesses in relation to accounting and auditing aspects, they can still bring about major threats [29]. That is, there is great potential to complement the accounting and auditing professions, or even go further and replace them altogether. Doshi et al.'s (2020) notions on the use of Artificial Intelligence in those business functions differed from the ideas of Bizarro and Dorian (2017).

Luo et al. (2018) saw that some issues which are related to the adoption of Artificial Intelligence technologies in the field of accounting and auditing include the lack of experience, lower returns when it comes to higher investments, and the lack of needed skills and capabilities of professionals, as technology is taking over the field even more [23]. A main factor derived from the use of Artificial Intelligence technologies in accounting and auditing is that there are constant changes in the laws and regulations in such practices, requiring continuous maintenance and updates of the systems of Artificial Intelligence adopted in the organization in order for them to be updated and in compliance with the changes of regulations [30]. An example here is the changes in tax policies and requirements, which resulted in the need to make major changes and updates in accounting and auditing systems in order to ensure their compliancy with the updated laws and regulations of governments.

Zemankova (2019) indicated that the application of Artificial Intelligence in accounting and auditing may lead to inequality of income, less need for labor, lower financial safety, etc. [31]. Also, the application of Artificial Intelligence carries risks of exploiting algorithm and the possibility of deception, having error or bias, or incorporating some form of human logic or emotional error. Makridakis (2017) aimed at assessing the role of humans in the future as Artificial Intelligence is taking over the tasks that humans perform [22]. He noted that there may be numerous negative drawbacks related to the application of Artificial Intelligence which include the increased rate of unemployment as technology is taking over some tasks and professions and the possibility of inequality of wealth.

Mohammad et al. (2020) indicated that the challenges that are associated with the adoption and use of Artificial Intelligence in accounting and auditing include the development of effective policies for the Artificial Intelligence systems, to have skilled and professional labor with specific skills and capabilities which may be hard to find and require more funds to have the right professionals, and the lack of motivation and commitment from Artificial Intelligence from managers and leaders [32].

Within the functions of accounting and auditing, Artificial Intelligence raises ethical and moral concerns [31]. With continuously changing practices and procedures in businesses, and the method in which fraud is committed, new forms of crimes are rising. This increases the need for ethical guidance with regard to the way in which Artificial Intelligence can be used to detect such crimes through updating the system and ensuring the right coding required to ensure that the system catches such acts and doesn't increase the risks and errors associated with the business functions.

3 Conclusion

1. Artificial intelligence has dramatically changed the capabilities of accounting and auditing.
2. For many years, accountants have used technology to improve their jobs while adding value to the businesses they work in, which is the profession's ultimate goal.
3. The use of artificial intelligence allows to change the brand and transform the quality of business processes.
4. Artificial intelligence is a valuable tool to simplify and modernize business processes and operations.
5. The number of accountants and auditors cannot be completely reduced by the application of artificial intelligence techniques in the workplace because it still requires the need for human creativity and judgment.
6. As the auditing profession moves away from the method of preparation and system entries and more towards more specialization, there is a need for both centralization and decentralization of tasks and processes.
7. In the field of accounting, and within the function of auditing, Artificial Intelligence will not completely replace accountants and auditors, rather it would shift their focus towards different tasks as computers and machines take over specific tasks of data gathering and entry to ensure their efficiency and add value to the organization.

8. Artificial Intelligence should be seen as a compliment to human intelligence as a means of benefiting the processes of accounting and auditing and supporting the tasks within them to have better functioning business tasks.
9. Artificial Intelligence is not to take place of all accountant and auditor jobs, but rather to enhance their work processes and decrease the time and effort required to do them.

References

1. FSB. Artificial Intelligence and machine learning in financial services Market developments and financial stability implications (2017). http://www.fsb.org/terms_conditions
2. Greenman, C.: Exploring the impact of artificial intelligence on the accounting profession. J. Res. Bus. Econ. Manag. (JRBEM) **8**(3), 1451–1454 (2017)
3. Yudkowsky, E.: Artificial Intelligence as a positive and negative factor in global risks. In: Miri Machine Intelligence Research Institute, pp. 308–345. Oxford University Press, New York (2008)
4. Lombardo, T.: The future evolution of consciousness, world future review. Sage J. **6**(3), 322–335 (2015)
5. Deloitte: EFMA: AI and you: perceptions of artificial intelligence from the EMEA financial services industry (2017). http://efma.com
6. Alex, H., Fogel, K., Wilbank, C., Benard, G., Serge, M.: AI, robotics and the future of jobs. Pew Research Centre (2014). http://www.pewinternet.org/2014/08/06/future_of_jobs/
7. Murungi, S., Kayigamba, C.: The impact of computerized accounting system on financial reporting in the ministry of local government of rwanda. J. Emerg. Trends Econ. Manag. Sci. (JETEMS) **6**(4), 261–265 (2015)
8. McGuigan, N., Ghio, A.: Art, Accounting and technology: unravelling the paradoxical "in-between". Meditari Account. Res. **27**(5), 789–804 (2019)
9. Tarmidi, M., Rozalan, A., Rasli, M., Roni, R., Alizan, N.: Artificial intelligence accounting system (ALIAS). Global Bus. Manag. Res. **10**(3), 1116 (2018)
10. Can, T., Türkyılmaz, M., Birol, B.: Impact of RPA technologies on accounting systems. Muhasebe Finansman Dergisi **82**, 32–45 (2019)
11. Hansen, J., McDonald, J., Stice, J.: Artificial intelligence and generalized qualitative-response models: an empirical test on two audit decision-making domains. Decis. Sci. **23**(3), 708 (1992)
12. Munoko, I., Brown-Liburd, H.L., Vasarhelyi, M.: The ethical implications of using artificial intelligence in auditing. J. Bus. Ethics **167**(2), 209–234 (2020). https://doi.org/10.1007/s10551-019-04407-1
13. Rezaee, Z., Sharbatoghlie, A., Elam, R., McMickle, P.: Continuous auditing: building automated auditing capability. Audit. J. Pract. Theory **21**, 147–163 (2002)
14. Longinus, O.: Artificial intelligence system: implication for proper record keeping in micro-finance banks in Nigeria. Int. J. Acad. Res. Account. Financ. Manag. Sci. **8**(1), 131–136 (2018)
15. Kwarbai, J., Omojoye, E.: Artificial intelligence and accounting profession. Babcock J. Account. Financ. **1**(1), 1–26 (2021)
16. Jooman, S.: The influence of artificial intelligence on the future of the internal auditing profession in South Africa (2019)
17. Mhlanga, D.: Industry 4.0 in finance: the impact of artificial intelligence (AI) on digital financial inclusion. Int. J. Financ. Stud. **8**(3), 45 (2020)

18. Dongre, N., Pandey, A., Gupta, O.: Artificial Intelligence in accounting: opportunities & challenges. J. Xi'an Univ. Archit. Technol. **XII**(V), 1858–1864 (2020)
19. Davenport, T., Ronanki, R.: Artificial intelligence for the real world. Harv. Bus. Rev. (HBR) **96**, 110–116 (2018)
20. Chukwuani, V., Egiyi, M.: Automation of accounting processes: impact of artificial intelligence. Int. J. Res. Innov. Soc. Sci. (IJRISS) **4**, 444–449 (2020)
21. Omoteso, K.: The application of artificial intelligence in auditing: looking back to the future. Expert Syst. Appl. **39**, 8490–8495 (2012)
22. Makridakis, S.: The forthcoming artificial intelligence (AI) revolution: its impact on society and firms. Futures **90**, 46–60 (2017)
23. Luo, J., Meng, Q., Cai, Y.: Analysis of the impact of artificial intelligence application on the development of accounting industry. Open J. Bus. Manag. **6**, 850–856 (2018)
24. Ucoglu, D.: Current machine learning applications in accounting and auditing. Pressacademia **12**, 1–7 (2020)
25. Zhang, Y., Xiong, F., Xie, Y., Fan, X., Gu, H.: The Impact of artificial intelligence and blockchain on the accounting profession. IEEE Access **8**, 110461–110477 (2020)
26. Cheyenne, W., Matthew, S.: AI: ovetrated or the future of accounting (2018)
27. Omoteso, K.: The application of artificial intelligence in auditing: looking back to the future. Expert Syst. Appl. **39**(9), 8490–8495 (2012)
28. Bizarro, P., Dorian, M.: Artificial intelligence: the future of auditing. Intern. Audit. **5**, 21–26 (2017)
29. Doshi, H.A., Balasingam, S., Arumugam, D.: Artificial intelligence as a paradoxical digital disruptor in the accounting profession: an empirical study amongst accountants. Int. J. Psychosoc. Rehabil. **24**, 873–885 (2020)
30. Huang, Z.: Discussion on the development of artificial intelligence in taxation. Am. J. Ind. Bus. Manag. **8**, 1817–1824 (2018)
31. Zemánková, A.: Artificial intelligence and blockchain in audit and accounting: literature review. WSEAS Trans. Bus. Econ. **16**, 568–581 (2019)
32. Mohammad, S., Hamad, A., Borgi, H., Thu, P., Sial, M., Alhadidi, A.: How artificial intelligence changes the future of accounting industry. Int. J. Econ. Bus. Adm. **8**, 478–488 (2020)

The Impact of Artificial Intelligence on the Human Resource Industry and the Process of Recruitment and Selection

Amal Khalifa Al Aamer[1], Allam Hamdan[2](✉), and Zaher Abusaq[3]

[1] The Gulf Downstream Association, Awali, Kingdom of Bahrain
[2] Ahlia University, Manama, Kingdom of Bahrain
allamh3@hotmail.com
[3] Jeddah College of Engineering, University of Business and Technology, Jeddah 21448, Kingdom of Saudi Arabia

Abstract. In the contemporary world, artificial intelligence (AI) is an industry that continues to transform human lives and has a profound impact on business in almost all spheres. Organizations are searching for bright, dynamic, and potential employees to remain competitive in this digital age. Managing the digital world and developing the business environment will require organizations to employ a suitable individual with an effective recruitment strategy. Hence, hiring skilled employees who are experienced and efficient in achieving the job objectives is a crucial part of any organization's recruitment strategy. Artificial Intelligence has a key role to play in recruiting, its ultimate goal is to allow computers to carry out the same work as normally performed by humans. Artificial intelligence functions and reacts as if it were a human leading with incredible speed and accuracy. This study aims at analyzing how Artificial Intelligence (AI) impacts the Human Resource Industry. The study sheds light on the way artificial intelligence is used during the recruiting and selection process. The study finds that artificial intelligence technology capabilities can significantly improve the Human Resource and recruitment process with potential benefits such as increased productivity, reduced costs, improved accuracy, reduced workload, and improved candidate experience.

Keywords: Artificial intelligence · Human resource · Natural language processing · Recruitment · Selection

1 Introduction

AI, also referred to as artificial intelligence is a term that was developed by the computer scientist and mathematician John McCarthy who pioneered the field of artificial intelligence in his published paper "Computing Machinery and Intelligence". He describes it as "Artificial intelligence is the science and engineering of making intelligent machines, especially intelligent computer programs". Which opened a whole new field called artificial intelligence. In simple words, artificial intelligence mimics human intelligence using computers that function in similar ways to human beings by learning, adapting, identifying, and resolving problems (McCarthy 1950).

B. Alareeni et al. (Eds.): ICBT 2022, LNNS 620, pp. 622–630, 2023.
https://doi.org/10.1007/978-3-031-26953-0_57

Organizations are utilizing AI technology to automate repetitive tasks and aid in rapid and more accurate strategic decisions by using predictive algorithms. Recently, a growing number of enterprises have shown interest in implementing AI systems for Human Resource practices, such as recruiting, screening, and selecting candidates. Artificial intelligence technologies are already being used by leading companies to make better decisions and give workers predictive analytics. Organizations equipped with artificial intelligence can successfully compete in the market, and they can also experience operational excellence.

A key aspect of organizational performance is how the organization's Human Resource Management, and manpower impact its results and performance. Therefore, one of the most challenging tasks of the human resource manager is to hire the right aspirant with the requisite skill set to match the job specification and meet the organizational goals and objectives, In order to plan the recruitment pool for new millennials that can fit into their organizations, human resource managers need to identify how they plan to retain and attract new talent and how to assess the strategies, functions, and conditions of their organizations. Artificial intelligence technologies make organizational competence and knowledge challenges more complex and requires a new approach to human resource management. The purpose of engaging humans and machines together in human resources is not to put people out of work, but rather to restructure difficult processes and lead to a radical shift in how organizations recruit people, learn from them, and develop them. It is artificial intelligence, which plays one of the most significant roles in transforming Human Resource functions and has had a positive impact on Human Resource employees, organizations, and people. As a result, the purpose of this research paper is to demonstrate the role of artificial intelligence in improving the efficiency of administrative systems in the human resource industry, and more specifically, is to answer the key question: what is the effect of artificial intelligence in the human resource field? (Garg and Sharma 2021). Therefore, analyzing how Artificial Intelligence affects the Human Resource industry and the recruitment and selection process in organizations as technology has increasingly been recognized as playing a role in the organizations as they deal with enormous amounts of information and data and must embrace technology like artificial intelligence to transform themselves digitally. This reason has led to the advent of AI in corporate management, which has not only transformed the way employees work but has also completely changed business models. Human Resources is no exception to this situation; it, too, must embrace innovative technology (A. Waheed et al. 2019).

2 Literature Overview

Human resources' responsibilities include accepting thousands of resumes, nominating a few, conducting individual interviews, employing the best candidate, and other chores linked to the employee, such as benefits, training, and managing affairs. On average, more than 500 requests for leave, medical insurance, pay information, evaluation, interview results, and other topics are received each day. Artificial intelligence is having an increasingly large impact on all aspects of life and work, including human resource management. Artificial intelligence techniques have made it possible to mimic human intellectual functions such as decision-making, profiling, learning, problem-solving and evaluation, allowing human resource managers to automate many of their tasks, such as screening the Curriculum Vitae-CV examination and the usage of conversation tools that are very interactive to answer employee inquires. These AI technology free up time for human resources managers to focus on more productive duties, such as motivating employees and enhancing their cultural and creative perspectives, by eliminating tedious manual processes that consume the majority of their working hours. This section includes an overview of the effect of artificial intelligence technologies on the human recurse industry and the future view of using AI in the selection process of recruiting. This research will cover the impact and outcomes of using artificial intelligence in the Human Resource Industry and recommendations on how to use artificial intelligence in the recruitment process, and how Humans and artificial intelligence can work simultaneously (Attfield 2020).

2.1 Human Resource Algorithms

Based on a BBC Research, it was found that AI has grown 20 percent per year on average for the past five years, Artificial intelligence is superior to human performance in its capability to do more than one job, advance the Interaction with employees, and quality of hire, However, to create an Artificial Intelligence system, a large amount of data must be gathered, which is tedious, costly, and expensive process. The best way to utilize artificial intelligence is to feed it with neutral data to reduce racial discrimination between employees and bias for any new job seekers, with AI, companies can also narrow their search for qualified candidates by deciphering the answers provided-or to the words and phrases on a resume. For example, looking at the choice of words can reveal any potential bias in the responses or even in the way that the company frames the questions to the candidates. Intelligent systems should be highly secure when dealing with information about both candidates and employees. Using predictive analytics in human resources means using an advanced set of analytical capabilities that includes machine learning, mining data, real-time logging, text analysis, and statistical methods. By creating formulas and algorithms that simulate the results, these analyses can reveal patterns in the data and predict what will happen in the future. All with the goal of allowing managers to make better decisions in order to reduce costs. A decision tree algorithm can also be used to predict business outcomes and candidate performance, which is an easy and ordinary method that consists of tree models consisting of decisions and their possible outcomes. The nodes represent the tests for a specific attribute, the branches represent their outcomes, and decision trees are used to predict business outcomes and candidate performance (Jing 2009).

2.1.1 Natural Language Processing (NLP)

In the field of human resources, the interaction between humans is the main focus of the use of artificial intelligence and for this reason, the use of natural language processing is the main target, which aims to automatically process (human) natural language in a written form by helping computers to interpret, understand and process human language, By using these algorithms, it allows us to handle huge amounts of text, audio, and other data, in order to perform Artificial intelligence can be used to transform human resource management techniques by automating tasks such as content classification, modeling, context extraction, sentiment analysis, as well as the translation of text-to-speech and speech-to-text. There are still very few organizations able to build predictive models because of a lack of access to HR data and the difficulty in using it. When applying NLP to human resource management, NLP can be employed for forecasting the future job performance of applicants through asynchronous video interviews, together with other AI technologies such as voice analysis, chatbots, and facial expressions. NLP's powers can be employed to automatically evaluate and translate text and implement a user-centered communication and information system (Beysolow 2018).

2.2 Recruitment and Selection Process in Artificial Intelligence

2.2.1 The Impact of Using Artificial Intelligence (AI) on Recruiters

The future of recruiting using artificial intelligence is augmented intelligence. Human intellect cannot be completely replaced with technology. An augmented intelligence approach calls for the creation of technologies that improve human aptitude and efficiency.

2.2.1.1 Hiring The Best Candidates

Quality in recruiting is one of the top KPIs in Human resources as data has become easier to collect, access, and analyze over the years. Artificial intelligence's promise for improving the quality of hiring lies in its ability to use data to standardize the matching between candidates' experience, knowledge, skills, and job specifications. It is predicted that this improvement in job matching will lead to happier, more productive employees who are less likely to leave.

2.2.1.2 Automating High-Volume Tasks Save Time

Screening resumes manually remains the most time-consuming part of recruiting, especially when the majority of resumes are unqualified. A recruiter's time for a single hire is estimated at 23 h for screening resumes and interviewing shortlisted candidates. Automated assessment of resumes, scheduling interviews with candidates, and automating the screening of resumes are some of the tasks recruiters can automate with artificial intelligence for recruiting. AI-powered technology can not only automate a part of your recruiting workflow but integrate seamlessly with your existing recruiting stack so that it does not disrupt your workflow. Automating these aspects of recruiting can reduce time-to-hire, which means you're less likely to lose the best candidates.

2.2.2 What is Artificial Intelligence (AI) for Recruiting Process?

By using an artificial intelligence algorithm, companies can collect more data about applicants, such as information on social media, past work records, and educational background based on the job description. Artificial intelligence may replace traditional methods of screening resumes and selecting a new candidates. As a result, it is possible to reduce the recruitment time by conducting searches that are unbiased or free from bias based on race, gender, or religious stereotypes. In addition to collecting data about candidates. Machines now scan the CVs of candidates and select those who are suitable for a personal interview for recruitment and qualification. It is particularly effective in the case of large firms with hundreds of employees on a yearly basis. AI uses the blind screening system for CVs to collect additional information such as work skills, culture, and personality data and sends an email with a test that includes questions in the same vein as an In-person interview to determine whether a candidate qualifies for an interview. As part of this technology, names are removed from profiles along with gender, race, age, and other personal information so that only qualifications are compared, which makes it impartial. Artificial intelligence can be used to develop a visual interview assessment tool that analyzes answers to interview questions, word choice, word frequency, eye movements, facial expressions, and other traits. Interview questions can be customized for each job identifying characteristics to increase the chances of the appropriate candidate being chosen for the right job. Some companies also use predictive models for future job performance to know the eligibility of a candidate for a job and the likelihood of him staying or leaving, and these models combine many prediction tools, such as intelligence tests and personality tests (Council 2018).

2.2.3 Impact of AI on Human Resource Employees

Chatbots have recently become popular in the field of human resources to reduce work pressure and time, increase employee independence, and reduce bureaucracy by allowing chatbots to handle both easy and complex activities at the same time, such as answering employee queries and dealing with routine issues. Oracle has introduced a chatbot for the human resources department that can build a virtual preview of an employee's inquiry card, containing his name, address, and contact information, in only a few seconds via SMS discussion. Chatbots are built with advanced natural language processing technology to provide accurate answers in the right direction, and a decision tree is used based on frequently asked questions so that employees can access information with fewer questions, and the return on investment of chatbots for the organization is expected to be quickly realized by reducing reliance on human resources administrators, as other long-term benefits include reduced attrition, and increase employee development, which is reflected in the strengthening of the organization's reputation. Artificial intelligence technology will play a key role in the future ongoing training of the employees and the transfer of talents from one generation to another, so an employee's path with learning and progress does not stop with just getting a job. Honeywell, for example, has developed tools that leverage virtual and augmented reality capabilities, as well as artificial intelligence, to monitor work experience and extract lessons learned for new employees (Marr 2018). Some employees may be wary of AI because they believe it will replace them at work, causing them to lose their jobs; nevertheless, the issue is more of a "reinforcement" than a "replacement." Batter works with technologies that monitor workflow, provide intelligent suggestions, and execute repetitive chores, as an example of how technology may help us do our jobs more efficiently and successfully. AI may absolutely be used to monitor employees at work, and organizations currently use techniques such as sound analysis to identify the amount of tension or anxiety with a primary focus on monitoring and assessing employee wellness. The presence of machines in factories and warehouses has become common, but what is new is their presence in offices, and among these robots are delivery robots (Segways) that can move through workplace corridors to deliver messages and parcels, as well as security monitoring robots (Gamma) and parking robots (ParkPlus). Each organization has its own set of goals and strategies for investing in AI. Understanding and being aware of these patterns allows individuals in charge of digital transformation efforts to see the possibilities. Then get to work on seizing it (FraiJ and László2021).

2.2.4 Potential Outcomes of AI in the Recruitment and Selection Process

Artificial intelligence saves time by storing records in a way that prevents the same event from being repeated. Spending enough time screening candidate resumes is the standard mode of recruitment. As a result, sifting through resumes is a time-consuming process. AI helps to find the best people for the job, it also focuses on competency-based candidates in order to match them with the suitable position and talent. The process of finding a suitable candidate for the company is completed in a professional manner and the use of recruiting agencies is decreased. AI tools assist in cost reduction and functions in such a way that it uses a large amount of data for recruiting and does unbiased screening and selection. As a result, quality candidates are hired, and Employees receive up-to-date information and prompt responses to their questions, which results in satisfied employees and increased employee engagement. It also aids in lowering staff turnover and rewards those who provide excellent service to the company. The hiring of candidates is done entirely by machines, with no human intervention. Therefore, unbiased screening and applicant selection are possible. AI software aids in the screening and selection of qualified candidates. It aids in the identification of candidates' skills, competencies, and characteristics that are relevant to the position being applied for so that only talented individuals are hired.

2.3 Defining a New Way of Recruiting

Research shows that 75% of hiring organizations do not provide feedback to unsuccessful candidates, while 18% of the rejected candidates stop doing business with the company that rejected them. Active talent searches are still at an early stage. As recruitment costs rise and scarce talent becomes harder to find, screen, and maintain, recruiters must change the rules of the game. Candidates must be treated as customers, and AI-driven People Analytics must make the hiring process - and the information it relies on - more accountable and human-centric.

Through the use of chatbots, the recruitment process can be accelerated, and the hiring cost lowered significantly. Sentiment analysis and computational linguistics have the potential to accelerate, multi-focus, neutralize, and measure the candidate's experience autonomously during the selection process. Candidates' expectations align with those of the teams they might join. Candidate engagement is measured (and predicted) based on the content choices they make on social media. Return on investment of a new recruit can be measured, not only in terms of the organization's ROI in hiring, training, and remuneration, but also in terms of the individual's capability to learn on the job, develop new skills, and his or her social contribution. Achieving digital IQ and EQ in the next phase of People Analytics.

By analyzing the data above, the algorithms develop the ability to understand what is called 'the unconscious level of information. In the future, autonomy-learning machines will be able to simulate human behavior by assembling and analyzing people's statements, mood changes, and intentions on social media and other sources of publicly available information. This allows employees' experiences to be validated on a day-to-day basis. Skills will be scanned at the corporate level every day. Data analysis can also provide important information about which employees are engaged and challenged (and which aren't) and if there are any indications that social cohesion is at risk. A strategic workforce planning approach like this helps to reduce employee turnover and gives a new dimension to strategic workforce planning. It helps you determine what skills and talents are needed to maintain balance in the workplace and find the right blend between man and machine.

2.4 AI Applications for Human Resources (HR) That Are Available Today

2.4.1 Chatterbot for Candidate Screening

An AI tool assists candidates in engaging themselves before or after applying for positions advertised by the business. Companies operating in the digital economy can connect with candidates using a chatterbot, which is an AI tool. A chat box assists the applicant by replying to questions, and an AI tool asks for comments and required candidate information.

2.4.2 Messaging Systems for Candidate Engagement

Job seekers use a variety of job sites to search for and apply for jobs, but only a small percentage of them return to the application. The AI technology uses automated e-mails or a messaging system to automate the candidate application process. As this automatic data may maintain in communication with the candidate, it is possible that the candidate may respond quickly.

2.4.3 Re-engagement System

Tracking systems for job applications are frequently closed when a job opening closes. However, employing AI technology allows targeting a specific candidate and establishing their level of interest in the position that was applied for, it also makes use of the engagement opportunity to keep track of candidates who apply for a new job opening.

2.4.4 New Hire Onboarding System

After the candidate has completed the application process and all other procedures, he or she will be invited to accept the offer. And once they've accepted, there's a gray area, which usually occurs two weeks before they start working for the organization they want to work for. The new hire onboarding system is said to be an orientation program, It is effective for new hires since it introduces the organization's policies, procedures, and cultures. All of these formal procedures may be answered by AI tools for candidates, and it can also assist new hires by providing knowledge and resources that lead to current programs.

2.5 Implementing AI Applications in the HR Industry

A Real-Life Examples

Univer deployed HireVue AI-driven assessments with facial and voice recognition software. The company uses HireVue proprietary algorithms that is using words, speech patterns, body language, tone, and facial expressions to determine which candidates are best suited for a particular job. A hiring chatbot created by Mya Systems streamlines recruiting processes at staffing agencies such as L'Oréal, Adecco, Hays, and Deloitte using conversational AI. The Mya hiring process guides candidates from the job search to the onboarding process. Powered by machine learning engineers and NLP engineers, Mya uses state-of-the-art natural language processing and understanding techniques. AI-driven solutions from HireScore can help companies hire and retain employees by integrating seamlessly with their existing HR systems to ensure compliance and security. In addition, HiredScore enables large companies to make better hiring decisions by analyzing how they hire candidates and providing unique insights to applicants so that recruiters can focus on those who are the best match for the job. As well, HiredScore circulates high-quality leads that were either rejected in previous processes or signed up to receive future job offers from the company. Recruiters and job seekers alike benefit from Wade & Wendy's AI-driven recruitment solutions. Wade & Wendy is powered by a proprietary recruiting conversation system, cutting-edge text parsing techniques, intelligent workflow automation, and a robust knowledge graph that includes conversational utterances, linguistic logic, and information about job seekers, candidates, and job positions. Hiretual provides a comprehensive set of AI-powered solutions for the majority of recruiting needs. With Hiretual, you can integrate more than 30 Applicant Tracking Systems (ATSs) in one seamless workflow, sync candidate activities, and manage duplicates all in one place. The solution is backed by smart business analytics and industry-standard security and compliance measures. Virtual recruiters and advanced analytics tools provide a safer and more transparent selection process, breaking ties with inconsistent and laborious methods and advancing to a modernized, digitalized HR process (Ribeiro 2020).

3 Conclusion and Future of AI in the HR Industry

It is undeniable that artificial intelligence plays a critical role in optimizing recruitment techniques. Artificial intelligence solutions relieve the load of tedious and time-consuming repetitive processes like sourcing and screening applicants. Such leverage will greatly reduce the cost of hiring while also boosting the quality of recruitment. Furthermore, Artificial intelligence will increase transparency in the hiring process, eliminate human biases, and improve job seeker impressions of employers, enhancing the image and brand of employers. Artificial intelligence's role in recruitment and selection will undoubtedly grow as a result of all of these potential benefits. Artificial intelligence (AI) is a technology that can function as intelligently as a human brain in a variety of settings. When compared to traditional recruitment approaches, it garners more attention and relevance in automating recruiting systems. All organizations must work on recruitment as a primary activity. Now, the recruitment sector is gaining traction by using a smart approach to hiring, namely, artificial intelligence-assisted hiring.

Moreover, several industries are paying attention to the changes in the recruitment process. AI technology has a significant impact on recruiting since it allows recruiters to align unstructured candidate data, create uniform profiles, and find and match skill sets required by the sector. Recruiters believe AI technology is competing with them for recruitment activities in today's world. However, it is human-built software that makes the process go more smoothly. To summarize, AI's role is a combination of humans and AI that leads to data upkeep, cost and time savings for enterprises, and access to the entire recruitment process (Gawdat 2022). AI will help us to do more in less time. Therefore, it is not AI versus humans, it is a combination of AI and Humans versus problems.

References

Garg, P.K., Sharma, L.: Artificial intelligence: challenges and future applications. In: Artificial Intelligence, pp. 229–245 (2021). https://doi.org/10.1201/9781003140351-22

Waheed, A., Miao, X., Waheed, S., Ahmad, N., Majeed, A.: How new HRM practices, organizational innovation, and innovative climate affect the innovation performance in the IT industry: a moderated mediation analysis. Sustainability **11**(3), 621 (2019). https://doi.org/10.3390/su11030621

Attfield, P.: How artificial intelligence is transforming the recruitment process. The globe and mail, 30 October 2020. https://www.theglobeandmail.com/business/article-how-artificial-intelligence-is-transforming-the-recruitment-process/. Accessed 26 April 2022

Jing, H.: Application of fuzzy data mining algorithm in performance evaluation of human resources. In: 2009 International Forum on Computer Science-Technology and Applications (2009). https://doi.org/10.1109/ifcsta.2009.90

Beysolow II, Taweh: What is natural language processing? In: Applied Natural Language Processing with Python, pp. 1–12. Apress, Berkeley, CA (2018). https://doi.org/10.1007/978-1-4842-3733-5_1

FraiJ, J.D., László, V.: Literature review: artificial intelligence impact on the recruitment process. Int. J. Eng. Manage. Sci. **6**(1), 108–119 (2021). https://doi.org/10.21791/ijems.2021.1.10

Council, F.C.: Council post: 10 ways artificial intelligence will change recruitment practices. Forbes, 10 August 2018. https://www.forbes.com/sites/forbescoachescouncil/2018/08/10/10-ways-artificial-intelligence-will-change-recruitment-practices/?sh=29ff7d573a2c. Accessed 26 Apr 2022

Marr, B.: The amazing ways honeywell is using virtual and augmented reality to transfer skills to millennials. Forbes, 7 March 2018. https://www.forbes.com/sites/bernardmarr/2018/03/07/the-amazing-ways-honeywell-is-using-virtual-and-augmented-reality-to-transfer-skills-to-millennials/?sh=62ca503d536a. Accessed 27 Apr 2022

Ribeiro, J.: 5 companies that are revolutionizing recruiting using artificial intelligence. Medium, 16 November 2020. https://medium.com/tech-cult-heartbeat/5-companies-that-are-revolutionizing-recruiting-using-artificial-intelligence-9a70986c7a7e#:~:text=Hiring%20chatbot%20Mya%20Systems%20uses,and%20up%20to%20the%20onboarding. Accessed 27 Apr 2022

Gawdat, M.: Scary Smart. Pan Macmillan, London (2022)

The Effectiveness of Applying Artificial Intelligence in Recruitment in Private Sectors

Abdulla Mohamed Husain Almajthoob[1], Allam Hamdan[2](✉), and Hanadi Hakami[3]

[1] College of Business and Finance, Manama, Bahrain
[2] Ahlia University, Manama, Bahrain
allamh3@hotmail.com
[3] Jeddah College of Engineering, University of Business and Technology, Jeddah 21448, Kingdom of Saudi Arabia

Abstract. Companies seek for improving quality in order to survive in the competitive market. To survive, you must search for the right candidates and hire them in the best-fit position at the right time. Due to the rapid development in the world, several methods (traditional methods) are no longer efficient. People and companies are looking forward deploying more efficient methods in recruiting candidates. Currently, the trend in recruitment is moving forward applying artificial intelligence. This paper focus on impact of Artificial intelligence (AI) on private sectors. Since this industry is introduced recently, various companies are interested in applying AI. But the dilemma is that the advantages and disadvantages of applying AI in HR is still not commonly known for all of them. Also, the long-term effects for this industry still not revealed.

Keywords: HRM · E-HRM · E-recruitment · Artificial intelligence · Digital recruitment · AI in recruitment · Artificial intelligence in recruitment

1 Introduction

What is the most important element when considering evaluating a corporate's value? The criteria of evaluating the resources and values of companies have changed since the beginning of the eighties of the last century [1]. Previously, evaluation depended on tangible resources such as lands, Factories and equipment with a rate ranging from 70% to 90%. In the beginning of the current century, intangible resources cover about more than 60% of average firms' value [1]. As a result of this change, recruitment turned out to be an extensive concern for HR and CEOs over the last years [1].

Due to the effects of the forth industrial revolution, since we live in the beginning of it, Human resource management was enforced to take a step forward into dealing with their concern which was the manpower that plays the most important role in the performance for organization. Hiring qualified candidates who his/her qualifications and skills match with strategy of the firm and the required position is considered to be the most challenging task overall human resource management activities. According to Edwin B.

B. Alareeni et al. (Eds.): ICBT 2022, LNNS 620, pp. 631–641, 2023.
https://doi.org/10.1007/978-3-031-26953-0_58

Flippo, "Recruitment is the process of searching of the candidates for employment and stimulating them to apply for jobs in the organization" [13].

Recruitment is a responsibility for human resource management (HRM) and it is accomplished through two aspects; inter and external. Internal recruitment. Internal recruitment occurs inside the firm by advertising the vacancy to the employees therefore employees apply for it then he/she will be transfered to that position or by giving a chance for an employee to get promoted for that vacancy. While external recruitment is basically attracting external candidates who looks forward to be hired to the vacancy by publishing required skill and experience through the public media. At the end, Human recourse manager decides whether a candidate fits for the job role or not [13].

At present, recruitment is mainly achieved by recruiters who sit and search for specific applicants through available sources. Recruiters are in charge of carrying out all steps required such as contacting candidates, rejecting applicants, conducting interviews, etc. Due to the time those steps take and limitation of human being, it is not easy to follow up with all tasks. In addition, human being constraints such as prejudices, time limits and pre-concepts represent obstacles for impartial and efficient recruitment. Therefore, firms are obligated to solve this issue since this lead them to lose the most desirable applicants which are more valuable resources for firms. Developing efficient methods becomes essential based on the previous reasons (roy setiawan 2021). As result, recruitment practices and methods has been changed by technology via using multimedia tools, online applicant tracking system and self-learning computing systems. Moreover, these developments in recruitment led to emerging the term artificial intelligence in recruitment [1].

1.1 Research Problem

The HR sector is rapidly developing. This rapid transformation is introducing new challenges to organizations and the whole industry. This fast development is raising new difficulties to companies as well as the whole sector. The organization has to deploy new solutions to meet the market's requirements as existing process are not efficient. New approaches offered by artificial intelligences can help organization to overcome these challenges.

1.2 Research Objectives

The main objective for this paper is to encourage companies and people to invest more in AI recruitment by

- To define E-HRM and Digital recruitment.
- To define artificial intelligence in general.
- To define artificial intelligence in recruitment.
- To show advantages and challenges in AI recruitment.

2 Literature Review

2.1 E-HRM

According to Sbhi, Hanan and Bzar [8] Human resources (HR) was introduced in 1960s as management regulation. Human resource management (HRM) can be defined as management of workforce [8]. Based on RJ Stone, A Cox, M Gavin [6] human resource management concentrates on how to manage the relationship between employee-employer. Human resource managements organizes and impede shortage in workforce of corporations via stimulating older worker to keep working until a large age and preparing younger workers to substitute the older workers which results in reducing retirement costs and work continuity [7]. In order for organization to fit with quick internal or external changes, organizations moved forward using HRM with help of technology tools which results in a term called HRIS [8]. HRIS stands for Human Resource information system which uses technology tools in order to achieve human resource management functions [8].

Thereafter, due to enhancements in human resource practices, Electronic Human Resource Management (E-HRM) has substituted HRIS. The term E-HRM was raised in 1990s means implying automating management and employee services [8]. E-HRM can be referred as 'Web-based HRM', 'Virtual HRM', 'electronic HRM', 'HRIT', 'computer-based HRM' and 'digital HRM' [8]. E-HRM can be expressed as "the application of computers and telecommunication devices to collect, store, retrieve and disseminate human resource date for business purposes" [8].

According to Al-kasasbeh, Omar and Halim [9], information technology is used by E-HRM in two approaches; first, Technology is essential to link users that are geographically distant and allow them to communicate regardless of where they are; even if they are in the same room or working in separate spots of the world. To clarify, technology serves HRM by establishing the connecting environment. According to that, e-HRM has the ability to improve the services given to HR department clients (both employees and management), increase efficiency and cost-effectiveness in the HR department, and allow HR to turn into a strategic partner that contributes to the fulfillment of organizational goals. Second, e-HRM is indeed a tool used to provide management solutions that contribute to human resources effectiveness, such as E-recruitment, appraisal, E-selection, E-performance, E-training, E-compensation and E-communication [9].

Additionally, Al-kasasbeh, Omar and Halim [9] explain how technology is involved in the solutions offered by E-HRM to HR department as the following: 1) E-recruitment: is used by to advertise vacancy and building application for applying to that vacancy. 2) E-selection: internet is useful in selection process especially for applicants who lives in a widely far spot. 3) E-Compensation: tools provide higher access to information for managers in which may result in increasing effectiveness in compensation. 4) E-training: technology provides material such as laptop, PC, E-books, etc. and media players. 5) E-performance appraisal: technology contribution in performance appraisable is tangible in two practices; monitoring employees' computer performance unobtrusively in an automated method that requires minimum input from individuals over their work performance and through writing and providing employees' performance feedback. 6) E-communication: which can be clearly noticed in using Emails as a tool of communication in more than 75% in corporative companies.

Performance expectancy, effort expectancy, and peer pressure are the deciding factors of e-HRM [10]. Al-haideri and Siam [10] define performance expectancy as the number of goals and objectives that a client is expecting to achieve via using the framework. Whilst, effort expectancy, the second E-HRM determining factors; the extent to which comfort is tied to framework utilization. Lastly, Peer pressure can be defined as "how much one perceives it important to use the new framework because others are doing" [10].

The benefits of using E-HRM is to ensure availability in HR functions for employees and managers [8]. As a result, E-HRM has been employed in a variety of industries, including the healthcare industry as a web-based electronic hospital system and education for learning purposes [8]. Using E-HRM solutions increases HR flexibility, strategy, cost-effectivity; it also improves decision-making, reduces response time, minimizes operational effort, increases productivity, and improves client service [8].

However, Although E-HRM has shown potential in development HR process, 90% of basic organizational information management system refused to use E-HRM in Phuket hotels while only 10% agreed [8]. The literature on e-HRM has revealed that the use of e-HRM is dependent on the use of technology as a base element, as well as the acceptance model of technology [10]. Based on Al-Haidari and Siam [10] a study has shown that if the use of the HRM system is aligned with the system's suggested motivation and its facilitation condition, it may be gradually coupled to the e-HRM system. Whilst There have been several studies on e-HRM, but it has yet to be determined if it is effective or not, various techniques can define the effectiveness of HRM in employee satisfaction and commitment [10].

2.2 Digital Recruitment

2.2.1 Digital Recruitment 1.0

In the mid-to-late 1990s, digital recruitment via the internet has created an enormous revolution for both applicants and employers [1]. Monstors.com; early digital recruitment portals; started in 1994 attracted thousands of recruiters to publish plentiful vacancy details due to the lower cost since the cost of printing and publishing job ads through newspapers was avoided [1]. The relationship between Monster.com applicants and jobs were directly potential meaning increasing in jobs offered in Monters.com led to increasing in attracted candidates. The more job seekers applied, the more job were listed [1].

The process was as simple as the recruiters explored the CVs in the portal after that they filtered and sorted them then they selected the job seekers who were best-fit for the company's perspective. Likewise, applicants saved time and efforts in reviewing printing job ads, contacting firms, mailing firms, and sending resumes [1]. Based on Rosoiu, O., & Popescu, C [11] online recruitment process is 70% faster than the traditional method. The internet also allowed businesses to reach out to thousands of potential workers through corporate websites. They-might-incorporate as much static and dynamic information as they thought would be beneficial [1]. In addition, digital recruitment removed borders between countries and allowed job seekers to apply for several jobs in several areas [11].

However, according to Campus, Arrazola and Hevia [12] the performance of job portals is was determined by four main attributes: 1) the amount of information gathered about candidates; 2) the amount of information available for job seekers in a portal; 3) the cost of portals; 4) the undetected quality of applicants that necessitates the firm's need to screen job candidates further. In fact, Person–environment (P–E) fit theories proposed that when potential applicants see a fit or match with the organization, they will have a good attitude toward recruiting material provided on job portals [12]. Job seekers matches the description in job portal with their skills to assess whether they fit for a company or not.

As a result of enormous outcomes job portals brought, companies abounded the old practice of recruitments (analog recruitment) and shifted toward online recruitment since it allowed them reaching various applicants [1]. Based on to Campus, Arrazola and Hevia [12], 90% of large companies in the United States of America use online recruitment in the last decade. The value digital recruitment offered for both recruiters and candidates during the decade led to establish new cooperation and new job boards proliferated [1]. For example, Monsters.com revenue in 2006 was $1.1 while in 1996 was 162.6 million [1].

2.2.2 Digital Recruitment 2.0

After one decade of the beginning of Digital Recruitment 1.0, Digital Recruitment 2.0 has been introduced to the public [1]. As mentioned by Bohmova [5] Digital Recruitment 2.0 can be defined as "a process of hiring new employees with the use of social media such as LinkedIn, Face book, Google +, Twitter and many other social media sites". There were two main advantages that digital recruitment 2.0 has over 1.0. First, it allow job seekers to explore several job platforms [1]. In other words, applicants have the ability to search for different jobs offered in multiple job portals without the need for visiting each one. Likewise, Companies may reach out to unique job seekers across all recruiting sites without having to post their offers on each one separately. Second, the appearance of digital professional and social network platforms [1].

With so many possibilities accessible in this day of plentiful Internet and social media, choosing which sourcing tool to use in the recruitment task becomes a vital decision for any firm or recruiter [4]. Social platforms benefit recruiters in two ways; 1) they are used to find and attract passive and semi passive job seekers [4, 5] 2) "posts per week" which means how the company interact with its followers in the platform [5]. In 2017, 93% of recruiters were using social media for recruiting purposes [4] compared to 80% in 2012 [5]. However, social media websites differ in popularity from geographical area to another [5].

LinkedIn was one of the first and most well-known professional networking platforms. LinkedIn, launched in 2003, allows users to endorse others in their network, and get endorsement from others in their networks, develop, exchange information and professional networks and communities of interest [1]. Statistics shows that in 2012 LinkedIn users reached up to 332 million member over 200 countries [5]. Whilst, in 2017 linked had 3 million job listing and 467 million users [4]. In fact, 95% of recruiters were using LinkedIn for recruiting processes [4]. Indeed, the popularity that LinkedIn gained because of the ability in LinkedIn to build professionals relations as it was being seen by the public [4].

On the other hand, Twitter and Facebook were as high use as LinkedIn since they were more general social media [4]. Members of Facebook may expand their network by 'friending' other individuals and keep them up to date by publishing images and videos [1]. While Twitter is considered as microblogging platform allowing members to communicate within 140 characters for each tweet [4]. Studies show that 66% of recruiters were using Facebook in 2014 while 52% for Twitter [4].

2.2.3 Digital Recruitment 3.0

Digital recruitment was introduced after 2.0 and the main advantage in this type of recruitment is inserting artificial intelligence in the process of recruitment [1] which will be reviewed in this paper.

2.3 Artificial Intelligence

Artificial intelligence (AI) is a constantly dynamic frontier of developing computing capabilities, not a technology or group of technologies [3]. AI does not have a single definition and can take on a variety of meanings according on its context, applications, and intelligence [15]. Further, AI may be described as a system that can comprehend and learn from external inputs in order to achieve certain goals by adapting to the circumstances [15]. Whilst, Kot [Kot] defines AI as a broad class of computer-based technology that supports various business tasks with human-like intelligence for prosperous higher intellectual process. In general, AI may be defined as a system that simulates typical humanistic behavior such as learning, speaking, and problem-solving, causing it to act similarly to an intelligent person [15]. According to the definition, it is clearly that AI not only provide solution for problems with fast analyze and response but also have the ability to mimic and increase human-like intelligence which include cognitive, emotional, and social competences [15].

The machine learning technologies that are at the core of contemporary AI have greater autonomy, deeper learning capacity, and are more inscrutable than any of the "intelligent" IT artifacts that have made before [3]. Current AI technologies, such as robotics and autonomous vehicles, facial detection, natural language processing, and various types of virtual agents, are being used in a broad range of problem areas [3]. An estimations on AI states that in 2020 more than half of enterprises were employing some kind of this new wave of technology and the applications are growing at an incredible rate [3]. These advancements are significant because AI has limitless potential for improving people's lives in a wide range of domains, including their homes, medical, education, career, entertainment, safety, and transportation. AI offers businesses unparalleled prospects for developing intelligent goods, developing unique service offerings, and developing new business opportunities and organizational structures [3].

Advanced technology such as Artificial intelligence has three principles in order to be understood; Combinations, recursiveness and phenomena [2]. First, combinations means machine learning depends on improved performance-price ratio of computer-processing technology, data storage and management. By combining these factors, it results in making AI a significant tool for producing products, service or platform [2]. Second, AI implementation structures modular architecture forming complicated network of technologies where each technology is enhanced independently according to it is goals [2]. In addition, AI is a set of technologies that are set of technologies. Changing or improvement in a segment of technology may lead to a conflict of other's objectives [2]. For Example, enhancements in a service recommendation engine leads to enhancements to its convenience, personalization and ease of use [2]. Lastly, third principles, Phenomena; today's AI has data-driven learning at its center. For instance, a type of machine learning algorithms are able to teach themselves how to recognize a dog or a cat if trained with pre labeled set depending on how big the data is without setting explicit rules for recognizing [2].

Furthermore, intelligence-based AI systems may be divided into three types: analytical, human-inspired, and humanized AI [15]. Analytical AI with cognitive intelligence has the ability to make future judgments based on previous data learning and analysis, which might be used in Virtual teaching fraud detection, assistance, photo identification, and other applications [15]. Human-inspired AI can recognize and assess human emotions such as anger, enthusiasm, and so on, which impacts their decision-making. Virtual recruiters, for example, recognize a candidate's emotions during the selection process [15]. Humanized AI combines all three cognitive, social, and emotional abilities.

Aside from the benefits of AI, there are other obstacles that must be addressed. Data problems, political, legal, and regulatory obstacles, and ethical issues are among them. As AI requires a big quantity of data, issues such as transparency, low quality, unavailability of data gathering format, data discontinuity and lack of data availability and so on pose important obstacles [15]. To protect people's privacy and safety, the government must develop a legal framework that achieve a balance between AI data and public privacy [15]. Because AI requires a massive quantity of data, rigid laws and regulations may stymie its application. As per the European GDPR (General data protection standards) enacted in 2018, businesses must take great care while handling and transacting personal data, which may affect the freedom of utilizing data in AI. It is critical to preserve ethical purpose while obtaining public data from social media or other private sources. AI ethical issues are caused by moral quandaries, discrimination and biased decisions made by AI, appropriateness, and compatibility between a person and a computer [15].

2.4 Artificial Intelligence in Recruitment

Applying Digital Recruitment 2.0 in large companies ended up in a huge number of accumulated CV's and choosing the best fit applicants created a challenge for the firm [15]. Not reviewing candidates' application was costly, but also there was a probability of human bias in selecting candidates [15]. Thus, the need to overcome these obstacles required applying AI digital solutions since they are more efficient than human-being abilities [15].

AI in recruiting became widely employed in enterprises in 2018 and has been a significant trend to this day [15]. AI technologies were used at almost every stage of the recruiting process, which revolutionized the recruitment market in a more novel way and gave a huge aid in selecting the top applicants from big pools of varied aspirants [15]. These tools were effective in a variety of ways, including producing job descriptions with appropriate terminology and language that is bias-free, gender-neutral, and targets a specific set of candidates [15]. For instance, L'Oreal used AI to eliminate gender bias language in their advertisements, allowing them to hire an equal number of male and female candidates [15].

Based on Geetha AI improve and develop recruitment process in 8 steps; 1) screening candidates: by using chatbot [13] or CV screening tool ATS (applicant tracking system) which assess candidates profile based on keyword specified by recruiter [15]. 2) Candidate engagement: AI tools response to aspirants via auto email generated or messaging system. 3) Re-engagement. 4) Post-offer acceptance. 5) New hire on-boarding: It is beneficial to new recruits since it presents the organization's policies, processes, and cultures. All of these formal procedures may be answered by AI tools for applicants, and it also assists new recruits with knowledge and resources that link to current programs. 6) Career development. 7) Employee relation. 8) Scheduling.

Moreover, AI improve recruitment process via several tools and systems such as chatbot [Ayes]. AI-powered chatbot are increasingly becoming popular in the recruiting market. These chatbots can connect with applicants, answer their questions 24 h a day, and provide real-time and personalized interaction via text message, email, social media, and other channels. AI-powered chatbots are trained and prepared by using human natural language in order to communicate with candidates like humans via using emotion, contextual words and shorthand [15]. Also, video chat analysis such as Affectiva, HireIQ and HireVue, is widely used during interviews to analyze candidates' characteristic and performance such as tone, words-used, emotions, cadence, etc. [15].

Ayesha [15] believes that by assigning the routine tasks of screening to AI, recruiters will be able to stay focused on more strategic and creative matters in their daily routines, while HR managers will switch their focus from operational functions to a management role, motivating and cultivating the potential of their teams. According to Hmoud and Laszlo [14], administrative routine jobs will be gradually replaced by clever AI technology, allowing recruiters and HR managers to focus more on strategic activities.

The advancement of AI offers potential options for recruiters to maximize talent acquisition by automating time-consuming repetitive operations such as sourcing and screening applications, to improve the recruiting process's quality, and to minimize biased – human deception. According to Ayesha [15], these benefits stem from AI's ability to process information and make decisions at volumes and speeds far exceeding human capacity, as well as the availability of AI-enabled recruiting tools and systems that overcome common cognitive biases that undermine the reliability and validity of human judgment in recruiting activities.

Also, AI tools significantly reduce time in recruitment process especially in screening. For example, Ideal AI solutions provider claims that enabling AI tools decreases time to hire by 62% (from 24 days to 9 days). In addition, Hilton Hotels decrease their time to hire by 88% (52 days to 5 days) after applying AI screening tools. Lastly, L'Ore'al deployed AI-enabled screening technologies, and the time it took to evaluate a resume was reduced by 90%, from 40 min to 4 min [1]. In fact, as consequence of reducing time in recruitment process with help of AI tools, hiring cost becomes cheaper for companies [15]. Since numbers of applicants are increasing and evaluating and screening process are becoming more complicated, a company would hire a huge number of recruiters in order to help the company choosing the right talented candidate. Thus, the company has no choice other than applying AI recruitment.

Bias recruitment represents a huge issue for efficiency in old-fashioned recruitment. Human biases are possible during the screening step of recruiting, however AI eliminates these prejudices and provides promising alternatives for acquiring right personnel. AI may eliminate human prejudices by using algorithms that disregard biases such as color, name, gender, school or university attended, and so on. On the other side, it filters candidates based on data provided to it such as qualification, abilities, and experience, among other things. Thus, AI is based on facts, and no emotions or sympathies may influence its evaluation [15].

In addition, AI recruitment can enhance decision making for human resource managers. Ayesha [15] states that Data analytics technologies are essential for making better decisions and forecasts about prospects. According to Geetha and Reddy [13] AI packages aid in the screening and selection of qualified candidates. It aids in identifying candidates' talents, competencies, and characteristics that are relevant to the position being applied for. As a consequence, a talented individual is hired.

However, in order to apply achieve impartial and bias-free recruitment, AI tools' algorithms must not only built and written in an efficient way but they must be transparent and available for inspection. Ayesh [15] says that if a company apply AI tools thus, it is necessary to demonstrate openness and transparency in how the algorithms are built and how they work to choose a candidate. Thus, in a scenario like this, a candidate will always feel justice why he/she were rejected. Ayesh also adds rejected applicants, who had a great experience when they were rejected, are more likely to be open to a future chance.

Besides the advantages that AI recruitment brings to human resource management, there server risks and challenges are raised in applying and moving forward to AI. One of the main challenges is lack of knowledge [15]. Since AI recruitment applied 2018, a lot of companies are in the early stages of implementing AI therefore many tools are still unknown to HR experts. If recruiters do not know how to use the system or how algorithms are built, they would doubt and reject the solution. Since AI tools are not error-free, recruiters must explain how fairness and efficient their tools to the candidates otherwise they may lose candidates' trust [15].

Additionally, among the most frequently mentioned and essential problems of employing AI in the recruiting process is ensuring data protection and taking adequate precautions against ethical concerns. Recruiters might obtain personal information that is not directly relevant to recruiting by using various technologies utilized in the recruitment process. Age, health, body image, gender, sexual orientation, and other factors, for example, can be utilized to classify applicants and even discriminate when possible. The collection of this additional information may raise ethical and privacy concerns [15].

According to Ayesha [15] some studies show that a decent number of candidates do not trust machines recruitment functionality. The applicants always believe that human touch is significant at any phase of the hiring process, and they are convenient interacting with a human during the interview. Not only candidates, but also HR recruiter's feels that AI recruitment tools is a threat for their job. Companies used to hire employees more than they fire which is not the situation in the current time [16]. However, AI tools cannot be left working alone without observation from humans [15].

Lastly, poor quality data and bias input represent challenges for HR managers. Since AI tools output is generated according the input [15], a poor input definitely results in an inappropriate output which has nothing to do with the machine's efficiency or the algorithm that runs it. For instance, Because of the biased input data given into the system in amazon, the computer trained itself to be prejudiced, causing it to punish female candidates despite having enough qualification [15].

3 Conclusion

Recruitment process is one of the main challenging and important practice in HR since worker represent more than 60% of companies' value [1]. Previously, recruiters had to advertise job vacancy through traditional ways such as newspapers, reviewing profiles and selecting candidates which was considered to a costly and long procedure. After that, a big leap was brought to HR recruitment process by moving to digital recruitment through websites and social media platforms such as monster.com, Facebook and LinkedIn [1]. According to huge numbers of candidates' profiles to be reviewed by recruiters which costs time and an effort, the need for more efficient method was emerged.

Artificial intelligence is a machine or tool that can act like human via learning by its self and generated results according to input from human. AI tools in recruitment such as ATS, Chatbot and video chat analysis can be developed and involved in every step of recruitment process [15]. In order for AI tools to be applied efficiently and impartial, they must be transparent for candidates and available for inspection. AI shows several benefits for HR managers in recruitment such as reducing cost, reducing time especially in screening CVs, bias-free recruitment, making HR managers more professionals, better decision making and making job applicants experience smooth and positive. In contrast, several challenges and risks must be considered when applying AI in recruitment such as lack of knowledge, bias-input, poor quality data, collaboration cost, losing jobs, machine trust and ethical issues like privacy [15].

Despite the risks and obstacles of AI recruiting, many firms have had success with their AI deployment. According to Ayesha [15], Okolie observed that enterprises benefitted from lower expenditures, more applications, and better applicant matching with individuals having a smoother application procedure, a diverse range of career possibilities, and, finally, a greater response rate from the organization to get feedback. She also claims that when Google launched the Cloud Jobs project, some of its clients, including Johnson & Johnson and FedEx, started using it to enhance interaction with potential applicants in their recruitment platforms, as well as to increase visibility and matching likelihood for job searchers. On the other hand, some companies faced issues in implementing AI tools such as Amazon [15].

Eventually, artificial intelligence leads to employee satisfaction and engagement. Furthermore, it contributes to a lower staff turnover and ensures that the organization receives good service. The applications in the recruiting process are promising, and the growing demand for these tools with new capabilities makes it even more so. Despite the fact that AI decrease number of jobs causing some people to be fired, the touch of humans will always be needed and therefore it will generate new types of jobs and business. Since this field is considerably recent, the number of conducted studies on AI recruitment are not sufficient. Thus, methods are not widely employed in the recruiting industry. So there are many things to learn in order to seamlessly integrate and adapt to these new technology.

References

1. Black, J.S., van Esch, P.: AI-enabled recruiting: what is it and how should a manager use it? Bus. Horiz. **63**(2), 215–226 (2020)
2. Gregory, R.W., Henfridsson, O., Kaganer, E., Kyriakou, H.: The role of artificial intelligence and data network effects for creating user value. Acad. Manag. Rev. **46**(3), 534–551 (2021)
3. Berente, N., Gu, B., Recker, J., Santhanam, R.: Managing artificial intelligence. MIS Q. **45**(3), 1433–1450 (2021)
4. Koch, T., Gerber, C., De Klerk, J.J.: The impact of social media on recruitment: are you LinkedIn? SA J. Hum. Resour. Manag. **16**(1), 1–14 (2018)
5. Bohmova, L.: The use of social media in the recruitment process. FAIMA Bus. Manag. J. **4**(2), 20 (2016)
6. Stone, R.J., Cox, A., Gavin, M.: Human Resource Management. Wiley, Hoboken (2020)
7. Pak, K., Kooij, D.T., De Lange, A.H., Van Veldhoven, M.J.: Human resource management and the ability, motivation and opportunity to continue working: a review of quantitative studies. Hum. Resour. Manag. Rev. **29**(3), 336–352 (2019)
8. Zeebaree, S.R., Shukur, H.M., Hussan, B.K.: Human resource management systems for enterprise organizations: a review. Period. Eng. Nat. Sci. **7**(2), 660–669 (2019)
9. Al-kasasbeh, A.M., Halim, M.A.S.A., Omar, K.: E-HRM, workforce agility and organizational performance: a review paper toward theoretical framework. Int. J. Appl. Bus. Econ. Res. **14**(15), 10671–10685 (2016)
10. Siam, M.R., Alhaderi, S.M.: The scope of e-HRM and its effectiveness. Polish J. Manag. Studies **19**, 353–362 (2019)
11. Rosoiu, O., Popescu, C.: E-recruiting platforms: features that influence the efficiency of online recruitment systems. Inform. Econ. **20**(2), 46 (2016)
12. Campos, R., Arrazola, M., de Hevia, J.: Finding the right employee online: determinants of internet recruitment in Spanish firms. Appl. Econ. **50**(1), 79–93 (2018)
13. Geetha, R., Bhanu, S.R.D.: Recruitment through artificial intelligence: a conceptual study. Int. J. Mech. Eng. Technol. **9**(7), 63–70 (2018)
14. Hmoud, B., Laszlo, V.: Will artificial intelligence take over human resources recruitment and selection. Netw. Intell. Stud. **7**(13), 21–30 (2019)
15. Brishti, J.K., Javed, A.: The viability of AI-based recruitment process: a systematic literature review (2020)
16. Cavaliere, L.P.L., et al.: The impact of E-recruitment and artificial intelligence (AI) tools on HR effectiveness: the case of high schools. Doctoral dissertation, Petra Christian University (2021)

Check for updates

The Impact of Artificial Intelligence on Financial Institutes Services During Crisis: A Review of the Literature

Eman Salem Abdulla[1], Allam Hamdan[2](✉), and Hatem Akeel[3]

[1] College of Business and Finance, Manama, Bahrain
[2] Ahlia University, Manama, Bahrain
allamh3@hotmail.com
[3] College of Business Administration, University of Business and Technology, Jeddah 21448, Kingdom of Saudi Arabia

Abstract. Artificial Intelligence (AI) is the general field that covers everything related to importing "intelligence" to computer systems in order performing tasks which require human intelligence, this is achieved by using algorithms that can detect patterns, generate insights from the data presented to them, to apply them to future decision-making processes and predictions, Artificial intelligence is a new concept of technological innovation where different technologies, processes and methods have been combined to create alternative solutions which are precise and to the point to enhance the economies as well as the competitive edge of the organization. The implications of the AI technology have been seen in various fields of life including medical automotive as well as financial industries. Financial institutes with crisis face challenges to deal with during to improve it quality and efficiency, so now days the financial services are under strain and challenges due to the wake up of the Covid-19 pandemic, Therefore, many financial institutes shift to implementing AI in its services to enhance it services, satisfy their customer and increase productivity.

Keywords: Artificial intelligence · Financial institutes services · Crisis · Covid-19

1 Introduction

Over the past year's technology has become a major in our daily life and we have been connected to technology in one way or another, because it provides us with the useful resources that put all the information we need at our fingertips, technology has entered into our personal life, business, industry, agriculture, education and medicine. A major part of this technology is Artificial intelligence as we start recently to hear it a lot and how it can help in different fields in our normal life (Dupont et al. 2018).

Artificial Intelligence (AI) is the general field that covers everything related to importing "intelligence" to computer systems in order performing tasks which require human intelligence, this is achieved by using algorithms that can detect patterns, generate

B. Alareeni et al. (Eds.): ICBT 2022, LNNS 620, pp. 642–655, 2023.
https://doi.org/10.1007/978-3-031-26953-0_59

insights from the data presented to them, to apply them to future decision-making processes and predictions (Fernandez 2019). AI is more about the ability to think critically and analysis data than it is about a specific shape or function.

The implications of the AI technology have been seen in various fields of life including medical automotive as well as financial industries.

AI could improve the quality of products and service for clients due to a broader and deeper analytical basis and information. In addition, AI could lead to higher efficiency and lower costs (Sion 2018), Therefore, most institutions seek to invest in the applications and tools of modern artificial intelligence.

Financial institutes with crisis face challenges to deal with during to improve it quality and efficiency, so now days the financial services are under strain and challenges due to the wake up of the Covid-19 pandemic (Billio and Varotto 2020). And in order to reduce the spread of the virus, precautions were taken such as social distance, in addition many countries took several plans responding to the pandemic for example staff absences, cancelation of travel or limitation of transportation and interruption of technology (WHO 2020). Therefore, many financial institutes shift to implementing AI in its services this will be explained furthermore in the literature review.

2 Related Theoretical Review

2.1 Artificial Intelligence (AI)

To start with through history there were three revolutions, the first industrial revolution started in 1784 with the steam engine, the second revolution was in 1870 with the electricity and the third revolution was in 1969 with computers and IT, but now days we are facing the fourth revolution is about the big data and network which artificial intelligence is a part of it, AI is a tool or type of computer science that use machines to help in finding solutions to problems that human faces (Nilsson 2013), not only solving problems AI can help in planning, reasoning and learning.

Artificial intelligence has four categories that are used in our world:

1. **Machine learning (ML):**
   This type of AI application design and develop algorithms and techniques that give computers the ability to learn, it uses statistics to find patterns within the data available (Numbers, Words, Pictures, Clicks and many (McCarthy 2004).
   ML can be found in financial sector, it can be found in banks, online shopping or social media (Fernandez 2019), this type of application plays an important role in making an efficient, smooth and safe experience.
2. **Deep Learning (DL):**
   This type of AI application help computers learns to simulate the way human think based on training and gaining experience from reading images, voices the same way human do (Meiring and Myburgh 2015).
   DL depends on processing a huge amount of data by passing it through deep neural networks (DNN) to help it to extract useful information for the data just like in neural network in human, the different between ML and DL is that deep learning characterized by different levels of algorithms that form artificial neural networks which have the ability to understand unregulated data (McCarthy 2004).

3. **Big Data:**
   The big data known as a large and diversify group of information that is growing very fast, include increasing in size, speed and amount. It simply means that the big data is the data stores everyday by the world as every individual considered as a data creator (Mubark 2019).
   Big data have the ability to help companies improve it operations and make faster decisions, through the enormous information provided, as the data is collected from different sources such as social media, electronic stores, mobile application, database, servers and even through surveys, invoices to purchase products and medical examinations. These data been collected and reviews so it can be processed and filtered then analyzed according to what company's needs (Meiring and Myburgh 2015).
   An example of implementing the big data in banking and financial sector is in predicting and dealing with cybercrimes operations, more in differentiate between fake and original credit cards by analyzing customers past data and knowing the movement of these cards, this will contribute to arising errors that face financial sector (Mardanghom and Sandal 2019).
4. **Text mining/Natural language processing (NLP):**
   It's a tool of AI that is concerned with the field of letting computers understanding human natural languages by analyzing huge amount of data extracted from human natural language, NLP can solve simple problems like answering a query on the internet to complex ones that require terabytes of data to train.
   NLP used in many programs that need to analyze text or audio data for example:

   - Search engines.
   - Social network development.
   - Chatbots
   - Spell check software.
   - Sort annoying and unsafe emails (Latimore 2018).

After explaining the artificial intelligence categories that is used in our world, we will define the AI types, AI have four types systems:

1. **Reactive machine:**
   This type of system has no memory mean they are the basic types of artificial intelligence systems and fully interactive systems, as they do not have the ability to form memories, nor do they have the ability to use past experiences to inform current decisions, as interactive machines perform basic tasks only. Machines that use this type respond to some inputs with some output, and their mechanism of action does not include any self-learning process.
   Deep Blue an example for this type of machine or device, as it is the supercomputer that plays chess from IBM, which defeated international defender Garry Kasparov in the late 1990s (Mardanghom and Sandal 2019).
2. **Limited Memory:**
   In this type, artificial intelligence has the ability to store data, or past forecasts, and use them to make better predictions in the future. And with limited memory,

engineering and building machine learning technologies becomes more complex. Among the examples of machines that use this type of artificial intelligence, we find that most of today's devices that rely on artificial intelligence use limited memory, including personal assistance applications such as Google Assistance, voice and picture recognition programs, and chatbots (Fernandez 2019).

3. **Theory of mind:**
   The third type of artificial intelligence is theory of mind, unlike the two types of artificial intelligence Reactive Machines and Limited Memory Machines, whose applications are widely implemented, the system based on theory of mind are still in the development phase.
   It represents the next level of progress for artificial intelligence, where we will be able to better understand the entities with which it interacts, by distinguishing the needs, emotions, beliefs, and intellectual processes of their own. Machines will therefore be more sophisticated, and the type will not only constitute perceptions of the world, but also of factors or entities (Kunwar 2019).
4. **Self-Aware:**
   However, this type is still not existing but in future, humans may finally develop a self-conscious AI. It is the same entity that we see in science fiction movies. This kind of AI may raise many hopes, but it also raises many concerns. The idea of a self-aware robot with a special and independent intelligence is troubling, because this means that humans must then negotiate with the machine that they created with their own hands, and the result of these negotiations gives way to many assumptions, expectations and imaginations (Buchanan 2019).

2.2 Financial Institutes Services

The term financial institutes mean all firms engaged in the business of dealing with finance such as loans, deposits, currency exchange, investment under the law and regulations of the countries, they are part of the financial market and financial system. The financial institutes might be public, privet and corporates, the service represents in the institutes are an intangible activity or business obtained by a beneficiary or individuals or machines through which the service is provided the level of satisfaction of this beneficiary is related to the level of performance of individuals and machines, and the banking service represents:

- An activity or work presented to the beneficiary by individuals, machines and devices.
- Its presentation may or may not be linked to the commodity, for example: The beneficiary's access to the information he needs through his meeting with the bank's workers makes this type of service linked to the person who provides it.
- The beneficiary's withdrawal of cash via an ATM is linked to the use of this cash machine.

The level of satisfaction of the beneficiary is related to the efficiency of the workers and the level of development of the used machines and devices, it helps to speed up the delivery and delivery of the service. Financial institutes provide for the beneficiary's many services in:

- Provision of funds such as banking services venture capital, asset financing, credit cards and trade financing.
- Managing investible funds: portfolio management.
- Risk financing such as insurance, project preparty services.
- Market operations such as stock market, money market and asset management.

2.3 Crisis

The concepts of crises and points of view are many, and there is a difference and variation from one author to another. It is difficult to know the real reasons for the emergence of the crisis due to the difference in time, place and circumstances. Therefore, we can identify causes that are outside the nature of human and it is difficult to control or stop them and there is no ability to predict their arrival or occurrence, such as floods, volcanoes, hurricanes, fire, earthquakes and diseases (Haleem and Javaid 2020).

During the month of December 2019, a new disease began to grab the attention of news and news channels around the world. Scientists quickly discovered that this disease is the Corona virus, it was a new virus for which there is no definitive treatment at the present time. Since then, public fears have increased in parallel with the massive media coverage and the spread of the virus to other parts of the world.

The coronavirus (COVID-19) is a primary health issue that started in Wuhan, china in December 2019, as the world health organization said its infectious disease caused by a newly discovered coronavirus.

Most people infected with the COVID-19 virus will experience mild to moderate respiratory illness and recover without requiring special treatment. Older people, and people with medical issues like cardiovascular disease, diabetes, chronic respiratory disease, and cancer are more likely to develop serious illness.

With no vaccine founded and in order to stop the spread of it many countries have started taking very strict measures, Countries and capitals imposed strict embargoes and lockdowns which affect economies (Haleem and Javaid 2020).

3 Literature Review

The literature review seeks to identify research areas related to the impact of using AI during crisis and the possible research gabs. Three questions were defined to formulate the basis of the research papers selection: The research of the impact of using AI during crisis in literatures can be divided into three categories: The development of AI in financial services, Impact of crisis on Finance Institutes and Implementing Artificial intelligence in Financial Institutes services during crisis. After searching for the relative topics under the defined research questions, a screening processing was conducted to select the papers under the desired scope.

3.1 The Development of AI in Financial Services

3.1.1 Anti-fraud and Risk

Now days with the increase of E-commerce, online fraud and increase in payment this leads financial institutes specially banks to find a way to enhance their services therefore,

AI tools is an option to improve their services and prevent frauds (Buchanan 2019), the ability of AI in understanding algorithms were very productive in detecting frauds, 70% of global financial institutes use AI in their services and 80% of their frauds were detected by using AI technology (Pothumsetty 2020), agreed to that (Zavadskaya 2017) he stated that the past few years payments was done by using mobile channels and as sequence fraud were increase very fast, therefore, using traditional detecting methods will be inaccurate, expansive and consuming time. So financial institutes shifted to use AI tools based on that some studies suggest that applying AI tools is to avoid misrepresentation, negligence, detecting illegal tax and any dangers of frauds (Kunwar 2019).

Detecting frauds start by using the algorithms to analysis characteristics data available in the database of the institutes said (Tadapaneni 2020), so it decreases the number of wrong rejections which will safely time.

Adding to that one of the advantages of AI is sending alert to investigate about threats by using fraud detecting system which banks uses (Pothumsetty 2020), (Buchanan 2019) further added that banks using one of the most effective application which is the fraud detecting to detect credit card frauds, they use algorithm to test and validate the data available regarding the credit card and do the comparison with the payment data than label which is fraud or non-fraud transactions so it can be alert and investigate about it, Such system have been use by Mastercard. (Suhel et al. 2020) explain that in 2016 Mastercard have launched the "Decision Intelligence" service, this tool uses algorithms to provide a proactive result to the issuer of the card through an intelligent analysis process. After that, this information will be included in the existing standard transaction then the credit card issuers can activate the Mastercard comprehensive tool, enabling real-time decision-making based on available and specially designed account data, including setting alert criteria and rejecting transactions which will revel fraud and non-fraud transactions. They continued that yearly 118 billion US dollar cost from decline transactions but by applying the decision intelligence services it will reduce operation cost which will reduce risk and increase revenues for retailer and banks, below a figure shows how Mastercard decision intelligence works.

3.1.2 Chatbots

One of the most used AI tools in the financial institutes is the chatbots, the recent years the demand of financial services has increased, therefore in order to serve clients in an efficient manner, the financial institutes start to implement chatbots (Bhasin 2019), this tool to help customers with their needs by providing answers right away to them around 24/7, with the ability to answer big number at the same time (Zavadskaya 2017).

(Latimore 2018) defined this tool as a technology that use algorithms interact with customers by text. It works by answering questions or requests, where it use AI to enter the databases available in the institutes to respond for customers request and answer it (Suhel et al. 2020), in addition (Patil and Kulkarni 2019) agreed that chatbots is conversational software where it use AI to create a platform that enable to answer certain type of questions requested by customers. Additionally, (Fernandez 2019) stated that using chatbots to answer client in banking services, investment and policies in term of insurance that will save time for the financial institutes and give the time to focus in fixing different issues, so they enhance customer support, because it helps in cards, payments, passwords, loans and dealing with frequent answers.

(Dupont et al. 2018) mentioned that the value of chatbots is to reduce cost while it will enhance customers experience and will increase revenues to the institutes, (Buchanan 2019) agreed that financial institutes can benefit from chatbot in being more engaged with their customers, because customers will reach the institute faster by chatbots answering their requests in 24/7 means they won't need weekends, holidays or leaves so whenever customer need an answer anytime they can ask it with no waiting time this will increase the brand loyalty and image,more to add (Bhasin 2019) stated that chatbots can save lots of cost because there is no need for hiring employees this will reduce cost of human resources and will be no cost for training them which will lead to saving in costs. (Mubark 2019) supported that chatbots save costs because its one-time investment just needs developing and maintenance over the time this will decrease the need of hiring and increase the institutes margins.

However (Valmohammadi and Beladpas 2014) argue that engaging of employee especially the front line staff may impact customer satisfaction and this was backup by (Kunwar 2019)who argue that in order the financial institutes provides high quality services the communication between the institute and the customer is very important, (Bhasin 2019) stated that financial institutes cant completely depend on chatbots, as they may not have sometime the answer for all questions and these answers needs human intervention, therefore some institutes will complete it with employees to solve unknow questions to increase its customer satisfaction (Bhatti 2019).

In some institutes the chatbot can be given a persona, means they might name it, avatar or personalize it to let the customers feel they are interacting with humans.

An example of implementing chatbots in banking is Erica a chatbot used in bank of America (Buchanan 2019) they launched Erica a chatbot and one of the most effective applications in the bank, it provides daily services to clients in:

- Paying bills, plan payments and deposits.
- Services in respond to account information like account number, transfers checking account details, sending money and loan details.
- View proactive insights
- Sending and receiving money.

He continued that Erica was built to enhance client services and have the ability to help them to understand all about mobile banking.

Bank of America has announced that since launching Erica in 2018 over 150,000 users tap on Erica insight each week.

More an Example implementing chatbot in financial institutes is in Lemonade, it is a B2C website offer property and casualty insurance to its customers, where they use chatbot to insure services and get paid from them (Kunwar 2019).

If we will look for examples there are many examples, but last is Cleo which is used in capital management, Cleo is a digital assistant that replace banking applications, it connects banks card to customers so they can manage their budgets and save money by the help of Cleo (Suhel et al. 2020).

3.1.3 Algorithmic Trading

Another AI application tool use in financial institutes is algorithmic trading or some call it stock trading, this tool use AI and implement it in a program to be used in trading that allow financial institutes to make better trading decisions by monitoring the results and news in the real time to discover ways to enable stock prices to rise or fall (Kunwar 2019), moreover (Buchanan 2019) added that algorithmic trading can analysis hundreds of data sources in one time to know about the changes in market conditions in term of price level and predict what happen, however (Tadapaneni 2020) mentions this system needs high level of organizing check to operate since it include flash crashes.

According to (Zavadskaya 2017) the use of algorithmic trading increased after introducing computers trading system to the financial market during the 1970s.

It use three types of strategies, first the performance based strategy where the investors will use it making many purchases as it focus on execution based plans, second the statistical arbitrage strategy in this strategy it will identify the cost different within different markets and can dispense, as for the last strategy it is the momentum investment strategy which focus on the business movements, this strategy helps traders to predict the market based on the actual actions (Pothumsetty 2020).

(Kunwar 2019) stated that the benefit of algorithmic trading are to increase the accuracy and reduce the chance of errors, in addition it will reduce human errors, more algorithmic trading allows trades to be done with the possible rate due to its accuracy and how it is able automatic and simultanly checking multiple markets conditions, (Buchanan 2019) agreed to that and reduce mistakes as there will be less human interaction this mean less emotion errors, more it will be executing faster with the best possible prices not to forget with the ability to check more than one market at the same time.

Regardless to it advantages and benefits (Reznik and Pankratova 2018) stated that there are some disadvantages needed to mention about the algorithmic trading, such as technical errors, the algorithmic trading is a complicated process the order made by computer not a server, which means any internet disconnect will lead to not updating the market with orders, this mean it needs for monitoring by employees to avoid any mechanical failure due to disconnect problems, power loss or any other technical errors.

3.1.4 Anti Money Laundering (AML)

First Money laundering means owing illegal money the purpose of it to be owned or used or invested and exchange in in illegal terms (Kunwar 2019), agreed to that (Buchanan 2019) It is an illegal operation that aims to make the funds produced from criminal activity - such as drug trafficking and terrorist financing, they have strategies and thoughts on how do they use it or invest it, this make it hard to track and monitor, therefore an Anti-money laundering (AML) system needed to help in detecting suspicious activity, (Pothumsetty 2020) stated banks are increasingly turning to artificial intelligence as a new weapon to aid in against financial crime, for example identifying fraudsters serving the system. But (Fernandez 2019) stated as financial institutions incorporate more AI into their anti-money laundering and compliance systems, criminals will undoubtedly look to AI systems likewise to cover their tracks. The result is a potential AI arms race.

(Mardanghom and Sandal 2019) stated that the ML activity overall is between 2% to %% from the world global GDP, with the resources available in the financial institutes to fight money laundering this approach is not enough in producing results.

Artificial intelligence can help banks they use two ways one based on using the big data and the other use AI capabilities in recognizing patterns, some banks prefer to use the data as it is based on the transactions and details information available about the customers with their business activities this will contribute to the effort of detecting money laundering (Vedapradha and Ravi 2018).

He continued that AML is a challenge that face financial institutes services, but the technology of AI can help in preventing the activity.

Impact of crisis on Finance Institutes:
As mention above regarding crisis that impact our world in this literature review, we will be focusing on a recent crisis which is covid-19 pandemic that is still affecting our world since December 2019. It starts to affect many sectors (social, economic, financial and health), during the pandemic there were decrease in bank transactions, decrease in payments and massive decrease in using ATM cash machines which affect the profit, although there was an increase on online services and increase in digitalization. The impact of Covid-19 also affected the employees, customers, liquidity and communications of the financial institutes.

3.1.5 Liquidity

Since the continues of spreading of covid-19 virus countries try to search for solutions to save the economy from the repercussions of this epidemic, therefor many countries rushed to follow policies that suit them, where some of them their central banks used the monetary policy to confront the crisis. (Melamedov 2020).

In order to support the level of liquidity that were affected (Liddy 2020) stated that IMF reveals that the majority of central banks and financial institutes restored to reducing monetary interest rate.

In addition, (Yerchuru and Chidambaram 2020) cleared that despite to the general agreement on monetary policy to confront the crisis, the use of it varied between lowering interest rate and employing open market operations, while others resorted to use the legal cash reserve ratio to increase the ability of banks to grant financing.

They continued countries that have high monetary reserves have follow incentive policies at the level of monetary policy to support it domestic economy demand, so it can enhance the liquidity of banking sector and encourage commercial banking to finance the private sector and small and medium entrepreneurs.

More to add many precautions Meuser and emergency preparations have been taken in the financial institutes to mitigate the effect of the spread of the virus on financial market, therefor some have approved to restructure or postponed the financial provided to client without additional fees, as well as providing the necessary financing to private sector clients who have lost their jobs (Billio and Varotto 2020).

In addition, (Kapan and Minoiu 2021) stated some countries have launched incent programs that aims to support the private sector and enabling it to play it role in promoting economic growth and empowering the sector, they continued some central banks have restored to set a temporary daily limit on cash withdrawal and deposit at banks branches this aims to reduce the risk of spreading of the virus. Supporting that (Liddy 2020) stated governments reduce mandatory reserves on bank deposits and made agreement with banks to payback for up to a year and offer financial packaging to set off the liquidity pounding.

3.1.6 Customers and Communications

With the spread of covid-19 virus countries tend to lock down financial institutes stop providing it services directly to the customers in their places, this let them experience a decrease in demand, therefor, many institutes created an alternative way to reach and communicate with their customers, they tend to put strategies and plans to ensure they deliver the services to everyone during the pandemic (Rumiyati and Syafarudin 2021).

Digital banking was one of the ways banks could communicate with their clients and customers, they had to make sure that customers are satisfied and feel comfortable completing their needs like old days (Ramasamy 2020), other institutes should take in consideration transferring to provide it services online like banks to satisfy its customers.

A survey made by Captech in March 2020 to understand the impact of covid-19 on customers in financial institutes to find new opportunities and ways to engage with their customers. The result of the survey shows that the impact of covid-19 on customers within ages 38 to 50 were higher, they are the most impacted in receiving financial services. More to add the pandemic impacted more customers with incomes between 75,000 to 99,999 dollars, but this didn't change their financial decision on receiving the services from their financial institutes.

In addition, the survey conducts the way customers preferred to receive their banking services during the pandemic, it shows almost 53% preferred completing their banking services via websites, where 44% prefer mobile application, even more the percentage of customers preferring to speak to an associate on phone or opening a meeting with them virtually have increased during the crisis.

This let many institutes forced to take in consideration finding new ways to communicate with their clients such like digital banking, because customers are satisfied with the services provided online during the covid-19 and are considering continuing and switch in receiving the services online after covid-19.

3.1.7 Employees

The one effect of Covid-19 that there were changes in the workplaces and worker (Liddy 2020), there been many changes will be explained below:

- Changes in workplace:
  The work routine has been changing rapidly in the covid-19 crisis, due to that and according to (Liddy 2020) there were acceleration to virtual and online work. A survey was done for 229 human resource department in a financial service showed that 80% of their employees work from home during the covid-19 crisis (Gartner 2020), this increases the productivity of the employees, on the other hand many employees face

many challenges especially if they were living with their family and they don't have enough space to attend the work virtually so they need to occupied other space.

- Management and virtual leadership:
  (Bapuji et al. 2020) stated that due to the covid-19 crisis millions of employees with different position require to work from home due to that leaders can work from home as well. In additional they stated that absence of physical attendance improves the relationship among the employees and the managers.
  According to (Stoker 2019) the leader style will be change in such crisis due to that leaders must be prepared especially in the assessment of the employees as they will not be able to monitor the staff directly. Moreover, according to (Vandenberghe et al. 2019) the lack of attendance face to face with the leaders will affect the staff learning opportunities from their leaders.
- Social Distance:
  As it was explained earlier working from home was one of the impacts of covid-19 as a consequence loneliness and loss of social interaction which affect the employee behavior and performance negatively. (Ozcelik and Barsade 2018), In additional online communication can cause misunderstanding.
  On the other hand, researchers show the social interaction and chatting between employees each other are essential for mental health. (Billio and Varotto 2020).

Therefore, governments stressed the importance of these institutes preservation on their employees because of the past crisis showed the importance of the pattern of intervention in the long run (Bapuji et al. 2020).

3.2 Implementing Artificial Intelligence in Financial Institutes Services During Crisis

Crisis face human long time ago, the world with every crisis try to invest it resources available in that time, now days the world is fighting against new crisis which is covid-19 virus, the virus explained above with no treatment available and a wat to stop the spreading of it, but not to forget we are in the era of the fourth revolution where it focus in technology, many institutes try to invest in resources related on technology specially in artificial intelligence, as with the lock down and services can't be provided in the institutes place therefore they have to let AI to help in continues the work and changing the way financial do.

Before covid-19 some financial institutes were able to adopt and use AI in their services, as (Anderson et al. 2021) mentioned that the financial institutes were the biggest sector in spending on AI tools in many services that help in engaging with clients, detect frauds and others as mentioned above.

Therefore, the use of AI during the pandemic have increased, but (Pugliese 2020) argued even though institutes implemented AI during covid-19, this crisis effected the performance of the institutes this led to a reduce in investing in new projects.

While (Gandham 2020) stated that with the lock down finance institutes came up with strategies to adopt and start to implement AI in their services that will help them provide services to satisfy their customers.

He continued that this pandemic will also faster the Neo banking which is a new concept of banking uses AI tools to analysis and understand the clients in term of their expanses, so it helps them to manage their investment, wealth and other.

The recent pandemic and the lack of research and information provided in implementing AI in financial institutes services drive us to the outcome of a webinar hosted by bank of England on the 10th of August 2020 with panel of experts in finance and AI and a number of journals.

Before Covid-19 AI were implemented in many financial institutes and using it during the pandemic have shifted the market for digital banking for example it have been working in the pattern and consumer behaviors which drives down the cost of spending, increase productivity and enhance client interface (Gharbawi), agreed to him (Goh) she stated that during Covid-19 people were concerned about the spreading of the virus and were quarantined so they tend to focus on digitalization and using contactless interactions which have increased the amount of online transactions and generated huge amount of data that will help in increasing the efficiency and effectiveness of the financial sector data base, therefore, AI have help in detecting frauds banks and have increase it level of activeness in using AI tools 40% more compering to it used before the pandemic.

As for (Moore) he stated that one of AI strength is that in the past financial institutes weren't convinced of using AI tools but now they start to immerge it quickly to the institutes to improve their clients experience, although the succussed of using AI depends on having a ready platform and processes in place.

He continued that one of the risks facing implementing AI during the pandemic is stretching the digital divided between banks had adopted to use AI and institutes which are behind, so banks are ahead will go further ahead than the other in behind.

Experts concluded the webinar saying not knowing how long the Covid-19 will last it will lead financial institutes to hold back on developing new AI applications that will let them reduce investment in new actions.

Existing studies have explained how financial institutes benefit from AI technology (Aberg and Khali 2018). On the other hand, the implementation of advance AI is new and previous research of using this technology during crisis is limited.

4 Conclusion

It's too early to know how AI will be implemented in financial institutes and its services during Covid-19 as the pandemic is still recent so overcoming the result will take time, much more employees will need to be more educated about AI tools and awareness to have the advantage in implementing it in solving issues caused because of the pandemic, this might lead in future to relaying on AI much more then these days, Although it can be concluded that implementing AI in financial institutes can enhance it productivity and efficiency in term of its services and can help the institutes to overcome the difficulty cause of Covid-19 pandemic, although many challenges and opportunities face implementing AI during and post Covid-19 pandemic. In addition, this literature review didn't focus on immediate actions as this crisis will leads to profound and lasting changes. Therefore, further studies needed to focus on actions in the short and long run that will stronger the financial institutes after the pandemic.

References

Anderson, J., Bholta, D., Gharbawi, M., Thew, O.: The impact of COVID-19 on artificail intelligence in banking (2021)

Bapuji, H., Ertug, G., Shaw, J.: Organization and societal economic ineuality: a review ans way forward. Acad. Manag. Ann. 60–91 (2020)

Bhasin, R.: Chatbots for financial inclusion. Int. J. Adv. Res. Dev. 14–19 (2019)

Bhatti, A.: Exploring the adoption of AI in the finance industry the case of chatbot in kenya (2019)

Bhatti, A.: Exploring the adoption of artificial intelligence in the finance industry the case of chatbot in kenya (2019)

Billio, M., Varotto, S.: A new world post covid 19 lessons for business, the finance industry and policy makers (2020)

Buchanan, B.: Artificial intelligence in finance (2019)

Dupont, L., Fliche, O., Yang, S.: Governance of Artificial Intelligence in Finance (2018)

Fernandez, A.: Artificial intelligence in finance services. Econ. Bull. Anal. Artic. (2019)

Gharbawi, M., Goh, L., Jain, C., Moore, A.: Artificial intelligence, financial services and the impact of Covid-19, Webinar – Bank of England

Gandham, S.: How data and AI utilisation is not the same in banking after covid-19 (2020)

Gartner, M.: Gartner HR survey reveals 41% of employees likely to work remotely at least some of time post coronavirus pandemic (2020)

Haleem, A., Javaid, M.: Effects of COVID-19 pandemic in daily life (2020)

Kapan, T., Minoiu, C.: Liquidity insurance Vs. Credit Provision Evidence from the covid 19 crisis (2021)

Kunwar, M.: Artificial intelligence in finance: understanding how automation and machine learning is transforming the financial industry (2019)

Latimore, D.: Artificail intelligence in banking (2018)

Liddy, J.: Six considerations in dealing with the impact of Covid-19: a view on the impact for financial institutes. KPMG (2020)

Mardanghom, R., Sandal, H.: Artificial intelligence in financail services (2019)

McCarthy, J.: What is arificail intelligence (2004)

Meiring, G., Myburgh, H.: A review of intelligent driving style analysis systems and related artificial intelligence algorithms (2015)

Melamedov, L.: Coronavirus (Covid-19) and the banking industry: impact and solutions (2020)

Mubark, M.: Alarming Influence of AI and chatbot in the banking and finance industry (2019)

Nilsson, N.J.: Principles of Artificial Intelligence. Morgan Kaufmann, Burlington (2013)

Ozcelik, H., Barsade, S.: No employee an island: workplace loneliness and employee performance. Acad. Manag. J. 2343–2366 (2018)

Patil, K., Kulkarni, M.: Artificail intelligence in financail services: customer chatbot advisor adoption. Int. J. Innov. Technol. Explor. Eng. 4296–4303 (2019)

Pothumsetty, R.: Implementation of artificial intelligence and machine learning in financial services. Int. J. Eng. Res. Adv. Technol. (2020)

Pugliese, A.: The impact of AI (2020)

Ramasamy, K.: Impact analysis in banking, insurance and financial services industry due to COVID-19 pandemic. Pramana Res. J. 19–29 (2020)

Reznik, N., Pankratova, L.: High frequency trade as a component of algorithmic trading: market consequences. National University of Life and Environmental Science (2018)

Rumiyati, R., Syafarudin, A.: The influence of service quality, marketing mix, on bank customer satisfaction in the era Covid-19. Ilomata Int. J. Tax Account. (IJTC) 84–96 (2021)

Sion, G.: Participating in a highly automated society: how artificial intelligence disrupts the job market. J. Self-Gov. Manag. Econ. (2018)

Stoker, G.: Tightening the leash after a threat: a multi level event study on leadership behavior following the financial crisis. Leadersh. Q. (2019)

Suhel, S., Shukla, V., Vyas, S., Mishra, V.: Conversation to Automation in banking through chatbot using artificail machine Intelligence language. In: Infocom Technologies and Optimization, pp. 611–618 (2020)

Tadapaneni, N.: Artificial intelligence in finance and investment. Int. J. Eng. Res. Adv. Technol. (2020)

Valmohammadi, C., Beladpas, M.: Customer relationship management and service quality a survay within the banking sector. Ind. Commer. Train. 77–83 (2014)

Vandenberghe, L., Bentein, A., Mignonac, R.: A dynamic model ofthe effects of feedback seeking behavior and organizational commitment on newcomer turnover. J. Manag. (2019)

Vedapradha, R., Ravi, H.: Application of artificial intelligence in investment banks. Rev. Econ. Bus. 131–136 (2018)

Yerchuru, S., Chidambaram, M.: The COVID-19 crisis: implications for the financial services industry (2020)

Zavadskaya, A.: Artifical intelleigence in finance: forcasting stock market returns using artificail neural network. Hanken school of economic (2017)

Marketing, E-commerce and Digitalization

Customer Resource Integration in Virtual Brand Communities: Conceptual Framework

Muhammad Dharma Tuah Putra Nasution(✉), Endang Sulistya Rini, Beby Karina Fawzeea Sembiring, and Amlys Syahputra Silalahi

Faculty of Economics and Business, Universitas Sumatera Utara, Medan, North Sumatera, Indonesia
dharmatuah@gmail.com, {endang.sulistya,beby}@usu.ac.id

Abstract. Although social media has been recognized to facilitate collaborative value creation, there are differences in perspectives on how to understand the process of integrating customer resources into the brand community on social media. To understand this perspective in depth, this study considers it necessary to identify the process of integration in virtual brand communities. Therefore, the purpose of this study is to explore and analyze the configuration of the resource integration process in the brand community on social media. A content analysis is used to draw the conceptual framework in this study. The proposed study assumed, first, how to configure the process of integrating customer resources into the virtual brand community on social media, second, the effectiveness of customer interaction during the integration process into the virtual brand community.

Keywords: Resource integration · Mutually beneficial interaction · Social media brand community · Customer social participation

1 Introduction

Although research on resource integration from the perspective of S-D Logic is in the early stages of development, integration of resources has been highlighted in various areas of marketing literature. Even if not explicitly stated, most studies characterize resource integration as a precondition for value creation (Edvardsson and Tronvoll 2013; Lusch and Webster 2011) and a requirement for service exchange (Vargo and Lusch 2008). There are suggestions from scholars to expand the concept of resource integration in order to construct a robust theory (Peters et al. 2014) and as practical approaches to configure the resource integration process (Korkman et al. 2010; Kleinaltenkamp et al. 2012). Regardless of how common resource integration is in business settings, theoretical constructions need a pragmatic approach to resource integration practices and integration process configurations (Kleinaltenkamp et al. 2012).

Kleinaltenkamp and colleagues assert that integrators of resources will use collaborative mechanisms to generate value that is defined phenomenologically (Kleinaltenkamp et al. 2012). As technological and Internet advancements facilitate collaboration among

B. Alareeni et al. (Eds.): ICBT 2022, LNNS 620, pp. 659–666, 2023.
https://doi.org/10.1007/978-3-031-26953-0_60

actors, research interest in resource integration processes and collaborative value creation mechanisms will expand. Interactive technology enables the application of a wide range of social interaction techniques and methods to stimulate customer participation and involves the corporate network in collaborating from idea generation to launch (McAlexander et al. 2003; Sawhney et al. 2005; Schau et al. 2009; Korkman et al. 2010).

Ironically, not every business is eager to integrate resources with their customers to collaboratively create value (Saarijärvi 2012), since this strategy could be a double-edged sword in some instances. Co-created service failures may result in negative confirmations and increased customer dissatisfaction (Edvardsson and Tronvoll 2013; Caridà et al. 2015). Edvardsson and his colleagues discuss that resource integration refers to the process in which actors integrate and utilize resources. The mechanism consists of phases in which actors engage cooperatively and collaboratively. The steps provide actors with experience and reciprocal behavioral outcomes (Caridà et al. 2015).

Existing work on resource integration is relatively abstract and presents few practical implications for practitioners to achieve its potential value. Therefore, the problems with this study highlight two points. First, the process of integrating resources into a brand community is highly reliant on its nature: virtual or non-virtual, commercial or social, including who handles the platform whether company or community, or even social media influencers. Second, customer interaction inside the brand community will result in a different integration process, depending on the context of usage. Thus, the focus of research is on the integration of resources through a theoretical framework and a practical lens. The specific objectives of this study include first identifying and exploring the activity of integrating resources into social media brand communities. Second, analyze the link between social participation and customer interaction in the resource integration process.

2 Literature Review

2.1 Resource Integration

Themes related to resource allocation and transfer among many customers in S-D Logic research are still in the early stages of development. This study is proposed to contribute to the development of the limited literature on the dynamic process of integrators (customers) who collaborate and integrate their resources to create value as explicit concepts (Caridà et al. 2015).

Although "recursive processes" cannot be ensured without the management of integrated resources (Caridà et al. 2019), studies on the comprehensive nature of integrated resources and the mechanisms of coordination and modification of resource integrators' activities in relation to one another that are considered insufficient (McColl-Kennedy et al. 2012; Saarijärvi 2012; Caridà et al. 2019).

In the S-D Logic perspective (Vargo and Lusch 2008), resource integration reflects a basic condition of pre-service exchange and collaborative value creation processes. The outcomes of activities and interactions in which resources are integrated and collectively created, as well as valued in use, will determine the value creation (Gummerus 2013; Laamanen and Skålén 2015). S-D Logic's focus on the collaborative dimension to create value will get the attention of business and social actors who utilize networks to communicate and share resources (Vargo and Lusch 2008; Laamanen and Skålén 2015). These capture the essence of resource mobilization and utilization (Kleinaltenkamp et al. 2012), which is thus important for the development of initiatives involving resource integration.

The integration of such resources occurs when actors allocate their resources into the mechanisms of other collaborators in line with their expectations and competencies. This will mean that there are social and cultural reasons for actors to become members of a network (Gummesson and Mele 2010). From such a point of view, this is an ongoing process that links a set of activities taken by an actor to a set of particular interactions among actors with specific resources (Payne et al. 2008; Peters et al. 2014; Ballantyne and Varey 2006).

Specifically, interactions enable collaborators to acquire more resources, thereby creating new resources that are transferred via integration (Vargo and Lusch 2011). In addition, collaborators will allocate their expertise and other resources to unite interconnected members of a business that share a common set of competencies via interaction and integration. These will provide value innovation in the form of a new value proposition and facilitate higher levels of collaborative value creation (Mele et al. 2010).

Two-way interaction and mutually beneficial utilization of resources are essential parts of the collaborative value creation mechanism (Vargo and Lusch 2008), which first and foremost requires actors to coordinate their resources dynamically. Löbler (2013) contends that resources do not only exist but must also become existent, and resources can be seen as dynamic concepts that are generated and reconfigured via the process of resource integration activity (Pels et al. 2009). The extent to which particular potential resources from one resource are utilized is reliant on the availability of other potential resources from other resources, and the extent to which resource utilization depends on the capability of beneficiaries to integrate all of them (Vargo and Lusch 2011). Resources therefore become existing objects or concepts until they are integrated via interaction during the implementation of such an activity (Löbler 2013).

2.2 Mutually Beneficial Interaction

Enterprise and customer relationships (B2B; B2C; C2B) have been generalized to relationships between actors (A2A) (Gummesson and Mele 2010). Also mentioned in the concept of "many-to-many marketing," C2C interactions are part of a network (A2A) whose position is outside the company (Gummesson 2008). Although research on C2C is generally considered information from word-of-mouth, the presence of social media has led to the expansion of new research on C2C interactions for collective value creation (Gummesson and Mele 2010).

There are two main stages that underlie the entire process of collective value creation. First, interactions are conducted to evaluate each other's resources, competencies, and processes that underlie the integrator's capability to contribute to value creation (Gummesson and Mele 2010). The integrators involved organize dialogue and transfer knowledge and other resources for organizational learning and resource creation and innovation. Second is the resource integration stage, in which integrators evaluate the operant resources they have and what they can do (Mele et al. 2010).

Resource integration is generalized to actor-to-actor interactions in which they link mutually beneficial resources. In the process of integration, resources that differ in quality and quantity need to be adjusted for one another. When procedures, understanding, and brand community engagement are aligned, collaborative value creation will occur.

2.3 Customer Social Participation

According to the literature, co-creation is meaningless without customer participation (Bharti et al. 2014). Therefore, customers will play an active part in the process of value generation. Customer participation is the extent to which customers contribute resources so long as they interact with the firm (Bharti et al. 2014; Chan et al. 2010) or with other customers in different settings (Pandey and Kumar 2020). The active participation of customers in activities and inputs beyond consumption, such as the search for and sharing of knowledge, will be a crucial element for the success of value-sharing (Yi and Gong 2013). Other studies demonstrate that customer participation may be defined as the extent to which customers exchange information, provide recommendations, and participate in collective decision-making that reflects the customers' efforts to produce collaborative creations (Chen and Chen 2017).

In the context of digital platforms, social media facilitates customer participation as an integral part of the co-creation process (Ramaswamy and Ozcan 2016), in which customers as individuals and members of a community can co-create value with other customers interconnected on social media platforms (Carlson et al. 2019). In the brand community on social media, for instance, customers share information about the brand, support decisions about the need for product development opportunities and enhanced brand experience, as well as produce content with other customers (Gensler et al. 2013).

Despite the literature highlighting customer participation in value creation that has validated its measurement (Yi and Gong 2013), in the context of social media, customer participation is still conceptually limited, and relatively few have empirical results (Khan 2017). Therefore, the study of Kamboj and Sarmah (2018) has attempted to validate customer participation by adopting the definition of customer social participation from Chae and Ko (2016), which they divided into three dimensions. Chae and Ko (2016) contend that customer social participation in social media indicates an effort to co-create values through interactive participation in the generation and delivery of services. Accordingly, in the framework of this study, the operational definition of customer social participation has been broadened and related to earlier research (Chae and Ko 2016). Customer social participation reflects an initiative to achieve collective value on social media through interactive participation during interaction and integration activities.

2.4 Brand Community on Social Media Platforms

Social media has been well documented in many significant events in human history, including the growth of enterprises and brands (Kaplan and Haenlein 2010). This evolution of ICT has encouraged social interaction between companies and their networks and led to the development of collaborative platforms (Camarinha-Matos 2009). For instance, a virtual brand community requires strategic and relational alignment as well as member participation in order to achieve collective value outcomes, and a social media brand community facilitates such customer participation by allowing them to share their knowledge, experience, and opinions regarding brands or services as well as the creation and collaborative exchange of content that can represent the process of collective value creation (Abeza et al. 2020). There is a potential that participation in virtual brand communities may enhance a company's chances of gaining access to resources (additional knowledge and ideas) provided by its members in order to generate new combinations of knowledge elements that are unique and valuable (Colurcio et al. 2012; Kristensson et al. 2004).

According to a managerial standpoint, the brand community is a catalyst for collaborative value creation (Skålén et al. 2015). In this model, members of the brand community act as both providers and beneficiaries to create values and benefits for every collaborator engaged inside the network ecosystem such as for themselves, the brand community, and the company (Pongsakornrungsilp and Schroeder 2011). Although collective value creation in the realm of virtual brand communities has been investigated (Schau et al. 2009; Skålén et al. 2015; Pongsakornrungsilp and Schroeder 2011; Rowley et al. 2007), it remains unclear how firms should be involved in collaborations, and this is necessary to make it operate. Empirical analysis reveals that business endeavors to interact with virtual communities are often ineffective (Schröder and Hölzle 2010), and it will become a challenge to build and sustain effective online brand communities.

With a pragmatic approach, academicians and scholars have sought to define the process of collaborative value creation in the brand community (Schau et al. 2009; Skålén et al. 2015; Russo-Spena and Mele 2012). By focusing on how things happen and what the consequences are, this practical approach has become a central approach to studying the process of collaborative value creation in brand communities (Schau et al. 2009; Skålén et al. 2015; Russo-Spena and Mele 2012).

Schau and colleagues provide a series of collective value-creation practices. There is a common perspective that connects these activities: behavior, appearance, and representation. It addresses three primary elements: procedure, understanding, and participation (Schau et al. 2009). Co-creation, as articulated by Russo and colleagues, reflects a collection of practices at each step of the innovation process and, more precisely, a sequence of activities undertaken by actors by integrating specialized resources (Russo-Spena and Mele 2012) (Fig. 1).

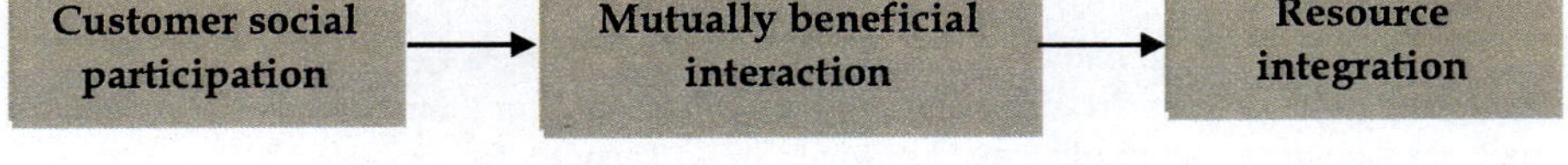

Fig. 1. Conceptual framework

The hypotheses developed and proposed are as follows.

H1: Customer social participation will have a positive and significant effect on mutually beneficial interactions in the brand community on social media.

H2: Mutually beneficial interactions will have a positive and significant effect on the integration of resources in the brand community on social media.

H3: Mutually beneficial interactions will mediate customer social participation relationships and resource integration in the brand community on social media.

3 Research Methods

This study used a content analysis which derived from various articles and past researches in order to develop a framework. A number of referred articles justify drawing and proposing the above concept.

4 Conclusion

This study aims to overcome the contextual gap in terms of the process of integrating resources into virtual brand communities. The study investigates the relationship of customer social participation as a resource allocator; mutually beneficial interaction as a medium to transfer resources; and resource integration or value-in-context. Overall, the study will provide a better understanding of the resource integration activities of actors in virtual brand communities.

This study's empirical results could be valuable for marketing professionals, academicians, and enterprises who collaborate with brand communities on social media. These insights will inform stakeholders on how to optimize for a more favorable virtual brand community and lead to brand co-creation. The practical contribution of research may provide brands engaged in the process of online interaction with appropriate ideas for establishing and managing their initiatives and enhancing brand equity. The authors will conduct empirical validation of the proposed framework as part of continued studies.

Acknowledgment. We would like to thank you for the Doctoral Dissertation Research Grant to the Ministry of Education and Culture and Technology Research of Indonesia in 2022.

References

Abeza, G., O'Reilly, N., Finch, D., Séguin, B., Nadeau, J.: The role of social media in the co-creation of value in relationship marketing: a multi-domain study. J. Strateg. Mark. **28**(6), 472–493 (2020)

Babin, B.J., Zikmund, W.G.: Exploring Marketing Research. Cengage Learning, Boston (2015)

Ballantyne, D., Varey, R.J.: Creating value-in-use through marketing interaction: the exchange logic of relating, communicating and knowing. Mark. Theory **6**(3), 335–348 (2006)

Bharti, K., Agrawal, R., Sharma, V.: What drives the customer of world's largest market to participate in value co-creation? Mark. Intell. Plan. **32**(4), 413–435 (2014)

Camarinha-Matos, L.M.: Collaborative networked organizations: status and trends in manufacturing. Annu. Rev. Control. **33**(2), 199–208 (2009)

Caridà, A., Edvardsson, B., Colurcio, M.: Modes of resource integration driving value co-creation: an empirical investigation in virtual brand communities (VBCs). In: Proceedings of Naples Forum on Service, pp. 9–12, June 2015

Caridà, A., Edvardsson, B., Colurcio, M.: Conceptualizing resource integration as an embedded process: matching, resourcing and valuing. Mark. Theory **19**(1), 65–84 (2019)

Carlson, J., Wyllie, J., Rahman, M.M., Voola, R.: Enhancing brand relationship performance through customer participation and value creation in social media brand communities. J. Retail. Consum. Serv. **50**, 333–341 (2019)
Chae, H., Ko, E.: Customer social participation in the social networking services and its impact upon the customer equity of global fashion brands. J. Bus. Res. **69**(9), 3804–3812 (2016)
Chan, K.W., Yim, C.K., Lam, S.S.: Is customer participation in value creation a double-edged sword? Evidence from professional financial services across cultures. J. Mark. **74**(3), 48–64 (2010)
Chen, C.C.V., Chen, C.J.: The role of customer participation for enhancing repurchase intention. Manag. Decis. **55**(3), 547–562 (2017)
Colurcio, M., Tregua, M., Melia, M., Caridà, A.: Social networks for outsourcing and developing a firm's creativity. In: Advances in the Human Side of Service Engineering, pp. 62–71. CRC Press (2012)
Edvardsson, B., Tronvoll, B.: A new conceptualization of service innovation grounded in S-D logic and service systems. Int. J. Qual. Serv. Sci. **5**(1), 19–31 (2013)
Gensler, S., Völckner, F., Liu-Thompkins, Y., Wiertz, C.: Managing brands in the social media environment. J. Interact. Mark. **27**(4), 242–256 (2013)
Gummerus, J.: Value creation processes and value outcomes in marketing theory: strangers or siblings? Mark. Theory **13**(1), 19–46 (2013)
Gummesson, E., Mele, C.: Marketing as value co-creation through network interaction and resource integration. J. Bus. Mark. Manag. **4**(4), 181–198 (2010)
Gummesson, E.: Extending the service-dominant logic: from customer centricity to balanced centricity. J. Acad. Mark. Sci. **36**(1), 15–17 (2008)
Hair, J.F., Jr., Sarstedt, M., Hopkins, L., Kuppelwieser, V.G.: Partial least squares structural equation modeling (PLS-SEM): an emerging tool in business research. Eur. Bus. Rev. **26**(2), 106–121 (2014)
Hair, J.F., Sarstedt, M., Ringle, C.M., Mena, J.A.: An assessment of the use of partial least squares structural equation modeling in marketing research. J. Acad. Mark. Sci. **40**(3), 414–433 (2012)
Kamboj, S., Sarmah, B.: Construction and validation of the customer social participation in brand communities scale. Internet Res. **28**(1), 46–73 (2018)
Kamboj, S., Sarmah, B., Gupta, S., Dwivedi, Y.: Examining branding co-creation in brand communities on social media: applying the paradigm of stimulus-organism-response. Int. J. Inf. Manage. **39**, 169–185 (2018)
Kaplan, A.M., Haenlein, M.: Users of the world, unite! The challenges and opportunities of social media. Bus. Horiz. **53**(1), 59–68 (2010)
Khan, M.L.: Social media engagement: what motivates user participation and consumption on YouTube? Comput. Hum. Behav. **66**, 236–247 (2017)
Kleinaltenkamp, M., Brodie, R.J., Frow, P., Hughes, T., Peters, L.D., Woratschek, H.: Resource integration. Mark. Theory **12**(2), 201–205 (2012)
Korkman, O., Storbacka, K., Harald, B.: Practices as markets: value co-creation in e-invoicing. Australas. Mark. J. (AMJ) **18**(4), 236–247 (2010)
Kristensson, P., Gustafsson, A., Archer, T.: Harnessing the creative potential among users. J. Prod. Innov. Manag. **21**(1), 4–14 (2004)
Laamanen, M., Skålén, P.: Collective–conflictual value co-creation: a strategic action field approach. Mark. Theory **15**(3), 381–400 (2015)
Löbler, H., Hahn, M.: Measuring value-in-context from a service-dominant logic's perspective. In: Malhotra, N.K. (ed.) Review of Marketing Research, vol. 10, pp. 255–282. Emerald Group Publishing Limited, Bingley (2013)
Löbler, H.: Service-dominant networks: an evolution from the service-dominant logic perspective. J. Serv. Manag. **24**(4), 420–434 (2013)

Lusch, R.F., Webster, F.E., Jr.: A stakeholder-unifying, cocreation philosophy for marketing. J. Macromark. **31**(2), 129–134 (2011)
Malhotra, N., Nunan, D., Birks, D.: Marketing Research: An Applied Approach. Pearson (2017)
McAlexander, J.H., Kim, S.K., Roberts, S.D.: Loyalty: the influences of satisfaction and brand community integration. J. Market. Theory Practice **11**(4), 1–11 (2003)
McColl-Kennedy, J.R., Vargo, S.L., Dagger, T.S., Sweeney, J.C., Kasteren, Y.V.: Health care customer value cocreation practice styles. J. Serv. Res. **15**(4), 370–389 (2012)
Mele, C., Spena, T.R., Colurcio, M.: Co-creating value innovation through resource integration. Int. J. Qual. Serv. Sci. **2**(1), 60–78 (2010)
Nambisan, S., Baron, R.A.: Virtual customer environments: testing a model of voluntary participation in value co-creation activities. J. Prod. Innov. Manag. **26**(4), 388–406 (2009)
Pandey, S., Kumar, D.: Customer-to-customer value co-creation in different service settings. J. Cetacean Res. Manag. **23**(1), 123–143 (2020)
Payne, A.F., Storbacka, K., Frow, P.: Managing the co-creation of value. J. Acad. Mark. Sci. **36**(1), 83–96 (2008)
Pels, J., Möller, K., Saren, M.: Do we really understand business marketing? Getting beyond the RM and BM matrimony. J. Bus. Ind. Mark. **24**(5/6), 322–336 (2009)
Peters, L.D., et al.: Theorizing about resource integration through service-dominant logic. Mark. Theory **14**(3), 249–268 (2014)
Pongsakornrungsilp, S., Schroeder, J.E.: Understanding value co-creation in a co-consuming brand community. Mark. Theory **11**(3), 303–324 (2011)
Ramaswamy, V., Ozcan, K.: Brand value co-creation in a digitalized world: an integrative framework and research implications. Int. J. Res. Mark. **33**(1), 93–106 (2016)
Rowley, J., Kupiec-Teahan, B., Leeming, E.: Customer community and co-creation: a case study. Mark. Intell. Plan. **25**(2), 136–146 (2007)
Russo-Spena, T., Mele, C.: "Five Co-s" in innovating: a practice-based view. J. Serv. Manag. **23**(4), 527–553 (2012)
Saarijärvi, H.: The mechanisms of value co-creation. J. Strateg. Mark. **20**(5), 381–391 (2012)
Sawhney, M., Verona, G., Prandelli, E.: Collaborating to create: the Internet as a platform for customer engagement in product innovation. J. Interact. Mark. **19**(4), 4–17 (2005)
Schau, H.J., Muñiz, A.M., Jr., Arnould, E.J.: How brand community practices create value. J. Mark. **73**(5), 30–51 (2009)
Schröder, A., Hölzle, K.: Virtual communities for innovation: influence factors and impact on company innovation. Creat. Innov. Manag. **19**(3), 257–268 (2010)
Skålén, P., Pace, S., Cova, B.: Firm-brand community value co-creation as alignment of practices. Eur. J. Mark. **49**(3/4), 596–620 (2015)
Vargo, S.L., Lusch, R.F.: Service-dominant logic: continuing the evolution. J. Acad. Mark. Sci. **36**(1), 1–10 (2008)
Vargo, S.L., Lusch, R.F.: It's all B2B… and beyond: toward a systems perspective of the market. Ind. Mark. Manage. **40**(2), 181–187 (2011)
Yi, Y., Gong, T.: Customer value co-creation behavior: scale development and validation. J. Bus. Res. **66**(9), 1279–1284 (2013)

Which E-Wom Dimensions are More Likely Leading to Impulsive Buying on Online Travel Agent?

Hana Ulinnuha[1](✉), Weldy Lim Wirya[1], and Anastasia Bergita Andriani[2]

[1] Tourism Department, Faculty of Digital Communication and Hotel and Tourism, Bina Nusantara University, Jakarta 11480, Indonesia
hana.ulinn@gmail.com
[2] PT Global Tiket Network, Jakarta, Indonesia

Abstract. To increase the purchase of a tourism product, tourism marketers can consider e-WOM marketing strategies. This study aims to contribute the development of e-WOM strategies, which persuades people to make impulse purchases of tourism products from Online Travel Agent (OTA). With a total of 161 respondents, the research employs a quantitative approach with Likert scale questionnaire. In this study, the simple linear regression equation yields $Y = 13.100 + 0.547X$. According to the coefficient of determination test results, e-WOM has a 47.9% effect on impulse purchases of tourism products. Furthermore, four dimensions, including frequency of accessing information, frequency of interaction, product variation, and transaction & website condition, have a more significant impact on impulsive tourism purchase in OTAs, which tourism product providers in OTAs should prioritize.

Keywords: e-WOM · Impulsive buying · Tourism

1 Introduction

Industry 4.0 produces many innovative technologies, including artificial intelligence, big data, robot, financial technology, e-commerce, and e-marketing. In both services and products industry, digital technology was applied and implemented to ease production and marketing management, including tourism industry [1, 2].

The tourism industry cannot be separated from technological development. Online travel agent (OTA) is one of the examples of available technology which impacts the tour and travel business. There are a lot of OTA (online travel agents) currently with various types of tourism products offered. OTA offers various ways provides the platform for tourism service business to do, for example, marketing activities. One feature that has a significant advantage not only for improving aftersales services but also helps build brand image and gain competitiveness is the review feature. Buyers and visitors can give comments after purchasing and experiencing [3].

According to We Are Social's research in their 2021 report, Indonesia has become one of the ten countries with the longest duration of smartphone usage per day (3.2 h). Seeing

B. Alareeni et al. (Eds.): ICBT 2022, LNNS 620, pp. 667–675, 2023.
https://doi.org/10.1007/978-3-031-26953-0_61

as netizens can do a variety of daily activities with a single smartphone, including leaving comments and reviews on online platforms, Indonesians have become large potential customer base for online tourism businesses. The testimony gathered in the OTA or any review platform is now referred to as "e-wom," or electronic word of mouth. Such a length of use, directly or indirectly, supports the presence of e-Wom and presumably affect the consumer's impulsive buying activity.

Therefore, in this study, we collected 161 samples from adults over the age of 20. Who had done an impulsive buying activity caused by e-Wom provided by an online travel agent.

2 Literature Review

2.1 E-WOM

E-WOM is defined as a negative or positive statement about a product or company made by actual, potential, or previous customers that is uploaded to social media or a location where people or institutions can access it online [1]. So far, e-wom has been identified as a behavioral influencer in impulse purchases [2–4]. However, the goal of this research is to look into this phenomenon in the tourism context. The e-wom dimension developed by Husnain in Table 1 [4], has been employed in this study to examine the e-wom occurrence.

Table 1. E-wom dimension

Intensity	Frequency of accessing various types of information from various online media
	Frequency of interaction with other users in various online media
	The number of reviews written by the users in various online media
Opinion valence	Positive comments from users in various online media based on the products, seller and services offered
	Negative comments from users in various online media based on the products, seller and services offered
	User recommendations in various online media based on the products, seller and services offered
Content	Information related to various tourism products offered
	Information related to tourism product quality offered
	Information related based on the price for a tourism product offered
	Information-related transaction security

2.2 Impulsive Buying

People are frequently compelled to buy right away, which leads to impulsive purchases with hardly any regard for the consequences. The circumstances are called impulsive buying. Many scholars describe impulsive buying as one that is made without prior planning because consumers are interested in price discounts, sales promotions, and the presentation of unique or exciting goods, all of which pique the consumer's interest in purchasing [5, 6]. Tourist Impulsive Buying has already been discussed and linked to augmented reality [5], time pressure [7, 8], and emotion [9]. This study, on the other hand, seeks to identify the tourist impulsive purchasing attribution on e-wom.

Based on [5], impulsive buying influenced by seven main dimensions, including Urge to Purchase; Positive Affect; In-store Browsing; Shopping Enjoyment; Time Available; Money Available; Impulsive Buying Tendency. Table 2 explains the dimensions used in this study.

Table 2. Impulsive buying dimension

Urge to purchase	An urge or desire that someone feels when buying something suddenly or spontaneous is strong, sometimes irresistible and tends to act unexpectedly
Positive affect	Influences are Affected by pre-perceived mood and affective disposition, coupled with reactions to encounters in the store's environment (e.g., desired items and sales encounters)
In-store browsing	A vital component of the impulse buying process is store shopping. The longer the consumer browses the store, the consumer will tend to find more stimuli, which will increase the likelihood of the consumer making an impulse purchase
Shopping enjoyment	Shopping pleasure refers to the satisfaction obtained from the shopping process. In this case, it relates to the context of shopping in a mall or shopping centre. Some research suggests that impulse buying can be a person's attempt to relieve depression or to cheer themselves up
Time available	The time available for individuals to do shopping activities. Because time pressure can reduce impulse buying
Money available	Availability of money refers to the number of extra funds or the budget that must be spent when shopping. Money becomes a facilitator or liaison for purchasing an object
Impulsive buying tendency	Tendency to experience a sudden urge to make an on-the-spot purchase or an urge to act on that impulse with little consideration or evaluation of the consequences

3 Research Methods

This study employs a quantitative scientific method. The purpose of this study is to determine the available effect of e-Wom on impulsive purchasing of tourism products, and thus the hypothesis developed was e-wom activity has a significant impact on impulse buying of tourism products (H1).

In this study, information was gathered by distributing questionnaires containing several questions about the existence of e-Wom and impulse buying to the people who experienced impulsive buying in OTA.

4 Results

4.1 Profile of Respondents

Based on the data that has been collected, the profiles of the respondents in this study are as follows:

Table 3. Profile of respondents

Gender	Male	57%
	Female	43%
Age	20–35 y.o	84%
	>35 y.o	16%
Education level	High School	16%
	Diploma	12%
	University	72%
Monthly revenue	5-million-rupiah	28%
	5–10-million-rupiah	32%
	>10-million-rupiah	40%
Occupancy	Student/College Student	44.72%
	Staff	26.71%
	University Lecturer	1.86%
	Teacher	1.86%
	Security	1.24%
	Housewife	0.62%
	Entrepreneur	22.98%

According to Table 3, the respondents in this study were mostly Generation Z and Millennial men with a university education. All respondents had purchased impulsively in the past and were asked to complete a questionnaire about tourism product exposure in OTAs and impulsive purchasing. Table 4 displays the results as follows:

Table 4. e-WOM means

	Dimensions	Means
Intensity	Frequency of accessing information	4.33
	Frequency of interaction	4.24
	Number of reviews	4.365
Opinion valence	Positive comment	4.445
	Negative comment	4.41
	Recommendations	4.39
Content	Product variations	4.41
	Product quality	4.49
	Prices offered	4.465
	Transaction and website	4.335

According to Table 4, while reading the online travel agent discussion reviews, respondents are more frequently exposed to the content of the product, specifically the product quality. People, on the other hand, tend to buy impulsively because they have impulsive buying tendency. This is explained further in Table 5. The tendency to buy on impulse is recognized as a manifestation of general impulsiveness [10].

Table 5. Impulsive buying means

Urge to purchase	4.275
Positive comment	4.325
Shop-shopping	4.41
Shopping	4.415
Availability of time	4.4
Availability of money	4.3
Impulsive buying tendency	4.435

4.2 Instrument Test

The validity and reliability test was used to evaluate the instrument.

Validity Test. The validity test is carried out by examining the value of r arithmetic > r table. The r table value used in this study is 0.1646 and here is the calculated r-value:

- Variable X has a r count of 0.306**–0.558**
- Variable Y has a r count of 0.383**–0.627**

As a result, the research instrument was declared valid in this study. Valid.

Reliability Test. The reliability test was carried out to determine whether or not the research instrument was reliable. The Cronbach Alpha value must be greater than 0.6 for this test to be performed. The Cronbach Alpha value for the X variable is 0.785, and the Y variable is 0.736, indicating that the research instrument in this study is reliable.

4.3 Classical Assumption Test

The classical assumption test was conducted to determine whether this study's analysis requirements are met. The following is the classical assumption test used:

Normality Test. The normality test in this study was carried out using the Kolmogorov-Smirnov method with a value of 0.088 so that it is greater than 0.05, and the data in this study can be stated to be normally distributed.

Linearity Test. The linearity test in this study resulted in the number 1.806, which was greater than 0.05, so the two variables in this study were stated to be linearly related.

Heteroscedasticity Test. The heteroscedasticity test in this study used the Glejser test method, and the test result was 0.251, which was greater than 0.05, indicating that there were no symptoms of heteroscedasticity and that the study could proceed.

Multicollinearity Test. The multicollinearity test yielded a tolerance value of 0.603–0.823 and a VIF value of 1.215–1.660, indicating that no multicollinearity symptoms existed.

4.4 Hypothesis Testing

Simple Linear Regression Analysis. The results of simple linear regression in this study are $Y = 13.100 + (0.547) X$ So that means that every time there is an increase of one unit in the X (e-Wom), it means that there is an increase in the value of 0.547 for impulsive buying activity. And without the e-Wom, the impulsive buying activity has a constant value of 13.100.

Coefficient of Determination. E-wom based on the coefficient of determination has affecting the impulsive buying activity for 47.9% and it means that 52.1% is affected by other variables not examined in this study.

T-test. If the value of t table is smaller than t count, then H0 is rejected and H1 is accepted and vice versa if t count is smaller than t table then H0 is accepted and H1 is rejected. In this study, the t-count value is 12.099 with the t-table value of 0.676. So, in this study H0 was rejected and H1 was accepted: e-wom activity affects impulsive tourism buying on OTA.

Table 6. T-Test for each dimension of e-wom & impulsive buying

Model	t	Sig.	Explanation
Constant	12.099	.000	Significant
Frequency of accessing information	2.030	.044	Significant
Frequency of interaction	2.566	.011	Significant
Number of reviews	1.381	.169	Not significant
Positive comment	1.154	.250	Not significant
Negative comment	1.314	.191	Not significant
Recommendations	1.643	.103	Not significant
Product variations	2.274	.024	Significant
Product quality	1.367	.174	Not significant
Price offered	.280	.780	Not significant
Transaction and website	2.993	.003	Significant

This study examines the use of Instagram social media more deeply by looking at the results of the t-test for each dimension studied.

Table 6 shows that four dimensions of e-wom are important to impulse buying: frequency of accessing information, frequency of interaction, product variation, and transaction & website condition. The other six dimensions, including the number of reviews, positive and negative comments, recommendations, product quality, and price offered, have no bearing on impulsive purchasing behavior.

5 Conlusion, Limitation, and Future Research

5.1 Conclusion

Based on the results of the research that has been done, the conclusions in this study are as follows:

1. E-wom has 47.9% effect towards impulsive buying activity on tourism products.
2. E-wom influences impulsive acquiring in OTA.

The ten dimensions of e-wom are certainly linked to impulsive buying.

However, only four e-wom dimensions have a significant impact on impulsive buying: frequency of information access, frequency of interaction, product variation, and transaction & website condition.

If there are many negative reviews or comments given at a low price offered, tourists will not be enticed to purchase a tourism product. However, if the number of reviews is filled with positive reviews at relatively high prices product, there would be the opportunities to invite tourists or consumers to purchase a tourism product.

Assume that the condition of a tourist product receives positive feedback. In that case, it will entice new people to buy a tourism product immediately, potentially increasing the frequency of other reviews. Finally, if the number of tourists purchasing a tourism product increases and there is a similarity in their experiences or opinions about a tourism product, the frequency of interaction may increase [11].

5.2 Suggestion

Researchers made the following recommendations based on the findings:

1. Among the ten dimensions, tourism product providers in OTAs should prioritize the four main dimensions that have the greatest impact on e-wom, which are frequency of accessing information, frequency of interaction, product variation, and transaction & website condition.
2. There is an opportunity for additional researchers to gain insight into 52.1% of other variables that can influence impulsive purchasing behavior.

References

1. Hennig-Thurau, T., Gwinner, K.P., Walsh, G., Gremler, D.: Electronic word-of-mouth via consumer-opinion platforms: what motivates consumers to articulate themselves on the internet? J. Interact. Mark. (2004). www.interscience.wiley.com. Accessed 06 Oct 2022
2. Astuti, S.R.T., Khasanah, I., Yoestini, Y.: Enhanced reader. Dipenogoro Int. J. Bus. (2020). Accessed 08 Oct 2022
3. Singh, S., Verma, H.: A study of E-WOM stimuli urging e-impulse buying. Int. J. Eng. Manag. **3**(1), (2017). https://www.researchgate.net/publication/340095198. Accessed 08 Oct 2022
4. Husnain, M., Qureshi, I., Fatima, T., Akhtar, W.: The impact of electronic word-of-mouth on online impulse buying behavior: the moderating role of big 5 personality traits. J. Acc. Mark. **5**(4), 1000190 (2016). Related papers: Perceived value, personality and behavioral intention of electronic brands customers in Kenya (kiprop kibos). Influence of electronic word of mouth (e-WOM) on brand credibility and Egyptian consumers' purchase intentions (Reham Elseidi). The mediating role of trust towards E-Wom on the relationship between big five personality characteristics and influence by E-Wom
5. Do, H.N., Shih, W., Ha, Q.A.: Effects of mobile augmented reality apps on impulse buying behavior: an investigation in the tourism field. Heliyon **6**(8), e04667 (2020)
6. Alavijeh, M.R.K., Golestani, M.: Investigating the effect of scarcity messages on motivation and impulsive buying behavior of tourists in booking online (moderating role of travel experience). Tourism Management Studies (2022)
7. Sohn, H.K., Lee, T.J.: Tourists' impulse buying behavior at duty-free shops: the moderating effects of time pressure and shopping involvement. J. Travel Tour. Mark. **34**(3), 341–56 (2017). https://www.tandfonline.com/doi/abs/10.1080/10548408.2016.1170650. Accessed 08 Oct 2022
8. Li, C., Wang, Y., Lv, X., Li, H.: To buy or not to buy? The effect of time scarcity and travel experience on tourists' impulse buying. Ann. Tour. Res. **1**(86), 103083 (2021)
9. Ahn, J., Kwon, J.: The role of trait and emotion in cruise customers' impulsive buying behavior: an empirical study. J. Strateg. Mark. **30**(3), 320–33 (2022). https://www.tandfonline.com/doi/abs/10.1080/0965254X.2020.1810743. Accessed 08 Oct 2022

10. Sharma, P., Marshall, R., Sivakumaran, B., Roger, M.: Impulse buying and variety seeking: a trait-correlates perspective (2009)
11. López, M., Sicilia, M.: Determinants of E-WOM influence: the role of consumers' internet experience. J. Theor. Appl. Electron. Commer. Res. **9**, 28–43 (2014). www.jtaer.comwww.jtaer.com. Accessed 13 Oct 2022

Consumer Response Model for Luxury Brands

Yossie Rossanty(✉), Endang Sulistya Rini, Beby Karina Fawzeea Sembiring, and Amlys Syahputra Silalahi

Faculty of Economics and Business, Universitas Sumatera Utara, Medan, North Sumatera, Indonesia
yosunpab@gmail.com, {endang.sulistya,beby}@usu.ac.id

Abstract. Even though the Covid-19 epidemic has affected the marketing of goods and services, this has not deterred people from purchasing luxury brands. Brand owners that wish to reach a large number of customers via social media marketing, including engaging and transacting on virtual platforms, have also discovered this. Social media marketing will assist customers in selecting things without their needing to physically visit the business. It must be acknowledged that purchasing things, particularly luxury items, demands accuracy hence, brand-owning corporations require their own marketing strategies. For instance, product photographs must be realistic so that they resemble actual items. Product specifics must be expanded upon and, most significantly, brand involvement must be increased. If customer participation in the brand is correctly achieved, then consumer reaction will be good and luxury brand sales will grow. This research is to assess marketing efforts on social media and luxury brand participation in relation to customer reactions when engaging and transacting on virtual platforms. A content analysis is used to draw the conceptual framework in this study. It is anticipated that the empirical contribution of this study will contribute to the advancement of science, particularly on the issue of consumer responses to marketing activities on social media.

Keywords: Social media marketing activities · Luxury brands · Consumer brand engagement consumer response

1 Introduction

Over the last three decades, luxury brands have generated one of the fastest-growing worldwide marketplaces, with an approximate value transaction of $1 trillion in 2018 (Altagamma 2019). Europe and the United States account for less than half of all sales. Nonetheless, recent data indicates that the economies of other countries, notably Asia, have been growing at a far quicker pace (Deloitte 2019). For instance, Chinese customers accounted for 90% of global market growth in 2019, while European sales increased by just 1% (Bain and Co. 2020).

Spending in this sector is predicted to grow over the next decades as a result of increased Chinese spending and the effect of millennials and Generation Z. According to Statista's 2021 Consumer Market Forecast, Asia, headed by mainland China, is expected

B. Alareeni et al. (Eds.): ICBT 2022, LNNS 620, pp. 676–681, 2023.
https://doi.org/10.1007/978-3-031-26953-0_62

to spend the most, followed by Europe, North America, South America, Africa, Australia, Oceania, and Australia (Statista Consumer Market Outlook 2021). Regardless of the decline in discretionary spending and the poor economic situation, demand has fallen drastically in 2020 caused by the COVID-19 outbreak. The worldwide market for luxury goods is expected to reach $382.6 billion by 2025. According to iprice analysis, the COVID-19 outbreak has not affected Indonesian shoppers' appetite to purchase online (Iprice 2021). The tendency suggests that fashion goods are the most popular category of luxury items, and Sirclo reports that digital transactions among Indonesian shoppers will reach ninety percent by 2025 (Sirclo 2021).

This study focuses on the key challenges associated with social media marketing for luxury goods, such as the personal motives for luxury consumption (Kemper et al. 2022); the intentions for customer participation on social media for premium brands; and the underlying motivations for customers' preference for luxury brands (Park and Ahn 2021; Javornik et al. 2021). Previous research suggests that luxury goods necessitate an emotional relationship between the consumer and the product, with the role of social media marketing-generated content potentially driving customer responses (Tan and Chen 2021; Zhao et al. 2022), though this has not been extensively tested.

Although the pandemic's transmission rate has decreased and activities are limited, online sales are still growing and online shopping remains popular. Since this is more convenient and lowers the requirement for needless movement, thus, based on previous studies, this study will investigate the correlation between social media marketing activity and customer-firm engagement on customer response with a focus on luxury brands post the COVID-19 pandemic. This study is consistent with the field of marketing research, which focuses on current and strategic issues.

2 Theoretical Framework and Hypotheses

2.1 Social Media Marketing Activities

Social media are characterized as internet-based interaction, virtual platforms, media, and applications to collaborate and share information (Dania and Griffin 2021; Li et al. 2021). During the past decade, firms have expanded their usage of social media for marketing reasons, which includes social media forms, social bookmarking, podcasts, and blogs. Ads and marketers use social media for a variety of goals, including brand building, customer relationship marketing, sales promotion, and advertising. The utilization of social media marketing for brand promotion has grown dramatically in recent years (Kumar and Nanda 2019).

Luxury brands are seen as a symbol of wealth and prestige. However, luxury brands will no longer be considered prestigious if everyone has them (Arrigo 2018). Digitalizing allows firms to produce brand content by promoting the history and values of brands to the target audience. Luxury, on the other hand, is typically associated with rarity, wisdom, and exclusivity; thus, the accessibility and infinite availability of content in the digital world appears to be at odds with the exclusivity of luxury brands (Arrigo 2018).

2.2 Consumer Brand Engagement

Customers' brand loyalty may go beyond purchases that are driven by a motivation factor. Customers' brand engagement as a behavioral expression since the majority of them demonstrate actual engagement activities such as consuming and sharing, which are often seen as essential characteristics of consumer and brand interactions on social media (Machado et al. 2019). Reading comments, looking at visuals related to a certain brand, and watching videos are all forms of consumption (Muntinga et al. 2011). These individuals are known as stalkers, and marketers appreciate them since their active consumption of marketing content indicates voyeuristic curiosity (Azar et al. 2016). The contribution refers to user-to-user and user-to-content interactions with companies, such as posting, reviewing, and sharing content with a specific brand purpose. Customers' consumption patterns and contributions are impacted by how a firm and its customers interact.

2.3 Consumer Response

Consumer responses may assist the company in a variety of ways, such as persuading customers to purchase the brand, enhancing the brand's perceived value, and fostering brand loyalty (Marbach et al. 2019). There are two key outcomes of consumer-brand interactions: brand purchase intention and willingness to pay a premium price (Hollebeek et al. 2014). In more detail, Hollebeek and colleagues implicitly explain that purchase intention refers to a consumer's desire to purchase a particular brand, while customers' willingness to pay refers to the readiness to spend a greater premium for the chosen brand than for competitor brands of similar size and quantity. Willingness to pay more shows the value of the brand and is a key part of the whole strategy. These dimensions are seen as a direct driver of actual behavior.

2.4 The Relationship Between Social Media Marketing Activity and Consumer Brand Engagement

Luxury brands are acquired for a multitude of reasons, including as a symbol of wealth, enjoyment, or as presents. Regardless of the reasons for acquiring luxury goods, brands serve as platforms for connecting consumers (Han et al. 2010). These social interactions may be classified depending on the kind and quality of the consumer's interactions and relationships (Zhu and Chen 2015). The majority of brand-related information consumption is motivated by pleasure, entertainment, and leisure rather than need. Consumers utilize social media platforms to find interesting discussion topics and current information.

The quantity of support that is also provided by social media platforms for customized services is related to the degree of customization (Godey et al. 2016). Customization of social media platforms has the potential to enhance brand loyalty and long-term user brand engagement (Martin and Todorov 2010), in addition to the simplicity with which ideas, thoughts, and emotions may be shared through social media (Kim et al. 2014).

Customers are provided with an engaging experience by generating and distributing sponsored entertainment content across many social media networks. Customers' capability to express their personality through online customization may play a pivotal role in enhancing consumer engagement. Moreover, consumers utilize social media to keep up with breaking news and trending topics. Even though the digital world gives marketers a lot of opportunities to connect with customers, not many use social media.

H1: Social media marketing activity has a positive and significant effect on consumer brand engagement.

2.5 The Relationship Between Social Media Marketing Activity and Consumer Response

Social media platforms provide brand-specific information that assists in the establishment of customer preferences and purchase intents (Naylor et al. 2012). When determining purchase intentions for a product or brand, social media users may evaluate product-related comments posted on social media sites. Several studies have been conducted to investigate the influence of social media marketing on customer responses such as brand preference, brand loyalty, and purchase intent, with contradicting findings indicating a positive relationship between social media marketing activity and purchase intent. The impact of social media marketing on customers more willingness to pay a premium price is positive.

H2: Social media marketing activity has a positive and significant effect on consumer response.

2.6 The Relationship Between Consumer Brand Engagement and Consumer Response

Customer-brand interactions have typically been examined in an attempt to elicit positive consumer responses such as consumer-based brand equity and brand loyalty. However, little attention has been paid to the relationship between brand consumer engagement, brand purchase intention, and willingness to pay a premium for a brand. Interactions with brand-related touch points increase brand acquisition intent (Algharabat 2018). Consumers who interact with their preferred brand through social media are more likely to have positive product responses compared to consumers who interact with their favorite brand through channels that do not enable customer interaction.

3 Research Methods

This study used a content analysis which derived from various articles and past researches in order to develop a framework. A number of referred articles justify drawing and proposing the above concept.

4 Conclusion

Consumer responses to luxury products may benefit the company in a number of ways, including generating goodwill, acquiring the brand, enhancing the perceived value of the brand, and fostering brand loyalty. Following the end of the COVID-19 outbreak, luxury brands are experimenting with online sales. Of course, not being able to directly view anything might impact consumer responses when selecting a product, particularly if the product is an expensive luxury brand product.

The capability of users to express their personality via website customization, according to the consumer response model, may play a vital role in enhancing customer engagement through customization. Consumers use social media to stay up to date on the latest news and popular topics. This is not the case, despite the fact that the Internet provides various options for marketers to engage customers with luxury products.

When assessing or developing purchase intents for a certain product or brand, social media users may leverage product-related comments made on social media sites. Social media marketing efforts increase customers' willingness to pay premium prices. Customer-brand interaction have typically been evaluated in an attempt to elicit positive consumer responses such as consumer-based brand equity and brand loyalty. Consumers who interact with their preferred brand on social media are more likely to have positive product responses compared to those who interact with brands that do not support social media channels.

References

Algharabat, R.S.: The role of telepresence and user engagement in co-creation value and purchase intention: online retail context. J. Internet Comm. **17**(1), 1–25 (2018)

Altagamma, B.C.G.: The true-luxury global consumer insight (2019)

Arrigo, E.: Social media marketing in luxury brands: a systematic literature review and implications for management research. Manag. Res. Rev. **41**(6), 657–679 (2018)

Azar, S.L., Machado, J.C., Vacas-de-Carvalho, L., Mendes, A.: Motivations to interact with brands on facebook-towards a typology of consumer–brand interactions. J. Brand Manag. **23**(2), 153–178 (2016)

Bain and Co. China's Unstoppable 2020 Luxury Market. Bain & Company, Inc. (2020). https://www.bain.cn/pdfs/202012160134321779.pdf

Dania, A., Griffin, L.L.: Using social network theory to explore a participatory action research collaboration through social media. Qual. Res. Sport Exerc. Health **13**(1), 41–58 (2021)

Deloitte. Global powers of luxury goods 2019: Bridging the gap between the old and the new. Touche Tohmatsu Limited (2019). https://www2.deloitte.com/global/en/pages/consumer-business/articles/gx-cb-global-powers-of-luxury-goods.html

Godey, B., et al.: Social media marketing efforts of luxury brands: Influence on brand equity and consumer behavior. J. Bus. Res. **69**(12), 5833–5841 (2016)

Hair, J.F., Black, W.C., Babin, B.J., Anderson, R.E., Tatham, R.: Multivariate Data Analysis. Cengage, Hampshire (2019)

Han, Y.J., Nunes, J.C., Drèze, X.: Signaling status with luxury goods: the role of brand prominence. J. Mark. **74**(4), 15–30 (2010)

Hollebeek, L.D., Glynn, M.S., Brodie, R.J.: Consumer brand engagement in social media: conceptualization, scale development and validation. J. Interact. Mark. **28**(2), 149–165 (2014)

Iprice. https://iprice.co.id/trend/insights/report-peta-persaingan-e-commerce-q3-2021/
Javornik, A., et al.: Strategic approaches to augmented reality deployment by luxury brands. J. Bus. Res. **136**, 284–292 (2021)
Johnson, P., Gill, J.: Research methods for managers, pp.1–288 (2010)
Kemper, J.A., Bai, X., Zhao, F., Chiew, T.M., Septianto, F., Seo, Y.: Sharing luxury possessions in the age of digital experience economy: consumption type and psychological entitlement. J. Bus. Res. **142**, 875–885 (2022)
Kim, E., Sung, Y., Kang, H.: Brand followers' retweeting behavior on twitter: how brand relationships influence brand electronic word-of-mouth. Comput. Hum. Behav. **37**, 18–25 (2014)
Kumar, V., Nanda, P.: Social media in higher education: a framework for continuous engagement. Int. J. Inf. Commun. Technol. Educ. (IJICTE) **15**(1), 97–108 (2019)
Li, F., Larimo, J., Leonidou, L.C.: Social media marketing strategy: definition, conceptualization, taxonomy, validation, and future agenda. J. Acad. Mark. Sci. **49**(1), 51–70 (2020). https://doi.org/10.1007/s11747-020-00733-3
Machado, J.C., Vacas-de-Carvalho, L., Azar, S.L., André, A.R., Dos Santos, B.P.: Brand gender and consumer-based brand equity on facebook: the mediating role of consumer-brand engagement and brand love. J. Bus. Res. **96**, 376–385 (2019)
Marbach, J., Lages, C., Nunan, D., Ekinci, Y.: Consumer engagement in online brand communities: the moderating role of personal values. Eur. J. Mark. **53**(9), 1671–1700 (2019)
Martin, K., Todorov, I.: How will digital platforms be harnessed in 2010, and how will they change the way people interact with brands? J. Interact. Advert. **10**(2), 61–66 (2010)
Muntinga, D.G., Moorman, M., Smit, E.G.: Introducing COBRAs: exploring motivations for brand-related social media use. Int. J. Advert. **30**(1), 13–46 (2011)
Naylor, R.W., Lamberton, C.P., West, P.M.: Beyond the "like" button: the impact of mere virtual presence on brand evaluations and purchase intentions in social media settings. J. Mark. **76**(6), 105–120 (2012)
Park, J., Ahn, J.: Editorial introduction: Luxury services focusing on marketing and management. J. Retail. Consum. Serv. **58**, 102257 (2021)
Saunders, M., Lewis, P., Thornhill, A.: Research Methods for Business Students. Pearson Education, Boston (2009)
Sekaran, U., Bougie, R.: Research methods for business: A skill building approach. John wiley & sons, Hoboken (2016)
Sirclo. https://www.sirclo.com/blog/sirclo-rilis-laporan-tren-perkembangan-industri-ecommerce-dan-harbolnas-di-indonesia-saat-pandemi/
Statista Consumer Market Outlook, 2021 In-depth Report: Luxury Goods (2021). https://www.statista.com/study/61582/in-depth-luxury/
Tan, S., Chen, W.: How marketer-generated content characteristics affect consumer engagement? empirical evidence from China's WeChat food marketing. Brit. Food J. **124**(1), 255–274 (2021)
Zhao, K., Zhang, P., Lee, H.M.: Understanding the impacts of user-and marketer-generated content on free digital content consumption. Decis. Support Syst. **154**, 113684 (2022)
Zhu, Y.Q., Chen, H.G.: Social media and human need satisfaction: implications for social media marketing. Bus. Horiz. **58**(3), 335–345 (2015)

Indian Cooperative Trade Platform (ICTP): A Grounded Model

A. J. Lakshmi[1], Abilash Unny[2], and M. P. Akhil[3](✉)

[1] Sree Narayana College, Chempazhanthy, India
[2] PricewaterhouseCoopers LLP, London, UK
[3] Alliance University, Bengaluru, Karnataka, India
akhil.mp@kristujayanti.com

Abstract. Information technology and communication technology have coupled themselves and, have become a disruptive combination which has affected all walks of human existence. Digital Transformation is used to highlight the changes made by information and communication technology. Cooperatives are proven to overcome the shortcomings of other forms of institutions in terms of bringing an inclusive idea, failure to address the larger social concerns and its inherent nature of being highly focused on social and economic attributes and are capable of bringing in technological shifts or even changes in business models through their own initiatives. However, the small and locally based cooperatives are still away from digitalisation. The study aims to analyse how the concept of digital transformation works in social enterprise as well as incorporate the concept of digital transformation in Cooperatives and propose a model of model known as the Indian Cooperative Trade Platform (ICTP). The study tried to analyse the fact that the development of information technology through social enterprise is a solution for the importunate circumstances of a deprived section of society. The requirement of digitalisation in the cooperative sector is essential in this competitive era. With the help of the right support and use of digitalisation, ICTP like a centralised platform can be developed and made functional so as to ensure the efficient trading of agricultural products and that's how to ensure the upliftment of deprived classes. The model will facilitate the attainment of SDGs as well if it is implemented and monitored systematically.

Keywords: Entrepreneurship · Collective social entrepreneurship · Cooperatives · Digital transformation · ICTP · Sustainability

1 Introduction

Historians who acknowledge the significant factors theory of societal growth consider information technology as the significant factor in the present societal changes (Wang et al. 2006).On the practical front, it is evident that the last few decades have brought substantial alterations to social movements and priorities (Wedel 2020). Information technology and communication technology have coupled themselves and, have become a disruptive combination and have affected all walks of human existence (Vadrot 2020).

B. Alareeni et al. (Eds.): ICBT 2022, LNNS 620, pp. 682–695, 2023.
https://doi.org/10.1007/978-3-031-26953-0_63

The second half of the 20th century introduced the concept of digital electronics and, at a fast pace, the same has side-lined the analogue version. Hoque and Alam (2010) affirm that digital electronics is the backbone of information and communication technology. The term 'Digital Transformation' is therefore used to highlight the changes made by information and communication technology. Digital transformation is defined as the incorporation of digital technology into all areas of activities resulting in fundamental changes in how everything operates (Mergel et al. 2019). It is a cultural shift in the way in which the traditional methods of transacting.

A paradigm shift in the way contemporary society measures the value created for communities and clients by business enterprises has triggered an inclusive growth strategic focus within them, resulting in the rise of social enterprises (Abbatiello et al. 2018). The area of social entrepreneurship was once portrayed as a phenomenon created to address urgent social challenges through innovation on a global scale (Osberg and Martin 2007). This has resulted in various definitions contributing to abstruseness about the subject (Nicholls 2010). Bornstein (2004) and Nicholls (2006) defined social entrepreneurship as a systemic social change model, while Austin et al. (2006) explained it as a hybrid partnership arrangement. To add further confusion, Alvord et al. (2004) and Yunus (2008) presented it as a structure for empowerment or political reform.

The advantages of cooperatives in ensuring inclusive sustainable development are very much on a higher pedestal (Sehlin et al. 2019). Cooperatives are the type of institutions that are preferred to overcome the shortcomings of other forms of institutions. These shortcomings are generally in the capacity of other forms of institutions in bringing an inclusive idea, failure to address the larger social concerns and its inherent nature of being highly focused on economic attributes (Franco *et al.* 2010). The success story of cooperatives in the end mass is usually connected with those cooperatives that have been created to overcome these aspects (Alavosius *et al.* 2009). Rather the utility of cooperatives for serving the underprivileged population is very much an indicator of the importance of cooperatives in the context of inclusive sustainable development (Castilla-Polo and Sánchez-Hernández 2020). These narratives stand good even in the Indian context. The statistical data shows that there is a larger presence of cooperatives in the rural landscape (Ghosh 2007).

Even though there are cooperatives, which are working on a pan-India basis or in one and more states or covering a large or full area of a particular state, still the majority are small cooperatives serving or confined to a small locality (Singh 2016; Majee and Hoyt 2011). Similarly, there are credit cooperatives that are working as well as banking organisations. These types of bigger cooperatives and large amounts of credit cooperatives have already got themselves introduced into the utilisation of information and communication technology to a great extent (Elee 2021; Sanfilippo and Chattopadhyay 2020). However, the remaining smaller cooperatives which are serving a limited number of members within a local community generally shy away from digitalisation (Ortmann and King 2007). Multiple reasons can be identified for this situation. However, more than the limiting circumstances, the need for digitalisation is highly relevant.

2 Research Question

2.1.1. Is it possible to address the trading issues of the farmers through social entrepreneurship?
2.1.2. Can digital transformation through cooperative social entrepreneurship be used as a tool to support rural agriculturists?
2.1.3. Whether a centralised (national level) Cooperative (trade) platform can be used to address the trading issues of the rural agriculturalists?

3 Research Objective

3.1.1. Practical Implication - instigating practical change

3.1.1.1. To analyse how the concept of digital transformation works in Cooperative social enterprise

3.1.2. Theoretical Implication – the advancing body of knowledge

3.1.2.1. To incorporate the concept of digital transformation in the Cooperatives

3.1.3. To contribute toward social entrepreneurship knowledge

3.1.3.1. To examine the relevance of cooperatives as a social enterprise and bring in the Indian Cooperative trade platform

4 Research Methodology

The paper is doctrinal in nature and reliance is only on data from secondary sources. The relied secondary materials include books, articles, reports etc. This paper is trying to analyse the fact that the development of information technology through a social enterprise is a solution for the importunate circumstances of a deprived section of society. Need for cooperatives to infuse the same by taking an example from the rural population of a country like India so that it will be highly potent.

5 Digitalisation

Digitalisation and digitisation are two conceptual words that are frequently used synonymously in a variety of literary works (Rachinger *et al.* 2019). The conversion of analogue data into digital form is referred to as digitisation (Ghobakhloo 2020). In contrast, the term "digitalisation" describes "the adoption or increase in usage of digital or computer technology by an organisation, industry, country, etc." Digitisation is a component of digitalisation, and realising digitalisation begins with digitisation. Digitalisation is the use of digital technologies to change a business model and provide new revenue and value-producing opportunities; it is the process of moving to a digital business (Kaur Narula and Rana 2017). By encouraging innovation, creating efficiency, and enhancing services, digital technologies have the ability to promote more equitable and sustainable growth (Reis *et al.* 2020).

6 Digital Transformation

Consumer expectations and behaviours have fundamentally changed as a result of the digital transformation and the ensuing business model innovation (Iivari et al. 2020). This has put pressure on established businesses and disrupted several markets. Customers may access dozens of media outlets, actively and easily interact with businesses and other customers, and navigate an ever-growing number of customer journey touchpoints, many of which are digital (Wang et al. 2006).

The demand for digital transformation is being driven by three main external factors. First, a rising number of supporting technologies have emerged since the advent of the World Wide Web and its widespread acceptance, which has bolstered the growth of e-commerce (Niranjanamurthy *et al.* 2012). Big data's pervasiveness and the emergence of new digital technologies like artificial intelligence (AI), blockchain, the internet of things (IoT), and robotics are expected to have profound repercussions on the corporate world (Saarikko et al. 2020).

Second, the competitive landscape is drastically shifting as a result of these new digital technologies. Technologies have changed the nature of competition in the retail, pushing revenues to relatively new internet companies (Kim and Srivastava 2007). The competition has expanded internationally, but it has also grown fiercer.

Third, as a result of the digital revolution, consumer behaviour is altering. Consumers are moving their business to online retailers, according to market data, and digital touch-points play a significant part in the customer experience that influences both online and offline sales (Kumar and Kumar 2011). The use of modern search and social media tools has increased customer connectivity, knowledge, empowerment, and activity.

The use of new digital technology has the potential to challenge established traditional corporate norms. Businesses that cannot adapt to these changes lose their appeal to clients and are likely to be replaced by businesses that do make use of such technologies.

7 Example of Revolutionary Changes Using Digitalisation

The huge potential of digital technology for transforming a traditional landscape stands exemplified by an initiative implemented by a corporate entity in India when it ventured into a business area which was highly dependent on agricultural products as inputs (Kraus *et al.* 2021; Li 2020). This was implemented by introducing a procuring mechanism which was directly accessible to the corporate entity as well as the agriculturists. An information and communication technology-based system was implemented by it wherein agriculturalists were provided with access to computers. This facility was provided at the traditional market locations with someone in charge and farmers being able to identify the requirements. Real-time information and customised knowledge enhance the farmers to take decisions and align the farm output with market demand and secure quality and productivity. The farmers benefited through enhanced farm productivity and higher farm-gate prices (Krishna *et al.* 2022). The corporate entity benefited from the lower cost of procurement despite giving higher prices to the farmers by eliminating cost in the supply chain that was not adding value (Abokyi *et al.* 2020). This attempt was highly successful and was subsequently the subject of multiple studies and the same is demonstrating the positive utilisation of digital technologies in bringing in revolutionary changes (Deichmann *et al.* 2016). It needs to be noted that the transformation was that of a traditional way of doing business and was making root-level changes to a community that did not even know about information and communication technology (Rijswijk *et al.* 2021).

In the above example, the corporate entity was benefiting itself and was also transforming the existing system and it also resulted in the benefit of farmers. This was an isolated example wherein at the behest of corporate entities lives of the rural population benefited (Mukerji 2020). The example discussed above stands referred to as 'e-Choupal' and was the initiative of ITC Limited (Dangi and Singh 2010). It was launched in June 2000 and is the largest initiative among all Internet-based interventions in rural India and reaches out to over 4 million farmers growing a range of crops like soya bean, coffee, wheat, rice, pulses, shrimp etc. in over 35000 villages through 6100 kiosk field (Bowonder et al. 2003).

8 Potential in Digitalisation and Digital Transformation in Cooperatives

There is a huge potential for cooperatives to utilise the benefits of digital transformation and to bring a positive change in the life of less prioritised categories field (Sehlin et al. 2019; Kuimov *et al.* 2019). Cooperatives are generally much sought after in sectors otherwise not a priority for other types of economic organisations. Cooperatives, in the Indian context, are generally focused on benefits for rural India (Singh 2016). Cooperatives as well as beneficiaries are generally slow-paced in adopting the latest technologies (Luo *et al.* 2022). There is a need to overcome this lethargy and there is a requirement to implement information and communication technology benefits in a thorough scale because a digital transformation is capable of making a huge impact for both cooperatives and the beneficiaries (Montegut-Salla et al. 2013).

The present era is considered the new era of industry 4.0 and at the same, the concept of digitalisation has altered the way entire things work (Ghobakhloo 2020). Digitalisation, in the context of society, has been a switch that has resulted in a total revamping of the entire system (Williams 2021). Digitalisation can also utilise information and communication technology for easing the way of doing something (Rachinger *et al.* 2019). The use of information and communication technology in the context of an organisation can be classified into three categories. It is initially informative, then communicative and lastly transactional (Joshi *et al.* 2013). In the first phase, the information about a particular organisation is made available to any interacting party. For example, providing information about a particular organisation. In the second phase, the interacting party is provided with a means of digitally communicating with an organisation. For example, clicking on a link on a website to send a message to the organisation. In the third stage, the interacting party can initiate and complete a transaction with the organisation. The ideal example can be transactions with e-commerce websites or banking websites. These aspects of digitalisation are from the perspective of the external interactions of the organisation (Slavinski and Todorović 2020). Before venturing into an external interaction through information and communication technology means an organisation needs to have an internal digitalisation (Rachinger *et al.* 2019). This process usually happens for the purpose of efficiency of the work within the organisation itself. The advantages of digitalisation can be categorised in multiple ways like a reduced work day, improved accuracy/fewer errors, increased revenues/collection, reduced cost of delivering services and improved documentation and reporting (Bardakci 2020; Schroeder and Ziaee Bigdeli 2018). The above-mentioned advantages are from a generic perspective and apply to all institutions. From the context of cooperatives, this may be verified based on the priorities.

The need for bringing in the utilisation of information and communication technology for the process of digitalisation cannot be expected in such a scenario (Saarikko et al. 2020). Lastly, the interest and inclinations of the stakeholders are reflected in the institution. However, the need for utilising information and communication technology is unavoidable because of the huge potential of the digitisation process that has been narrated earlier (Yang and Gu 2021; De' et al. 2020). This needs to be attempted and implemented by taking a high degree of initiative.

Some cooperatives are as good as any corporate entities in their capacity. Such cooperatives are capable of bringing in technological shifts or even changes in business models through their own initiatives (Boevsky and Kostenarov 2020). On the other hand, cooperatives do not have the capacity to identify and implement such technological features. This in capacity may be due to a lack of awareness, technical competence, insufficiency of capital, etc. (Purbasari et al. 2022). In such a scenario, there is a need to create some means to ensure the inclusion of technology. This can be achieved by a governmental initiative or by the formation of an apex cooperative tasked exclusively with bringing in the process of digitalisation (Yang *et al.* 2021). Between these two extremes, there may be some cooperatives that may require some sort of initiation into the introduction of technology. With respect to the first category, there is not much necessity for any inducement of bringing in digitalisation (Camargo Benavides and Ehrenhard 2021). Said category of cooperatives is capable of bringing in any changes by its own means just like any other institution. Switching to technology is a regular affair for it because of the availability of all sorts of means for conceiving and implementing the same (Alavosius *et al.* 2009).

9 A Framework for the Indian Cooperative Trade Platform (ICTP)

The benefits of digitalisation from the context of cooperatives can be dyed in the wool by developing an Indian Cooperative Trade platform. This trade platform will be a cooperative type establishment and uses a digital platform or an application for facilitating trade transactions. The membership of this platform provides both sellers (members of different cooperatives signed up to this platform) and buyers (also organisations with buyer membership interested in purchasing seller's products) a virtual space to do their trade. The membership fees collected (a larger proportion from buyers than sellers) are used for the maintenance and development of any future roadmap features of this platform making it self-sufficient and any extra income after expenses will be used for pre-agreed social welfare projects which will directly benefit the rural agriculturist community. In the context of the Indian Cooperative trade platform, it is not the digitalisation of existing cooperatives but the cooperative framework by utilising a digital model which provides services and the sale of goods (Fig. 1).

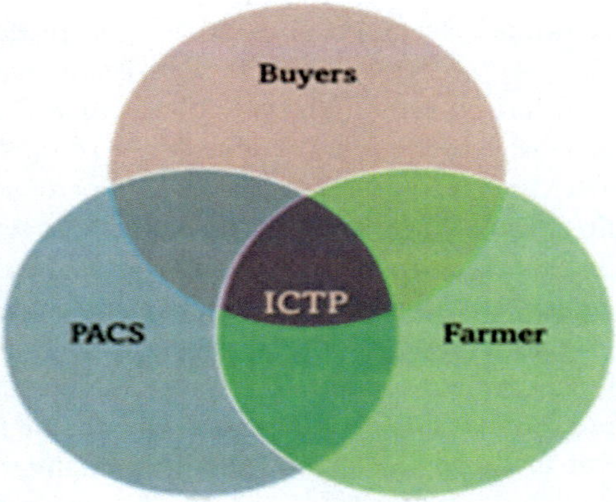

Fig. 1. Indian cooperative trade platform framework

9.1 Functioning of the Indian Cooperative Trade Platform (ICTP)

An Indian cooperative Trade platform can be developed only with the support and assistance of the apex or national levels organisations such as the National Cooperative Union of India (NCUI), National Cooperative Development Corporation (NCDC), National Bank for Agricultural and Rural Development (NABARD) and the Central or State Governments. The idea behind this trade platform is to digitalise the trading transactions of the farmers so that they will get good prices for their produce. It can be started as a trial platform.

9.1.1 Primary Agricultural Credit Cooperative Societies (PACS)

The Primary Agricultural Credit Societies to be referred to as PACS moving forward play a vital role in supporting farmers by providing agricultural loans and function as the backbone of the rural credit structure in the State (Asodiya *et al.* 2014). PACS play a major role in bringing rural and low-income populations to the banking system. Apart from banking functions, some cooperatives also play a major role in promoting Agricultural operations including the marketing of Agricultural products and are actively involved in the socio-economic development of their entire area of operations (Vanlalmuana and Laldinliana 2020; Devi Sekhar and Vijayan 2021).

The proposed model, suggests that PACS needs to play a more active role and facilitate a collection of agricultural products from the agriculturalists who are their members. They also need to carry out the quality and quantity checks for each product including their grading. After this process, the PACS will prepare and submit a product list showing all the details of the product including quality, grading, quantity and reserve price information with recommended auction timeline via the PACS portal for listing the product for auction. ICTP team will review the product listing information and then handle the auction process. The monies of this transaction are normally held in an ESCROW-type account handled by ICTP. After finalising the sale the PACS will arrange the delivery of goods for the buyers and update the PACS portal. The buyers will submit a Goods received note (GRN) ideally as soon as they receive the goods or within 24 h of the delivery. The monies of this transaction are released to the PACS account as soon as the GRN is submitted on the ICTP platform. PACS will transfer the monies to the seller after retaining a small charge pre-agreed with their members for the services rendered.

The high-Level benefits to PACS using the ICTP are getting a customisable reusable centralised product repository, having the ability to set a reserve price, getting access to product market rate data for reserve price setting, getting access to product demand forecast leveraging on sale data from the ICTP platform and finally the ability to access a secure way to complete these transactions through the platform.

9.1.2 Farmers

The rural agriculturalist is the one who produces agricultural products like wheat, paddy, cereals etc. the agriculturalists have to see to it that they produce the best quality products so that they could fetch high prices for the same. These agricultural products have to be given to the primary agricultural credit society. The further process is done by the PACS. Through this process, the farmers can acquire the best prices for their produce. High-Level Benefits are the ability to receive a market price or above for products, the ability to view and track sale progress, ability to understand market demand.

9.1.3 ICTP (Indian Cooperative Trade Platform)

Indian Cooperative Trade Platform is a platform where buyers and sellers (including PACS) can trade agricultural produce on a digital trade platform. As mentioned before when the PACS have submitted the product details via their portal ICPT team will verify this information and manage the auction process end to end including the listing of products for auction including product delivery arrangement, providing notifications and relevant updates to buyers, PACS and sellers and also handle any related financial transactions. The high-level benefits of ICTP are product database access for Buyers, PACS and RA, product description with quality and grading, a customised portal for Buyers, PACS and RA, a centralised communication channel for all platform users, ability to develop sales and market data for the forecast.

9.1.4 Buyers

The buyers can be individuals, member cooperatives, companies and governments. As mentioned above they will enter into a membership agreement with ICTP to have the ability to access the information on the platform and purchase required agricultural produce. By undergoing this process the buyers could get the best quality products at the best price. High-Level Benefits for buyers using the ICTP platform are the option to register early bids, availability of a live bidding portal, and end-to-end buyer support by the ICTP team.

To explain the functioning of the Indian Cooperative Trade platform a High -level benefit Model is detailed below (Fig. 2):

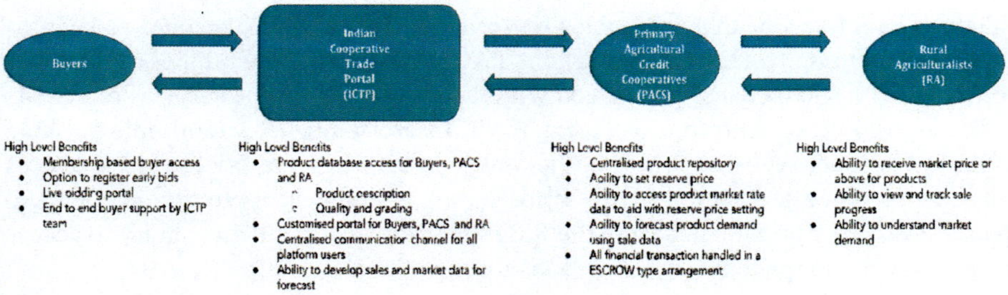

Fig. 2. High-level benefit model of ICTP

9.2 Indian Cooperative Trade Platform - A Grounded Process Model

While analysing the research question, Can a centralised Cooperative platform be used to address the trading issues of the Rural Agriculturalists? It reflects on an interpretation of what constitutes a fact. The ontology of the same is associated with the nature of reality, but in this context, there are multiple realities, which are enumerated below and the research question is addressed with the help of a process model, which is explained in detail as follows (Fig. 3):

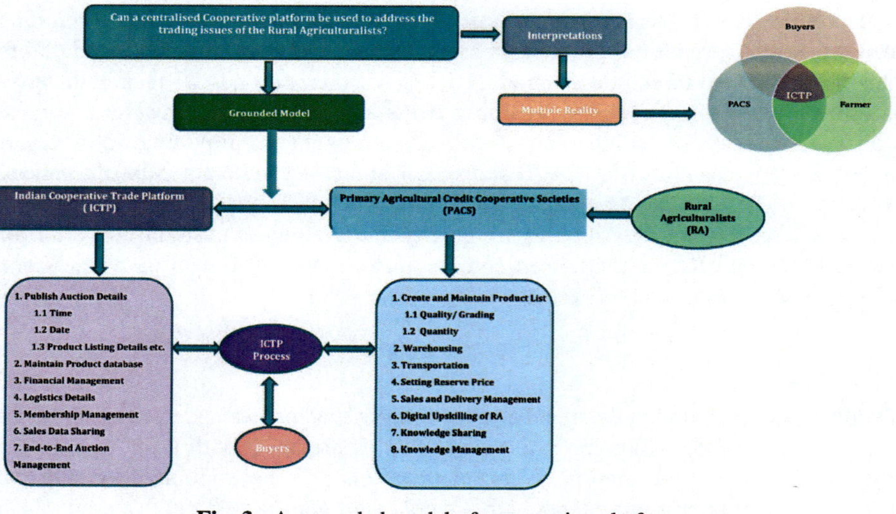

Fig. 3. A grounded model of cooperative platform

9.3 Area of Operation

The area of operation is confined to the village, panchayath or municipality where exists a PACS in which the farmers have a membership. Normally the farmers will be members of the PACS where they have agricultural land.

9.4 Membership

PACS will be a member of this centralised platform so that the members of these PACS can do trading transactions through the same.

9.5 Capital

Capital contribution of this centralised platform has to be through the central and state government, National Cooperative Union of India(NCUI), National Cooperative Development Corporation(NCDC), National Bank for Agriculture and Rural Development (NABARD), International Cooperative Alliance (ICA) or other successful cooperatives like Indian Farmers Fertiliser Cooperative Limited (IFFCO), Anand Milk Union Limited (AMUL), Uralungal Labour Contract Cooperative Society (ULCCS) etc. other than these they can collect membership fees from the PACS.

9.6 Management and Administration

The management and administration of this trade platform have to be vested with trained personnel who processes the qualities of true co-operators, who are aware of the values and principles of cooperatives and who also have management skills.

10 Aims of ICTP

- Creation of a common trade platform for agricultural produce
- Addressing the principle of cooperation among cooperatives
- The digitalisation of the cooperative agricultural sector
- Developing a database for the agricultural produce
- Creating benefits for the farmers
- Avoiding the intermediaries of trading channels
- Forecasting the demand and supply
- Planning the trading in advance
- Transparent transactions

11 Conclusion

With respect to the cooperatives, it is evident that there is a requirement for driving digital transformation in this sector. This requires considerable support and financial backing from the government, apex and international organisations. There are two schools of thought: one is of the opinion that there is a need for verifying the necessity of bringing in digitalisation by analysing the requirements of individual cooperatives. Similarly, there is yet another school of thought which concludes that digitalisation should not be avoided or delayed because the same is a necessary attribute of efficiency. Anyhow there is a convergence of both of these schools on one point there is a need for a feasibility verification of digitalisation. Thus apart from those cooperatives that are having means for bringing in digitalisation, there is definitely a requirement for verifying the digitalisation of other coops. With the help of the right support and use of digitalisation a centralised platform i.e., ICTP can be developed and made functional so as to ensure the efficient trading of agricultural products. Hence ICTP will help rural agriculturalists from getting exploited by intermediaries and fetch the best price for their products not only at the domestic level but also internationally. On top of this, the model will help to attain the Sustainable Development Goals, particularly SDGs 1, 2, 9 and 12 - No Poverty, Zero Hunger, Industry Innovation and Infrastructure, Responsible consumption and Production. The success of the proposed model can be ensured only through proper execution and continuous monitoring. Ultimately it is not the idea, it is the implementation that counts.

References

Abbatiello, A., Agarwal, D., Bersin, J., Lahiri, G., Schwartz, J., Volini, E.: The Rise of the Social Enterprise - Deloitte Global Human Capital Trends Survey 2018 (2018). https://www2.deloitte.com/content/dam/Deloitte/at/Documents/human-capital/at-2018-deloitte-human-capital-trends.pdf. Accessed 1 Jan 2021

Bharadwaj, A., El Sawy, O.A., Pavlou, P.A., Venkatraman, N.: Digital business strategy: toward a next generation of insights. MIS Q. **37**(2), 471–482 (2013)

Abokyi, E., et al.: The impact of output price support on smallholder farmers' income: evidence from maize farmers in Ghana. Heliyon **6**(9), e05013 (2020). https://doi.org/10.1016/J.HELIYON.2020.E05013

Alavosius, M., et al.: Use of a cooperative to interlock contingencies and balance the commonwealth. J. Organ. Behav. Manag. **29**(2), 193–211 (2009). https://doi.org/10.1080/01608060902874575

Alvord, S., Brown, L., Letts, C.: Social entrepreneurship and societal transformation: an exploratory study. J. Appl. Behav. Sci. **40**(3), 260–283 (2004)

Asodiya, P.S., et al.: Impact of primary agricultural cooperative societies's on farmer's economy of Panchmahal district of middle Gujarat, India. J. Agric. Sci. **6**(6), 195 (2014). https://doi.org/10.5539/JAS.V6N6P195

Austin, J., Gutierrez, R., Ogliastri, E.: Effective management of social enterprises: lessons from business and civil society organizations in Ibero–America. Harvard University, Cambridge (2006)

Bardakci, D.H.: Benefits of digitalization in international logistics sector. Int. J. Soc. Sci. Econ. Res. **05**(06), 1476–1489 (2020). https://doi.org/10.46609/IJSSER.2020.V05I06.009

Boevsky, I.Y., Kostenarov, K.D.: The potential of digitalization for increasing members' satisfaction of Bulgarian cooperatives. In: IOP Conference Series: Materials Science and Engineering (2020). https://doi.org/10.1088/1757-899X/940/1/012073

Bornstein, D.: How to Change the World: Social Entrepreneurs and the Power of New Ideas. Oxford University Press, Oxford (2004)

Bowonder, B., Gupta, V., Singh, A.: Developing a rural market e-hub the case study of e-Choupal experience of ITC (2003)

Camargo Benavides, A.F., Ehrenhard, M.: Rediscovering the cooperative enterprise: a systematic review of current topics and avenues for future research. Voluntas **32**(5), 964–978 (2021). https://doi.org/10.1007/s11266-021-00328-8

Castilla-Polo, F., Sánchez-Hernández, M.I.: Cooperatives and sustainable development: a multilevel approach based on intangible assets. Sustainability **12**(10) (2020). https://doi.org/10.3390/su12104099

Dangi, N., Singh, H.: e-Choupal: hope or hype? Am. J. Econ. Bus. Adm. **2**(2), 179–184 (2010). https://doi.org/10.3844/AJEBASP.2010.179.184

De', R., Pandey, N., Pal, A.: Impact of digital surge during Covid-19 pandemic: a viewpoint on research and practice. Int. J. Inf. Manag. **55**, 102171 (2020). https://doi.org/10.1016/J.IJINFOMGT.2020.102171

Deichmann, U., et al.: Will digital technologies transform agriculture in developing countries? (2016). https://doi.org/10.1596/1813-9450-7669

Devaux, A., Torero, M., Donovan, J., Horton, D.: Agricultural innovation and inclusive value-chain development: a review. J. Agribus. Dev. Emerg. Econ. **8**, 99–123 (2018)

Elee, E.G.: Impact of information technology on cooperative services delivery in Nigeria. UNIZIK J. Educ. Res. Policy Stud. **2**, 247–259 (2021). https://unijerps.org/index.php/unijerps/article/view/35. Accessed 11 Sept 2022

Devi Sekhar, R., Vijayan, D.S.: Productive efficiency of primary agricultural credit societies in Kerala, India. Nat. Volatiles Essent. Oils **8**(5), 8032–8040 (2021)

Franco, R.C., et al.: Cooperatives, history and theories of. In: Anheier, H.K., Toepler, S. (eds.) International Encyclopedia of Civil Society, pp. 563–570. Springer, New York (2010). https://doi.org/10.1007/978-0-387-93996-4_102

Ghobakhloo, M.: Industry 4.0, digitization, and opportunities for sustainability. J. Clean. Prod. **252**, 119869 (2020). https://doi.org/10.1016/J.JCLEPRO.2019.119869

Ghosh, A.K.: Cooperative movement and rural development in India. Soc. Change **37**(3), 14–32 (2007). https://doi.org/10.1177/004908570703700302

Iivari, N., Sharma, S., Ventä-Olkkonen, L.: Digital transformation of everyday life – how COVID-19 pandemic transformed the basic education of the young generation and why information management research should care? Int. J. Inf. Manag. **55**, 102183 (2020). https://doi.org/10.1016/J.IJINFOMGT.2020.102183

Joshi, A., et al.: The role of information and communication technology in community outreach, academic and research collaboration, and education and support services (IT-CARES). Perspect. Health Inf. Manag. **10**(Fall) (2013). /pmc/articles/PMC3797554/. Accessed 10 Sept 2022

Kaur Narula, T., Rana, S.: Digitalization in India. Int. J. Emerg. Technol. (Spec. Issue NCETST-2017) **8**(1), 298–303 (2017). www.researchtrend.net. Accessed 11 Sept 2022

Kim, Y.A., Srivastava, J.: Impact of social influence in e-commerce decision making. In: ACM International Conference Proceeding Series, vol. 258, pp. 293–302 (2007). https://doi.org/10.1145/1282100.1282157

Kraus, S., et al.: Digital transformation: an overview of the current state of the art of research **11**(3) (2021). https://doi.org/10.1177/21582440211047576

Krishna, V.V., et al.: Conservation agriculture benefits Indian farmers, but technology targeting needed for greater impacts. Front. Agron. **4**, 24 (2022). https://doi.org/10.3389/FAGRO.2022.772732/BIBTEX

Kuimov, V.V., et al.: Digital transformations in the development of cooperative network interactions. In: ACM International Conference Proceeding Series (2019). https://doi.org/10.1145/3372177.3373345

Kumar, R., Kumar, M.: E-commerce: living and working in an interconnected world (2011). Undefined

Singh, S.K.: Problems and prospects of the cooperative movement in India under the globalization regime. Int. J. Indian Psychol. **3**(4), 59 (2016)

Li, F.: The digital transformation of business models in the creative industries: a holistic framework and emerging trends. Technovation **92–93**, 102012 (2020). https://doi.org/10.1016/J.TECHNOVATION.2017.12.004

Luo, L., et al.: Research on the influence of education of farmers' cooperatives on the adoption of green prevention and control technologies by members: evidence from rural China. Int. J. Environ. Res. Public Health **19**(10) (2022). https://doi.org/10.3390/IJERPH19106255

Niranjanamurthy, M., et al.: E-commerce and M-commerce: issues and recommended screening. Int. J. Mark. Technol. **2**(8) (2012). http://citeseerx.ist.psu.edu/viewdoc/summary?doi=10.1.1.375.2360. Accessed 2 Sept 2021

Majee, W., Hoyt, A.: Cooperatives and community development: a perspective on the use of cooperatives in development. J. Community Pract. **19**(1), 48–61 (2011). https://doi.org/10.1080/10705422.2011.550260

Hoque, S.M., Alam, S.M.: The role of information and communication technologies (ICTs) in delivering higher education-a case of Bangladesh. Int. Educ. Stud. **3**(2) (2010). www.col.org/ODLIntro/introODL.htm. Accessed 11 Sept 2022

Mergel, I., Edelmann, N., Haug, N.: Defining digital transformation: results from expert interviews. Gov. Inf. Q. **36**(4), 101385 (2019). https://doi.org/10.1016/J.GIQ.2019.06.002

Montegut-Salla, Y., Cristóbal-Fransi, E., Gómez-Adillón, M.J.: Understanding the situation and factors of ICT adoption in agricultural cooperatives. J. Electron. Commer. Organ. **11**(3), 1–26 (2013). https://doi.org/10.4018/JECO.2013070101

Mukerji, M.: Re-examining strategic and developmental implications of e-Choupal, India. Electron. J. Inf. Syst. Developing Countries **86**(4), e12132 (2020). https://doi.org/10.1002/ISD2.12132

Nicholls, A.: The legitimacy of social entrepreneurship: reflexive isomorphism in a pre–paradigmatic field. SAGE J. (2010). https://doi.org/10.1111/j.1540-6520.2010.00397.x. Accessed 6 Jan 2021

Nicholls, A.: Introduction: the meanings of social entrepreneurship. In: Nicholls, A. (ed.) Social Entrepreneurship: New Paradigms of Sustainable Social Change, pp. 1–36. Oxford University Press, Oxford (2006)

Ortmann, G.F., King, R.P.: Agricultural cooperatives I: history, theory and problems. Agrekon **46**(1), 18–46 (2007). https://doi.org/10.1080/03031853.2007.9523760

Osberg, S., Martin, R.: Social entrepreneurship: the case for definition. Stanford Soc. Innov. Rev. Spring **5**, 28–39 (2007)

Purbasari, R., Raharja, S.J., Raharja, S.J.: Digital transformation in cooperative business processes: a study on cooperatives in the greater Bandung area. Inovbiz: Jurnal Inovasi Bisnis **10**(1), 16–22 (2022). https://doi.org/10.35314/INOVBIZ.V10I1.2367

Rachinger, M., et al.: Digitalization and its influence on business model innovation. J. Manuf. Technol. Manag. **30**(8), 1143–1160 (2019). https://doi.org/10.1108/JMTM-01-2018-0020/FULL/PDF

Reis, J., Amorim, M., Melão, N., Cohen, Y., Rodrigues, M.: Digitalization: a literature review and research agenda. In: Anisic, Z., Lalic, B., Gracanin, D. (eds.) IJCIEOM 2019. LNMIE, pp. 443–456. Springer, Cham (2020). https://doi.org/10.1007/978-3-030-43616-2_47

Rijswijk, K., et al.: Digital transformation of agriculture and rural areas: a socio-cyber-physical system framework to support responsibilisation. J. Rural Stud. **85**, 79–90 (2021). https://doi.org/10.1016/J.JRURSTUD.2021.05.003

Saarikko, T., Westergren, U.H., Blomquist, T.: Digital transformation: five recommendations for the digitally conscious firm. Bus. Horiz. **63**(6), 825–839 (2020). https://doi.org/10.1016/J.BUSHOR.2020.07.005

Sanfilippo, M., Chattopadhyay, T.: Sociotechnical cooperatives: the impact of technology on cooperative organizations. SSRN Electron. J. (2020). https://doi.org/10.2139/SSRN.3749545

Schroeder, A., Ziaee Bigdeli, A.: Capturing the benefits of digitalization and service innovation: a business network perspective. Acad. Manag. Proc. **2018**(1), 18491 (2018). https://doi.org/10.5465/AMBPP.2018.18491ABSTRACT

Sehlin, D., Truedsson, M., Cronemyr, P.: A conceptual cooperative model designed for processes, digitalisation and innovation. Int. J. Qual. Serv. Sci. **11**(4), 504–522 (2019). https://doi.org/10.1108/IJQSS-02-2019-0028

Slavinski, T., Todorović, M.: The impact of digitalisation on the organisational capability changes – evidence from Serbia (2020). https://doi.org/10.2991/SENET-19.2019.41

Vanlalmuana, M., Laldinliana, D.: Structure and financial performance of primary agricultural credit cooperative societies in Mizoram. Int. J. Adv. Sci. Hub **2**, 202–206 (2020). https://doi.org/10.47392/IRJASH.2020.120

Vadrot, A.B.M.: Re-thinking the conditions for social change and innovation. Innov. Eur. J. Soc. Sci. Res. **33**(1), 1–3 (2020). https://doi.org/10.1080/13511610.2020.1713455

Wang, Y.-S., Lin, H.-H., Luarn, P.: Predicting consumer intention to use mobile service. Inf. Syst. J. **16**(2), 157–179 (2006). https://doi.org/10.1111/J.1365-2575.2006.00213.X

Wedel, M.: Social change and innovation for times of crises. Innov. Eur. J. Soc. Sci. Res. **33**(3), 277–279 (2020). https://doi.org/10.1080/13511610.2020.1789313

Williams, L.D.: Concepts of digital economy and industry 4.0 in intelligent and information systems. Int. J. Intell. Netw. **2**, 122–129 (2021). https://doi.org/10.1016/J.IJIN.2021.09.002

Yang, F., Gu, S.: Industry 4.0, a revolution that requires technology and national strategies. Complex Intell. Syst. **7**(3), 1311–1325 (2021). https://doi.org/10.1007/S40747-020-00267-9/TABLES/2

Yang, W., et al.: How could policies facilitate digital transformation of innovation ecosystem: a multiagent model. Complexity (2021). https://doi.org/10.1155/2021/8835067

Yunus, M.: Creating a world without poverty: social business and the future of capitalism. Public Affairs, New York (2008)

The Influence of Instagram Social Media on Participant Interest in MICE Tourism (Case Study: Bina Nusantara University Students)

Fithria Khairina Damanik(✉) and Nabila Fidy Thyssen

Tourism Department, Faculty of Digital Communication, Hotel and Tourism, Bina Nusantara University, Jakarta 11480, Indonesia
fithria.damanik@binus.edu

Abstract. This research is aiming to analyze the influence of Instagram social media on the participation's interest in MICE (Meeting, Incentive, Conference, and Exhibition) events among Bina Nusantara University students. This research is inspired by the lack of interest to participate in MICE events among the students. This research uses a quantitative approach with simple linear regression analysis using SPSS (Statistical Program for Social Science) to reach its purposes. The analysis is based on a questionnaire to 396 students among various majors in Bina Nusantara University. The results of the simple linear regression test and hypothesis testing indicate that there is an influence of Instagram social media on to student's participation interest in MICE events with the regression value of 0.96. Furthermore, the collaboration dimension from social media influence variable got the highest value with the total of 3.36. Meanwhile, in the participation interest on MICE event, the desire dimension got the highest value with the total of 3.29.

Keywords: Instagram social media · MICE tourism · Participant interest

1 Introduction

The data show that 4% of Indonesia national income is coming from tourism industry [1]. Another data show that 55% of Indonesia income is produced in Java Island with total contribution of DKI Jakarta Province is reaching 17.23% [2]. However, related to tourism industry in DKI Jakarta, the total contribution from this industry is only 9.20% in year of 2020 [2]. Refers to its potentials, this number can be maximized more especially due to the position of Jakarta as the capital city of Indonesia.

Therefore, special attention needs to be addressed to develop Jakarta's tourism. One of the focuses is through the concept of MICE (Meeting, Incentive, Conference, and Events) as stated in The Medium-term Development Plan (RPJMD) of DKI Jakarta. MICE activities are believed to be a sub sector in tourism which can be beneficial for various stakeholders in the industry, such as travel agent, event organizer, accommodation, food and beverage, and local small medium enterprises. This capital city of Indonesia for sure has a lot of potential in MICE for its complete infrastructure and

B. Alareeni et al. (Eds.): ICBT 2022, LNNS 620, pp. 696–707, 2023.
https://doi.org/10.1007/978-3-031-26953-0_64

accessibility. However, the data show that MICE is on the second lowest rank of the tourist travel purposes with the decreasing trend, the score is only 0.5% [1].

In this digital era, there is a change in the communication behavior of the millennial generation (25–34 years old) and generation Z (15–24 years old). These changes can be seen in everyday life, such as the lack of face-to-face interaction due to the presence of gadgets [3]. The digital era of the tourism industry provides convenience in the form of distributing information that can reach a wide network area and can be accessed by many people so that the search for information on services or tourism services can be more effective and efficient. The form of technological progress and other information is the existence of social media. The use of social media to support tourism development is also based on the high number of social media user in Indonesia [4].

Based on the data, social media users are reaching 4.54 billion out of the total of 7.75 billion people of worldwide populations [4]. Meanwhile, in Indonesia, the internet user is reaching 175.4 million people or 64% or the total population with 160 million people are social media users (Youtube, Facebook and Instagram [4]. This research focuses on one platform of social media which is Instagram with 85 million users with 40% advertising potentials [4].

Based on that background, the author has a research question on how the influence of social media Instagram to the participation's interest on MICE tourism is, with case study on students at Bina Nusantara University. The group of students is chosen as the population of sampling because another data shows that people on the age of 15–24 is the category of people that infrequently doing tourism activity [1]. Therefore, the aim of this research is to analyze the influence of Instagram social media on participation interest in MICE tourism. This research hopefully will give suggestions on social media optimalization to increase participation's interest in MICE tourism among the students. The aim of the research will be answered in the following structure of paper containing of introduction, literature review, research method, findings, and conclusions.

2 Literature Review

MICE is defined as convention tourism, with the limitation of convention services business, travel incentives, and exhibition. MICE is tourism activities combining between business and leisure with the activity including meeting, incentive, convention, and exhibition that involving big amount of people [5]. Another definition stated that MICE is a tourism activity where entrepreneurs and professionals gathered in a place to do discussion related to topic or problems [6] MICE is a tourism product that can be sold to associations, institutions, corporation, or big company on a local, national, and international level. This industry needs collaboration with various stakeholders to provide the best service for their market. MICE industry has positive as the increase of economic and income of the community [5].

Within this digital era, internet use cannot be denied as one of the daily needs of people worldwide. One of its implementations is the use of social media that can be utilized to promote various industries including tourism. To show indicators in using social media, this research uses 4C theory [7]. Including Context, Communication, Collaboration, and Connection.

1. Context: "How we frame our stories"
   Context is how users form a story or message (information) into valuable and qualified content.
2. Communication: "The practice of sharing our story as well as listening, responding, and growing"
   Communication is a form of content sharing including listening, responding, and growing together. By doing content sharing through social media, communication can be happening between organizer and social media user through photo or caption sharing.
3. Collaboration: "Working together to make things better and more efficient and effective"
   Collaboration is a cooperation between organizer and information user through social media to deliver the message more effectively and efficiently.
4. Connection: "The relationships we forge and maintain"
   Connection can be defined to maintain relationship between organizer and social media user that has been built virtually through social media. Maintaining relationships needs to be done sustainably to make the user feel closer to the account promoting product and services.

Meanwhile for the participation variable, the indicator used is the AIDA model. AIDA stands for Awareness, Interest, Desire, and Action. The intention to participate happened after someone got the information related to an event. AIDA model used to measure the participation interest; it has a hierarchy of somebody's response after seeing an advertisement [8]. Here is the explanation of each indicator in AIDA model.

1. Awareness
   Awareness is the first hierarchy on the model, where it shows the awareness coming as a response to product or service. In this stage, the information provider introduces their product or services by giving various information to the market.
2. Interest
   Advertising of a product or service is intended to create a market interest to know more about it. Hence, people can evaluate whether the product or service is valuable for market expectation.
3. Desire
   Desire is the third hierarchy to show the intention of the customer to buy a product or use a service,
4. Action
   Action is the last hierarchy where consumers are ready to do the transaction to fulfill their needs and wants for the product or service.

3 Research Methods

The research was conducted based on a purpose to see if Instagram social media has the influence on participation interest of MICE event with case study on Bina Nusantara University students. Therefore, these following hypotheses were proposed.

H0: Instagram social media has no influence on the participation interest of MICE events among BINUS University students.
H1: Instagram social media has influence on the participation interest of MICE events among BINUS University students.

This research uses the sampling method by counting the total of Bina Nusantara University students who are assumed as social media active users. The total population is 45,925 people. The sample needed is counted based on this following formula.

$$n = \frac{N}{N.d^2 + 1}$$

Description:
n = sample needed
d = Standard of distribution with a = 5%
N = total population
Therefore, based on that formula, here is the sample needed for this research:

$$n = \frac{45,925}{45,925x0.0025^2 + 1}$$

$$n = \frac{45,925}{115 + 1} = 395.905 = 396$$

Hence, the minimum requirement for respondents is 396 people.

The sample of this research will be given an online questionnaire containing several questions related to respondent's basic profiles and several statements about reaction after seeing promotion on social media based on 4C (Context, Communication, Collaboration, and Connection), and AIDA model. The valid answer is only the answer from the respondents who actively using Instagram social media. The following table shows the connection between variables and indicators that proceed as a statement in questionnaire for the respondents (Table 1).

Table 1. Connection between social media variables, indicators, and statements.

Indicator of social media	Definition	Question/statement
Context	Content packaging through choices and variations of photos and informative captions	1. The MICE event content on Instagram is quite interesting 2. Content related to MICE events contains informative captions

(*continued*)

Table 1. (*continued*)

Indicator of social media	Definition	Question/statement
Communication	Interaction between account owners and other users	3. MICE event Instagram account managers need to interact with followers/audiences to improve two-way communication (engagement) 4. The MICE event Instagram account manager is quite interactive
Collaboration	Contribution and engagement of other users through comments and like buttons	5. Audience plays an active role in giving likes and comments on MICE event content 6. MICE event content on Instagram gets a lot of likes and comments in every upload
Connection	The use of hashtags to make it easier for users to connect quickly and easily	7. The use of hashtags (#/hashtag) is an effective and efficient way to find information on MICE events 8. Through the hashtag /hashtag feature, it makes it easier for me to choose the type of MICE event that I will participate in. Likert

Meanwhile, for the second variable of participation interest based on AIDA model, the statement given to the respondent based on the indicator is on the following table (Table 2).

Table 2. Connection between participation interest variables, indicators, and statements.

Indicator of participation interest (AIDA)	Definition	Question/statement
Awareness	Is a form of attention and awareness of the existence of a product or service	9. Through Instagram it helps me to find out information related to MICE events 10. Marketing of MICE event content is often found through Instagram

(*continued*)

Table 2. (*continued*)

Indicator of participation interest (AIDA)	Definition	Question/statement
Interest	The emergence of a citizen's interest in a product or service, which is introduced by an account	11. Social media Instagram became a platform that made me interested in participating in MICE events 12. MICE events on Instagram provide content that makes the audience interested in participating
Desire	Desire that arises because of the exchange process that is interesting for netizens	13. Uploading MICE events encourages me to know more about the activities and benefits that I will get 14. Uploading the MICE event on Instagram became a consideration for me to participate in the event
Action	The act of buying or participating in an activity	15. I chose to participate in the MICE event because of the upload on Instagram 16. I plan to participate in MICE events again in the future

The statements that are asked to the respondents will be measured using the Likert Scale. The answer is divided into 5 levels of opinions from 1–5 representing Very disagree until very agree level. The answer will be analyzed with a statistical test using SPSS. Since the questionnaire is given online through google form, the result can be downloaded directly into excel form and imported to SPSS.

4 Findings

The questionnaires were distributed to 396 respondents of Bina Nusantara University students through online questionnaires. The answer shows the composition of respondents are 46,70% or 185 students of male respondents and 53,3% or 211 students of female respondents. The respondents came from various departments in Bina Nusantara. Then, the next question is about their usage of Instagram, non-instagram users' respondents will be excluded from the analysis.

The first step of the analysis is checking the validity and readability test of the instrument. The validity test is aiming to test the accuracy of the measuring instrument or questionnaire in taking measurements, while the reliability test is aiming to test whether the information used is reliable or consistent (Table 3).

Table 3. Validity score of participant's interest variable.

Variable	Number	R (Count)	The value of R table	Value of Sig	Decision
The social media influence (X)	X1	.671	0.098	.000	Valid
	X2	.702	0.098	.000	Valid
	X3	.657	0.098	.000	Valid
	X4	.741	0.098	.000	Valid
	X5	.844	0.098	.000	Valid
	X6	.695	0.098	.000	Valid
	X7	.844	0.098	.000	Valid
The participation interest of MICE events (Y)	X8	.844	0.098	.000	Valid
	Y1	.691	0.098	.000	Valid
	Y2	.762	0.098	.000	Valid
	Y3	.582	0.098	.000	Valid
	Y4	.643	0.098	.000	Valid
	Y5	.748	0.098	.000	Valid
	Y6	.766	0.098	.000	Valid
	Y7	.691	0.098	.000	Valid
	Y8	.762	0.098	.000	Valid

Based on the table, the items used as research instruments on the variables of Instagram's social media influence and participation interest, the score of sig value is under 0.005 which means all indicators are valid (Table 4).

Table 4. Reliability score of participant's interest variable.

Reliability statistics		
Cronbach's alpha	R table	N of items
.889	.098	8

The next analysis is to see the reliability score for each variable. The Cronbach's Alpha value shows the number of 0.889, which is bigger than the r table value of 0.098. It can be said that the student participation interest variable (Y) is reliable and the answers from respondents are consistent and reliable (Table 5).

Table 5. Reliability score of social media influence variable.

Reliability statistics		
Cronbach's alpha	R table	N of items
.874	.098	8

Meanwhile, the reliability check for social media influence variable shows Cronbach's Alpha value of 0.874 which is bigger than the r table value of 0.098, which means that the social media influence variable (X) is reliable and the answers from respondents are consistent and reliable (Table 6 and 7).

Table 6. Descriptive analysis of social media influence's indicator

Dimensions	No	Question/statement	N	*Min*	*Max*	*Mean*	*Std. Deviation*
'Context	X1	The MICE event content on Instagram is quite interesting	396	1	5	3.27	1.27
	X2	Content related to MICE events contains informative captions	396	1	5	3.26	1.26
Communication	X3	MICE event Instagram account managers need to interact with followers/audiences to improve two-way communication (engagement)	396	1	5	3.26	1.26
	X4	The MICE event Instagram account manager is quite interactive	396	1	5	3.26	1.24
Collaboration	X5	Audience plays an active role in giving likes and comments on MICE event content	396	1	5	3.36	1.26

(*continued*)

Table 6. (*continued*)

Dimensions	No	Question/statement	N	*Min*	*Max*	*Mean*	*Std. Deviation*
	X6	MICE event content on Instagram gets a lot of likes and comments in every upload	396	1	5	3.19	1.31
Connection	X7	The use of hashtags (#/hashtag) is an effective and efficient way to find information on MICE events	396	1	5	3.36	1.26
	X8	Through the hashtag /hashtag feature, it makes it easier for me to choose the type of MICE event that I will participate in. Likert	396	1	5	3.36	1.26
Summary			396	8	40	26.32	10.12

Table 7. Descriptive analysis of Participation Interest's indicator

Dimensions	No	Question/statement	N	*Min*	*Max*	*Mean*	*Std. Deviation*
Awareness	Y1	Through Instagram it helps me to find out information related to MICE events	396	1	5	3.07	1.25
	Y2	Marketing of MICE event content is often found through Instagram	396	1	5	2.86	1.29
Interest	Y3	Social media Instagram became a platform that made me interested in participating in MICE events	396	1	5	3.02	1.26

(*continued*)

Table 7. (*continued*)

Dimensions	No	Question/statement	N	*Min*	*Max*	*Mean*	*Std. Deviation*
	Y4	MICE events on Instagram provide content that makes the audience interested in participating	396	1	5	3.01	1.23
Desire	Y5	Uploading MICE events encourages me to know more about the activities and benefits that I will get	396	1	5	3.24	1.38
	Y6	Uploading the MICE event on Instagram became a consideration for me to participate in the event	396	1	5	3.34	1.43
Action	Y7	I chose to participate in the MICE event because of the upload on Instagram	396	1	5	2.87	1.28
	Y8	I plan to participate in MICE events again in the future	396	1	5	3.01	1.23
Summary			396	8	396	8	40

The next analysis is simple linear regression to see whether the X variable influences Y variable or not (Table 8).

Table 8. Simple linear regression analysis

Model		Coefficients*				
		Unstandardized coefficients		Standardized coefficients		
		B	Std. Error	Beta	t	Sig.
1	(Constant) Influence of Instagram social media	21.875	1.325		16.512	0.000
a.	Dependent Variable: Interest of Student	0.096	0.048	0.099	2.008	0.045

Equation based on Linear Regression.

Participant Interest = 21.875 + 0.0096 (social media).

Based on the above equation, the constant value is 21.875 states that if there is no increase in the value of the Instagram Social Media Influence variable (X), then the value of Student Participation Interest (Y) is 21.875. The regression coefficient (b) is 0.096, which means that each addition (because of the + sign) one value of the Instagram Social Media Influence variable (X) will give an increase of 0.096 on the Student Participation Interest variable (Y). Based on the results of these calculations, the authors conclude that the Instagram Social Media Influence variable (X) has a positive (unidirectional) influence on the Student Participation Interest variable (Y) (Table 9).

Table 9. Coefficient and correlation

Correlations			
Model		Influence of social media	Dependent variable: interest of student
Influence of social media	Pearson Correlation	1	0.99*
	Sig. (2-tailed)		0.48
	N	396	396
Dependent variable: participation interest	Pearson Correlation	0.499*	1
	Sig. (2-tailed)	0.048	
	N	396	396

If the number of Pearson correlation values is included in the decision rule, then the value is 0.099 > 0.098 which means there is a correlation between the X and Y variables. Then, this correlation analysis shows r(count) which is positive, which means that if the value of the Instagram social media influence (X) increases, the value of Participation Interest on MICE events (Y) will also increase, and vice versa (Table 10).

Table 10. T test

Model			t	Sig.
1	(Constant) Influence of social media		16.512	0.000
a.	Dependent Variable: Participation Interest		2.008	0.045

With a significant level of 5% or 0.05. The basis for decision making is as follows:

- If the value of Sig. > 0.05 and the value of t (count) < value of t(table) then H0 is accepted.

- If the value of Sig. < 0.05 and the value of t (count) > the value of t(table) then H1 is accepted.

The value of t (count) for the X variable = 2.008, while the value of t (table) is calculated by (0.05/2) = 0.0025 and df N−1 (396–1) = 395. The results obtained for t (table) is 1.962. Hence, based on the table, the value of Sig. The influence of social media Instagram (X) is 0.045 < 0.05 and the value of t (count) is 2.008 > t (table) 1.962 with a regression coefficient value of 0.096, then H0 is rejected and H1 is accepted. This means that the influence of Instagram social media (X) has a significant positive effect on Student Participation Interest (Y). This explains that the higher Instagram social media (X) value, the higher Participation Interest on MICE tourism (Y) will be.

5 Conclusion

Based on the research, it can be concluded that the Social Media Effect of Instagram (X) has a significant and positive effect on Student Participation Interest (Y) with the case study of Bina Nusantara University Students. In other words, the higher value of Instagram social media influence (X), the higher the Participation Interest (Y) on MICE events will be. Therefore, further researchers can also use other social media platforms such as YouTube, Tiktok, Twitter, Facebook, or the official website of tourism. On the other hand, the research can also be about specific accounts which provide information about MICE events. Each platform of social media will show different characters, different kinds of content, and various generations of users. Further research will show another type of result which is expected to enlarge the knowledge of tourism development, especially in terms of destination marketing.

References

1. Badan Pusat Statistik. Laporan Perekonomian Indonesia 2020, Badan Pusat Statistik, Jakarta (2020)
2. Badan Pusat Statistik. BPS Kuartal II-2021, Badan Pusat Statistik, Jakarta (2021)
3. Zis, S.F., Effendi, N., Roem, E.R.: Perubahan Perilaku Komunikasi Generasi Milenial dan Generasi Z di Era Digital. Satwika: Kajian Ilmu Budaya dan Sosial, pp. 69–87 (2021)
4. WeAreSocial and Hootsuite, Digital 2021: Global Overview Report, WeAreSocial & Hootsuite, 2021 (2021)
5. Desinathini, MICE, Jakarta: UNPAM PRESS (2019)
6. Indrajaya, T.: Potensi Industri MICE (Meeting, Incentive, Conference and Exhibition) di Kota Tangerang Selaran. Jurnal Ilmiah Widya (2015)
7. Arief, G.M., Millianyani, H.: Pengaruh Sosial Media Marketing Melalui Instagram Terhadap Minat Beli Konsume Sugar Tribe (2015)
8. Rehman, F.U., Nawaz, T., Ilyas, M., Hyder, S.: A comparative analysis of mobile and email marketing using AIDA model. J. Basic Appl. Sci. Res. **4**, 38–49 (2014)

User's Continuance Intention Towards Digital Payments: An Integrated Tripod Model DOI, TAM, TCT

A. Pushpa[1], C. Nagadeepa[1], K. P. Jaheer Mukthar[1](✉), Hober Huaranga-Toledo[2], Laura Nivin-Vargas[3], and Matha Guerra-Muñoz[4]

[1] Kristu Jayanti College Autonomous, Bengaluru, India
jaheer@kristujayanti.com
[2] Universidad Nacional Mayor de San Marcos, Lima, Peru
[3] Universidad Nacional Santiago Antunez de Mayolo, Huaraz, Peru
[4] Universidad Popular del Cesar, Valledupar, Colombia

Abstract. Technology advancement have revolutionalised the financial service sector. Digital payment have motivating paperless, faceless and cash less transactions strengthening the country's economic growth. Access to financial services is deemed as one among the key factors to socioeconomic resilience in the pandemic period. Pandemic has catalysed the access and usage of financial services across the world transforming the way people made and received payments or borrowed and saved. Experts predicted that epidemic has accelerated adoption of digital platforms at rapid speed and repetitive usage could lead to continuance usage, but there is lack of study in discussing the motivators that leads to user's continuance intention. Hence, the study aims to undertake an empirical analysis and synthesise the users continuance usage of digital payments post pandemic with integration of tripod theories, namely technology acceptance model (TAM), Technology continuance model (TCT) and Diffusion of Innovation (DOI). Factors like perceived usefulness, perceived ease of use, confirmation, Compatibility, and trialability as the functions of satisfaction and continuance intention were considered. The study followed a quantitative research design approach, using a survey method data was collected from 250 respondents in Bengaluru. The research model was tested using SEM analysis. The findings of the study reveal that all the factors have a positive association with the user's intention to continue using digital payments.

Keywords: Continuance intention · Digital payments · SEM analysis

1 Introduction

Fintech is a fast-shepherding industry equipped with technology and swift processes, it is synthesis of finance and technology. Fintech has empowered consumers to access advanced financial services like online savings and investments, online payments, mobile payments, Peer to peer lending (P2P), Government to people (G2P), budgeting and financial planning, crowdfunding, Xie et al. (2021); smart contracts, robo advisors, e-aggregators, block chain technology, cloud computing, big data, etc., D Acunto et al.

B. Alareeni et al. (Eds.): ICBT 2022, LNNS 620, pp. 708–717, 2023.
https://doi.org/10.1007/978-3-031-26953-0_65

(2019); Belanche et al. (2019). It is supplemented with a constellation of technologies that include internet, mobile networks, cloud computing, data analytics, big data, and artificial intelligence (AI), Horner and Cunnane (2017). Internet and Mobile Association of India (IAMA)-Kantar, based on ICUBE 2021 reported that usage of internet in rural India grew to 37% in 2021 from 31% in 2020, but urban India internet usage remained 66% and 69% since 2019. The concept of digital payment in India started flouring through the launch of "Digital India" mission in the year 2016 motivating paperless, faceless and cash less transactions and strengthening the country's economic growth. Access to financial services is deemed as one among the key factors to socioeconomic resilience in the pandemic period, Al Nawayseh (2020); Karusala et al. (2019). It is a game changer that ensured a comprehensive economic growth during the pandemic. Pandemic had a negative impact on most industries, every other industry experienced a slump in the growth, but fintech boomed as covid protocols like curtailed physical movement, lockdowns, social distancing, etc., encouraged contactless transactions. The pandemic has given a push to the adoption of digital payments, as it created an instantaneous need for both online and offline transactions. According to Pwc study, 2021 the global cashless payments increased by 42%. Pandemic has catalysed the acceptance and usage of financial services across the world transforming the way people made and received payments or borrowed and saved.

Digital payments enable electronic mode of making payments, facilitating users transfer funds amid transaction accounts through electronic accounts, traditional banking accounts or any other payment instruments. Digital payments are based on mobile wallets which replicate the physical wallets in mobile devices like smart phones and tablets, Mumtaza et al. (2020); Kaur et al. (2020), provide provisions to transfer from person to business transfers (online or offline purchases), government to person, person to person and other forms of payment Bills /fees/penalty and many more, Nofie Iman (2018). With this backdrop, the study aims to undertake an empirical analysis and synthesise the attitude of users for continuance usage of digital payments post pandemic. Considering this discussion, the present study deems it suitable to take on a review of available literature to understand the results proved, identify the limitations and the research design suitable for the research before further empirical research and is presented as background and hypothesis development for proposed model in the Sect. 2; Sect. 3 reveals the methodology used in conducting the research; Sect. 4 discusses analysis of data and discussion; Sect. 5 unveils the conclusion, implications followed by directions for future research.

2 Background and Hypothesis Development

Technology acceptance model (TAM) descents from TRA (Theory of reasoned action) proposed by Fishbein et al. (1975). TAM, Davis (1989) one of the robust and most dominant theories of technology/innovation acceptance, Venkatesh et al. (2000). The theory established the causal association between perceived usefulness (PU), perceived ease of use (PEOU) with user's intention to use. Researchers have always proposed additional variables to TAM strengthening TAM. Extended TAM 2 incorporated wide range of external factors like social and cognitive factors, Venkatesh (2000). The Diffusion innovation theory, Rogers (2003) represents as one of the most powerful theory in the literature of innovation adoption and used by most researchers to find insights on various technology contexts. Technology continuance theory (TCT) probes continuance intention of users towards technology/innovation; it encompasses factors embedded in TAM, ECM and cognitive model (COG), Liao et al. (2009); Expectation confirmation Model (ECM), Bhattacherjee (2001), evaluates three important determinants of technology continuance intention like confirmation, the perceived usefulness and the satisfaction.

2.1 Hypothesis Development for the Proposed Model

Using the theoretical background of tripod models namely, TAM, DOI and TCT, the study framed a research model that recognises some drivers and inhibitors of digital payment continuance intention. The basic notion hypothised in the proposed research model is that digital payment user's continuance intention is determined in association with the perceived usefulness, the perceived ease of use, confirmation, Compatibility, and trialability, as the functions of satisfaction and continuance intention. Firstly, TAM elements namely, perceived ease of use and perceived usefulness are incorporated in the integrated model. Secondly, confirmation, satisfaction and continuance intention factors of TCT are featured. Finally compatibility and trialability factors of DOI are integrated.

Hypothesis Development

Confirmation is described as "realisation of expected benefits", (Bhattacherjee 2001). There are numerous studies that examined the association between the confirmation and satisfaction of technology user, Hoehle et al., 2012; researchers like Zhou et al. (2018); Foroughi et al. (2019) established the association between confirmation and user's satisfaction. Likewise, studies by hoelhle et al. (2012); Albuhisi and Abdallah (2018), recognised significantly influence of confirmation and perceived usefulness on continuance intention of users. Hence, analysing the relationship between confirmation and satisfaction is relevant Perceived usefulness (PU) refers to the subjective assessment of users in understanding if the technology or system enhance job performance, Davis et al. (1989). It describes the magnitude to which the user trust that the technology could be a driving force in achieving his/her goals. PU affects the user's perceptions on satisfaction during acceptance or post acceptance phase, Bhattacherjee (2001b). Previous studies have reliably explained the positive association between perceived usefulness and satisfaction, Chen et al. (2013); Tam C et al. (2018); Dai et al. (2020). Perceived ease of use (PEOU) is the degree of conviction that users experience in using a technology effortless, Davis (1989). Studies relevant to Information technology unveils that perceived ease of use (PEOU) as a critical influencer on user's satisfaction, Shin et al. (2011); Joo et al. (2018). Hence, examining the perceived usefulness and perceived ease of use and its association with satisfaction is relevant. Trialability defines the degree to which the people deliberate that before adopting innovation they need to experience. With reference to the present study trialability implies how users view digital payments, this is highly influenced by observability, Y H Lee (2007). Studies in the area of TAM and DOI has significant effect on PU on the observability, M M Yang (2007); L Y Huang (2004). Lin et al. 2017) supported empirically trailability and users intention. There seems to be scarcity in the literature integrating this concept, so the Traibility factor was included in the study. Compatibility envisages the fact that the technology/innovation is attuned with social/cultural values and beliefs, previously introduced ideas and with client needs.

Studies opined that innovations that exhibit a higher compatibility and enables user's social and personal prominence are more likely to be acknowledged, Zilkepli et al. (2015). Covid-19 context digital payments have gained wide acceptance and adopted as it perceived compatibility where social distancing was highly encouraged, Sharma D et al. (2021). Earlier studies have deliberated the association between compatibility and behavioural intention, Lin T T C et al. (2017); but there are no studies incorporating the association of compatibility on satisfaction of users that leads to continuance usage. Hence, the study included Compatibility as inhibitor in the study. Satisfaction refers to the cumulative feelings due to multiple reactions in different contexts, San Martin et al. (2013). If users are not satisfied with the technology/system they discontinue its usage. Studies have revealed that satisfaction is one the key determinant of continuance usage, Zhou (2014); Lee et al. (2015). Based on the past literature, the study predicts a positive impact of satisfaction on user's continuance usage. The prevailing literature on user's continuance intention to use is evidenced that it is mainly influenced by satisfaction, Chen (2012). Wang et al. (2019) established the presence of significant impact between satisfaction and users intention to continue usage. With this back ground seven Hypothesis were framed and tested.

H1: Confirmation has a positive impact on satisfaction of users.

H2: Perceived usefulness has a positive impact on satisfaction of users.

H3: Perceived ease of use has a positive impact on satisfaction.

H4: Trailability has a positive impact on satisfaction.

H5: Compatibility has positive impact on satisfaction.

H6: Satisfaction has a positive impact on continuance intention.

2.2 Proposed Model

The casual relationship among the constructs is integrated into the proposed model and presented in the Fig. 1.

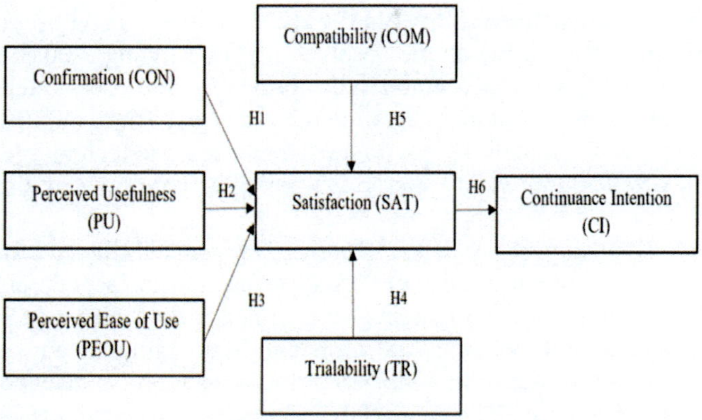

Fig. 1. A Tripod model of digital payment acceptance and continuance intention

3 Methodology

This study employed an exploratory one, used a survey method to test the hypothesis formulated in the previous section, the subsequent section discusses about the questionnaire development and data collection. The questionnaire entails two sections, the first part measures demographic aspects of the respondents, and the second part measures constructs using a five-point Likert's scale, anchored at 1 as strongly disagree and 5 as strongly agree to test theoretical model proposed. The constructs were adapted into digital payment continuance intention context. Four items measuring each perceived usefulness and perceived ease of use were adapted from Samar et al. (2017), three items measuring confirmation was adapted from Bhattacherjee (2001), Two items measuring compatibility and three items measuring trailability was included in the questionnaire; Three items measuring continuance adoption was adapted from Ghani et al. (2017).

As the research conducts is a multivariate analysis, the sampling size was derived at least 5 to 10 times the number of constructs, as suggested by Sugiyono (2018). The sample was chosen in the area of Bengaluru city, a sample of 280 respondents were provided with questionnaire through Google forms and social media platforms like WhatsApp and Instagram. 260 samples were received out of which 10 samples were rejected due to missing values and 250 samples were used for study. The demographic analysis in terms of gender and age showed that about 62.03% of them were male and 37.97% were female. Majority of them were in the age group of 21 and 30 (43.4%), 19.3% were below 20, and 37.3% of the respondents were in the age group above 30.

4 Data Analysis and Discussion

The hypothesis set were tested in three stages, firstly using SPSS 24.0 the instruments was tested for reliability to measure. Secondly, CFA was measured to check the indicator loadings for convergent and discriminant validity. Finally, SEM evaluated the coefficients of determination (R2), the path coefficients to arrive at the results and suggest implications. The Cronbach alpha estimated the reliability of the measures, and the values for the constructs gratified the cut-off of 0.70, J C Nunnally et al. (1884); Gefen et al. (2000) satisfying the adequate levels of reliability.

CFA and SEM Analysis

SEM analysis a multivariate statistical tool, that accommodates multiple interrelated dependence among the constructs in a model, J F Hair et al. (1998); J Anderson (1998) is used in the study. All the constructs of the model's factor loading signposts that all the loading are significant to their respective factors with p values < 0.05 at 5% LoS. The proposed model fit indices bared an overall fit and is statistically and theoretically fit. The fit measures valued are presented in the below table (Table 1).

Table 1. Results of model fit indices

$\chi 2$	P	GFI (Goodness of fit index)	AGFI (Adjusted Goodness of fit index)	CFI (Comparative fit index)	IFI (Incremental fit index)	RMR (Root mean square residual)
232.764	< 0.05	0.870	0.850	0.918	0.913	0.052

The Fig. 2 represents the SEM analysis integrating TAM, DOI and TCT with the result. The seven hypothesis relationship between the constructs are tested and the results of the relationship is significantly supported.

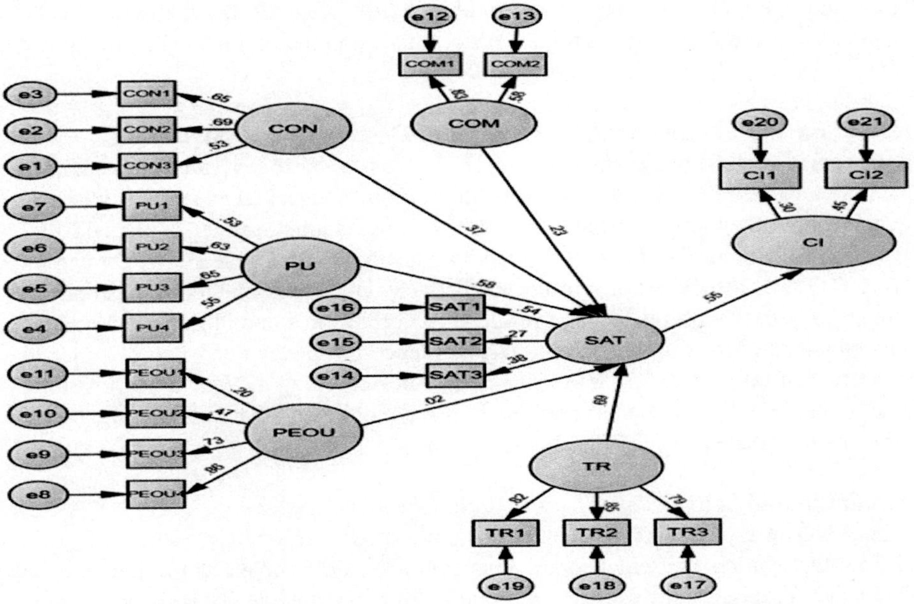

Fig. 2. SEM-Tripod model of digital-payment acceptance and continuance intention

5 Result Discussion

Confirmation and Satisfaction Towards Digital-Payment System
The one of the features in e-payment system Confirmation has an influence towards customer satisfaction with β value 0.37 and p is less than 0.05 of e-payment system. The respondent's perceived experience regarding using of e-payments was better option for them during the Covid-19, which was more than what they expected, the features of e-payments such as cash back offer, redeemable coupons, extended warranties, freebies, and discount offers meet the respondent's expectation and confirms their satisfaction. This result support the study by various researchers (Rahi 2020; Foroughi, 2019) proved that confirmation affects user's satisfaction. The majority of e-wallet applications offer a user-friendly design that makes it easy for users to understand how each feature works (Puspitasari 2021). These characteristics validate users' expectations of an electronic wallet, which impact user attitudes, perceived app benefits, and general contentment.

Perceived Usefulness and Satisfaction Towards Digital-Payment System
Customers' perceived usefulness of e-payment system shows a positive influence towards customer satisfaction with β value 0.58 and p is less than 0.05 of e-payment system. There are many useful attributes of e-payment system satisfies the respondents. It helps them to track their spending through e-payments, it generates the financial report and further helps them to manage their personal finance very effectively. In line with earlier studies, the researcher found that perceived usefulness significantly influenced users' satisfaction. Further, it suggests that consumers' continued usage of digital wallets is determined by their attitude toward them and their perception of their utility, Daneji (2019). According to earlier research, perceived utility and PEOU are the one of the most significant predictors of a respondent's attitude and intention to engage in a given behaviour, Davis (1989); Vijayasarathy (2004).

Perceived Ease of Use and Satisfaction Towards Digital-Payment System
The perceived ease of use e-payment system influences customer satisfaction with β value 0.37 and the $p < 0.05$. Various benefits of e-payments such as very easy to use, how convenient they are, and how easily they can be handled, influence respondents' satisfaction with them. This finding conflicts with the findings of a research by Daragmeh et al. (2021) who discovered insignificance between customers' perceived ease of use and their happiness with the use of e-wallets. According to some of the other studies, perceived ease of use is one of the primary influences on behaviour when it comes to the adoption of information systems (Davis, 1989). According to Velicia-Martin et al., COVID-19 affected respondents' decision to use a mobile app that showed whether or not they had contact with COVID-19-infected people.

Compatibility and Satisfaction Towards Digital-Payment System
Compatibility of e-payments, influences customer satisfaction with β value 0.23 ($p < 0.05$). As, the features in e-payment systems are makes the people comfortable in their daily financial transactions, further it is suitable for their lifestyle. These kind of comfortable features influences the respondents towards the satisfaction of e-payments. The

result of the study lines are supported by a research (Sebetci, 2018) on customer satisfaction through technology compatibility, which is considered that one of the important factors influencing overall satisfaction, was technology compatibility.

Trialability and Satisfaction Towards Digital Payment System
Trialability of e-payments influences customer satisfaction with β value of 0.69 ($p < 0.05$). COVID-19, pushed the people to use e-payment system in their day to day life. The trialability features such as "respondents' wish to e-payments option for a certain period of time", "to test some features of e-payment system during the trial period" and "Based on trial, it helps the respondent to decide whether to use the option or not" influenced them to accept and satisfy with the new technology. The result of the study is in the line with the earlier study (Puspitasari, 2021), proved that increased trialability increases consumer satisfaction. A fully functional application is available in the trial version for the customer to test before deployment. It is anticipated that the user would be satisfied with the e-wallet programme when he or she chooses to use it.

Satisfaction on Usage of Digital Payments and Continuance Intention
One of the key ideas in accepting any technology is satisfaction, which is described as the reaction of users to a technology. Users' intentions to continue using any technology are influenced by their contentment with prior usage and the technology's perceived value. The result of the path satisfaction towards e-payments influences the customer to show continuance intention towards its usage with the β value of 0.55 ($p < 0.05$) of e-payment system. Most of them started using this e-payments and made it as their daily habit. This finding was substantiated by a research provided by Basak and Calisir (2015).

6 Conclusion

The present study aimed to investigate the user's continuance intention towards digital payments with integration of tripod models namely, TAM, TCT and DOI. A schematic review of literature suggested existence of several studies on initial adoption of technology/innovation from various contexts. Post-adoption behaviour is yet to be tested. Thus, the study anticipated a cohesive model to explore the continuance intention among the users towards the digital payments. The new proposed model was verified with the SEM analysis. The study verdict suggests user's continuance intention is jointly anticipated by perceived usefulness, perceived ease of use, confirmation, compatibility, and trialability positively influenced satisfaction, leading the users intention. However, the study has few limitations, the study outcomes cannot be generalised for external validity as the results were derived from a single study and specific region because of differences in culture, a research in future determining if verdict of the study could be generalised is suggested. Also, future studies can include user's personal characteristics and feelings like innovative marketing that could impact user's continuance intention. A follow up study could be conducted in future with moderating variables to analyse the moderating or mediating effect.

References

Basak, E., Calisir, F.: An empirical study on factors affecting continuance intention of using facebook. Comput. Hum. Behav. **48**, 181–189 (2015)

Bhattacherjee, A.: Understanding information systems continuance: an expectation-confirmation model. MIS Q. **25**(3), 351–370 (2001). https://doi.org/10.2307/3250921

D'Acunto, F., Prabhala, N., Rossi, A.G.: The promises and pitfalls of roboadvising. Rev. Finan. Stud. **32**(5), 1984–2020 (2019)

Daneji, A.A., Ayub, A.F.M., Khambari, M.N.M.: The effects of perceived usefulness, confirmation and satisfaction on continuance intention in using massive open online course (MOOC). Knowl. Manag. E-Learning **11**(2), 201–214 (2019)

Daragmeh, A., Sági, J., Zeman, Z.: Continuous intention to use e-wallet in the context of the COVID-19 pandemic: integrating the health belief model (HBM) and technology continuous theory (TCT). J. Open Innovation: Technol. Market Complex. **7**, 132 (2021). https://doi.org/10.3390/joitmc7020132

Davis, F.D.: Perceived usefulness, perceived ease of use, and user acceptance of information technology. MIS Q. **13**(3), 319–340 (1989)

Ghani, M., Rahi, S., Yasin, N., Alnaser, F.: Adoption of internet banking: extending the role of technology acceptance model (TAM) with e-customer service and customer satisfaction **35** (2017). https://doi.org/10.5829/idosi.wasj.2017.1918.1929

Foroughi, B., Iranmanesh, M., Hyun, S.S.: Understanding the determinants of mobile banking continuance usage intention. J. Enterp. Inf. Manag. (2019)

Hoehle, H., Huff, S., Goode, S.: The role of continuous trust in information systems continuance. J. Comput. Inf. Syst. **52**(4), 1–9 (2012)

Horner, S., Cunnane, P.: Value of Fintech (2017)

Kaur, H., Singh, T., Arya, Y.K., Mittal, S.: Physical fitness and exercise during the COVID-19 pandemic: a qualitative enquiry. Front. Psychol. **11**, 590172 (2020). https://doi.org/10.3389/fpsyg.2020.590172

Huang,L.Y.: A study about the key factors affecting users to accept Chunghwa telecom's multimedia on demand. M.S. Thesis, Dept. Inf. Syst., Nat. Sun Yat-Sen Univ. (2004)

Lin, T.T.C., Bautista, J.R.: Understanding the Relationships between mHealth Apps' Characteristics, Trialability, and mHealth Literacy. J. Health Commun. **22**, 346–354 (2017). https://doi.org/10.1080/10810730.2017.1296508

Yang,M.M.: An exploratory study on consumers' behavioral intention of usage of third generation mobile value-added services. M.S. Thesis, Dept. Inf. Syst., Nat. Cheng Kung Univ. Tainan City, Taiwan (2007)

Fishbein, M., Ajzen, I.: Belief, Attitude, Intention and Behaviour: An Introduction to Theory and Research. Addison-Wesley, California (1975)

Mumtaza, Q.M.H.M., Intishar, S., Amaliya, S., Rosabella, Y., Hammad, J.A.H.: Worldwide mobile wallet: a futuristic cashless system. Bull. Soc. Inf. Theor. Appl. **4**(2), 70–75 (2020)

Iman,N.: Is mobile payment still relevant in the fintech era? Electron. Commer. Res. Appl. **30**, 72–82 (2018). https://doi.org/10.1016/j.elerap.2018.05.009. ISSN 1567-4223

Puspitasari, I., Wiambodo, A.N.R., Soeparman, P.: The impact of expectation confirmation, technology compatibility, and customer's acceptance on e-wallet continuance intention. In: AIP Conference Proceedings, vol. 2329, no. 1, p. 050012. AIP Publishing LLC, February 2021

Rahi, S., Khan, M.M., Alghizzawi, M.: Extension of technology continuance theory (TCT) with task technology fit (TTF) in the context of internet banking user continuance intention. Int. J. Q. Reliab. Manage. (2020)

Rogers E.M.: Diffusion of innovations, 5th Edition. Simon and Schuster (2003). ISBN 978-0-7432-5823-4

Park, S.Y.: An Analysis of the technology acceptance model in understanding university students' behavioral intention to use e-learning. Educ. Technol. Soc. **12**(3), 150–162 (2009)

Sebetci, Ö.: Enhancing end-user satisfaction through technology compatibility: an assessment on health information system. Health Policy Technol. **7**(3), 265–274 (2018)

Sharma D., Nawab A.Z.B., Alam M.: Computational intelligence methods in COVID-19: surveillance, prevention, prediction and diagnosis. In: Integrating M-Health with IoMT to Counter COVID-19, pp. 373–396. Springer, Singapore (2021)

Tsai, H., Lee, Y.-P., Ruangkanjanases, A.: Understanding the effects of antecedents on continuance intention to gather food safety information on websites. Front. Psychol. **11**, 579322 (2020). https://doi.org/10.3389/fpsyg.2020.579322

Venkatesh, V.: Determinants of perceived ease of use: integrating control, intrinsic motivation, and emotion into the technology acceptance model. Inf. Syst. Res. **11**(4), 342–365 (2000)

Venkatesh, V., Morris, M.G., Davis, G.B., Davis, F.D.: User acceptance of information technology: toward a unified view. MIS Q. **27**(3), 425–478 (2003). https://doi.org/10.2307/30036540

Velicia-Martin, F., Cabrera-Sanchez, J.-P., Gil-Cordero, E., Palos-Sanchez, P.R.: Researching COVID-19 tracing app acceptance: incorporating theory from the technological acceptance model. PeerJ Comput. Sci. **7**, e316 (2021)

Vijayasarathy, L.R.: Predicting consumer intentions to use on-line shopping: the case for an augmented technology acceptance model. Inf. Manag. **41**, 747–762 (2004)

Xie, J., et al.: Understanding FinTech platform adoption: impacts of perceived value and perceived risk. J. Theor. Appl. Electron. Commer. Res. **16**(5), 1893–1911 (2021)

Zhang, Y., et al.: Factors influencing patients' intentions to use diabetes management apps based on an extended unified theory of acceptance and use of technology model: web based survey. J. Med. Internet Res. **21**(8), e15023 (2019). https://doi.org/10.2196/15023

Zhou, W., Tsiga, Z., Li, B., Zheng, S., Jiang, S.: What influence users' e-finance continuance intention? The moderating role of trust. Ind. Manag. Data Syst. **118**(8), 1647–1670 (2018). https://doi.org/10.1108/imds-12-2017-0602

A Study on Cosmetics and Women Consumers: Government Protective Measures and Exploitative Practices

Syed Kazim[1], K. P. Jaheer Mukthar[2](✉), Robert Jamanca-Anaya[3], Cilenny Cayotopa-Ylatoma[4], Sandra Mory-Guarnizo[4], and Liset Silva-Gonzales[4]

[1] CMS Business School, JAIN (Deemed-to-be University), Bengaluru, India
[2] Kristu Jayanti College Autonomous, Bengaluru, India
jaheer@kristujayanti.com
[3] Universidad Nacional Santiago Antunez de Mayolo, Huaraz, Peru
[4] Universidad Señor de Sipán, Chiclayo, Peru

Abstract. Women have been using cosmetics for ages. Women for ages want to look good, beautiful, and presentable. Today cosmetics products have become a part and parcel of every urban woman. When the woman goes to work, her purchasing power increases, and therefore the consumption rate of cosmetics also increases. As women are the major consumers of cosmetics, the study is only focused to study the behavior of women when it comes to buying and using cosmetics. As per a report published by research and markets, the Indian beauty and personal care market is estimated to be USD 24.53 billion in 2022 and is expected to reach USD 33.33 billion by 2027, growing at a CAGR of 6.32%. Taking this into consideration a study is conducted with special reference to women. The objective of the study is to analyze the purchase behavior, the role of government in regulating the cosmetics industry, and the extent to which unethical activities prevail in the industry.

Keywords: Cosmetics · Women · Government · Exploitative practices

1 Introduction

The cosmetics industry came into existence to create beauty products for women. One of the first cosmetics products in India was a fairness cream which was called Afghan Snow, which was launched in 1919. The product was manufactured by ES Patanwala. The cream was named after King Zahir of Afghanistan, who felt that it reminded him of the snow from his homeland. Since then, there has been no looking back for the cosmetics industry. After India's independence, the cosmetics industry grew significantly and also led to the creation of many new products.

The cosmetics industry actually gained popularity after India implemented Liberalization, Privatization, and Globalization (LPG) in 1991, and especially after Aishwarya Rai won Miss World in the year 1994. This win made the Indian women feel that the

B. Alareeni et al. (Eds.): ICBT 2022, LNNS 620, pp. 718–732, 2023.
https://doi.org/10.1007/978-3-031-26953-0_66

world appreciated Indian beauty. After this many foreign players also entered the Indian cosmetics market and from then on it has only grown. Another reason for the growth of cosmetics in India is due to the purchasing power of women, which increased as the number of women going to work increased gradually over the years. Cosmetics marketing is estimated at 5,000 crores in India.

Over the years, the concept of because and consciousness to look good has grown to such an extent that even men want to use cosmetics products. As exclusive products were not created for men, they would use products that were created for women. But, of late, companies like Emami and others have come forward to create separate beauty and fairness products only for men, in order to meet the market demands.

Some of the major products which are sold in the cosmetics industry are talcum powder, shampoo, eyeliner, foundation cream, face wash cream, peel-out creams, fairness creams, fairness oil, nail polish, lipstick, antiperspirants and deodorants, hair dye, hair oil, and body massage cream.

One of the issues which were recently highlighted and discussed recently was about famous personalities becoming brand ambassadors of fairness creams, as it promotes discrimination and racism. The issue was in the news when South Indian actress Sai Pallavi declined a 2 crore offer to endorse a fairness cream. Some of the other actors and actresses who have refused to endorse fairness creams are Kangana Ranaut, Anushka Sharma, Priyanka Chopra Jonas, Swara Bhaskar, Kalki Koechlin, Sushant Singh Rajput, Abhay Deol, Ranbir Kapoor, and Randeep Hooda.

In order to understand the cosmetics industry, we also need to understand the role played by the government, various acts and organizations, and the administration, who are playing a role to regulate cosmetics companies and their products. It is important to understand the legal framework of healthcare in India, the drug and cosmetics act 1940, the amendments of the drug and cosmetics act 1940, the amendments in the drug and cosmetics act in 1986, and the role played by the central drug control administration, the central standard control organization (CDSCO) and the state drug control administration (SDCA).

Over the years, the Advertisement Standard Council of India (ASCI), a private organization has been raising objections over the unethical and inapposite advertisements which are created by various companies.

A research gap was found after the review of the literature. Not much study was done highlighting the ethical and unethical aspects of the study and the role of the government to control the cosmetics industry with respect to production and promotional strategies. The paper is divided into 5 parts. Introduction, literature review, research methodology, analysis and interpretation, and conclusion.

Objectives of Study

1. To assess the use of cosmetics products among women from various economic backgrounds
2. To identify the factors which influence the purchase of cosmetics among women
3. To analyze the current measure taken by the government toward the cosmetics industry
4. To highlight the unethical practices adopted by cosmetics companies and advertisement agencies

5. To identify the factors which are responsible for adopting unethical practices with respect to women consumers
6. To offer suggestions and recommendations to women cosmetics consumers.

2 Literature Review

There exists a strong criticism that beauty is linked with success. Cosmetics companies have strived to refine beauty with the help of cosmetics. The cosmetics companies have only shown that they would look beautiful if they use cosmetics products. As these companies want to sell their cosmetic products, they have only promoted the positive side of the products and have never highlighted the negative impact it can create on human skin in the long run. As also all human skins are not the same, there are changes that some cosmetics products would not be suitable for many people (Martin 1999).

Movies have played a significant role in popularizing various cosmetics products and to create a link between femineity and cosmetics. Over the years, the companies have also adopted aggressive marketing and promotional strategies to promote all their beauty products to the general public. This promotion campaign has given milage and popularity to various cosmetics products (Jennifer 1999).

Various body care products, clothes, and jewellery are used as adornments for women, which adds to their beauty. In order to be distinguished in society, women feel that they will have to use all these products. Various cosmetics advertisements create an idea for women, which they have to follow, if they have to be accepted, noticed, and called beautiful in society. These companies redefined beauty and strive to make it a social norm through their external communication (Joan 2001).

Women who consume cosmetics products regularly and are very conscious about beauty products, tend to be unhappy with their body and their self-image. These types of women tend to spend more money on cosmetics products so that they can look beautiful and could be acceptable in society (2001).

3 Research Methodology

Area of Study
Bengaluru City is taken into consideration for the study.

Sample Design

Sources of Data: Data was collected from both primary and secondary sources. Primary data was collected by using simple random sampling technique.

Data Analysis
Tools such as one-way ANOVA, two-way ANOVA, binomial test, chi-square test, cluster analysis, factor analysis, and mann-whitney U test were employed.

4 Analysis and Interpretation

Hypothesis Testing

Age-Wise Distribution
H_0: Age does not have a relationship with awareness about the use of health issues due to the using use of cosmetics.
H_1: Age has a relationship with awareness about the use of health issues due to the use of cosmetics (Table 1).

Table 1. Relationship of age with awareness about the use of health issues due to the using use of cosmetics

Age	Number of respondents	Percentage
15 to 25	142	37.9
26 to 35	122	32.6
36 to 45	64	17.1
46 and above	46	12.4
Total	374	100
Chi-square value: 25.375	Sig.: 0.000	

Source: Primary Data

The significant value of the Chi-square is less than 0.05. By this, the null hypothesis is rejected and the alternative hypothesis is accepted. This shows that age has a relationship with awareness about the use of health issues due to the use of cosmetics. From the table, it can also be observed that the level of awareness is high in the age group of 15 to 25 years and 26 to 35 years. As they get older, the lack of awareness reduces.

Education-Wise Distribution
H_0: Education does not have a relationship with awareness about the use of health issues due to use of cosmetics.
H_1: Education has a relationship with awareness about the use of health issues due to the use of cosmetics (Table 2).

Table 2. Relationship of education awareness about the use of health issues due to use of cosmetics

Age	Number of respondents	Percentage
No education	34	9.0
Up to +2	52	13.9
Graduation	154	41.2
Working	110	29.4
Others	24	6.5
Total	374	100
Chi-square value: 76.471	Sig.: 0.000	

Source: Primary Data

The significant value of the Chi-square is less than 0.05. Thus, the null hypothesis is rejected and the alternative hypothesis is accepted. From this, we can conclude that education has a relationship with awareness about the use of health issues due to the use of cosmetics. Respondents who are graduates and working professionals are more aware when compared to other respondents.

Age-Wise Distribution

H_0: There is no relationship between the age of women and the level of awareness of women about the practice of reading the text on the packaging.

H_1: There exists a relationship between the age of women and the level of awareness of women about the practice of reading the text on the packaging (Table 3).

Table 3. Relationship of age of women with the level of awareness of women about the practice of reading the text on the packaging

Age	Number of respondents	Percentage
15 to 25	104	34.7
26 to 35	102	34.0
36 to 45	58	19.3
46 and above	36	12.0
Total	300	100
Chi-square value: 38.236	Sig.: 0.000	

The significance value of the chi-square is less than 0.05. Thus, the null hypothesis is rejected and the alternative hypothesis is accepted. There exists a relationship between the age of women and the level of awareness of women about the practice of reading the text on the packaging. The majority of the respondents were between the age group of

15 to 25 years and 26 to 35 years are more concerned about reading the content on the packet of the product.

Education Wise Distribution

H_0: There is no influence of education on the awareness level of women about the practice of reading the text on the packaging.

H_1: There is an influence of education on the awareness level of women about the practice of reading the text on the packaging (Table 4).

Table 4. Relationship of education the awareness level of women about the practice of reading the text on the packaging

Age	Number of respondents	Percentage
No education	26	8.7
Up to +2	44	14.7
Graduation	130	43.3
Working	84	28.0
Others	16	5.3
Total	300	100
Chi-square value: 65.308	Sig.: 0.000	

The significance value of chi-square is less than 0.05. Thus, the null hypothesis is rejected and the alternative hypothesis is accepted. There is an influence of education on the awareness level of women about the practice of reading the text on the packaging. Among the respondents, the graduated and the working professionals are more cautious and serious about the text mentioned on the packaging when compared to the other respondents.

Age-Wise Distribution

H_0: There is no influence of age on the level of awareness of women about patch testing.

H_1: There is an influence of age on the level of awareness of women about patch testing (Table 5).

Table 5. Influence of age on the level of awareness of women about patch testing

Age	Number of respondents	Percentage
15 to 25 years	126	34.8
26 to 35 years	116	32.0
36 to 45 years	62	17.1
46 and above	58	16.1
Total	362	100
Chi-square value: 40.244	Sig.: 0.000	

Source: Primary Data

The significance value of the chi-square is less than 0.05. In this scenario, the null hypothesis is rejected, and the alternative hypothesis is accepted. Thus, there is an influence of age on the level of awareness of women about patch testing. Even in this case, the women in the age group of 15 to 25 years and 26 to 35 years are more prone to conducting the patch test when compared to the other respondents.

Education-Wise Distribution

H_0: There is no association between education and level of awareness among women consumers about patch testing.

H_1: There is an association between education and level of awareness among women consumers about patch testing (Table 6).

Table 6. Education and level of awareness among women consumers about patch testing

Age	Number of respondents	Percentage
No education	26	7.2
Up to +2	52	14.4
Graduation	156	43.1
Working	104	28.7
Others	25	6.6
Total	362	100
Chi-square value: 64.528	Sig.: 0.000	

Source: Primary Data

The significance value of chi-square is less than 0.05. Thus, the null hypothesis is rejected and the alternative hypothesis is accepted. Thus, there is an association between education and the level of awareness among women consumers. The respondents who are graduates and who are working professionals are more aware about the patch test when compared to the other respondents.

Factor Analysis

Factor Analysis was performed to understand various factors which encourage manufacturers and advertisers to indulge in exploitative practices (Table 7).

Table 7. Factor analysis

Component	Initial eigen values			Rotation sums loadings		
	Total	% of variance	Cumulative %	Total	% of variance	Cumulative %
1	2.919	18.72	18.72	2.625	16.86	16.75
2	.926	12.09	30.81	1.937	12.28	28.92
3	1.520	9.463	40.28	1.548	9.692	38.50
4	1.333	8.146	48.42	1.383	8.589	46.98
5	1.177	7.108	55.53	1.259	7.763	54.64
6	1.112	6.675	62.21	1.247	7.681	62.21
7	.943	6.217	74.25			
8	.985	5.829	79.56			
9	.953	5.614	84.64			
10	.827	4.770	89.06			
11	.774	4.424	93.43			
12	.767	4.371	96.11			
13	.514	2.685	98.12			
14	.411	2.002	100.00			
15	.393	1.880				

Source: Primary Data

From the table, we can learn that only six components contribute to 62% of the variance (Table 8).

Table 8. Factors leading to exploitation

I	The lack of awareness among women consumers	Value
1	The lack of awareness about testing in patches	.833
2	Not reading the text on the packaging	.792
3	Unaware of chemicals being mixed in cosmetics products	.783
4	Unaware of the health issues of using cosmetics products	.732
II	The lack of awareness about legal action	Value
1	Unaware of the consumer forums and organizations	.889
2	Unaware of the consumer court for legal action	.888
3	Do not want to go into the legal procedure as it is expensive and would consume a lot of time	.486
III	The behavior of consumers towards Ssolving a grievance	Value
1	Taking the initial step to take the legal action	.721
2	Complaining to the cosmetics company about the problem	.720
3	Contact local consumer forums	.579
IV	The response of consumers	Value
1	Not able the get the required know-how and support to face the consumer exploitation	.789
2	Do not want to take any kind of action against the cosmetics company	.464

Source: Primary Data

The above table displays four major factors from twelve statements. These statements are those statements that encourage manufacturers and companies to indulge in unethical and exploitative practices. The first factor is the Lack of Awareness among Women Consumers. The second factor is the Lack of Awareness of Legal Action. The third factor is the Behavior of Consumers Towards Solving a Grievance. The fourth factor is the Response of Consumers.

The Lack of Awareness Among Women Consumers

The factor basically indicated the lack of awareness about testing in patches, not reading the text on the packaging, being unaware of chemicals being mixed in cosmetics products, and being unaware of the health issues of using cosmetics products.

The Lack of Awareness About Legal Action

The factor basically indicated the unawareness of the consumer forums and organizations, the unawareness of the consumer court for legal action and not wanting to go into the legal procedure as it is expensive and would consume a lot of time.

The Behavior of Consumers Towards Solving a Grievance

The factor basically indicated taking the initial step to take legal action, complaining to the cosmetics company about the problem, and contacting the local consumer forums.

The Response of Consumers

The factor basically indicated the consumers were not able to get the required know-how and support to face the exploitation and did not want to take any kind of action against the cosmetics company.

Mann-Whitney Test

In order to perform the cluster analysis, the respondents were divided into 2 groups, namely, first on the basis of awareness of the legal procedure and second on the basis of aspects relating to the usage of cosmetics. The first group consists of 436 respondents and the second group is comprised of 262 respondents. There is a high degree of lack of awareness among women consumers, which encourages cosmetics companies to indulge in exploitative practices. Further research is an attempt to check the difference or the similarity which exists between these two groups (Tables 9 and 10).

Table 9. Analysis of two groups of women

		Cluster 1		Cluster 2		Z	Asymp. N % N % Z Sig (2-tailed)
		N	%	N	%		
Used patch testing	No	428	98.17	30	11.45	16.491	0.000
	Yes	8	1.83	232	88.55		
Total		436	100	262	100		
Used to read the contents on the package	No	402	92.20	40	15.27	14.420	0.000
	Yes	34	7.80	222	84.73		
Total		436	100	262	100		

(*continued*)

Table 9. *(continued)*

		Cluster 1		Cluster 2		Z	Asymp. N % N % Z Sig (2-tailed)
		N	%	N	%		
No serious thought on health concern and issues, especially on the skin	No opinion/Disagree	166	38.07	58	22.14	−3.084	0.002
	Agree	270	61.93	204	77.86		
Total		436	100	262	100		
Chemicals and natural raw material are used in cosmetics	No opinion/Disagree	380	87.16	68	25.95	11.530	0.000
	Agree	56	12.84	194	74.05		
Total		436	100	262	100		
Not have any kind of information or support to fight exploitation from the cosmetics companies	No opinion/Disagree	176	40.37	74	28.24	−2.284	0.022
	Agree	260	59.63	194	71.76		
Total		436	100	262	100		
Cosmetics have side effects on health in the long run	No opinion/Disagree	370	84.86	64	32.06	10.002	0.000
	Agree	66	15.14	178	67.94		
Total		436	100	262	100		
Case should be filed in case of any ethical violation	No opinion/Disagree	436	75.69	262	62.60	−2.600	0.009
	Agree	330	24.31	164	37.40		
Total		436	100	262	100		
Complaints should be on the company manufacturing the cosmetics products	No opinion/Disagree	408	93.58	214	81.68	−3.450	0.001
	Agree	28	6.42	48	18.32		
Total		436	100	262	100		

(continued)

Table 9. (*continued*)

		Cluster 1		Cluster 2		Z	Asymp. N % N % Z Sig (2-tailed)
		N	%	N	%		
Unaware of the legal products and address of the consumer court	No opinion/Disagree	164	37.61	150	57.25	−3.566	0.000
	Agree	272	62.39	112	42.75		
Total		436	100	262	100		

Table 10. Analysis of the opinions of the two groups of surveyed women who do not differ significantly

		Cluster 1		Cluster 2		Z	Asymp. N % N % Z Sig (2-tailed)
		N	%	N	%		
Legal process is expenses and it also time consuming	No opinion/Disagree	146	33.49	114	43.51	−1.873	0.061
	Agree	290	66.51	148	56.49		
Total		436	100	262	100		
Not taking any kind of action	No opinion/Disagree	110	25.23	56	21.37	−0.818	0.413
	Agree	326	74.77	206	78.63		
Total		436	100	262	100		
Aware about the process of testing and grading the cosmetics	No	404	92.66	242	92.37	−0.101	0.919
	Yes	32	7.34	20	7.63		
Total		436	100	262	100		

5 Suggestions, Discussion, and Conclusion

We are living in a consumeristic society and women contribute more than men in this society where people want to consume more and more. This is the reason why women are the main focus of the study. Cosmetics are something used by both men and women, but the industry is more inclined to sell cosmetics products to the women, as they are more concerned about their beauty and their looks.

In order offer suggestions, it can be offered to various stakeholders. Some of the major stakeholders who were kept in mind while conducting the study were women consumers, the Government, Directorate of Drug Control, the Bureau of Indian Standards Institution, and the Consumer Disputes Redressal Agencies.

With respect to women consumers, they should follow the world-famous formula 'let the buyer beware'. Women consumers should be cautious about what they are buying and what they are using. Instead of blaming the companies, government or advertising agencies, they need to do a little bit of research before consuming the products. The product ingredients might contain some technical terms and chemical terms, but the help of experts in the field can be taken to get a better understanding. The study has also revealed that the consumers in the age group of 15 to 25 years lack awareness about the products which they are buying and using.

It is also necessary for the consumers to have a comprehensive understanding of the impact of the product in the long run, as some products have a negative impact on the skin if it is used constantly. Other than the product the consumers should also be cautious about the advertisements, as many advertisements have misleading and exaggerated claims. For example, a product in the advertisement claiming that the skin colour would change in a span of 14 days is a misleading and exaggerated claim.

Women are also very much engaged in household activities. Thus, they can also put effort to learn how cosmetics products can be prepared only with the help of natural ingredients. Today in the era of information and knowledge explosion, information could be collected from various online sources or they could also get the information from the elders, who would have been using this technique for a long time. Among the various cosmetic products, the most consumed are fairness creams, deodorants, antiperspirants, and fairness creams.

We need to remember that customer is the king, and if the customer's attitude and perception change, the companies would also be forced to change their products and advertisements. Thus, the concept of 'customer is the king' should not be taken in a narrow sense but in a much broader sense.

With respect to the Government, they can take a number of steps to create a better and more ethical society. Firstly, the government should consider health as the top priority and should ensure that no compromise would be done when it comes to the health of the citizens. Secondly, the government should make strict laws with respect to any kind of unethical or exploitative practices, which can be with respect to the product or with respect to the advertisements of the products. Thirdly, the government should have a well-organized process to approve and give permission to any new kind of cosmetic products and take all measures to ensure the product is safe. Fourthly, the government should also strengthen organizations like the Advertisement Standard Council of India (ASCI), as they are tirelessly working to ensure that companies use lawful and ethical means to promote their products.

With respect to the Directorate of Drug Control, it works based on the Drugs and Cosmetics Act 1940 and also the Cosmetics Act 1962. Various changes and amendments should be done with respect to banning various kinds of chemicals which are used in these cosmetics products and to have strict laws where all the ingredients of the product are clearly mentioned on the packet of the product. In the long run, the government to set up testing laboratories.

If the product has some harmful chemicals and is permitted by the government, then the cosmetic companies should be asked to give a visible disclaimer on the product, like how a disclaimer is given on cigarette packets. The Directorate of Drug Control should have cost-effective testing methods, where the sample of the products can be tested on a regular basis, this would always make the manufacturers cautious about what they are doing. They can also provide licenses to various beauty parlors as they use various cosmetics on their consumers.

With respect to the Bureau of Indian Standards Institution, which is a body with strong powers to control various aspects of business and product quality. The organization should come up with a compulsory certification scheme, BIS Act, and lay down standards for testing. All this should be done by creating a hazel-free process or it might discourage cosmetic companies to do business and launching new and innovative products. Various European countries have string rules for all kinds of cosmetics products, due to which they end up avoiding harmful products to the customers.

Generally, when products are exported to various developed countries, they are required to meet various product standards and requirements. Similar standards can also be employed. The organization should also create awareness about various certifications and hallmarks to the general public. Various hallmarks like ISI, BIS, etc.

With respect to the Consumer Disputes Redressal Agencies, they will have to work tirelessly to ensure that justice is given to the consumer and that their issues and disputes are addressed and solved as quickly as possible. In order to encourage people, free legal aid should be provided which would ultimately help consumers to voice their opinion and to create a good and feasible market for the customers.

On the other hand, the agencies should make the legal process very simple and easy, so that the consumers will be easily able to fulfill and complete all the formalities and procedures at ease. Processes like this would instill fear among the manufacturers and they would think twice before breaking the law, misinterpreting the law, or committing some unethical practice.

The study also has great scope for further study. A similar study can also be conducted on men. A study can be conducted in different metropolitan cities other than Bengaluru. A study can be conducted in tire 2 cities as well as tire 2 cities are moving towards urbanization and their income is also growing gradually. Conducting the study in tire 2 cities will give a good insight to companies on the behavior and thought process of the customers and would help organizations to come up with strategies accordingly.

References

Jennifer, M.: Casting race: a history of make-up technology in the United States film industry. **60**(4), 913–919 (1999)

Joan, L.S.: Designing women; Studies in the representation of femineity in Roman society. Roman Republic Roman Empire **62**(04), 15–22 (2001)

Pollay, R.W., Mittal, B.: Here's the beef: factors determinants and segments in consumer criticism of advertising. J. Mark. **57**(3), 99–114 (1993)

Raju, J.K.: The glamourous world of advertising. Indian J. Mark. **8**(3), 15-20 (1996)

Rao, N.D.: Consumerism in India - emerging from its teething troubles. Indian J. Mark. **32**(1), 3–5 (2002)

Rayudu, C.S.: Consumer movement in India. Indian J. Mark. **16**(1), 21–28 (1988)

Robert, O.: Consumerism: its goals, organisations and future. J. Mark. **34**(2), 55–60 (1970)

Tangade, S.F., Basavaraj, C.S.: Awareness and perception of educated consumers about consumer protection laws. Indian J. Mark. **4**(1), 3–9 (2004)

Sadar, A., Fulsele, P.: An introspection of consumer movement in India. Indian J. Mark. **5**(2), 24–29 (2004)

Seshaiah, K., Raju, C.K.: Consumer satisfaction is a myth. Indian J. Mark. **20**(1), 32–38 (1990)

Shivakumar, K., Sujatha, R.: Cosmetics retailing- an empirical study. Indian J. Mark. **33**(1), 20–28 (2003)

Gundala, R.R., Chavali, K.: Ethical aspects in the advertising of fairness creams. ICFAI J. Mark. **8**(2), 60–68 (2005)

Rajasekar, N.: Fairness creams - a study on market trends & brand preferences. Indian J. Mark. **32**(1), 9–11 (2002)

https://www.globenewswire.com/en/news-release/2022/04/15/2423222/28124/en/India-Beauty-and-Personal-Care-Market-is-Projected-to-Register-a-CAGR-of-6-32-in-Terms-of-Value-Between-20222027.html#:~:text=India's%20Beauty%20and%20Personal%20Care,at%20a%20CAGR%20of%206.32%25.&text=India's%20Beauty%20and%20Personal%20Care%20Market%20is%20segmented%20based%20on,%2C%20Distribution%20Channel%2C%20and%20Category. Accessed 12 Aug 2022

https://economictimes.indiatimes.com/journey-of-fairness-creams-advertising-in-india/articleshow/30997189.cms. Accessed 13 Aug 2022

The Influence of Key Antecedents on Attitude and Revisit Intention: Evidence from Visitors of Homestay in Kundasang, Sabah, Malaysia

Syarifah Hanum Ali[1](✉), Kamaliah Sulimat[2], and Nor Azma Rahlin[1]

[1] Faculty of Business, Economics, and Accountancy, Universiti Malaysia Sabah, 88400 Kota Kinabalu, Sabah, Malaysia
syarifahhanum.ali@ums.edu.my

[2] Politeknik Kota Kinabalu, No. 4, Jalan Politeknik KKIP Barat Kota Kinabalu Industrial Park, 88460 Kota Kinabalu, Sabah, Malaysia

Abstract. The study on revisit intention has recently attracted significant attention from scholars and practitioners. This study attempts to contribute to the body of knowledge in marketing by focusing on issues relating to the factors that influence the revisit intention. Specifically, this study empirically investigates the influence of perceived authenticity, perceived value, perceived risk, electronic word-of-mouth (EWOM), and price sensitivity towards revisiting intention to the homestay in Kundasang, Sabah. The mediating effect of the attitude toward staying in a homestay on the relationship between variables was tested. Data collected from 178 homestay visitors were analysed using the partial least squares-structural equation modelling (PLS-SEM) approach supported by SmartPLS. The results reveal that perceived authenticity, risk, and EWOM significantly influence the revisit intention, whereas perceived value and price sensitivity have no significant effect. Furthermore, it was found that the attitude mediated the relationship between perceived value, price sensitivity, and revisit intention.

Keywords: Perceived authenticity · Perceived value · Perceived risk · Electronics Word-of-Mouth (EWOM) · Price sensitivity · Revisit intention

1 Introduction

Revisit intention has been an essential topic in service quality. As a consequence of satisfaction, revisit intention or repurchase intention remains an essential topic and has gained interest among researchers, academicians, and practitioners, particularly in marketing and tourism. Researchers have highlighted several important issues regarding revisiting behavior intention [1, 2]. Repeat visitation is crucial in the accommodation provider, particularly in hotels and the homestay industry. Homestay operators depend on repeat visitors to maintain their income as accommodation providers. Particularly, repeat visitors generated more significant profit due to the longer duration of stay and lesser cost to serve them [3, 4]. Moreover, loyalty is created by the loyal visitors from

B. Alareeni et al. (Eds.): ICBT 2022, LNNS 620, pp. 733–742, 2023.
https://doi.org/10.1007/978-3-031-26953-0_67

the repeat visit, which will spread their experiences to potential customers [5]. However, many certified homestay operators faced difficulties sustaining the operation due to decreased visitors [6].

The structure of the article:

1. Introduction
2. Literature Review
3. Methodology
4. Result
5. Discussion

    5.1. The Influence of Antecedences and Attitude on Revisit Intention
    5.2 The Mediating Effect of the Attitude

6. Contribution of the study
7. Conclusion
8. References

2 Literature Review

2.1 Revisit Intention

One of the influential attributes of revisit intention is the number of previous visits [4]. Understanding the underlying concept of revisit intention is crucial since repeat visitors generate profit for the service provider due to the longer duration of stay and lesser cost to serve them [3; 4]. Moreover, loyalty is created by the loyal visitors from the repeat visit, which will spread their experiences to other potential customers [5]. Moreover, understanding the concept of revisit and repurchase intention is vital as the repurchase intention is influenced by service quality [7]. Specifically, as a consumer evaluates perceived quality [8], service quality significantly influences behaviour such as repurchase intention [7]. Service quality is crucial for competitiveness and has been widely studied and applied in many industries since the introduction of service quality theory (SERVQUAL) by [9]. Five dimensions are determined to measure service quality: tangibles, reliability, responsiveness, assurance, and empathy. According to [10], service quality looks at how consumers' needs are met, and whether services fulfilled their expectations. In tourism, service quality refers to the visitor's judgement about the service's overall superiority or excellence [11].

2.2 Customers' Attitudes

Customers' satisfaction positively affects how they feel about the products and services they buy, and those good feelings make them more likely to buy again [12–14]. Higher customer satisfaction leads to more positive customer attitudes and loyal customers [15, 16]. [17] say that tourists feel about a destination change based on the overall quality of services and how happy they are with the destination. Several studies [12, 13, 18] have also found that customers' high attachment to quality assessment makes them happier with the company's offers. It makes customers feel good about the company's products or services and more likely to repurchase them. [16] found that how consumers see a product's social and functional benefits affect how satisfied they are with it and that satisfaction affects how they feel about the product or service.

2.3 Perceived Authenticity

Authenticity refers to social construction that is not fixed and adjustable based on different evaluators' interpretations and perceptions of various contexts such as place, situation, person, or object [19]. The concept of authenticity has gained attention among researchers, particularly in the tourism area [20], since the seminal work on authenticity was studied by [21]. However, the concept introduced by [21] was problematic [22]. According to [22], authenticity in tourism study consists of two types, namely 'objective related authenticity' and 'activity related authenticity. With objective authenticity and constructive objectivity dimensions, objective authenticity refers to natural and genuine things [23]. The activity-related authenticity refers to the existential authenticity of human nature [23]. For the visitors of homestay and Airbnb, the local living experience, which is in existential authenticity, is the main attraction [24, 25].

2.4 Perceived Value

Various definitions of value have been introduced by scholars [26–29], who defined value for money as the relationship between the price and the quality rate. At the same time, [26] described perceived value as the difference between perceived quality and sacrifice associated with price. [29] explained that the definition of value differs among consumers, such as low price, benefits received, service they paid for or what they received after they paid. According to [27], the perceived value is defined overall assessment of the service's net worth based on benefits and cost or sacrifice. To sum up, [28] explained that perceived value represents the trade-off between what the consumers obtain, such as utility, benefits, and quality, and what the consumer gives up, such as price and sacrifice.

2.5 Perceived Risk

Perceived risk is the uncertainty and consequences usually associated with bad outcomes [30]. [31] found four types of risks: natural disaster risk, physical risk, political risk, and performance risk. They also found that people are less likely to return if they think the risk is high. [2] studied how likely people were to return to a dangerous place. They found that perceived physical risk did not significantly affect how the site was seen, but it directly affected how likely people were to return. In the case of Airbnb, perceived risk is essential for deciding whether to return because guests share the space with others, like the hosts or guests [25]. So, perceived risk is always a sign of a bad thing when it comes to a person's desire to return to a place [32].

2.6 Electronic Word-of-Mouth Marketing (EWOM)

In recent years, academicians and practitioners from various disciplines have long recognized the importance of WOM marketing. [33] Word-of-mouth (WOM) advertising is "an oral, person-to-person communication between a receiver and a communicator whom the receiver perceives as non-commercial, regarding a brand, product, or service". Traditionally, the expression of word-of-mouth was mainly applied to verbal communication among friends and family. However, rapid technological development shows the convenience of sharing processes about products or brands [34]. Therefore, it is common nowadays to extend the term to electronic communication between producers and customers or between customers themselves [35]. The role of WOM as the most influential in advertising tools [35–37] and cost-effectiveness are the reasons that lead to the selection of WOM marketing among researchers. The information provided by the customer-to-customer is perceived to be more trustworthy and exciting than the information provided by the firm [38].

2.7 Price Sensitivity

Usually, the price of a product is how much money people have to pay for it. So, consumers know how much they must give up in exchange for what they already have. The income effect is something that many people have noticed: if everything else stays the same, consumers are willing to spend more if they have much money. [39] said, "People with more money spend lots of money". A quick look at the Consumer Expenditures report from the U.S. Bureau of Labor Statistics, for example, shows that as average income goes up, so do spending levels on almost every type of product. Also, [39] said, "American households have more money to spend on whatever they want," so they buy more expensive luxury goods.

2.8 Underlying Theory

The Theory of Planned Behavior (TPB) is related to the degree to which an individual has a favourable or unfavourable evaluation of the behaviour of interest. It involves a consideration of the consequences of performing the behaviour [40]. Over the years, TPB received substantial attention from researchers in marketing to determine the outcomes of behaviour. Based on the TPB, several key antecedents of attitude were adopted in the current study, such as Perceived Authenticity, Perceived Value, EWoM, and Perceived Risk. TPB's underlying theory best explains the relationship between antecedent constructs on Attitude and Revisit Intention and homestay' revisit Intention. The dependent construct is the "revisit intention" while the mediator is attitude. The independent-mediator-dependent framework is applied to the current research. The independent constructs for the current research are perceived authenticity, perceived value, perceived risk, EWOM and price sensitivity.

2.9 Hypotheses Development

H1: Perceived authenticity, perceived value, perceived risk, EWOM and price sensitivity influences the homestay's revisit intention.

H1a:Perceived authenticity positively influence the homestay's revisit intention.
H1b:Perceived value positively influence the homestay's revisit intention.
H1c:Perceived risk negatively influence the homestay's revisit intention.
H1d: EWOM positively influence the homestay's revisit intention.
H1e:Price sensitivity positively influences the homestay's revisit intention
H2: Attitude mediated the relationship between perceived authenticity, perceived value, perceived risk, EWOM and price sensitivity and homestay's revisit intention.
H2a:Attitude mediated the relationship between perceived authenticity and revisit intention.
H2b: Attitude mediated the relationship between perceived value and revisit intention.
H2c: Attitude mediated the relationship between perceived risk and revisit intention.
H2d:Attitude mediated the relationship between EWOM and revisit intention.
H2e:Attitude mediated the relationship between price sensitivity and revisit intention.

3 Methodology

The response rate for the current research is 178 respondents. The target sample for the research is 149. Ninety-eight respondents answered printed questionnaires, and 80 responded received from Google Forms. They were dominated by the respondents living in Sabah (97.2%) and the remaining living outside Sabah (2.8%). Most of the respondent is female, 36 female (74.7%), while 25.3% are male. In terms of marital status, single dominated the responded sample (69.7%), while the remaining 29.2% were married, and 1.1 per cent whether widowed, divorced or separated. Regarding the age, 48.9% are 20-year-old, and below, 20.2% are between 21- to 30-year-old, 19.1% are between 31- to 40-year-old, 10.1 per cent are between 41- to 50-year-old, and 1.7% is 51-year-old and above. The ethnic group was dominated by Bumiputera Sabah (81.5), followed by Malay (8.4%), and the remaining were Chinese (3.4%), Bumiputera Sarawak (1.7%), Indian (1.1%) and others (3.9%).

This study tested the five hypotheses by employing path analysis in the SmartPLS. The capability to predict the structural models and the relationship between the constructs was tested by performing the path analysis.

4 Results

The R square represents explained variance and the f square for effect size [41]. The SmartPLS model and results yielded by the PLS algorithm are presented in Figure 1. The results found that all the independent constructs (perceived authenticity, perceived value, perceived risk, EWOM and price sensitivity) explain 46.4 per cent of the variance in revisit intention. In the disciplines such as consumer behaviour, the R square value of 0.20 is considered high [41].

4.1 Results of Hypothesis Testing (Independent Constructs and Revisit Intention)

In examining the results of hypotheses testing, the minimum bootstrapping with 5000 samples suggested by [41] was used. The results Model Resolution by SmartPLS using PLS algorithm indicated that perceived authenticity ($\beta = 0.113, p > 0.05$), perceived risk ($\beta = 0.014, p > 0.05$) and EWOM ($\beta = 0.033, p > 0.05$) not significantly influence the visitors revisit intentions to homestay in Kundasang. Hence, H1a, H1c and H1d were not supported. However, perceived value ($\beta = 0.347, p < 0.05$) and price sensitivity ($\beta = 0.326, p < 0.05$) have significant relationship with visitors' revisit intentions to homestay; H1b and H1e were supported.

4.2 Results of Mediating Role of Attitude

Table 1. Results of hypothesis testing (mediator)

Relationship	Indirect Effect Beta	Std. Error	t- values	P Values	0.025	0.975	Results
PA - ATT - RI	0.071	0.042	1.68	0.093	−0.011	0.151	Not Supported
PV - ATT - RI	0.219	0.061	3.575	0.000	0.113	0.351	Supported
PR - ATT - RI	0.009	0.037	0.239	0.811	−0.053	0.093	Not Supported
E-WOM - ATT - RI	0.021	0.039	0.528	0.598	−0.058	0.092	Not Supported
PS - ATT - RI	0.205	0.045	4.53	0.000	0.127	0.301	Supported

PA = Perceived Authenticity; PV = Perceived Value; PR = Perceived Risk; EWOM = Electronic Word-of-Mouth Marketing; PS = Price Sensitivity; ATT = Attitude; RI = Revisit Intention

Table 1 illustrate the result of hypothesis testing for the mediator. The results revealed that the attitude mediated the relationship between perceived value and revisit intention and between price sensitivity and revisit intention. Hence, hypotheses H2b and H2e were supported. However, attitude did not mediate the relationship between perceived authenticity, perceived risk, and EWOM towards revisiting intention. Therefore, H2a, H2c and H2d were not supported.

5 Discussion of Results

5.1 The Influence of Antecedences and Attitude on Revisit Intention

The results indicate that the perceived authenticity is not significantly influencing the revisit intention. The possible explanation for the result is that the visitors seek the authentic local experience while visiting the homestay in Kundasang. The result is also in line with previous studies [25, 32] and [41]. In addition, the finding is contradicted by [42], which indicates that perceived authenticity increases the visitor's perceived value and does not influence the revisit intention. This study found that the perceived value influences the revisit intention of the homestay in Kundasang. The result is quite surprising due to the value for money and the consistent quality of the homestay. The finding parallels [25] and [41] found that perceived value significantly influences the revisit intention. The result of the study reveals that the revisit intention does not significantly influence the perceived risk.

Regarding this result, the possible reason for the finding is that the visitors spent most of their time long hours in the homestay. Usually, the visitors visit Kundasang to visit the attractive places around Kundasang and Ranau. The EWOM does not significantly influence the revisit intention to homestay in Kundasang. The result has not confirmed that WOM marketing is the most influential type of advertising in promotional strategy [35], which affects the purchase decision [43]. In the context of homestay in Kundasang, internet searching found that it is common for homestay operators to have an online platform to advertise their homestay. The sensitivity level to price is significantly increasing the homestay visitors' revisit intention. The result is similar to the study by [25] that indicated that price sensitivity influences the repurchased intention in the context of Airbnb. In addition, the results also contrast with the finding by [43] that economic benefits influence collaborative consumption in the accommodation sector. e sector of accommodation.

5.2 The Mediating Effect of the Attitude

The initial model examines the role of attitude as the mediator in the relationship between antecedents and revisits intention. The result is in line with [44] suggested that other than satisfaction, attitude is a mediator in the relationship between antecedents and revisit intention.

6 Contribution of the Study

This study led to several helpful contributions to many parties. This study's empirical results help homestay operators figure out relevant factors in enhancing visitor attitude and increasing revisit intention in the study model. Thus, homestay operators must concentrate on three antecedents: perceived value, price sensitivity, and attitudes, since they significantly influence visitors' intentions to revisit homestay at Kundasang, Sabah, Malaysia. With attitude exerting a more significant impact, homestay operators could use advertisements and promotions to draw visitors' attention to homestay at Kundasang, Sabah, Malaysia, as this may influence their revisit intention homestay in the future. Lastly, by looking into three significant factors to the revisit intention, related ministries and agencies can plan appropriate development programmes and training for homestays, primarily at Kundasang, Sabah. The results of this study also provided helpful ideas to related ministers and agencies to help homestay operators improve their service, understand homestay visitors' revisit intention behaviour and become more competitive.

7 Conclusion

The results of this study provide a better knowledge of the factors influencing revisit intention from developing nation markets. Additionally, the study guided homestay operators on the various behavioural aspects under TPB factors with their visitors over time, with a stronger emphasis on increasing revisit behaviours. Indeed, this study support TPB, which is that attitude is one of the most critical factors of intention. Therefore, homestay operators must acquire knowledge and awareness of the factors that drive the intention to revisit and revisit behaviour on homestay at Kundasang, Sabah, Malaysia, to guarantee that this niche industry can survive and sustain. To do this, stakeholders must focus on three antecedence factors from this study that can help them accelerate growth and increase revisit intention that benefit them, their visitors, and the country.

References

1. Som, A.P.M., Marzuki, A., Yousefi, M., Khalifeh, A.N.A.: Factors influencing visitors' revisit behavioral intentions: a case study of Sabah, Malaysia. Int. J. Mark. Stud. **4**(4), 39–50 (2012)
2. Ajzen, I.: The theory of planned behavior. Organ. Behav. Hum. Decis. Process. **50**(2), 179–211 (1991)
3. Arndt, J.: Role of product-related conversations in the diffusion of new product. J. Mark. Res. **4**(August), 291–295 (1967)
4. Baur, R.A.: Consumer behavior as risk taking. In: Hancock, R.S. (ed.) Dynamic Marketing for a Changing World. American Marketing Association, Chicago (1960)
5. Bigne, J.E., Sanchez, M.I., Sanchez, J.: The role of variety seeking in short and long run revisit intentions in holiday destinations. Tour. Manag. **22**(6), 607–616 (2001)
6. Blazevic, V., et al.: Beyond traditional word-of-mouth: an expanded model of customer-driven influence. J. Serv. Manag. **24**(3), 294–313 (2013)
7. Chen, S.: Symptoms of stress related to the characteristics of customer service in warehouse superstores. J. Retail. Cons. Serv. **19**(2), 202–210 (2012)
8. Chew, E.Y., Jahari, S.A.: Destination image as a mediator between perceived risks and revisit intention: a case of post-disaster Japan. Tour. Manag. **40**, 382–393 (2014)
9. Chhabra, D., Healy, R., Sills, E.: Staged authenticity and heritage tourism. Ann. Tour. Res. **30**(3), 702–719 (2003)
10. Chiu, C.M., Wang, E.T., Fang, Y.H., Huang, H.Y.: Understanding customers' repeat purchase intentions in B2C e-commerce: the role of utilitarian value, hedonic value, and perceived risk. Inf. Syst. J. **24**(1), 85–114 (2014)
11. Choi, Y.H., Choo, H.J.: Effects of Chinese consumers' relationship benefits and satisfaction on attitudes toward foreign fashion brands: the moderating role of country of salesperson. J. Retail. Cons. Serv. **28**, 99–106 (2016)
12. Dodds, W.B., Monroe, K.B.: The effect of brand and price information on subjective product evaluations. ACR North Am. Adv. **12**(1), 85–90 (1985)
13. Ekinci, Y., Dawes, P.L., Massey, G.R.: An extended model of the antecedents and consequences of consumer satisfaction for hospitality services. Eur. J. Mark. **42**(1/2), 35–68 (2008)
14. Fuchs, M., Weiermair, K.: New perspectives of satisfaction research in tourism destination. Tour. Rev. **58**(3), 6–14 (2003)
15. Grayson, K., Martinec, R.: Consumer perceptions of iconicity and indexicality and their influence on assessments of authentic market offerings. J. Cons. Res. **31**(2), 296–312 (2004)

16. Guttentag, D.: Airbnb: disruptive innovation and the rise of an informal tourism accommodation sector. Curr. Issues Tour. **18**(12), 1–26 (2013)
17. Hair, J.F., Ringle, G.T., Sarstedt, C.M.: A Primer on Partial Least Squares Structural Equation Modelling (PLS-SEM), 2nd edn. Sage, Thousand Oaks (2017)
18. Harris, L.C., Goode, M.M.: The four levels of loyalty and the pivotal role of trust: a study of online service dynamics. J. Retail. **80**(2), 139–158 (2004)
19. Hellier, P.K., Geursen, G.M., Carr, R.A., Rickard, J.A.: Customers repurchase intention: a general structural equation model. Eur. J. Mark. **37**(11/12), 1762–1800 (2003)
20. Johnson, M.S., Garbarino, E., Sivadas, E.: Influences of customer differences of loyalty, perceived risk and category experience on customer satisfaction ratings. Int. J. Mark. Res. **48**(5), 601–622 (2006)
21. Kashyap, R., Bojanic, D.C.: A structural analysis of value, quality, and price perceptions of business and leisure travelers. J. Travel Res. **39**(1), 45–51 (2000)
22. Kotler, P., Keller, K.L., Ancarani, F., Costabile, M.: Marketing Management, 14th edn. Pearson, Boston (2014)
23. Liang, L.J., Choi, C.H., Joppe, M.: Understanding repurchase intention of airbnb consumers: perceived authenticity, electronic word-of-mouth, and price sensitivity. J. Travel Tour. Mark. **35**(1), 73–89 (2018)
24. Litvin, S.W., Goldsmith, R.E., Pan, B.: Electronic word-of-mouth in hospitality and tourism management. Tour. Manag. **29**(3), 458–468 (2008)
25. MacCannell, D.: Staged authenticity: arrangements of social space in tourist settings. Am. J. Sociol. **79**(3), 589–603 (1973)
26. Kasim, M.M., Kayat, K., Ramli, R., Ramli, R.: Sustainability criteria for the Malaysia homestay program. Int. Rev. Manag. Mark. **6**(S7), 250–255 (2016)
27. Olsen, S.O.: Comparative evaluation and the relationship between quality, satisfaction, and repurchase loyalty. J. Acad. Mark. Sci. **30**(3), 240–249 (2002)
28. Opermann, M.: Tourism destination loyalty. J. Travel Res. **39**, 78–84 (2000)
29. Parasuraman, A., Zeithaml, V.A., Berry, L.L.: Servqual. J. Retail. **64**(1), 12–40 (1998)
30. Petrick, J.F.: Are loyal visitors desired visitors. Tour. Manag. **25**, 463–470 (2005)
31. Silverstein, M.J., Fiske, N., Butman, J.: Trading Up: Why Consumers Want New Luxury Goods and How Companies Create Them. Portfolio, New York (2005)
32. Smith, D., Menon, S., Sivakumar, K.: Online peer and editorial recommendations, trust, and choice in virtual market. J. Interact. Mark. **19**(3), 15–37 (2005)
33. Steiner, C.J., Reisinger, Y.: Understanding existential authenticity. Ann. Tour. Res. **33**(2), 299–318 (2006)
34. Stephen, A.T., Lehmann, D.R.: Why do People Transmit Word-of-Mouth? The Effect of Recipient and Relationship Characteristics on Transmission Behaviors. Columbia, New York (2009)
35. Tian-Cole, S., Cromptіon, J.: A conceptualization of the relationships between service quality and visitor satisfaction, and their links to destination selection. Leisure Stud. **22**(1), 65–80 (2003)
36. Trusov, M., Bucklin, R.E., Pauwels, K.: Effects of word-of-mouth versus traditional marketing: findings from internet social networking site. J. Mark. **73**(5), 90–102 (2009)
37. Tussyadiah, I.P., Pesonen, J.: Impacts of peer-to-peer accommodation use on travel patterns. J. Travel Res. **55**, 1–19 (2015)
38. Um, S., Chon, K., Ro, Y.: Antecedents of revisit intention. Ann. Tour. Res. **33**(4), 1141–1158 (2006)
39. Van Doom, J., Lemon, K.N., Mass, S., Pick, D., Pimer, P., Verhoef, P.C.: Customer engagement behavior: theoretical foundations and research directions. J. Serv. Res. **13**(3), 253–266 (2010)
40. Wang, N.: Rethinking authenticity in tourism experience. Ann. Tour. Res. **26**(2), 349–370 (1999)

41. Wilkie, W.: Consumer Behavior. John Wiley & Sons, New York (1994)
42. Wu, W.Y., Chang, M.L.: The role of risk attitude on online shopping: experience, customer satisfaction, and repurchase intention. Social Behav. Pers. **35**(4), 453–468 (2007)
43. Zeithaml, V.A.: Consumer perceptions of price, quality, and value: a means- end model and synthesis of evidence. J. Mark. **52**(3), 2–22 (1988)
44. Zhu, F., Zhang, X.: Impact of online consumer reviews on sales: the moderating role of product and consumer characteristics. J. Mark. **74**(2), 133–148 (2010)

A Literature Review on Digital Human Resources Management Towards Digital Skills and Employee Performance

Reno Candra Sangaji, Alldila Nadhira Ayu Setyaning(✉), and Endy Gunanto Marsasi

Universitas Islam Indonesia, Yogyakarta, Indonesia
dilanadhira@uii.ac.id

Abstract. Digitalisation in business processes has become very popular, even though it is difficult for companies to implement. Digital transformation and digital disruption of Human Resource Management triggered the concept of Digital HRM. This article aims to clarify the general understanding of phenomena related to the relationship between Digital HRM, Digital Skills, and Employee Performance. The concept of Digital HRM brings an evolutionary advancement for the future. Human Resource demands digital skills that produce and distribute ideas and information. Thus, understanding the factors behind the different levels of various digital skills is very important. Digitalisation grades differently among HR, contributing to an organisation's level of success. They include information, communication, collaboration, critical thinking, creativity, and problem-solving. Our literature review finds that employees who can access information systems through the digital process are considered to have a successful performance. Human resource information systems are helpful in HR processes and strategic organisational development tools. In this case, technology acts as a trigger to better share technology-based facilities, services, and decision analysis.

Keywords: Digital human resources management · Digital skills · Employee performance

1 Introduction

Rapid technological advances have an impact on all aspects. One of the visible impacts is on the way an organisation works. The digital revolution affects all levels of the organisation, one of which is the pressure on employees to adapt to a rapidly changing environment and increasing digital technology [1]. Digitalisation has become a key in the business world and is recognised as something organisations must embrace to survive [2].

Almost all manual work is replaced by intelligent computer programs such as artificial intelligence (AI), besides being able to access large amounts of information commonly known as big data [3, 4]. This change is called the digitisation process, and

B. Alareeni et al. (Eds.): ICBT 2022, LNNS 620, pp. 743–750, 2023.
https://doi.org/10.1007/978-3-031-26953-0_68

digitisation is everything that can be digitised [5]. Digitalisation changes how individuals within an organisation communicate and interact as humans and, in turn, how organisations operate [6].

The development of information technology has played a vital role in the evolution of HRM [7]. In the business world, digital technology is changing every aspect of HRM, from attracting & recruiting new employees, training and developing, performance appraisal and compensation and rewards [8]. Currently, research in the management field has not been done much in digital human resources, mainly focusing on digitalisation for marketing and business, such as customer satisfaction, purchasing behaviour and customer relationship management. There are several reasons why this article was written: First, an organisation is a phenomenon in organisations that is common and irrelevant, not only for HRM but for all parts of the organisation [9, 10]. Second is the dependence between the digitalisation of organisations and Digital HRM [11, 12]. It is, therefore, appropriate to conceptualise the digitisation of HRM along with the general digitisation of an organisation to consider these dependencies. Third, the existing literature on widespread digitalisation is more developed than on digital HRM. It is thus appropriate to take advantage of the general insights available in clarifying Digital HRM.

2 Literature Review

2.1 Digital Human Resource Management (HRM) on Employee Performance

To get a deeper understanding of what digital HRM is, previously we must know what HRM is in general and the definition of HRM itself, according to Huselid [13]. The best areas of HRM include recruitment and selection, socialisation, job design, training and development, participation, performance appraisal, employee rewards and job security. Bardin and Suderlan also describe HRM as a practice related to the relationship between employees and their work organisation [14]. On the other hand, HRM can be explained as HR management so the organisation can develop.

According to Mediteti [15], digital HRM manages all HRM work through technology, applications, and the internet, so innovation needs to be brought into HRM practices to attract, reward, and evaluate employees. Today, social or virtual media is acceptable for retaining valuable employees.

In addition, Digital HRM has an impact on the way employees work in conducting training. Research conducted by Iwu [16] found that most employees agree that digital HRM will improve employee performance. Other studies that have been conducted have studied the impact of digitalisation on human resource development, talent management, and performance in the workplace [17]. Research conducted by Tripathi & Kushwaha [18] suggests that organisations put digitalisation forward in their HRM practices as it has become very significant in recent times. Recent studies by Fedorova [19] show that digitising HRM processes makes it possible to remove many routine tasks, reduce the risk of human error and solve critical problems. To the description above, it is hoped that if the application of digital HRM, in this case, is to monitor online work in real-time so that managers are right in conducting performance appraisals, the effect on employee performance will improve.

P1: Digital HRM affects employee performance.

2.2 Digital Human Resource Management (HRM) on Digital Skills

Digital technology has radically changed the lives of individuals and societies, work styles, and how organisations operate their organisations and businesses. In this case, an example is the work style with technology related to employee training and development. With the existence of Digital HRM for training management for employees, it can be accessed anywhere with virtual classrooms and assess progress through systems or portals.

Research conducted by Anthony [20] explained that technology enables improvement through cloud-based services, analytics that allow decisions, and talent management across the organisation. In addition, research by Nawaz & Gomes [21] found that HR information systems help improve HR skills and act as strategic tools for employee and organisational development. Specifically, digital HR strategies can be understood as (the idea of) a blend of human resources and technology aimed at generating sustainable corporate advantages when they are valuable, rare, inimitable and exploitable. (Barney, 1991). As briefly indicated, there are a variety of recognised theoretical approaches that are directly suited to explaining and establishing digital HRM.

With digitalisation in the field of HRM, employees will be more informed, more involved and more alert because they can handle and perform their duties anywhere and anytime, so they will bring feedback in real-time to management [15]. On the other hand, employees will use mobile devices to enable employees to do their jobs more efficiently. In addition, digitalisation will bring innovations, collaborations, and strategies to organisations. So that employees who have digital skills will become global employees because they will share and bring their knowledge and ideas to an international level [15]. Referring to the existing description, digitisation is a benchmark of employee performance behaviour in influencing skills in the digital field, starting from using mobile devices that can make it easier to complete a job, such as online and virtual simulation training. Affect employee skills in terms of using digital media to be efficient and effective.

P2: Digital HRM affects digital skills.

2.3 Digital Skills on Employee Performance

In the millennial era, the current generation is considered a digital workforce. They have devices connected through the internet and web-based service applications attached to their daily lives. Thus, current employees need to have digital skills to engage with digital employees and be able to work more effectively and efficiently [22]. Digital skills have various elements, including digital skills in information, communication, critical thinking, creative digital, and problem-solving digital skills.

Information management skills concern the ability to maintain the information, as workers must be able to manage documents, files, emails, and other digital forms of communication as part of their work activities [23]. Employees need skills to do filing in the right place and to be consistent in naming digital files. Skills in digital information include assessing the usefulness, relevance, and reliability of retrieved digital information [24]. In addition, employees must develop information skills to check whether the information they find is correct and up-to-date.

Employees must have skills in digital communication. This skill means the good competence to send information online and present this information to others [25]. Computerised transmission is becoming very common in this era of globalisation [26]. This allows employees to improve digital skills in a collaborative way to work effectively and passionately in teams to achieve common goals and assume shared responsibility for completing a task. The work will be performed faster than expected because a group does it [27]. Digital collaboration skills are needed to identify each member's specific function based on their expertise [28]. This capability supports doing the same document work, together or not.

An employee must be able to think critically and appropriately assess digital information and communication, consider sharing perspectives and decide whether the content is supported by proven arguments or reasons [29]. Critical thinking is the essential skill in distinguishing wrong, outdated, outdated information and communication and involves the skills to provide evidence and argument [30].

Digital skills must also be creative and use digital tools, and employees must be able to use online digital platforms to give clever turns to existing processes [31]. This can increase employee creativity, using digital communication information technology to provide more experience and complete tasks creatively [32]. Besides that, employees are expected to solve problems digitally, and information and communication technology can be used to analyse complex issues [33] [34]. In solving a problem, it is necessary to evaluate problem-solving skills in terms of flexibility and effectiveness.

To the description above, digital skills, in this case, including digital skills in information, communication, Critical-thinking digital skills, Creative digital skills, and problem-solving digital skills, must be possessed by employees so that work can be completed effectively and efficiently. So that if digital skills are improved, employee performance will increase.

P3: Digital Skills affect employee performance (Fig. 1).

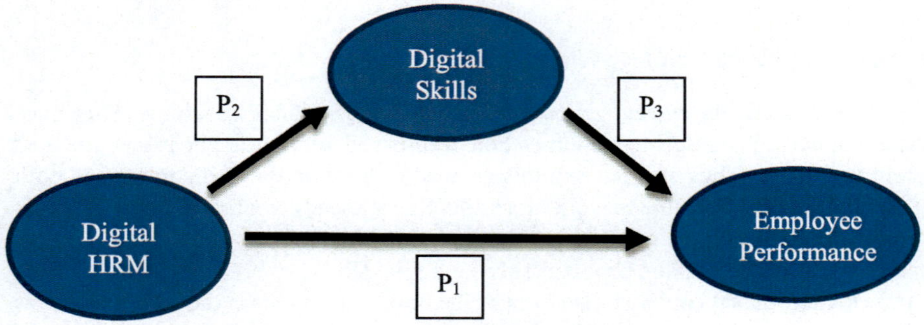

Fig. 1. Proposed research model

3 Discussion

The ability of HRM in technology shows that the effectiveness of advanced information technology does not only depend on the structure of the technology but can also be influenced by other structures, such as the internal context, especially the quality of human resources. The application of Digital HRM to organisations and individuals; shows the immediate effect that occurs if the intra-organizational context regarding the impact of digitisation is often neglected. Digital HRM is more often focused on factors such as employee acceptance of technology, HR professional analytical skills, or assisting with purchasing. The level of HRM capability is defined as the ability of an organisation to manage and develop its workforce. The Digital Capability of HRM can be thought of as an intra-organizational context that can act to drive the complete Digitalization implementation process. High Digital HRM capabilities equip the workforce to support strategic business plans. They show that the organisation can improve competency-based processes in carrying out tasks, impacting digital HRM practices. Experts suggest that when HRM capabilities are high, Digital HRM practices make it easier to provide efficient services that better meet line managers' needs, promoting social networking between HR managers and line managers. This helps to realise complementary functional HR practices. This process in the area of Digital HRM promotes the difference, consistency, and consensus perception of employees about the HRM system, which further enables the effectiveness of HRM. In summary, we suggest that digital HRM practices impact HRM effectiveness through the internal consistency of HR practices and the external social networks of HR managers and line managers. This impact's positive or negative effect depends on the maturity of the organisation's HRM system.

Digital skills require that one of the skills focused on is digital information management skills: skills to search, evaluate, and manage digital information. The amount of online information and the proliferation of databases make the effective and efficient use of search engines essential. Information management skills involve the ability to maintain information where workers must be able to manage documents, files, emails, and other digital forms. In this case, the data is part of their work activities. HR professionals need the skills to store files correctly, be consistent in naming digital files and organise digital files through a hierarchical folder structure. Information evaluation skills include assessing digital information's usefulness, relevance, and reliability. Workers need skills to check whether the information found is correct and up-to-date. Another skill demand is digital communication management skills, i.e. the skills to send information online and to contemplate the best way to present this information to users.

As a result of globalisation and technological developments, computers have become prevalent in today's workplace. Experts point out that communication skills are the ability to form interpersonal impressions and obtain satisfactory results from online interactions. Workers must be able to articulate themselves clearly through various online media. The skill of choosing the right location to send a message and carefully considering the content is essential to convey the message and achieving what is desired from online interactions. Closely related to this component are communication network skills and the ability to mobilise online contacts to achieve specific goals, such as increasing brand awareness, facilitating resource mobilisation, or generating business. Finally, content-sharing skills communication is the ability to share content online, from status updates, photos, and videos to writing comments and blogs.

The pattern of successful performance can be seen from the number of employees who can access information about training programs from any location or attend training in virtual classrooms. HR information systems through digital processes are helpful in HR processes and also act as strategic tools for organisational development—the whole organisation. Performance can be related to the success rate of communication between superiors and inferiors in the organisational structure, namely communicating electronically and all information being disseminated in video/audio or via mobile devices. In addition, employee performance management is also carried out and inspected digitally. Next is management's effort to increase employee engagement or management level that digitally enables employees and management. Companies will use mobile devices to allow employees to do their jobs better and more efficiently. Employees will be more informed, engaged and alert because they can handle and perform tasks anywhere and anytime, bringing honest feedback. -time to management. What was previously impossible can bring about broad organisational change through collaboration and new corporate strategies. Employees will become global employees because they will share and bring their knowledge and ideas to a worldwide level.

4 Conclusion

The digital transformation of human resource management has completely changed the traditional form of work. The introduction of artificial intelligence in the process of human resources becomes the problem-solving and automation of several functions in work. The correlation between technological processes (digitalisation and automated work contexts) and those involving methods in human resources (recruitment, selection, performance) is very pronounced.

Radical changes rapidly increase the need for an advanced workforce in their use and adaptation, making the relationship between digitalisation and human resources a productive force resistant to intense competition and frequent changes. The activity process in the Human Re has undergone significant changes that continue today. The recruitment and selection process has changed; digital platforms, international connections, globalisation, social networks and many other aspects have influenced training, and digital forms of communication have shown their importance even in crisis and pandemic situations. So the state's role must be present, primarily to support small and medium enterprises (MSMEs) so that all participants in the process of realising infrastructure related to the readiness of instruments and operators in the digitisation process. Digitisation as a process enters an entirely different phase and is soon expected to belong to all parties in the use of science, where information technology systems are applied. In other words, it is the use of the latest technology. In an ever-changing world where technology changes, so are people, varying requirements, and consumer wants. With ever-increasing demand, there is hope for everything, and innovation is always part of a challenge in the global dimension.

References

1. Goldstein, J.: Digital technology demand is transforming HR. Work. Solut. Rev. **6**(1), 28–29 (2015)
2. Deloitte, B.: Global Human Capital Trends 2016, the New Organization: Different by Design. Deloitte University Press Chicago, IL (2016)
3. Shah, N., Irani, Z., Sharif, A.M.: Big data in an HR context: exploring organizational change readiness, employee attitudes and behaviors. J. Bus. Res. **70**, 366–378 (2017)
4. Makridakis, S.: The forthcoming Artificial Intelligence (AI) revolution: its impact on society and firms. Futures **90**, 46–60 (2017)
5. Andersson, J.: Digital Transformation, Moderna affärssystem (2017)
6. Larkin, J.: HR digital disruption: the biggest wave of transformation in decades. Strat. HR Rev. **16**(2), 55–59 (2017)
7. Kavanagh, M.J., Thite, M., Johnson, R.D.: The future of HRIS. Emerg. Trends HRM IT (2009)
8. Nawaz, N.: A comprehensive literature review of the digital HR research filed. In: Information and Knowledge Management, vol. 7, no. 4 (2017)
9. Gebayew, C., Hardini, I.R., Panjaitan, G.H.A., Kurniawan, N.B.: A systematic literature review on digital transformation. In: 2018 International Conference on Information Technology Systems and Innovation (ICITSI), pp. 260–265 (2018)
10. Bohnsack, R., Hanelt, A., Marz, D., Marante, C.: Same, same, but different!? a systematic review of the literature on digital transformation. Acad. Manag. Proc. **2018**(1), 16262 (2018)
11. Amladi, P.: HR's guide to the digital transformation: ten digital economy use cases for transforming human resources in manufacturing. Strat. HR Rev. **16**(2), 66–70 (2017)
12. Bondarouk, T., Ruël, H., Parry, E.: Electronic HRM in the Smart Era. Emerald Publishing Limited, Bingley (2017)
13. Huselid, M.A.: The impact of human resource management practices on turnover, productivity, and corporate financial performance. Acad. Manag. J. **38**(3), 635–672 (1995)
14. Bredin, K., Söderlund, J.: Human Resource Management in Project-Based organizations: The HR Quadriad Framework. Springer, Heidelberg (2011)
15. Saini, S.: Digital HRM and its effective implementation: an empirical study. Int. J. Manag. Stud. **2**(7), 62–66 (2018)
16. Iwu, C.G.: Effects of the use of electronic human resource management (E-HRM) within human resource management (HRM) functions at universities. Acta Univ. Danubius. Adm. **8**(1) (2016)
17. Betchoo, N.K.: Digital transformation and its impact on human resource management: a case analysis of two unrelated businesses in the Mauritian public service. In: 2016 IEEE International Conference on Emerging Technologies and Innovative Business Practices for the Transformation of Societies (EmergiTech), pp. 147–152 (2016)
18. Tripathi, R.T., Singh, P.K.: A study on innovative practices in digital human resource management. In: Proceedings of the National Conference "Digital transformation of business in India: Opportunities and challenges", Dehradun, pp. 1–13 (2017)
19. Fedorova, A., Zarubina, A., Pikulina, Y., Moskovskikh, A., Balandina, T., Gafurova, T.: Digitalization of the human resource management: Russian companies case. In: International Conference on Education, Social Sciences and Humanities, vol. 12271230 (2019)
20. Anthony, L.: AntConc (Version 3.4. 3)[Computer Software]. Tokyo, Japan: Waseda University (2014)
21. Nawaz, N., Gomes, A.M.: Human resource information system: a review of previous studies. J. Manag. Res. **9**(3), 92 (2017)

22. Aggarwal, V., Sharon, S.D.: Digital human resource management. Gyan Manag. **11**(2), 23–27 (2017)
23. Hwang, Y., Kettinger, W.J., Yi, M.Y.: Personal information management effectiveness of knowledge workers: conceptual development and empirical validation. Eur. J. Inf. Syst. **24**(6), 588–606 (2015)
24. Powell, E., Wang-Hall, J., Bannister, J.A., Colera, E., Lopez, F.G.: Attachment security and social comparisons as predictors of Pinterest users' body image concerns. Comput. Human Behav. **83**, 221–229 (2018)
25. Ananiadou, K., Claro, M.: 21st century skills and competences for new millennium learners in OECD countries. OECD education working papers, no. 41.. OECD Publication (2009)
26. Schulze, J., Schultze, M., West, S.G., Krumm, S.: The knowledge, skills, abilities, and other characteristics required for face-to-face versus computer-mediated communication: Similar or distinct constructs? J. Bus. Psychol. **32**(3), 283–300 (2017)
27. Mishra, P., Kereluik, K.: What 21st century learning? a review and a synthesis. In: Society for Information Technology & Teacher Education International Conference, pp. 3301–3312 (2011)
28. van Damme, M., et al.: Verzonken technologie. ICT en de arbeidsmarkt, vol. 114. Sociaal en Cultureel Planbu (2005)
29. Higgins, S.: Critical thinking for 21st-century education: a cyber-tooth curriculum? Prospects **44**(4), 559–574 (2014)
30. Wickstrom, A.: Influencing gender relations through child parent relationship therapy. Int. J. Play Ther. **19**(2), 79 (2010)
31. Oei, A.C., Patterson, M.D.: Playing a puzzle video game with changing requirements improves executive functions. Comput. Human Behav. **37**, 216–228 (2014)
32. Oldham, G.R., Da Silva, N.: The impact of digital technology on the generation and implementation of creative ideas in the workplace. Comput. Human Behav. **42**, 5–11 (2015)
33. Neubert, J.C., Mainert, J., Kretzschmar, A., Greiff, S.: The assessment of 21st century skills in industrial and organizational psychology: complex and collaborative problem solving. Ind. Organ. Psychol. **8**(2), 238–268 (2015)
34. Greiff, S., Funke, J.: Interactive problem solving: exploring the potential of minimal complex systems (2017)

Digital Transformation During the Pandemic Performed by SMEs in ASEAN Countries: A Review of Empirical Studies

Arif Hartono(✉), Ratna Roostika, and Baziedy Aditya Darmawan

Department of Management, Faculty of Business and Economics, Universitas Islam Indonesia, Yogyakarta, Indonesia
arif.hartono@uii.ac.id

Abstract. SMEs have been the backbone of numerous economies all over the world. But since Covid-19 pandemic hit many economies, it changed practically every part and component of SMEs. In response to the pandemic, one transition has been accelerated, namely digital transformation. Although previous studies on digital transformation performed by SMEs in ASEAN countries have been conducted, currently, there is no single empirical study that compare digital transformation conducted by SMEs across ASEAN countries. Therefore, we discuss and reflect on the implementation of digital transformation by SMEs in ASEAN countries by reviewing existing empirical studies on SMEs' digital transformation published between 2020–2022 are reviewed.

Keywords: Digital transformation · Covid-19 · SMEs · ASEAN

1 Introduction

The crisis was brought on by Covid-19, which changed practically every part and component of SMEs such as the way they operate, access information, communicate, make choices, purchase or sell items, and retrain personnel, all must be performed digitally. In addition, the Covid-19 crisis's weakest link is the economy's SMEs. How resilient must SMEs to be endure in a pervasive condition of uncertainty under Covid-19 crisis? In response to the COVID-19 epidemic that has affected the world since the end of 2019, one transition has been accelerated: digital transformation.

Technology advancements promote changes in organizations and society as a result of digital transformation. These changes include corporate procedures, modifications to the way the company interacts with its clients, workers, and customers, and alterations to market conditions are all part of this. The business model that is the product of digital transformation Innovations have significantly altered customer expectations and behavior, suppressed established businesses, and disrupted a number of marketplaces [1].

Although previous studies on digital transformation performed by SMEs in ASEAN countries have been conducted, to the best our knowledge, there is no single empirical

B. Alareeni et al. (Eds.): ICBT 2022, LNNS 620, pp. 751–755, 2023.
https://doi.org/10.1007/978-3-031-26953-0_69

study that compare digital transformation conducted by SMEs across ASEAN countries, existing studies currently are dominated by conceptual studies that focus on a single country. Therefore, an empirical study that compare digital transformation across SMEs in ASEAN countries that use primary data, remain unexplored. It is expected that this study as the early-stage prior further study on digital transformation comparison among SMEs in ASEAN countries to be conducted.

1.1 Objective and Structure of the Research

This study aims to analyze digital transformation performed by SMEs in ASEAN countries in time of the pandemic by reviewing published empirical studies during 2020–2022. This study is divided into five sections i.e. (1) Introduction, that consists of objective of the study, theoretical issue and research gap, (2) Literature Review, that covers issues on the reason for performing digital transformation and digital transformation pathways, (3) Findings on digital transformation pathways during the pandemic, and (4) Discussion and Implication for future research.

2 Literature Review

2.1 Summary of the Reviewed Empirical Studies

Table 1 presents summary of empirical studies on digital transformation among SMEs in ASEAN countries. Only studies that used primary data are included, while conceptual studies are excluded. Majority of existing studies tend to focus on conceptual review on digital transformation, therefore only limited number of the studies were reviewed in this study.

2.2 The Urgency for Digital Transformation

The digital transformation has been hastened through Covid-19 for businesses throughout the world where workers are required to work remotely and use a variety of technologies to do their tasks. The challenge is that SMEs should be able to have the necessary organizational skills, culture, and talent to achieve digital transformation. The literature presents proof that the strategic deployment of digital transformation can boost performance, productivity, and competitiveness [13].

A part from Covid-19, other factors that also drive digital transformation are mainly from external [1]. First, A growing number of supporting technologies have emerged since the invention of the World Wide Web and its widespread adoption, which has bolstered the growth of e-commerce. Second, the competition is drastically shifting as a result of these new digital technologies. Technologies have changed the competitive landscape in retail, pushing revenues to relatively new internet companies. Third, the digital revolution is causing changes in consumer behavior. Consumers are moving their purchases to online retailers, according to market research, and digital touchpoints are crucial in the customer journey, which impacts both online and offline sales.

Table 1. Empirical studies on digital transformation studies among SMEs in ASEAN countries

Authors	Countries	Methods	Digital transformation pathways
[2]	Indonesia	*Qualitative*	Adjusting the business model transformation with the support of digital technologies
[3]	Indonesia	Quantitative	Implementing digital marketing (e.g., mobile application and social media)
[4]	Indonesia	Quantitative	Implementing digital technology i.e., financial technology
[5]	Indonesia	Quantitative	Implementing digital marketing (e.g., mobile application and social media)
[6]	Indonesia	Quantitative	Implementing digital marketing (e.g., mobile application and social media)
[7]	Indonesia	Quantitative	Implementing digital marketing (e.g., mobile application and social media)
[8]	Malaysia	*Qualitative*	Implementing digital marketing (e.g., mobile application and social media)
[9]	Philippines	*Qualitative*	Digital entrepreneurship
[10]	Thailand	Quantitative	Implementing digital marketing (e.g., mobile application and social media)
[11]	Thailand	*Qualitative*	Implementing digital technology i.e., cloud computing
[12]	Vietnam	Quantitative	E-commerce adoption

3 Findings on Digital Transformation Pathways During the Pandemic

3.1 Adjusting the Business Model

According to [14] different digital transformation pathways depend on SMEs' contextual factors and they divided into three paths. First, SMEs with a high level of digital maturity that respond to the pandemic by quickening the shift toward digitalized enterprises are the first to come to mind. Second, SMEs with low levels of digital maturity and financial concerns who choose to digitize solely the sales function. The third category of SMEs has very little digital literacy but is well-supported by social capital and they tend to partnering with other SMEs that have outstanding digital skills.

3.2 Jump into Digital Marketing

Previous studies such as [2, 4–6, 9] reveal that many SMEs implement digital marketing as the easiest way to cope with crisis during the pandemic. It turns out that digital marketing has positive impact on financial performance as well as substantial impact on the long-term viability of businesses.

3.3 Implementing Digital Technology

The third pathway in relation with SMEs' digital transformation is the implementation of digital technology such as financial technology [4] and cloud accounting [11] in time of the pandemic. The usage of fintech is related to digital transformation on financial behavior. The findings demonstrate that the Covid-19 epidemic has an impact on SMEs' (old behavioral intention) use of fintech, and thus strengthens the influence on new behavior in utilizing fintech (new behavioral intention) [4]. While cloud accounting help SMEs to increase their efficiency, financial organization, and flexibility by implementing cloud-based accounting [11].

3.4 E-Commerce Implementation

The last emerged digital transformation from the previous studies is the adoption of e-commerce [12]. The study shows that technological perceived compatibility, followed by management support and outside pressure, has the biggest impact on the adoption of e-commerce. In contrast, external support has no effect on the uptake of e-commerce during the pandemic.

4 Discussion and Implication for the Future Research

This study aims to review previous studies on digital transformation conducted by SMEs across ASEAN countries during Covid-19 that hit many economies severely. This study is important to be conducted because it can be used as the initial effort to support further study on digital transformation comparison among SMEs in ASEAN countries that currently has not been studied. This study also provides practical contribution for SMEs' owner and/or managers to support digital transformation decision making.

Lastly limitation of the study needs to be acknowledged. First, only limited number of previous empirical studies are reviewed, as a result, insights on digital pathways is also limited. Second, most of the reviewed studies are not published in the top rank academic journals, hence, its affect the quality of the reviewed journals. Therefore, future studies must address such issues.

Acknowledgement. This work was supported by Indonesian Ministry of Education, Culture, Research, and Technology (Proposal ID: 6120ff2f-e964-42c0-bef7-104c24a33ae3).

References

1. Verhoef, P.C., et al.: Digital transformation: a multidisciplinary reflection and research agenda. J Bus Res. **122**, 889–901 (2021). https://doi.org/10.1016/j.jbusres.2019.09.022
2. Priyono, A., Moin, A., Putri, V.N.A.O.: Identifying digital transformation paths in the business model of smes during the covid-19 pandemic. J. Open Innov.: Technol. Mark. Complexity. **6**, 1–22 (2020). https://doi.org/10.3390/joitmc6040104
3. Effendi, M.I., Sugandini, D., Istanto, Y.: Social media adoption in SMEs impacted by COVID-19: the TOE model*. J. Asian Finan. Econ. Bus. **7**, 915–925 (2020). https://doi.org/10.13106/jafeb.2020.vol7.no11.915

4. Wiyono, G., Kirana, K.C.: Digital transformation of SMEs financial behavior in the new normal era. Jurnal Keuangan dan Perbankan **25** (2021). https://doi.org/10.26905/jkdp.v25i1.4954
5. Purba, M.I., Simanjutak, D.C.Y., Malau, Y.N., Sholihat, W., Ahmadi, E.A.: The effect of digital marketing and e-commerce on financial performance and business sustainability of MSMEs during COVID-19 pandemic in Indonesia. Int. J. Data Netw. Sci. **5**, 275–282 (2021). https://doi.org/10.5267/j.ijdns.2021.6.006
6. Patma, T.S., Wardana, L.W., Wibowo, A., Narmaditya, B.S.: The shifting of business activities during the COVID-19 pandemic: does social media marketing matter? J. Asian Finan. Econ. Bus. **7**, 283–292 (2020). https://doi.org/10.13106/JAFEB.2020.VOL7.NO12.283
7. Patma, T.S., Wardana, L.W., Wibowo, A., Narmaditya, B.S., Akbarina, F.: The impact of social media marketing for Indonesian SMEs sustainability: lesson from Covid-19 pandemic. Cogent Bus. Manag. **8** (2021). https://doi.org/10.1080/23311975.2021.1953679
8. Fabeil, N.F., Pazim, K.H., Langgat, J.: The impact of Covid-19 pandemic crisis on micro-enterprises: entrepreneurs' perspective on business continuity and recovery strategy. J. Econ. Bus. **3** (2020). https://doi.org/10.31014/aior.1992.03.02.241
9. Cueto, L.J., Frisnedi, A.F.D., Collera, R.B., Batac, K.I.T., Agaton, C.B.: Digital innovations in MSMEs during economic disruptions: experiences and challenges of young entrepreneurs. Adm. Sci. **12** (2022). https://doi.org/10.3390/admsci12010008
10. Yawised, K., Apasrawirote, D., Padgate, U.: Enhancing SMEs' leader and business resilience towards digital marketing engagement during the COVID-19 pandemic (2021)
11. Sastararuji, D., Hoonsopon, D., Pitchayadol, P., Chiwamit, P.: Cloud accounting adoption in Thai SMEs amid the COVID-19 pandemic: an explanatory case study. J Innov Entrep. **11** (2022). https://doi.org/10.1186/s13731-022-00234-3
12. Hoang, T.D.L., Nguyen, H.K., Nguyen, H.T.: Towards an economic recovery after the COVID-19 pandemic: empirical study on electronic commerce adoption of small and medium enterprises in Vietnam. Manag. Mark. **16**, 47–68 (2021). https://doi.org/10.2478/mmcks-2021-0004
13. Papadopoulos, T., Baltas, K.N., Balta, M.E.: The use of digital technologies by small and medium enterprises during COVID-19: implications for theory and practice. Int. J. Inf. Manag. **55** (2020). https://doi.org/10.1016/j.ijinfomgt.2020.102192

Modern Challenges of Payment Systems' Efficient Functioning

Kvasnytska Raisa(✉), Forkun Iryna, and Gordeeva Tetyana

Khmelnytskyi National University, 11, Instytuts'ka str., Khmelnytskyi 29016, Ukraine
rstepanivna@gmail.com, gordeevat_2004@ukr.net

Abstract. The paper substantiates that the basis of the effective functioning of banking institutions in conditions of uncertainty is the identification and management of risks inherent in banking activity. Therefore, the purpose of the article is to substantiate the essence of payment systems, their role in ensuring the acceleration of monetary turnover both at the level of a sovereign country and at the international level; and also to study the international experience of central banks and the approaches of experts of the World Bank and the International Monetary Fund regarding the oversight of payment systems in order to ensure their effective functioning. The article defines that the sphere of payment systems belongs to the risky spheres of activity, which are characterized by such types of risks as legal, payment, operational, systemic and financial risks. At the same time, it is emphasized that cyber risk is the component of operational risk and cyber risk is considered as the risk of the implementation of cyber threats against information resources and/or information infrastructure, as well as the consequences of such events. Considering that modern payment systems are intermediaries in the funds' movement, and, therefore, they are at risk of negative influences, challenges, threats and dangers, that can lead to damaging the national interests of states, it is proposed to consider cyber risk as a separate type of risk of the payment systems functioning.

Keywords: Payment system · International payment system · Risks of payment systems · Information security of payments

1 Introduction

Various payment systems are an important component of the economic and financial infrastructure of the countries of the world. Ukraine is no exception in this regard, in recent years its economy and banking system have undergone radical changes due to the development of market relations in the national economy. Nowadays, the state of domestic and international payments in the country is characterized by rapid growth in the volume of monetary turnover, a decrease in the likelihood of risks, and an increase in the security of payments through payment systems and money transfer systems. Thus, it is necessary to study the essence, varieties and features of the functioning of payment systems in the contemporary changing conditions of the development of economic relations between the subjects of the economy at the national and international levels.

B. Alareeni et al. (Eds.): ICBT 2022, LNNS 620, pp. 756–765, 2023.
https://doi.org/10.1007/978-3-031-26953-0_70

2 Relevant Research

The strategic development of digitization of payment operations between economic entities has received great attention at the government level of many leading countries of the world (the USA, China, Japan, Canada, Australia, etc.). Issues of the development of the information (digital) economy are constant in the areas of research of international organizations, such as the World Bank, the OECD, the European Commission, etc. Consideration of the functionality of various payment and monetary systems, and the study of their advantages, safety and development opportunities are given attention in the works of many economists, namely: P. Adamyk, O. Vovchak, O. Jusov, I. Kravchenko, T. Kokkola, A. Saadi, B. Summers, R., J. Spindler, N. Trusova, I. Chkan and others. and others. In the mentioned works thorough research on the functioning and development of payment systems was conducted, also the shortcomings of electronic payments were highlighted and attention was focused on their low security. The authors pay considerable attention to the issue of the level of security of payment systems and their transformation under the influence of innovative development and the latest technologies. However, existing and potential threats to the security of payment systems create the possibility of a violation of the regular mode of their operation (including disruption and/or blocking of the system, and/or unauthorized management of its resources), endanger the safety (security) of electronic information resources, which requires additional of research in this direction.

3 Obtained Research Results

Settlement operations and payments occupy an important place in the market economy of any country, enabling the movement of money between economic entities, as well as, in the international economic area. After all, each day economic entities carry out a large number of operations on the exchange of goods, services and financial assets, which, in turn, are mediated by cash payments and transfers. Under the conditions of developing the new type of economy, an important guarantee for the stable functioning of the economy of both sovereign countries and the world economy as a whole is the active use of payment systems, which have recently received rapid development in Ukraine as well. It is the expansion of the types of payment systems, and the spectrum of their functionality, that has a significant impact on the development of the country's payment infrastructure, and the emergence of new types of payment instruments, which, depending on their specific types, can stimulate the emergence of new markets, the growth of national economies, increase their competitiveness, reduce unemployment, improvement of the standard of living.

So, what is a payment system? It is worth noting that a unified approach to the concept of "payment system" has not been developed either at the level of normative definition in the international spheres of functioning of these systems or at the level of scientific circles. The Committee on Payments and Market Infrastructures (CPMI) of the Bank for International Settlements (BIS) defines a payment system as "a set of tools, procedures and rules for transferring funds between participants or among participants; the system includes participants and the organization that is its operator. The basis of the payment system, as a rule, is an agreement between the participants and the operator, and the transfer of funds takes place in accordance with the operational infrastructure agreed upon by them" [1] The representative of the European Central Bank, Tom Kokkola, noted that the concept of a "payment system" can be considered similar to the concept of a "funds transfer system", which means "an official agreement based on a private contract or legislation with several members, general rules and standardized mechanisms for transfer, clearing, netting and/or repayment of monetary obligations arising between its members" [2]. The Law of Ukraine "On Payment Services" defines a payment system as "a system for performing payment transactions with formal and standardized agreements and general rules regarding processing, clearing and/or settlement between participants of the payment system" [3]. It should be noted that in Ukraine, as in the whole world, there is a tendency towards a rapid increase in the volume of cashless settlements and payments. Money transfers in Ukraine are carried out using domestic (a payment system in which the payment organization is a resident and which carries out its activities and ensures the transfer of funds exclusively within the borders of Ukraine) and international payment systems (a payment system in which the payment organization can be a resident, as well as a non-resident and which carries out its activities on the territory of two or more countries and ensures the transfer of funds within this payment system, including from one country to another). As of January 1, 2022, 54 payment systems were operating in Ukraine, including [4]: 2 - state payment systems (System of Electronic Payments (SEP) and National Payment System "PROSTIR", which includes 78 banks-participants); 9 - payment systems, the payment organization of which is a bank, that includes 78 banks-participants. The largest number of participants (51 participants) has the international payment system "Welsend" JSC UKRGAZBANK (Ukraine); 21 - payment system, the payment organization of which is a non-bank institution, which has 157 participants. The largest number of participants (26 participants) has the international funds' transfer system "AVERS No. 1" (Ukraine); 6 - international card payment systems, which have 119 participants. The largest number of participants (55 participants) has the international card payment system "MasterCard" (USA); 10 - international money transfer systems, which have 71 participants. International funds transfer systems "RIA" and "MoneyGram" (USA) have the largest number of participants (18 participants); 6 – intrabank payment systems.

Note that today the National Bank of Ukraine in partnership with the SWIFT company is executing a project to implement the ISO 20022 international message exchange standard in the payment infrastructure of Ukraine. The transition to using international standards for the exchange of financial messages in the payment infrastructure of Ukraine will enable: to harmonize the Ukrainian payment space with the global one; to expand the details of payments with additional information; to increase the level of service and efficiency of payments; to enrich the functionality of payment instruments for the benefit of banks and their clients; to increase the level of payment automation [5]. Sure thing, the operation sphere of payment systems belongs to the risky spheres of activity, which is explained by the presence of a significant number of connections between participants, the volume and size of operations performed in these systems, high mobility and promptness of calculations, the rapid development of the latest technologies, the development of systems of remote banking services, which in turn creates potential opportunities for disrupting the normal flow of payments, which can immediately cause huge negative consequences for the entire payment system. Risk level management involves conducting risk analysis in the process of strategic, tactical and operational planning and allows obtaining a qualitative and quantitative assessment of possible risks [6, p. 47]. Thus, the National Bank of Ukraine has developed Regulations on Procedure for Payment Infrastructure Oversight in Ukraine [7], Methodological recommendations for risk management in payment systems [8] and Methodological recommendations on operational risk management (including cyber risk and business continuity) and ensuring the storage of customer information by payment infrastructure facilities [9], which take into account the best international experience of central banks and the approaches of experts from the missions of the World Bank and the International Monetary Fund on payment systems oversight. Among the main risks in the banking and non-banking spheres is operational risk, that is, the risk of loss of profit due to errors in the implementation of daily traditional financial transactions [10, p. 261].

There is the following typification of the payment systems' risks: 1) legal risk - the risk of the absence of legal regulation, changes or unforeseen application of the provisions of the legislation of Ukraine, which may lead to losses of the oversight object; 2) financial risks that include: credit risk - the risk that the oversight object will not be able to fulfil its financial obligations; liquidity risk - the risk that the oversight object will not have enough funds to properly fulfil its financial obligations, but it will be able to fulfil them at another time in the future; general commercial risk - the risk of a deterioration in the financial condition of the oversight object due to a decrease in its income or an increase in expenses, as a result of which expenses exceed income and lead to losses, which are covered by capital; depository risk - the risk of loss of financial assets of the oversight object; investment risk - the risk of loss or unavailability of financial assets of the oversight object, arising as a result of their investment; 3) payment risk - the risk that payments in the payment system will not be carried out properly; 4) operational risk - the risk that deficiencies in information systems or internal processes, human error, operational failures due to external events will lead to a reduction, deterioration or cessation of the provision of services by the oversight facility. As part of operational risk, cyber risk is considered - the risk of implementation of cyber threats against information resources and infrastructure; 5) systemic risk - the risk that the inability of one of the participants of the payment system and the operator of the payment infrastructure services to fulfil its obligations or a violation of the continuity of the activity of the payment system itself will lead to a disruption of the activities of the participants of the payment system, other institutions or functioning of the financial system as a whole.

Simultaneously, the normative documents of the National Bank of Ukraine define cyber risk as a component of operational risk. However, given the fact that contemporary payment systems are intermediaries in the movement of funds, and therefore they are at risk of negative influences, challenges, threats and dangers that can lead to damaging the national interests of the State, cyber risk should be considered as a separate type of payment system risks.

To prevent and avoid cyber risks, the Central bank of the State should increase the level of information security and cyber protection in money transfers. In order to strengthen the reliability and efficiency of payment systems, the regulator should establish precise requirements for payment market participants regarding: building a system of information protection and ensuring cyber security; preparedness for possible cyber-attacks, security, the ability to detect cyber-attacks, react and absorb them, ensuring adaptability and the possibility of their recovery [11] and the procedure for detecting cyber-attacks that reduce the reliability of the functioning of payment systems.

It is worth noting that the question of cyber security and payment fraud is more topical than ever all over the world today, and Ukraine is no exception. Thus, the National Bank of Ukraine notes an increase in the activity of fraudsters during 2021. The "most popular" method of fraud with payment cards in Ukraine, as well as in the world, traditionally remains social engineering, thanks to the application of which people themselves transfer money to fraudsters or reveal their card data to them. Therefore, in order to intensify the fight against such incidents at the macro level in 2022, the Central Bank of Ukraine, together with the Cyber Police Department of the National Police of Ukraine, as well as with the support of the International Finance Corporation (IFC) in partnership with the Swiss State Secretariat for Economic Affairs (SECO), The Good Governance Fund of Great Britain and the Ministry of Digital Transformation of Ukraine, launched a nationwide information campaign on payment security called "Fraud Goodbye". The company's goal is to improve citizens' awareness of cyber hygiene, promote the formation of a culture of safe behaviour in the virtual space, as well as, to learn the basic safety rules of non-cash payments in various payment systems [12]. Banking institutions are one of the main subjects of the creation of payment systems, their payment organizations and participants. Banks, in the process of making payments, transfers and settlements through the functionality of a certain payment system, are faced with the need to protect their own and client information and ensure cyber security. Therefore, the issue of the information security policy of banking institutions is not new, it is periodically highlighted in the scientific works of scientists during the last decades since it was during them that the informatization of banking activity was active, but this issue is gaining special relevance in our time, which has coincided with the digitization boom services. The bank's information security policy is understood as a set of legal and moral and ethical norms, rules, administrative, organizational measures and technical, software and cryptographic means aimed at protecting the bank's information infrastructure from accidental and intentional interference in the process of its functioning. Information security policy is a set of requirements, rules, restrictions, and recommendations that regulate the order of information activities in the organization and are aimed at achieving and maintaining the state of information security of the organization [13].

The modern practice of finding and implementing new approaches to ensuring the information security of banks is characterized by a certain diversity. Thus, to analyze modern banking practices regarding their provision of information security in Ukraine, the following selection of banks was made: two banks were selected from the leaders in terms of assets within each group, allocated following the decision of the NBU of February 5, 2021 No. 40. Such banks are: banks with a state share: JSC CB PRIVATBANK and JSC OSCHADBANK; banks of foreign banking groups: JSC RAIFFEISEN BANK and JSC UKRSIBBANK; banks with private capital: JSC UNIVERSAL BANK and

JSC TASCOMBANK. These banks offer customers a wide range of payment and settlement services because they are participating in many international payment systems and urgent money transfer systems (Table 1).

Such a variety of systems of urgent money transfers and payment services leads to an increase in the probability of realizing the risk of breach of information security during calculations with their help. At the same time, problems with a violation of information security should be considered from two sides: 1) on the part of the client, who can perform part of banking operations independently thanks to remote service and e-banking technologies; 2) on the part of a bank employee who intentionally or negligently may violate the information security policy introduced by the bank.

Table 1. Banks-participants in payment systems

Payment systems	JSC CB PRIVAT BANK	JSC OSCHAD BANK	JSC RAIFFEISEN BANK	JSC UKRSIB BANK	JSC UNIVERSAL BANK	JSC TASCOM BANK
Master Card	+	+	+	+	+	+
Visa	+	+	+	+	+	+
China UnionPay	+	–	–	–	–	–
NPS PROSTIR	+	+	+	–	–	+

The first direction of ensuring information security involves increasing the financial literacy of clients. According to the second direction of ensuring and preventing information security incidents, banks develop such an internal normative document as the "Information Security Policy", which is published on the official website of each bank. Thus, the "Information Security Policy" is a document that formulates and expresses the position of the bank's management regarding information security, as well as defines the main principles and tasks of ensuring information security in the bank [14]. The main goals of assuring information security are to ensure such quality characteristics of information as confidentiality, integrity, availability and accountability. A glossary is also provided to avoid differences in the interpretation of the terms and concepts used in the event of information security incidents.Note that the banks from our sample have in common that the type of ownership to which the bank belongs has a significant impact on the information security policy.

Thus, according to the first proposed criterion, namely the normative and legal one, JSC CB PRIVATBANK and JSC OSCHADBANK, which are banks with a state share, provide a detailed list of laws of Ukraine ("On information", "On information protection in information and telecommunications systems", "On protection of personal data", etc.), normative legal acts of the NBU on information security (in particular, "On approval of the Regulation on the organization of measures to ensure information security in the banking system of Ukraine"), standards ISO/IEC 27002:2015 and payment card industry data security standards PCI DSS [14, 15]. JSC RAIFFEISEN BANK and JSC UKRSIBBANK, as banks belonging to foreign banking groups, characterize their policy, which regulates the information security management system, as meeting the requirements of Ukrainian legislation, standards, resolutions and regulatory acts of the NBU on information security. At the same time, these banks additionally refer to the regulatory documents of the international banking groups to which they belong. Thus, it is emphasized that JSC RAIFFEISEN BANK's policy is based on the regulatory document of the RBI Group REG-2016-0065 Group IT Security [16]; in JSC UKRSIBBANK on the BNP Paribas Group Information Security Policy and CyberSecurity Program 2020 [17].

Banks with private capital – JSC UNIVERSAL BANK and JSC TASCOMBANK – note only in general that the bank's policy complies with the legislation of Ukraine and the normative legal acts of the National Bank of Ukraine, which relate to information security [18]. According to the second, organizational criterion, the availability of the Information Security Management System (ISMS), through which the bank's information security management is organized thanks to the consolidation of human, methodological, intellectual and software and technical resources, is common [19]. However, certain peculiarities can be seen in the ISMS levels and their hierarchy. Thus, JSC CB PRIVATBANK and JSC OSCHADBANK have a common feature – an extensive multi-level ISMS. Therefore, the ISMS Committee on implementation, maintenance and control of the information security management system was established and operates continuously at JSC OSCHADBANK. As part of the information security management system, the participants of information security in this bank are: Management of the bank; ISMS Committee; The administrator of the information resource/system; Information protection administrator; Heads of branches of the bank; Owners of information assets; Heads of independent structural units; All employees of the bank and users of the bank's information assets; Partners, suppliers, and other counterparties.

Also, the "Information Security Policy" of JSC OSCHADBANK describes in detail the rights and duties of all information security participants. In JSC CB PRIVATBANK, the ISMS is somewhat more compact. The bank has a collective management subdivision, which is empowered to implement and operate the ISMS, a person responsible for the bank's information security – a Chief information security officer (CISO) – has been appointed, and an Information Security Unit has been created, which reports directly to the CISO. JSC RAIFFEISEN BANK has three levels of information security management: the Bank's Management Board, the Information Security Unit and information technology units, which in their activities are guided by the requirements of Raiffeisen Bank International Group regulatory documents and information security regulatory documents. In JSC UKRSIBBANK, in accordance with the standards of the BNP Paribas Group, at the initiative of the bank's top management and the bank's information security control unit, the Committee for Information Security and IT Risks was established and operates regularly. Each employee of the bank or employee of the counterparty (within contractual obligations) participates in maintaining the high level of information security of the bank. Thus, it can be concluded that the banks of foreign banking groups strictly adhere to corporate standards in the management of information security.

Information security management in banks with private capital – JSC UNIVERSAL BANK and JSC TASCOMBANK – is carried out through the bank's ISMS, which is aimed at protecting the information resources and assets of the banks from external and internal threats and threats associated with intentional and unintentional actions of its employees or third parties.

The next of the information security criteria we highlighted, the technical criterion can be characterized both through the provision of information confidentiality, protection against malicious software, backup and recovery of information, license purity, physical security, etc. and through access management, rights and authority management, management of authorization methods. Most of the mentioned above are not detailed by the banks in the "Information Security Policy", as protected by banking and commercial secrecy. Only JSC CB PRIVATBANK notes that in the bank's information systems, which directly provide automation of banking activities, it is prohibited to combine the following powers within the same function (role): development and support (administration), development and operation, support (administration) and operation, execution operations in such systems and further control over their implementation. The bank uses the standards, documents and guidelines of the Open web application security project (OWASP) to develop secure web applications [15].

It should be noted separately, that all banks from the sample adhere to the principle of providing a minimum level of rights when providing access to the bank's information systems (including access by privileged users). Thus, JSC OSCHADBANK defines the essence of this principle as "a minimum of rights – this is the access of bank employees and users of information systems to the information resources of the bank's computer network should be organized in such a way as to grant only those powers necessary for the performance of official tasks" [14].

Banks with a state share – JSC CB PRIVATBANK and JSC OSCHADBANK – declare in their "Information Security Policies" that "bank employees are responsible for non-compliance with information security requirements established by the bank's internal documents and current legislation" [14, 15].

In the "Information Security Policy" of JSC RAIFFEISEN BANK, it is specified that "employees are personally responsible for… in particular, the preservation of banking secrecy, personal data of clients and other confidential information of the bank, maintaining the appropriate level of Information Security in the performance of their official duties" [16].

JSC UKRSIBBANK in the "Information Security Policy" separately highlights the section "Response to information security incidents", which prescribes the actions of the bank's customers and counterparties in the event of such a situation regarding their assets with the aim of eliminating it or clarifying the causes of its occurrence [17].

There is a peculiarity in ensuring information security at JSC UNIVERSAL BANK is the Policies of "Clean Desktop" (regarding workplaces, paper media, removable electronic media) and "Clean Screen" (regarding automated workplaces to reduce the risk of unauthorized access) to ensure additional protection of information in the bank's premises. On the desktop, there are only documents that are currently necessary for the performance of functional duties. Information carriers with limited access must be stored in an orderly form in specialized storage locations [18].

JSC TASCOMBANK emphasizes that "at the moment of hiring, all employees must be familiarized with the bank's information security policy under their signature and provide an obligation to observe confidentiality". Also, only in the information security policy of this bank is there a point "Model of threats and Model of the infringer" [19]. This point provides a description of current threats to information security and a description of the possible actions of the violator, which is based on an analysis of the type of intruder, the level of his right, knowledge, and theoretical and practical capabilities. Each type of infringer category, infringer motive, infringer qualification, tools and access level of the infringer is assigned a certain level of potential threat. This ranking makes it possible to effectively reduce the probability of realizing the risk of breaching the bank's information security.

So, summing up the documentation of the information security policy of the banks included in the sample, we can say that ensuring information security banks use standardized provisions, which are based on global standards and rules and domestic legislation and regulation of the NBU. Certain features may be due to belonging to different ownership groups, corporate culture, etc. We consider it appropriate to also focus on the above-mentioned peculiarities of the information security policies of JSC TASCOMBANK, namely the description of the "Model of threats and Model of the infringer" with the ranking of types of threats and their inherent risk levels, and JSC UNIVERSAL BANK, namely the implementation of the "Clean Desktop" and "Clean Screen" rules, as their wider implementation in the practice of ensuring information security of most domestic banks will increase the effectiveness of risk management and the level of protection against cyber threats.

4 Conclusions

Based on the results of the research, it can be concluded that an important guarantee for the stable functioning of the economy of both sovereign countries and the global economy is the active use of the payment systems functionality, which is an organizationally formed set of system participants and the relations between them regarding the transfer of funds based on legislation norms at the level of sovereign countries or at the international level. The presence of a significant number of connections between the participants of payment systems, the volume and size of operations performed in these systems, high mobility and promptness of calculations, the rapid development of the latest technologies, and the development of remote banking service systems are determining factors of the riskiness of the payment systems sphere. One of the risks that may arise as a result of the realization of cyber threats to information resources and/or information infrastructure, and may also be a consequence of such events, is the cyber risks. In order to prevent, detect, respond, absorb cyber risk, and ensure adaptability and the ability to restore the payment system, the central bank of the state should increase the level of information security and cyber protection in the field of funds transfer. The information security policy of banks must be constantly supplemented and changed in accordance with the specified set of criteria for assuring information security. At the same time, the main goals of assuring information security are ensuring such quality characteristics of information (criteria for ensuring information security), such as confidentiality, integrity, availability, and accountability. Timely identification of cyber risk is an important component in ensuring the overall security of the payment system functioning. In order to effectively manage cyber risks, it is advisable to improve approaches to the development of the information security policy of payment organizations of payment systems. Therefore, the subject of further research should be the development of effective methods of cyber risk identification and assessment of its impact on the payment systems' efficient functioning.

References

1. Principles for financial market infrastructures/CPSS-IOSCO. https://www.bis.org/cpmi/publ/d101a.pdf. Accessed 28 Sep 2022
2. Kokkola, T.: The payment system Payments, securities and derivatives, and the role of the Eurosystem, no. 369, ECB (2010)
3. The Verkhovna Rada of Ukraine, On payment services. https://zakon.rada.gov.ua/laws/show/1591-20#Text. Accessed 2022/09/28
4. Information from the Register of payment systems, settlement systems, participants of these systems and payment infrastructure service operators of the NBU. https://bit.ly/3fmoBzm. Accessed 20 Sep 2022
5. ISO 20022 implementation in the payment infrastructure of Ukraine. https://bank.gov.ua. Accessed 23 Sep 2022
6. Kvasnytska, R., Dotsenko, I., Matviychuk, L., Oliinyk, L.: Organization of management of moral risks of banking activity in a modern business environment. In: Alareeni, B., Hamdan, A. (eds.) Sustainable Finance, Digitalization and the Role of Technology. ICBT 2021. LNNS, vol. 487, pp. 243–249. Springer, Cham (2023). https://doi.org/10.1007/978-3-031-08084-5_18
7. National Bank of Ukraine, Regulations on Procedure for Payment Infrastructure Oversight in Ukraine. https://zakon.rada.gov.ua/laws/show/1132-2010-%D0%BF. Accessed 29 Sep 2022
8. Methodological recommendations for risk management in payment systems. https://bit.ly/3RmxKFu. Accessed 29 Sep 2022
9. Methodological recommendations on operational risk management (including cyber risk and business continuity) and ensuring the storage of customer information by payment infrastructure facilities. https://bit.ly/3rhGHW6. Accessed 29 Sep 2022
10. Trusova, N.V., Chkan, I.O.: Payment systems in Ukraine and the risks of their functioning. Bus. Inform. **1**, 257–263 (2021)
11. Lysenko, S., Bobrovnikova, K., Gaj, P., Sochor, T., Forkun, I.: Resilient computer systems development for cyberattacks resistance. In: CEUR Workshop Proceedings, no. 2853, pp. 353–361 (2021). http://ceur-ws.org/Vol-2853/short41.pdf
12. The information campaign of the National Bank on payment security #Fraud Goodbye is launched. https://cutt.ly/BVVLyCe. Accessed 29 Sep 2022
13. Information security policy. https://bit.ly/3Cnkf47. Accessed 29 Sep 2022
14. JSC OSCHADBANK. https://cutt.ly/kVVLPnG. Accessed 29 Sep 2022
15. JSC CB PRIVATBANK». https://cutt.ly/TVVXhc8. Accessed 29 Sep 2022
16. JSC RAIFFEISEN BANK. https://cutt.ly/3VVL4pC. Accessed 29 Sep 2022
17. JSC UKRSIBBANK. https://cutt.ly/OVVZTWO. Accessed 29 Sep 2022
18. JSC UNIVERSAL BANK. https://ppt-online.org/931513. Accessed 29 Sep 2022
19. JSC TASCOMBANK. https://cutt.ly/eVVZ7bm. Accessed 29 Sep 2022

Instagram Book Review Codebook: A Content Analysis of Book Reviews by Bookstagrammers on Instagram

Harshita Singh(✉) and Ginu George

CHRIST (Deemed to be University), Bangalore 560029, India
harshita.singh@res.christuniversity.in,
ginu.george@christuniversity.in

Abstract. The growing number of Social Media Influencers (SMIs) include a niche of influencers known as 'Bookstagrammer'. Bookstagrammers are literary book influencers on Instagram; Their book reviews are now replacing those that expert journalists and columnists wrote. Bookstagrammers are successfully grabbing readers' attention on Instagram with their online book-review style. This trend of book reviewing is now influencing the purchase intention of many readers online and impact the sales of the publishing industry. The present study attempts to develop a codebook by exploring and highlighting the key attributes of the book reviews posted by Bookstagrammers by conducting a content analysis on the Instagram posts. The blend of existing literature with the content analysis of the posts helps understand how these features create engagement in the book review posted.

Keywords: Bookstagrammer · Social media influencers · Instagram · Marketing · Book reviews

1 Introduction

1.1 Book Reviewing in the Digital Era

The meaning of review originated in 15th century France and came from the French word '*revoir*'. The prefix *re-* (again) added to *voir* (to see) and signifies looking again at any material. Random House defined book reviews as "an evaluation, analysis or critique of a newly published book by a critic, reporter or another person in a newspaper or magazine" (Lin et al. 2005). Literary criticism is "a part of literary theory that provides contemporary interpretation, commenting, evaluation and appraisal of current literature pieces" (Stanková 2021). Margaret Atwood, a writer, advocated three objectives of book reviews: providing the target reader with an accurate and detailed introduction to a book, attracting readers who were initially apathetic or indifferent to a book, and pointing out the author's problems that require deeper thinking (Nischik 2015). Earlier, only selected columnists and journalists could write book reviews for publications and journals. The same book reviewing concept has now gone digital. Internet book reviews are defined

B. Alareeni et al. (Eds.): ICBT 2022, LNNS 620, pp. 766–775, 2023.
https://doi.org/10.1007/978-3-031-26953-0_71

as all public reviews of published books by readers on the Websites of bookstores, publishers, or private individuals (Lin et al. 2005). Many readers read book reviews to know if the book interests them and then make their purchase decision (Stanková 2021). Online book reviews are electronic Word-Of-Mouth (eWOM) sources regarding books that influences consumers' purchasing strategies (Voss et al. 2003). Positive Internet book reviews are capable of inducing interest in consumers towards books and contribute to the ability of consumers to find suitable books (Lin et al. 2005). Chevalier & Mayzlin, in their study, show that customers use reviews to screen their purchases and know if the book is the right fit for them (Chevalier et al. 2006). Blogs, e-commerce websites, or specific book reviewing platforms like 'Goodreads' are good sources of online book reviews; Instagram is the latest.

1.2 Social Media Influencers: Bookstagrammers

Instagram is a free social media application that enables users to upload and share content via videos and photos. it is the largest and fastest-growing visual-based social media network, with the number of active users increasing day by day (Rietveld et al. 2020). Driven by user-generated content and Instagram is highly influential in many settings, from purchasing/selling behaviour, entrepreneurship, and political issues to venture capitalism (Greenwood and Gopal 2015). As Instagram and other social media platforms grew popular, they led to a growing community of social media influencers. Social media influencers have built a reputation for their knowledge and expertise on a specific topic. They regularly post pictures and videos on their preferred social media platform to generate a large following on their social media handles. Each influencer has a niche of their own and attract those who share the same interest. (Araujo et al. 2017). This influencer marketplace can be split into four categories namely mega-influencers, macro-influencers, micro-influencers and nano-influencers (Campbell and Farrell 2020). Based on the study of Table 1 shows the distinct categories of influencers that are based not only on follower counts but also perceived authenticity, accessibility, expertise, and cultural capital (Association of National Advertisers 2018).

Among these growing influencers is a niche of influencers known as Bookstagrammers. They influence the marketplace and the interaction between publishers and readers by being a platform where readers gather and discuss about books (Lo 2020). Bookstagram is a community of book readers on Instagram, and those who post book reviews are 'Bookstagrammers.' Bookstagrammers share book reviews, current reads, new book releases, etc., and create aesthetically pleasing photos of the books and pair them with their content. For books, literary critics like Bookstagrammers generally provide "product tests" in the form of book reviews. As a result of their broad reach and public attention, makes them an opinion leader who, reinforces by word-of-mouth effect. This can have a critical impact on a book's sales success (Clement et al. 2007). Habitual readers often make the final purchase decision after reading book reviews (Saima and Khan 2020). This helps them know if the book is relevant for them and what other readers have to say about it (Mohammad Alghamdi and Ihshaish 2021).

Table 1. Types of influencers

Influencer	Scope	Follower count
Nano influencers	Nano influencers are just starting their influencer careers; therefore the majority of their followers are friends, acquaintances, and locals	0–10k followers
Micro influencers	Micro influencers are successful enough to make a living as influencers, but they are lower in size and scope than macro influencers. Their audience is typically more regionally focused relative to their geographic basis, and the majority of their income is generated by affiliate-link schemes or occasional brand partnerships	10k–100k followers
Macro influencers	Macro influencers are influencers who have yet to gain celebrity but nevertheless are extremely successful	100k–1m followers
Mega influencers	Mega influencers, like celebrity influencers, are those who have seen a huge increase in their social media following. They have gained a celebrity status based on their field of expertise	1m + followers

1.3 Objectives

A book review has a promotion effect for writers and book publishers that is comparable to the publisher's marketing or public relations (PR) initiatives. The book is brought to the attention of readers, having a good impact on book's success (Kamakura et al. 2006). To our knowledge, there is very little academic literature available on the attributes of a book review in today's digital era. To bridge this gap, this study has developed a comprehensive codebook for multiple researchers to use in their study on online book reviews and its impact on readers. The attributes of Bookstagrammers book reviews give an insight on the current trend of book reviewing. Based on the existing literature, this study also aims to highlight the desired attribute outcomes.

1.4 Methodology

To obtain the sample frame of Bookstagrammers on Instagram, the search term "Bookstagrammer" was entered into the Instagram search. The content is filtered in the second phase based on two conditions. Firstly, posts with the word 'book review' mentioned in the caption are selected for the study. Secondly, posts of those Bookstagrammers who came under the category of micro-influencers. The audience prefer micro-influencers, who harness greater authenticity and trust and are often more connected to their followers' needs and interests. They find micro influencers' recommendations more genuine than those made by more prominent celebrities, whom they may view as more prone to "sell out" (Campbell and Farrell 2020). This study was conducted between August 15 to September 15, 2022, to ensure the content is recent and updated with the trend. The post count of these micro-influencers came up to 118 posts. All the posts are analyzed using

Content analysis to develop the codebook structure with the support of existing literature. The comprehensive codebook contains eight codes that reflect the functionality of each code.

2 Attribute Codebook and Themes

One of the key elements in a qualitative study is the systematic coding of text. Codes are building blocks for theory or model building and is defined as "tags or labels for assigning units of meaning to the descriptive or inferential information compiled during a study" (Miles and Huberman 1994). Codes are allocated to data chunks, typically phrases, sentences, or paragraphs tied to a particular context or situation, to guarantee that labels are meaningful. All the codes are put together to make a codebook. The codebook is essential to help analyse qualitative research as it provides a formalized operation of the codes (DeCuir-Gunby et al. 2011). This study focuses on creating one such codebook that highlights the attributes of the book review on Instagram. There is no specific style of writing a book review on Instagram; the Bookstagrammers have made book reviewing formats that appeal to their audience. As Instagram is a visual platform, the first thing that attracts the viewers' attention is a photograph (Ramos-serrano and Martínez-garcía 2016), Bookstagrammers click aesthetic pictures of books to post on Instagram. The vivid visual cues arouse interest and catches the follower's attention towards the post (Lee and Shin 2014)(Ramos-serrano & Martínez-garcía 2016). Bookstagrammers make their profile look visually appealing and the picture content is accompanied with well-balanced and meaningful book review that serves the readers (Novotna et al. 2021) readers looking for book recommendations on Instagram. Table 2 below shows the attribute codes and themes that are developed based on the sample frame of 118 posts with the post count and the percentages. Table 3 highlights the total and average number of likes and comments on the 118 book reviews posts selected for this study.

Table 2. Total post count and percentages of attribute codes and themes

Attribute codes and themes	Post count	Percentage
• Review length (Average word count)	250.30	N/A
• Emoticons	102	86.44%
• Star Rating	76	64.40%
• Book Quote	32	27.12%
• Reader Experience	105	88.98%
• Story Plot	107	90.68%
• Spoiler Alert	01	0.85%
• QOTD (Question Of The Day)	61	51.69%

Table 3. Likes and Comments on the Bookstagrammers' book review post

Post Engagement	Total	Average
Likes	171253	1452.30
Comments	5415	45.89

2.1 Review Length

Research has said that review depth and length predict review helpfulness (Chua and Banerjee 2014), and per Instagram guidelines, the caption only has a limit of 2,200 characters. Hence, Bookstagrammers must keep the 2,200 characters limit in mind when writing a book review on Instagram. As Instagram truncates the caption to 125 characters, the viewer can only see the first line of the caption. It becomes crucial to make the opening line catchy enough for readers (Lo 2020). Therefore, it is desirable to have a good blend of review depth, length, helpful content, and a catchy opening line (Chua and Banerjee 2014). The posts studied for this paper showed that the word count ranged from 160 words to 331 words. All the reviews put together the average word count of the book reviews posted by the Bookstagrammers came up to 250 words.

2.2 Emoticons

Emojis are "picture characters" developed for mobile phones in Japan in the late 1990s. They have recently gained popularity around the globe in text messaging and social media due to the proliferation of smartphones that support emoji character input and display. The advent of emojis has caused a significant shift in how people write online by potentially replacing these user-defined language options with predefined pictorial icons. Expressing emotions in words is now a part of writing as contexts, intentions, and emotional states are challenging to represent in short messages. Hence emoticons play a vital role here (Pavalanathan and Eisenstein 2015). Various emoticons are used to compensate for the lack of facial cues to express and share feelings visually and keep the readers involved with their content (Li et al. 2019) (Sari 2018). Out of all the 118 book reviews, it is observed that 86.4% of the book review had emoticons used in the caption to keep the reader engaged with their content.

2.3 Star Rating

Numerical ratings are essential in the early stages of search and awareness (Hu et al. 2014). Product reviewers give star ratings or create their unique rating scale when writing a product review. Review websites, mostly, require their users to rank products or services out of the scale of 5, denoted as star rating. These star ratings are known to serve as valid cues for the content of a lengthy product review (Lak and Turetken 2014). Rating scales make it easier for users to consider a particular product for purchase (Chen 2008 64% of the Bookstagrammer book reviews have given the books a star rating in the review. These rating scales were in the form of star emoticons or numerals. Out of these 64% of posts used a rating scale, 92% had an emoticon star rating, and the remaining 8% had a numerical rating scale.

2.4 Book Quote

Informative appeals can be considered an overt persuasion attempt to influence the readers' purchase intention (Rietveld et al. 2020). A practice among the Bookstagrammers is sharing a quote from the book in their review. Sharing a sneak peek of the book helps grab the readers' attention. However, only 27% of the book review posts have shared a quote in their book reviews. The Bookstagrammers shared these book quotes in two ways: visual and text. From the visual context, they share the book quotes as a picture in the carousel post format. Such images serve as a visual cue for the readers and make it easy to reshare the relatable post on their Instagram profiles. Of the 27% posts, 53% shared a book quote in a visual form, and the remaining 47% shared it in the form of text. The book quote is written in the caption within the dialog quotes or italics format to make it stand apart from the book review.

2.5 Reader Experience

Content focusing on user experience and aspirational values has more engagement power (Balan 2017). Book reviews allow readers to share their post-purchase reading experiences with other readers. These reviews are perceived as more authentic and credible than marketer-generated information (Chua and Banerjee 2014). Hence, Book reviews by Bookstagrammer appear genuine and personal to the readers when they share their reading experiences. 89% of the book review posts by the Bookstagrammers focused on their reader experience and the book review. Since the Bookstagrammers are genuine readers, their followers believe their reviews be honest.

2.6 Story Plot

Bookstagrammers often include a brief story plot in their book reviews without revealing any crucial part of the story. This crisp content often encourages the readers to read the complete review before making their final decision. 91% of the book reviews had the story's plot shared in the content, while the remaining 9% of posts only shared the reader experience with the review.

2.7 Spoiler Alert

When readers see a review involving spoilers (undesirable descriptions of the stories), they may not be interested in reading the book as they now know the crucial part of the story (Ikeda et al. 2010). Hence, Bookstagrammers avoid giving out any spoilers in their review. Readers who haven't yet read the book want to avoid spoilers when reading a book review. In some instances, when a post is about book spoilers, Bookstagrammers caution their readers about it. Such was the case with only one book review out of the 118 book reviews where the Bookstagrammer had included a spoiler in the content and explicitly wrote 'spoiler alert' above the spoiler content.

2.8 QOTD

A QOTD is an acronym for 'Question of the Day', and the influencers often use these questions to help boost engagement on their posts. Towards the end of a book review, many Bookstagrammers were seen writing QOTD followed by a question for the readers. 52% of the book review posts of Bookstagrammers had QOTD mentioned in them, along with various reader-centric questions.

3 Attribute Outcomes

All the attributes listed in the codebook give book reviews a more structured format, and the creativity of the Bookstagrammer becomes the unique selling proposition to the readers. Bookstagrammers leverage these attributes to impact the reader's purchase intention. Based on existing literature the desired attribute outcomes into two broad categories:

3.1 Reader Engagement

As a social media platform, Instagram has higher interaction, co-creation, and engagement usage than other social media platforms (Coelho et al. 2016) (Valentini et al. 2018). Bookstagrammers adopt various book reviews patters to create engagement on their posts. On Instagram, viewer engagement is measured by Likes, Shares, Comments, and Following. We quantify the number of readers who appreciated the post through likes. This form of engagement is a solid social signal on Instagram that demonstrates that users liked the photos or videos posted by the Bookstagrammer. Post comments provide a deeper level of engagement in which followers contribute to the post through their opinions and discussions about the content (Bakhshi et al. 2014). Putting together all the 118 posts, Table 3 shows that they had an average of 1451 likes and 46 comments as the reader engagement. Readers who view the Instagram posts of the Bookstagrammers engage in various ways in the comments section; They tag fellow readers, share their opinion and share decision to read that book. The number of followers, in turn, represents the degree of approval of the profile, acting as a metric of engagement, recognition, and reach. Often influencers are recognized and categorized based on their follower count. Followers appreciate influencers posting relevant, reliable, and enjoyable content (Balaban et al. 2020). Sharing represents the extent to which users exchange, distribute and receive content (Virtanen et al. 2017). Sharing the post of a Bookstagrammer helps the influencer understand what readers have in common with them and identify what type of content readers prefer sharing. This form of peer-to-peer sharing of information.

3.2 Influence Reader's Purchase Intention

Researchers have established that online consumer reviews (OCRs) influence consumer attitudes and behaviors, including information adoption (Filieri and Mcleay 2014), product considerations and choice, brand awareness of as well as attitudes toward goods and services and purchase intentions (Senecal and Nantel 2004). This element of eWOM in online book reviewing influences the reader's purchase intention (Voss et al. 2003). When readers look around to buy books, they look at the book cover, title, synopsis, price, discounts, etc., but book recommendations significantly influence the book choices of the readers (Leitão et al. 2018). Bookstagram has given book reviews a visual add-on. The Bookstagrammer's informative posts are considered an overt persuasion attempt that influences the readers' purchase intention (Rietveld et al. 2020). Studies have stated that Photo styles and influencers' suitability are positively associated with photo perceptions and purchase intention (Primasiwi et al. 2021). Reader look for a ratings in the book review to make their purchase decision (Chen 2008). The story plot discussed by the Bookstagrammers in review help readers know if the book is the right fit for them, making it easy for the readers to choose their next book (Ikeda et al. 2010).. Since the people reviewing books are genuine readers, their followers believe the reviews be honest.

4 Limitations and Future Research

Some limitations of the study must be acknowledged. This qualitative research requires more time and interaction between the researcher and the subject. As the focus was on Instagram, we conducted a content analysis on the Bookstagrammers' book review posts and studied the available academic literature to establish the relation between them. Since there is a lack of scholarly literature on book reviewing activities on Instagram and their influence on the publishing industry ecosystem, future research can empirically test the code book developed in the current study and develop theories.

5 Conclusion

This paper contributes to the emerging trend of book reviewing on Instagram and developed a codebook highlighting the attributes of the book reviews written by the Bookstagrammers. Publishing industry relies on book reviews as their marketing and promotion tool to drive their sales (Clement et al. 2007)(Kaur and Singh 2021). In this digital era, Bookstagrammers' book reviews are gaining popularity among the reading community making them the influencers and opinion leaders. The existing literature helps to establish reader engagement and influence the reader's purchase intention for the book. Bookstagrammers book reviews have served as a promotion tool as many authors and their books have received attention from the readers, and considered the book for their next purchase.

References

Araujo, T., Neijens, P., Vliegenthart, R.: Getting the word out on Twitter : the role of influentials, information brokers and strong ties in building word-of-mouth for brands. Int. J. Advert. **36**(3), 496–513 (2017). https://doi.org/10.1080/02650487.2016.1173765

Association of National Advertisers. Survey Report: How ANA Members Are Using Influencer Marketing 34 (2018). https://www.ana.net/getfile/26389

Bakhshi, S., Shamma, D.A., Gilbert, E.: Faces engage us : photos with faces attract more likes and comments on instagram, pp. 965–974 (2014)

Balaban, D.C., Culic, L., Mustăþea, M., Pavelea, A., Culic, L.: What determines young people to follow influencers? The role of perceived information quality and trustworthiness on users' following intentions. Rom. J. Commun. Public Relat. **22**(3), 5–19 (2020)

Balan, C.: Nike on instagram : themes of branded content and their engagement power, pp. 13–18 (2017)

Campbell, C., Farrell, J.R.: More than meets the eye: the functional components underlying influencer marketing. Bus. Horiz. **63**(4), 469–479 (2020). https://doi.org/10.1016/j.bushor.2020.03.003

Chen, Y.: Herd behavior in purchasing books online. Comput. Hum. Behav. **24**, 1977–1992 (2008). https://doi.org/10.1016/j.chb.2007.08.004

Chevalier, J.A., Mayzlin, D., Chevalier, J.A., Mayzlin, D.: The effect of word of mouth on sales: online book reviews. J. Mark. Res. **43**(3), 345–354 (2006)

Chua, A.Y., Banerjee, S.: understanding review helpfulness as a function of reviewer reputation , review rating , and review depth (2014). https://doi.org/10.1002/asi.23180

Clement, M., Proppe, D., Rott, A.: Do critics make bestsellers? Opinion leaders and the success of books. J. Media Econ. **20**(2), 77–105 (2007). https://doi.org/10.1080/08997760701193720

Coelho, R.L.F., de Oliveira, D.S., de Almeida, M.I.S.: Does social media matter for post typology? Impact of post content on Facebook and Instagram metrics. Online Inf. Rev. **40**(4), 458–471 (2016). https://doi.org/10.1108/OIR-06-2015-0176

DeCuir-Gunby, J.T., Marshall, P.L., McCulloch, A.W.: Developing and using a codebook for the analysis of interview data: an example from a professional development research project. Field Methods **23**(2), 136–155 (2011). https://doi.org/10.1177/1525822X10388468

Filieri, R., Mcleay, F.: E-WOM and accommodation: an analysis of the factors that influence travelers' adoption of information from online reviews. J. Travel Res. **53**, 44–57 (2014). https://doi.org/10.1177/0047287513481274

Greenwood, B.N., Gopal, A.: Tigerblood: newspapers, blogs, and the founding information technology firms. Inf. Syst. Res. **26**(4), 812–828 (2015)

Hu, N., Koh, N.S., Reddy, S.K.: Institutional Knowledge at Singapore Management University Ratings Lead you to the Product , Reviews Help you Clinch it ? The Dynamics and Impact of Online Review Sentiments on Products Sales Ratings Lead You To The Product , Reviews Help You Clinch It?, pp. 42–53 (2014)

Ikeda, K., Hijikata, Y., Nishida, S.: Proposal of deleting plots from the reviews to, pp. 346–347 (2010)

Kamakura, W.A., Basuroy, S., Boatwright, P.: Is silence golden? An inquiry into the meaning of silence in professional product evaluations. Quant. Mark. Econ. **4**(2), 119–141 (2006). https://doi.org/10.1007/s11129-006-3181-x

Kaur, K., Singh, T.: Impact of online consumer reviews on amazon books sales: empirical evidence from India. J. Theor. Appl. Electron. Commer. Res. **16**(7), 2793–2807 (2021). https://doi.org/10.3390/jtaer16070153

Lak, P., Turetken, O.: Star ratings versus sentiment analysis--a comparison of explicit and implicit measures of opinions. In: Proceedings of the Annual Hawaii International Conference on System Sciences, pp. 796–805 (2014). https://doi.org/10.1109/HICSS.2014.106

Lee, E., Shin, S.Y.: Computers in human behavior when do consumers buy online product reviews? Effects of review quality, product type, and reviewer's photo. Comput. Hum. Behav. **31**, 356–366 (2014). https://doi.org/10.1016/j.chb.2013.10.050

Leitão, L., Amaro, S., Henriques, C., Fonseca, P.: Journal of retailing and consumer services do consumers judge a book by its cover? A study of the factors that influence the purchasing of books. J. Retail. Consum. Serv. **42**, 88–97 (2018). https://doi.org/10.1016/j.jretconser.2018.01.015

Li, M., Chng, E., Chong, A.Y.L., See, S.: An empirical analysis of emoji usage on Twitter. Ind. Manag. data Syst. **119**(8), 1748–1763 (2019). https://doi.org/10.1108/IMDS-01-2019-0001

Lin, T.M.Y., Luarn, P., Huang, Y.K.: Effect of internet book reviews on purchase intention: a focus group study. J. Acad. Librariansh. **31**(5), 461–468 (2005)

Lo, E.Y.: How social media, movies, and TV shows interacts with young adult literature from 2015 to 2019. Publishing Res. Q. **36**(4), 611–618 (2020). https://doi.org/10.1007/s12109-020-09756-8

Alghamdi, A.M., Ihshaish, H.: The use and impact of goodreads rating and reviews, for readers of Arabic books. Int. J. Bus. Inf. Syst. **37**, 442–466 (2021)

Novotna, A., Matula, K., Kociánová, V., Vojtěch, S.: Lessons Learned from Bookstagrammers for Library Promotion and Promotion of Readership : Qualitative Study DigitalCommons @ University of Nebraska - Lincoln Lessons Learned from Bookstagrammers for Library Promotion and Promotion of Readership : Qualitati. Library Philosophy and Practice, December (2021)

Nischik, R.M.: "The writer, the reader, and the book": Margaret Atwood on reviewing in conversation with Reingard M. Nischik. Am. Rev. Can. Stud. **45**(4), 522–531 (2015). https://doi.org/10.1080/02722011.2015.1123419

Miles, M.B., Huberman, A.M.: Qualitative Data Analysis: An Expanded Sourcebook, 2nd edn. SAGE, Thousand Oaks (1994)

Pavalanathan, U., Eisenstein, J.: Emoticons vs. emojis on Twitter: a causal inference approach (2015)

Primasiwi, C., Irawan, M.I., Ambarwati, R.: Key Performance Indicators for Influencer Marketing on Instagram, vol. 175, pp. 154–163 (2021)

Ramos-serrano, M., Martínez-garcía, Á.: Personal style bloggers: the most popular visual composition principles and themes on Instagram. Observatorio **10**(89), 109 (2016)

Rietveld, R., Dolen, W.V., Mazloom, M., Worring, M.: ScienceDirect what you feel is what you like influence of message appeals on customer engagement on Instagram. J. Interact. Mark. **49**(20), 53 (2020). https://doi.org/10.1016/j.intmar.2019.06.003

Saima, Khan, M.A.: Effect of social media influencer marketing on consumers' purchase intention and the mediating role of credibility. J. Promot. Manag.**27**(4), 503–523(2020)https://doi.org/10.1080/10496491.2020.1851847

Sari, P.M.:. Semiotic interpretation of emoticon used by followers on instagram (2018)

Senecal, S., Nantel, J.: The influence of online product recommendations on consumers' online the influence of online product recommendations on consumers' online choices, November 2017 (2004). https://doi.org/10.1016/j.jretai.2004.04.001

Stanková, M.: Online literary criticism: when reader becomes critic. Eur. J. Media Art Photogr. **9**(1), 128–135 (2021)

Valentini, C., Romenti, S., Murtarelli, G., Pizzetti, M.: Digital visual engagement: influencing purchase intentions on Instagram. J. Commun. Manag. **24**(362), 381 (2018). https://doi.org/10.1108/JCOM-01-2018-0005

Virtanen, H., Björk, P., Sjöström, E.: Follow for follow: marketing of a start-up company on Instagram. J. Small Bus. Enterp. Dev. **24**, 468–484 (2017). https://doi.org/10.1108/JSBED-12-2016-0202

Voss, K.E., Spangenberg, E.R., Grohmann, B., Voss, K.E., Spangenberg, E.R., Grohmann, B.: Measuring the hedonic and utilitarian dimensions of consumer attitude. J. Mark. Res. **40**(3), 310–320 (2003)

Digital Technologies and Small-Scale Rural Farmers in Malaysia

Herwina Rosnan[1](✉) and Norzayana Yusof[2]

[1] Arshad Ayub Graduate Business School, Universiti Teknologi MARA, Shah Alam, Malaysia
herwina@uitm.edu.my
[2] Graduate School of Business, HELP University, Kuala Lumpur, Malaysia

Abstract. Malaysia is moving toward a digital economy as outlined in Malaysia Digital Economy Blueprint. In agriculture, digital technologies like the Internet of Things (IoT), Artificial Intelligence (AI), robotics, and drone have been introduced in large-scale farming. These technologies enable smart farming, which makes farming activities more efficient, productive, and sustainable. However, the digital divide can pose a challenge to the digitalization process. Rural areas are associated with limited access to technological infrastructure, low levels of digital skills, and technology literacy that may hinder the adoption of digital technologies. Hence, the main objective of this study is to gain an in-depth understanding of the effect of digital technology on small-scale rural farmers. Data were collected through interviews and a total of fifteen small-scaled farmers participated in the study. The findings show that the adoption of digital technologies contributes to efficiency but it does not improve the well-being and income of small-scale farmers in rural areas. The implication highlights the necessity for policymakers to consider different models for the inclusion of small-scale rural farmers in the digitalization process.

Keywords: Digital technologies · Small-scale farmers · Agriculture sector

1 Introduction

Digital technologies have become an integral part of human life. Digital technologies are expected to improve businesses, transform processes, create business value and generate better revenue leading to achieving a digital economy. The application of digital technologies is referred to as digitalization. The journey toward the adoption and application of digital technologies is known as digital transformation. Digital transformation involves all aspects of change required which include business processes, business models, data, technologies, and most importantly human aspects. The application of digital technology will drive greater business efficiency in this sector which includes higher crop yield, reduce demand for farmlands, and more efficient farming. It is also expected that the attractiveness of the agriculture sector will reduce workers' migration to urban areas. Agriculture is not only important for a nation's food security but it has the potential to reduce poverty by raising the income of the rural population, especially in developing countries and the world's poor.

B. Alareeni et al. (Eds.): ICBT 2022, LNNS 620, pp. 776–783, 2023.
https://doi.org/10.1007/978-3-031-26953-0_72

Under Malaysia's new vision of Shared Prosperity Vision 2030, smart and high-value agriculture has been identified as a key economic growth activity. Malaysia's digitalization journey has started in 1996 with the development of the Multimedia Super Corridor. Pursuant to the effort, Malaysia's Digital Economy Blueprint was formulated to improve the country's readiness to embrace the digital economy. It is a foundation to drive digitalization across the nation. Its three objectives are to encourage industry players to become creators, users, and adopters of innovative business models, harness human capital to thrive in the digital economy, and nurture an integrated ecosystem that allows society to embrace the digital economy (EPU 2021).

The COVID-19 pandemic has accelerated the adoption and acceptance of technologies (Maalsen and Dowling 2020) which is a good indication of the growth of the digital economy. The vision to advance the digital economy is consistent with the Shared Prosperity Vision 2030, which aims to provide fair and equitable economic development among all levels of society by 2030.

Malaysia is committed to achieving a digital economy as outlined in Malaysia Digital Economy Blueprint. In agriculture, digital technologies like the Internet of Things (IoT), Artificial Intelligence (AI), robotics, and drone have been introduced in large-scale farming. These technologies enable smart farming, which makes farming activities more efficient, productive, and sustainable. However, the digital divide can pose a challenge to the digitalization process. Rural areas are associated with limited access to technological infrastructure, low levels of digital skills, and technology literacy that may hinder the adoption and application of digital technologies. An effort is required to ensure large-scale, and small-scale farmers benefit in the emerging digital society. Despite various initiatives to bring digitalization into agriculture, there is a lack of study on agriculture digitalization and small-scale farmers in Malaysia. The main objective of the current study is to fill the knowledge gap in understanding the effect of digital adoption and application on small-scale farmers. The study seeks to answer the main research question: How does the adoption of digital technologies affect small-scale rural farmers?

The paper begins with the introduction of digital technologies and the agriculture sector in Malaysia followed by a review of the literature on agriculture digitalization. The methodology was explained justifying the adoption of the qualitative research method to gain an understanding of the current phenomenon. Findings and discussion are presented followed by the conclusion and recommendation. The paper contributes to the understanding of how small-scale farmers are affected by digital technologies. The limitation of the study is related to the adoption of a qualitative research method which does not seek generalization of the findings.

2 Literature Review

The new technologies are transforming the agriculture sector into a more productive and profitable sector. As reported by the United States Department of Agriculture (USDA) Precision Agriculture technologies increased net returns and operating profits (Schimmelpfennig 2016). Mushi et al. (2022) stated that in agriculture, the advanced digital technologies mostly used by large-scale farmers contribute significantly to sustainable agriculture. Particularly, when tackling the environmental sustainability issue in farming, digital technologies are increasingly used to maintain the sustainability of farm production.

The studies on technological innovation are extensive. The literature discusses various aspects of the role of technology which include the reasons and purpose of its development, how technology brings benefits or posed challenges, what impact it brings to the sector, and other issues. Based on the actor-network theory, it offers a view from the perspective of multiple agencies' roles (Shove et al. 2012). It views technological change or innovation beyond merely the development of the technology itself but the inter-plays among different actors. It requires the understanding of the role of people's behavior, the structure of the economy and the institutional structure comprises people's skills and knowledge, the market, the organization forms, consumer preference, and policy goals.

The new digital technologies have been deemed a solution to accelerate the agricultural transformation and contribute to sustainable rural development, particularly for underprivileged countries with limited national capacities (Canton 2021). In the context of these countries, the process of digitalization can contribute to the development of the agriculture sector and communities (WEF Report 2018 and OECD Report 2019). Consequently, the report stated that it can also contribute to realizing UN Sustainable Development Goals (SDGs) in rural areas which encompass SDG Goals number 1 (No Poverty), 2 (Zero Hunger), and 13 (Climate Action). However, according to some scholars, the expected positive outcome of digital technologies is overestimated. There are many factors influencing the digital transformation of the agriculture sector which is affected the by vision, hope, and imagination of actors (Lajoie-O'Malley et al. 2020). In addition, rural communities struggle with several problems such as difficulty in reaching markets, aging population, depopulation, migration, and remoteness, which can also negatively affect the adoption of digital technologies for sustainable food production. According to a market analysis by Grand View Research, the factors that would facilitate the adoption of sustainable farming technologies include better education and training of farmers, sharing of information, easy availability of financial resources, and increasing consumer demand for food (Grand View Research 2019). Furthermore, the size of the farm also affects the success of the technological application. Mushi et al. (2022) highlighted the inequality that exists in the application of digital technologies between large-scale and small-scale farmers. Farmers will only invest in technology with potential economies of scale as profit margin increases with farm size.

According to Salemink et al. (2017) initiatives to promote digital transformation in rural agriculture are also dependent on digital skill levels, access to private investment, trust in technologies, and the willingness of the farmers. For these reasons, policies can contribute to an increase in social exclusion for fragile actors, such as elders or low-educated people, or forms of dependency by digital providers that control both technologies and data ownership.

Sustainable agriculture is one of the main reasons for promoting digital technologies in farming. The case of smart farming, for example, capitalizes on the advancement of data management as data is the key element in modern agriculture to help farmers with critical farm supervisory. Agriculture data attained through sensors maximize productivity and increase efficiency which can circumvent the waste of resources and the pollution of the environment. This type of data-driven agriculture coupled with robotic solutions with artificial intelligence (AI) techniques, is a possible solution for future agriculture sustainability. However, smart farming requires a relatively heavy investment that deters small-scale rural farmers from embracing it. This is one of the factors that differentiate large-scale farms and small-scale farms (Mushi et al. 2022).

3 Methodology

This study is an exploratory study that adopts a qualitative research approach where the interview is the primary source of data collection. The focus is to gain an in-depth understanding of how the adoption and application of digital technologies affect small-scale rural farmers. Data were collected through interviews with rural farmers in selected states in Malaysia. The sampling technique in qualitative research is not drawn statistically but purposive as the aim of the study is to gain an in-depth understanding rather than to achieve representativeness. Informants were selected based on the purposive sampling technique. Among the informants' selection criteria to participate in the study are (1) small-scale farmers and (2) those who have adopted digital technologies in running their farming activities. A semi-structured interview protocol was prepared to guide the data collection process. A total of fifteen small-scale farmers participated in the study. They were selected from various agriculture sectors namely oil palm plantation, rice plantation, livestock farming, vegetable plantation, and rubber plantation. Table 1 stated the details of the informants.

Table 1. The Informants.

Agriculture sector	Number of informants	Location (State in Malaysia)
Oil palm plantation	3	Johor
Rice plantation	2	Selangor
Livestock farming	2	Perak
Vegetable plantation	4	Selangor
Rubber plantation	4	Pahang

The method for data analysis for this study is using thematic analysis. Thematic analysis is one of the common methods for analyzing qualitative data. The process involves identifying, analyzing, and interpreting patterns of meaning within qualitative data. The interview data was first transcribed and coded, followed by finding the pattern or themes, reviewing the pattern, and finally drawing the conclusion (Braun and Clarke 2006).

4 Findings and Discussion

In general, based on MyDigital Report (EPU 2021) digital adoption among industries in Malaysia is still in the infancy stage. Compared to the neighboring countries, industrial development showing slow progression. In the context of the agriculture sector, the government formulated two main initiatives namely (1) to promote smart farming adoption through a centralized open data platform amongst industry players, and (2) to create more local digital platforms to enable access to the 'Farm to Table' digital marketplace. Table 2 outlined the government initiatives, outcomes, and targets.

Table 2. The sectoral initiative by the government in the agriculture sector.

Initiatives	Outcomes	Targets
Promote smart farming adoption through a centralized open data platform amongst industry players	Increased digital adoption and generated new business models by accessing the open data platform and identifying specific cost-cutting measures	1. To have machine-readable data, with access through API 2. Contribute to the creation of at least 5,000 start-ups by 2025 3. Increase in digital adoption rate across businesses
Create more local digital platforms to enable access to the 'Farm to Table' digital marketplace	Increased participation in the digital marketplace and sales of farmers	1. Increase in digital adoption rate across businesses 2. Contribute to the creation of at least 5,000 start-ups by 2025 3. Contribute to a 30% uplift in labor productivity across all sectors

(Source: Malaysia Digital Economy Blueprint).

This research sought to investigate the potential and actual effect of digital technologies on rural farmers. The scope of the impact encompasses economic, environmental, and social impact. In general, how rural farmers are affected by digital technologies differs between sectors. The sophistication of digital technologies used is also different. In terms of digital technology adoption, the motivation to adopt technology was partly due to the role of government initiatives which include financial and non-financial assistance. The majority of the farmers believed that technologies contribute to farm efficiency. The application of technology among rural farmers in this study mostly are at an early stage. For example, rice plantations were exposed to artificial intelligence (AI) namely drone technology to man farming activities. In oil palm plantations, an in-field sensor that is able to read soil conditions, and plant status helps the farmer to use the resources like fertilizer and pesticides more efficiently. Despite the positive expected outcome from the application of digital technologies, according to the farmers, the benefits were overestimated. In the rice plantation, for example, the farmers pointed out that digital technologies ease the work and reduce the dependence on manual labor. But, the cost of the technology is exorbitant, and maintaining the drone requires additional cost. The technologies were able to slightly increase yield but they also increased the cost of using and maintaining the technology. As a result, there is no increase in income.

The main challenges faced by rural farmers are due to the aging population with limited ability to adapt to new digital skills and knowledge. The case of youth migration to urban areas looking for a more promising career is indeed a global phenomenon (FAO 2014). Compared to other sectors and careers, the agriculture sector is deemed less attractive for young people with relatively higher education than their parents. As a result, the agriculture sector has to depend on aging smallholders who are less likely to adopt new technologies. Consistent with the view of Lajoie-O'Malley et al. (2020) that rural communities struggle with several problems such as difficulty in reaching markets, an aging population, and the migration of the younger generation to urban areas.

Since most farmers are elderly, it is not easy to learn and adapt to new technology. There is a tendency that rural farmers to end up as passive technology users rather than creators. The study inferred that rural farmers are dependent on digital technology providers. The finding is inconsistent with Malaysia's digital economy aspirations to turn industry players to become creators and integrate themselves into the ecosystem that allows society to embrace the digital economy. Hence, enticing young people to participate in the rural agriculture sector is critical. A previous study by Abdullah et al. (2012) on the inclination of rural youth to participate in agriculture indicate that even though most rural youths have a positive perception of the agriculture sector the actual number of youths participating in the sector was low.

In addition, participation in marketing, sales, and services are not visible in the oil palm, rubber, and rice plantation sector since farmers still depend on intermediaries or distributors. Hence, the expected benefits from web-based technologies like a digital marketplace that enable the farmers to access the market and increase their bargaining power are not evident.

From an environmental sustainability point of view, digital technology supposedly would be able to help farmers manage resources efficiently. The Internet of Things (IoT) can help promote better farm management and digital automated waste management can minimize or eliminate potential groundwater and soil pollution. However, that is not the case in the case of poultry farming under study. The study found that the sophistication of the technology application influences the positive effects of digital technology. For example, for many years, poultry farming in Perak state has been under criticism for polluting the environment with the continual issue of fly infestation (The Star, Sept 14, 2021). Although technologies for better farm management are available, the farmers seem to avoid investing in these technologies. Data from the interview revealed that cost is the main concern for farm operators to adopt digital technologies. The study deduced that the current way of doing business is more economical than investing in digital technologies. The study supports the findings by Salemink et al. (2017) that the willingness of the farmers to embrace digital technology is among the factor of technology adoption.

The study also discovered that rural farmers' attitude toward upholding environmentally sustainable farming is still low. One of the reasons is probably due to a lack of awareness of the importance of protecting the environment. The motivation to adopt digital technologies is mostly due to the expected increase in income. With regard to the impact of digital technologies on environmental sustainability and climate change, most of the informants admit that have not thought about the connection between these two issues. Albeit they admit that to a certain degree, digital technologies allow them to manage all the farm activities more effectively.

5 Conclusion and Recommendations

The findings show that the adoption of digital technologies does help to ease the work. However, informants claimed that the technologies had not improved much of their well-being. Most farmers lamented that even though digital technology like mobile applications and drones contribute to better efficiency and ease in farm management, it does not improve their income. The main challenges are the skills required to use the new technologies and the cost of maintaining those technologies. Furthermore, the adoption of digital technologies was primarily in response to the government initiative to support the nation's digitalization process. Digital technologies are expected to enable farmers to operate more efficiently, optimize resources, expand revenue, and, most importantly, improve their living and well-being. Unfortunately, the new technology used, such as drones, has become a burden. Moreover, most rural farmers are an aging population with limited capability to keep abreast of new technologies.

The implication of this study highlights the necessity for policymakers to consider different models for the inclusion of small-scale rural farmers in the digitalization process. Instilling an innovative culture through training and awareness programs might be necessary to promote digital technology adoption. Furthermore, the issue of the attractiveness of the agriculture sector among the younger generation needs to be tackled. Nevertheless, the study on how digital technology adoption affects rural small-scale farmers may be premature as the government is targeting to achieve increased technology adoption by 2025.

The Food and Agriculture Organization of United Nations (FAO) believes that given a supportive environment young people are able to venture into the agriculture sector innovatively. Providing educational or training programs that motivate young people on the potential of making a living in rural areas, offering farming knowledge and skills, improving rural infrastructure related to ICT, and facilitating access to credit are among the interventions that governments and/or philanthropies can contribute.

Acknowledgments. The authors would like to thank the Ministry of Higher Education Malaysia for the financial support through the Fundamental Research Grant Scheme (FRGS), File No: 600-IRMI/FRGS 5/3 (380/2019), and the Research Management Centre (RMC), Universiti Teknologi MARA, Shah Alam for managing the fund.

Paper Contribution to Related Field of Study. The study contributes to the understanding of the challenges and benefits experienced by small-scale rural farmers. Looking from the perspective of actor-network theory, the findings revealed the role of multiple actors that influence the success of digital technology adoption. The understanding of inter-play between different actors is critical as a factor that could foster or hamper digital technological innovation.

Notwithstanding, this research seeks to gain an in-depth understanding of the current phenomenon, hence the main limitation of the study is that the findings cannot be generalized to the larger population.

References

Abdullah, F.A., Samah, B.A., Othman, J.: Inclination towards agriculture among rural youth in Malaysia. Econ. Plan. **11**(773.3), 3–6 (2012)

Braun, V., Clarke, V.: Using thematic analysis in psychology. Qual. Res. Psychol. **3**(2), 77–101 (2006)

Canton, H.: Food and Agriculture Organization of the United Nations—FAO. In The Europa Directory of International Organizations 2021, pp. 297–305. Routledge, Abingdon (2021)

Economic Planning Unit. Malaysia Digital Economy Blueprint. Economic Planning Unit, Prime Minister's Department

FAO, C. IFAD. Youth and agriculture: key challenges and concrete solutions. IFAD, Rome (2014)

Grand View Research. Precision Farming Market Analysis. Estimates and Trend Analysis, pp 1–58. Grand View Research Inc., San Francisco (2019)

Aqilah, I.: Fly infestation taking a toll on Manjung folk. The Star (2021). https://www.thestar.com.my/metro/metro-news/2021/09/14/fly-infestation-taking-a-toll-on-manjung-folk

Lajoie-O'Malley, A., Bronson, K., van der Burg, S., Klerkx, L.: The future (s) of digital agriculture and sustainable food systems: an analysis of high-level policy documents. Ecosyst. Serv. **45**, 101–183 (2020)

Maalsen, S., Dowling, R.: Covid-19 and the accelerating smart home. Big Data Soc. **7**(2), 2053951720938073 (2020)

Organisation for Economic Co-operation and Development. Digital Opportunities for Better Agricultural Policies. OECD Publishing (2019)

Salemink, K., Strijker, D., Bosworth, G.: Rural development in the digital age: a systematic literature review on unequal ICT availability, adoption, and use in rural areas. J. Rural. Stud. **54**, 360–371 (2017)

Schimmelpfennig, D.: Farm profits and adoption of precision agriculture (No. 1477-2016-121190). USDA 2016, 217, 1–46 (2016)

Shove, E., Pantzar, M., Watson, M.: The Dynamics of Social Practice: Everyday Life and How it Changes. Sage, Thousands Oaks (2012)

World Economic Forum (WEF). Innovation with a Purpose: The Role of Technology Innovation in Accelerating Food Systems Transformation; WEF, Geneva, Switzerland (2018)

Third Coffee Wave - Factors Influencing Consumers' Coffee Purchase Decision in Shah Alam

Arlinah Abd Rashid(✉), Azlina Hanif, Ammar Ahmad, Muhammad Salihin Jaafar, and Nadia Kamilah Hamdan

Arshad Ayub Graduate Business School, Universiti Teknologi MARA, 40450 Shah Alam, Malaysia
arlinah@uitm.edu.my

Abstract. The rising interest in coffee consumption has caught the attention of many local businesses, especially small and medium-sized businesses (SMEs) in Malaysia. Thus, there is an urgent need for understanding the consumers' behaviours and needs, particularly in their purchase decision of coffee. This research aims to study the factors that influence the purchase decision of coffee in Shah Alam. An online survey was distributed to 308 respondents to gather customers' opinion on the factors that influence their purchase decision of coffee. The result of the study reveals that price, digital marketing and job performance are the significant predictors, with digital marketing as the most influential factor. The findings suggest that business owners should focus more on digital marketing and pay attention to advertise the coffee as a means to improve job performance. Based on the variables presented in the study, the business owners particularly in the coffee industry able to gain competitive advantages to boost the sales and to sustain in a highly competitive market. From the angle of emphasizing digital marketing, putting a good price point for products and location of promotion in focus for enhancing job performance among coffee drinkers would increase consumer purchase decision towards their coffee at their cafes.

Keywords: Marketing mix · Purchase decision · Consumer behavior

1 Introduction

Drinking coffee has become part of societal norms all around the world including Malaysia. With increased urbanization and hectic lifestyles, more and more people are demanding for coffee. The rise of the middle class, explosion of social media and hedonistic lifestyle, particularly in urban areas such as Klang Valley, has also influenced coffee drinkers to visit cafes. Based on a study conducted by Md Saleh et al. (2021), new generation of Malaysians have contributed to the expansion of independent cafés who seek out experiences outside mainstream culture, and food and beverage has evolved into a culinary fashion experience.

B. Alareeni et al. (Eds.): ICBT 2022, LNNS 620, pp. 784–793, 2023.
https://doi.org/10.1007/978-3-031-26953-0_73

Apart from the traditional coffee shop also known as *kopitiam*, coffee chains and independent cafes are paving their ways into the Malaysia market. This results in the growth of cafes offering competitive products in terms of quality, variety and taste. Consequently, this makes it more challenging for independent cafes to retain customers. With a myriad of coffee options in the market, implementing efficient marketing mix tactics based on the 4Ps is critical for convincing customers to make a buy and repeat purchase. Essentially, this study is conducted to understand the factors that influence consumer purchase decisions, specifically the marketing mix, which includes taste, price, café atmosphere, digital marketing and job performance at local independent cafés. The following section of this paper will discuss previous studies conducted on the marketing mix. This is followed by the methodology section, findings and analysis, and finally the conclusion and recommendation.

2 Literature Review

Marketing mix is generally used as a conceptual framework to identify choices that fulfill consumers' needs. It acts as controllable variables that affect the buyers' opinions on particular products or services. Furthermore, it is composed of combination of tactics by the organization to fully understand its business desire to effectively serve its products and services to the targeted audience (Kotler 2000, as cited in Thabit and Raewft 2018). The marketing mix components commonly used to measure consumers' purchase decision include taste, price, atmosphere, digital marketing and job performance.

2.1 Taste

Survey conducted in food industry suggested that taste is the main reason of food choices (Ellison et al. 2021). Taste and quality are the most important motivations that explain the local consumer's purchase (Cranfield et al. 2012; Costanigro et al. 2014, as cited in Trentinaglia De Daverio et al. 2021). Furthermore, the unique taste of local food will motivate people to satisfy their hunger (Coskun and Norman 2021). Samoggia and Riedel (2018) mentioned the key influences of consumers' coffee consumption are the coffee's qualities, taste, and smell. According to Urwin et al. (2019), consumers in Canada value coffee taste over price. Ramírez-Correa et al. (2020) explained the qualities and characteristics of coffee is reflected with its unique features and distinctive taste. These studies provide insight that taste and quality of coffee are important criterias for coffee consumption.

2.2 Price

According to Darmawan (2018), price has a significant impact on the purchasing decision process. Price plays an important role on purchase decision (Shah 2020). Recent study by Qalati et al. (2019) found that price has an influence on consumer buying behaviour across all industries and sectors. The study of sustainability of the food sector by Grunert (2011) mentioned that the high price will bring a negative effect to the purchasing behaviour. Sepúlveda et al. (2016) investigated the purchase behaviour of specialty coffee and concluded that there is a direct relationship of price with customer buying decisions. Mony and Be (2021) further explained that the main factor that will affect the position of the businesses' competition is the price of the products or services. These studies have indicated that price is highly associated with the customers buying behaviour and decision.

2.3 Atmosphere

Physical atmosphere in terms of retail setting is vital in influencing customers' behaviours (Helmefalk and Hultén 2017). Nguyen and Leblanc (2002) defined the physical surroundings as the place where service and products are delivered. Han and Ryu (2009) explained that the crucial aspects in the context of restaurants are layout, ambient, décor and artifacts. According to Samoggia and Riedel (2018), lifestyle experience and the retails' atmosphere are the other crucial factors consumers visit the coffeehouses, as well as the taste of the beverage. Mony and Be (2021) stated that the physical environment influences the consumption behaviour of Cambodian customers of Café Amazon. The good physical environment leads to customers' positive experiences to a restaurant in terms of emotion, intention, satisfaction, and loyalty (Han and Ryu 2009). This concludes that the role of the café's overall atmosphere would be dominant in influencing customers' behaviour.

2.4 Digital Marketing

Digital Marketing is another important element influencing customer purchase decision. Dwivedi et al. (2021) state that cognition, emotion, experience, and personality used in social media marketing influence the process of decision making made by the consumers. According to Samoggia and Riedel (2018), consumers are highly accessible toward information regarding the effects of coffee consumption on television, newspapers, websites, and other social media. The local coffee shops find it more convenient to use the online store as a platform for coffee shops to sell their products (Chen and Demirci 2019). This concludes that effective digital marketing will increase the sales due to positive influence on customer purchase decisions resulting from the growth of digital channels.

2.5 Job Performance

Another reason for drinking coffee is its benefit in terms of improving productivity and job performance. According to Fogaça et al. (2018), job performance is defined as all types of behaviours employees at work. Stroebaek (2013) mentioned that coffee breaks help to improve workplace's productivity. Further study by Waber et al. (2013) found that there is an overall increase in the workplace's productivity by increasing mental's alertness and motor speed, reduce sleepiness, and show a longer reaction time compared to non- coffee drinker. However, there are studies suggest that coffee consumption has no or mild impact on work performance and does not link to increasing of mental alertness or mental performance (Rogers et al. 2012).

The main research design consists of gathering responses about the perception towards coffee's taste, price, cafe's atmosphere, its digital marketing, and job performance. The aspects of the study were illustrated into independent variables and dependent variable. The conceptual framework provides an overview of the variable's association and hypothesis of each variable is established. The following Fig. 1 demonstrates the conceptual framework of the study.

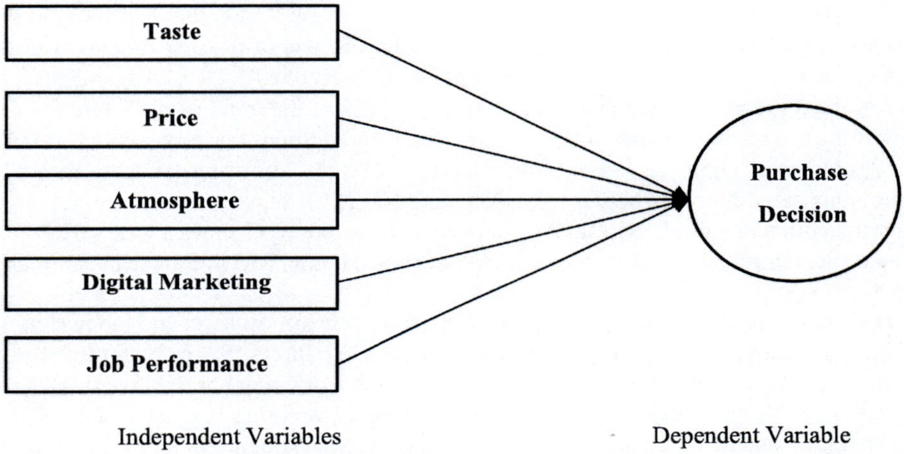

Fig. 1. Conceptual framework

The selected elements do not embrace all factors influencing coffee's purchase decision but are representative enough to demonstrate the relationships between these two variables. Apart from these five variables in the study, the other variables that could be taken into considerations are the coffee packaging (Waheed et al. 2018), branding (Song et al. 2019), cultural and geographical context (Samoggia and Riedel 2018) and IT services (Song et al. 2019).

3 Research Method

This study was conducted from May 2022 till June 2022 in Shah Alam, Selangor. Using convenience sampling, a total of 308 responses were collected from five independent local cafes. The distributed questionnaire contained seven sections to gather information on demographics, taste, price, atmosphere, digital marketing, job performance, and purchase decision variables. Each item is measured using a Five-Likert scale.

The method of analysis is quantitative data analysis. Thus, the gathered information is analysed using the SPSS software where the descriptive statistics, correlation analysis and multiple regression analysis were conducted. Reliability analysis was first conducted before the other methods were used. This was to ensure that the 30-item instruments in the questionnaire have good internal consistency. The descriptive statistics would be generated next to determine the mean and standard deviation for each variable. Pearson Correlation analysis would be carried out to determine the strength of association between the variables. Lastly, the multiple regression analysis is conducted to examine the relationship between dependent and independent variables. Through these methods, the objectives of the study would be met.

4 Findings and Analysis

Findings from the demographic section indicates that 53.3% of the respondents were female. The highest age group were those from 25–35 years old (45.1% of respondents). Most of the respondents were married accounting 77.9% of the respondents. More than 50% of the respondents work in the private sector. In addition, the respondents were mostly holders of Diploma or Bachelor's degree (74%). 31.2% of respondents earn a monthly household income between RM4851 and RM7110. When asked how often do you drink coffee in a week, 34.7% of respondents answered 3 to 5 times a week. 61.7% of the respondents claimed that they spent between RM10 and RM30 for each visit to a café, for coffee.

The means and standard deviation for taste, price, café atmosphere, and job performance were analyzed using descriptive statistics. Table 1 shows that cafe atmosphere has the highest mean of 3.86. On the other hand, digital marketing has the lowest mean score of 3.27. Nevertheless, the value for the mean of all variables is greater than 3.00 demonstrating that respondents agreed to all items in the questionnaire. As for the standard deviation, Marketing registered the highest standard deviation of 0.94 while taste has the lowest standard deviation of 0.583.

Table 1. Descriptive analysis

	Minimum	Maximum	Mean	Std deviation	Rating (Based on Mean)
Taste	3	5	3.86	0.583	2
Price	1	5	3.35	0.797	4
Atmosphere	2	5	4.34	0.656	1
Marketing	1	5	3.27	0.94	5
Job Performance	1	5	3.79	0.833	3
Purchase	1	5	3.32	0.859	

All the variables were analysed using Pearson correlation. The data were extracted and presented in Table 2. Price and digital marketing have a positive and strong relationship with purchase decision. Atmosphere and job performance have a negative and weak relationship with purchase decision. The values range from the lowest score of –.13 to the highest score of .87. As for taste, there is a negative, very weak relationship between taste and purchase decision.

Table 2. Summary of Pearson correlation analysis

	1	2	3	4	5	6
1. Taste	–					
2. Price	–.13*	–				
3. Atmosphere	–.09	.25**	–			
4. Digital Marketing	–.12*	.86**	.29**	–		
5. Job Performance	–.20**	.22**	.18**	.23**	–	
6. Purchase Decision	–.13*	.84**	.30**	.30**	.87**	.27**

** p < .001, * p < .05

Multiple regression was performed to test the relationship among variables in the study framework using the 0.05 significance level.

Table 3. Model summary of multiple linear regression analysis

	Standardized Coefficients β	P Value
Constant		
Taste	–.006	.810
Price	.329	.000
Atmosphere	.042	.132
Digital marketing	.561	.000
Job performance	.056	.044

Note: R2 = .793, ** p < .001, * p < .05

Based on Table 3, the model is statistically significant [F (5,302) = 231.041, p < .001]. The R-squared value shows 79.3% of the variation in purchase decision can be explained by variations in the independent variables. The remaining 20.7% is due to other predictor variables. Table 3 also shows that price and digital marketing have a strong relationship on consumers' purchase decision, with standardized beta value of 0.329 and 0.561 respectively. The results are consistent with the findings by Sepúlveda et al. (2016) who found that there was a direct relationship between price and customer buying decisions towards specialty coffee. The findings showed that price is one of important predictors in consumers' purchase decision in the context of coffee. This might be because the price of a cup of coffee at the independent café is slightly higher from other restaurants. The price could reflect the quality and dining experience as customers expects value when purchasing the coffee. As the customers pay at higher price, the quality of the coffee should match the value because higher price with substandard quality would prohibit repurchase of the product. Therefore, price of a coffee is expected to bring a good quality and value to customers.

In relation to **digital marketing**, the emergence of online marketing through social media has influence customers when viewing the products. The local coffee shops are becoming more convenient by emerging platforms, following the same trend to build online stores (Wu et al. 2017). The use of digital marketing has been a central importance to reach potential customers as well as to update new products. The strong relationship between digital marketing and purchase decision found in this study can be explained by the use of smartphones as people tend to connect with social media every day. This makes digital marketing an advantage to reach those target market faster as compared to conventional marketing.

The standardized beta value for **job performance** is 0.056, which influence the relationship between job performance and consumer purchase decisions, despite being a weaker predictor than price and digital marketing. This research demonstrates that coffee drinkers have a sense of engagement, comfort, self-fulfillment, and pleasure, which influences their job performance. This is because coffee has an energizing effect and customers often consume two to three cups each day (Nomisma 2018). This is consistent with previous findings showing coffee drinkers experience significant enjoyment and fulfillment from their coffee consumption, which boosts their mood and pleasant emotions (Nehlig 2010).

In contrast, the results indicated that taste and atmosphere had a weak relationship with the purchasing decisions of consumers. This may be due to the respondent profile of this study who are mostly female and may not enjoy the taste of coffee but consumes it routinely. This is corroborated by Samoggia et al. (2020) who found that women do not enjoy the flavor of coffee, with habit and a sense of alertness being the other major determinants of coffee use. In addition, previous research has demonstrated that consumers take coffee because it is a ritual or a daily habit (Agoston et al. 2017), a childhood tradition, and a component of breakfast or snacking (Sousa et al. 2016).

5 Conclusion and Recommendation

This study shows that a large percentage of millennials in Malaysia drink coffee frequently (more than three to five times a week), contributing to the country's rising coffee consumption. The coffee market is expected to expand at a rapid rate in the coming years, making it crucial for businesses to keep tabs on the sector.

The empirical findings showed that some key factors like price, digital marketing, and job performance have a higher influence on consumer decisions than taste and café ambiance. Price is obviously a major consideration, and it is reasonable to expect it to reflect the coffee's value and quality. The vast majority of respondents are willing to pay a premium for their coffee, which could be explained by the fact that they are high income earners. However, many of them are also regular customers of international coffee chains like Starbucks and Coffee Beans & Tea Leaves, so the value proposition offered by local independent cafés is crucial to their purchasing decisions. Even more so, the influence of one's coworkers and immediate social circle can be felt in one's digital marketing efforts and in one's performance on the job.

These variables should not be overlooked in which this study should be replicated and receive attention in future research. New findings are required to highlight the debate and resolve the issue. Ufer et al. (2019) mentioned that consumer choices are becoming increasingly complex, particularly for specialty products. Specialty products carry extra dimensions as compared to non-specialty products and requires careful attention. Overall, the current study investigated marketing mix elements in relation to purchase decision. This research makes useful contribution by incorporating important elements in predicting consumers' purchase decision in the context of independent café. Understanding the marketing mix elements may assist in critical goals of obtaining repeat purchase and ultimately, increase the profitability of the business.

This study provides a series of significant findings; however, there are few limitations. First, the study uses convenience sampling and may not be the best representative of the target population. Second, the data were collected from respondents at five local independent cafes. This might limit the generalizability of the results to cafes in other areas. Hence, future research can consider using probability sampling and including more premises to generate better responses and results.

References

Md Saleh, M.F., Abd. Halim, N., Farid, A.-N.M.: Food tourism motivation and customer satisfaction on hipster café in Johor Bahru, Malaysia. J. Tour. Hosp. Environ. Manag. **6**(26), 155–162 (2021)

Kotler, P.: Marketing Management: The Millenium Edition, 10th edn. Prentice Hall, Upper Saddle River (2000)

Thabit, T., Raewf, M.: The evaluation of marketing mix elements: a case study. Int. J. Soc. Sci. Educ. Stud. **4**(4), 100–109 (2018)

Ellison, B., McFadden, B., Rickard, B.J., Wilson, N.L.: Examining food purchase behavior and food values during the COVID-19 pandemic. Appl. Econ. Perspect. Policy **43**(1), 58–72 (2021)

Cranfield, J., Henson, S., Blandon, J.: The effect of attitudinal and sociodemographic factors on the likelihood of buying locally produced food. Agribusiness **28**, 205–221 (2012)

Costanigro, M., Kroll, S., Thilmany, D., Bunning, M.: Is it love for local/organic or hate for conventional? Asymmetric effects of information and taste on label preferences in an experimental auction. Food Qual. Prefer. **31**, 94–105 (2014)

Trentinaglia De Daverio, M.T., Mancuso, T., Peri, M., Baldi, L.: How does consumers' care for origin shape their behavioural gap for environmentally friendly products? Sustainability **13**(1), 190 (2021)

Coskun, G., Norman, W.: The influence of impulsiveness on local food purchase behavior in a tourism context. Tour. Int. Interdiscip. J. **69**(1), 7–18 (2021)

Samoggia, A., Riedel, B.: Coffee consumption and purchasing behavior review: insights for further research. Appetite **129**, 70–81 (2018)

Urwin, R., Kesa, H., Joao, E.S.: The rise of specialty coffee: an investigation into the consumers of specialty coffee in Gauteng. Afr. J. Hosp. Tour. Leis. **8**, 1–17 (2019)

Ramírez-Correa, P., Rondan-Cataluna, F.J., Moulaz, M.T., Arenas-Gaitan, J.: Purchase intention of specialty coffee. Sustainability **12**(4), 1329 (2020)

Darmawan, M.D.: The effect of price, product quality, promotion, social factor, brand image on purchase decision process of loop product on youth segment (Case Study of Pt Telekomunikasi Selular). In: Proceeding of International Seminar & Conference on Learning Organization (2018)

Shah, B.: Consumer's buying behaviour of motorcycles in Janakpurdham. J. Manag. **3**(1), 22–34 (2020)

Qalati, S., Yuan, L., Iqbal, S., Hussain, R., Ali, S.: Impact of price on customer satisfaction; mediating role of consumer buying behaviour in telecom sector. Int. J. Res. **6**(4), 150–165 (2019)

Grunert, K.G.: Sustainability in the food sector: a consumer behaviour perspective. Int. J. Food Syst. Dyn. **2**(3), 207–218 (2011)

Sepúlveda, W.S., Chekmam, L., Maza, M.T., Mancilla, N.O.: Consumers' preference for the origin and quality attributes associated with production of specialty coffees: results from a cross-cultural study. Food Res. Int. (Ottawa, Ont.) **89**, 997–1003 (2016)

Mony, A.V., Be, B.: Factors Influencing Customer Behaviors To Purchase Coffee At Cafe Amazon In Phnom Penh (2021)

Helmefalk, M., Hultén, B.: Multi-sensory congruent cues in designing retail store atmosphere: effects on shoppers' emotions and purchase behavior. J. Retail. Consum. Serv. **38**, 1–11 (2017)

Nguyen, N., Leblanc, G.: Contact personnel, physical environment and the perceived corporate image of intangible services by new clients. Int. J. Serv. Ind. Manag. **13**(3), 242–262 (2002)

Han, H., Ryu, K.: The roles of the physical environment, price perception, and customer satisfaction in determining customer loyalty in the restaurant industry. J. Hosp. Tour. Res. **33**(4), 487–510 (2009)

Dwivedi, Y.K., et al.: Setting the future of digital & social media marketing research: perspectives and research propositions. Int. J. Inf. Manag. **59**, 102168 (2021)

Chen, C.W., Demirci, S.: Factors affecting mobile shoppers' continuation intention of coffee shop online store: a perspective on consumer tolerance. Int. J. Electron. Commer. Stud. **10**(2), 203–238 (2019)

Fogaça, N., Rego, M.C.B., Melo, M.C.C., Armond, L.P., Coelho, F.A., Jr.: Job performance analysis: scientific studies in the main journals of management and psychology from 2006 to 2015. Perform. Improv. Q. **30**(4), 231–247 (2018)

Stroebaek, P.S.: Let's have a cup of coffee! Coffee and coping communities at work: let's have a cup of coffee! Symb. Interact. **36**(4), 381–397 (2013)

Waber, B.N., Olguin, D., Kim, T., Pentland, A.: Productivity through coffee breaks: changing social networks by changing break structure. Massachusetts Institute of Technology (2013)

Rogers, P.J., Heatherley, S.V., Mullings, E.L., Smith, E.J.: Faster but not smarter: effects of caffeine and caffeine withdrawal on alertness and performance. Psychopharmacology **226**(2), 229–240 (2012)

Waheed, S., Khan, M.M., Ahmad, N.: Product packaging and consumer purchase intentions. Mark. Forces **13**(2), 97–114 (2018)

Song, H., Wang, J., Han, H.: Effect of image, satisfaction, trust, love, and respect on loyalty formation for name-brand coffee shops. Int. J. Hosp. Manag. **79**, 50–59 (2019)

Wu, J., Shu-Hua, C., Ping, K.L.: Why should i pay? Exploring the determinants influencing smartphone users' intentions to download paid app. Telemat. Inform. **34**(5), 645–654 (2017)

Nomisma. Coffee Monitor Nomisma-Datalytics: 260 Euro la Spesa Media Annua Degl Italiani perilCaffè (2018). https://nomisma.it/wp-content/uploads/2019/11/COFFEE_MONITOR_NOMISMA.pdf

Nehlig, A.: Is caffeine a cognitive enhancer? J. Alzheimer's Dis. **20**(1), S85–S94 (2010)

Samoggia, A., Del Prete, M., Argenti, C.: Functional needs, emotions, and perceptions of coffee consumers and non-consumers. Sustainability **12**(14), 5694 (2020)

Ágoston, C., Urbán, R., Király, O., Griffiths, M.D., Rogers, P.J., Demetrovics, Z.: Why do you drink caffeine? The development of the motives for caffeine consumption questionnaire (MCCQ) and its relationship with gender, age and the types of caffeinated beverages. Int. J. Ment. Health Addict. **16**(4), 981–999 (2017). https://doi.org/10.1007/s11469-017-9822-3

Sousa, A.G., Machado, L.M., Silva, E.F.D., Costa, T.H.M.D.: Personal characteristics of coffee consumers and non-consumers, reasons and preferences for foods eaten with coffee among adults from the Federal District, Brazil. Food Sci. Technol. **36**(3), 432–438 (2016)
Ufer, D., Lin, W., Ortega, D.L.: Personality traits and preferences for specialty coffee: results from a coffee shop field experiment. Food Res. Int. (Ottawa, Ont.) **125**, 108504 (2019)

Evaluation of Reliability and Validity of Instruments for Digital Government Competency Framework for Omani Public Sector Administrators: Acceptance Study

Juma Al-Mahrezi(✉), Nur Azaliah Abu Bakar, and Nilam Nur Amir Sjarif

Razak Faculty of Technology and Informatics, Universiti Teknologi Malaysia, Jalan Sultan Yahya Petra, 54100 Kuala Lumpur, Malaysia
jumaphd2016@gmail.com, {azaliah,nilamnur}@utm.my

Abstract. Globally, governments invest in transformation projects as a component of digital transformation is referred to as Digital Government (formerly known as e-Government). Therefore, this process needs people, technology, and a set of strategies. Government staff must therefore be prepared with pertinent digital skills that are currently not thoroughly studied if the Digital Government is to succeed. This research aimed to evaluate the face and content validity and analysis of the result of the pilot study of the new instrument for "digital government competency framework for Omani public sector managers." Seven experts in the IT and non-IT fields participated in face validity, and 110 were used as a sample for the pilot study. The suggestions and comments were taken into account. The judging evidence of the instrument is being evaluated for content validity by an expert panel of eight academicians. In the final instrument, only items with a Content Validity Index (CVI) higher than 0.80 were used. The final instrument has 75 multiple-choice questions on a 5-point Likert scale. Results support the validation of this 75-item questionnaire; hence could be further researched on construct validity.

Keywords: Digital government · Cronbach's alpha · Public sector · Content validity · Face validity · Pilot study · Construct validity

1 Introduction

Currently, The competencies necessary to make the digital transformation successful differ from those from past years. To be successful in attempts to transform the organization through digital technology both now and in the future, organizations' administrators must identify the skills needed by employees in ICT relative to other departments. In a survey conducted by Gerald C. Kane, Doug Palmer [1], among the most important technological talents, Global CIO skills, creativity, emotional intelligence, and learning skills are in high demand and, over the next three years, supply should outpace demand.

B. Alareeni et al. (Eds.): ICBT 2022, LNNS 620, pp. 794–805, 2023.
https://doi.org/10.1007/978-3-031-26953-0_74

Finding technology experts with the necessary abilities to complete the ICT project group is, according to [2] the primary obstacle to carrying out a digital initiative successfully. Also, research of the Australian business sector by Gekara, Snell [3], Supervisors, technicians, and business people professional classes show strong demand for digital capabilities. Additionally, the digital environment needs to be more widely understood. This condition is a significant flaw in the developing digital economy. The future of business is probably going to be led by emerging technology. As a result, the workers will need an advanced level of general digital abilities and specialized technology in specialized applications [3].

In contrast, all administration on organizationalional is working harder to implement the expectations for digital systems. Consequently, it needs a higher level of technical proficiency [3]. Managers at public enterprises have unique challenges since they have to adopt a digital attitude and reevaluate every procedure [4]. Similar problems with management-level digital government capabilities and skills may arise with Oman's Digital Government plan. Workforce demands and competencies were impacted by changes in organizational processes and conventional business models for running and providing goods and services. The researchers of this study spoke with researchers in the Omani Public Sector to gain their thoughts to confirm that these concerns with digital competency are equally pertinent to the setting of the Omani Public Sector. Five experts contributed to this discussion [5]. Furthermore, Al-Kalbani [6] highlighted that "government loyalty, knowledge and training, organization assurance, audit and monitoring, legal and social forces, technology capability, system integration, and technology compatibility and reliability are essential components for effective compliance with information security in the Digital Government of Oman."

Interestingly, while the prior study has focused on the many aspects of digital government, it has given less consideration to the competencies of digital government. According to the problem background review, there is a lack of research on public sector managers' digital government competencies.

1.1 Research Gap

Previous research on digital government competencies was conducted from an organizational viewpoint rather than an employee's. Additionally, there has been an unavailable digital government competency framework for public sector managers and a lack of research on digital government competencies in the Omani public sector. As a result, this study is aware that this topic has not been thoroughly researched and that for Digital Government projects to be successful, government employees must be trained with essential digital skills. To fill this gap, this research examines public employees' digital government competencies and suggests a public sector digital government competency framework.

The pilot study and research instrument validation is an essential analysis process to consider a good instrument [7]. Validation is "measuring what is intended to be measured" [8]. According to Taherdoost [9], there are four main types of validity, namely, face validity, construct validity, content validity, and criterion validity. This study only focused on the two most commonly used validity techniques, which are face validity and content validity. In addition, this paper will analyze the results of the pilot study.

Using those tools will help identify digital government competencies in the Omani public sector, whereby the reliability and validity of the instrument have important implications for the various stakeholders in the public sector. Therefore, this study aims to examine the instrument's validity and reliability and analyses the result of a pilot study of digital government competencies in the Omani public sector.

2 Methodology

This paper developed an instrument and validated the instrument using the methodology used by [10]. Th which is an expert or one-way translation. Consequently, this research has made use of expert translator services in Oman. e four essential parts of this method are the development of the instrument, questquestioniers'translation, instrument validation of instruments, and pilot research. The design and development in the first phase are based on literature reviews, exploratory research, and relevant reports. Based on the thematic analysis results, theories, including the Technology Organization Environment Theory (TOE) and the Human Capital Theory (HCT), were examined as the basis for choosing the variables in this study. Next, a translation procedure into local languages is one method for assisting study participants in providing accurate responses. The instrument's validity is established in the third step by following procedures, which are content validation and face validation. Conducting a pilot study to finish the validity and reliability procedure of the instrument is the last step in assessing the reliability of this study.

2.1 Instrument Development

The questionnaire has two main sections; A-Profile of the Respondent and B-Information on digital government competencies. Based on preliminary studies in the design and construction of this study model, a total of 15 constructs were consolidated into one proposed model [11]. The initial questionnaire consists of 75 items of measurement. This research used the Likert Scale for measuring instrument items. of the survey. A point Likert scale "1 = Strongly Disagree, 2 = Disagree, 3 = Neutral, 4 = Agree, 5 = Strongly Agree" were being using for measurement items. The items of measurement are adapted from the various previous study.

2.2 Translate the Survey

To make the survey more understandable to participants, it was translated from English into Arabic. This research utilizes the approach recommended by [12], which is "an expert or one-way translation". Consequently, this research has made use of expert translator services in Oman and then validated the translated questioners' items by academic experts from the IT field.

2.3 Face Validation of the Survey

Face validation is "the degree to which a measure appears to be related to a specific construct in the judgment of non-experts such as test-takers and legal system representatives" [9]. The dichotomous scale can be used with the categorical options of "Yes" and "No," representing the advantageous and disadvantageous items, respectively, to assess the face validity. According to Masuwai [7], The procedural suggestion is evaluated by two or more independent judges. Therefore, seven respondents were invited to the study and performed face validity procedures. The respondents include government employees, university students, and academicians.

2.4 Content Validation of the Survey by Experts

Expert judgment intends to ensure the measurement items correctly represent the construct and that each item measures what it intends to measure. In this study, the expert and academics will perform the content validity test to validate the instrument suggested by McKenzie [13]. A study by Kennedy [14] proposed that a board of 5 to 10 professionals is adequate to assess measurement items. Eight academics and experts were therefore selected for the content validation test based on their expertise, academic background, area of interest, experience, and skills in developing survey instruments, statistical analysis, and digital government competencies. The experts graded each item using one of three measures to determine its relevance and clarity. 1 = Not relevant or not clear; 2 = Relevant or not clear but needs some work; 3 = Very relevant or very clear [15, 16]. Additionally, the experts are requested to offer suggestions or comments on any Construct measurement.

The validity of the survey items is measured quantitatively using the Content Validity Ratio (CVR) and Content Validity Index (CVI) calculations [9]. CVR is "an item's statistic indicating the usefulness of item measurement to be accepted or rejected. CVR and CVI offer practicality in terms of time and cost, and also, it is quick and easy to perform" [17]. Besides, CVI is flexible as it requires a minimum of three experts. Using Lawshe [15], Each measurement item's CVR was measured using the CVR calculation, which is described as follows:

The value Ne is the number of experts indicating "relevant" (score of 2 and 3), and the value N is the total number of experts. Because there were eight experts, a minimum CVR of 0.75 is required to approve the measurement items [15].

2.5 Pilot Study

The purpose of the pilot study is to test the validity of the questionnaire, the feasibility of instruments, and the syntax of the questions with a small number of targeted respondents. Small-scale or A trial run analysis is performed before applying comprehensive research. This is a way of pre-testing the appropriateness of the specific research instrument [18]. It allows the researcher to revise the confusing questions before collecting the actual data, both reliability and validity used in the instrumented assessment. Reliability analysis of a pilot study was carried out utilizing Cronbach's alpha. Four reliability cut-off points have been proposed by Hinton et al. [19] including excellent reliability (0.90 and above), high reliability (0.70–0.90), moderate reliability (0.50–0.70), and low reliability (below 0.50). (0.50 and below). While reliability is essential for the study, it is insufficient without validity [9]. 110 people participated in the pilot study for the current investigation. The initial purpose is to develop the standard of the questionnaires. Secondly, assessing respondents' level of understanding and clarifying the actual survey being performed. (Kaiser-Meyer-Olkin [KMO] and Bartlett's test of sphericity are used for each scale to determine whether the data set is sufficient for factor analysis.

3 Results

Based on the findings and suggestions from content validittion and face validition and, the survy have been improved.

Face Validity of the Questionnaire
Participants offered some critical feedback on the examined instrument along with some positive suggestions. The need for improved language, reduced items, appropriate sentence structure, and double-barrel inquiries are some of the solutions.

Expert Content Validity of the Questionnaire
According to expert assessments, every construct is accepted as being a component of this study model. However, because of the computation done using the CVR approach, several items were deleted. The validition of the survey achieved at 96% of CVI, as shown in Table 1's CVR values for each item, CVI value for each construct, and all survey validity.

Table 1. Content Validity Index (CVI) of the survey instrument

Construct	No of initial item	Total accepted items (CVR > 0.56)	CVI
Data science competency	6	6	0.96
Information security	5	4	0.85
Management competency	4	4	1
Soft skills	5	3	0.83
Digital leadership competency	7	7	1
Digital creativity and innovation	3	3	1
Digital literacy	7	5	0.88
Vocational training and education	4	4	1

(continued)

Table 1. (*continued*)

Construct	No of initial item	Total accepted items (CVR > 0.56)	CVI
Social and ethical responsibility	5	5	1
Relationships and engagement	5	5	1
Digital talents development	3	3	1
Change management and digital transformation	7	6	0.89
Digital platform usability	6	6	1
Attitudes	4	4	1
Digital government competency	3	3	1
Overall CVI			0.96%

3.1 Pilot Study

The results of the reliability analysis using Cronbach's alpha for the pilot research are displayed in Table 2. The inter-correction among the constructs' items is highly reliable and acceptable for the actual data collection method, as shown by the fact that Cronbach's alpha values for all constructs are greater than 0.7 [20].

Table 2. Reliability Analysis Result of a pilot study testing

Construct	No of items	Cronbach's alpha	Result
Data science competency	6	0.808	Accepted
Information security competency	4	0.757	Accepted
Management competency	4	.8920	Accepted
Soft skills	4	0.742	Accepted
Digital leadership competency	7	0.763	Accepted
Digital creativity and innovation	3	0.934	Accepted
Digital literacy	5	0.805	Accepted
Vocational training and education	4	0.845	Accepted
Social and ethical responsibility	5	0.920	Accepted
Relationships and engagement	5	0.880	Accepted
Digital talents development	3	0.754	Accepted
Change management and Digital transformation	6	0.911	Accepted
Digital platform usability	6	0.803	Accepted

(*continued*)

Table 2. (*continued*)

Construct	No of items	Cronbach's alpha	Result
Attitudes	4	0.757	Accepted
Digital government competency	3	0.715	Accepted

The Exploratory Factor Analysis (EFA), Principal Component Factor Analysis, and Varimax Rotation were used to assess the scales' validity and unidimensionality. For each scale, the Kaiser-Meyer-Olkin [KMO] and Bartlett's sphericity tests were used to determine its suitability for factor analysis. to determine if the data set is suitable for factor analysis, the KMO and Bartlett's sphericity test are examined. The descriptive statistics of the analysis outcome adopting EFA from the pilot study test data set are shown in Table 3. The EFA with Keyser-Meyser-Olkin (KMO), Bartlett's test, and descriptive statistics show the scale's excellent internal quality. The adequate sample size in the prior investigation indicates the KMO value of 0.780 [21].

Table 3. Results of JMO and Bartlett's test

Kaiser-Meyer-Olkin Measure of Sampling Adequacy		**0.780**
Bartlett's Test of Sphericity	**Approx. Chi-Square**	1224.997
	df	105
	Sig.	.000

3.2 Descriptive Statistics

The value of the coefficient of variation (CV) for each item is also shown in Table 4, demonstrating the consistency of the respondents. According to [22], a CV value that approaches 0 is preferable since it indicates that the data items are more consistent.

Table 4. Descriptive statistics

Variable & Item	Mean	Std. deviation	CVI
Data science competency			
DSC-1	3.36	1.002	0.2982143

(*continued*)

Table 4. (*continued*)

Variable & Item	Mean	Std. deviation	CVI
DSC-2	3.41	0.97	0.2844575
DSC-3	3.49	0.993	0.2845272
DSC-4	3.58	0.961	0.2684358
DSC-5	3.67	0.949	0.2585831
DSC-6	3.24	1.049	0.3237654
Information Security Competency			
ISC-1	3.24	0.888	0.2740741
ISC-2	2.77	0.974	0.3516245
ISC-3	3.7	0.841	0.2272973
ISC-4	2.4	1.167	0.48625
Management Competency			
MC1	2.9	0.867	0.2989655
MC2	2.89	0.98	0.3391003
MC3	3.6	0.988	0.2744444
MC4	3.11	1.103	0.3546624
Soft Skills			
SOS1	3.69	0.875	0.2371274
SOS2	3.3	1	0.3030303
SOS3	3.12	0.955	0.3060897
SOS4	3.05	0.956	0.3134426
Digital leadership competency			
DLC1	3.39	1.041	0.3070796
DLC2	3.56	0.904	0.2539326
DLC3	3.56	0.852	0.2393258
DLC4	3.59	0.881	0.2454039
DLC5	2.91	1.105	0.3797251
DLC6	2.94	0.96	0.3265306
DLC7	4.04	1.211	0.2997525
Digital Creativity and Innovation			
DCI1	3.33	1.05	0.3153153
DCI2	3.45	1.055	0.3057971
DCI3	3.34	1.086	0.3251497

(*continued*)

Table 4. (*continued*)

Variable & Item	Mean	Std. deviation	CVI
Digital literacy			
DL1	3.41	0.98	0.28739
DL2	3.88	0.713	0.1837629
DL3	4.04	0.62	0.1534653
DL4	4.12	0.763	0.1851942
DL5	4.32	0.777	0.1798611
Vocational training and education			
VTE1	2.7	1.231	0.4559259
VTE2	2.89	1.07	0.3702422
VTE3	3.03	1.192	0.3933993
VTE4	2.89	1.229	0.4252595
Social and ethical responsibility			
SER1	3.04	0.985	0.3240132
SER2	3.06	0.96	0.3137255
SER3	3.6	0.988	0.2744444
SER4	3.16	1.027	0.325
SER5	3.3	1.208	0.3660606
Relationships and engagement			
RE1	3.26	0.809	0.2481595
RE2	3.43	1	0.2915452
RE3	3.36	0.955	0.2842262
RE4	3.48	0.906	0.2603448
RE5	3.42	0.902	0.2637427
Digital talents development			
DTD1	3.16	1.154	0.3651899
DTD2	3.04	1.022	0.3361842
DTD3	2.84	1.138	0.4007042
Change management and Digital transformation			
CMDT1	3.23	0.983	0.3043344
CMDT2	3.6	0.859	0.2386111
CMDT3	3.39	1.015	0.29941
CMDT4	3.51	0.993	0.282906

(*continued*)

Table 4. (*continued*)

Variable & Item	Mean	Std. deviation	CVI
CMDT5	3.45	0.934	0.2707246
CMDT6	3.81	0.904	0.2372703
Digital Platform Usability			
DPU1	3.78	0.861	0.2277778
DPU2	4.05	0.722	0.1782716
DPU3	3.6	1.033	0.2869444
DPU4	3.6	1.015	0.2819444
DPU5	3.23	1.046	0.323839
DPU6	3.56	1.169	0.3283708
Attitudes			
AT1	3.03	0.943	0.3112211
AT2	3.27	0.898	0.2746177
AT3	3.46	0.864	0.249711
AT4	3.85	1.033	0.2683117
Digital Government Competency			
DGC1	3.05	0.887	0.2908197
DGC2	3.37	0.844	0.2504451
DGC3	3.95	0.892	0.2258228

4 Discussion

This study validated the face and content validity of the questionnaires and pilot study analyses used to evaluate the digital government competency framework of Omani public sector managers. The survey needs to solve certain issues shown by face validity. The focus is placed, among other things, on the clarity of the language and sentence structure. The judgment is viewed as appropriate. The CVI employed in this research shows that 96% of the survey instrument's validity was achieved. The CVI scored above the predicted minimum CVI of 0.80 [17]. Only 15 constructs and 75 items were selected as the final instrument by descriptive analysis using EFA, with internal reliability Cronbach's alpha for all structures and items above 0.7 as advised.

5 Conclusion

This study has gone through various stages of evaluating the validity and reliability of each suggested variable by creating questions based on prior research and expert confirmation. The critical and crucial components of each variable were identified via CVR and CVI analyses. To evaluate the purification and EFA scale, the data gathered during the pilot study were analyzed. In this investigation, Cronbach's alpha > 0.7 is employed. The new instruments have been determined to demonstrate an appropriate and acceptable measuring performance required for a future descriptive study to evaluate the Omani public sector managers' competency with the digital government. The validity and reliability test for this survey seems to be sufficient. Therefore the following steps and the rest of the actual data gathering and analysis can be established.

6 Study's Implications

The findings will be helpful because the Omani government has to enhance and raise the performance of the Digital Government. Using the suggested framework, the government can update its plans and policies to enhance public sector managers' digital government management capabilities. Any government, including Oman, will benefit from measuring and recognizing the necessary competencies. The digital government competencies framework will assist the organization in determining priorities and establishing performance benchmarks for all departments and agencies. Executives and staff will be guided by understanding their desired behaviors and abilities and how to achieve them. Furthermore, Oman Vision 2040 is supported by this research. It includes ten key indicators, including national competencies with dynamic capabilities and skills to compete locally and globally.

7 Limitations and Future Work

In order to identify digital government competencies in the Omani public sector, a future quantitative study will need to demonstrate that this new instrument has the required and appropriate measurement performance. However, this survey's face and content validity is considered adequate. Therefore, it may be further established for future actions and for conducting the complete actual data analysis using SPSS and Smart PLS.

Acknowledgment. This work is financially supported by Universiti Teknologi Malaysia UTM SPACE under Grant Number R.K130000.7756.4J574.

References

1. Kane, G.C., et al.: Aligning the organization for its digital future (2016). https://www2.deloitte.com/us/en/insights/topics/emerging-technologies/mit-smr-deloitte-digital-transformation-strategy.html. Accessed 12 Nov 2019
2. Half, R.: Staffing Digital Projects: Not as Straightforward as It Sounds (2017). https://www.roberthalf.com/blog/management-tips/staffing-digital-projects-not-as-straightforward-as-it-sounds. Accessed 11 Dec 2019
3. Gekara, V., et al.: Skilling the Australian workforce for the digital economy. Research Report (2019)
4. Mergel, I.: Competencies for the Digital Transformation of Public Administrations (2020). https://www.co-val.eu/blog/2020/04/08/digital-transformation-of-public-administrations/

5. Malanda, D.: Digital skills in the public sector: asystematic literature review. In: Digital Innovation and Transformation Conference (2019)
6. Al-Kalbani, A.: A compliance based framework for information security in e-government in Oman (2017)
7. Masuwai, A.M., Saad, N.S.: Evaluating the face and content validity of a Teaching and Learning Guiding Principles Instrument (TLGPI): a perspective study of Malaysian teacher educators. Geografia-Malaysian J. Soc. Space **12**(3) (2017)
8. DiStefano, C., Hess, B.: Using confirmatory factor analysis for construct validation: an empirical review. J. Psychoeduc. Assess. **23**(3), 225–241 (2005)
9. Taherdoost, H.: Validity and reliability of the research instrument; how to test the validation of a questionnaire/survey in a research. How to test the validation of a questionnaire/survey in a research (2016)
10. Rodrigues, I.B., et al.: Development and validation of a new tool to measure the facilitators, barriers and preferences to exercise in people with osteoporosis. BMC Musculoskelet. Disord. **18**(1), 1–9 (2017)
11. Al-Mahrezi, J., Bakar, N.A.A., Sjarif, N.N.A.: Digital government competency for omani public sector managers: a conceptual framework. In: Saeed, F., Mohammed, F., Al-Nahari, A. (eds.) IRICT 2020. LNDECT, vol. 72, pp. 1009–1020. Springer, Cham (2021). https://doi.org/10.1007/978-3-030-70713-2_90
12. Dhamani, K., Richter, M.: Translation of research instruments: research processes, pitfalls and challenges. Afr. J. Nurs. Midwifery **13**(1), 3–13 (2011)
13. McKenzie, J.F., et al.: Establishing content validity: using qualitative and quantitative steps. Am. J. Health Behav. (1999)
14. Kennedy, L.G., et al.: Validity and reliability of a food skills questionnaire. J. Nutr. Educ. Behav. **51**(7), 857–864 (2019)
15. Lawshe, C.H.: A quantitative approach to content validity. Pers. Psychol. **28**(4), 563–575 (1975)
16. Yaghmaei, F.: Content validity and its estimation (2003)
17. Tojib, D.R., Sugianto, L.-F.: Content validity of instruments in IS research. J. Inf. Technol. Theory Appl. (JITTA) **8**(3), 5 (2006)
18. Alsanea, M.: Factors affecting the adoption of cloud computing in Saudi Arabia's government sector. Goldsmiths, University of London (2015)
19. Perry Hinton, D., et al.: SPSS Explained. Routledge (2004)
20. Hult, G.T.M., et al.: Addressing endogeneity in international marketing applications of partial least squares structural equation modeling. J. Int. Mark. **26**(3), 1–21 (2018)
21. Williams, B., Onsman, A., Brown, T.: Exploratory factor analysis: a five-step guide for novices. Australas. J. Paramedicine **8**(3) (2010)
22. Florida, R.: The economic geography of talent. Ann. Assoc. Am. Geogr. **92**(4), 743–755 (2002)

Factors that Impact a Company's Digitalization and Employee Skills: The Case of Saudi Aramco

Sarah Al Buainain, Yousif Abdelrahim(✉), and Aliah Zafer

College of Business Administration, Prince Mohammad Bin Fahd University, Dhahran, Kingdom of Saudi Arabia
{yabdelrahim,azafer}@pmu.edu.sa

Abstract. With the effect of rapid advances in communication and technology, it has become a necessity for businesses to keep up with the times. In this process, it is necessary to carry out digitalization-based transformation activities such as revising the ways of doing business in businesses and designing new business models. In addition to the feeling of having to comply with the transformation, the main goal should be for the managers who manage the transformation to combine digitalization with creativity and innovation, and to try to create and present the technology of the future in advance. In the digital world, which is developing faster than ever with the development of the digital age, businesses are trying to transform rapidly, while the adoption of digital technologies continues to progress rapidly. The objective of this research is to study the impact of company size, Six Sigma, and online staff training of Saudi Aramco on its digitalization (i.e., digital transformation), which in turn influences its employee skills. The author supported the argument from the literature review and asked seventy female and male Aramco employees about their opinions on the impact of their company size, Six Sigma, and online training on the company's digitalization and their skills. The author posits three propositions based on the argument and respondents' opinions.

Keywords: Digital transformation · Employee skills · Company size · Online training

1 Introduction

Digital transformation refers to the digital technology adoption by an organization or a company. It is a universal phenomenon, and the main goal of this process is to modify or create new customer experiences, and culture, to improve value, efficiency, and innovation to meet the requirements of the market and the changing businesses. Every company has its digital transformation for this reason it is difficult to set one definition that aligns with all the organizations. However, it is mainly considered as the digital technology integration in all business areas which has an impact on how the business delivers and operate to the customers (Ko et al. 2022). Customer satisfaction and happiness are critical for organizations to acquire new customers and retain existing customers. In other words, organizations are always in the race of getting a high market share to increase

B. Alareeni et al. (Eds.): ICBT 2022, LNNS 620, pp. 806–816, 2023.
https://doi.org/10.1007/978-3-031-26953-0_75

sales, revues, and therefore, profits. No organization will survive without happy and satisfied customers; Hence, digital technology integration is vital to organizations in the world of globalization and the internet.

Before KSA, since technological development has evolved the government of Saudi Arabia has evolved too. This was done by facilitating several services to the Saudi companies by replacing the traditional way with a technological process due to its effectiveness and better quality. Moreover, KSA has gained the Government Leadership Award due to its high support of technological advancement and investment in development goals which had assisted in facing several crises for both private and public sectors (Singhdong et al. 2021). For instance, during the covid-19 pandemic, most Saudi companies were able to survive and cope with this crisis as the employees were used to digital solutions but of course, the dependence on technology had raised during this period. One of the main pillars of the 2030 Vision is a digital transformation that will change how the organization is communicating, processing, experiencing, and productivity. In addition, this is considered as revolutionizing act as businesses take advantage of these technologies in order to make them more scalable, accessible, and efficient. This is essential to be adopted by all organizations because business needs to evolve in order to align with the technological development that is happening and taking place globally. Each and every business need to decide how to implement new tech and how to transform it (Pousttchi et al. 2019). The enterprise digital transformation has three key areas which are customer experience in order to understand them better and to enhance customer growth. Secondly, the operational process by leveraging automation and digitization to enhance the internal process, enable digital tools for employees and monitor performance by collecting data as this is beneficial in making business strategic decisions. Thirdly, the business model by offering augmenting physical with services and digital tools, global shared services by using technology, and introducing digital products (Wolf et al. 2018).

Studies of digital transformation have shown that the increasing competitive positioning of successful firms does not depend solely on the technologies they adopt, but more importantly, on the strategies developed by their leaders. While the building blocks of a digital transformation strategy for managers are known, clearly defined rules on how to approach digital transformation and implement a well-defined digital transformation strategy are lacking. In this study, the concept and scope of digital transformation, the reasons for realizing digital transformation, the difficulties encountered and the solution proposals, the explanation of digital transformation, and business strategies used in Saudi Aramco have been examined.

As discussed in the previous earlier, digital transformation emerges with the blending of personal and corporate information technology environments and covers the transformational impact of new digital technologies such as social, mobile, analytics, cloud, and the internet of things in businesses. Digital transformation requires the transition of manufacturing from more labor-intensive processes to mechanical processes based on information technology. The point to be considered is not only the acquisition of information but also the interpretation and use of this information. Digital transformation involves the application of digital technologies to change core business transactions, products, processes, organizational structures, and management concepts. Increasing sales and productivity, creating value, and providing innovations in customer interaction can be shown among the benefits of digital transformation. Because of the development of digital technologies and the impact of digitalization on all business processes, managers will need to understand how the transformation will affect their operations. In this process, it is important how the managers will follow the path to realize the digital transformation, what kind of strategies they will include, and how they will evaluate the situation of the business. Digital transformation, which affects the ways of doing business, designs, models, and processes in order to improve the performance and activities of businesses, requires the capabilities that companies should have for data collection, processing, and analysis. The dynamic capabilities approach developed by the company is expressed as high-level capabilities that enhance an organization's ability to adapt its tangible and intangible resources to a changing business environment. From this point of view, it can be said that dynamic capabilities are the basis for the realization of digital transformation in businesses.

The dynamic capabilities of businesses are of great importance in the realization of digital transformation in an increasingly competitive environment. It will be easier for businesses that want to gain a competitive advantage in a constantly changing business environment, to choose appropriate internal and external resources to create different dynamic capabilities and to realize digital transformation as a result of having these resources. These resources could be Sigma Six, company size, and online staff training as explained in the literature review.

Fig. 1. The theoretical framework for the relationship between a company size, six sigma, online staff training, digital transformation and employee skills

2 Literature Review and Proposition Development

(See Fig. 1)

2.1 Factors Influence Digital Transformation

Digital transformation refers to the digital technology adoption by an organization or a company. It is a universal phenomenon, and the main goal of this process is to modify or create new customer experiences, and culture, to improve value, efficiency, and innovation to meet the requirements of the market and the changing businesses. Every company has its digital transformation for this reason it is difficult to set one definition that aligns with all the organizations. However, it is mainly considered as the digital technology integration in all business areas which has an impact on how the business delivers and operate to the customers (Ko et al. 2022). Customer satisfaction and happiness are critical for organizations to acquire new customers and retain existing customers.

Before KSA, since technological development has evolved the government of Saudi Arabia has evolved too. This was done by facilitating several services to the Saudi companies by replacing the traditional way with a technological process due to its effectiveness and better quality. Moreover, KSA has gained the Government Leadership Award due to its high support of technological advancement and investment in development goals which had assisted in facing several crises for both private and public sectors (Singhdong et al. 2021). For instance, during the covid-19 pandemic, most Saudi companies were able to survive and cope with this crisis as the employees were used to digital solutions but of course, the dependence on technology had raised during this period. One of the main pillars of the 2030 Vision is a digital transformation that will change how the organization is communicating, processing, experiencing, and productivity. In addition, this is considered as revolutionizing act as businesses take advantage of these technologies to make them more scalable, accessible, and efficient. There is some digital transformation that is in the nascent stages and others have been decades ongoing on this transformation. This is essential to be adopted by all organizations because business needs to evolve in order to align with the technological development that is happening and taking place globally. Each and every business need to decide how to implement new tech and how to transform it (Pousttchi et al. 2019). The enterprise digital transformation has three key areas which are customer experience in order to understand them better and to enhance customer growth. Secondly, the operational process by leveraging automation and digitization to enhance the internal process, enable digital tools for employees and monitor performance by collecting data as this is beneficial in making business strategic decisions. Thirdly, the business model by offering augmenting physical with services and digital tools, global shared services by using technology, and introducing digital products (Wolf et al. 2018).

Studies of digital transformation have shown that the increasing competitive positioning of successful firms does not depend solely on the technologies they adopt, but more importantly, on the strategies developed by their leaders. While the building blocks of a digital transformation strategy for managers are known, clearly defined rules on how to approach digital transformation and implement a well-defined digital transformation strategy are lacking. In this study, the concept and scope of digital transformation, the reasons for realizing digital transformation, the difficulties encountered and the solution proposals, the explanation of digital transformation, and business strategies used in Saudi Aramco have been examined.

Digital transformation emerges with the blending of personal and corporate information technology environments and covers the transformational impact of new digital technologies such as social, mobile, analytics, cloud, and the internet of things in businesses. Digital transformation requires the transition of manufacturing from more labor-intensive processes to mechanical processes based on information technology. The point to be considered is not only the acquisition of information but also the interpretation and use of this information. Digital transformation involves the application of digital technologies to change core business transactions, products, processes, organizational structures, and management concepts. Increasing sales and productivity, creating value, and providing innovations in customer interaction can be shown among the benefits of digital transformation. Because of the development of digital technologies and the impact of digitalization on all business processes, managers will need to understand how the transformation will affect their operations. In this process, it is important how the managers will follow the path to realize the digital transformation, what kind of strategies they will include, and how they will evaluate the situation of the business. Digital transformation, which affects the ways of doing business, designs, models, and processes in order to improve the performance and activities of businesses, requires the capabilities that companies should have for data collection, processing, and analysis. The dynamic capabilities approach developed by the company is expressed as high-level capabilities that enhance an organization's ability to adapt its tangible and intangible resources to a changing business environment.

2.2 Employee Skills

Digital transformation requires several important skills that should be possessed by the employees to work effectively. Some of these skills are digital and data security, leadership quality, machine learning, big data analysis, and mobility management. In the digital era, it is essential for upskilling employees and to ensure that they have the right skills that align with the objectives and needs of the organization. This is essential as it plays an essential role in the growth, productivity, and innovation of the business. A lack of these skills can impact the organization negatively. For this reason, organizations have to focus on managing skills at a time when everything is becoming digitized at a quick rate (Ekmeil and Abumandil 2020). High-skilled employees are needed to avoid the gap between the needs and skills of the organization.

This can be achieved by providing regular training to the employee while preparing for the digital transformation workplace. There are four main steps that the companies should focus on which are evaluating the skills of the employees to understand the current employees' digital skills that they have. This can include CRM experience, software knowledge, proprietary systems, or Microsoft suite expertise. Secondly, to reveal talent gaps to understand the need for outside talent to handle digital transformation or not. Thirdly, to offer training through higher education institutions or professional associations. Lastly, promoting within the organization as it is less expensive (Singhdong et al. 2021).

2.3 Effectiveness of Digital Transformation

Digital transformation is considered as effective and important for all organizations to follow as it offers higher opportunities and facilities on how to conduct business professionally. However, most businesses cannot complete independently on the digital transformation and seek products, service providers, and consultants to help them to migrate and overhaul practices that are long-standing in the digital environment. This process has various benefits for both the business and the customers (Singhdong et al. 2021). Digital transformation is a more powerful tool in growing business profits and productivity than previously. Replacing manual processes with digital makes customers achieve more with the resources and time provided. This technological leap offers an opportunity for innovation, creativity, and growth by reducing the ongoing expenses (Tarutė et al. 2018).

However, there are some businesses that are resistant to taking this step and embracing technologies as they believe change is disruptive. This is because, they don't have the patience, willingness, and capability to master this process. This process requires enhancing additional value or process and there is a need for new costs that may not be interesting for others to take this risk. But, strategies of change management should take place here as the digital transformation will not work effectively without broad awareness and high visibility. This transformation is effective because it reduces labor costs while increasing productivity (Pousttchi, et al. 2019). One of the most impactful and beneficial ways to transform any business is to use technology.

For instance, the money and time spend by enterprises to update and train new employees with the help of digital resources can be a powerful tool in finishing this quickly. This reveals that with the right tools the productivity level is increased, and costs are decreased. Moreover, it enhances competitive advantage as it drives innovation. This is because deciding not to embrace these transformations reveals that the business is not interested in upgrading and developing its services which will result in negative consequences. Thirdly, it enhances customer experience such as live chat, email, social media, mobile apps, etc. This improves the communication level with the organizations and customers by offering a great experience (Wolf et al. 2018).

2.4 Digital Transformation in Saudi Aramco

In 2022, the digital strategy vision of Saudi Aramco is spearheading digital innovation, maximizing shareholders, and digitalized energy corporations to be the world's leading in energy globally. Whereas, its mission is to innovate with technology, maximize localization, enhance the digital workforce, and improve margin, and revenue diversification (Aramco 2022).

There are four categories of digitalization capabilities and levels in Saudi Aramco. Firstly, connected that includes data visualization, intelligent leak detection, chain management information Hubs, and remotely field-controlled vehicles. Secondly, collaborative which encompasses 3D printing, edge computing, demand-driven optimization, AR/VR, fleet management, field mobility, global logistics visibility, and integrated 3D visualization asset. Thirdly, predictive includes digital twin, machine learning, predictive analysis, image analytics, asset performance management, and production surveillance systems. Fourthly is autonomous which focuses on autonomous field vehicles. Saudi Aramco has adopted the digital transformational program to accelerate and optimize its projects that aligns with its objectives. They encourage and support the workers to create digital value-driven solutions.

In addition, their coordinated approach is to enhance sustainability, efficiency, and safety by benefitting the business. Moreover, the digital revolution of Saudi Aramco consists of six flagship programs which are; operational, digital workforce, compliance, finance, sustainability, and supply chain (Aramco 2022).

2.5 Factors Affecting Digital Transformation

2.5.1 The Company Size

Digital transformation is considered more effective for large corporations due to the challenges they may face in time and finishing projects (Singhdong et al. 2021). With the help of this transformation, the business will be able to finish the tasks within a short time, will save money, and will gain a positive reputation in the market. Of course, it is also powerful for small businesses but its demand is higher for larger corporations (Wolf et al. 2018). Following the above-mentioned discussion, the author argues that company size is more likely to affect its digital transformation and posits proposition I (P1): P1: A company's size will positively influence its digital transformation; such that larger companies will have more digital transformation.

2.5.2 Six Sigma

As digital transformation is impacting both external and internal interaction of the business, questions have been raised about the traditional frameworks of TQM such as Six Sigma, Lean, and Lean Six Sigma as an integration of both of them. In a certain situation in the technology space, it is not ideal as it shows a high focus on the incremental effect and the ability of the digital movement's disruptive nature. However, digital transformation and Lean Sigma are not exclusive mutually. The concentration of Lean is on the value through the non-value add and waste elimination processes. Whereas, the focus of Six Sigma is on variations and defect reduction. The innovation incorporation into functions of the business requires an efficient operational process (Tay and Loh 2022). The organizational change strategy and development of technology with the Lean Sigma implementation framework as a key focus assist in the success of the digital transformation and any digital strategy. Successful technology implementation in an organization requires a practical, standardized, and strong operational process to provide continuous improvement and facilitate innovation. The framework of the Lean Sigma application can assist in the development of sustainable organizations due to its high concentration on enhancing and redesigning the operational process (Ekmeil and Abumandil 2020). Hence, the author argues that Six Sigma helps a company's digital transformation, and therefore, employee's skills and posits proposition 2 (P2): P2: A company's Six Sigma Six influences its digital transformation, and therefore, employees' skills.

2.5.3 Staff Training

The digitalized employee training core element is the learning management system (LMS). This shows that digital transformation has a high influence on the training of employees because it assists in unlocking new opportunities and making content learning more available due to the interactive learning tools that are provided. Digital transformation offers numerous benefits such as it allows learning materials to deliver to trainees more effectively, organizing communication and collaboration between trainees and trainers, helping in tracking profess, assessing training results, and more. It is suitable for online learning and hybrid learning by combining e-learning and traditional classroom. This transformation offers a great opportunity for the employees to learn and develop their skills due to the high facilities it offers (Pousttchi et al. 2019). With the help of this transformation, they can undergo training at home, the workplace, or any other place. This is because mobile applications provide access at any time to training materials and have higher capabilities. The post-training phase can also be done by utilizing mobile apps to brush the skills of the employees. For instance, applications for audio simulation can assist the employees to communicate through simulation with customers of real-life dialogues. There are other extended reality technologies that are effective for employee development and training such as virtual reality, augmented reality, and mixed reality (Oh et al. 2022). Following the above-mentioned line of discussion, the author believes that online training helps companies improve digital transformation, and therefore, the employees' skills and posits proposition3 (P3): P3: online training positively improves digital transformation, and therefore, the employees' skills.

2.5.4 Technology

Technology is an essential factor that supports the need for digital transformation and organizational digitization. However, there is no single technology or application that enables transformation. There are various critical digital transformation technologies for digitalization. For instance, cloud computing offers faster access to the organization to its new functionalities, software, and updates at all times from anywhere aside from data storage. In addition, mobile platforms help in completing work everywhere and at any time (Singhdong et al. 2021).

Automation is also powerful technology that deploys bots for repetitive tasks and handles mundane more accurately and faster than humans such as RPA. Emerging transformational technology helps businesses to work more efficiently, move faster, and to implement new services and products through social media, edge computing, IoT, virtual reality and augmented reality, and blockchain. Digital technologies are used in the digital transformation process to make it more successful and effective (Wolf et al. 2018). Therefore, the author argues that technology in an organization enhances digital transformation, and therefore employees' skills and posits proposition 4 (P4): P4: Technology in companies enhances the company's digitalization and its employees' skills.

3 Robust Study

In this robust study, the author tries to add more support to the argument by asking eleven questions to seventy male and female employees at Saudi Aramco via online self-administered survey. The survey questions were designed to get Saudi Aramco employees' opinions on the impact of the company size on digitalization and employeees' skills. Then, the impact of Six Sigma on the company digitalization and the employees' skills. Finally, whether or not online training has helped their company improve digitalization and employee skills.

4 The Survey Results

The survey results have gathered a population of around 70 respondents from Saudi Aramco. The following are the results of the survey questionnaire of 11 questions (Table

1). The respondents ages from 29–38 (45.7%), 18–28 (44.3%), 39–48 (8.6%), and 49–58 (1.4%). The respondents' departments include the Human Resources department (30.4%), business administration (23.2%), the engineering department (17.4%), finance (13%), and IT department (11.6%).

Table 1. The responses of 70 aramco employees asked about their opinions about the impact of digital transformation on their company and their skills.

Questions	Responses with "Yes Answer"	Respondents with "No Answer"
Digital transformation influenced my work positively	85.7%	11.4%
Digital transformation has influenced my skills positively	92.8%	7.2%
There is strong relationship between the company size and digital transformation	95.6%	4.4%
Technology has assisted the success and efficiency of the company	98.6%	1.4%
Digital transformation has been implemented in all areas in our company	38.2%	8.8%
Have you encountered any issues during the implementation of digital transformation	2.4%	4.4%
Digital transformation can change my life in a good way	78.6%	2.8%
My company usually trains employees to be familiar with digital transformation	75.7%	5.7%

5 Conclusions

The survey results show that digital transformation in Saudi Aramco has a great impact on the employee's skills, positively and as well as there is a strong relationship between Saudi Aramco's company size and digital transformation. It is safe to say that with the effects of the digital economy and digital age in today's businesses, the new search is to be "digital". In many industries, companies are dedicating resources towards becoming digit companies to bring new products to digital technology to streamline their business processes and access more information.

The potential impacts of digital transformation cover all industries. Organizations need to continue existing services and develop strategies to manage the change in the analog-to-digital mix. Therefore, businesses must establish an overall development direction and implement digital innovations, while remaining active in evaluating and optimizing existing business opportunities. To achieve digital transformation, they must develop their ability to continually learn as an indispensable prerequisite for coping with change and successfully innovating. The effects of digitalization will be felt in different dimensions between developed, developing, and underdeveloped countries, because the effects of digitalization have sociological, cultural, and social dimensions in addition to their economic dimensions. For example, in developed organizations, such as Saudi Aramco, that manage information, and production technology, and are highly digitized, fewer human resources are needed due to increased productivity. With similar logic, human resource demands for operational purposes will continue for a while in countries that cannot manage information, produce technology and have low levels of digitization. When it comes to the global dimension, it can be said that the need for human resources (labor force) will decrease if digitalization increases with this speed.

At the same time, it should be emphasized that together with human cognitive abilities, it is the basic element of digital transformation. New technologies are the foundation of digital transformation. Depending on their readiness, these technologies can be integrated into companies step by step.

References

Aramco. How Aramco's digital transformation is shaping the workplace of the future. Aramco **2**(1), 1 (2022). https://www.aramco.com/en/magazine/elements/2022/aramco-digital-transformation-shaping-future-workplace

Ekmeil, F., Abumandil, M.: Implementation of digital transformation on six sigma with total quality management case study in the Palestinian healthcare sector during coronavirus disease (COVID-19). Seoul J. Bus. **2**(1), 1–12 (2020). https://www.researchgate.net/publication/347485700_Implementation_of_Digital_Transformation_on_Six_Sigma_With_Total_Quality_Management_Case_Study_In_The_Palestinian_Healthcare_Sector_During_Coronavirus_Disease_COVID-19

Ko, A., Fehér, P., Kovacs, T., Mitev, A., Szabó, Z.: Influencing factors of digital transformation: management or IT is the driving force? Int. J. Innov. Sci. **14**(1), 1–20 (2022). https://doi.org/10.1108/IJIS-01-2021-0007

Oh, K., Kho, H., Choi, Y., Lee, S.: Determinants for successful digital transformation. MDPI **14**(1), 1–14 (2022). https://doi.org/10.3390/su14031215

Pousttchi, K., Gleiss, A., Buzzi, B., Kohlhagen, M.: Technology impact types for digital transformation. In: Conference on Business Informatics (CBI), vol. 3, no. 2, pp. 1–11. (2019). https://www.researchgate.net/publication/335362246_Technology_Impact_Types_for_Digital_Transformation

Singhdong, P., Suthiwartnarueput, K., Pornchaiwiseskul, P.: Factors influencing digital transformation of logistics service providers: a case study in Thailand. J. Asian Finan. Econ. Bus. **8**(5), 241–251 (2021). https://www.koreascience.or.kr/article/JAKO202112748674943.page

Tarutė, A., Duobienė, J., Klovienė, L., Vitkauskaitė, E., Varaniūtė, V.: Identifying factors affecting digital transformation of SMEs. In: Proceedings of the 18th International Conference on Electronic Business, vol. 2, no. 6, pp. 373–381 (2018). http://iceb.johogo.com/proceedings/2018/ICEB2018_paper_04_full.pdf

Tay, H., Loh, H.: Digital transformations and supply chain management: a Lean Six Sigma perspective. J. Asia Bus. Stud. **16**(2), 340–353 (2022). https://doi.org/10.1108/JABS-10-2020-0415

Wolf, M., Semm, A., Erfurth, C.: Digital Transformation in Companies –Challenges and Success Factors. In: Innovations for Community Services, vol. 3(2), pp. 178–193 (2018). https://www.researchgate.net/publication/325572599_Digital_Transformation_in_Companies_-_Challenges_and_Success_Factors

Malaysian Student's Attitude Towards Organic Food Buying Behaviour

Mohamed Bilal Basha[1(✉)], Lawal Yesufu[1], Saheed Busari[2], Gail AlHafidh[3], and Fatima Sultan Khalfan Helis Alali[4]

[1] Business Department, Higher College of Technology, Sharjah Women's College, Sharjah, UAE
mbilal@hct.ac.ae
[2] Business Department, Islamic University, Kuala Lumpur, Malaysia
[3] General Studies, Higher College of Technology, Sharjah Women's College, Sharjah, UAE
[4] Higher College of Technology, Sharjah Women's College, Sharjah, UAE

Abstract. In many developing countries export earnings account for a large percentage of the country Gross Domestic Product (GDP), government revenues and expenditures and public investment. High dependency on a limited number of commodities for export increases the economy vulnerability to price fluctuations in the global markets. Not only instability in export earnings reduces the ability of the economy to finance development, but also increases uncertainty about future growth. Economic diversification underscores the importance of productivity growth to balance development and reduce dependency on international markets.

The new economy, driven by digital technologies could help developing countries diversify output and foster economic growth. Building capacity for digital development enhances the country capabilities to promote innovation, create knowledge and disseminate information. Digital networks increase communication as well as allow people, regions and nations to collaborate and share information aiming at fostering growth and sustaining development. For developing countries, external knowledge enhances the economy readiness to diversify output, create employment opportunities and improve global competitiveness.

Keywords: Purchasing attitudes and intentions · Student buying behaviour · Organic foods

1 Introduction

The global interest in organic food and products is a growing trend. The increased concern about climate change and lifestyle has led to consumers questioning their purchase choices, especially young people. Customers are becoming increasingly aware of some of the negative aspects of food production to date, for example; genetically modified seeds; the use of harmful pesticides; soil quality depletion due to modern intense farming methods and the heavy carbon footprint that worldwide deliveries generate. Organic products, on the other hand, are considered to be both beneficial to the environment and

B. Alareeni et al. (Eds.): ICBT 2022, LNNS 620, pp. 817–830, 2023.
https://doi.org/10.1007/978-3-031-26953-0_76

better for our health. The advantages include: "high nutritional value (higher contents of vitamins, minerals, essential fatty acids, proteins, etc.); high biological quality; high technological quality (when stored they maintain their quality for a longer period of time, they are more suitable for processing and distribution, they require less energy to be processed)", (Golijan and Dimitrijevic 2018).

In Malaysia, the focus of this study, the supply of locally grown organic food is limited in both volume and variety such that suppliers are unable to keep up with the growing demand, especially amongst young people. Much of the organic food sold in Malaysia is therefore imported and thus makes organic food more expensive than local produce and yet the market is still growing (Somasundram et al. 2016).

This study explores a specific sector of that growing market: university students in Kuala Lumpur. As young adults, the choices made by University students reflect the potential for future market changes as their purchase decisions can point to untapped market potential and, as such, this unique study, though seemingly narrow in scope, will enhance and shed light on the current understanding of organic market forces for the benefit of stakeholders and interested parties, globally.

This research differs from previous studies that have separately identified the factors influencing consumer behaviour towards organic products. In past papers, a constant variable is identified and tracked as it influences other dependent variables driving the trend in consumer intent for organic products. The researchers discovered a gap in the literature wherein the whole picture was not being represented. This study, therefore, aims to fill that gap by exploring the full range of factors such as; health, safety, awareness, environmental concerns, past-experience, taste, availability of products and religious beliefs to generate a deeper and more holistic understanding of organic food buying behaviour.

This important study aims to explore this gap in knowledge regarding the motivations behind the purchase intentions of this particular group by examining the consumer behaviour of university students, Generation Y, in their response to organic products. The objective of this study is, therefore, to explore how certain variables such as availability, knowledge government policy and religious intent have influenced the purchasing behaviour and consumption switch towards organic products among university students in Malaysia. The paper will proceed with the following sections: Literature review; Sampling Procedure and Data Collection (including measurement Instruments); Analysis; Findings & Implication of the Study and then, lastly the Conclusion.

2 Literature Review

There are several recent studies relating to consumer behaviour towards organic products as interest from the major stakeholders worldwide increases and data-backed research is required to inform producer targets, distribution channels and retail catalogues. However, to date, most of the studies focus on a broad range of demographics with few focussing solely on the 'millennials'. This study will explore 6 hypotheses to try to shed light on how health, environment, safety, price and taste influence consumer behaviour on organic products in Malaysia.

2.1 Health and Lifestyle

Health and lifestyle has been found to be one of the major factors that influences consumers towards choosing organic food (Basha 2014a). It is a well-established fact that there is a positive link between the health benefits of a product and the purchase intention of a consumer. Lian (2017), investigates the factors that motivate consumer purchase of organic products and identifies past experiences of the use of natural products on health as a consumer motivation value. The study employed structural equation modelling to analyse data gathered from 421 respondents on their consumption of organic products, specifically in Klang Valley, Malaysia. Even though the research focused on consumers in Klang Valley, the percentage of respondents that were classified as university students was not known (Francis et al. 2003; Basha et al. 2021). This study therefore aims to focus specifically on Generation Y university students to narrow the focus and gain more clarity. Based on the extant research and discussion, hypothesis 1 was generated as follows: H 1: Health and lifestyle have no significant effects on consumer behaviour towards organic food.

2.2 Environmental Consciousness

In recent years, consumers have become much more environmentally-aware and of the impact of the actions of humans on the natural world around us. Governments, world-wide, are discussing strategies to help control climate change, the increasing temperatures and environmental related health issues. In response, consumers too are now showing more concern about the food and its production impact on the land and its environment (Basha et al. (2015a); Basha (2014b)).

Several studies has proven the positive connection between the environmental concern and consumer positive attitude towards organic products. Hossain and Lim (2016), examine the effect of consumer attitude towards environmental protection and knowledge of environmental issues, which affects actual purchase behaviour. The research showed that a positive attitude towards environmental protection and education has a significance influence on consumers' actual purchase behaviour, using a sample of consumers in Penang, Malaysia. The use of Mall intercept technique and questionnaires confirmed that there are three components of consumer attitudes toward environmental protection: the individual, government, industries responsibility and financial responsibility (Hossain and Lim 2016). Their findings showed that the increase in the level of awareness about the benefit of consuming natural products had triggered a significant increase in demand for organic products fuelled by the belief that the deterioration in the environment is because of the decrease in the production of the natural product. Based on the extant research and discussion, hypothesis 2 was generated as follows: H 2: Environmental Consciousness has no significant effect on consumer behaviour towards organic foods.

2.3 Government Support and Policy

The recent increase in the demand for organic products has triggered the interest of the governments in making policies towards promoting the importance of organic food to the people. Moreover, government policies have been made towards promoting the efforts of the farmers in producing organic products. The government of Malaysia is a role model in this regard providing sustainable activities such as awareness of the benefits of producing and consuming organic products and providing soft loans to organic product farmers in the country (Chen and Chai 2010).

Furthermore, the Malaysian government has engaged in a series of campaigns through the media on the importance of producing and consuming organic products. This campaign is one of the Malaysian green, sustainable development goals by introducing an eco-labels activity to increase consumer awareness (Lotter 2003). Also, the Malaysian government has introduced the Malaysian Farm Accreditation scheme under the auspices of the Department of Agriculture (DOA) with an environmental condition of producing high quality, safe and suitable products for human consumption (Hossain and Lim 2016). Impact studies have further proved that consumer attitude toward the organic food consumption are influenced by the role of government and industry financial roles towards environmental protection (Yesufu 2016). This shows government policy and support has a positive and significant impacts on consumers' actual purchase behaviour, although environmental awareness has no significance. Based on the extant research and discussion, hypothesis 3 was generated as follows: H 3: Government support & policy has no significant influence on consumer behaviour towards organic foods.

2.4 Convenience and Price Consciousness

The price of organic food and its price premiums over conventional food have been of significant influence in consumer behaviour switch. The consumer behaviour towards the natural product is driven by the active, cognitive and intentional processes (Holt-Giménez et al. 2012). The price elasticity, price perceptions, price knowledge and consumer willingness to pay are significant factors that motivate consumers to purchase behaviour of organic products (Rödiger and Hamm 2015; Shamsuddin et al. 2020). Although, different factors also affect consumer willingness to purchase organic products. Two studies, Basha et al. (2015b) and Krystallis and Chryssohoidis (2005) used factor analysis to determine the factors that motivate consumers to pay for different organic products. Moreover, the statistical result shows that the differences in prices of the organic product has a significant impact on the willingness of consumers' to pay as compared to conventional products which remain non-significant (El Alfy et al. 2017). Based on the extant research and discussion, hypothesis 4 was generated as follows: H 4: Convenience and price has no significant effects on consumer behaviour towards organic foods.

2.5 Religious Intent to Consume Organic Food

Religion is a significant factor that determines the buying intent of people in the changing and globalized world. Religious intent has become a fundamental ethos that specifies a large number of people (Porter 2013). Every religion is taught to be fair with humans and animals. Therefore, it preaches our duty towards protecting flora and fauna. Due to ill-treatment in the animal factory, consumers move towards free-range or organic chicken as a sign of protest towards the ill-treatment. Consumers are reluctant to purchase the pesticide and substance use in the chicken or plants, which is subject to unsafe and non-halal unhygienic food. Almost, all the religion preaches the consumer to avoid consuming the food, which is not safer for their health (Shaharudin et al (2010). However, this practice is not commonly followed and seen as necessary among the majority of the group. Based on the extant research and discussion, hypothesis 5 was generated as follows: H5: Perceived beliefs/religious intent has no significant influence on consumer behaviour towards organic foods.

2.6 Subjective Norms

Subjective norms are significant social factors that influence consumer purchase intention of certain products and services. The extent of social interaction from peer groups, family members, colleagues will determine the influence of subjective norms in a given society (Basha and Lal 2019; shamsudin et al (2018); Basha et al. (2015a); Lal et al. (2019). Highly interactive social behaviour tends to have significant effect on its members particularly in consumption switch (Thøgersen et al. 2016). Consumer behaviour to purchase certain products is a psychological process which involves recognition of needs and how to resolve those needs. Available information about consumer needs has influence on the buying decision. Hence, subjective norms are a significant social order which gives consumers information, influences its interpretations, plan and eventually influences the consumer choice on needs among on available varieties (Basha et al. 2015b). Consumer plan behaviour is a significant process in the purchase of apparel products. The perception of an individual about the opinions of others will determine the level of subjective norms. A consumer that is intransigent will likely not be influenced by other people's buying behaviour or relative information from a social group (Thøgersen et al. 2016). Meanwhile, Maloney et al. (2014), suggested that subjective norms are an external purchasing social factor that have direct influence on consumer purchasing intention of apparel products. Based on the extant research and discussion, hypothesis 6 was generated as follows: H 6: Subjective norms have no significant influence on consumer behaviour towards organic foods.

3 Research Methodology

The primary aim of this study is to identify the six key determinants outlined in the literature review. These determinants have shown positive influence towards the organic food buyers. However, these determinants have not been tested collectively in any other studies in Malaysia, particularly Kuala Lumpur. Therefore, these important variables are used in this study to understand the consumer attitude towards organic food. Moreover, all the extant studies have given less importance to understanding university students as a specific focus group. Consequently, this study was mainly developed to identify the student attitude and perception towards organic food. The primary reason for selecting the Kuala Lumpur student is due to metropolitan city environment and the affluent lifestyle enjoyed by many of the students.

3.1 Sampling Procedure and Data Collection

This research explores how university students purchase intent of organic products is influenced by availability, knowledge, religious intent, subjective norms and government policy. The research involved a sample of 329 student participants in Kuala Lumpur, Malaysia. The study collected data via a structured questionnaire survey that was distributed in an international Islamic University library and through online students' contacts. However, due to the small size of the sample, the results have limited representativeness and should be used as pre-test results for a broader study.

3.2 Measurement Instrument

The questionnaire consisted of five sections. Apart from the first section which records respondents' demographics information, the remaining four sections measure the four study variables (availability, knowledge government policy and religious intent). The study adapts a 5-point Likert scale (1 = strongly agree to 5 = strongly disagree).

3.3 Data Analysis and Findings

To validate the sample, data was collected from 500 students using random sampling. Questionnaires were distributed and collected from the students using random sampling in the university. Initially, the students were asked about their involvement in the purchasing decision of household items and some primary questions on organic food to know their understanding of organic food. The questionnaire are only given to those students who positively indicated that they are involved in the household purchase and have a knowledge of organic food. Out of (500) five hundred students, approximately 329 effectively responded to the questionnaire which was 65% approximately. The remaining returned questionnaires either partially filled or the respondent showed no interest in completing the data. Thus, the incomplete answer and the questionnaire were removed. Some of the questionnaires were also removed where responses were totally negative or positive attitude towards all the questions.

The table below shows the breakdown of the demographic factors of the participants in the study in Kuala Lumpur. The male buyers represent 72% of the participants and female buyers are 38%. In addition, the educational background of the organic food buyers is divided into the following categories with the percentage of respondents. The bachelor's degree holders 35.5%, followed by postgraduate degree 45.3%, doctorate degree 17.6% professional/technical 9% and foundation students 6%. The collected data from the respondents was analysed using SPSS software program. The reason for selecting this software is due to the nature of the data set. SPSS is the most common software used to understand the consumer attitude and perception through the analysis of multi-regression, Cronbach alpha and Anova test and EFA test to validate the data. These various analysis methods determine the factors affecting the purchasing decision of the chosen sample. See the Table 1 below.

Table 1. Demographic

Demographic	Frequency	Percent
Male	204	62.0
Female	125	38.0
Education		
Secondary/High School	29	8.8
Bachelors Degree	88	26.7
Postgraduate Degree	149	45.3
Doctorate Degree	58	17.6
Professional/Technical	3	.9
Foundation	2	.6
Age		
Less than 20	22	6.7
20 to 29	154	46.8
30 to 39	111	33.7
40 to 49	41	12.5
60 and above	1	.3

3.4 Reliability, Factor and Correlation Test Analysis

Cronbach's alpha measure of internal consistency (Cronbach 1951) was used to measure the construct in order to test if the alpha value was greater than 0.7 (Nunnally 1978). The purpose of this test to identify how close the questionnaire fit to the intent study. Any questionnaires of below 0.7 were removed from the intent study. See Table 2 below.

Table 2. Reliability analysis

Variable	No. of items	Cronbach's alpha coefficient
HCLS	5	0.764
EC	5	0.749
GSP	5	0.769
CP	5	0.876
SN	5	0.832
ATT	5	0.730
Religious Intent toward consumption	5	0.858
Purchase intention	5	0.725

All the thirty questionnaires were further tested to validate the data using multivariate statistical procedure of factor analysis, data reduction, data summarization, the Kaiser-Meyer-Olkin Measure of Sampling Adequacy (KMO) and Bartlett for analysing the test results (Tobias and Carlson 1969). In analysing this particular data set, Kaiser (1974) suggested that a minimum KMO value of 0.5 was imperative for conducting factor analysis with such data. The results of this reliability test showed the alpha scores of all six variables are higher than the set KMO and Bartlett's test. Thus, the result indicated that almost all the variables are higher than the set values of limit of 0.5. Therefore, the inference from these results indicated that the testing instrument designed for this study was validated to use for the study sample. See Table 3 below.

Table 3. KMO and Bartlett's Test

Kaiser-Meyer-Olkin Measure of Sampling Adequacy		.752
Bartlett's Test of Sphericity	Approx. Chi-Square	4479.569
	df	435
	Sig	.000

In order to understand the strength of the association between the determinants, the Pearson bivariate correlation analysis was performance. This analysis interprets the relationship strength between the two variables. Thus, the findings will help to know which factors is highly independent and correlated with the attitude (Pallant 2001).

The majority of the independent variables were highly correlated with consumer purchase intentions, as they are at the 0.01 level. All of the above variables showed a strong inter-relationship among each other. Interestingly, the results of this correlation showed that the variables of Health and Lifestyle (HLS) and Purchase Intention (PI) had a relatively negative relationship, whereas the Environmental Care (EC), Subjective Norms (SN), Government policy and support (GSP) Convenience and price (CP) and Religion (Rel) had distinct positive relationships with the independent variable of purchase intentions.

3.5 Multi Regression and Anova Analysis

To further check the prediction of independent variables on dependent variable, multiple regression analysis was tested. The test is a more sophisticated extension of correlation and was used to explore the predictive ability of a set of independent variables on one dependent variable (Pallant 2001). To identify and validate the interactive effect of one variable on another multiple regression analysis was used. The result are outlined as Table 5. For this study, multiple regression analysis (standard regression) was used to test the hypotheses. This particular test was adopted in order to test the hypothesis and determine the extent of the interactive effects of the independent variables on the dependent variable. The results are outlined in Table 4.

The results from Table 5 summarize the testing of the six determinants. To summarize, the testing of the hypotheses between the determinants of REL, GSP, HCLS, EC, CP

Table 4. Summary between Determinants (REL, GSP, HCLS, EC, CP and SN) with Purchase Intention –

Variance		R	R square	Adjusted R square	Std. error of the estimate
Dimension0	REL, GSP, HCLS, EC, CP	.371[a]	.138	.124	.57879
SN	SN	.221[a]	.049	.046	.62734

and SN – and the purchase intentions of the consumers shows that the value of the R squares indicated that there were approximately 13.8% of variance in the purchase intentions of consumers, in terms of the determinant variables of REL, GSP, HCLS, EC and CP. The findings showed that these determinants have important contributions in understanding the consumer purchase intentions for organic foods. Further, the value of the subjective norms (SN) R Squares indicated that less than ten percent, which is 4.9% of variance towards the Purchase Intention (PI) can be explained by the predictor variable of Attitude (AT). The value of the findings is less than 10%. However, the finding cannot be undermined as it still has some effect on purchase intention. These results showed that positive attitude towards the determinant will leads to purchase intention. This result shows that consumers are more likely to purchase the product when they have positive attitude.

Table 5. ANOVA – Determinant and purchase intention.

Model		Sum of squares	df	Mean Square	F	Sig.
REL, GSP, HCLS, EC and CP	Regression	15.980	5	3.196	9.541	.000[a]
Subjective norms (SN)	Regression	6.635	1	6.635	16.858	.000[a]

To further test the validity of the model, ANOVA test is conducted to check the fit of the variable. The test is used to check the determinant fit within the model. The model should has less than 0.05 to validate the model. Remarkably, the level of significance of the determinants towards consumer purchasing intentions was found to be less than 0.5 (actual level was 0.00), which indicated validity of this model for further testing of beta co-efficiency. In examining the results of the ANOVA test on the Predictors (REL, GSP, HCLS, EC, CP and SN) against the consumer Purchase Intention (PI), results indicated that the F-test was significant ($F = 9.541$, $p < 0.05$), and SN ($F = 16.858$, $p < 0.05$) signifying that this regression model was indeed valid. This validity indicated a significant relationship between this set of predictors and the dependent variable.

3.6 Analyses of Coefficients on Determinants

A continuance of the ANOVA test, coefficient test was conducted to check the contribution of each of the determinants of the model using Standardized β Coefficients. In construing the results for this test, the higher the value in this test specifies that the factor has a larger effect on the independent variable (Purchase Intention – PI). The values of the β Coefficients are shown in Table 6.

Table 6. Coefficients – Predictors (REL, GSP, HCLS, EC, CP and SN) and Purchase Intention (PI) Coefficients[a]

Model		Unstandardized coefficients		Standardized coefficients	t	Sig.
		B	Std. error	Beta		
1	(Constant)	2.069	.310		6.671	.000
	HCLS	.016	.046	.020	.355	.723
	EC	.141	.053	.150	2.635	.009
	GSP	−.001	.058	−.001	−.012	.991
	CP	.188	.043	.262	4.399	.000
	REL	.103	.039	.144	2.631	.009
	SN	.223	.054	.221	4.106	.000

Source: Author generated from the data.
a. Dependent Variable: ATT.
Source: Author generated from data analyses.
Note: The Dependent Variable used for Tests is Purchase Intention (PI).

Inferring from the results of this test, the determinants which contributed significantly towards the dependent variable of the purchase intention of the respondents were the three determinant variables of: Environment Concern ($\beta = 0.141$, $p = 0.009$); convenience and price ($\beta = 0.188$, $p = 0.000$) religion ($\beta = 0.103$, $p = 0.009$) and Subjective Norms ($\beta = 0.223$, $p = 0.000$). Furthermore, the determinants of Health and Life style ($\beta = 0.016$, $p = 0.723$) Government support and policy ($\beta = -0.001$, $p = 0.991$) was seen as not significantly contributing to the dependent variable of the purchase intentions towards purchasing organic foods.

Results of Hypothesis One

From the above analyses of Table 7, the hypothesis on Health and life style **is accepted**, and consequently, the results of this study contradict with the findings of the existing research. In this case, this result shows that consumers in Kuala Lumpur are not considering their health and life style while purchase an organic food. Therefore, the study did not support health and safety as an influencing factors towards the purchase of organic foods.

Results of Hypothesis Two

From the above analyses, the hypothesis on Environmental concern **is rejected**, and consequently, the results of this study upkeep with the majority of the findings of the existing research. In this case, this result shows that consumers in the Kuala Lumpur are distressed about the environment while purchase an organic food. Therefore, the study backings the environmental concern as an influencing factor towards the purchase of organic foods.

Results of Hypothesis Three

From the above analyses, the hypothesis on Government support and policy **is accepted,** and consequently, the results of this study challenge with the findings of the existing research. In this case, this result shows that consumers in the Kuala Lumpur are not concerned with the government support and policy while purchasing an organic food. Therefore, the study did not support the government support and policy as an influencing factors towards the purchasing of organic foods. The findings is rather scarce and surprising. This could be result of lack of consumer trust on organic food production method and the government policies.

Results of Hypothesis Four

From the above analyses, the hypothesis on convenience and price **is rejected**, and consequently, the results of this study backing with the majority of the findings of the existing research. On this case, this result shows that consumers in the Kuala Lumpur are not concerned with the convenience and price while purchasing organic food. On this note, the consumer is willing to pay a higher price and willing to travel to purchase the organic food. The interference of price and convenience doesn't deter the consumer from purchasing organic food, as consumers believe that having organic food will bring well-being for everyone, including the flora and fauna. Thus, consumers show a positive attitude and willingness towards the purchase of organic foods.

Results of Hypothesis Five

From the above analyses, the hypothesis on religion **is rejected**, and consequently, the results of this study support with the findings of the existing research. In this case, this result shows that consumers in the Kuala Lumpur are concerned with their religion while purchasing organic food. Consequently, the study supports religion as an influencing factor towards the purchasing of organic foods.

Results of Hypothesis Six

From the above analyses, the hypothesis on religion **is rejected**, and therefore, the results of this study support with the findings with the majority of the existing research. On this case, this result shows that consumers in the Kuala Lumpur are highly concerned with the subjective norms whilst purchasing organic food. Consequently, the study supports the subjective norms as an influencing factors towards the purchasing of organic foods. Malaysian is closely tied community lifestyle. Leaders, community head and social environmental plays an important role shaping the people's behaviour. Thus, the result shows that people in Kuala Lumpur are highly influenced by subjective norms.

4 Conclusion

The purpose of the study is to understand the determinant key factors and its influence on consumer purchase intention, specifically amongst young university students. The six determinants were chosen based on the extant research literature review and their influence toward purchase intention in Kuala Lumpur. Findings of the study shows that four key determinants environmental concern (EC), Convenience and price (CP), Religion (REL) and Subjective norms (SN) significant influence on purchase intention. Health and lifestyle failed to influence consumer purchase intention; this contradicts other findings globally such as Bellows et al. (2008) in their USA study and Maruyama and Trung (2007) in their Vietnam study. This needs further investigation to strengthen the reasoning for consumers not considering organic food as a health and lifestyle option.

Furthermore, government support and policy (GSP) failed to show any positive contribution towards the purchase intention. These findings will allow the policy makers, government agencies, organic retail managers to understand the millennial perception towards organic food. Moreover, this study provides all the stakeholders with the adequate information and indication of the consumer distress towards the environmental issues, pressure from the society and religion factors has positive contribution towards the purchase of organic food. It is in the industry's favour that consumers who have a positive attitude towards organic food are less concerned about the price and hindrance of availability of organic food.

5 Managerial Implication

The policymakers and the managers must understand the consumer attitude and educate the consumer about the benefits of organic food. Policymakers should bring more weekly markets (sandai) of organic food near to the residential areas so the products can be purchased directly by the consumer and at a cheaper price. This will solve the consumer hindrance towards finding organic food outlets. The policymakers should also educate the consumer about the benefit to the environment and the associated flora and fauna of having organically produced food. Additionally, the managers of supermarkets should provide training to their employees so that they will encourage consumers to purchase organic food by promoting the health benefits. Finally, the government should help build trust among the consumers by showcasing the process of organic food agriculture and its regress quality check throughout farming, distribution, storage, until the product reaches the retail market. This transparency will ensure increased consumer trust and help to cultivate and maintain a positive attitude towards organic food.

Further research is imperative if these results are to be generalised universally. However, these current findings suggest that the demand for organic produce is increasing substantially among young people in Malaysia and that this is likely to be the case in Asia generally. In western countries, factors such as health and lifestyle are documented as the big "pull" factors towards organic produce, whereas the results of this paper reflect a different set of motivations. An understanding of these regional variations such as seen in this vignette case study, reported and supported by empirical evidence in the global context, are crucial to a fuller comprehension of the forces at work in the burgeoning organic markets, worldwide.

References

Al-Hyari, K., Alnsour, M., Al-Weshah, G., Haffar, M.: Religious beliefs and consumer behaviour: from loyalty to boycotts. J. Islamic Mark. **3**(2), 155–174 (2012). https://doi.org/10.1108/17590831211232564

Basha, M.:. Driving factors of purchase intention towards organic food: a conceptual study. In: 4th International Conference Engineering Technology and Technopreneurship (ICE2T) (2014). https://doi.org/10.1109/ICE2T.2014.7006211

Basha, M., Mason, C., Shamsudin, M., Hussain, H., Salem, M.: Consumers attitude towards Organic food. Procedia Econ. Finan. **31**, 444–452 (2015). https://doi.org/10.1016/S2212-5671(15)01219-8

Basha, M.B., Ghafar, A., Wahid, F., Alhafid, G., Al Shaer, E., Shamsudin, M.F.: Consumer buying behaviour towards organic food–a case of UAE. Trans. Mark. J. **9**(1) (2021)

Basha, M., Mason, C., Shamsudin, M., Hussain, H., Salem, M., Azlan, A.: Consumer acceptance towards Organic food. Glob. J. Interdisc. Social Sci. **4**(3), 29–32 (2015)

Basha, M., Lal, D.: Indian consumers' attitudes towards purchasing organically produced foods: an empirical study. J. Clean. Prod. **215**, 99–111 (2018). https://doi.org/10.1016/j.jclepro.2018.12.098

Basha, M.B.: Driving factors of purchase intention towards organic food: a conceptual study. In: 2014 4th International Conference on Engineering Technology and Technopreneuship (ICE2T) , pp. 19–22 IEEE (2014)

Bellows, A., Onyango, B., Diamond, A.H.: Understanding consumer interest in organics: production vs. purchasing behaviour. J. Agric. Food Ind. Organ. **6**(1), 1169–1169 (2008). https://doi.org/10.2202/1542-0485.1169

Basha, Mohamed Bilal, Lal, David: Indian consumers' attitudes towards purchasing organically produced foods: an empirical study. J. Clean. Prod. **215**, 99–111 (2019). https://doi.org/10.1016/j.jclepro.2018.12.098

Cronbach, L.J.: Coefficient alpha and the internal structure of tests. Psychometrika **16**(3), 297–334 (1951)

El Alfy, S., G6mez, J.M., Ivanov, D.: Exploring instructors' technology readiness, attitudes and behavioral intentions towards e-learning technologies in Egypt and United Arab Emirates. Educ. Inf. Technol. **22**(5), 2605–2627 (2016). https://doi.org/10.1007/s10639-016-9562-1

Francis, C., et al.: Agroecology: the ecology of food systems. J. Sustain. Agric. **22**(3), 99–118 (2003). https://doi.org/10.1300/J064v22n03_010

Golijan, J., Dimitrijevic, B.: Global organic food market. Acta Agriculturae Serbica **XXIII**(46), 125–140 (2018). https://doi.org/10.5937/AASer1846125G

Holt-Giménez, Eric, Shattuck, Annie, Altieri, Miguel, Herren, Hans, Gliessman, Steve: We already grow enough food for 10 billion people … and still can't end hunger. J. Sustain. Agric. **36**(6), 595–598 (2012). https://doi.org/10.1080/10440046.2012.695331

Hossain, M., Lim, P.: Consumers' buying behavior towards organic foods: evidence from the emerging market. Malay. Inst. Manag. **51**(2), 7–25 (2016)

Kaiser, H.F.: An index of factorial simplicity. Psychometrika **39**, 31–36 (1974)

Krystallis, A., Chryssohoidis, G.M.: Consumers' willingness to pay for Organic food. Brit. Food J. **107**(5), 320–343 (2005). https://doi.org/10.1108/00070700510596901

Lal, D., Basha. M., Ivanov, D.: Indian consumer buying behaviour towards Organic foods: Empirical evidence. In: Proceedings of the 2019 2nd International Conference on Computers in Management and Business (ICCMB 2019), Cambridge, U.K, pp. 14–18 (2019). ISBN 978-1-4503-6168-2. https://doi.org/10.1145/3328886.3328889

Lotter, D.W.: Organic agriculture. J. Sustain. Agric. **21**(4), 59–128 (2003). https://doi.org/10.1300/J064v21n04.06

Maruyama, M., Trung, L.V.: Supermarkets in Vietnam: opportunities and obstacles. Asian Econ. J. **21**(1), 19–46 (2007). https://doi.org/10.1111/j.1467-8381.2007.00245.x

Basha, M.B., Ramesh, K.: Consumer attitude towards organic food in Trichy -South India. Int. J. Mark. Human Res. Manag. **5**(5), 1–10 (2014)

Shamsudin, M.F., Basha, M.B., Saida, K., Kadir, B.: A study on the buying behavior of consumers toward cosmetic products in large retail outlets with special reference to Chennai City. Int. J. Psychol. Rehabil. **24**(7), 4188–4208 (2020). https://doi.org/10.37200/IJPR/V24I7/PR270414

Pallant, J.: SPSS Survival Manual: A Step by Step Guide to Data Analysis Using SPSS for Windows Version 10. Open University Press, Buckingham (2001)

Porter, C.: The religion of consumption and Christian neighbour love. Dissertations. 731. Loyola University Chicago (2013). https://ecommons.luc.edu/luc_diss/731

Reiche, P., Ghemawat, P.: National cultural differences and multinational business (2017). https://www.semanticscholar.org/paper/National-Cultural-Differences-and-Multinational-Ghemawat-Reiche/dbb70b0e14146d1a5a6c4c85bb815c25557849fa

Rödiger, M., Hamm, U.: How are organic food prices affecting consumer behaviour? a review. Food Qual. Prefer. **43**, 10–20 (2015). https://doi.org/10.1016/j.foodqual.2015.02.002

Shamsudin, M.F., Basha, M.B., Mason, C., Hussain, H.I., Salem, M.A.: Environmental concern towards purchase intention of organic food product. Natl. Acad. Manag. Staff Cult. Arts Herald **1**, 938–942 (2018)

Shaharudin, M.R., Pani, J.J., Mansor, S.W., Elias, S.J., Sadek, D.M.: Purchase intention of organic food in kedah, malaysia: a religious overview. Int. J. Mark. Stud. **2**(1), 96–103 (2010)

Sharma, N., Singhvi, R.: Consumer perception and behaviour towards organic food: a systematic review of literature. J. Pharmac. Phytochem. **7**(2), 2152–2155 (2018)

Somasundram, C., Razali, Z., Santhirasegaram, V.: A review on organic food production in Malaysia. Horticulturae **2**(12), 2152–2155 (2016). https://doi.org/10.3390/horticulturae2030012

Tobias, S., Carlson, J.E.: Brief report: Bartlett's test of sphericity and chance findings on factor analysis. Multivariate Behav. Res. **4**(3), 375–377 (1969). https://doi.org/10.1207/s15327906mbr0403_8

Thøgersen, J., Zhou, Y., Huang, G.: How stable is the value basis for organic food consumption in China? J. Clean. Prod. **134**, 214–224 (2016). https://doi.org/10.1016/jclepro.2015.06.036

Yesufu, L.: The effect of Human Resource practices on the psychological contract of academics. Unpublished PhD thesis. University of Bath (2016)

Employee Productivity in the Service Industry: Does Human Resource Quality Matters?

Sharifah Rahama Amirul, Khairul Hanim Pazim, Rasid Mail, Jakaria Dasan, and Sharifah Milda Amirul(✉)

Universiti Malaysia Sabah, 88400 Kota Kinabalu, Sabah, Malaysia
sma@ums.edu.my

Abstract. Relying on quantitative perspectives to determine employee productivity is highly subjective and is regarded as a bias for the service industry. The conventional employee productivity measure, which is derived from an economic position based on the output and input relation, is no longer feasible for the service industry. The human component plays a much larger role in this sector's operations. This is due to the fact that the service process is highly dependent on the quality of the workforce, which also has a propensity to be more customer intensive. Employee productivity is not a new concept but there is still a huge gap in addressing the importance of human resource quality (HRQ) as a measure of employee or labour productivity and the significance of placing a value on human resources has been the subject of much debate in the existing literature. In this paper, we focused on the significance of HRQ and the reasons why it is relevant to the service business. Within the scope of this paper, a conceptual framework of human resource quality as an important component for the measurement of employee productivity in the service sector is proposed.

Keywords: Human resource quality · Employee productivity · Qualitative productivity measure

1 Introduction

Customer centric industry, market-oriented product and service competitiveness realizes the importance of products and service quality. Therefore, continues improvement for customers' value for money is necessary. In that notion, Nilda et al. (2014) advocated that the improvement in products and service quality must also be in line with the enhancement in employee performance. Furthermore, considerable body of evidence suggest that work quality improve productivity (Dong 2019; Royuela and Suriñach 2013) through greater human resource quality (HRQ). Indeed, it is important to measure the HRQ, in this context, as emphasised by Kaplan and Norton (1996) "what gets measured gets done". However, little attention has been given to the importance of HRQ in capturing employee productivity in service industry. This may be due to the lack of qualitative employee productivity measurement in existing literature because employee productivity measurement is dominated by productivity measured in manufacturing than service industry.

B. Alareeni et al. (Eds.): ICBT 2022, LNNS 620, pp. 831–839, 2023.
https://doi.org/10.1007/978-3-031-26953-0_77

In addition to that, definition of employee productivity is also overshadowed by economic and quantitative view which simply conceptualised on individual output per unit, per hour spent, a real gross domestic product (GDP) per hour worked or employee output per hour worked (Asia Productivity Organisation 2012; OECD 2001; 2005; Dean and Kunze 1992; Gulezian and Samelian 2003; Spring, Singapore 2011; Thomas and Sanvido 2000) and value added (OECD 2001; Spring Singapore 2011). These definitions are not conscientiously measure employee productivity in the service industry which is more complicated as it is not merely calculating the quantity aspect but also quality (Agya and Singh 2014; Rutkauskas and Paulavičiene 2005). Another economic or quantitative definition or measurement to employee productivity is a total factor productivity (TFP), which typically measured as the ratio of aggregate output (e.g., GDP) to aggregate inputs (Sickles and Zelenyuk 2019).

The assumption of TFP implied that employees and (average) hours are perfect substitutes. Dixon and Shepherd (2010) argued that this assumption is not supported by either economic theory or the empirical evidence neither that it incorporates an appropriate or sensitive specification of the employee services input in the production function. The component of human resource is relatively higher in-service sector as compared to manufacturing sector. As such HRQ is more critical to attain process improvement for the former. In that setting, customer-employee interaction and relationship are very crucial in the service sector to attract and retain the customer where customer satisfaction is the key element to build loyalty (Castañeda 2011; Cheng et al. 2011; June et. al 2000) and repurchase intention (Huddleston et al. 2009).

2 Service Versus Manufacturing Processes

To understand the importance of HRQ in service industry, it is noting worth to comprehend the different processes or operation between service and manufacturing industries. All manufacturers are focusing its operation on transforming resources into finished product, which are tangible and generally standardized product. Service industry on the other hand provides intangible product such as banking, entertainment, or education. Unlike the manufacturer, service provider is not generally standardized but customized to satisfy the specific needs of a customer (Saylor Academy, n.d). Agya and Singh Sekhon (2014) further explained the different between manufacturing and service industry as shown in Fig. 1. Agya and Singh Sekhon (2014) indicated that manufacturing-based disregard the impact of services on customers and stakeholders, but process in services on the other hand include customer inputs and extend beyond the manufacturing-based production process.

Meanwhile in Fig. 2, Agya and Singh Sekhon (2014) extend their explanation on the service-based production process. According to Agya and Singh Sekhon (2014) service outcome is determined and measured by stakeholder satisfaction and is dependent on the quality and quantity of input resources and the transformation process type adopted by a firm. The technological readiness and readiness to value co-creation play a key role in facilitating service delivery. Efficiency and effectiveness in this sense are determined by the quantity and quality of inputs, respectively. This is also supported by Rutkauskas and Paulavičienė (2005) who claimed that both the quantity and quality aspects must be considered together to provide a joint impact on the total productivity of the service organizations. Taken from Grönroos and Ojasalo (2004), and Agya and Singh Sekhon (2014) productivity concepts, the customer aspect cannot be separated from services productivity, hence the economic definition for productivity is not a comprehensive measure in the service industry but must be accompanied by qualitative aspect.

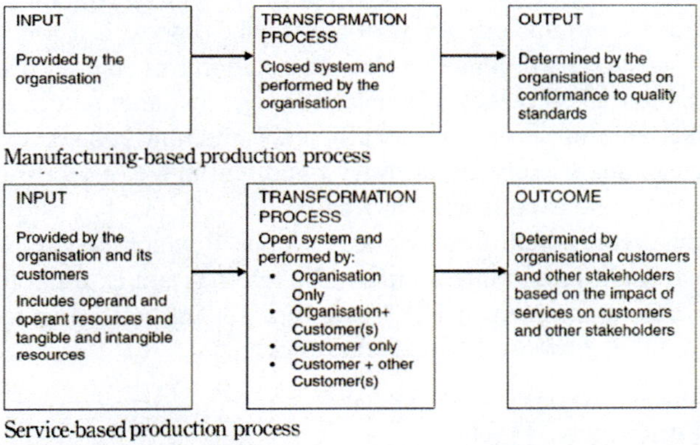

Fig. 1. The difference between manufacturing and services-based production process

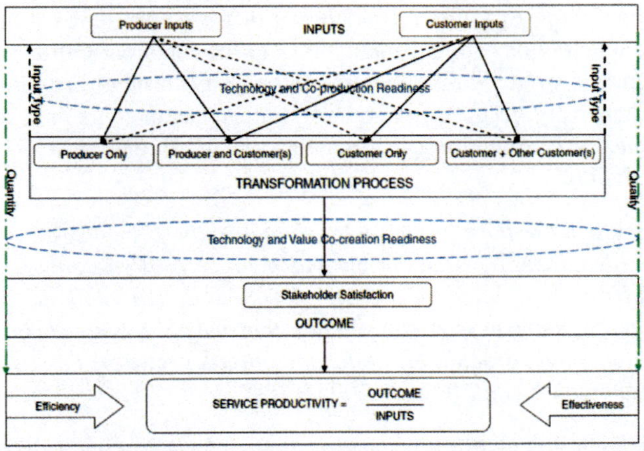

Fig. 2. Schematic framework service production process and service productivity relationship. Source: Agya and Singh Sekhon (2014)

3 Human Resource Quality in Attaining Employee Productivity

Employee productivity which is measured from economic or quantitative definitions are straightforward conventional measure of employee inputs. With a diverse workforce, quantitative LP measure is an effective and convenient measure of employee inputs (Li 2013). However, it is not comprehensive measure in service industry because employee hour is precisely for efficiency evidence, but it does not reflect employee quality about how effectively the service operation as a whole (Li 2013; Grönroos and Ojasalo 2004). Grönroos and Ojasalo (2004) exemplified that the more efficiently the service organization uses its own resources as input into the processes and the better the organization can educate and guide customers to give process-supporting inputs to produce a given amount of output, the better the internal efficiency of the service process is. From the provider's point of view, how customers produce services in isolation from the service provider has no direct effect on internal efficiency. However, through customer perceptions of service quality it has a decisive impact on external efficiency and, thus, on service productivity (p. 417–418). Hence, quality services are very important in the service industry. Therefore, it is crucial to innate excellence HRQ particularly to satisfy customer needs. As such, in this study, a conceptual framework is proposed which detailed the HRQ that is divided into two main important elements: the internal and external features of the employees for greater attainment of HRQ (see Fig. 3). The following section provide further discussion on the proposed HRQ framework.

4 External Feature of HRQ

The external feature of HRQ represents by employee competency (LC) which are significant aspect in assessing employees' productivity (Jabar et al. 2011; Shamsuddin et al. 2018). The seminal work of Ollila (2013) stated that organizations can improve productivity by qualitatively developing inputs used in the function. Ollila (2013) further described that change in qualitative factors may signify through benefiting the competence comprehensively because maintaining human competence is a central channel to add productivity. According to Organization for Economic Cooperation and Development (OECD),

> *"A competency is more than just knowledge and skills. It involves the ability to meet complex demands, by drawing on and mobilizing psychosocial resources (including skills and attitudes) in a particular context. For example, the ability to communicate effectively is a competency that may draw on an individual's knowledge of language, practical IT skills and attitudes towards those with whom he or she is communicating"* (OECD 2005).

Scholars like Dubois et al. (2004), and Griffiths and Washington (2015) also defined LC as a specific set of behaviour or characteristics that individuals have and use in an appropriate and consistent ways to achieve exceptional performance. These characteristics reflect unique combination of knowledge, skills, aspects of self-image, social motives, traits, thought patterns, mind-sets, and ways of thinking, feeling, and acting. It can be justified into two terms. First are the characteristics of a person who meets the minimum requirements, secondly is tied to productivity improvement and competitive advantage (Dubois et al. 2004). Griffiths and Washington (2015) added that competence is manifested in explicit behaviours (what you do and how you show up) and performance (decisions, actions, and results); while intent and potential are part of the much more complex, and largely unseen, world of values, traits, and motivations (the O in KSAOs).

The existing literature recognised that employee competency (LC) substantially improve productivity (Dubois et al. 2004; Griffiths and Washington 2015; Ollila 2013). Competencies are prevalent method used to classify ideal employees and have become a fundamental part of talent management systems across organisations (Griffiths and Washington 2015). The improvement of productivity does not mean working faster or increasing the amount of work, but more realistic working based on justified insights (Griffiths and Washington 2015).

5 Internal Feature of HRQ

The second element is the internal feature of HRQ that captured employee well-being. Healthy workforce has become a paramount investment made by organization because empirical evidence continues to demonstrate that wellness has a positive impact on worker productivity (Envick 2012). High levels of well-being at work is good for the employee and the organization. It means lower sickness-absence levels, better retention and more satisfied customers (Robertson and Cooper 2011). Beaton et al. (2009) proposed that worker productivity is a combination of absenteeism (off from work) and presenteeism (reduced levels of productivity while at work).

Envick (2012) asserted that wellness is generally defined as the quality and state of being healthy in body and mind. Therefore, this paper emphasised on two important aspects of employee well-being. First is psychological well-being (mind) and the second one is physical well-being (body). According to (Bloom et al. 2015; Johnson et al. 2018) psychological well-being is the ability to handle stress, emotion, maintaining positive attitudes and sense of purpose. While physical well-being is defined as absence of sickness which consists of the ability to perform physical activities and carry out social roles that are not hindered by physical limitations and experiences of bodily pain, and biological health indicators (Capio et al. 2014).

A study conducted by Envick (2012) stated that numerous studies revealed a direct positive relationship between psychological well-being and worker productivity (Cropanzano and Wright 2001; Guthrie and Wright 2005; Wright and Bonett 2007; Wright et al. 2002). Therefore, Envick (2012) extends these studies by proving that physical wellness as an important precursor to psychological well-being. Therefore, both of physical and psychological aspects are compliment to one another, which affecting the well-being of employees and then productivity.

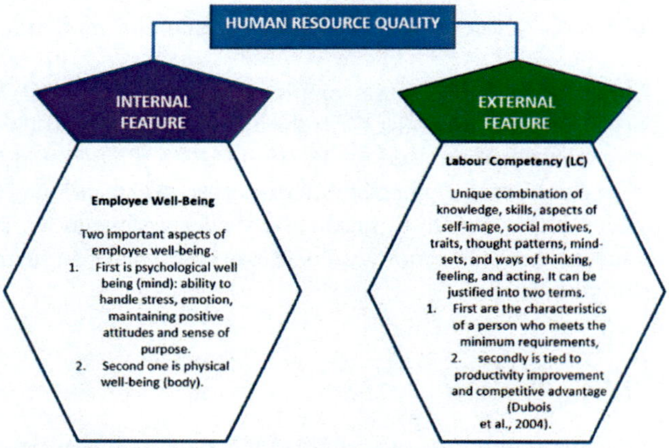

Fig. 3. Proposed Human Resource Quality (HRQ)

6 Discussion and Conclusion

In measuring employee productivity for service industry, HRQ should take into account the employees' length of service (and other demographic factors) as a regulating factor. Ample research has shown that measuring employee productivity in the service industry is difficult to quantify. Studies have shown that productivity in the service industry is not feasible to be measured economically and quantitatively such as ways employee productivity is determined in the manufacturing industry. In the service industry, the human resource quality (HRQ) emerges as a distinct element in measuring productivity. This is held true since HRQ is highly related to the satisfaction felt by the end receivers or users of the service provided. As for example, in the hotel service industry, the hospitality delivered by the hotel employees has been pointed out as one of the major factors of why repetitive visits to the same hotel or accommodation take place (Dasan et al. 2016). It is observed that high star-rated hotels still rely on their senior employees to be the frontline personnel particularly at the areas where first impression matters, for instance, at the entrance and the receptionist counter. Here, they less favoured much younger employees. This gap of years provides the employees with the much-preferred attitudes that are embedded on the way hospitality is shown. Consequently, not only the travellers come for frequent visits, but also disseminate information on the hospitality of the hotel, either verbally or virtually. These observations imply how the human relation factor (Castañeda 2011; Cheng et al. 2011; June et al. 2000) in the form of external (competency) and internal (productivity) of HRQ, serves as the main construct for measuring employee productivity. Importantly, this factor should be considered that the effect of the construct measure is continuous, rather than a merely onetime impact.

The inter cross elements of internal features of well-being and external feature of competency lead to the identification of HRQ. Based on the internal features of HRQ, the balance between the psychological state and the physical well-being explains the readiness of individual employee to be presentable at workplace (Johnson et al. 2018). A clear state of mind with bright attitude enables employees to adopt the concept of working smart rather than working hard. Such work approach enables employees to prioritize tasks and responsibilities at hands leading to a more efficient and productive work performance. Moreover, smart working is linked to enhancing creativity and being innovative that are shown to be directly related to productivity (Gastaldi et al. 2014). Meanwhile, external feature of competency equips employees with required knowledge, skills, and necessary preparations to face challenges. Competency prompts employees to display implicit attitudes prior to executing explicit behavior (Buckwalter 2018). In other words, implicit attitudes can control explicit behavior from being exaggerated or appearing overdoing. Nevertheless, neither internal feature nor external feature could stand alone in ascertaining productivity. Instead, it is the mix of both features (external and internal) that constitute HRQ. Yet, these features represent quality measures and objectifying the level of competency within these constructs remains undefined. For physical well-being, investigating the effect of presenteeism on productivity in the service industry should contribute additional understanding to the existing body of knowledge in this focused theme. For future considerations, human resource quality as a (an) employee productivity measures can be made comprehensively in by including other factors such as job satisfaction, customer satisfaction, employee engagement and performance goals. A new set of employee qualitative measures can be developed in the future research to fill in existing gap of employees' productivity studies in the service industry.

Acknowledgement. Sponsored by Ministry of Higher Education Malaysia FRGS/1/2017/SS03/UMS/03/1 (FRG460–2017).

References

Agya, V.A., Singh Sekhon, H.: Service production process: implications for service productivity. Int. J. Product. Perform. Manag. **63**(8), 1012–1030 (2014)

Asian Productivity Organization. APO Productivity Databook. In Asian Productivity Organization, Tokyo (2012)

Beaton, D., et al.: Measuring worker productivity: frameworks and measures. J. Rheumatol. **36**(9), 2100–2109 (2009). https://doi.org/10.3899/jrheum.090366

Bloom, N., Liang, J., Roberts, J., Ying, Z.J.: Does working from home work? evidence from a Chinese experiment. Q. J. Econ. **130**(1), 165–218 (2015). https://doi.org/10.1093/qje/qju032. Advance

Buckwalter, W.: Implicit attitudes and the ability argument. Philos. Stud. **176**(11), 2961–2990 (2018). https://doi.org/10.1007/s11098-018-1159-7

Capio, C.M., Sit, C.H., Abernethy, A.B.: Physical well-being. In: Michalos, A.C. (ed.) Encyclopedia of Quality of Life and Well-Being Research. Springer, Dordrecht (2014). https://doi.org/10.1007/978-94-007-0753-5_2166

Castañeda, J.A.: Relationship between customer satisfaction and loyalty on the internet. J. Bus. Psychol. **26**(3), 371–383 (2011). https://doi.org/10.1007/s10869-010-9196-z

Cheng, C.C., Chiu, S., Hu, H., Chang, Y.: A study on exploring the relationship between customer satisfaction and loyalty in the fast food industry: with relationship inertia as a mediator. Afr. J. Bus. Manag. **5**(13), 5118–5126 (2011). https://doi.org/10.5897/AJBM10.870

Cropanzano, R., Wright, T.A.: When a "Happy" worker is really a "Productive" worker: a review and further refinement of the happy-productive worker thesis. Consult. Psychol. J. **53**(3), 182–199 (2001). https://doi.org/10.1037/1061-4087.53.3.182

Dasan, J., Chan, K.L.J., Shaari, S.C.: Gen-Y's aura impact on HR new model and practices of hospitality sector. Int. J. Res. Tour. Hosp. **2**(1), 1–7 (2016). https://doi.org/10.20431/2455-0043.0201001

Dean, E.R., Kunze, K.: Productivity measurement in service industries. In: Griliches, Z. (ed.) Output Measurement in the Service Sectors, pp. 73–107. University of Chicago Press (1992)

Dixon, R., Shepherd, D.: Models of labour services and estimates of total factor productivity. Appl. Econ. **42**(28), 3629–3634 (2010). https://doi.org/10.1080/00036840802314549

Dong, S.: Indonesia: enhancing productivity through quality jobs. Asian-Pac. Econ. Literat. **33**(1), 131–132 (2019). https://doi.org/10.1111/apel.12260

Dubois, D., Rothwell, W., Stern, D., Kemp, L.: Competency-Based Human Resource Management. Davies-Black Publishing, Palo Alto (2004)

Envick, B.: Investing in a healthy workforce: the impact of physical wellness on psychological well-being and the critical implications for worker performance. Acad. Health Care Manag. J. **8**(1/2), 21–32 (2012)

Gastaldi, L., Corso, M., Neirotti, P., Aaolucci, E.: Smart working: rethinking work practices to leverage employees' innovation potential. In: 5th International Continuous Innovation Network (CINet) Conference "Operating Innovation – Innovating Operations." Budapest, Hungary (2014)

Griffiths, B., Washington, E.: Competencies at Work: Providing a Common Language for Talent Management, 1st edn. Business Expert Press, New York (2015)

Grönroos, C., Ojasalo, K.: Service productivity - towards a conceptualization of the transformation of inputs into economic results in services. J. Bus. Res. **57**(4), 414–423 (2004). https://doi.org/10.1016/S0148-2963(02)00275-8

Gulezian, R., Samelian, F.: Baseline determination in construction labor productivity-loss claims. J. Manag. Eng. **19**(4), 160–165 (2003). https://doi.org/10.1061/(ASCE)0742-597X(2003)19:4(160)

Guthrie, J.P., Wright, P.M.: human resource management and labor productivity: does industry matter ? University of Texas at arlington. Acad. Manag. J. **48**(1), 135–145 (2005)

Huddleston, P., Whipple, J., Mattick, R.N., Lee, S.J.: Customer satisfaction in food retailing: comparing specialty and conventional grocery stores. Int. J. Retail Distrib. Manag. **37**(1), 63–80 (2009). https://doi.org/10.1108/09590550910927162

Jabar, M.A., Sidi, F., Selamat, M.H., Ghani, A.A.A., Ibrahim, H., Baharom, S.: Capturing tacit knowledge for assessing employees' competency and productivity. Am. J. Econ. Bus. Adm. **3**(2), 358–362 (2011)

Johnson, S., Robertson, I., Cooper, C.L.: Well-Being: Productivity and happiness at Work, 2nd edn. Palgrave Macmillan, Cham, Switzerland (2018)

June, M., Schultz, D.E., Bailey, S.E.: Customer/brand loyalty in an interactive marketplace. J. Advert. Res. **40**(3), 41–52 (2000)

Kaplan, R.S., Norton, D.P.: The balance scorecard: translating strategy into action. Harvard Business Review Press, Boston (1996). ISSBN 10: 0875846513

Nilda, T.T.P., Mohd Yusof, S., Hasan, A., Darma, H.S.: A structural equation model for evaluating the relationship between total quality management and employee productivity. Int. J. Qual. Reliabil. Manag. **34**(8), 1138–1151 (2014)

OECD. Measuring Productivity: Measuring the aggregate and industry-level productivity growth. In: OECD Manual (2001). https://doi.org/10.1787/9789264194519-en

OECD. The definition and selection of key competencies (2005)

Ollila, S.: Productivity in public welfare services is changing: the standpoint of strategic competence-based management. Social Work Public Health **28**(6), 566–574 (2013). https://doi.org/10.1080/19371918.2013.791524

Robertson, I., Cooper, C.: Well-Being: Productivity and Happiness at Work. Palgrave Macmillan (2011). https://doi.org/10.1057/9780230306738

Royuela, V., Suriñach, J.: Quality of work and aggregate productivity. Soc. Indic. Res. **113**(1), 37–66 (2013). https://doi.org/10.1007/s11205-012-0081-1

Rutkauskas, J., Paulavičienė, E.: Concept of productivity in service sector. Eng. Econ. **3**(3), 1–34 (2005)

Shamsuddin, Z., Mahmood, S., Ghazali, P.L., Salleh, F., Nawi, F.A.M.: Indicators for cooperative performance measurement. Int. J. Acad. Res. Bus. Soc. Sci. **8**(12), 577–585 (2018)

Sickles, R.C., Zelenyuk, V.: Measurement of Productivity and Efficiency (2019). https://doi.org/10.1017/9781139565981

SPRING Singapore. A Guide to Integrated Management of Productivity Activities (IMPACT). Singapore (2011)

Thomas, H., Sanvido, V.E.: Quantification of losses caused by labor inefficiencies: where is the elusive measured mile? J. Constr. Law Bus. **1**(3), 1–14 (2000)

Wright, T.A., Bonett, D.G.: Job satisfaction and psychological well-being as nonadditive predictors of workplace turnove. J. Manag. **33**(2), 141–160 (2007). https://doi.org/10.1177/0149206306297582

Wright, T.A., Larwood, L., Denney, P.J.: The different "faces" of hapiness-unhappines in organizational research: emotional exhaustion, positive affectivity, negative affectivity and psychologigkal well-being as correlates of job performance. J. Bus. Manag. **8**(2), 109–112 (2002)

The Role of Digital HRM: Contribution to the Improvement of Business Sustainability

Retno Purwani Setyaningrum[1] (✉) and Muafi Muafi[2]

[1] Universitas Pelita Bangsa, Cikarang, Indonesia
retno.purwani.setyaningrum@pelitabangsa.ac.id
[2] Department of Management, Business and Economics Faculty, Universitas Islam Indonesia, Yogyakarta, Indonesia

Abstract. The role of Digital Human Resources Management (DHRM) in the era of globalization, especially in the COVID-19 pandemic, has made a major contribution to sustainable business performance. This is interesting to study considering that the DHRM work process will take place through mobile, electronic media, social media via the internet, and also with the help of IT (information technology). This research aims to analyze the role of Digital Human Resource Management (DHRM) in contributing to the improvement of sustainable business performance in companies in DKI Jakarta. The importance of the role of DHRM is investigated because DHRM is able to do human work through software and several applications, which are supported by the internet network. Digitalization in HRM will enable companies to operate more efficiently and relevantly in the future.

This type of research is qualitative which involved managers working in oil companies and transportation companies in DKI Jakarta who used DHRM in the companies where they worked. This study analyzed the data using the triangulation method through documentation, interviews and direct observation in the field with case studies. The results of the study explain that several digital HRM practices have been carried out in several companies, but other practical activities have not been carried out optimally. This is because the support from the system and the digitization of business processes that are included in HR practices are not yet optimal. However, the company realizes that DHRM is able to improve business performance in a sustainable manner.

Keywords: Digital Human Resources Management (DHRM) · Business performance · Environmental performance and social performance

1 Introduction

In the current era of globalization and the COVID-19 pandemic, technological developments have improved the organization's digital performance. With digitalization, disruptive changes are not only occurring at the company's operational level, but also at the environmental and social levels. This is the reason that shows that during the last two

B. Alareeni et al. (Eds.): ICBT 2022, LNNS 620, pp. 840–848, 2023.
https://doi.org/10.1007/978-3-031-26953-0_78

decades research on DHRM has become a growing concern, with various topics investigated in the literature [1]. Digitalization is becoming one of the dominant aspects in the business world. It must be admitted that without digitalization, companies and their employees will be outdated [2, 3]. The acceleration of DHRM is increasingly being intensified by most organizations around the world as a consequence of globalization, especially in the era of the COVID-19 pandemic. DHRM is inseparable from sustainable business development. Sustainability is understood as an effort to maintain a balance between economic, environmental and social factors [4].

Today, the digital shift has been forced by external circumstances, but internal impacts also need to be considered. If organizations and leaders better understand the mechanisms behind digitization, it will be easier for organizations in building long-term strategies that are more aware and that can make a positive contribution to social and environmental change [5–7]. Just like every other part of the organization, DHRM transformation is critical where technology disruption is becoming a norm [8]. HRM teams that still rely on Excel sheets to collect and interpret data need to turn to digitization. This is because it is not only a waste of time, effort and cost, but also the accuracy factor in analyzing is relatively low. The DHRM work process will take place through mobile, electronic media, social media via the internet, and also with the help of information technology [9]. In simple terms, DHRM transformation requires implementing human resource tools, practices and strategies with the use of digitization, so that it can improve employee experience and optimize operational performance. In addition, this is also due to the need for data-based automated practices. A successful DHRM will not only optimize HR processes, but if done correctly, it can have a holistic effect in overall and sustainable organizational improvement [9]. This research wants to know how far the impact of DHRM can affect the improvement of sustainable business performance.

2 Literature Review

2.1 The Role of DHRM in Business Sustainability

Changing human resource practices from conventional to digital-based practices is very time consuming. All of which require patience and confidence that the decision to implement DHRM will succeed. Top management support is needed to prevent the organization from returning to its original practice. If companies want to be successful in this 21st century, companies must start to determine the position of human resource activities today. [10] argued that human resource management policies and practices are activities carried out by organizations effectively to achieve various goals. In theoretical studies, [11] put forward three theoretical perspectives on human resource management practices and policies, including;

1. The structural functionalism perspective, which states that the human resources department and all its activities are the result of organizational growth and/or the need to carry out activities carried out by a specialist.
2. The strategic contingency perspective, which views human resource management as a creative form of increasingly critical external pressures

3. The perspective of human resource management strategy, where human resource management activities are designed to support the company's strategic objectives in an integral way

The development of the implementation of human resource management practices is currently directed at the management and utilization of strategic human resources to achieve organizational goals. Consequently, managers in companies should pay more attention to the management of human resources. The management of human resources must be carried out effectively and efficiently to help achieve organizational goals by improving organizational performance.

In general, human resource management will be related to recruitment, selection, development, compensation, retention, evaluating and promoting personnel in the organization [10]. Added by [12, 13] which stated that human resource management is the life of an organization that focuses on the management and effective use of people to increase productive contributions to the organization through strategic, ethical and accountable ways. [14] defines that human resource practices are specific actions used to attract, motivate, or retrain the workforce for general human resource activities (planning, staffing, appraisal, compensation, training and development, and determining and maintain effective working relationships. The activities involved in human resource management are numerous and varied, which can be grouped into six policy areas. The six policy areas are;

1. Organizational design, including; human resource planning, job analysis, job design, work groups and information systems
2. Staffing, including; withdrawal/interview/hire, approval action, promotion/separation/transfer, outlacement service, induction/orientation and labor selection method,
3. Communication and Public Relations, including; personnel records/reports/information systems, workforce communication, suggestion system, and personnel research,
4. Performance management, including; management assessment/management by objective, productivity/expansion program and consumer-oriented performance appraisal,
5. Fulfillment, benefits and reward system, including; safety programs, health/clinical services, complaint procedures, compensation administration, Equal Employment Opportunity Commission compliance, wage/salary administration, insurance administration, non-employment compensation administration, pension/profit sharing programs, labor relations/collective bargaining,
6. Manpower and organizational development, including; management development, career planning, manpower assistance/counseling programs, skills training, layoff readiness programs and attitude surveys.

These six human resource practice activities are very difficult to judge as good or bad because they are not easy to implement. This is because the company depends on the characteristics of existing policies and procedures.

In the era of globalization and pandemic-19, almost all organizations rely heavily on Internet-based performance. It will be very helpful for the HRM Department to try to understand and analyze the relationship between HRM and information technology. Adaptation of digital technology will strike the perfect balance between efficiency and innovative aspects of any organization [9]. In the era of globalization and the Covid-19 pandemic, digitalization and technology have penetrated human life to the technology needed to perform routine and even mundane tasks. DHRM will help organizations through optimization of Social, Mobile, Analytics and Cloud (SMAC) technology, so that digitization in HRM will make it more efficient and relevant in the future [9, 15].

Several studies that examine the influence of HR practices on business performance have also been carried out by many researchers [16, 17]. HRM practice is expected to involve initiatives and ideas from management to be environmentally friendly so that it can reduce costs [18], can reduce the impact on the environment [19] and can be used as capital to manage the internal and external environment in their business operations, including social relations with the community [20, 21]. Companies must be able to translate their social goals into practice, namely having the company's social mission into reality and aligning it with the interests of the community [22]. Some of these researchers generally underline that digital human resource practices can have an influence on sustainable performance (business performance, environmental performance and social performance).

3 Method

This research approach used case studies by analyzing one particular object, namely companies that had used DHRM even though it had not been fully implemented in HRM Practices. This study places more emphasis on exploration to answer the question "How was DHRM done and why was DHRM done?" This is useful to provide an overview of the knowledge of DHRM practices that are implemented and their contribution in improving sustainable business. The events studied in this research are the extent to which digital HRM practices are in improving sustainable business performance [23–25]. This research was conducted at an oil company and liquid product transportation company in DKI Jakarta, Indonesia. The research participants are managers in developing companies that had implemented DHRM in their offices, but had not fully implemented it in every HR practice. This is because in certain practices sometimes the manager still uses a manual system.

This type of research is descriptive qualitative using primary and secondary data. Primary data was obtained from 3 managers who understood the company's internal conditions and the need for using digitalization within the company. These three participants happened to be HRD managers as users of the digitalization system in the company. Researchers conducted written interviews and direct interviews with respondents and made direct visits to company locations. 70% of DHRM had been implemented in the company, but in certain practices, for example e-recruitment, had not fully used the digitalization system, considering the trust and psychological factors in recruitment. More emphasis was placed on character recognition, ability and honesty and trustworthiness of the applicant who was already known by people in the company. In addition,

HRD staff sometimes still made special notes either through word or excel in personnel administration.

The researchers then conducted direct interviews by coming directly to the location of the companies. The goal is to make it easier to accommodate the answers submitted and to have participants provided high responses and participation. That was needed becasue the participants were used as key participants. This research conducted validity testing by searching for references and consulting with experts. Furthermore, regarding the reliability, the researchers carried out careful field observations and re-checking process with the participants. The research team also used and conducted structured interviews which were directed at the digital role of HRM practices in improving sustainable business performance. Furthermore, in testing the reliability of the study, triangulation was carried out. The way to test the reliability was by conducting interviews with other managers and staff in the field as well as the local community [26–28].

4 Research Result and Discussion

This study focuses on the contribution of the role of digitizing HRM in maintaining business sustainability in the era of globalization and the COVID-19 pandemic. Three key participants concluded the need for careful use of DHRM. The main thing that managers actually have to do is implementing DHRM in its entirety and explore it further, whether it has been implemented according to existing procedures. The ability to carry out digitization had not been fully mastered by the staff under these managers. Besides that, there is still dependence on the owner in making decisions for recruitment. Therefore, in the future, it is hoped that a system can work for humans, so that humans can control digitization well, without any fear of using digitalization. Likewise, there should be an even distribution of digitalization capabilities among all directors, managerial and staff, so that the company will be able to maintain its business (business, environmental and social performance).

A new McKinsey Global survey of executives statds that companies have accelerated the digitization of customer interactions and supply chains and their internal operations by three to four years [29]. Almost all respondents said that the company has provided at least a temporary solution to meet the many new demands on the organization, and much faster than managers thought before the Covid-19 pandemic crisis. What's more, the respondents expect that the board of directors will invest in digitalization, in order to be able to sustain the business.

The results of interviews with key participants indicate that they are more familiar with the term information technology system in HR than digital human resources management. In more detail, the results of the interview can be written as follows:

Respondent X: Participant 1

"So far we have done DHRM to improve green practices in business. However, considering that the company is still family-run and the ability to digitize is not evenly distributed, even the directors themselves still don't understand digitization, managerial staff have not used DHRM to its full potential".

"According to managers, the digitalization of e-recruitment interviewed exists, but its implementation has not been well implemented. This is because recruitment is still more from insiders, partners who know more about the abilities of prospective employees. Meanwhile, E-training does not yet exist, because employees are still limited, and training is held as needed. E-training has not been properly scheduled. DHRM emphasizes more on the payroll system, attendance, performance appraisal, and employee data. Managers are proposing to the board of directors for a complete DHRM that facilitates HR administration and improves HR performance".

"We now rarely hold regular meetings to discuss the digitalization system in the company because currently the company is more focused on company development through finding new clients for liquid transportation companies, meanwhile, oil companies are focusing on finding new investors to develop their companies."

"The Board of Directors has not focused on the administrative system in the company but is more focused on finding clients and investors. In addition, the directors often hold meetings outside the office. But we think that the directors have worked hard to advance the company, so our managerial team must continue to support the directors by working better and maximizing the managerial and staff capabilities. We have to work hard, because the directors have worked hard, so, we, who work in this company, also need to work hard to be able to participate in growing the company. Nevertheless, we still think and believe that DHRM will enable us to work more efficiently, be more on a green operation basis, and focus on increasing benefits to the community."

Respondent Y: Participant 2

"We actually want to maximize the DHRM system within the company as much as possible, but sometimes the family business system is still applied in the company where we work. However, the board of directors is very open to input from management. However, because the directors are busy developing the company, they sometimes pay less attention to the use of DHRM in the company, especially to the green environment which has not been a priority for the organization."

"We propose a DHRM that uses data and analytics to measure progress at every stage of the employee life cycle, from recruitment to learning and development to retention and offboarding. This is done with the aim of the company being able to turn quickly when the given strategy does not work. Besides that, we also participate in monitoring the company's business development so that the company can survive and even get the right investors so that the company can become bigger. In the end we will have increased welfare and feel comfortable working in this company because the directors are quite close to the employees."

Respondent Z: Participant 3

"Our managers are in talks to propose to the board of directors that DHRM utilize social media, such as recruiting using social media facilities and including social media in learning strategies. That includes consumerizing HR and reimagining HR self-service as a truly user-friendly experience for employees. But we still support the current board of directors who are very busy. Even with the zoom meeting, the meeting of the board of directors to find investors and develop the company can be carried out at any time. We as the management team are happy to support the company's vision and mission, although currently the implementation of DHRM has not been fully implemented in the company where we work. However, the board of directors is very concerned about business sustainability and managerial also actively participates in the development of the company. We even feel that this company belongs to us too because of our closeness to the board of directors."

"The most important thing in the implementation of DHRM is the effectiveness and efficiency in managing human resources. This can be done by increasing employee productivity through DHRM and DHRM is made cloud-based so that data is safe and easily accessible".

5 Conclusions, Limitations and Implications

Based on the results of the research presented, it can be concluded that DHRM was highly considered by the managerial staff at the research site and had been carried out even though it is only 70%. Managers realized that DHRM was important for the company but the directors and staff did not fully have the desire to practice the company's digitalization system as a whole. This is due to the uneven knowledge of digitization capabilities. Therefore, there is a need for digitalization training which is attended by directors, managers and staff. Likewise, recruitment that currently does not utilize E-Recruitment owned by the company is expected to be able to utilize social media so that it can enable companies to get qualified employees in accordance with company expectations. Social media can also be used for learning and e-training, and it is hoped this will be included as a special training agenda.

The company also has not completely focused on the company's internal green environment, because the company is currently focusing on company development. This study has limitations, namely the number of participants involved is only three people. This may allow this study to be less useful for general conclusions. Researchers conducted in-depth interviews, even saw the condition of companies that are all busy with their respective responsibilities, so that there was no right time to take part in digitalization training. From the results of interviews with managers, and literature reviews from previous studies, the authors conclude some of the company's advantages when utilizing DHRM facilities, as follows:

1. DHRM is not only for HRM administration, attendance and payroll systems, but DHRM can improve the HRM process by using analytics to determine what is and is not aligned with the HRM strategy.
2. The duplicative system used can be streamlined, for example HR and the learning system, into one portal so that it can be more efficient and effective
3. Using a mobile-first strategy can help provide a better employee experience for job candidates and workers by making HR self-service and other processes easier to use, which in turn can increase engagement and retention.
4. Taking an agile approach to strategy, processes, and people, where projects are launched quickly and iteratively, can lower the risk of widespread strategy and technology failure that cannot be easily changed.
5. By using design thinking throughout technology implementation, HR can ensure that employees truly adopt digitization. If that doesn't work, design thinking can help HR leaders and professionals quickly change direction.

6. Cloud Human Capital Management and talent management systems can streamline processes across the employee lifecycle, lower costs, and enable integration of modern tools and technologies.

The implication of this research is to provide views on the importance of practicing HRM digitization for companies that are developing. In the era of globalization, especially during the COVID-19 pandemic, organizations need to pay attention to the readiness of industry 5.0 and the company's competitive advantage. Organizations must implement a first mover in the field of DHRM, so that it can be something interesting but must still consider the costs and risks that will be borne. This must really be taken into account carefully and thoroughly in order to remain profitable.

References

1. Kraus, S., Durst, S., Ferreira, J.J., Veiga, P., Kailer, N., Weinmann, A.: Digital transformation in business and management research: an overview of the current status quo. Int. J. Inf. Manag. **63**, 102466 (2022). https://doi.org/10.1016/j.ijinfomgt.2021.102466
2. Schallmo, D.R.A., Williams, C.A.: History of Digital Transformation, no. January, pp. 3–8 (2018). https://doi.org/10.1007/978-3-319-72844-5_2
3. Kontić, L., Vidicki, Đ: Strategy for digital organization: testing a measurement tool for digital transformation. Strateg. Manag. **23**(2), 29–35 (2018). https://doi.org/10.5937/straman1801029k
4. Kuzior, A., Kettler, K., Rąb, Ł.: Digitalization of work and human resources processes as a way to create a sustainable and ethical organization. Energies **15**(1) (2022). https://doi.org/10.3390/en15010172
5. Giudici, E., Dettori, A., Caboni, F.: Challenges of humanistic management education in the digital era. In: Aguado, R., Eizaguirre, A. (eds.) Virtuous Cycles in Humanistic Management. CMS, pp. 21–35. Springer, Cham (2020). https://doi.org/10.1007/978-3-030-29426-7_2
6. Ramdani, L., Boudinar, C.: Human resources management in the era of digital transformation in algeria: challenges and issues. In: 2021 1st International Conference on Cyber Management and Engineering, CyMaEn 2021 (2021). https://doi.org/10.1109/CyMaEn50288.2021.9497296
7. Karwehl, L.J., Kauffeld, S.: Traditional and new ways in competence management: application of HR analytics in competence management. Grup. Interaktion. Organ. Zeitschrift fur Angew. Organ. **52**(1), 7–24 (2021). https://doi.org/10.1007/s11612-021-00548-y
8. Kane, G.C., Palmer, D., Nguyen, A.P., Kiron, D., Buckley, N.: Strategy, not technology, drives digital transformation. MIT Sloan Manag. Rev. Deloitte (57181), 27 (2015). http://sloanreview.mit.edu/projects/strategy-drives-digital-transformation/
9. Varadaraj, D.A., Wadi, B.MAl.: A study on contribution of digital human resource management towards organizational performance. Int. J. Manag. Sci. Bus. Adm. **7**(5), 43–51 (2021). https://doi.org/10.18775/ijmsba.1849-5664-5419.2014.75.1004
10. Bernadin, H.J., Russel, E.: Human Resources Management: An Experiental Approach, 2nd edn. McGraw-Hill Companies, Inc. (1998)
11. Tsui, A.S.: Defining the activities and effectiveness of the human resource department: a multiple constituency approach. Hum. Resour. Manag. **26**(1), 35–69 (1987). https://doi.org/10.1002/hrm.3930260104
12. Walters, R.W.: HRM in perspective. In: Tracey, D.W.R. (ed.). Human Resources Management and Development Handbook, pp. 3–11. AMACOM, New York (1985)

13. Jr Werther, K.W.B.: Human Resources Management and Personnel Management, 5th edn. Irwin McGraw Hill, USA (1996)
14. Schuler, V.L., Huber, R.S.: Personnel and Human Resources Management, 5 edn. West Publishing Company, Minneapolish (1993)
15. Huang, Y., Qiu, H., Wang, J.: Digital technology and economics impacts of covid-19: experiences of the People's Republic of China. ADBI Work. Pap. Ser. **1276**, 75 (2021)
16. Arqawi, S., et al.: Green human resource management practices among palestinian manufacturing firms- an exploratory study. J. Resour. Dev. Manag. **52**(2019), 62–69 (2019). https://doi.org/10.7176/JRDM
17. Jackson, M., Renwick, S.E., Jabbour, D.W.S., Müller-Camen, C.J.C.: State-of-the-art and future directions for green human resourcemanagement: Introduction to the special issue, ZeitschriftfürPersonalforschung (ZfP). Rainer HamppVerlag, Mering **25**(2), 99–116 (2011)
18. Rohilla, J.: An innovative approach of green human resource management: practices in the organization. Glob. J. Res. Anal. **6**(6), 174–176 (2017)
19. Masri, H.A., Jaaron, A.A.M.: Assessing green human resources management practices in Palestinian manufacturing context: an empirical study. J. Clean. Prod. **143**, 474–489 (2017). https://doi.org/10.1016/j.jclepro.2016.12.087
20. Das, S., Singh, R.K.: Green hrm and organizational sustainability: an empirical review. Researchgate.Net, no. October (2016). https://www.researchgate.net/profile/Sudhir-Chandra-Das/publication/320686237_Green_HRM_and_Organizational_Sustainability_An_Empirical_Review/links/59f45ca1458515547c2083e7/Green-HRM-and-Organizational-Sustainability-An-Empirical-Review.pdf
21. Yusoff, Y.M., Nejati, M., Kee, D.M.H., Amran, A.: Linking green human resource management practices to environmental performance in hotel industry. Glob. Bus. Rev. **21**(3), 663–680 (2020). https://doi.org/10.1177/0972150918779294
22. Stachowicz-Stanusch, A.: Corporate Social Performance: Reflecting on the Past and Investing in the Future. IAP Information Age Publishing, United State of America (2016)
23. Yin, R.K.: Case Study Research: Design and Methods. SAGE, Thousands Oaks (2009)
24. Eisenhardt, K.M., Eisenhardt, K.M.: Linked references are available on JSTOR for this article: agency theory : an assessment and review. Acad. Manag. **14**(1), 57–74 (2018)
25. Stake, R.: The Art of Case Study Research, p. 175 (1995)
26. Carter, N.: The use of triangulation in qualitative research. Number 5/September 2014 **41**(5), 545–547 (1969)
27. Egon, G.G., Lincoln, Y.S.: Competing Paradigms in Qualitative Research (1994)
28. Whyte, W.F.: Participatory Action Research. Sage, Newbury Park (1991)
29. LaBerge, L., O'Toole, C., Schneider, J., Smaje, K.: COVID-19 digital transformation & technology. McKinsey, no. October, p. 6 (2020). https://www.mckinsey.com/business-functions/strategy-and-corporate-finance/our-insights/how-covid-19-has-pushed-companies-over-the-technology-tipping-point-and-transformed-business-forever

Internal Control System on Using Digital Banking Applications and Services in Jordanian Banks During the Corona Virus Pandemic

Reem Oqab Al-Khasawneh[1](✉) and Satih Razouk[2]

[1] Al-Balqa' Applied University (BAU), Salt, Jordan
dr.reemokab@bau.edu.jo
[2] Aleppo University, Aleppo, Syria

Abstract. A study has aimed at identifying the extent of effectiveness and ability of internal control to supervise digital banking applications and services which have been highly used in Jordanian banks due to the Corona.

A descriptive analytical approach has been used due to the nature of the study; a questionnaire has been designed as a tool of data collection and distributed to study sample, composed of internal auditors, chief accountants, and financial managers in Jordanian commercial and Islamic banks. 96 questionnaires have been distributed. Statistical package for the social sciences (SPSS).

The study has found that digital banking applications and services used by Jordanian bank have contributed to the achievement of internal control system's goals. In addition, it has indicated that digital banking applications and systems, used by Jordanian banks, are characterized by information properties which make internal control system more effective and efficient during the Corona pandemic. Moreover, the study has shown that public or regulatory administrative control procedures have achieved the effectiveness of digital banking applications and services used by Jordanian banks in the light of the Corona pandemic; and the procedures and systems of data security and safety have highly affected the effectiveness of digital banking applications.

Keywords: Internal control · Digital services · Internal auditor · Jordanian banks · The Corona Virus Pandemic

JEL Classification: M42

1 Introduction

Due to the Corona virus pandemic, many economics of world countries have been besieged; like other economic sectors, financial markets and banking institutes have suffered financial damages and losses because of outbreak of Corona virus and closures. Thus, a need for electronic financial transactions has emerged; many banks have provided these electronic service to avoid gatherings of customers and comply with the requirements and conditions of public safety. Accordingly, electronic banking transactions have become a necessity and shall be developed in a manner that meet all financial

B. Alareeni et al. (Eds.): ICBT 2022, LNNS 620, pp. 849–865, 2023.
https://doi.org/10.1007/978-3-031-26953-0_79

services provided to banks' customers. Then, the banks shall be shifted to digital dealing in order to continuously provide their financial services and meet customers' services when they request via their smart electronic devices; the data on financial transactions are secured and confidential.

Despite of various advantages, electronic banking transactions are fraught with risks because e-transactions are closely connected to data security so that customers' balances can be tampered; or innovative e-transfer and paying processes can be conducted through customers' accounts. Thus, this has contributed to the development of internal control systems and methods; and attention is paid to data safety and security in addition to privacy and reliability; raising the levels of performance, productivity and speed in the field of digital banking services is taken into account.

The importance of the study stems from the need to identify the development of control procedures in a way that is in line with the development of digital banking applications in light of the Corona pandemic. It is important because it demonstrates the need to provide important scientific means in order to protect and ensure the accuracy and ability of internal control systems to follow the accounting systems used in these applications, and to process transactions. In addition, the study is important because it ensures the safety of the accounting treatment of digital banking applications when practicing different types of applications and electronic services, especially with the multiplicity of risks arising from the use of digital banking services and the scientific dimension of its effects; moreover, the importance of this study comes from identifying the recent trends in electronic information security to overcome these risks, and to provide protection means that ensure the continuity of electronic banking applications, and thus success in performing the supervisory work.

The world witnesses an accelerated development of the field of providing banking services in the light of digital economy. This, therefore, requires similar attention and development of the field of internal control in order to ensure the effectiveness of these services, maintain the ability to meet management and customers' needs and secure the infrastructure for supporting digital banking services and user data. In addition, the confidentiality of personal and financial data shall be protected and shall not be penetrated. It is also important to study the changes of controls in digital banking services environment. Undoubtedly, amended laws and regulations which are in line with the nature of digital services are required. This study focuses on the awareness of internal control effectiveness through the use of digital banking services and applications in the light of the Corona Virus Pandemic in Jordanian banks. Accordingly, problem of study can be formulated into the following main question:

What extent is internal control effective in the light of the use of digital banking applications and services in Jordanian banks during the Corona Pandemic?

2 Literature Review

The (Mansour 2021) study, it aimed to demonstrate the importance of cyber security through its impact on internal control and the value of the economic unit by adopting the framework of information technology governance (COBIT5). to eight axes that reflect the requirements of the research, through the use of Google form forms, and a study has concluded that there is acceptance and agreement on the existence of a relationship between the dimensions and requirements of cyber security (operations and procedures, cyber risks, confidentiality, and privacy protection, logical security, strategy) on modern frameworks of control The study recommended the need for the economic unit to adopt effective means for continuous evaluation of internal control to maintain information security by adopting modern frameworks for internal control COBIT5 and through the integration of procedures and characteristics in light of modern frameworks to avoid means of penetration of electronic systems and attempts to manipulate their information. Study (Zhu and Song 2021) The study demonstrated the role of internal control systems in preventing risks when serious emergencies occur, thus ensuring the efficient performance of business organizations during crises. In light of the rapid spread of the virus, COVID-19 in China, this study provides a test of the ability of the internal control system to mitigate the effects of the epidemic on the performance of business organizations. The study concluded that business organizations that have a high-quality internal control system achieve better financial performance during the epidemic, and that the difference in the nature of the company's activity, the degree of computerization, the extent of the virus's spread in an area, and the percentage of government ownership in the company affects the role and efficiency of the internal control systems and has a more role importance in the performance of companies. In a study (Kawa Wali and Bnar K. Darwish 2021) The study demonstrated the role of how electronic accounting services enhance efficiency in financial institutions by evaluating the quality of the electronic accounting system used in banks operating in the Kurdistan region of Iraq. And its role in achieving the quality of financial reports, and the study concluded that electronic accounting services have a significant impact on the efficiency of the work of financial institutions as a result of the impact of the internal control system. It showed that there are indicators of some elements of the internal control system that show this effect, such as risk assessment and control activities. The study recommended the need for a further evaluation of internal control to determine and analyze the effectiveness of controls in financial institutions in the Kurdistan Region-Iraq. And (Al-Khasawneh –Bshayreh 2019) study has aimed at identifying the impact of developing an infrastructure (i.e. material and human resources, software, network and communications) on the effectiveness of accounting information systems. A descriptive approach was used; the questionnaires were distributed to employees at various administrative and accounting levels of bank. The study found that developments of information technology infrastructure (i.e. material and human resources, software, network and communications) had a very high impact on the effectiveness of accounting information systems; and development of information security had also a very high impact on the effectiveness of accounting information system. The study (Manar Al-Ghanimi and Hamza Al-Zubaidi 2016), shows the impact of the development of electronic banking on the internal control systems in a group of Iraqi banks and how electronic banking abolished the paper-based use of transactions and concealment of the workflow and concluded that there is a simple weakness and difficulty in the conduct of control operations In addition to the lack of knowledge of the employees on the latest developments in the means and systems, the study recommended the need to strengthen and improve the control environment in banks as it is the general framework of the bank and work on holding training courses for employees working in the internal control department in the bank to keep pace with the rapid development In the field of computer and electronic services provided by Iraq banks, Study (Aroussi Hayat 2015) This study aimed to reveal the impact of internal control on the quality of electronic banking services, especially in light of the developments taking place in the global economy. The quality of electronic banking services in commercial banks in Algiers, and that the bank has a specific and written accounting system. Management informs all employees of their roles within the framework of internal control processes in a clear and written manner. - There is effective control over the financial and accounting performance in the bank, the employees in the supervisory bodies enjoy scientific qualification and experience, and banks are keen to provide all information related to any new services. The most important thing recommended by the study is the need to develop an effective emergency plan to ensure the progress of work and reduce the possibility of failure of electronic devices and systems. This study is comprehensive and complementary to previous and recent studies in Jordan, as we dealt with internal control over digital transactions and banking applications in light of the Corona pandemic, as it covered topics that were not addressed.

From the above, hypotheses were formulated as follows

1. Digital banking applications and services in Jordanian banks achieve the purposes of internal control in the light of the Corona Pandemic.
2. Digital banking applications and services in Jordanian banks have information characteristics which make internal control system more effective and efficient during the Corona Pandemic.
3. Public or regulatory administrative control procedures achieve the effectiveness of digital banking applications and services in Jordanian banks during the Corona Pandemic
4. Procedures and programs of information security and safety affect the effectiveness of digital banking applications and services in Jordanian banks during the Corona pandemic.

3 Materials and Method

Methodology of Study. This research is quantitative, and the study used the deductive approach through which data was collected by questionnaire to evaluate the theoretical data about the impact of the Corona pandemic on internal control systems in light of the use of digital services and electronic banking applications.

Population and Sample of Study. The population is the banks in Jordan. The sample of the study is consists of internal auditors, chief accountants and financial managers working in 24 Jordanian banks, distributing as follows: 13 Jordanian commercial banks, 8 foreign banks and 3 Islamic banks. Sample study is limited to Jordanian commercial and Islamic banks. Accordingly, 96 questionnaires have been distributed as follows; 6 questionnaires have been distributed for each bank to be responded 2 internal auditors, 2 employees of computer department (information technology) and 2 employees of digital banking services; 86 questionnaires which are valid for analysing have been retrieved.

Questionnaire Design. A survey method was chosen for data collection. The research tool was a questionnaire as mentioned previously, and a five-level Likert scale was used to answer the questionnaire questions and test hypotheses, and the results were interpreted as follows: 1–1.79 indicated that there was no effect. 1.8–2.59 indicated that the impact of the epidemic was weak; 2.60–3.39 indicated a moderate impact of the pandemic; 3.40–4.19 indicated the high impact of the pandemic; As for 4.20–5 it indicates a very high impact. The respondents' answers were analyzed using the SPSS statistical program.

Reliability of Study Tool. To ensure the reliability of the study tool, Cronbach's alpha coefficient was calculated for the internal consistency test. Table 1 shows that Cronbach's alpha lies in the range 0.95 to 0.97″ This indicates very good correlation strength and proves that the choice of questions is appropriate for the purpose of the questionnair.

Table 1. Cronbach's alpha for the study hypotheses

Field	Internal consistency
Purposes of internal control system in the light of providing digital banking services and applications	0.95
The features of digital banking applications and their role in achieving the effectiveness of internal control system	0.96
The general control procedures in digital banking and applications	0.95
The procedures of controlling security and safety of operations and processing of digital banking applications and services	0.97

Analysis Data Tools: All primary data collected were analysed using SPSS. Descriptive statistics were deployed showing arithmetic means and Arithmetic means and significance probability (Sig.) data statistical significance within 95% confidence level.

4 Results and Findings

4.1 Result Analysis and Hypotheses Testing

First Hypothesis Testing

The following table shows the arithmetic mean and standard deviation of the first hypothesis paragraphs.

The table indicates that all items are statistically significant at the level of significance, the arithmetic mean of all clauses was above 4 (total score of 5). This indicates that the responders strongly concur with the information's main points. Accordingly, "Banks provide digital banking applications with the means to detect errors, fraud, and abuse" has the greatest acceptance rating. In general, the axis' arithmetic mean is 4.81, and

Table 2. Purposes of internal control system in the light of providing digital banking services

Statement	Arithmetic mean	Sig.	Response orientation
Internal control system applied within banking applications can protect balances, files and information of the users from misuse	4.78	0.000*	Very high impact
Applied internal control system achieves validity, credibility and accuracy of data in digital banking transactions	4.84	*0.000	Very high impact
Internal control system explains to the users the policies and procedures which must be adhered to when activating the digital services	4.80	*0.000	Very high impact
Internal control system helps achieve production efficiency and bank return and cost reduction	4.80	*0.000	Very high impact
Banks provide with digital banking applications means of discovering errors, fraud and misuse	4.85	*0.000	Very high impact
Total	4.81	*0.000	Very high impact

the probability of significance (Sig.) is 0.000. As a result, at the significance level, this domain is statistically significant. Additionally, this domain's typical response rate is greater than the average ability score of 3.This means that respondents agree with the paragraphs of this domain. As a result, the hypothesis that "digital banking applications and services in Jordanian banks achieve internal control purposes in light of the Corona pandemic" was accepted.

Second Hypothesis

The following table shows the arithmetic mean and standard deviation of the second hypothesis paragraphs.

According to Table 3, every paragraph is statistically significant at the level of significance. Each paragraph's arithmetic mean has been greater than 4; it often ranges between 4.5 and 4.91. It indicates that the sample respondents strongly concur with the paragraphs' substance. The first statement, "Banking accounting processing is handled at blazing speed through digital apps," has therefore obtained the greatest level of approval.

The averages for the arithmetic means of the axes is 4.80, and the significance probability (Sig.) is 0.000. Consequently,

This section has statistical significance at a certain level. Additionally, this field's average response rate is higher than the average ability score, which is 3. It indicates that respondents concur with the statements made in this subject; in other words, the study population is persuaded by the features of available digital banking applications that make internal control systems more effective and efficient.

Table 3. The features of digital banking applications and their role in achieving the effectiveness of internal control system

Statement	Arithmetic mean	Sig.	Response orientation
Through digital applications, banking accounting processing is performed at blazing speed	4.91	*0.000	Very high impact
Accounting system of digital applications conducts many pre-programmed supervisory tests in order to ensure that required conditions of digital financial transactions are available	4.89	*0.000	Very high impact
Digital banking applications provide the users who are different at speed with the information required for decision making process	4.90	*0.000	Very high impact
Digital banking applications perform effectively accounting processes	4.90	*0.000	Very high impact
Through digital banking applications, accounting system classifies the information saved in the system in order to be valid for decision-making process	4.69	*0.000	Very high impact
Through e-banking applications, accounting data is recorded and saved in accordance with generally applicable accounting principles and rules	4.89	*0.000	Very high impact
Digital banking applications have a power to detect cases of fraud, tampering and intentional and unintentional error	4.50	*0.000	Very high impact
Digital banking applications provide highly accurate results, except for the error the customer may do during data entry process	4.75	*0.000	Very high impact
Total	4.80	*0.000	Very high impact

As a result, the hypothesis stating 'Digital banking applications and services in Jordanian banks have information characteristics which make internal control system more effective and efficient during the Corona Pandemic, accepted.

Third Hypothesis

The following table shows the arithmetic mean and standard deviation of the hypothesis paragraphs.

Table 4 indicates the following:

Table 4. The availability of general (organizational) control procedures in digital banking and applications

	Statement	Arithmetic mean	Sig.	Response orientation
Administrative control procedures During The Corona Virus Pandemic				
1	Digital and electronic banking duties and services are separated in order to reduce employees' fraud	4.77	*0.000	Very high impact
2	Dual control procedures of functions such as retrieval of encryption file or large transfer process via internet are ensured	4.70	*0.000	Very high impact
3	Administrative procedures and instructions, allowing to deport unauthorized employees to access the system, are available	4.70	*0.000	Very high impact
4	Procedures of settling electronic and digital banking transactions are available	4.70	*0.000	Very high impact
5	Suspicious activities are reviewed and frauds are detected, in addition, targeted review of amounts and sizes of transactions is performed in order to prevent unintentional errors	4.66	*0.000	Very high impact
6	Documents registering work and processing method of applications and sites are maintained by the administration	4.73	*0.000	Very high impact
7	Errors are checked and customers are instructed in order prevent unintentional errors	4.74	*0.000	Very high impact
8	It is ensured that alternative channels are available in order to approve account activity or continuous changes of maintenance and develop plans and strategies of downtime	4.75	*0.000	Very high impact

(continued)

Table 4. (*continued*)

	Statement	Arithmetic mean	Sig.	Response orientation
9	Fees of performed transactions via electronic sites and application are determined and customers are informed of these fees	4.72	*0.000	Very high impact
10	Amounts of collected fees and discounts are reviewed in order to ensure that customers are not charged for more or less transaction fees	4.70	*0.000	Very high impact
	Total	4.70	*0.000	Very high impact
Compliance control procedures and policies and their role in achieving the efficiency and effectiveness of digital banking applications and services During The Corona Virus Pandemic				
1	Official names of electronic applications and websites are clearly identified	4.60	*0.000	Very high impact
2	Privacy and security policies of customers are properly disclosed while using digital services	4.70	*0.000	Very high impact
3	Digital banking services are advertised and disclosed in accordance with laws issued by the Central Bank and electronic systems including electronic signature			Very high impact
	Total	4.65	*0.000	Very high impact
Supervising and controlling the third party During The Corona Virus Pandemic				
1	Control procedures are placed in order to secure internal network and follow up the data and detect the weaknesses	4.66	*0.000	Very high impact
2	Existing emergency plans for service providers is ensured in order to ensure the continuity of providing remote services	4.68	0.000*	Very high impact

(*continued*)

Table 4. (*continued*)

	Statement	Arithmetic mean	Sig.	Response orientation
3	A follow up report shall be issued in order to monitor service interruption, and follow –up by the bank to find out the reasons and solves	4.86	* 0.000	Very high impact
4	A report of security accidents illustrating a volume of rejected attempts and resetting password in addition to virus attacks is available	4.78	*0.000	Very high impact
5	Passwords are used in order to ensure that only authorized persons are allowed to access to the system	4.80	*0.000	Very high impact
	Total	4.76	*0.000	Very high impact
Controls of business continuity During The Corona Virus Pandemic				
1	An analysis of effect of service interruption on digital banking service and the minimum level that can be provided shall be conducted	4.50	*0.000	Very high impact
2	Time required for retrieving the services shall be determined	4.55	*0.000	Very high impact
3	Business continuity plans shall be updated in order to with any interruption	4.65	*0.000	very high impact
4	Customer communication plans shall be developed before interrupting	4.55	*0.000	Very high impact
5	Abilities to resume the work shall be periodically tested in order to determine the possibility of achieving the goals	4.65	*0.000	Very high impact
6	Business continuity plans shall be accepted by the third party	4.50	*0.000	Very high impact
	Total	4.50	*0.000	Very high impact

(*continued*)

Table 4. (*continued*)

	Statement	Arithmetic mean	Sig.	Response orientation
Data safety During The Corona Virus Pandemic				
1	Processes of developing the test of digital applications and sites of IP addresses are financially separated	4.65	*0.000	Very high impact
2	Tasks of operation teams and database support are separated	4.55	*0.000	Very high impact
3	Application support and database administrator jobs are separated in the application and sites	4.60	*0.000	Very high impact
4	Live data is not used in structure test	4.70	*0.000	Very high impact
5	Confidential customer data such as debit or credit card is encrypted in database	4.75	*0.000	Very high impact
	Total	4.65	*0.000	Very high impact
	Total fields	4.66	*0.000	Very high impact

- Arithmetic means of all first axis articles pertaining to administrative control procedures were higher than 4: the averages fell between 4.65 and 4.77. Additionally, this field's average response rate is higher than the average ability score, which is 3. At a certain threshold of significance, every paragraph is statistically significant. It denotes that a large majority of sample respondents strongly concur with the points made in the paragraphs relating to administrative control methods. Additionally, the axis paragraphs' total arithmetic mean was 4.70.
- All second axis paragraphs that discuss compliance control policies and processes and how they contribute to the effectiveness and efficiency of digital banking apps and services have arithmetic means that are greater than 4; on average, these means fall between 4.60 and 4.70. The typical response rate in this area is more than the typical ability score, which is 3. As a result, every paragraph is statistically significant at some level. It indicates that a high level of agreement exists among sample respondents with the statements made in the paragraphs under this axis, which is titled "Compliance Control Procedures and Policies and Their Role in Achieving the Efficiency and Effectiveness of Digital Banking Applications and Services." Additionally, the cumulative arithmetic mean of the axis paragraphs is 4.65.

The third axis's arithmetic means for all of the paragraphs that deal with overseeing and managing the third party were all higher than normal, falling between 4.80 and 4.86. When compared to the average ability score, which is 3, this field's average response

Table 5. Procedures of controlling security and safety of operations and processing of digital banking applications and services during the Corona Pandemic

	Statement	Arithmetic mean	Sig.	Response orientation
Supervising the security of electronic applications and sites of bank during the Corona Pandemic				
1	Ceilings of processes performed through banking application or electronic sites comply with banking business rules such as transaction limits or ceilings	4.60	*0.000	Very high impact
2	Bank obtains publication certification 'Service oriented architecture (SOA)'	4.77	*0.000	Very high impact
3	Data between bank and the concerned parties is encrypted	4.78	*0.000	Very high impact
4	Authentication control such as passwords, security questions and answers, and authentication token is available	4.75	*0.000	Very high impact
	Total	4.73	*0.000	Very high impact
Security of operating systems during the Corona Pandemic				
5	It is ensured that application and site servers are efficiently running	4.50	*0.000	Very high impact
6	Servers of banking applications and electronic sites and servers of database are separated for security purposes	4.55	*0.000	Very high impact
7	Security requirements are provided in order to review the extent of adequacy of servers to meet the requirements of operating systems	4.66	*0.000	Very high impact
8	A multi-layer wall is placed in order to achieve adequate security at various levels of network, thereby preventing the penetration	4.55	*0.000	Very high impact
	Total	4.60	*0.000	Very high impact

(*continued*)

Table 5. (*continued*)

	Statement	Arithmetic mean	Sig.	Response orientation
Security of database during the Corona Pandemic				
9	Rules of access to database shall be consistent with bank policies related to passwords such as length and expiration	4.65	*0.000	Very high impact
10	Super users shall be identified in database in order to ensure that their access is authorized and required for their jobs	4.60	*0.000	Very high impact
11	The users shall be given uniquely identifiable IDs and passwords in order to access to database, thereby ensuring the accountability and non-repudiation	4.50	*0.000	Very high impact
12	Database transactions ensure that data base is confidential, safe and available	4.65	*0.000	Very high impact
13	Users' access to database is regularly reviewed in order to ensure its adequacy and relevant exceptions are subjected to management actions	4.75	*0.000	Very high impact
14	Database backups are scheduled	4.70	*0.000	Very high impact
	Total paragraphs	4.64	*0.000	Very high impact
	Total paragraphs of field	4.67	*0.000	Very high impact

rate is higher. Therefore, at a level of significance, every paragraph is statistically significant. This indicates that the sample respondents agree with the ideas expressed in the paragraphs under the heading "supervising and managing the third party." The sum of the arithmetic means of the paragraphs on the axis was 4.76.

- Arithmetic means of all fourth axis articles relating to business continuity controls were higher than averages, falling between 4.55 and 4.65. When compared to the average ability score, which is 3, this field's average response rate is higher. Therefore, at a level of significance, every paragraph is statistically significant. This indicates that the sample respondents concur with the statements made in the paragraphs under the heading "Controls of business continuity." Additionally, the cumulative arithmetic mean of the axis paragraphs is 4.56.
- All fifth axis articles relating to data security have arithmetic means higher than 4; the averages fall between 4.55 and 4.75. When compared to the average ability

score, which is 3, this field's average response rate is higher. Therefore, at a level of significance, every paragraph is statistically significant. This indicates that the sample respondents agree with the statements made in the paragraphs relating to data safety. Additionally, the cumulative arithmetic mean of the axis paragraphs is 4.56.

- The average rate of response to this field is higher than the average ability score, which is 3, and the average arithmetic mean of the axis is 4.66. Probability of significance (Sig.) is equal to 0.000. As a result, this field has statistical significance at a certain level. Moreover. The hypothesis, "Public or regulatory administrative control processes achieve the effectiveness of digital banking apps and services in Jordanian banks during the Corona Pandemic," is accepted as a consequence of the respondents' agreement with the axis paragraphs relating to this hypothesis.

Fourth Hypothesis
Statistical tests have been used to test fourth hypothesis and find out whether average response has reached average ability which is 3; Table 4 illustrates the results.

- The averages for the arithmetic means of the axes' paragraphs have ranged between 4.50 and 4.78. Compared to the average ability score, which is 3, the average rate of reaction is higher. Additionally, every paragraph is statistically significant at a certain level. This indicates that the sample respondents agree with the statements made in the axis' paragraphs about monitoring the security of bank websites and electronic apps as well as the security of operating systems and databases.
- Arithmetic mean of axis equals 4.67, and the average rate of response to this field is higher than the average ability score which is 3. Accordingly, It means that the respondents agree to the paragraphs of axis related to this hypothesis; As a result, the hypothesis stating 'Procedures and programs of information security and safety affect the effectiveness of digital banking applications and services in Jordanian banks during the Corona pandemic', is accepted.

5 Conclusion and Recommendations

Conclusion
The study results as follows:

- Digital banking applications and services have achieved to a very high degree the goals of internal control system; statement related to availability of means of detecting errors, fraud and misuse in digital banking applications digital banking used by the banks, has received the highest degree of approval of relevant questions. Accordingly, banking applications and services shall include all means achieving the goals of internal control system.
- Digital banking applications and services shall have information characteristics which make internal control system more effective and efficient to a very high degree during the Corona pandemic; statement, stating that banking accounting processing is quickly

performed by using digital applications, has received the highest degree of approval of hypothesis questions.

- General or regulatory procedures of administrative control, which include administrative control procedures, compliance control policies and procedures, third-party control and supervision procedures in addition to controls of business continuity and data safety, have achieved to a very high degree an effectiveness of digital banking applications and services the Jordanian banks use in the light of the Corona pandemic. The following procedures are available to a very high degree:
- Procedures and programs of information security and safety which include control over the security of electronic applications and sites, operation systems and database, have highly affected the effectiveness of digital banking applications and services the Jordanian banks use in the light of the Corona pandemic.
- Jordanian banks pay an attention to develop particularly digital banking services and provide appropriate controls which contribute to the increase of customer willingness to use them.

Recommendations. The study has recommended the following:

- Jordanian banks shall keep up with all developments of electronic banking applications and services, and shall develop appropriate control systems and means.
- Internal auditors shall pay attention to control over the availability of these systems and means and they shall test control means in order to identify the extent of their efficiency and effectiveness.
- Central bank shall pay attention to provide general instructions organizing general and special control procedures which must be available while using digital applications and services.
- Availability of a secure environment is one of factors encouraging the customers to deal with digital applications. Hence, the banks shall pay attention to controls of data security and confidentiality.
- Legislative and legal controls related to digital applications and services shall be available in order to protect the customer; in addition, they shall not bear the customer a full responsibility of protecting password and user name.

Limitations of Study. The Banking sector in Jordan is considered one of the sectors that is witnessing a great and rapid development in the use of electronic systems before the Corona pandemic, and for those in light of the pandemic, banks were able to provide banking services and seemed to transform the technology by encouraging customers to use electronic applications on phone devices, and this study covered the period The current period from 2020 until the beginning of 2021, which may witness the sector developing more in the use of electronic and digital means in banking services.

References

Al-Buhaisi, E., Al-Sharif, H.: Risks of electronic accounting information systems: applied study on banks working in Gaza strip. J. Islamic Univ. **16** (2008)

Al-Hadithy, E.S.N.: Evaluation of Internal Control System in Institutions that Use Computer: A Field Study on Financial and Banking institutions in The Hashemite Kingdom of Jordan, Unpublished Master Thesis, University of Jordan, Amman (2008)

Al-Hakim, S.: Impact of possibility of control over automated accounting information systems. In: Public Institutions With Economic Nature By Inspectors of Central Financial Supervision Agency. Damascus Univ. J. Educ. Legal Sci. **26** (2010)

Al-Matarneh, G.F.: Contemporary Audit – Theoretical Aspect. Dar Al-Masera For Publishing and Distribution, Amman (2009)

Burtuon, R.N.: Discussion of information technology. Related activities of internal auditors. J. Inf. Syst. (2000)

Chow, C.W., Kato, Y.: The use of organization controls and their effects on data manipulation and management and management myopia: a Japan vs. U.SA comparison. Account. Organ. Soc. **21** (2006)

Edwards, D., Oxner, T.: Internal Auditing in the Banking Industry Bank Accounting & Finance. (Euromoney Publications PLC) (2001)

Edmon, G., Edmon, T.: Extent of effectiveness of accounting information systems in Iraqi private commercial banks for the point of view of management. Unpublished Master Thesis, Middle East University, May 2010

Hamada, R.: Impact of general controls of electronic accounting information systems on increase of accounting information reliability. Damascus Univ. J. Educ. Legal Sci. **26** (2010). Unpublished Master Thesis

Al-Khasawneh, R.O.H.: Study and evaluation of government electronic accounting information systems - a field study in the Hashemite Kingdom of Jordan. Res. J. Financ. Account. (2012). https://www.iiste.org/Journals/index.php/RJFA/article/view/1808

Al-Khasawneh, R.O.H., Al-Oqool, M.A.: Impact of the development of ICT infrastructure and security on the effectiveness of accounting information in the Jordanian banking sector. Int. Bus. Res. Arch. **12**(12) (2019)

Al-Khasawneh, R.O.H.: Impact of Corona Pandemic (COVID-19) on external audit on Jordanian banks. Acad. Account. Financ. Stud. J. **25**(1) (2021). https://www.abacademies.org/articles/Impact-of-Corona-Pandemic-Covid-19-on-External-Audit-on-Jordanian-Banks-1528-2635-25-1-642.pdf

Al- Khasawneh, R.O.H.: Challenges facing external auditor while auditing banking accounting systems in the light of the use of digital technologies of fourth industrial revolution in Jordan. Am. J. Ind. Bus. Manag. **12**(4) (2022)

Hurt Hill, R.: Accounting Information Systems: Basic Concepts & Current Issues, south-western, Thomson, London (2007)

Ratcliffe, T.A., Munter, P.: Information technology, internal control, and financial statement audits. CPA J. (2002)

Abdel Razaq, Q.: Analysis and design of accounting information systems, Dar Al- Thaqafa for Printing and Distribution, Amman (2004)

Stephen, M., Seken, M.: Accounting Information System For Decision Making, Translated by Saed Kamal Al-Din and Hajjaj, Ahmad, Dar Al-Marech For Publishing, Saudi Arabia, Al-Riyadh, (2002)

Al-Ghanimi, M., Al-Zubaidi, H.: The evolution of electronic banking and the impact of internal control on electronic banking operations (Applied Research on a Sample of Iraqi Private Banks). J. Univ. Coll. Herit. **1**(20), 361–388 (2016)

Laroussi, H.: Internal control and its impact on the quality of electronic banking services by application on a sample of banks in Algiers, Master's Thesis, Mohamed Boudiaf University of Msila (2015)

Mansour, A.: The impact of cyber security on internal control and its reflection on the economic unit - An exploratory study of the opinions of a sample of auditors and accountants in the ministry of higher education and scientific research. J. Adm. Econ. Issue **127**, 223–238 (2021)
Wali, K., Darwish, B.K.: Electronic accounting services and their effect on enhancing the efficiency of financial institutions Babylon. Univ. J. **29**(1), 21–39 (2021)
The Canadian Public Accountability Board (CPAB) (2020) COVID-19 Insights: Understanding internal control in the audit. https://www.cpab-ccrc.ca/
Zhu, P., Song, J.: The role of internal control in firms' coping with the impact of the COVID-19 pandemic: evidence from China. Sustainability **13**(11), 62–94 (2021). https://doi.org/10.3390/su13116294

Author Index

B. Alareeni et al. (Eds.): ICBT 2022, LNNS 620, pp. 867–870, 2023.
https://doi.org/10.1007/978-3-031-26953-0